PESTICIDE PROFILES

Toxicity, Environmental Impact, and Fate

PESTICIDE PROFILES

Toxicity, Environmental Impact, and Fate

Edited by

Michael A. Kamrin

Institute for Environmental Toxicology
Michigan State University
East Lansing, Michigan

LEWIS PUBLISHERS

Boca Raton   New York

Library of Congress Cataloging-in-Publication Data

Kamrin, Michael A.
    Pesticide profiles : toxicity, environmental impact, and fate / Michael A. Kamrin.
       p.   cm.
    Includes bibliographical references and index.
    ISBN 0-56670-190-2
       1.Pesticides—Toxicity. 2. Pesticides—Environmental aspects. I. Title.
    RA1270.P4K285 1997
    3638.17′92—dc20   96-34362
                                                                  CIP

No claim to original U.S. Government works
International Standard Book Number 0-56670-190-2
Library of Congress Card Number 96-34362
Printed in the United States of America   2 3 4 5 6 7 8 9 0
Printed on acid-free paper

Preface

This book has its genesis in the efforts of a large number of individuals over a long period of time. It is the outgrowth of an effort started a number of years ago to provide clearly understandable information about the toxicology and environmental chemistry of pesticides to those who use and may be affected by these compounds. The people most responsible for these initial activities are Dr. James Witt and Dr. Frank Dost, both from Oregon State University.

About a decade ago, these pioneers were joined by other scientists at the University of California at Davis, Cornell University, and Michigan State University to form a consortium known as EXTOXNET. Together this group planned the coordinated production of Pesticide Information Profiles (PIPs) (summarizing environmental toxicology and chemistry information) and Toxicology Information Briefs (TIBs) (providing brief overviews of important environmental toxicology and chemistry concepts). This consortium was able to obtain a small amount of funding that originated in the U.S. Environmental Protection Agency (Office of Pesticides and Toxic Substances) and was administered by Extension Service-USDA through the National Pesticide Applicator Training Program. Combined with large contributions of time and effort by the participants, these funds were used to produce 100 profiles and about 15 briefs that were published in 1989.

After a brief hiatus, new funding for this effort was identified in the USDA National Pesticide Impact Assessment Program and this funding has continued to the present. Nancy Ragsdale and Dennis Kopp of USDA-NAPIAP have been instrumental in supporting this program and facilitating its efforts. With the additional funding, the original PIPs and TIBs were revised and an additional 85 PIPs and about 5 TIBs were produced. A second collection of materials, called the *Pesticide Information Notebook,* was published in 1994 with many of the additional PIPs and TIBs included.

Because these materials were produced at four different universities over a long period of time, and involved a large group of contributors, a number of inconsistencies arose in the information provided and the way this information was described. This book represents an almost 2-year effort to update and standardize the PIPs and TIBs and to include additional information so that an even more valuable resource is the result.

As suggested previously, this book could not have been written without the essential contributions of faculty, staff, and students at the four universities involved in EXTOXNET. At the University of California at Davis, the effort is headed by Arthur Craigmill, with the assistance of Sandy Ogletree and Loreen Kleinschmidt. At Cornell University, the work is supervised by Don Rutz and William G. Smith, with the current assistance of Eric Harrington and previous help from Linda A. Seyler. Previous faculty contributors from Cornell include Kean S. Goh, James W. Gillett, Christopher F. Wilkinson, Barbara Hotchkiss, and Ann T. Lemley. The Oregon State University faculty who have succeeded Jim Witt and Frank Dost are Terry Miller and

Jeff Jenkins, and they have been assisted by Pat Thomson. In addition, another faculty member, Sheldon Wagner, has been a long-term contributor to the effort.

At Michigan State University, the EXTOXNET program is headed by the editor of this book, Michael Kamrin, and he has been fortunate to have a large number of talented students contribute their efforts to this project. They include Suanne Miller, Monica Schmitt, Jon Allan, Susan Pigg, Renee Pionk, and Matthew Moon. However, the person most responsible for bringing together all the components needed to produce this book is Bradley Aaron, M.S. In addition, a number of staff at the Institute for Environmental Toxicology have been instrumental in many tasks required for the successful completion of this project. They include Mary Robinson, Carole Abel, Carol Fischer, Darla Conley, Paul Groll, and Richard Davis.

Last, but not least, a large number of scientists in academia and industry volunteered their time to review drafts of the PIPs and TIBs, and their comments have been invaluable in producing an accurate and readable final product. Many of these scientists are members of the Society for Environmental Toxicology and Chemistry, which assisted in soliciting a large number of experts as reviewers. Although the names of the reviewers are not listed here, each of them has the deep gratitude of the editor and the EXTOXNET members.

In a large project of this kind, it is impossible to identify everyone who has contributed. The editor hopes that those who have not been mentioned will still take pride in being associated with this project and in the publication of this volume.

Michael A. Kamrin, Ph.D.

The Editor

Michael A. Kamrin, Ph.D., is a Professor in the Institute for Environmental Toxicology and the Department of Resource Development at Michigan State University. Dr. Kamrin obtained his training at Cornell University, where he received a B.A. in Chemistry in 1960, and at Yale University, where he received an M.S. in Biophysical Chemistry in 1962 and a Ph.D. degree in Biophysical Chemistry in 1965. He served as an Assistant Professor at Michigan State University from 1967 to 1972; an Associate Professor from 1972 to 1979; and a Professor from 1979 to the present. He joined the Institute for Environmental Toxicology in 1982 and the Department of Resource Development in 1990. Dr. Kamrin was also appointed a Docent at the University of Turku (Finland) in 1996.

Dr. Kamrin is a member of the American Association for the Advancement of Science, the Society of Toxicology, the Society of Environmental Toxicology and Chemistry, the Society for Risk Analysis, the American Chemical Society, and the honorary society Sigma Xi. He has served as president of the Michigan Chapter of the Society of Toxicology and the Central Great Lakes Chapter of the Society of Environmental Toxicology and Chemistry. He received the Public Communications Award of the Society of Toxicology and the University of Turku (Finland) Memorial Medal. He has been the recipient of many research and outreach grants from the U.S. Department of Agriculture, the National Institute of Environmental Health Sciences, and the U.S. Environmental Protection Agency. Dr. Kamrin is the author of more than 50 scientific articles and outreach publications on environmental toxicology. He is sole or primary author of two books and co-editor of two other books. His major interests are risk assessment and risk communication, especially with respect to environmental contaminants.

Contents

Section III: Basic concepts in toxicology and environmental chemistry

Section IV: Appendices

Comprehensive Figure List

Chapter 1
There are no figures in Chapter 1.

Chapter 2

Chapter 3

Chapter 4

Chapter 5

Cumulative Table List

Section I

Introduction to pesticide profiles and basic concept papers

chapter one

Introduction and profile description

1.1 Introduction

The purpose of this book is to provide user-friendly summaries of the environmental, toxicological and chemical properties of pesticides for a variety of audiences, both technical and nontechnical. To accomplish this, data about each active ingredient have been gathered from a wide variety of sources and are provided to the reader in each profile in both qualitative and quantitative formats.

To assist the reader in understanding exactly what information is provided in each subsection of the profile and to appreciate what the quantitative equivalents are to the qualitative statements in each of these summary documents, a detailed description of the categories in each profile is provided in the following section of this chapter. This description should assist the user in interpreting any part of an individual profile that may not be clear.

The actual experimental results (e.g., specific effects at particular doses) are presented for those with backgrounds in environmental toxicology and chemistry. Lay people or those with other backgrounds are provided with general summary statements regarding potential impacts of the compound on humans, wildlife, and the environment.

This introductory chapter is followed by the section of the book containing the pesticide profiles. This second section is divided into chapters containing summaries of pesticides in the same or similar chemical groups. Thus, all of the organophosphates are in the same chapter. The rationale for this is that structural similarities among members of a given group are usually related to similarities in biological activity and environmental behavior.

Based on this concept, the beginning of each chapter in this section is devoted to a general profile describing the common properties of the pesticides in that chemical class. The purpose of this section is to assist the user in accessing at least some information about an active ingredient even if the book does not contain a profile specifically devoted to it. The reader does, however, need to know the chemical group to which the pesticide belongs in order to choose the correct chapter to examine.

At the end of each profile are sections dealing with physical constants and exposure guidelines. These will be most useful to those with the relevant background but not to the lay reader. The physical constants are quantitative descriptors of properties (e.g., water solubility) that affect the behavior of the compound in both the inanimate and animate environment. The exposure guidelines can be compared to measured or calculated environmental exposures to determine if established exposure limits have been exceeded. Also, for readers who would like additional information, references are provided to support the experimental results, physical constants, and exposure values.

The final chapter in this second part of the book contains profiles for pesticides that fall into chemical groups in which relatively few members are used as pesticides. In this chapter, the profiles are divided by how the pesticides are used: e.g., herbicide, insecticide, etc. Since this chapter does not contain pesticides grouped by class, there is no generic profile at the beginning of the chapter.

To complement the second part of the book, the third section of this volume includes short summaries of important environmental toxicology and chemistry concepts that can be used to better understand different aspects of the pesticide profiles. For example, toxicological properties such as carcinogenicity and dermal toxicity are described. As another, a description of the movement of pesticides in the environment is provided.

The appendices are designed to enhance the value of this volume even further. The first appendix lists a large number of trade names and the active ingredient(s) that correspond to each of these products. The list includes trade names that are no longer used because old products are consistently found in significant quantities on farms, in business establishments, and in homes. Given the enormous number of past and present trade names, this is not an exhaustive listing but should contain the most commonly used products.

The second appendix provides a conversion table so that the readers can visualize how the quantities described in metric units compare to units with which they may be more familiar. The last appendix is a glossary of terms commonly used throughout the book. The reader should be able to find the definition of any term that is new to them in this section of the book.

1.2   Description of profile categories

Each profile is divided into the following sections.

Trade or other names
Regulatory status
Introduction
Toxicological effects
Ecological effects
Environmental fate
Physical properties
Exposure guidelines
Basic manufacturer
References

These sections are described in more detail below.

Trade or other names

This section lists the most common trade and other names of products that contain the pesticide active ingredient.

Regulatory status

This section indicates whether the active ingredient is a General Use Pesticide (GUP) or a Restricted Use Pesticide (RUP) according to U.S. Environmental Protection Agency (EPA) guidelines, and its EPA toxicity class. According to EPA guidelines, GUPs may be sold to the public for general, unrestricted use. RUPs, on the other hand, are for retail sale to and use by only certified applicators or persons under their direct supervision. In some cases, all products containing a given active ingredient are RUPs; in others, only some formulations or products are RUPs, and the rest are for general use. In cases where one (or more) of the products or formulations containing the pesticide active ingredient is (are) classified as RUP(s), the active ingredient is generally referred to as a RUP.

The SIGNAL WORD (in capital letters) is also included in this section. The signal word is based on the EPA toxicity class, and must be included by law on labels for products containing the active ingredient. The EPA toxicity class is based on the likely acute toxicity to humans (as determined in animal tests), potential for eye or skin damage to applicators, ecological effects or potential to contaminate ground or surface water. Table 1.1 shows the toxicity classes, the corresponding qualitative descriptors, and the criteria for each class. In certain cases, different products or formulations containing the same active ingredient will be in different EPA toxicity classes based on toxicity test results. These differences may be due to the proportion of the active ingredient in the final product or the type of formulation or product.

Table 1.1 EPA Toxicity Class and Signal Word in Relation
to Acute Toxicity and Skin/Eye Irritation

EPA toxicit class	Toxicity rating	Signal word	Characteristic acute toxicity in experimental animals
I	Highly toxic	DANGER — POISON	Oral LD_{50}: 0–50 mg/kg Dermal LD_{50}: 0–200 mg/kg Inhalation LC_{50}: 0–0.2 mg/L Skin/eye irritation: severe
II	Moderately toxic	WARNING	Oral LD_{50}: >50–500 mg/kg Dermal LD_{50}: >200–2000 mg/kg Inhalation LC_{50}: >0.2–2.0 mg/L Skin/eye irritation: moderate
III	Slightly toxic	CAUTION	Oral LD_{50}: 500–5000 mg/kg Dermal LD_{50}: >2000–20,000 mg/kg Inhalation LC_{50}: >2.0–20 mg/L Skin/eye irritation: slight
IV	Practically nontoxic	None required	Oral LD_{50}: >5000 mg/kg Dermal LD_{50}: >20,000 mg/kg Inhalation LC_{50}: >20 mg/L Skin/eye irritation: none

Introduction

This section describes the pesticide's chemical family, its most common applications and uses, and the formulations in which it is available. Common formulation types are liquids, wettable powder, emulsifiable concentrates, and dusts. There may be a brief explanation of how the pesticide works to control the pest, or in what circumstances it may be applied (e.g., pre- or post-emergence in the case of herbicides). Also contained in this section may be information on the compatibility of the active ingredient with other pesticides with which it may commonly be mixed. In general, the information presented pertains to the technical grade of the compound unless otherwise noted.

Toxicological effects

Acute toxicity

Included in the acute effects section are those effects that have been noted in test animals or in cases of accidental human exposure. Acute toxic effects are those caused by exposures over brief periods of time, such as several minutes, hours, or 1 day. Exposure may occur by the oral route (ingestion), dermal route (skin contact), or inhalation.

The most commonly used measures of acute ingestion and skin toxicity are the oral and dermal doses at which 50% of the test animals die within 14 days after exposure to the test substance. These are also known as the oral median lethal dose (or oral LD_{50}) and the dermal median lethal dose (dermal LD_{50}), respectively. This dose is expressed as the amount of the pesticide active ingredient (or in some cases, formulated product) applied (in milligrams) per kilogram of test animal body weight.

The acute toxicity via inhalation is assessed as the airborne concentration of the test substance that is lethal to 50% of the test animals over a specific exposure time (in many cases, 4 hours). This is often called the median lethal concentration, and is abbreviated LC_{50}. It is expressed in terms of the amount of pesticide or formulation (in milligrams) in a given volume of air (either in liters, L, or cubic meters, m^3). It is important to remember that the LC_{50} refers to the airborne concentration of the substance to which the animal is exposed, not the dose to the animal.

Chronic toxicity

The chronic toxicity section summarizes the data from animal studies of pesticide effects, as well as data from human epidemiological studies where available. Chronic effects refer to those effects that occur due to multiple or continuous exposures for extended periods of time, such as several weeks, months, or years. In many cases, there will be a long delay (latency period) between the time of the initial exposure(s) and the onset of health problems. In addition to the general chronic effects identified in human epidemiology and animal studies, data on reproductive, teratogenic, mutagenic (and genotoxic), and carcinogenic effects are presented in separate subsections. In the chronic toxicity section, and in all of its subsections, special attention is paid to the dose levels at which effects were observed in animal studies and their relevance to human health.

Reproductive effects

This section summarizes the results of studies of the effects of pesticide exposure on the ability of test animals to successfully produce normal numbers of healthy offspring. Female and/or male animals are fed a range of doses of the pesticide and then observations are made to determine whether they are fertile and produce live offspring, often through several generations.

Teratogenic effects

This section summarizes studies of the effects of pesticide exposure on the viability and development of test animal offspring. Parent animals receive various doses of the pesticide. Their offspring are examined for the presence of structural birth defects and abnormal development.

Mutagenic effects

This section summarizes the results of tests to examine whether pesticide exposures result in changes in gene structure or function in biological systems. Tests may be performed on bacteria, isolated animal or human cell cultures, as well as in experimental animals to determine if the pesticide causes changes in the chemical structure (mutation) of genetic material (DNA), aberrations in chromosome structure, alterations in the synthesis and repair of DNA, or proper gene expression.

Carcinogenic effects

This section summarizes results of long-term animal studies, generally extending over a lifetime, performed to determine the cancer-causing potential of the pesticide. Effects such as abnormal cell growth, cellular changes related to cancer, and tumor formation are commonly monitored and reported.

Organ toxicity

This section describes the organs or organ systems that are primarily affected by chronic exposure to the pesticide.

Fate in humans and animals

The section on the fate of a pesticide in humans or animals includes information regarding the absorption of the pesticide, its distribution throughout the body, and how easily it may be degraded and eliminated from the body. Where available, data are presented on the route(s) of elimination, the biological half-life of the pesticide (how long it will take for one half the amount present in the system to be degraded and/or eliminated), and the principal breakdown products of the pesticide.

Ecological effects

Effects on birds

This section summarizes the toxicity of acute oral (single administration) or subchronic dietary (approximately 8 days) pesticide exposure on birds. Toxicity is measured by the median lethal oral dose, the oral LD_{50}, or the median lethal dietary concentration, or dietary LC_{50}. These refer, respectively, to the dose or concentration at which 50% of the test birds die. Here again, it is important to remember that the LD_{50} is expressed in terms of the amount of pesticide or formulation per kilogram of the test bird, whereas the LC_{50} represents a concentration, in this case the concentration of the pesticide in the test bird diet. To avoid confusion, the oral LD_{50} is expressed in units of milligrams per kilogram, and the dietary LC_{50} is expressed as parts per million (ppm). The toxicity of pesticide active ingredient is qualitatively classified according to the toxicity categories shown in Table 1.2.

Effects on aquatic organisms

This section summarizes results of acute toxicity testing on fish and other aquatic species. Results are reported as the median lethal concentration (LC_{50}), or the water concentration that kills 50% of the exposed individuals (either fish or other aquatic species) within a specific time period (typically 48 or 96 hours). The toxicity categories

Table 1.2 Categories of Ecotoxicity

Toxicity category	Birds acute oral LD$_{50}$ (mg/kg)	Bird dietary LC$_{50}$ (ppm)	Fish water LC$_{50}$ (mg/L)
Very highly toxic	<10	<50	<0.1
Highly toxic	10–50	50–500	0.1–1
Moderately toxic	>50–500	>500–1000	>1–10
Slightly toxic	>500–2000	>1000–5000	>10–100
Practically nontoxic	>2000	>5000	>100

for aquatic organisms are also included in Table 1.2. The toxicity to fish or other aquatic organisms may be influenced by factors such as pH, temperature, total dissolved oxygen, and total water hardness. Where available, the bioconcentration factor for fish or other aquatic organisms is also included.

Effects on other organisms (non-target species)

Some pesticides, especially if applied incorrectly or at excessive rates, affect organisms other than the intended pest; these are referred to as non-target species. In some cases, these species may be economically important (e.g., bees) or ecologically important (e.g., earthworms or non-target plants). This section summarizes the results of studies on the effects of the pesticide on these non-target organisms, especially bees.

Environmental fate

Breakdown in soil and groundwater

This section summarizes the results of studies on pesticide persistence, soil binding, mobility, and microbial degradation, as well as the potential for pesticides or harmful degradates to leach into groundwater. Data describing the frequency of detection and concentration of the pesticide in soil and groundwater samples are included if available. The soil persistence of the pesticide is qualitatively classified on the basis of observed field half-life as shown in Table 1.3.

Pesticide behavior in a specific soil/groundwater environment will depend on many factors, including the soil type, temperature, vapor pressure, the amounts of sunlight and rain, the water solubility of the pesticide, the type of soil, and the rate of application. Thus, information presented in each profile should serve as a means of comparison between pesticides and should not be intrepreted in an absolute fashion.

Table 1.3 Pesticide Persistence in Soils

Persistence categories	Time required for 50% of the applied pesticide to degrade at or near the soil surface (half-life)	Examples of pesticide classes (there are several exceptions in each class)
Low persistence	Less than 30 days	Carbamate and organophosphate herbicides
Moderate persistence	30–100 days	Triazine, urea, amide, and phenoxy herbicides
High persistence	Greater than 100 days	Chlorinated hydrocarbon insecticides

Breakdown in water

Information on pesticide behavior in the surface water environment is summarized here. This section describes how easily the pesticide dissolves and by which pathways it is broken down or eliminated from surface waters. Such factors as water pH, sunlight, and available oxygen may affect the persistence of the pesticide in water. If available, this section may include results of studies on the concentrations of the pesticide found in surface water.

Breakdown in vegetation

This section summarizes data on how various plants take up, distribute, and process or eliminate the pesticide. For example, the pesticide may damage plant leaves or stems, or it may be degraded as it travels within the plant.

Physical properties

This section describes the chemical and physical characteristics of the pesticide active ingredient. This includes the physical appearance of the pure or technical grade chemical, Chemical Abstracts Service number, molecular weight, solubilities in water and various other solvents (see Table 1.4), melting point, and vapor pressure. Also included where available are the octanol–water partition coefficient (K_{ow}) and the soil adsorption coefficient.

Table 1.4 Categories of Solubility

Rating	Solubility at 20–30°C (room temperature)
Insoluble	Less than 1 mg/L
Slightly soluble	1–100 mg/L
Soluble	100–10,000 mg/L
Very soluble	Greater than 10,000 mg/L

The soil adsorption coefficient depends on several soil properties, including organic matter content, organic matter type, particle size distribution, clay mineral composition, and pH. Where available, the reported soil adsorption coefficient is the K_{oc}, which takes into account the soil organic matter content. Where this is not available, the unadjusted observed soil adsorption coefficient, K_d, is reported. The values for these variables are experimentally determined, and reported values may in many cases consist of the most representative value.

Exposure guidelines

The exposure guidelines are the maximum acceptable doses or levels of exposure. Exposure guidelines have been developed by different agencies and organizations such as the EPA, the U.S. Occupational Safety and Health Administration (OSHA), and the American Council of Government Industrial Hygienists (ACGIH).

EPA guidelines and standards

Guidelines developed by the EPA include reference dose (RfD) and the Health Advisory level (HA). The Reference Dose (RfD) is defined as the dose of the administered chemical that during an entire lifetime, appears to be without appreciable risk. RfDs are generally based on data from animal studies that are adjusted for scientific uncertainties to determine levels which are unlikely to result in harmful effects in human populations. A margin of safety for both adults and children is built into these values.

Health Advisory levels (HAs) refer to drinking water contaminant levels that would not be anticipated to cause adverse health effects over a given exposure period. A margin of safety to account for scientific uncertainty is built into these values, again using animal data as a foundation. Drinking water Health Advisory levels are developed for exposures of 1-day, 10-days, long-term (about 7 years), and lifetime duration.

Health Advisories do not carry the force of law. In cases where a legally enforceable drinking water standard has been established, it is presented in the profile instead of the HA.

The maximum contaminant level (MCL) is the standard established under the Safe Drinking Water Act for the maximum permissible level of a contaminant in drinking water that is delivered to the users of a public water system. In cases where a contaminant is a known or probable human carcinogen, a zero-tolerance approach is used, and the MCL is determined by the limitations of quantitative chemical analysis.

OSHA standards

Under the Occupational Safety and Health Act (OSHAct), OSHA has established acceptable levels of exposure to airborne contaminants in the workplace environment. These are known as Permissible Exposure Limits (PELs), which have the force of law in occupational situations where the OSHAct is applicable. PELs are intended for application in the occupational environment, and assume that exposure occurs 8 hours a day for 40 hours a week. OSHA PELs are developed using evidence from human health studies as well as animal studies, allowing for scientific uncertainty.

Where the PEL is not available, the ACGIH Threshold Limit Value (TLV) is presented.

ACGIH guidelines

The ACGIH has established maximum workplace concentrations of air contaminants to which nearly all workers may be repeatedly exposed without adverse effect. These are known as the ACGIH Threshold Limit Values (TLVs). Like OSHA PELs, TLVs assume exposure periods of 8 hours within a 40-hour work week. ACGIH also publishes Short Term Exposure Limits (STELs), to be exceeded not more than four times in a day, with at least 1 hour between successive excursions.

Other guidelines

A guideline used by both the U.S. Food and Drug Administration and the World Health Organization is the Acceptable Daily Intake (ADI). The ADI is obtained in a similar fashion to the RfD using animal study data, and incorporating scientific uncertainty. There may be some differences between the ADI and RfD because of the data used and/or the amount of uncertainty that is thought to exist by different agencies.

Basic manufacturer

The name, address, and phone number of the primary manufacturer of the pesticide are listed here. The emergency phone number of the primary manufacturer is included if available. The company listed here may be the firm registered with the U.S. EPA to process and market the product in the United States and not the actual producer.

1.3 Disclaimer

The information provided in this book does not replace or supersede the information on the pesticide product labeling or other regulatory requirements.

Section II

Pesticide profiles

chapter two

Pyrethroids and other botanicals

2.1   Class overview and general description

Background

Pyrethroids and other botanical pesticides are grouped together not because they have similar toxicological properties, like the organophosphates or carbamates, but because they all have biological origins. Pesticides may be extracted from naturally occurring materials or they can be synthesized in commercial laboratories. The most common natural materials come from plants. Some plants have developed, over long periods of time, substances that deter plant-eating insects. The plants use these compounds as protection from insects. Researchers have isolated a number of compounds from plants that are effective insecticides or fungicides. Development of these botanical pesticides is one of the fastest growing areas in the pesticide industry.

The World Health Organization, in 1967, stated that "all of the most poisonous materials so far known are, in fact, of natural origin" (1). Whether this is still true or not is unimportant; however, the comment points out that just because something is found in nature does not mean that it is harmless (1). Pesticides extracted from plants or synthesized to mimic compounds found in plants are powerful and effective control agents.

Plant-derived pesticides fall into several different broad classifications. By far the largest group of such pesticides consists of pyrethrum and its related synthetic compounds, pyrethroids. Pyrethrum comes from chrysanthemums. There are currently over 20 pyrethroids and they constitute the single largest group of natural insect chemical control agents in the world (2).

Another class of botanicals is the rotenoids. They are found in several plants in the bean family. Rotenone is one of six compounds in this group. It is used as a fish poison. A third important category of botanicals, the nicotinoids, comes from tobacco and several other plants. Nicotine is the most commonly used compound in the group. The fourth group of botanicals includes the compounds strychnine and scilliroside. These and several other compounds in the group are used to control rodents.

The latter three groups of botanicals will not be discussed in detail here, though they will be briefly described at the end of the overview in this chapter. Only the pyrethroids will be discussed at length in this chapter.

The generalized structure of the pyrethroids is shown in Figure 2.1, and the various pyrethroids are listed in Table 2.1.

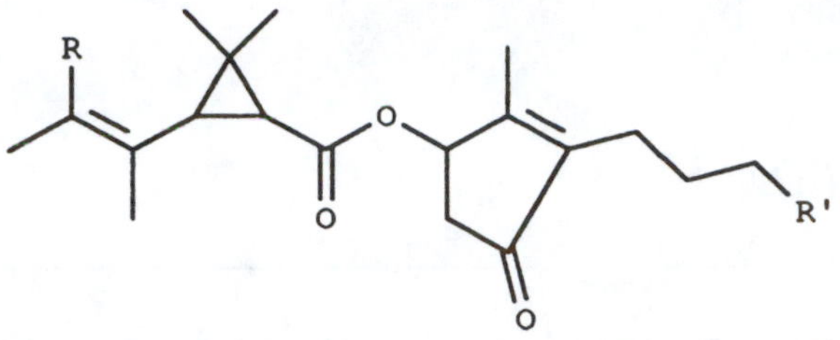

Figure 2.1 Generic pyrethroid structure.

Table 2.1 Pyrethroids

Allethrin*	Fenfluthrin
Barthrin	Fenvalerate
Bioallethrin	Flucythrinate*
Bioresmethrin	Fluorocyphenothrin
Cismethrin	Fluvalinate*
Cyphenothrin	Kadethrin
Cyflurthin	Permethrin*
Cyhalothrin	Pyrethrin I
Cypermethrin*	Pyrethrin II
Deltamethrin	Resmethrin*
Esfenvalerate*	Tetramethrin
Fenproponate	

Note: * indicates that a profile for this compound
is included in this chapter.

Uses of pyrethroids and other botanicals

Currently, there are no estimates of the national or regional use patterns of botanically derived pesticides in the United States.

Pyrethrum and pyrethroid mechanisms of toxicity

This group of pesticides interferes with the balance of sodium ions in the nerve junctions of target and non-target organisms, rendering them inactive (2). The pesticide provides a "knockdown" dose to insects but is generally not strong enough to kill them. Compounds in this group are toxic to insects but not as toxic to mammals because of their ability to break down the pyrethroids into metabolite compounds that are easily excreted.

Pyrethrins and pyrethroids are commonly combined with other insecticides to enhance their efficacy against insects. The synthetic pyrethroids are more potent than the natural compounds.

Acute toxicity

Pyrethroids are slightly to moderately toxic to animals. The LD_{50} of allethrin is 1100 mg/kg in male rats and 370 mg/kg in mice (2). Cypermethrin is moderately toxic, with an LD_{50} of 187 to 326 mg/kg in male rats (2).

Exposure to high doses of pyrethroids can be fatal. There was one instance when a man died after eating a meal cooked in 10% cypermethrin concentrate mistakenly used for cooking oil. He had symptoms of nausea that progressed to stomach pains, to diarrhea, to convulsions, to unconsciousness, and then to coma. Death then occurred due to respiratory failure (2).

For humans, dermal contact with large amounts of these compounds, such as deltamethrin, may result in numbness, burning and itching of the skin, and intoxication (2).

Chronic toxicity

Long-term studies of pyrethroids showed that several of them can cause liver effects (2). In a chronic feeding study with rats, doses of 125 mg/kg/day of resmethrin produced some pathological liver changes in addition to increased liver weights (3). Rats fed large doses of pyrethrins over two years showed liver damage (2).

Reproductive effects

In laboratory animal studies on about half of the pyrethroids, there were no significant reproductive effects. However, in experiments with two pyrethroids, permethrin and resmethrin, there were some reproductive effects. High oral doses of 250 mg/kg/day of permethrin during days 6 to 15 of pregnancy reduced the fertility of female rats (2). A three-generation rat study using resmethrin showed slight increases in premature stillbirths and a decrease in pup weight at the 25-mg/kg/day dose (2).

Therefore, botanical compounds at low doses are unlikely to cause reproductive effects in humans.

Teratogenic effects

Available evidence suggests that pyrethroids will be unlikely to cause teratogenic effects. No developmental effects were seen in the offspring of rats given doses of allethrin as high as 195 mg/kg/day (2). Also, no birth defects were observed in the offspring of rabbits given doses of resmethrin as high as 100 mg/kg/day (4).

Mutagenic effects

Laboratory animal experiments with various pyrethroids (compounds such as esfenvalerate, permethrin, and resmethrin) showed no mutagenic effects (5,6).

Carcinogenic effects

The majority of pyrethroids have been shown through animal experiments to be unlikely to produce carcinogenic effects (2,3).

However, there is one exception: the U.S. Environmental Protection Agency has classified cypermethrin as a possible human carcinogen since there was evidence of some benign lung tumors in female mice exposed to very high doses of this pyrethnoid (8).

Organ toxicity

As stated earlier in the chronic toxicity section, at high doses, pyrethroids may cause adverse effects on the central nervous system (CNS) and also adverse changes in the liver (9).

Fate in humans and animals

Pyrethroid compounds are metabolized and excreted rapidly in animals. For example, rats eliminated over 99% of cypermethrin within a few hours (2). Also, when an oral dose of 10 mg/kg resmethrin was given to laying hens, 90% of the dose was eliminated via urine and feces within 24 hours (10).

Humans also eliminate pyrethroid compounds such as cypermethrin relatively quickly. The "urinary excretion of cypermethrin metabolites was complete 48 hours after the last of five daily tracer doses of 1.5 mg" (2).

Ecological effects

Effects on birds

The majority of the pyrethroid compounds are practically nontoxic to birds. The LD_{50} of allethrin in mallard ducks is 2000 mg/kg and the LD_{50} of cypermethrin in mallard ducks is 4640 mg/kg (7,8).

Effects on aquatic organisms

Pyrethroid compounds are very highly toxic to fish and aquatic invertebrates. The LC_{50} of bioallethrin in coho salmon is 0.0026 mg/L, and the 96-hour LC_{50} of cypermethrin in rainbow trout is 0.00082 mg/L (7). Also, the acute LC_{50} for *Daphnia magna*, a small freshwater crustacean, is 0.0002 mg/L (8). In addition, pyrethroid compounds tend to bioaccumulate in fish. A bioconcentration factor of 1200× was determined in a flow-through study investigating the accumulation of cypermethrin in rainbow trout (8). In another study, the bioaccumulation factor of esfenvalerate in rainbow trout was found to be about 400× (11).

Effects on other organisms (non-target species)

The majority of pyrethroids are highly toxic to bees (8,12).

Environmental fate

Breakdown in soil and groundwater

Pyrethroid compounds have a strong tendency to adsorb to soil particles and are moderately persistent. For example, permethrin has a half-life of 3 to 6 weeks (13) and esfenvalerate has a half-life ranging from 15 days to 3 months (11). However, one of the pyrethroids, cypermethrin, has low persistence in sunlight inasmuch as it photodegrades rapidly with a half-life of 8 to 16 days (8). Because of the pyrethroid compounds' tendency to bind to soil, they are unlikely to cause groundwater contamination.

Breakdown in water

Pyrethroid compounds are relatively insoluble in water. They readily adsorb to sediment and soils and, therefore, concentrations of pyrethroids in silty water decrease rapidly.

Breakdown in vegetation

For many of the pyrethroid compounds, there is no information available regarding their fate in vegetation. In one study, a 4.5 ml/100 L solution of Cymbush 250

EC (cypermethrin) was applied to strawberry plants. Results revealed that 40% of the applied cypermethrin remained after 1 day, 12% remained after three days, and 0.5% remained after 7 days (14). In another study, permethrin was neither phytotoxic nor poisonous to most plants when used directly. However, some injury occurred on certain ornamental plants (15).

General properties of other botanicals

Rotenoids

The rotenoids are extracted mainly from two species, Derris and Lonchocarpus, members of the bean family. These compounds are found in over 65 species of legumes. Rotenoids are found in leaves, stems, roots, and seeds of these plants. Parts of the plant may contain up to 40% of the compound (2).

Rotenoids are highly toxic to fish and some insects; however, they are only moderately toxic to mammals (16). Refer to the rotenone profile for further information.

Nicotinoids

Three compounds make up the bulk of this category: nicotine, nornicotine, and anabasine. Nicotine is by far the most widely used of the group and is also the most potent. The active compounds in this group have been extracted from five different families of plants (2). Nicotine is extracted with steam or water from the different plants, though most commonly from tobacco. It is a moderately to highly toxic compound with an acute oral LD_{50} of 50 to 60 mg/kg (17). The compound affects the nerve's ability to function properly, much like the organophosphate and carbamate insecticides, resulting in similar symptoms. Nicotine can cause twitching and difficulties with breathing, convulsions, and death. The use of nicotinoids has been declining. They are being replaced with equally effective synthetic insecticides. Additionally, they are nonpersistent, and are ineffective in cold weather (16).

Botanical rodenticides

Three compounds are found in this group: strychnine, scilliroside, and ricin. Strychnine comes from the plant *Strychnos nux-vomica*, scilliroside comes from the red squill bulb, and ricin comes from castor beans.

2.2   *Individual profiles*

2.2.1   *Allethrin*

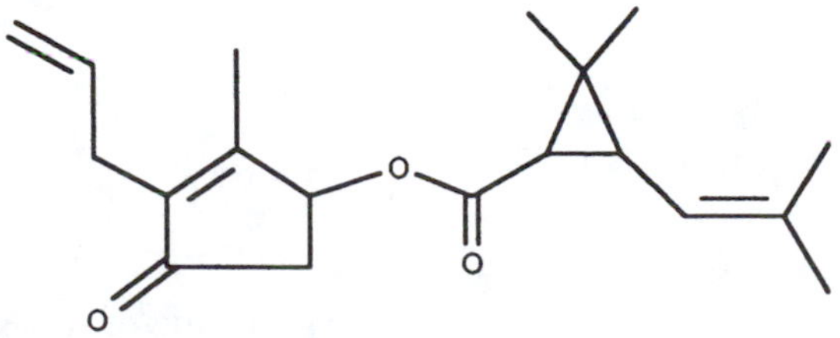

Figure 2.2 Allethrin.

Trade or other names

Trade names for allethrin include Alleviate, Pynamin, d-allethrin, d-cisallethrin, Bioallethrin, Esbiothrin, Pyresin, Pyrexcel, Pyrocide, and *trans*-allethrin.

Regulatory status

Pesticides containing allethrin are toxicity class III — slightly toxic, and bear the Signal Word CAUTION on the product label. Containers of technical grade d-*trans*-allethrin bear the Signal Word WARNING. Allethrin is a General Use Pesticide (GUP).

Introduction

Allethrin is a nonsystemic insecticide that is used almost exclusively in homes and gardens for control of flies and mosquitoes, and in combination with other pesticides to control flying or crawling insects. Another structural form, the d-*trans*-isomer of allethrin, is more toxic to insects and is used to control crawling insects in homes and restaurants. It is often used to control parasites living within animal systems. It is available as mosquito coils, mats, oil formulations, and as an aerosol spray.

Allethrin is a pyrethroid, a synthetic compound that duplicates the activity of the pyrethrin plant. It has stomach and respiratory action and paralyzes insects before killing them. Unless stated otherwise, information in this profile refers to unpurified allethrin.

Toxicological effects

Acute toxicity

Allethrin is slightly toxic by dermal absorption and ingestion. Short-term dermal exposure to allethrin may cause itching, burning, tingling, numbness, or a feeling of warmth, but not dermatitis (2). Exposure to large doses by any route may lead to nausea, vomiting, diarrhea, hyperexcitability, incoordination, tremors, convulsive twitching, convulsions, bloody tears, incontinence, muscular paralysis, prostration, and coma.

Allethrin is a central nervous system stimulant. Heavy respiratory exposure caused incoordination and loss of bladder control in mice and rats (2).

The toxicity of allethrin varies with the amounts of different isomers present. The oral LD_{50} for allethrin is 1100 mg/kg in male rats, 685 mg/kg in female rats, 480 mg/kg in mice, and 4290 mg/kg in rabbits (7,12). For d-allethrin, the oral LD_{50} is 1320 mg/kg in rats. The dermal LD_{50} is greater than 2500 mg/kg in rats (12,19).

Chronic toxicity

Reproductive effects
No data are currently available.

Teratogenic effects
No developmental defects were seen in the offspring of rats given doses as high as 195 mg/kg/day (7).

Mutagenic effects

Allethrin has been found to be mutagenic under certain conditions in strains of the bacterium *Salmonella typhinurium* (18). However, tests of d-allethrin (bioallethrin) for DNA damage and mutation were negative (19). Thus, allethrin appears to have little or no mutagenic activity.

Carcinogenic effects

Rats fed very high doses of d-allethrin for 2 years did not develop cancer (19).

Organ toxicity

Rats fed 75 mg/kg/day of d-allethrin exhibited decreased body weight gain, increased liver weights, and, in females only, increased levels of serum liver enzymes (19).

A 6-month study with dogs fed d-allethrin (bioallethrin) showed effects on the liver at 5 mg/kg/day (19). A dose of 50 mg/kg/day allethrin for 2 years produced no detectable effect in dogs (19).

Fate in humans and animals

Following oral administration, allethrin is readily absorbed and metabolized in mammalian systems to less toxic compounds that may be more easily eliminated by the body (12).

Ecological effects

Effects on birds

Allethrin is practically nontoxic to birds. The oral LD_{50} for allethrin is 2030 mg/kg in bobwhite quail and greater than 2000 mg/kg in mallards (12,19). The dietary LD_{50} is 5620 ppm for d-allethrin in mallards and bobwhite quail (19).

Effects on aquatic organisms

The pyrethroid insecticides, including allethrin, are toxic to fish. Fish sensitivity to the pyrethroids may be explained by their relatively slow metabolism and elimination of these compounds. The half-lives for elimination of several pyrethroids by trout are all greater than 48 hours (20,21). Generally, the lethality of pyrethroids to fish increases with increasing ability to dissolve in fat (22). The LC_{50} (96-hour) for channel catfish is 30 mg/L.

The toxicity of allethrin compounds ranges from an LC_{50} of 0.0026 mg/L for d-allethrin in coho salmon to an LC_{50} of 0.08 mg/L for s-bioallethrin in fathead minnows (19).

Effects on other organisms (non-target species)

Allethrin is slightly toxic to bees (19). Its LD_{50} is 0.003 to 0.009 mg per bee (19).

Environmental fate

Breakdown in soil and groundwater

No data are currently available.

Breakdown in water

In pond waters and in laboratory degradation studies, pyrethroid concentrations decrease rapidly due to sorption to sediment, suspended particles and plants. Microbial and photodegradation also occur (19).

Breakdown in Vegetation

No data are currently available.

Physical properties

Allethrin is a yellow to amber colored viscous liquid with a mild or slightly aromatic odor (12). It may decompose when exposed to heat or light.

> Chemical name: 2-methyl-4-oxo-3-(2-propenyl)-2-cyclopenten-1-yl 2,2-dimethyl-3-(2-methyl-l-propenyl)cyclopropanecarboxylate (12)
> CAS #: 584-79-2 (allethrin); 42534-61-2 (d-allethrin)
> Molecular weight: 302.4 (12)
> Solubility in water: Allethrin and d-allethrin are insoluble in water (12)
> Solubility in other solvents: alcohol v.s.; hexane v.s.; xylene v.s., petroleum ether v.s. (12).
> Melting point: 4°C (12)
> Vapor pressure: 16 mPa @ 30°C (12)
> Partition coefficient (octanol/water): 91,300 (12)
> Adsorption coefficient: Not available

Exposure guidelines

> ADI: Not available
> HA: Not available
> RfD: Not available
> PEL: Not available

Basic manufacturer

> Sumitomo Chemical Co., Ltd.
> 5–33, Kitahama 4-chome
> Chuo-ku Osaka 541 Japan
> Telephone:　81-6-220-3693
> FAX:　　　　81-6-220-3492
> Telex:　　　522–7541 SUMIKA J

2.2.2　*Cypermethrin*

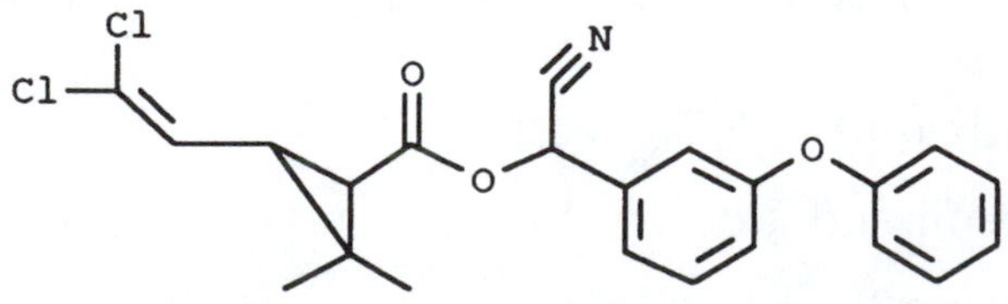

Figure 2.3 Cypermethrin.

Trade or other names

Trade names include Ammo, Arrivo, Barricade, Basathrin, CCN52, Cymbush, Cymperator, Cynoff, Cypercopal, Cyperguard 25EC, Cyperhard Tech, Cyperkill, Cypermar, Demon, Flectron, Fligene CI, Folcord, Kafil Super, NRDC 149, Polytrin, PP 383, Ripcord, Siperin, Stockade, and Super.

Regulatory status

Many products containing cypermethrin are classified as Restricted Use Pesticides (RUPs) by the EPA because of cypermethrin's toxicity to fish. RUPs may be purchased and used only by certified applicators. Cypermethrin is classified toxicity class II — moderately toxic. Some formulations are toxicity class III — slightly toxic. Pesticides containing cypermethrin bear the Signal Word WARNING or CAUTION on the product label, depending on the particular formulation.

Introduction

Cypermethrin is a synthetic pyrethroid insecticide used to control many pests, including moth pests of cotton, fruit, and vegetable crops. It is also used for crack, crevice, and spot treatment to control insect pests in stores, warehouses, industrial buildings, houses, apartment buildings, greenhouses, laboratories, and on ships, railcars, buses, trucks, and aircraft. It may also be used in non-food areas in schools, nursing homes, hospitals, restaurants, and hotels, in food processing plants, and as a barrier treatment insect repellent for horses. Technical cypermethrin is a mixture of eight different isomers, each of which may have its own chemical and biological properties. Cypermethrin is light stable. It is available as an emulsifiable concentrate or wettable powder.

Toxicological effects

Acute toxicity

Cypermethrin is a moderately toxic material by dermal absorption or ingestion (2,8). Symptoms of high dermal exposure include numbness, tingling, itching, burning sensation, loss of bladder control, incoordination, seizures, and possible death (2,8). Pyrethroids like cypermethrin may adversely affect the central nervous system (2,8). Symptoms of high-dose ingestion include nausea, prolonged vomiting, stomach pains, and diarrhea that progresses to convulsions, unconsciousness, and coma. Cypermethrin is a slight skin or eye irritant, and may cause allergic skin reactions (8).

The oral LD_{50} for cypermethrin in rats is 250 mg/kg (in corn oil) or 4123 mg/kg (in water) (2,8). The EPA reports an oral LD_{50} of 187 to 326 mg/kg in male rats and 150 to 500 mg/kg in female rats (8). The oral LD_{50} varies from 367 to 2000 mg/kg in female rats, and from 82 to 779 mg/kg in mice, depending on the ratio of *cis/trans-*isomers present (2). This wide variation in toxicity may reflect different mixtures of isomers in the materials tested. The dermal LD_{50} in rats is 1600 mg/kg, and in rabbits is greater than 2000 mg/kg (2,8).

Chronic toxicity

Reproductive effects

No adverse effects on reproduction were observed in a three-generation study with rats given doses of 37.5 mg/kg/day, the highest dose tested (8).

Teratogenic effects

Cypermethrin is not teratogenic (2). No birth defects were observed in the offspring of rats given doses as high as 70 mg/kg/day nor in the offspring of rabbits given doses as high as 30 mg/kg/day (8).

Mutagenic effects

Cypermethrin is not mutagenic, but tests with very high doses on mice caused a temporary increase in the number of bone marrow cells with micronuclei. Other tests for mutagenic effects in human, bacterial, and hamster cell cultures and in live mice have been negative (2).

Carcinogenic effects

EPA has classified cypermethrin as a possible human carcinogen because available information is inconclusive. It caused benign lung tumors in female mice at the highest dose tested (229 mg/kg/day); however, no tumors occurred in rats given high doses of up to 75 mg/kg/day (8).

Organ toxicity

Pyrethroids like cypermethrin may cause adverse effects on the central nervous system. Rats fed high doses (37.5 mg/kg) of the *cis*-isomer of cypermethrin for 5 weeks exhibited severe motor incoordination, while 20 to 30% of rats fed 85 mg/kg died 4 to 17 days after treatment began (2). Long-term feeding studies have shown increased liver and kidney weights and adverse changes in liver tissues in test animals (8).

Pathological changes in the cortex of the thymus, liver, adrenal glands, lungs, and skin were observed in rabbits repeatedly fed high doses of cypermethrin (23).

Fate in humans and animals

In humans, urinary excretion of cypermethrin metabolites was complete 48 hours after the last of five doses of 1.5 mg/kg/day (2). Studies in rats have shown that cypermethrin is rapidly metabolized by hydroxylation and cleavage, with over 99% being eliminated within hours. The remaining 1% becomes stored in body fat. This portion is eliminated slowly, with a half-life of 18 days for the *cis*-isomer and 3.4 days for the *trans*-isomer (2).

Ecological effects

Effects on birds

Cypermethrin is practically nontoxic to birds. Its acute oral LD_{50} in mallard ducks is greater than 4640 mg/kg (8). The dietary LC_{50} in mallards and bobwhite quail is greater than 20,000 ppm (8). No adverse reproductive effects occurred in mallards or bobwhite quail given 50 ppm, the highest dose tested (8).

Effects on aquatic organisms

Cypermethrin is very highly toxic to fish and aquatic invertebrates. The LC_{50} (96-hour) for cypermethrin in rainbow trout is 0.0082 mg/L, and in bluegill sunfish is 0.0018 mg/L (20). Its acute LC_{50} in *Daphnia magna*, a small freshwater crustacean, is 0.0002 mg/L (20).

Cypermethrin is metabolized and eliminated significantly more slowly by fish than by mammals or birds, which may explain this compound's higher toxicity in fish compared to other organisms (20).

The half-lives for elimination of several pyrethroids by trout are all greater than 48 hours, while elimination half-lives in birds and mammals range from 6 to 12 hours (20,23).

The bioconcentration factor for cypermethrin in rainbow trout was 1200 times the ambient water concentration, indicating that there is a moderate potential to accumulate in aquatic organisms (8). Elimination of half of the accumulated amount of the compound took nearly 8 days. After 14 days, 70 to 80% of the material had been eliminated from the organisms (8).

Effects on other organisms (non-target species)

Cypermethrin is highly toxic to bees (8,24).

Environmental fate

Breakdown in soil and groundwater

Cypermethrin has a moderate persistence in soils. Under laboratory conditions, cypermethrin degrades more rapidly on sandy clay and sandy loam soils than on clay soils, and more rapidly in soils low in organic material (8). In aerobic conditions, its soil half-life is 4 days to 8 weeks (8,12,25). When applied to a sandy soil under laboratory conditions, its half-life is 2.5 weeks (26). Cypermethrin is more persistent under anaerobic conditions (8).

It photodegrades rapidly with a half-life of 8 to 16 days. Cypermethrin is also subject to microbial degradation under aerobic conditions (8).

Cypermethrin is not soluble in water and has a strong tendency to adsorb to soil particles. It is therefore unlikely to cause groundwater contamination (12).

Breakdown in water

In neutral or acid aqueous solution, cypermethrin hydrolyzes slowly, with hydrolysis being more rapid at pH 9 (basic solution). Under normal environmental temperatures and pH, cypermethrin is stable to hydrolysis with a half-life of greater than 50 days and to photodegradation with a half-life of greater than 100 days (8).

In pond waters and in laboratory degradation studies, pyrethroid concentrations decrease rapidly due to sorption to sediment, suspended particles and plants. Microbial degradation and photodegradation also occur (22,27).

Breakdown in vegetation

When applied to strawberry plants, 40% of the applied cypermethrin remained after 1 day, 12% remained after 3 days, and 0.5% remained after 7 days, with a light rain occurring on day 3 (14).

When cypermethrin was applied to wheat, residues on the wheat were 4 ppm immediately after spraying and declined to 0.2 ppm 27 days later. No cypermethrin was detected in the grain. Similar residue loss patterns have been observed on treated lettuce and celery crops (28).

Physical properties

Pure isomers of cypermethrin form colorless crystals. When mixed isomers are present, cypermethrin is a viscous semi-solid or a viscous, yellow liquid (2,12).

Chemical name: (R,S)-alpha-cyano-3-phenoxybenzyl(1RS)-*cis,trans*-3-(2,2-dichlorovinyl)-2,2-dimethylcyclopropane-carboxylate (12)
CAS #: 52315-07-8
Molecular weight: 416.3 (12)
Solubility in water: 0.01 mg/L @ 20°C; insoluble in water (12)
Solubility in other solvents: methanol v.s.; acetone v.s.; xylene v.s. (12)
Melting point: 60–80°C (pure isomers) (12,2)
Vapor pressure: 5.1×10^{-7} nPa @ 70°C (12)
Partition coefficient (octanol/water): 4,000,000 (25)
Adsorption coefficient: 100,000 (25)

Exposure guidelines

ADI: 0.05 mg/kg/day (29)
MCL: Not available
RfD: 0.01 mg/kg/day (30)
PEL: Not available

Basic manufacturer

Zeneca Ag Products
1800 Concord Pike
Wilmington, DE 19897
Telephone: 800-759-4500
Emergency: 800-759-2500

2.2.3   *Esfenvalerate*

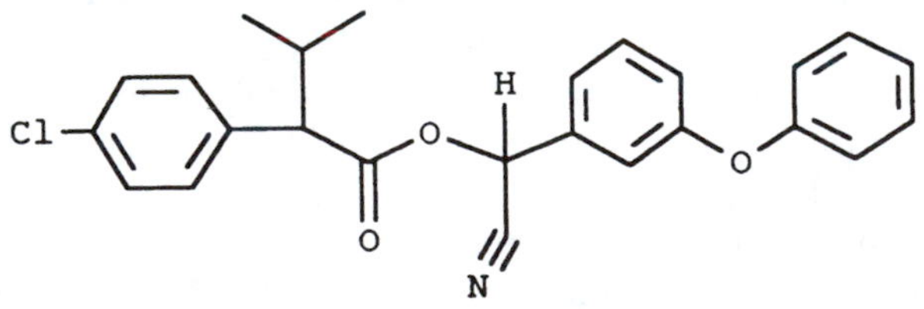

Figure 2.4 Esfenvalerate.

Trade or other names

Trade names for the older fenvalerate compounds include Ectrin, Pydrin, Sanmarton, Sumifly, Sumiflower, and Sumitick. Trade names for the new product, esfenvalerate, include Asana XL, Halmark, and Sumi-alfa. The compound may also be listed as S-fenvalerate.

Regulatory status

Most products containing esfenvalerate are General Use Pesticides (GUPs). The emulsified concentrate formulation is a Restricted Use Pesticide (RUP) because of possible adverse effects in aquatic organisms. Esfenvalerate is a moderately toxic pesticide in EPA toxicity class II; products containing it must contain the Signal Word WARNING on the label.

Introduction

Esfenvalerate is a synthetic pyrethroid insecticide used on a wide range of pests such as moths, flies, beetles, and other insects. It is used on vegetable crops, tree fruit, and nut crops. It may be mixed with a wide variety of other types of pesticides such as carbamate compounds or organophosphates. Esfenvalerate has replaced the naturally occurring compound fenvalerate (to which it is almost identical) for use in the U.S.

Much of the data for fenvalerate is applicable to the pesticide esfenvalerate because the two compounds contain the same components. The only differences in the two products are the relative proportions of the four separate constituents (isomers). Esfenvalerate has become the preferred compound because it requires lower applications rates than fenvalerate, is less chronically toxic, and is a more powerful insecticide. The compound contains a much higher percentage of the one insecticidally active isomer (84% for esfenvalerate and 22% for fenvalerate).

Toxicological effects

Acute toxicity

Esfenvalerate is a moderately toxic compound via the oral route. The reported oral LD_{50} of esfenvalerate is 458 mg/kg in rats. It is slightly toxic via the dermal route, with a reported dermal LD_{50} of 2500 mg/kg in rabbits. It is practically nontoxic via inhalation, with a reported inhalation LC_{50} of greater than 2.93 mg/L in rats (2,12).

Because esfenvalerate is a relatively new compound it has little usage history. The bulk of the evidence related to acute poisonings in humans due to esfenvalerate comes from incidents in India. Nearly 600 individual cases of poisoning were reported between 1982 and 1988. These cases were due to improper handling of the pesticide. Acute toxic effects were observed in workers and among the general public. Symptoms of acute poisoning included dizziness, burning, and itching (which was worsened by sweating and washing). Severe cases of direct contact caused blurred vision, tightness in the chest, and convulsions (2). The changes appear to be reversible.

In rats, high acute exposure to esfenvalerate produced muscle incoordination, tremors, convulsions, nerve damage, and weight loss. The compound may produce nausea, vomiting, headache, and temporary nervous system effects such as weakness, tremors, and incoordination at acute exposure levels in humans. Esfenvalerate is a strong eye irritant, producing tearing or blurring of vision.

Chronic toxicity

Rats fed fenvalerate at concentrations of approximately 12.5 mg/kg/day for 2 years had no compound-related changes in the blood or urine (12). In other studies significant reduction in body weight was the main adverse effect seen in both rats and mice of both sexes.

Reproductive effects

In a three-generation rat study, low doses (up to 12.5 mg/kg/day) of fenvalerate produced no toxicity in the fetus. Some maternal toxicity was noted in the second generation at the higher dose. When pregnant mice and rabbits were fed low dietary levels of fenvalerate (2.5 mg/kg/day) on days 6 to 15 of gestation, there was maternal toxicity in both species. It seems that during pregnancy, the females are more sensitive to fenvalerate than they would otherwise be, even though the toxicity is not reflected in any effect on the fetus (5). There are no specific data available for esfenvalerate, but it is not expected to cause reproductive effects at low doses.

Teratogenic effects

Esfenvalerate did not produce any birth defects in offspring at low dietary doses (2). It appears the pesticide would not pose a teratogenic threat to humans at expected exposure levels.

Mutagenic effects

Esfenvalerate shows no mutagenic effects. Numerous tests in hamsters, mice, and rats show no signs of mutagenic activity associated with this compound (2). It is likely that it poses no mutagenic risk to humans.

Carcinogenic effects

A rat study of fenvalerate conducted over a wide range of doses of up to 75 mg/kg, for 2 years, resulted in no evidence of cancer.

Mice fed diets containing small amounts of fenvalerate for 2 years showed no adverse effects (2). It appears that fenvalerate does not cause cancer.

Organ toxicity

Studies to date have not shown any dose-related effects on internal organs of test animals or in human populations (2).

Fate in humans and animals

Cows treated with 0.1 or 0.5 g fenvalerate on their skin had 0.03 to 0.06% of the applied chemical in the milk. When the cows received fenvalerate orally at very low levels, about 0.50% of the dose appeared in the milk.

Fenvalerate does not appear to be metabolized by bovine rumen, but it is degraded further down the digestive tract (32). This happens rapidly with less than 0.02% of the parent compound found in the urine and 20% of the major metabolite present. Higher concentrations of the parent compound are present in the feces. In the rat, fenvalerate is rapidly broken down and almost completely eliminated within several days. One study indicated mammals eliminated 96% in the feces in 6 to 14 days (32).

While the data presented here are for fenvalerate, esfenvalerate behaves in the same manner (33).

Ecological effects

Effects on birds

Esfenvalerate is slightly toxic to birds. Oral LD_{50} values for the compound are 1312 mg/kg in bobwhite quail and greater than 2250 mg/kg in mallard ducks (12).

Effects on aquatic organisms

Based on laboratory studies, fish are very sensitive to esfenvalerate. It has a 96-hour LC_{50} of 0.0003 mg/L in bluegill, 0.0003 mg/L in rainbow trout, 0.001 mg/L in carp, and 0.0002 mg/L in killfish (5). The LC_{50} in *Daphnia magna*, an aquatic invertebrate, is 0.001 mg/L. The pesticide is very highly toxic to these species. Water turbidity, such as would be found in the field, tends to reduce the toxicity of this compound (5).

Bioaccumulation factors in rainbow trout are about 400 times the background (ambient water concentration of the pesticide) levels (5).

Effects on other organisms (non-target species)

Esenvalerate is highly toxic to bees. The compound tends to repel bees for a day or two after application, causing bee visitations to drop during that time (5). Since most intoxicated bees die in the field before they can return to contaminate the hive, the brood is not exposed except by direct spray. Dried spray residues are not expected to pose a significant threat to bees (5).

Environmental fate

Breakdown in soil and groundwater

Under field conditions, esfenvalerate is moderately persistent with a half-life ranging from about 15 days to 3 months, depending on soil type (25). In a soil laboratory study, 17% of the applied chemical was lost in 90 days.

Esfenvalerate and its breakdown products are relatively immobile in soil and thus pose little risk to groundwater (11). The compound's ability to bind to soil increases with increasing organic matter. It is very insoluble in water. Fenvalerate has not been found in over 100 tested groundwater supplies (34).

Breakdown in water

Esfenvalerate will break down in water to one half of the original amount (half-life) in about 21 days due to sunlight (11).

Breakdown in vegetation

The parent compound is the residue most often found on foliage. In a series of trials in Canada where 16 varieties of fruits and vegetables were grown under typical outdoor conditions, no degradation products or metabolites were found (tests were sensitive to 0.05 mg/kg). Sampling was from 1 to 112 days after application. The half-life of esfenvalerate on plant surfaces is 2 to 4 weeks (31).

Physical properties

The pure compound exists as colorless crystals; the technical product as an amber liquid (12).

> Chemical name: (S)-alpha-cyano-3-phenoxybenzyl(S)-2-(-4-chlorophenyl)-3-methylbutyrate (12)
> CAS #: 66230-04-4
> Molecular weight: 419.9 (1)
> Solubility in water: <0.3 mg/L at 25°C, insoluble in water (12)
> Solubility in other solvents: v.s. in hexane, acetone, chloroform, and methanol @ 20°C (12)

Melting point: 59–60°C (12)
Vapor pressure: 0.067 mPa @ 25°C (12)
Partition coefficient (octanol/water): 1,660,000 (12)
Adsorption coefficient: 5300 (25)

Exposure guidelines

ADI: Not available
HA: Not available
RfD: Not available
PEL: Not available

Basic manufacturer

DuPont Agricultural Products
Walker's Mill, Barley Mill Plaza
P.O. Box 80038
Wilmington, DE 19880-0038
Telephone: 800-441-7515
Emergency: 800-441-3637

2.2.4 *Flucythrinate*

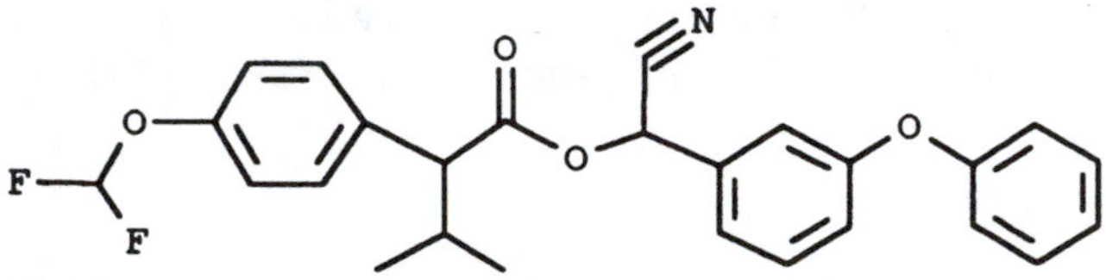

Figure 2.5 Flucythrinate.

Trade or other names

Trade names include AASTAR, AC 222705, Cybolt, Fuching Jujr, OMS 2007, and Pay-Off.

Regulatory status

Flucythrinate is a highly toxic pesticide in EPA toxicity class I. Pesticides containing flucythrinate must bear the Signal Word DANGER on the product label because of its high oral toxicity and potential to cause eye irritation. For some applications flucythrinate may be classified as a Restricted Use Pesticide (RUP) by the EPA. RUPs may be purchased and used only by certified applicators.

Introduction

Flucythrinate is a synthetic pyrethroid used to control insect pests in apples, cabbage, field corn, head lettuce, and pears, and to control *Heliothis* spp. in cotton, which is its primary use. It is available in emulsifiable concentrate, water-dispersible granules, and wettable powder formulations. The AASTAR product also contains phorate, and it may also be found in formulation with chlorpyrifos, dimethoate, methomyl, or phenthoate. *Unless otherwise stated, the information presented here refers to the technical product.*

Toxicological effects

Acute toxicity

Flucythrinate is highly toxic via the oral route. The oral LD_{50} for technical flucythrinate is 81 mg/kg in male rats (35), 67 mg/kg in female rats, and 76 mg/kg in mice (12).

Flucythrinate is moderately toxic via the dermal route with a reported dermal LD_{50} in rabbits of greater than 1000 mg/kg and in guinea pigs of greater than 2000 mg/kg (36).

Flucythrinate can cause mild to severe skin irritation (35). It failed to produce allergic reactions in guinea pigs (2,36). Skin application on human volunteers caused more severe paresthesia (i.e., abnormal sensations such as burning or tingling) on the earlobes than on the forearms. This condition lasted for approximately 24 hours after application on earlobes and 4 to 5 hours on forearms (36). When 13.8 mg/cm^2 was applied to the forearms of volunteers, paresthesia appeared 4 to 5 hours later and lasted for 3 days (36). Flucythrinate may also cause extreme eye irritation (36).

Flucythrinate is moderately toxic via the inhalation route. The 4-hour inhalation LC_{50} for technical flucythrinate in rats is 4.85 mg/L (12).

Chronic toxicity

Rats fed 15 mg/kg/day for 28 days showed severe motor symptoms (involuntary muscular movement), and rats fed 7.5 mg/kg/day showed moderate motor symptoms. In both cases, symptoms disappeared within 48 hours after resumption of a normal diet (2). Symptoms exhibited by animals in other short-term feeding studies include vomiting, diarrhea, incoordination, excessive salivation and urination, and hypersensitivity (2).

No adverse effects were observed when rats and dogs were fed flucythrinate for 90 days at doses of up to 3 mg/kg/day for rats and 3.75 mg/kg/day for dogs (35). Dogs fed 7.5 mg/kg/day for 2 years exhibited vomiting and decreased body weight gain. Rats fed 6 mg/kg/day for 2 years also exhibited decreased body weight gain (2).

Reproductive effects

In a three-generation reproductive study, rats given 1.5, 3, or 6 mg/kg/day showed reduced parental and pup weights, and decreased pup survival. Reduced litter size occurred at 3 and 6 mg/kg/day (2,35). It is unlikely that reproductive effects due to flucythrinate will be seen under normal circumstances in humans.

Teratogenic effects

Flucythrinate does not cause birth defects (12). No teratological effects were observed in studies with rats or rabbits (12). The highest dose tested was 8.0 mg/kg/day for rats and 60 mg/kg/day for rabbits (35).

Mutagenic effects

Ames tests using several strains of bacteria exposed to concentrations as high as 1000 µg/plate and a rat dominant-lethal test at up to 10.0 mg/kg showed no evidence that flucythrinate causes mutations (2,35).

Carcinogenic effects

No tumor formation was observed in mice or rats fed doses of up to 6 mg/kg for 24 months (11).

Organ toxicity

Pyrethroids primarily affect the central nervous system. Long-term, high-dose feeding studies have shown liver and kidney effects.

Fate in humans and animals

Flucythrinate is rapidly metabolized in mammals (12). When flucythrinate was administered orally to rats, 15 to 24% was eliminated in the urine, and 37 to 65.6% was eliminated in the feces during the first 24 hours. Within 8 days, 95.8 to 100% of the dose was eliminated in either urine or feces. A large amount of the chemical recovered in the feces was unaltered flucythrinate, suggesting that this portion passed through the gut without being absorbed into the bloodstream (36). Metabolites of flucythrinate are considered to be of no toxicological significance (11).

Ecological effects

Effects on birds

Pyrethroids are practically nontoxic to bird species. The oral LD_{50} for flucythrinate in the mallard duck is greater than 2510 mg/kg and in the bobwhite quail is 2708 mg/kg, and the 8-day dietary LC_{50} values are 4885 and 3443 ppm in mallard ducks and bobwhite quail, respectively (12).

Effects on aquatic organisms

The pyrethroid insecticides are very highly toxic to fish with 96-hour LC_{50} values generally below 0.001 mg/L in many species tested (12,20). Fish sensitivity to the pyrethroids may be explained by their relatively slow metabolism and elimination of these compounds. The half-lives for elimination of several pyrethroids by trout are all greater than 48 hours, while elimination half-lives for birds and mammals range from 6 to 12 hours (21). Flucythrinate accumulated in the edible tissues of bluegill sunfish to 487 times the concentration in surrounding waters (37).

Effects on other organisms (non-target species)

The compound is highly toxic to bees, with a reported topical application (contact) LD_{50} of 0.078 μg per bee (12).

Environmental fate

Breakdown in soil and groundwater

Flucythrinate is of low to moderate persistence with reported field half-lives of 21 to 60 days (25). Observed persistence will vary according to soil type and other variables. Most residues will be found within the top 3 inches of soil (37). It is nearly insoluble in water and has a very strong tendency to bind to soil particles (25). Therefore, it is unlikely to be mobile or to contaminate groundwater.

Breakdown in water

In pond waters and in laboratory degradation studies, pyrethroid concentrations decrease rapidly due to sorption to sediment, suspended particles, and plants. Microbial degradation and photodegradation also occur (37).

Breakdown in vegetation

No data are currently available.

Physical properties

Flucythrinate is a dark amber, viscous liquid with a faint odor.

Chemical name: (RS)-alpha-cyano-3-phenoxybenzyl-(S)-2-(4-difluoromethoxyphenyl)-3-methylbutyrate (12)
CAS #: 70124-77-5
Molecular weight: 451.4 (12)
Solubility in water: 0.5 mg/L @ 21°C (12), insoluble in water
Solubility in other solvents: s. in acetone, xylene, isopropanol, and most organic solvents (12)
Melting point: Not available
Vapor pressure: 0.0012 mPa @ 25°C (12)
Partition coefficient (octanol/water): 120 (12)
Adsorption coefficient: 100,000 (25)

Exposure guidelines

ADI: 0.02 mg/kg/day (29)
HA: Not available
RfD: Not available
PEL: Not available

Basic manufacturer

American Cyanamid Co.
One Cyanamid Plaza
Wayne, NJ 07470–8426
Telephone: 201-831-2000
Emergency: 201-835-3100

2.2.5 *Fluvalinate*

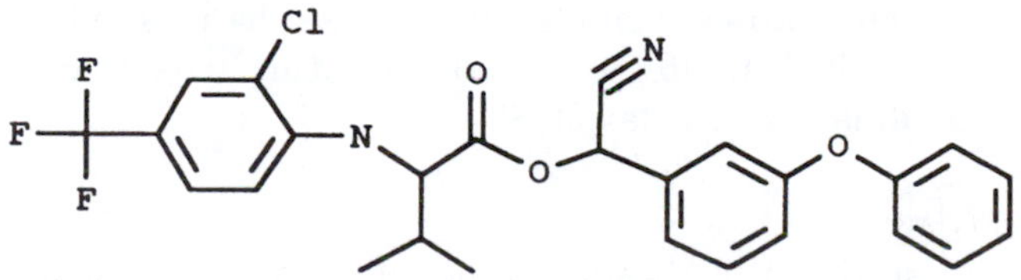

Figure 2.6 Fluvalinate.

Trade or other names

Trade names include Apistan, Klartan, Mavrik, Mavrik Aqua Flow, Spur, Tau-fluvalinate, and Yardex.

Regulatory status

Fluvalinate is a moderately toxic compound in EPA toxicity class II. Some formulations may have the capacity to cause corrosion of the eyes. Pesticides containing fluvalinate must bear the Signal Word DANGER on the product label. Fluvalinate is classified as a Restricted Use Pesticide (RUP) because of its high toxicity to fish and aquatic invertebrates. RUPs may be purchased and used only by certified applicators.

Introduction

Fluvalinate is a synthetic pyrethroid that is used as a broad-spectrum insecticide against moths, beetles and other insect pests on cotton, cereal, grape, potato, fruit tree, vegetable and plantation crops, fleas, and turf and ornamental insects. It is available in emulsifiable concentrates, suspensions, and flowable formulations.

Toxicological effects

Acute toxicity

Fluvalinate is moderately toxic via the oral route, with a reported LD_{50} in rats of 261 to 281 mg/kg (2,38). The product Mavrik 2E is less toxic via the oral route with a reported oral LD_{50} in rats of 1050 to 1110 mg/kg (38).

It is slightly to practically nontoxic via the dermal route, with a reported dermal LD_{50} for technical fluvalinate in rats and rabbits of greater than 20,000 mg/kg and greater than 2000 mg/kg, respectively (12). The dermal LD_{50} for Mavrik 2E in rabbits is greater than 2100 mg/kg (38). Fluvalinate is moderately irritating to the eye and it is a mild skin irritant (37). Fluvalinate does not cause allergic skin reactions (38). Some formulated products, including Mavrik 2E, can cause skin irritation and are corrosive to the eyes (39).

Some formulations of fluvalinate are practically nontoxic via inhalation; the reported 4-hour airborne LC_{50} for the formulated emulsifiable concentrate was greater than 5.1 mg/L in rats (12). Workers exposed to fluvalinate have reported coughing, sneezing, throat irritation, itching or burning sensations on the arms or face with or without a rash, headache, and nausea (38).

Chronic toxicity

A 90-day study with rats fed 3 mg/kg/day and a 6-month study with dogs fed 5 mg/kg/day both showed no adverse effects (38). In other studies, effects observed were increased liver and kidney weights and adverse changes in liver tissues in test animals (38). No neurological effects were observed in hens given doses of 20,000 mg/kg/day of fluvalinate for 21 days (38).

Reproductive effects

A reproductive study with rats showed no effects on offspring at 1 mg/kg. Toxic effects in fetuses occurred at 12.5 and 25 mg/kg/day, the highest doses tested (38). Based on these data, it is unlikely that reproductive effects would be seen in humans under normal circumstances.

Teratogenic effects

No birth defects were detected in the offspring of rats fed 50 mg/kg/day nor in the offspring of rabbits fed 125 mg/kg/day (38). Based on these data, it is unlikely that teratogenic effects would be seen in humans.

Mutagenic effects

Fluvalinate is not reported to have mutagenic activity (38).

Carcinogenic effects

No tumors were observed in mice given doses of up to 20 mg/kg/day, nor in rats given doses as high as 2.5 mg/kg/day for over 2 years (38).

Organ toxicity

Pyrethroids may cause adverse effects on the central nervous system, liver, and kidneys.

Fate in humans and animals

The elimination half-lives for mammals ranges from 6 to 12 hours (12). Following oral administration, approximately 40 to 50% is absorbed into the system, and more than 95% of the administered fluvalinate is excreted via urine and feces within 4 days (12). Thus, it has a low potential for bioaccumulation.

Ecological effects

Effects on birds

Fluvalinate is slightly toxic to birds. The acute oral LD_{50} for fluvalinate in bobwhite quail is greater than 2510 mg/kg (12). The dietary LC_{50} for fluvalinate in mallard ducks and bobwhite quail is greater than 5620 ppm (12,38).

Effects on aquatic organisms

Fluvalinate is very highly toxic to fish. The 96-hour LC_{50} for fluvalinate in bluegill sunfish is 0.9 µg/L, in rainbow trout is 2.9 µg/L, and in carp is 2.9 µg/L (12). Its 48-hour LC_{50} in *Daphnia magna*, a small freshwater crustacean, is 74 µg/L, and in mysid shrimp is 2.9 µg/L (38). The bioconcentration factor for fluvalinate in whole fish is 360 times the ambient water concentration, indicating a low to moderate potential to accumulate in aquatic organisms (38).

Effects on other organisms (non-target species)

Fluvalinate was not toxic to honeybees exposed to residues left on cotton leaves after application of ultralow volume (ULV) and emulsifiable concentrate (EC) formulations (24).

Environmental fate

Breakdown in soil and groundwater

Fluvalinate is of low persistence, with reported soil half-lives of 6 to 8 days (12,25). In sandy loam, sandy clay and clay soils, fluvalinate degrades under aerobic conditions with half-lives of 4 to 8 days. Under anaerobic conditions in sandy loam, its half-life may be 15 days (38). Fluvalinate is nearly insoluble in water and it has a strong tendency to bind to soil particles (25). It is therefore unlikely to contaminate groundwater; however, metabolites of fluvalinate may leach (12,38). Applications of less than 0.1 pound active ingredient per acre will decrease the potential for groundwater contamination (38). Photodegradation of fluvalinate does not occur on soil (38).

Breakdown in water

In water, fluvalinate is subject to photodegradation with a half-life of up to 1 day. Photodegradation yields anilino acid and 3-phenoxy benzoic acid (38). Fluvalinate is stable to hydrolysis under normal environmental temperatures and pH (38). In pond waters and in laboratory degradation studies, pyrethroid concentrations decrease rapidly due to sorption to sediment, suspended particles, and plants. Microbial and photodegradation also occur (22).

Breakdown in vegetation

No information was found.

Physical properties

Fluvalinate is a viscous, yellow oil (12).

Chemical name: (RS)-alpha-cyano-3-phenoxybenzyl N-(2-chloro-a,a,a-trifluoro-*p*-tolyl)-D-valinate (12)
CAS #: 102851-06-9
Molecular weight: 502.93 (12)
Solubility in water: 0.002 mg/L (12), insoluble in water
Solubility in other solvents: v.s. in organic solvents and aromatic hydrocarbons; s.s. in hexane (12)
Melting point: Not available
Vapor pressure: <0.013 mPa @ 25°C (12)
Partition coefficient (octanol/water): 7000 (12)
Adsorption coefficient: 1,000,000 (25)

Exposure guidelines

ADI: Not available
HA: Not available
RfD: 0.01 mg/kg/day (30)
PEL: Not available

Basic manufacturer

Sandoz Agro, Inc.
1300 E. Touhy Ave.
Des Plaines, IL 60018
Telephone: 708-699-1616
Emergency: 708-699-1616

2.2.6   Permethrin

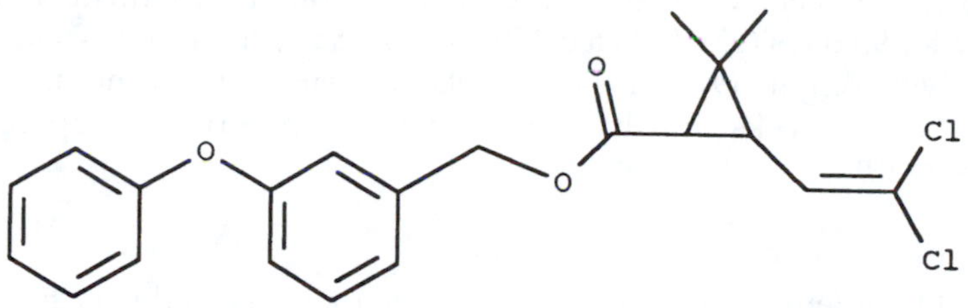

Figure 2.7 Permethrin.

Trade or other names

Trade names include Ambush, BW-21-Z, Cellutec, Dragnet, Ectiban, Eksmin, Exmin, FMC 33297, Indothrin, Kafil, Kestrel, NRDC 143, Pounce, PP 557, Pramex, Qamlin, and Torpedo.

Regulatory status

Permethrin is a moderately to practically nontoxic pesticide in EPA toxicity class II or III, depending on the formulation. Formulations are placed in class II due to their potential to cause eye and skin irritation. Products containing permethrin must bear the Signal Word WARNING or CAUTION, depending on the toxicity of the particular formulation.

All products for agricultural use (except livestock and premises uses) are Restricted Use Pesticides (RUPs) because of their possible adverse effects on aquatic organisms. RUPs may be purchased and used only by certified applicators.

Introduction

Permethrin is a broad-spectrum synthetic pyrethroid insecticide used against a variety of pests on nut, fruit, vegetable, cotton, ornamental, mushroom, potato, and cereal crops. It is used in greenhouses, home gardens, and for termite control. It also controls animal ectoparasites, biting flies, and cockroaches. It may cause a mite buildup by reducing mite predator populations. Permethrin is available in dusts, emulsifiable concentrates, smokes, ULV (ultra-low volume), and wettable powder formulations.

Toxicological effects

Acute toxicity

Permethrin is moderately to practically nontoxic via the oral route, with a reported LD_{50} for technical permethrin in rats of 430 to 4000 mg/kg (12). Via the dermal route, it is slightly toxic, with a reported dermal LD_{50} in rats of over 4000 mg/kg, and in rabbits of greater than 2000 mg/kg (2,12). Permethrin caused mild irritation of both the intact and abraded skin of rabbits. It also caused conjunctivitis when it was applied to the eyes (9). The 4-hour inhalation LC_{50} for rats was greater than 23.5 mg/L, indicating practically no inhalation toxicity. The toxicity of permethrin is dependent on the ratio of the isomers present, the *cis*-isomer being more toxic (12).

Chronic toxicity

No adverse effects were observed in dogs fed permethrin at doses of 5 mg/kg/day for 90 days (15). Rats fed 150 mg/kg/day for 6 months showed a slight increase in liver weights (9). Very low levels of permethrin in the diet of chickens (0.1 ppm for 3 to 6 weeks after hatching) have been reported to suppress immune system activity (9).

Reproductive effects

The fertility of female rats was affected when they received very high oral doses of 250 mg/kg/day permethrin during the 6th to 15th days of pregnancy (25). It is not likely that reproductive effects will be seen in humans under normal circumstances.

Teratogenic effects

Permethrin is reported to show no teratogenic activity (9).

Mutagenic effects

Permethrin is reported to show no mutagenic activity (9).

Carcinogenic effects

The evidence regarding the carcinogenicity of permethrin is inconclusive.

Organ toxicity

Permethrin is suspected of causing enlargement of the liver and nerve damage (9). Effects on the immune system have been noted in animal studies.

Fate in humans and animals

Permethrin is efficiently metabolized by mammalian livers (40). Breakdown products, or "metabolites," of permethrin are quickly excreted and do not persist significantly in body tissues (41). When permethrin is administered orally to rats, it is rapidly metabolized and almost completely eliminated from the body in a few days. Only 3 to 6% of the original dose was excreted unchanged in the feces of experimental animals (41). Permethrin may persist in fatty tissues, with half-lives of 4 to 5 days in brain and body fat (9). Permethrin does not block, or inhibit, cholinesterase enzymes (40).

Ecological effects

Effects on birds

Permethrin is practically nontoxic to birds (12). The oral LD_{50} for the permethrin formulation, Pramex, is greater than 9900 mg/kg in mallard ducks, greater than 13,500 mg/kg in pheasants, and greater than 15,500 mg/kg in Japanese quail (41).

Effects on aquatic organisms

Aquatic ecosystems are particularly vulnerable to the impact of permethrin. A fragile balance exists between the quality and quantity of insects and other invertebrates that serve as fish food (41).

The 48-hour LC_{50} for rainbow trout is 0.0125 mg/L for 24 hours, and 0.0054 mg/L for 48 hours (12). The 48-hour LC_{50} in bluegill sunfish and salmon is 0.0018 mg/L (12). As a group, synthetic pyrethroids were toxic to all estuarine species tested. They had a 96-hour LC_{50} of less than or equal to 0.0078 mg/L for these species (42).

The bioconcentration factor for permethrin in bluefish is 715 times the concentrations in water and is 703 in catfish. This indicates that the compound has a low to moderate potential to accumulate in these organisms.

Effects on other organisms (non-target species)

Permethrin is extremely toxic to bees. Severe losses may be expected if bees are present at treatment time, or within a day thereafter (2,43). Permethrin is also toxic to wildlife (9). It should not be applied, or allowed to drift, to crops or weeds in which active foraging takes place (12).

Environmental fate

Breakdown in soil and groundwater

Permethrin is of low to moderate persistence in the soil environment, with reported half-lives of 30 to 38 days (12,25). Permethrin is readily broken down, or degraded, in most soils except organic types. Soil microorganisms play a large role in the degradation of permethrin in the soil. The addition of nutrients to soil may increase the degradation of permethrin. It has been observed that the availability of sodium and phosphorous decreases when permethrin is added to the soil (44).

Permethrin is tightly bound by soils, especially by organic matter. Very little leaching of permethrin has been reported (45). It is not very mobile in a wide range of soil types (41). Because permethrin binds very strongly to soil particles and is nearly insoluble in water, it is not expected to leach or to contaminate groundwater.

Breakdown in water

The results of one study near estuarine areas showed that permethrin had a half-life of less than 2.5 days. When exposed to sunlight, the half-life was 4.6 days (13). Permethrin degrades rapidly in water, although it can persist in sediments (15,45). There was a gradual loss of toxicity after permethrin aged for 48 hours in sunlight at 0.05 mg/L in water (45).

Breakdown in vegetation

Permethrin is not phytotoxic, or poisonous, to most plants when it is used as directed. Some injury has occurred on certain ornamental plants. No incompatibility has been observed with permethrin on cultivated plants. Treated apples, grapes, and cereal grains contain less than one mg/kg of permethrin at harvest time (12).

Physical properties

Permethrin is an odorless, colorless crystalline solid or a viscous liquid that is pale brown (12).

Chemical name: 3-phenoxybenzyl(1RS)-*cis,trans*-3-(2,2-dichlorovinyl)-2,2-dimethylcyclopropanecarboxylate (12)

CAS #: 52645-53-1
Molecular weight: 391.3 (12)
Water solubility: ca. 0.2 mg/L @ 20°C (12), insoluble in water
Solubility in other solvents: s. in most organic solvents except ethylene glycol (12)
Melting point: 34–35°C (12)
Vapor pressure: 0.045 mPa @ 25°C (12)
Partition coefficient (octanol/water): 1,260,000 (12)
Adsorption coefficient: 100,000 (26)

Exposure guidelines

ADI: 0.05 mg/kg/day (29)
HA: Not available
RfD: 0.05 mg/kg/day (30)
PEL: Not available

Basic manufacturer

Zeneca Ag Products
1800 Concord Pike
Wilmington, DE 19897
Telephone: 800-759-4500
Emergency: 800-759-2500

2.2.7   *Resmethrin*

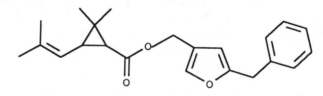

Figure 2.8 Resmethrin.

Trade or other names

Trade names include Chryson, Crossfire, Derringer, FMC 17370, Isathrine, NRDC 104, Pynosect, Raid Flying Insect Killer, Respond, Scourge, Sun-Bugger #4, SPB-1382, Synthrin, Syntox, Vectrin, and Whitmire PT-110.

Regulatory status

Resmethrin is a slightly toxic to practically nontoxic compound in EPA toxicity class III. Products containing resmethrin must bear the Signal Word CAUTION on the label. All products containing resmethrin for pest control at or near aquatic sites are classified as Restricted Use Pesticides (RUP) by the EPA because of potential fish toxicity. RUPs may be purchased and used only by certified applicators.

Introduction

Resmethrin is a synthetic pyrethroid used for control of flying and crawling insects in homes, greenhouses, indoor landscapes, mushroom houses, industrial sites, stored product insects, and for mosquito control. It is also used for fabric protection, pet sprays, and shampoos, and it is applied to horses or in horse stables. Technical resmethrin is a mixture of its two main isomers (molecules with the same chemical formula but slightly different configurations); a typical blend is 20 to 30% of the (1RS)-*cis*-isomer and 70 to 80% of the (1RS)-*trans*-isomer.

Toxicological effects

Acute toxicity

Resmethrin is slightly to practically nontoxic by ingestion. The oral LD_{50} for technical resmethrin in rats is variously reported as greater than 2500 mg/kg or 1244 mg/kg (3,12). Resmethrin is only slightly toxic through the dermal route as well. The reported dermal LD_{50} values for technical resmethrin are: greater than 3000 mg/kg in rats, greater than 2500 mg/kg in rabbits, and greater than 5000 mg/kg in mice (3,12). It is slightly toxic via inhalation, with a 4-hour inhalation LC_{50} for resmethrin of greater than 9.49 mg/L (3).

Symptoms of exposure by any route may include incoordination, twitching, loss of bladder control, and seizures (12). Dermal exposure may lead to local numbness, itching, burning, and tingling sensations near the site of exposure. Resmethrin is reported to be nonirritating to the skin and eyes of test animals and not to cause skin sensitization in guinea pigs (3).

Chronic toxicity

In a chronic feeding study with rats, 25 mg/kg/day (the lowest dose tested) caused liver enlargement. At 125 mg/kg/day, there were pathological liver changes in addition to increased liver weights. Doses of 250 mg/kg/day caused increased thyroid weight and thyroid cysts (3). In another study over 90 days, doses of 150 mg/kg/day did not produce any adverse effects in exposed rats (12). Increased liver weights occurred in dogs fed 30 mg/kg/day for 180 days. No effects were observed in dogs in this study at dose rates of 10 mg/kg/day (3).

In a 90-day inhalation study with rats, 0.1 mg/L, the lowest dose tested, produced behavioral changes, decreased blood glucose levels in males, and decreased body weights and increased serum urea levels in females (3).

Resmethrin was not neurotoxic to rats at doses of 62.5 mg/kg/day for 32 weeks, 250 mg/kg/day for 30 days, or 632 mg/kg/day for 7 days (4). It is unlikely that chronic effects will be seen in humans under normal circumstances.

Reproductive effects

A three-generation study with rats showed a slight increase in premature still-births and a decrease in pup weight at 25 mg/kg, the lowest dose tested (4). Since these doses are much higher than expected human exposures, it is unlikely such effects will occur in humans.

Teratogenic effects

No birth defects were observed in the offspring of rabbits given doses as high as 100 mg/kg/day (4). Skeletal aberrations were seen in the offspring of rats given doses higher than 40 mg/kg/day (3). No teratogenic effects were observed in mice at dose levels of 50 mg/kg/day over an unspecified period (12). It is unlikely that teratogenic effects will be seen in humans under normal circumstances.

Mutagenic effects

Resmethrin was not mutagenic in a test performed with the bacterium, *Salmonella typhimurium* (6).

Carcinogenic effects

No evidence of tumor formation was observed in a 2-year rat feeding study with doses as high as 250 mg/kg/day, nor in an 85-week study with mice given doses as high as 50 mg/kg/day (3,4).

Organ toxicity

Pyrethroids may cause adverse effects on the central nervous system. Long-term feeding studies have shown increased liver and kidney weights and adverse changes in liver tissues in test animals (12).

Fate in humans and animals

Resmethrin is quickly eliminated by chickens. When oral doses of 10 mg/kg resmethrin were given to laying hens, 90% of the dose was eliminated in urine and feces within 24 hours (46). In another study with hens given the same treatment, residues were low in hens sacrificed 12 hours after the treatment, with the highest levels found in the liver and kidneys. Low levels were found in the hens' eggs, with levels peaking 1 day after treatment in the whites and 4 to 5 days after treatment in the yolks (47).

Ecological effects

Effects on birds

Resmethrin is practically nontoxic to birds. Its LD_{50} in California quail is greater than 2000 mg/kg (3). In Japanese quail, the 5-day dietary LC_{50} is greater than 5000 ppm (48).

Effects on aquatic organisms

Resmethrin is very highly toxic to fish, with 96-hour LC_{50} values generally at or below 1 µg/L (0.001 mg/L) for most species tested. The LC_{50} for resmethrin in mosquito fish is 7 µg/L (49). The LC_{50} for resmethrin synergized with piperonyl butoxide in red swamp crawfish, *Procambarus clarkii*, is 0.00082 µg/L (48). The LC_{50} in bluegill sunfish is 0.75 to 2.6 µg/L, and 0.28 to 2.4 µg/L in rainbow trout (3). Other reported 96-hour LC_{50} values are 1.8 µg/L in coho salmon, 1.7 µg/L in lake trout, 3.0 µg/L in fathead minnow, 16.6 µg/L in channel catfish, and 1.7 µg/L in bluegill sunfish (50).

Fish sensitivity to the pyrethroids may be explained by their relatively slow metabolism and elimination of these compounds. The half-lives for elimination of several pyrethroids by trout are all greater than 48 hours, while elimination half-lives for birds and mammals range from 6 to 12 hours (20).

Effects on other organisms (non-target species)

Resmethrin is highly toxic to bees, with an LD_{50} of 0.063 µg per bee (3).

Environmental fate

Breakdown in soil and groundwater

Resmethrin is of low to moderate persistence in the soil environment. Its half-life has been estimated at 30 days (51). Observed half-lives will depend on many site-specific variables. In aerobic Kentucky loamy sand, the compound showed a half-life of nearly 200 days. Degradation end-products reported for resmethrin are chrysanthemic acid, benzaldehyde, benzyl alcohol, benzoic acid, phenylacetic acid, and various esters (52).

Resmethrin is tightly bound to soil and would not be expected to be mobile or to contaminate groundwater, especially in light of its extremely low solubility in water (51).

Breakdown in water

Resmethrin may enter surface waters through particulate run-off or misapplication. In pond waters and in laboratory degradation studies, pyrethroid concentrations decrease rapidly due to sorption to sediment, suspended particles and plants. Microbial and photodegradation also occur (22). The half-life in water is 36.5 days.

Breakdown in vegetation

No information was found.

Physical properties

Resmethrin is a waxy, off-white to tan solid with an odor characteristic of chrysanthemums (12).

Chemical name: 5-benzyl-3-furylmethyl (1RS)-*cis,trans*-2,2-dimethyl-3-(2-methylprop-1-enyl)cyclopropanecarboxylate (12)
CAS #: 10453-86-8
Molecular weight: 338.45 (12)
Solubility in water: <1 mg/L at 30°C (12), insoluble in water
Solubility in other solvents: s. in hexane, kerosene, xylene, methylene chloride, isopropyl alcohol, and aromatic petroleum hydrocarbons; m.s. in methanol (12)
Melting point: 43–48°C (12)
Vapor pressure: 0.0015 mPa @ 30°C (12)
Partition coefficient (octanol/water): Not available
Adsorption coefficient: 100,000 (51)

Exposure guidelines

ADI: Not available
HA: Not available
RfD: 0.03 mg/kg/day (30)
PEL: Not available

Basic manufacturer

Roussel Uclaf Corp.
95 Chestnut Ridge Road
Montvale, NJ 07645
Telephone: 201-307-9700

2.2.8   Rotenone

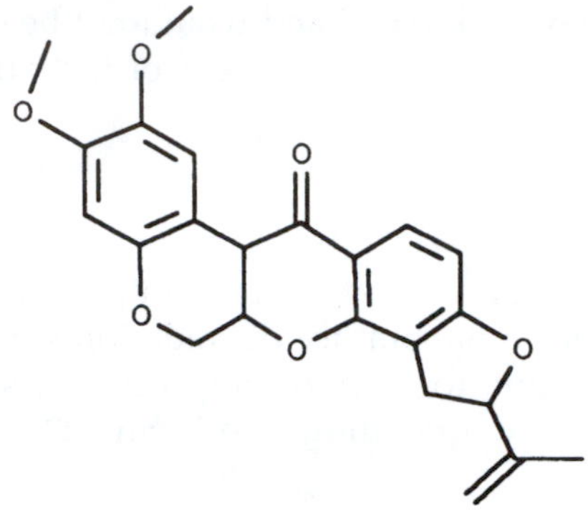

Figure 2.9 Rotenone.

Trade or other names

Trade names for products containing rotenone include Chem-Fish, Cuberol, Fish Tox, Noxfire, Rotacide, Sinid, and Tox-R. It is also marketed as Curex Flea Duster, Derrin, Cenol Garden Dust, Chem-Mite, Cibe Extract, and Green Cross Warble Powder. The compound may be used in formulations with other pesticides such as carbaryl, lindane, thiram, piperonyl butoxide, pyrethrins, and quassia.

Regulatory status

Rotenone is a General Use Pesticide (GUP), but uses on cranberries and for fish control are restricted uses. It is EPA toxicity class I or III — highly toxic or slightly toxic, depending on formulation. Rotenone, when formulated as an emulsified concentrate, is highly toxic and carries the Signal Word DANGER on its label. Other forms are slightly toxic and require the Signal Word CAUTION instead.

Introduction

Rotenone is a selective, nonspecific botanical insecticide with some acaricidal properties. Rotenone is used in home gardens for insect control, for lice and tick control on pets, and for fish eradications as part of water body management. Rotenone is a rotenoid plant extract obtained from such species as barbasco, cub, haiari, nekoe, and timbo. These plants are members of the pea (Leguminosae) family. Rotenone-containing extracts are taken from the roots, seeds, and leaves of the various plants. Formulations include crystalline preparations (approximately 95% pure), emulsified solutions (approximately 50% pure), and dusts (approximately 0.75 to 5% pure). *This profile refers to the crystalline preparation unless otherwise noted.*

Toxicological effects

Acute toxicity

Local effects on the body include conjunctivitis, dermatitis, sore throat, and congestion. Ingestion produces effects ranging from mild irritation to vomiting. Inhalation of high doses can cause increased respiration followed by depression and convulsions. The compound can cause a mild rash in humans and is a strong eye irritant to rabbits (2,8).

The oral LD_{50} of rotenone ranges from 132 to 1500 mg/kg in rats. The reported LD_{50} of rotenone in white mice is 350 mg/kg (12).

A spray of 5% rotenone in water was fatal to a 100-pound pig when exposed to 250 cubic centimeters (mL) of the airborne mixture (2).

In rats and dogs exposed to rotenone in dust form, the inhalation fatal dose was uniformly smaller than the oral fatal dose (2).

Rotenone is believed to be moderately toxic to humans with an oral lethal dose estimated from 300 to 500 mg/kg (2). Human fatalities are rare, perhaps because rotenone is usually sold in low concentrations (1 to 5% formulation) and because its irritating action causes prompt vomiting.

The mean particle size of the powder determines the inhalation toxicity. Rotenone may be more toxic when inhaled than when ingested (2), especially if the mean particle size is very small and particles can enter the deep regions of the lungs.

Chronic toxicity

Growth retardation and vomiting resulted from chronic exposures of rats and dogs. Rats fed diets containing rotenone at doses up to 2.5 mg/kg for 2 years developed no pathological changes that could be attributed to rotenone (53).

Dogs fed doses of rotenone up to 50 mg/kg/day for 28 days experienced vomiting and excessive salivation, but no decreased weight gain (53). Dogs fed rotenone for 6 months at doses up to 10 mg/kg/day had reduced food consumption and therefore reduced weight gain. At the highest dose, blood chemistry was adversely affected (53), possibly due to gastointestinal lesions and chronic bleeding. Examination of 35 tissue types revealed only one type of lesion that might have been associated with exposure to the test chemical: lesions of the GI tract (53).

Reproductive effects

Pregnant rats fed 10 mg/kg/day on days 6 through 15 of gestation experienced decreased fecundity, increased fetal resorption, and lower birth weight (23,53). Very high maternal mortality was seen at this dose. The 2.5-mg/kg/day dose produced no observable maternal toxicity or adverse effect on fetal development (2). Fetotoxicity and failure of offspring are reported in guinea pigs at doses of 4.5 and 9.0 mg/kg/day for an unspecified period (2). Thus, reproductive effects seem unlikely in humans at expected exposures.

Teratogenic effects

Pregnant rats fed 5 mg/kg/day produced a significant number of young with skeletal deformities (23,53). The effects were not observed at the 10-mg/kg/day level, so the data do not provide convincing evidence of teratogenicity (2,23) because the effects do not appear to be dose related. Thus, the evidence for teratogenicity is inconclusive.

Mutagenic effects

The compound was determined to be nonmutagenic to bacteria and yeast and in treated mice and rats. However, it was shown to cause mutations in some cultured mouse cells (2,23). In summary, the data regarding the mutagenicity of rotenone are inconclusive (4).

Carcinogenic effects

Studies in rats and hamsters have provided limited evidence for carcinogenic activity of rotenone. No evidence of carcinogenic activity was seen in hamsters at oral doses as high as 120 mg/kg/day for a period of 18 months (54). Studies of two species of rats evidenced no statistically significant cancerous changes in any organ site, including mammary glands, at oral doses of up to 75 mg/kg/day for 18 months (54).

Significant increases in mammary tumors have been reported in albino rats with intraperitoneal doses of 1.7 mg/kg/day for 42 days (54), and in Wistar rats at approximately 1.5 mg/kg/day orally for 8 to 12 months (54). In the latter study, however, higher dose rates (3.75 and 7.5 mg/kg/day) over the same period did not produce increased tumors (54).

Thus, the evidence for carcinogenicity is inconclusive.

Organ toxicity

Chronic exposure may produce changes in the liver and kidneys, as indicated by the animal studies cited above.

Fate in humans and animals

Absorption in the stomach and intestines is relatively slow and incomplete, although fats and oils promote its uptake. The liver breaks down the compound fairly effectively (2). Animal studies indicate that possible metabolites are carbon dioxide and a more water-soluble compound that can be excreted in the urine (2). Studies indicated that approximately 20% of the applied oral dose (and probably most of the absorbed dose) may be eliminated from animal systems within 24 hours (2).

Ecological effects

Effects on birds

Rotenone is slightly toxic to wildfowl. The LD_{50} values for rotenone in mallards and pheasants are greater than 2000 mg/kg and 1680 mg/kg, respectively (55). A dietary LC_{50} of 4500 to 7000 ppm is reported in Japanese quail (48).

Effects on aquatic organisms

Since rotenone is used as a fish toxin (piscicide), it follows that it is very highly toxic to fish. Reported 96-hour LC_{50} values were 0.031 mg/L in rainbow trout, 0.0026 mg/L in channel catfish, and 0.023 mg/L in bluegill for the 44% pure formulation (50). Aquatic invertebrates have a wide range of sensitivity to rotenone with 48-hour LC_{50} values ranging from 0.002 to 100 mg/L (50).

The compound is not expected to accumulate appreciably in aquatic organisms. The bioconcentration factor for rotenone in the sunfish is 181 times the ambient water concentration. In addition the highly toxic nature of this substance to aquatic organisms means that there is little survival of the organisms that accumulate the compound.

Effects on other organisms (non-target species)

The compound is not toxic to bees. However, it is toxic to bees when used in combination with pyrethrum (12).

Environmental fate

Breakdown in soil, water, and groundwater

Rotenone is rapidly broken down in soil and in water. The half-life in both of these environments is between 1 and 3 days (51). It does not readily leach from soil (51), and is not expected to be a groundwater pollutant.

Rotenone breaks down readily by exposure to sunlight (12). Nearly all of the toxicity of the compound is lost in 5 to 6 days of spring sunlight or 2 to 3 days of summer sunlight.

Breakdown in vegetation

Rotenone is a highly active but short-lived photosensitizer. This means that an organism consuming the compound develops a strong sensitivity to the sun for a short time. A number of photodecomposition products are formed when bean leaves are exposed to light. It is also sensitive to heat with, much of the rotenone quickly lost at high temperatures.

Physical properties

Chemical name: (2R, 6aS, 12aS)-1,2,6,6a,12,12a-hexahydro-2-isopropenyl-8,9-dimethoxychromeno[3,4-b]furo[2,3-h]chromen-6-one
CAS #: 83-79-4
Molecular weight: 394.43 (12)
Solubility in water: 15 mg/L @ 100°C (12), slightly soluble in water
Solubility in other solvents: s. in acetone, carbon disulfide and chloroform; s.s. in alcohols and carbon tetrachloride (12)
Melting point: 163°C (12)
Vapor pressure: <1 mPa @ 20°C (12)
Partition coefficient (octanol/water): Not available
Adsorption coefficient: 10,000 (51)

Exposure guidelines

ADI: Not available
HA: Not available
RfD: 0.004 mg/kg/day (30)
PEL: 5 mg/m³ (8-hour) (23)

Basic manufacturer

Fairfield American Corp.
238 Wilson Ave
Newark, NJ 07105
Telephone: 201-589-0263

2.2.9   *Ryania*

Trade or other names

Trade names of this product are Natur Gro R-50, Natur Gro Triple Plus, Ryanicide, and Ryan 50.

Regulatory status

Ryania is a General Use Pesticide (GUP). It is classified as EPA toxicity class III — slightly toxic. Products containing this pesticide are labeled with the Signal Word CAUTION.

Introduction

Ryania is a botanical insecticide made from the ground stems of *Ryania speciosa*, a native plant of tropical America. The principal alkaloid in this stem extract is ryanodine, which makes up approximately 0.2% of the product. Ryania is highly toxic to the fruit moth, coddling moth, corn earworm, European corn borer, and citrus thrip, but it is ineffective against the cabbage maggot, cauliflower worm, and boll weevil. Ryania is a complex mixture of many compounds; thus, no single structure would represent it.

Toxicological effects

Acute toxicity

Symptoms of poisoning include vomiting, weakness, and diarrhea (23). Rigidity of the muscles and depression of the central nervous system can lead to coma and death from respiratory failure at high doses (23).

The reported oral LD_{50} values for ryania in rats are 1200 mg/kg and 750 mg/kg (56). Reported LD_{50} values in guinea pigs and rabbits are 2500 mg/kg and 650 mg/kg, respectively (57). Dogs appear to be much more susceptible, with an LD_{50} of 150 mg/kg (23). The observed toxicity is apparently due to the ryanodine alkaloid, which is 500 to 700 times more toxic than the crude powder (23). A 2- to 5-mg/kg dose of ryanodine elicited symptoms in the frog and mouse (58). The dermal LD_{50} in rats is above 2000 mg/kg.

Inhalation data reported semiquantitatively indicate that ryania may be of low toxicity via inhalation (57).

Chronic toxicity

Rats and guinea pigs were fed diets containing 1% ryania powder (approximately 600 mg/kg/day for rats and 361 mg/kg/day for guinea pigs) during a 5-month period (57) and showed no symptoms. Rats fed 5% diets (approximately 2700 mg/kg/day) experienced severe weight loss and 100% mortality within 25 days (58).

Reproductive effects

No data are currently available.

Teratogenic effects

No data are currently available.

Mutagenic effects
No data are currently available.

Carcinogenic effects
No data are currently available.

Organ toxicity
Very large doses of ryania powder (from 1200 to 2700 mg/kg/day) in the diets of rats produced hemorrhages in the pancreas and intestines. They also produced lung complications in the test animals (57). Ryania mainly exerts its effects on striated (skeletal) muscle tissue and cardiac smooth muscle, although evidence for effects on other smooth muscles (e.g., bladder) is mounting (59,60).

Fate in humans and animals
No data are currently available.

Ecological effects

Effects on birds

Ryania is moderately toxic to birds and wildfowl. Some LD_{50} values include: wild birds, 1.78 mg/kg; pigeons, 2.31 mg/kg; and quail, 13.3 mg/kg (23). Chickens fed for 6 months on diets of 1% ryania showed no symptoms and did not have any evidence of cumulative effects (23).

Effects on aquatic organisms

It is considered moderately toxic to fish. Reported 96-hour LC_{50} values are 3.2 mg/L in rainbow trout and 18.5 mg/L in bluegill (23).

Effects on other organisms (non-target species)

No data are currently available.

Environmental fate

Breakdown in soil and groundwater

No data are currently available.

Breakdown in water

The toxicologically active portion of ryania, ryanodine, is water soluble but stable during storage and stable upon exposure to light.

Breakdown in vegetation

Ryanodine is not considered to be poisonous to plants (58). The major degradation product of ryanodine is anhydroxyanodine (58).

Physical properties

Ryanodine is a very stable solid at room temperature (12).

Chemical name: Ryania
CAS #: 15662-33-6
Molecular weight: 493.54 (12)
Water solubility: 100–10,000 mg/L (12), s. in water
Solubility in other solvents: s. in methanol, acetone, ether, and chloroform; i.s. in benzene and petroleum ether (12)
Melting point: decomposes @ 219–220°C (12)
Vapor pressure: Not available
Partition coefficient (octanol/water): Not available
Adsorption coefficient: Not available

Exposure guidelines

ADI: Not available
HA: Not available
RfD: Not available
PEL: Not available

Basic manufacturer

AgriSystems International
125 W. Seventh St.
Wind Gap, PA 18091
Telephone: 610-863-6700
Emergency: 610-863-8050

References

(1) World Health Organization. *Safe Use of Pesticides in Public Health. (WHO Tech. Rep. Ser. No. 356).* Geneva, Switzerland, 1967.

(2) Ray, D. E. Pesticides derived from plants and other organisms. In *Handbook of Pesticide Toxicology.* Hayes, W. J., Jr. and Laws, E. R., Jr., Eds. Academic Press, New York, 1991.

(3) U.S. Environmental Protection Agency. *Pesticide Fact Sheet Number 193: Resmethrin.* Office of Pesticides and Toxic Substances, Washington, D.C., 1988.

(4) U.S. Environmental Protection Agency. Tolerances in food administered by EPA: resmethrin. *Fed. Reg.* 48, 36246–36247, 1983.

(5) E. I. DuPont de Nemours Corp., Asana XL Technical Bulletin, Wilmington, DE, 1989.

(6) Herrera, A. and Laborda, E. Mutagenic activity of synthetic pyrethroids in *Salmonella typhimurium. Mutagenesis* 3(6), 503–514, 1988.

(7) U.S. Environmental Protection Agency. *Pesticide Fact Sheet Number 158: Allethrin Stereoisomers.* Office of Pesticides and Toxic Substances, Washington, D.C., 1988.

(8) U.S. Environmental Protection Agency. *Pesticide Fact Sheet Number 199: Cypermethrin.* Office of Pesticides and Toxic Substances, Washington, D.C., 1989.

(9) Hallenbeck, W. H. and Cunningham-Burns, K. M. *Pesticides and Human Health.* Springer-Verlag, New York, 1985.

(10) Christopher, R. J. Metabolism of *cis*-and *trans*-resmethrin in laying hens. *J. Agric. Food Chem.* 37(3), 800–808, 1989.

(11) Walker, M. H. and Keith, L. H. *EPA's Pesticide Fact Sheet Database.* Lewis, Boca Raton, FL, 1992.

(12) Kidd, H. and James, D. R., Eds. *The Agrochemicals Handbook,* 3rd ed. Royal Society of Chemistry Information Services, Cambridge, U.K., 1991 (as updated).

(13) Kaufman, D. D., Hayes, S. C., Jordan, E. G., and Kayser, A. J. Permethrin degradation in soil and microbial cultures. In *ACS Symposium. Ser. No. 42: Synthetic Pyrethroids.* Elliot, M., Ed. American Chemical Society, Philadelphia, PA, 1977.

(14) Belanger, A. A field study of four insecticides used in strawberry protection. *J. Environ. Sci. Health,* Part B, 25(5), 615–625, 1990.

(15) Thomson, W. T. Insecticides. In *Agricultural Chemicals: Book I.* Thomson Publications, Fresno, CA, 1985.

(16) Cremlyn, R. J. *Agrochemicals: Preparation and Mode of Action.* John Wiley & Sons, New York, 1991.

(17) Ecobichon, D. J. Toxic effects of pesticides. In *Casarett and Doull's Toxicology,* 4th ed. Amdur, M. O., Doull, J. and Klaassen, C. D., Eds. Pergamon Press, New York, 1991.

(18) Matsuoka, A., Hayashi, M., and Ishidate, M. Chromosomal aberration test on 29 chemicals combined with S9 mix in vitro. *Mutat. Res.* 66, 277–290, 1979.

(19) World Health Organization. *Allethrins: Allethrin, D-allethrin, Bioallethrin, S-bioallethrin-Environmental Health Criteria 87.* International Programme on Chemical Safety, Geneva, Switzerland, 1989.

(20) Bradbury, S. P. and Coats, J. R. Toxicokinetics and toxicodynamics of pyrethroid insecticides in fish. *Environ. Toxicol. Chem.* 8, 373–380, 1989.

(21) Haya, K. Toxicity of pyrethroid insecticides to fish. *Environ. Toxicol. Chem.* 8, 381–391, 1989.

(22) Muir, D. C. G., Rawn, G. P., Townsend, B. E., and Lockhart, W. L. Bioconcentration of cypermethrin, deltamethrin, fenvalerate and permethrin by *Chironomus tentans* larvae in sediment and water. *Environ. Toxicol. Chem.* 4, 51–61, 1985.

(23) U.S. National Library of Medicine. *Hazardous Substances Data Bank.* Bethesda, MD, 1995.

(24) Waller, G. D. Pyrethroid residues and toxicity to honeybees of selected pyrethroid formulations applied to cotton in Arizona. *Econ. Entomol.* 81(4), 1022–1026, 1988.

(25) Wauchope, R. D., Buttler, T. M., Hornsby, A. G., Augustijn-Beckers, P. W. M., and Burt, J. P. SCS/ARS/CES pesticide properties database for environmental decisionmaking. *Rev. Environ. Contam. Toxicol.* 123, 1–157, 1992.

(26) Harris, C. R. Laboratory studies on the persistence and behavior in soil of four pyrethroid insecticides. *Can. Entomol.* 113, 685–694, 1981.

(27) Agnihotri, N. P. Persistence of some synthetic pyrethroid insecticides in soil, water and sediment. I. *J. Entomol. Res.* 10(2), 147–151, 1986

(28) Westcott, N. D. and Reichle, R. A. Persistence of deltamethrin and cypermethrin on wheat and sweet clover. *J. Environ. Sci.,* Part B, 22(1), 91–101, 1987.

(29) Lu, F. C. A review of the acceptable daily intakes of pesticides assessed by the World Health Organization. *Regul. Toxicol. Pharmacol.* 21, 351–364, 1995.

(30) U.S. Environmental Protection Agency. *Integrated Risk Information System,* Washington, D.C., 1995.

(31) Food and Agriculture Organization of the United Nations. *Pesticide Residues in Food — 1981. FAO Plant Production and Protection Paper 42.* FAO, Geneva, Switzerland, 1981.

(32) Wsyolek, P. C., LaFaunce, N. A., Wachs, T., and Link, D. P. Studies of possible bovine urinary excretion and rumen decomposition of fenvalerate insecticide and a metabolite. *Bull. Environ. Contam. Toxicol.* 26, 262–266, 1981.

(33) Shell Chemical Company. *Pydrin insecticide. Technical Manual,* ADP82-015, 1982.

(34) Williams, W. M., Holden, P. H., Parsons, D. W., and Lorbe, M. N. *Pesticides in Ground Water Data Base. 1988 Interim Report.* U.S. Environmental Protection Angency, Washington, D.C., 1988.

(35) U.S. Environmental Protection Agency. Pesticide tolerance on an agricultural commodity: flucythrinate. *Fed. Reg.* 50, 21050–21051, 1985.

(36) Mehler, L. N. *Assessment of Human Exposure to Flucythrinate,* HS-1510. California Department of Food and Agriculture, Sacramento, CA, 1989.

(37) U.S. Environmental Protection Agency. *Pesticide Environmental Fate One-Line Summary: Esfenvalerate.* Environmental Fate and Effects Division, Washington, D.C., 1989.

(38) U.S. Environmental Protection Agency. *Pesticide Fact Sheet Number 86: Fluvalinate.* Office of Pesticides and Toxic Substances, Washington, D.C., 1986.

(39) Maddy, K.T. *A Study of Fluvalinate Dislodgeable Degradation Rates on Orange Foliage in Tulare County in California During May 1983.* California Department of Food and Agriculture, Sacramento, CA, 1984.

(40) Morgan, D. P. *Recognition and Management of Pesticide Poisonings,* 3rd ed. U.S. Environmental Protection Agency. Washington, D.C., 1982.

(41) Penick Corporation. *Technical Information Sheet: Pramex (Permethrin) Synthetic Pyrethroid Insecticide.* Lyndhurst, NJ, 1979.

(42) Schimmel, S. C. Acute toxicity, bioconcentration, and persistence of AC 222,705, benthiocarb, chlorpyrifos, fenvalerate, methyl parathion, and permethrin in the estuarine environment. *J. Agric. Food Chem.* 31(2), 399–407, 1983.

(43) Morse, R. A. Bee poisoning. 1988 New York state pesticide recommendations. *49th Annu. Pest Control Conf.* Cornell University, Ithaca, NY, 1987.

(44) World Health Organization. *Permethrin Environmental Health Criteria 94.* International Programme on Chemical Safety, Geneva, Switzerland, 1990.

(45) Wagenet, L. P. *A Review of Physical-chemical Parameters Related to the Soil and Groundwater Fate of Selected Pesticides in N.Y. State. Report #30.* Cornell University Agricultural Experiment Station, Ithaca, NY, 1985.

(46) World Health Organization. *Resmethrins — Resmethrin, Bioresmethrin, Cisresmethrin. Environmental Health Criteria 92.* International Programme on Chemical Safety, Geneva, Switzerland, 1989.

(47) Christopher, R. J. Distribution and depletion of carbon-14 resmethrin isomers administered orally to laying hens. *Pestic. Sci.* 16(4), 378–382, 1985.

(48) Hill, E. F. and Camardese, M. B. *Lethal Dietary Toxicities of Environmental Contaminants and Pesticides to Coturnix, Technical Report 2.* U.S. Department of the Interior, Fish and Wildlife Service, Washington, D.C., 1986.

(49) Tietze N. S. Acute toxicity of mosquitocidal compounds to young mosquitofish (*Gambusia affinis*). *J. Am. Mosq. Control Assoc.* 7(2), 290–293, 1991.

(50) Johnson, W. W. and Finley, M. T. *Handbook of Acute Toxicity of Chemicals to Fish and Aquatic Invertebrates. Resource Publication 137.* U.S. Department of the Interior, Fish and Wildlife Service, Washington, D.C., 1980.

(51) Augustijn-Beckers, P. W. M., Hornsby, A. G., and Wauchope, R. D. SCS/ARS/CES Pesticide properties database for environmental decisionmaking II. Additional compounds. *Rev. Environ. Contam. Toxicol.* 137, 1–82, 1994.

(52) Penick Corp. *Technical Information Sheet: SBP-1382 (Resmethrin).* Pesticide Technology Department, Orange, NJ, 1976.

(53) National Research Council. *Drinking Water and Health,* Vol. 5. National Academy Press, Washington, D.C., 1983.

(54) National Toxicology Program. *Toxicology and Carcinogenesis Studies of Rotenone in F344/N Rats and B6C3F Mice. (Report No. 320).* National Institute of Health, Bethesda, MD, 1984.

(55) Hudson, R. H., Tucker, R. K., and Haegele, M. A. *Handbook of Toxicity of Pesticides to Wildlife, Resource Publication 153.* U.S. Department of the Interior, Fish and Wildlife Service, Washington, D.C., 1984.

(56) Soloway, S. B. Naturally occurring insecticides. *Environ. Health Perspect.* 14, 109–116, 1976.

(57) Kuna, S. and Heal, R. E. Toxicological and pharmacological studies on the powdered stem of ryania speciosa, a plant insecticide. *J. Pharmacol. Exp. Ther.* 93, 407–413, 1948.

(58) Cusida, J. E., Pessah, I. N., Seibert, J., and Waterhouse, A. L. Ryania insecticide: chemistry, biochemistry and toxicology. In *Pesticide Science and and Biotechnology.* Greenlagh, R. and Roberts, T. R., Eds. Blackwell Publishers, Oxford, U.K., 1987.

(59) Zhang, Z. D., Kwan, C. Y., and David, E. E. Subcellular-membrane characterization of [3H]ryanodine-binding sites. *Biochem. J.* 290(1), 259–261, 1990.

(60) Gong, C., Zderic, S. A., and Levin, R. M. Ontogeny of the ryanodine receptor in rabbit urinary bladder smooth muscle. *Mol. Cell. Biochem.* 137(2), 169–182, 1994.

chapter three

Carbamates

3.1 Class overview and general description

Background

Carbamates were developed into commercial pesticides in the 1950s. Although members of this family of chemicals are effective as insecticides, herbicides, and fungicides, they are most commonly used as insecticides. Perhaps the most widely applied carbamate insecticide is carbaryl, which is especially utilized for lawns and gardens. There are approximately 25 other carbamate compounds currently in use as pesticides or pharmaceutical agents.

The carbamates have replaced chlorinated hydrocarbons as the pesticides of choice for controlling pests on agricultural crops such as citrus, forage crops, and cotton; on garden vegetables and lawns; and in the home. They control a wide variety of invertebrates; e.g., mites, spiders, and earthworms. The carbamates are generally applied to the soil and taken up by the plant.

As a group, carbamates are colorless and odorless compounds which are crystalline at normal temperatures (1). They are stable when exposed to air, light, or heat during storage. Carbamates are derivatives of carbamic acid and have the general structure given in Figure 3.1 (2). Table 3.1 lists the carbamates.

Carbamate usage

In 1987, herbicide production in the U.S. was estimated to be 623 million pounds (3). It was estimated that in 1987, carbamates represented 12.6% of all herbicides used. Assuming that all herbicides produced were applied, overall carbamate usage for 1987 was approximately 78.5 million pounds (4).

Mode of action and toxicology

Carbamates were originally identified in extracts of the calabar bean that grows in West Africa. These extracts contained a compound called physostigmine, which is a methylcarbamate ester (5). Many of the effects of carbamates have been documented from the study of this naturally occurring compound and its derivatives (5).

The primary way that insecticidal carbamates work on both target and non-target species is through the inhibition of the enzyme acetylcholinesterase. Acetylcholine (ACh) is a substance that transmits a nerve impulse from a nerve cell to a specific receptor such as another nerve cell or a muscle cell. Acetylcholine, in essence, acts as a chemical switch. When it is present (produced by the nerve cell), it turns the

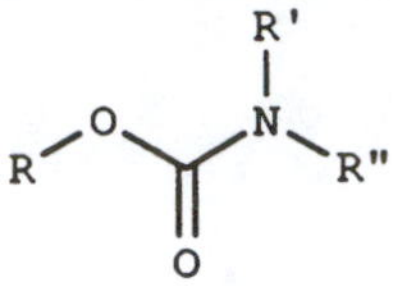

Figure 3.1 Generic carbamate structure, where R is an alcohol (OH), oxime (C=NOH), or, phenol ring, and R′ is either a hydrogen (H) or methyl group (CH$_3$).

Table 3.1 Carbamates

Aldicarb*	Isoprocarb
Bendiocarb*	Methiocarb
Carbaryl*	Methomyl*
Carbofuran*	Oxamyl*
Carbosulfan	Promecarb
Chlorpropham*	Propoxur*
Fenoxycarb*	Thiodiocarb

Note: * indicates that a profile for this compound is included in this chapter.

nerve impulse on. When it is absent, the nerve impulse is discontinued. The nerve transmission ends when the enzyme acetylcholinesterase breaks down the ACh into choline and acetic acid. Without the action of this enzyme, ACh builds up at the junction of the nerve cell and the receptor site, and the nerve impulse continues. Carbamate insecticides block (or inhibit) the ability of this enzyme, acetylcholinesterase, to break down the ACh and end the nerve impulse.

Carbamate inhibition of acetylcholinesterase is a reversible process. Estimates of the recovery time in humans range from immediate up to 4 days, depending on the dose, the specific pesticide, and the method of exposure (2,5). The breakdown of carbamate compounds within an organism is a complex process and is dependent on the specific pesticide structure (1).

Acute toxicity

The onset and severity of symptoms of carbamate pesticide poisoning are dose related. Ingestion of a carbamate insecticide at low doses can cause excessive salivation and an increase in the rate of breathing within 30 minutes. At higher doses, this is followed by excessive tearing, urination, uncontrollable defecation, nausea, and vomiting (5). At the highest doses, symptoms can include those listed above, along with violent intestinal movements, muscle spasms, and convulsions. Death has occurred in a few instances, usually due to respiratory failure resulting from paralysis of the respiratory muscles (5).

Chronic toxicity

While the insecticidal carbamates produce the typical symptoms of cholinesterase inhibition, they do not appear to induce a delayed neurotoxic reaction similar to that seen with some organophosphorus compounds. In separate neurotoxic studies of many of the carbamate compounds listed above, various acute doses of the compounds were administered to rats that were equivalent to half and up to several times the LD$_{50}$ values for those rats. Each of the study results showed no behavioral evidence of neurotoxicity during the 22-day observation period (5,6).

Reproductive effects

Many carbamates have been tested for reproductive effects in a variety of mammalian species. In three-generation rat studies with various carbamates, when low dietary doses below 50 mg/kg/day were given, there were no adverse effects on parental fertility, litter size, or lactation. However, at higher doses of 200 mg/kg or above, there were signs of reduction in parental food consumption, in growth, in lactation, in litter size, and in growth of the pups.

There is no evidence that carbamate insecticides in common use cause reproductive toxicity (5).

Teratogenic effects

Teratogenic studies have been conducted on the majority of the carbamate pesticides with no significant evidence of birth defects (7). Therefore, carbamate compounds are unlikely to cause birth defects in humans.

Mutagenic effects

There have been numerous gene mutation assays performed on bacteria, such as *Salmonella typhimurium* and *Escherichia coli* for many carbamate compounds. The results showed no mutagenic effects (5).

Tests for primary DNA damage for the above bacteria have also indicated negative results (5). However, weak mutagenicity has been reported for several of these carbamates, though only at highly toxic doses (8).

Therefore, exposure to carbamate compounds is unlikely to cause mutagenic effects in humans.

Carcinogenic effects

Carbamate compounds have been studied via 2-year dietary rat and mice studies. From these studies, no evidence has been found that the carbamates are carcinogenic (5).

Organ toxicity

During a 2-year dietary dose study, dogs given bendiocarb at 12 mg/kg/day had depressed acetylcholinesterase activity. There were no changes in organ weights, nor any other adverse effects (5).

In another 2-year dietary dose study, male rats fed 0.75 mg/kg/day of oxamyl had decreased weights of the heart, testes, and adrenal glands (9).

Therefore, chronic exposure to carbamate compounds may cause adverse effects on organs or acetylcholinesterase levels. These effects are unlikely to occur in humans at expected exposure levels.

Fate in humans and animals

Rats eliminate carbamate compounds rapidly via the urine. Some organisms oxidize the carbamate compounds into sulfoxides, which are then further oxidized before excretion. Most metabolites are excreted within 24 hours of exposure and therefore carbamate residues do not accumulate in animals (5).

Ecological effects

Effects on birds

Carbamate compounds are highly toxic to many types of birds, such as red-winged blackbirds (LD_{50}, 1.78 mg/kg for aldicarb) and ring-necked pheasants (LD_{50},

5.34 mg/kg for aldicarb) (10,11). It has been reported that one granule of carbofuran was sufficient to kill a small bird (10).

Effects on aquatic organisms

Carbamate compounds range from highly toxic to moderately toxic for fish such as bluegill sunfish (LC_{50} of 0.9 mg/L for methomyl), rainbow trout (LC_{50} of 1.6 mg/L for fenoxycarb), and bass (LC_{50} of 10 mg/L for chlorpropham) (10).

Effects on other organisms (non-target species)

Most carbamate compounds are highly toxic to bees (10). Also, bendiocarb has been observed to reduce earthworm populations (12).

Environmental fate

Breakdown in soil and groundwater

Carbamates, like the organophosphates, are generally not persistent in the environment (13). They usually remain active for a few hours to a few months in soils and crops and are rarely found in plants beyond the growing season. The rate of degradation in soil depends on moisture, pH, and organic content of the soil. The higher the organic content, the greater the binding to soil and thus the greater the persistence. Also, the higher the soil acidity, the longer it takes for carbamates to be degraded. In general, breakdown occurs rapidly enough so that groundwater contamination is unlikely (13).

For example, carbaryl has a half-life of 7 days in aerobic soil and 28 days in nonaerobic soil (14).

Breakdown in water

Carbamate compounds degrade through chemical hydrolysis and microbial processes and are unlikely to bioaccumulate in aquatic systems. They do not accumulate in water. In alkaline water and sunlight, carbamate compounds will decompose more rapidly (14).

Breakdown in vegetation

Carbamate compounds are absorbed and translocated through plants and treated crops. In most cases, carbamates will break down quickly in plants and the residues in plants will last for approximately 1 to 2 weeks (10).

3.2   Individual Profiles

3.2.1.   Aldicarb

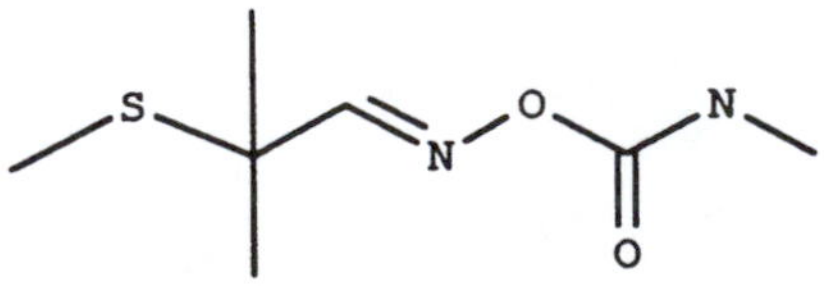

Figure 3.2 Aldicarb.

Trade or other names

Aldicarb is also called aldicarbe. Trade names include Temik, ENT 27093, OMS 771, and UC 21149.

Regulatory status

Aldicarb is a Restricted Use Pesticide (RUP) in the U.S. RUPs may be purchased and used only by certified applicators. Aldicarb is rated toxicity class I — highly toxic. Products containing aldicarb bear the signal word DANGER — POISON on the label.

Introduction

Aldicarb, a member of the carbamate class of chemicals, is an extremely toxic systemic insecticide. Used to control mites, nematodes, and aphids, it is applied directly to the soil. It is used widely on cotton, peanut, and soybean crops. In the mid-1980s, there were highly publicized incidents in which misapplication of aldicarb contaminated cucumbers and watermelons and led to adverse effects in people. In 1990, the manufacturer of Temik (aldicarb) announced a voluntary halt on its sale for use on potatoes because of concerns about groundwater contamination.

Toxicological effects

Acute toxicity

The primary route of human exposure to aldicarb is consumption of contaminated food and of water from contaminated wells (14). Occupational exposure to high levels of aldicarb is due to product handling, and most cases of aldicarb poisoning occur from loading and application of the pesticide.

Atypical of carbamates in general, aldicarb is extremely toxic through both the oral and dermal routes (5). Absorption from the gut is rapid and almost complete. When administered in oil or other organic solvents, aldicarb is absorbed rapidly through the skin. Its skin toxicity is roughly 1000 times that of other carbamates (5). In humans, the onset of symptoms is rapid, 15 minutes to 3 hours. Symptoms disappear in 4 to 12 hours (15).

Aldicarb's LD_{50} in rats, mice, guinea pigs, and rabbits ranges from 0.5 to 1.5 mg/kg when administered in liquid or oil form. The toxicities of the dry granules are distinctly lower (LD_{50} = 7.0 mg/kg), though still highly toxic (5,15).

Aldicarb is a cholinesterase inhibitor and can thus result in a variety of symptoms, including weakness, blurred vision, headache, nausea, tearing, sweating, and tremors. Very high doses can result in death due to paralysis of the respiratory system.

Chronic toxicity

There is very little evidence of chronic effects from aldicarb exposure. Rats and dogs fed low doses of aldicarb for 2 years showed no significant adverse effects (15). One epidemiological study suggested a possible link between low-level exposure and immunological abnormalities (15). The results of this study have been widely disputed.

Reproductive effects

Aldicarb administered to pregnant rats at very low levels (0.001 to 0.1 mg/kg/day) depressed acetylcholinesterase activity more in the fetuses than in the mother. The aldicarb was also retained in the mother's body for longer periods than in nonpregnant rats (16). A three-generation study at doses of 0.05 and 0.10 mg/kg/day produced no significant toxic effects; and in another study, a dose of 0.70 mg/kg/day produced no adverse effects (5). Thus, reproductive effects in humans are unlikely at expected exposure levels.

Teratogenic effects

No teratogenic effects have been found in rats exposed to aldicarb (5).

Mutagenic effects

Studies show that aldicarb is not mutagenic (15).

Carcinogenic effects

Studies indicate that aldicarb is not carcinogenic (15).

Organ toxicity

Aldicarb's effects are seen primarily on the nervous system.

Fate in animals and humans

Aldicarb is metabolized and quickly excreted. Rats and cows eliminate 80 to 90% of a dose of aldicarb within 24 hours. Elimination is mainly through urine, but some aldicarb leaves by way of the lungs in expired air and some through milk in cows (5).

Ecological effects

Effects on birds

Aldicarb is very highly toxic to birds. The greatest exposure to the organisms is through the ingestion of unincorporated granules and contaminated earthworms (17). The LD_{50} values of the compound range from 1.78 mg/kg in the red-winged blackbird to 5.34 mg/kg in the ring-necked pheasant.

Effects on aquatic organisms

Aldicarb is moderately toxic to fish. The LC_{50} (96-hour) is 8.8 mg/L in rainbow trout and 1.5 mg/L in bluegill sunfish (10). Bioconcentration in aquatic species is low (14,17,18).

Effects on other organisms (non-target species)

Aldicarb is not toxic to bees, even when applied directly (10).

Environmental fate

Breakdown in soil and groundwater

Aldicarb is moderately persistent in soil (10). Moisture and pH have important impacts on the rate of breakdown. Aldicarb is very soluble and mobile in soil (8).

Aldicarb movement is most serious for sandy or sandy loam soils. It has been found in wells in over 25 countries and in 12 states at concentrations above the drinking water Maximum Contaminant Level (14).

Breakdown in water

The half-life of aldicarb in water is from 1 day to a few months (14). In pond water, aldicarb is broken down rapidly and has a half-life of 5 to 10 days. It is degraded by bacteria, sunlight, and reactions with water. Because of its rapid degradation rate, levels in surface water may be lower than those in groundwater (5).

Breakdown in vegetation

In plants, aldicarb is rapidly converted to sulfoxide and more slowly to the sulfone compound. Citrus trees treated with 18 g per tree had the highest residue levels in the leaves (8). Residues of aldicarb also have been reported in sugar beets and grape leaves and fruit (14). In 1991, the manufacturer called for a halt to aldicarb use on bananas because of the elevated levels of residue found.

Physical properties

Aldicarb is a white crystalline solid. It is formulated as a granular mix (10 to 15% active ingredient) because of its toxicity. It is not compatible with alkaline materials and is noncorrosive to metals and plastics.

Chemical name: 2-methyl-2-(methylthio)propionaldehyde O-methylcarbamoyloxime (10)
CAS #: 116-06-3
Molecular weight: 190.27 (5)
Solubility in water: 6000 mg/L @ room temperature (10)
Solubility in other solvents: s. in acetone, xylene, ethyl ether, toluene, and other organic solvents (10)
Melting point: 99–100°C (10)
Vapor pressure: 13 mPa @ 20°C (10)
Partition coefficient (octanol/water): 1.13 (8)
Adsorption coefficient: 30 (13)

Exposure guidelines

ADI: 0.003 mg/kg/day (10)
MCL: 0.003 mg/L (19)
RfD: 0.001 mg/kg/day (20)
PEL: Not available

Basic manufacturer

Rhone-Poulenc Ag. Co.
P.O. Box 12014
2 T.W. Alexander Park, NC 27709
Telephone: 919-549-2000
Emergency: 800-334-7577

3.2.2   Bendiocarb

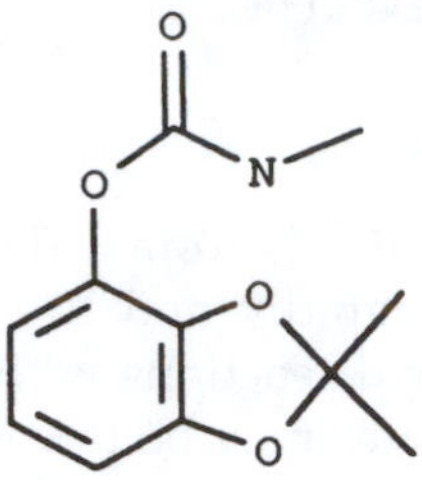

Figure 3.3 Bendiocarb.

Trade or other names

Trade names include Ficam, Dycarb, Garvox, Multamat, Multimet, Niomil, Rotate, Seedox, Tattoo, and Turcam.

Regulatory status

Most formulations of bendiocarb are classified as General Use Pesticides (GUP), with the exception of Turcam and Turcam 2.5 G, which are classified as Restricted Use Pesticides (RUPs). RUPs may be purchased and used only by certified applicators. Bendiocarb is toxicity class II — moderately toxic. Products containing bendiocarb bear the Signal Word WARNING.

Introduction

Bendiocarb is a carbamate insecticide. It is effective against a wide range of nuisance and disease vector insects. It is used to control mosquitoes, flies, wasps, ants, fleas, cockroaches, silverfish, ticks, and other pests in homes, industrial plants, and food storage sites. In agriculture, it is used against a variety of insects, especially those in the soil. Bendiocarb is also used as a seed treatment on sugar beets and maize and against snails and slugs. Pesticides containing bendiocarb are formulated as dusts, granules, ultra-low volume sprays, and as wettable powders.

Toxicological effects

Acute toxicity

Bendiocarb is moderately toxic if it is ingested or if it is absorbed through the skin (5). Absorption through the skin is the most likely route of exposure. It is a mild irritant to the skin and eyes (5).

Like other carbamate insecticides, bendiocarb is a reversible inhibitor of cholinesterase, an essential nervous system enzyme. Symptoms of bendiocarb poisoning include weakness, blurred vision, headache, nausea, abdominal cramps, chest discomfort, constriction of pupils, sweating, muscle tremors, and decreased pulse. If there is severe poisoning, symptoms of twitching, giddiness, confusion, muscle incoordination, slurred speech, low blood pressure, heart irregularities, and loss of reflexes may also be experienced. Death can result from discontinued breathing, paralysis of muscles of the respiratory system, intense constriction of the openings of the lung, or all three (5).

In one case of exposure while applying bendiocarb, the victim experienced symptoms of severe headache, vomiting, and excessive salivation, and his cholinesterase level was depressed by 63%. He recovered from these symptoms in less than 3 hours with no medical treatment, and his cholinesterase level returned to normal within 24 hours. In another case, poisoning occurred when an applicator who was not wearing protective equipment attempted to clean contaminated equipment. The victim experienced nausea, vomiting, incoordination, pain in his arms, hands, and legs, muscle spasms, and breathing difficulty. These symptoms abated within 2 hours after decontamination and treatment with atropine. The victim fully recovered by the following day (5).

The oral LD_{50} for bendiocarb is 34 to 156 mg/kg in rats, 35 to 40 mg/kg in rabbits, and 35 mg/kg in guinea pigs. The dermal LD_{50} is 566 mg/kg in rats (5). The acute inhalation LC_{50} (4-hour) is 0.55 mg/L air in rats (10).

Chronic toxicity

A 2-year study with rats fed high doses of 10 mg/kg/day showed a wide range of changes in organ weights, blood, and urine characteristics, as well as an increased incidence of stomach and eye lesions (5).

Reproductive effects

In a three-generation study with rats, fertility and reproduction were not affected by bendiocarb at dietary doses of up to 12.5 mg/kg/day. Very high prenatal and postnatal doses of 40 mg/kg/day were toxic to rat dams and reduced pup weight and survival rates. No effects were seen at 20 mg/kg/day (5). Thus, no reproductive effects are likely in humans at expected exposure levels.

Teratogenic effects

No teratogenic effects were seen in the offspring of rats given 4 mg/kg/day or in rabbits given 5 mg/kg/day of bendiocarb during gestation (5).

Mutagenic effects

Numerous studies show that bendiocarb is not mutagenic (5).

Carcinogenic effects

Bendiocarb was not carcinogenic in 2-year studies of rats and mice (5).

Organ toxicity

No changes in organ weight or harmful effects in tissues were observed in a 2-year dietary study of dogs fed doses of up to 12.5 mg/kg/day despite elevated serum cholesterol and decreased levels of calcium in the bloodstream (5).

Fate in humans and animals

Bendiocarb is absorbed through all the normal routes of exposure, but dermal absorption is especially rapid. Carbamates generally are excreted rapidly and do not accumulate in mammalian tissue. If exposure does not continue, cholinesterase inhibition and its symptoms reverse rapidly. In nonfatal cases, the illness generally lasts less than 24 hours (21). Within two days after feeding doses of up to 10 mg/kg of bendiocarb to rats, 89 to 90% of the dose was eliminated in the urine, 2 to 6% was exhaled, and another 2 to 6% was eliminated in the feces. This same pattern of elimination was observed in a human subject given an oral dose of bendiocarb (5).

Ecological effects

Effects on birds

Bendiocarb is moderately toxic to birds. The LD_{50} in mallard ducks is 3.1 mg/kg, and in quail is 19 mg/kg (22).

Effects on acquatic organisms

Bendiocarb is moderately to highly toxic to fish. The LC_{50} (96-hour) for bendiocarb in rainbow trout is 1.55 mg/L (10).

Effects on other organisms (non-target species)

Earthworm populations under turf are severely affected by bendiocarb (21). It is toxic to bees. The LD_{50} is 0.0001 mg per bee.

Environmental fate

Breakdown in soil and groundwater

The half-life of bendiocarb varies with soil type from less than 1 week to up to 4 weeks (10,23). It has a low soil persistence.

Breakdown in water

Bendiocarb is degraded in solution by the chemical action of water (hydrolysis). It does not accumulate in water.

Breakdown in vegetation

Bendiocarb is not toxic to plants when used as directed (10).

Physical properties

Bendiocarb is an odorless, white crystalline solid. It is stable under normal temperatures and pressures, but should not be mixed with alkaline preparations. Thermal decomposition products may include toxic oxides of nitrogen. It is non-corrosive.

Chemical name: 2,3-isopropylidenedioxyphenyl methylcarbamate (10)
CAS #: 22781-23-3
Molecular weight: 223.23 (10)
Solubility in water: 40 mg/L @ 20°C (10)
Solubility in other solvents: acetone v.s.; benzene s.; chloroform s.; dioxane v.s.; ethanol s.; hexane v.s. (10)
Melting point: 129–130°C (10)
Vapor pressure: 0.66 mPa @ 25°C (10)
Partition coefficient (octanol/water): 50 (10)
Adsorption coefficient: 570 (13)

Exposure guidelines

ADI: 0.004 mg/kg/day (10)
MCL: Not available
RfD: 0.0013 mg/kg/day (20)
PEL: Not available

Basic manufacturer

Roussel Uclaf Corp.
95 Chestnut Ridge Road
Martvale, NJ 07645
Telephone: 201-307-9700

3.2.3   Carbaryl

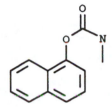

Figure 3.4 Carbaryl.

Trade or other names

Product names include Adios, Bugmaster, Carbamec, Carbamine, Crunch, Denapon, Dicarbam, Hexavin, Karbaspray, Nac, Ravyon, Septene, Sevin, Tercyl, Tornado, Thinsec, Tricarnam, and Union Carbide 7744.

Regulatory status

Carbaryl is a General Use Pesticide (GUP). However, various formulations vary widely in toxicity. For example, it is categorized as toxicity class I — highly toxic for Tercyl; toxicity class II — moderately toxic for Sevin 803; and toxicity class III — slightly toxic for some other products. Products containing carbaryl may bear the Signal Word DANGER — POISON, WARNING, or CAUTION, depending on the product formulation.

Introduction

Carbaryl is a wide-spectrum carbamate insecticide that controls over 100 species of insects on citrus fruit, cotton, forests, lawns, nuts, ornamentals, shade trees, and other crops, as well as on poultry, livestock, and pets. It is also used as a molluscicide and an acaricide. Carbaryl works whether it is ingested into the stomach of the pest or absorbed through direct contact. It is available as bait, dusts, wettable powders, granules, dispersions, and suspensions.

Toxicological effects

Acute toxicity

Carbaryl is moderately to very toxic. It can produce adverse effects in humans by skin contact, inhalation, or ingestion. The symptoms of acute toxicity are typical of the other carbamates. Direct contact of the skin or eyes with moderate levels of this pesticide can cause burns. Inhalation or ingestion of very large amounts can be toxic to the nervous and respiratory systems, resulting in nausea, stomach cramps, diarrhea, and excessive salivation. Other symptoms at high doses include sweating, blurring of vision, incoordination, and convulsions. The only documented fatality from carbaryl was through intentional ingestion.

The oral LD_{50} of carbaryl ranges from 250 to 850 mg/kg in rats, and from 100 to 650 mg/kg in mice (8,24). The inhalation LC_{50} in rats is greater than 200 mg/L (24). Low doses can cause minor skin and eye irritation in rabbits, a species in which carbaryl's dermal LD_{50} has been measured at greater than 2000 mg/kg (8).

Chronic toxicity

Reproductive effects

No reproductive or fetal effects were observed during a long-term study of rats fed high doses of carbaryl (8).

Teratogenic effects

The evidence for teratogenic effects due to chronic exposure is minimal in test amimals. Birth defects in rabbit and guinea pig offspring occurred only at dosage levels that were highly toxic to the mother (25).

Mutagenic effects

Carbaryl has been shown to affect cell division and chromosomes in rats (24). However, numerous studies indicate that carbaryl poses only a slight mutagenic risk (8,26). There is a possibility that carbaryl may react in the human stomach to form a more mutagenic compound, but this has not been demonstrated. In sum, the evidence suggests that carbaryl is unlikely to be mutagenic to humans (26,27).

Carcinogenic effects

Technical-grade carbaryl has not caused tumors in long-term and lifetime studies of mice and rats. Rats were administered high daily doses of the pesticide for 2 years, and mice for 18 months, with no signs of carcinogenicity (28). While N-nitrosocarbaryl, a possible by-product, has been shown to be carcinogenic in rats at high doses, this product has not been detected. Thus, the evidence indicates that carbaryl is unlikely to be carcinogenic to humans (29).

Organ toxicity

Ingestion of carbaryl affects the lungs, kidneys, and liver. Inhalation will also affect the lungs (5,30). Nerve damage can occur after administration of high doses for 50 days in rats and pigs (18). Several studies indicate that carbaryl can affect the immune system in animals and insects.

Male volunteers who consumed low doses of carbaryl for 6 weeks did not show symptoms, but tests indicated slight changes in their body chemistry (8). A 2-year study with rats revealed no effects at or below a dose of 10 mg/kg/day (25).

Fate in humans and animals

Most animals, including humans, readily break down carbaryl and rapidly excrete it in the urine and feces. Workers occupationally exposed by inhalation to carbaryl dust excreted 74% of the inhaled dose in the urine in the form of a breakdown product (24). The metabolism of up to 85% of carbaryl occurs within 24 hours after administration (24).

Ecological effects

Effects on birds

Carbaryl is practically nontoxic to wild bird species. The LD_{50} values are greater than 2000 mg/kg in mallards and pheasants, 2230 mg/kg in quail, and 1000 to 3000 mg/kg in pigeons (10).

Effects on aquatic organisms

Carbaryl is moderately toxic to aquatic organisms, such as rainbow trout (LC_{50} of 1.3 mg/L), and bluegill (LC_{50} of 10 mg/L) (10). Some accumulation of carbaryl can occur in catfish, crawfish, and snails, as well as in algae and duckweed. Residue levels in fish were 140-fold greater than the concentration of carbaryl in water. In general, due to its rapid metabolism and rapid degradation, carbaryl should not pose a significant bioaccumulation risk in alkaline waters. However, under conditions below neutrality, it may be significant (10).

Effects on other organisms (non-target species)

Carbaryl is lethal to many non-target insects, including bees and beneficial insects (10).

Environmental fate

Breakdown in soil and groundwater

Carbaryl has a low persistence in soil. Degradation of carbaryl in the soil is mostly due to sunlight and bacterial action. It is bound by organic matter and can be transported in soil runoff. Carbaryl has a half-life of 7 to 14 days in sandy loam soil and 14 to 28 days in clay loam soil. Carbaryl has been detected in groundwater in three separate cases in California (14).

Breakdown in water

In surface water, carbaryl is broken down by bacteria and through hydrolysis. Evaporation is very slow. Carbaryl has a half-life of about 10 days at neutral pH. The half-life varies greatly with water acidity (14).

Breakdown in vegetation

Degradation of carbaryl in crops occurs by hydrolysis inside the plants. It has a short residual life of less than 2 weeks. The metabolites of carbaryl have lower toxicity to humans than carbaryl itself. The breakdown of this substance is strongly dependent on acidity and temperature (8).

Physical properties

Carbaryl is a solid that varies from colorless to white or gray, depending on the purity of the compound. The crystals are odorless. Carbaryl is stable to heat, light, and acids. It is not stable under alkaline conditions. It is noncorrosive to metals, packaging materials, and application equipment.

Chemical name: 1-naphthyl methylcarbamate (10)
CAS #: 63-25-2
Molecular weight: 201.23 (10)
Solubility in water: 40 mg/L @ 30°C (10)
Solubility in other solvents: dimethylformamide v.s.; acetone s.; dimethyl sulfoxide v.s.; cyclohexanone s. (10)
Melting point: 142°C (10)
Vapor pressure: <5.3 mPa @ 25°C (10)
Partition coefficient (octanol/water): Not available
Adsorption coefficient: 300 (10)

Exposure guidelines

ADI: 0.01 mg/kg/day (10)
HA: 0.7 mg/L (lifetime) (24)
RfD: 0.1 mg/kg/day (20)
PEL: 5 mg/m^3 (8-hour) (31)

Basic manufacturer

Rhone-Poulenc Ag. Co.
P.O. Box 12014
T.W. Alexander Dr.
Research Triangle Park, NC 27709
Telephone: 919-549-2000
Emergency: 800-334-7577

3.2.4   *Carbofuran*

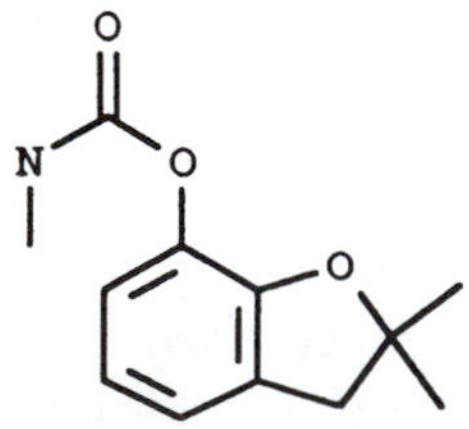

Figure 3.5 Carbofuran.

Trade or other names

Trade names include Furadan, Bay 70143, Carbodan, Carbosip, Chinufor, Curaterr, D 1221, ENT 27164, Furacarb, Kenafuran, Pillarfuron, Rampart, Nex, and Yaltox.

Regulatory status

Following a Special Review, the EPA initiated a ban on all granular formulations of carbofuran that became effective on September 1, 1994. Before 1991, 80% of the total usage of carbofuran was in granular formulations. The ban was established to protect birds and is not related to human health concerns. Bird kills have occurred when birds ingested carbofuran granules, which resemble grain seeds, and when predatory or scavenging birds ingested small birds or mammals that had eaten carbofuran pellets. There is no ban on liquid formulations of carbofuran.

Liquid formulations of carbofuran are classified as Restricted Use Pesticides (RUPs) because of their acute oral and inhalation toxicity to humans. Granular formulations are also classified as RUPs, but for a different reason: their toxicity to birds. Liquid formulations bear the Signal Word WARNING. Granular formulations bear the Signal Word DANGER. Formulations of carbofuran are in toxicity class I — highly toxic; or toxicity class II — moderately toxic.

Introduction

Carbofuran is a broad-spectrum carbamate pesticide that kills insects, mites, and nematodes on contact or after ingestion. It is used against soil and foliar pests of field, fruit, vegetable, and forest crops. Carbofuran is available in liquid and granular formulations, but as stated above, the granular form is banned in the U.S.

Toxicological effects

Acute toxicity

Carbofuran is highly toxic by inhalation and ingestion and moderately toxic by dermal absorption (5).

As with other carbamate compounds, carbofuran's cholinesterase-inhibiting effect is short term and reversible (5). Symptoms of carbofuran poisoning include nausea, vomiting, abdominal cramps, sweating, diarrhea, excessive salivation, weakness, imbalance, blurring of vision, breathing difficulty, increased blood pressure, and incontinence. Death may result at high doses from respiratory system failure associated with carbofuran exposure (5). Complete recovery from an acute poisoning by carbofuran, with no long-term health effects, is possible if exposure ceases and the victim has time to regain a normal level of cholinesterase and to recover from symptoms (5).

The oral LD_{50} is 5 to 13 mg/kg in rats, 2 mg/kg in mice, and 19 mg/kg in dogs. The dermal LD_{50} is >1000 mg/kg in rabbits (5). The LC_{50} (4-hour) for inhalation of carbofuran is 0.043 to 0.053 mg/L in guinea pigs (10).

Chronic toxicity

Rats given very high doses (5 mg/kg/day) for 2 years showed decreases in weight. Similar tests with mice gave the same results (5). Prolonged or repeated exposure to carbofuran may cause the same effects as an acute exposure (5).

Reproductive effects

Consuming high doses over long periods of time caused damage to testes in dogs, but carbofuran did not have any reproductive effects on rats or mice (5).

Available studies indicate carbofuran is unlikely to cause reproductive effects in humans at expected exposure levels.

Teratogenic effects

Studies indicate carbofuran is not teratogenic. No significant teratogenic effects have been found in offspring of rats given carbofuran (3 mg/kg/day) on days 5 to 19 of gestation. No effects were found in offspring of mice given as much as 1 mg/kg/day throughout gestation. In rabbits, up to 1 mg/kg/day on days 6 to 18 of gestation was not teratogenic (5).

Mutagenic effects

Weak or no mutagenic effects have been reported in animals and bacteria. Carbofuran is most likely nonmutagenic (5).

Carcinogenic effects

Data from animal studies indicate that carbofuran does not pose a risk of cancer to humans (5).

Organ toxicity

Carbofuran causes cholinesterase inhibition in both humans and animals, affecting nervous system function.

Fate in humans and animals

Carbofuran is poorly absorbed through the skin (32). It is metabolized in the liver and eventually excreted in the urine. The half-life in the body is from 6 to 12 hours. Less than 1% of a dose will be excreted in a mother's milk. It does not accumulate in tissue (5).

Ecological effects

Effects on birds

Carbofuran is highly toxic to birds. One granule is sufficient to kill a small bird. Bird kills have occurred when birds ingested carbofuran granules, which resemble grain seeds in size and shape, or when predatory or scavenging birds have ingested small birds or mammals that had eaten carbofuran pellets (33). Red-shouldered hawks have been poisoned after eating prey from carbofuran-treated fields (17).

The LD_{50} is 0.238 mg/kg in fulvous ducks, 0.48 to 0.51 mg/kg in mallard ducks, 12 mg/kg in bobwhite quail, and 4.15 mg/kg in pheasant (17). The LD_{50} is 25 to 39 mg/kg in chickens consuming carbofuran as a powder (10). The LC_{50} (96-hour) in Japanese quail is 746 ppm (34).

Effects on aquatic organisms

Carbofuran is highly toxic to many fish. The LD_{50} (96-hour) is 0.38 mg/L in rainbow trout and 0.24 mg/L in bluegill sunfish (10).

The compound has a low potential to accumulate in aquatic organisms. The bioconcentration factor ranges from 10 in snails to over 100 in fish (14).

Effects on other organisms (non-target species)

Carbofuran is toxic to bees except in the granular formulation (10).

Environmental fate

Breakdown in soil and groundwater

Carbofuran is soluble in water and is moderately persistent in soil. Its half-life is 30 to 120 days. In soil, carbofuran is degraded by chemical hydrolysis and microbial processes. Hydrolysis occurs more rapidly in alkaline soils (14). Carbofuran breaks down in sunlight.

Carbofuran has a high potential for groundwater contamination (14). Carbofuran is mobile to very mobile in sandy loam, silty clay, and silty loam soils; moderately mobile in silty clay loam soils; and only slightly mobile in muck soils. Small amounts of carbofuran have been detected (1 to 5 ppb) in water table aquifers beneath sandy soils in New York and Wisconsin (14).

Breakdown in water

In water, carbofuran is subject to degradation by chemical hydrolysis under alkaline conditions. Photodegradation and aquatic microbes may also contribute to degradation. The hydrolysis half-lives of carbofuran in water at 25°C are 690, 8.2, and 1.0 weeks at pH values of 6.0., 7.0, and 8.0, respectively. Carbofuran does not volatilize from water, nor does it adsorb to sediment or suspended particles (14).

Breakdown in vegetation

The half-life of carbofuran on crops is about 4 days when applied to roots, and longer than 4 days if applied to the leaves (8).

Physical properties

Carbofuran is an odorless, white crystalline solid. Heat breakdown can release toxic fumes. Fires, and the runoff from fire control, may produce irritating or poisonous gases. Closed spaces (storage, etc.) should be aired before entering.

Chemical name: 2,3-dihydro-2,2-dimethylbenzofuran-7yl methylcarbamate (10)
CAS #: 1563-66-2
Molecular weight: 221.25 (10)
Solubility in water: 320 mg/L @ 25°C (10)
Solubility in other solvents: acetone v.s.; acetonitrile v.s.; benzene v.s.; cyclohexone v.s. (5)
Melting point: 153–154°C (10)
Vapor pressure: 2.7 mPa @ 33°C (10)
Partition coefficient (octanol/water): 17–26 (10)
Adsorption coefficient: 22 (13)

Exposure guidelines

ADI: 0.01 mg/kg/day (10)
MCL: 0.04 mg/L (19)
RfD: 0.005 mg/kg/day (20)
TLV: 0.1 mg/m^3 (8-hour) (31)

Basic manufacturer

FMC Corporation
Agricultural Chemicals Group
1735 Market Street
Philadelphia, PA 19103
Telephone: 215-299-6661
Emergency: 800-331-3148

3.2.5   Chlorpropham

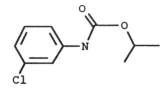

Figure 3.6 Chlorpropham.

Trade or other names

Trade names include Beet-Kleen, Bud Nip, Chloro IPC, CIPC, Furloe, Sprout Nip, Spud-Nic, Taterpex, Triherbide-CIPC, and Unicrop CIPC.

Regulatory status

Chlorpropham is a General Use Pesticide (GUP). The U.S. Environmental Protection Agency classifies it as toxicity class III — slightly toxic. Products containing it bear the Signal Word CAUTION.

Introduction

Chlorpropham, a member of the carbamate family of chemicals, is a plant growth regulator. It is used for pre-emergence control of grass weeds in alfalfa, beans, blueberries, cane berries, carrots, cranberries, ladino clover, garlic, seed grass, onions, spinach, sugar beets, tomatoes, safflower, soybeans, gladioli, and woody nursery stock. It is also used to inhibit potato sprouting and for sucker control in tobacco. It works by inhibiting root growth and photosynthesis. Chlorpropham is available as granules, dustable powder, and emulsifiable concentrate.

Toxicological effects

Acute toxicity

Chlorpropham may cause irritation of the eyes or skin (8). Symptoms of poisoning in laboratory animals exposed to high doses have included listlessness, incoordination, nose bleeds, protruding eyes, bloody tears, difficulty in breathing, prostration, inability to urinate, high fevers, and death. Autopsies of animals have shown inflammation of the stomach and intestinal lining, congestion of the brain, lungs, and other organs, and degenerative changes in the kidneys and liver (8).

The oral LD_{50} for chlorpropham in rats ranges from 5000 to 7500 mg/kg (10). In rabbits, the LD_{50} is 5000 mg/kg (10). The 4-hour inhalation LC_{50} in rats is greater than 32 mg/L (35).

Chronic toxicity

Reproductive effects

In reproductive studies with rats, 500 mg/kg/day, the highest dose tested, produced no adverse effects (4). The evidence suggests that chlorpropham does not cause reproductive effects, even at high doses.

Teratogenic effects

No birth defects occurred in a three-generation study of rats exposed to chlorpropham (35). No teratogenic effects were observed in rabbits at doses up to 250 mg/kg/day (31).

Mutagenic effects

Mutagenicity studies indicate that chlorpropham is either nonmutagenic or slightly mutagenic (8).

Carcinogenic effects

No evidence of carcinogenicity was found in one study of rats fed chlorpropham. This suggests that the pesticide is not carcinogenic, but does not provide conclusive evidence (8).

Organ toxicity

Chronic exposure of laboratory animals has caused retarded growth, increased liver, kidney, and spleen weights, and lesions of the spleen (4).

Fate in humans and animals

No data are currently available.

Ecological effects

Effects on birds

Chlorpropham is practically nontoxic to waterfowl (4). Its LD_{50} in mallards is greater than 2000 mg/kg (10).

Effects on aquatic organisms

Chlorpropham is moderately toxic to cold- and warmwater, freshwater fish. The LC_{50} (48-hour) for chlorpropham is 12 mg/L in bluegill sunfish (10). The LC_{50} is 10 mg/L in bass (10).

Effects on other organisms (non-target species)

Chlorpropham has no effect on streptomyces bacteria or on nitrifying bacteria. It is not dangerous to bees when used as recommended (10).

Environmental fate

Breakdown in soil and groundwater

Chlorpropham is moderately persistent in soil. It is subject to degradation by soil microbes. Photodegradation and volatilization do not readily occur. Soil half-lives of 65 days at 15°C to 30 days at 29°C have been reported (13).

Chlorpropham has some potential to contaminate groundwater because it is soluble in water and it has only a moderate tendency to adsorb to soil particles (13). However, chlorpropham adsorbs strongly to organic matter, so it is unlikely to leach through soils with high organic matter content (13).

Breakdown in water

Chlorpropham breaks down very slowly by reaction with water. At pH 4, 7, and 9 at 40°C, about 90% of the chlorpropham remained in a solution maintained in the dark for 32 days (41).

Breakdown in vegetation

Chlorpropham is absorbed by the roots of susceptible grass seedlings and transported throughout the plant. It is absorbed more slowly by leaves (35).

Physical properties

Technical chlorpropham is a white to light brown crystalline solid. It is stable under normal temperatures and pressures, but poses a slight fire hazard if exposed to heat or flame, and is a fire and explosion hazard in the presence of strong oxidizers. It may burn but will not readily ignite. Thermal decomposition may release highly toxic fumes of phosgene, toxic and corrosive fumes of chlorides, and oxides of carbon.

Chemical name: isopropyl 3-chlorophenylcarbamate
CAS #: 101-21-3
Molecular weight: 213.67 (10)
Solubility in water: 89 mg/L @ 25°C (10)
Solubility in other solvents: benzene v.s.; xylene v.s.; chloroform v.s.; ketones v.s.; acetone v.s.; methanol v.s.; esters v.s.; minerals and oils s. (10)
Melting point: 41.4°C (10)
Vapor pressure: 1.33 mPa @ 25°C (13)
Partition coefficient (octanol/water): Not available
Adsorption coefficient: 400 (estimated) (13)

Exposure guidelines

ADI: Not available
MCL: Not available
RfD: 0.2 mg/kg/day (20)
PEL: Not available

Basic manufacturer

ELF Atochem North America, Inc.
2000 Market Street
Philadelphia, PA 19103-3222
Telephone: 215-419-7219
Emergency: 800-523-0900

3.2.6   *Fenoxycarb*

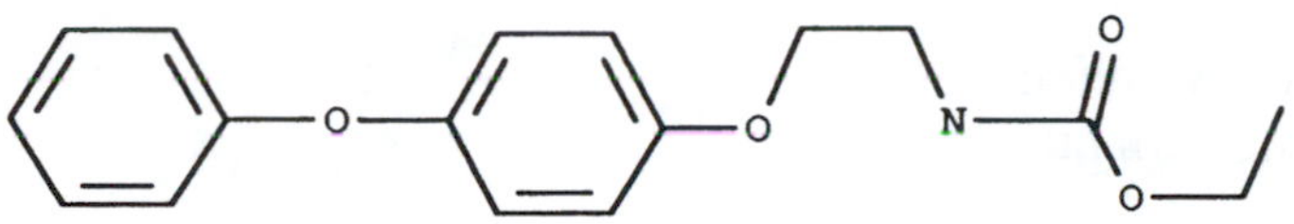

Figure 3.7 Fenoxycarb.

Trade or other names

Trade names for products containing fenoxycarb include Comply, Insegar, Logic, Pictyl, Torus, and Varikill.

Regulatory status

Fenoxycarb is a practically nontoxic pesticide in EPA toxicity class IV. It is a General Use Pesticide (GUP), and labels for products containing it must bear the Signal Word CAUTION.

Introduction

Fenoxycarb is a nonneurotoxic carbamate insect growth regulator used to control a wide variety of insect pests. It is used as a fire ant bait and for flea, mosquito, and cockroach control, and can also be used to control butterflies, moths, beetles, and scale and sucking insects on olives, vines, cotton, and fruit. It is also used to control these pests on stored products, and is often formulated as a grit or corncob bait.

Fenoxycarb blocks the ability of an insect to change into the adult stage from the juvenile stage (metamorphosis). It also interferes with larval molting, the periodic shedding or molting of the old exoskeleton and production of a new, larger one.

Toxicological effects

Acute toxicity

Fenoxycarb is practically nontoxic to mammals via the oral route. The oral LD_{50} of the compound is greater than 10,000 mg/kg in rats (10). It is slightly to practically nontoxic via the dermal route, with a reported dermal LD_{50} in the rat of greater than 2000 mg/kg (10). When 2000 mg/kg were administered to the animals' skin, rats exhibited labored breathing, curved body position, and diarrhea, but no deaths occurred (3). Fenoxycarb is not a skin sensitizer in guinea pigs and causes only minimal eye irritation when applied to rabbits (10). The inhalation toxicity of fenoxycarb is moderate, with an acute inhalation LC_{50} in rats of greater than 0.480 mg/L (10).

Chronic toxicity

Rats fed very low doses of fenoxycarb for a year had no compound-related effects at or below the 10-mg/kg/day dose. Dogs fed the compound at doses of 15.9 mg/kg/day or below for $1\frac{1}{2}$ months experienced no adverse effects (15). Similar results were noted for the compound in mice (1.4 mg/kg) and in rats (0.8 mg/kg/day) over a longer period (36).

Reproductive effects
No data are currently available.

Teratogenic effects
With an unknown species, there were no teratogenic effects observed at doses up to 300 mg/kg/day (36).

Mutagenic effects
According to the Environmental Protection Agency, fenoxycarb is not mutagenic (36).

Carcinogenic effects
No data are currently available.

Organ toxicity
The liver is the primary organ affected by fenoxycarb in long-term animal studies (36).

Fate in humans and animals
About 90% of a dose of fenoxycarb fed to rats was excreted within 96 hours. No residues were found in the animals' organs (36).

Ecological effects

Effects on birds

Fenoxycarb is practically nontoxic to birds (3). The compound has LD_{50} values greater than 3000 and 7000 mg/kg in mallard ducks and bobwhite quail, respectively (10,36). The dietary LC_{50} value for bobwhite quail is about 11,000 ppm (36).

Effects on aquatic organisms

Fenoxycarb is considered moderately to highly toxic to fish, with LC_{50} values ranging from 1.6 mg/L in rainbow trout to 10.3 mg/L in carp (10). Fenoxycarb is also considered highly toxic to the aquatic invertebrate *Daphnia* and affects growth and reproduction after chronic exposures to >1.6 ng/L (36). However, when fenoxycarb was applied at rates ranging from 0.015 to 0.03 lb/acre to ponds, the compound had no effect on a number of different invertebrates, including cladocerans, copepods, ostracods, and mayfly nymphs (38).

In one study, bluegill sunfish accumulated 20 times the amount of the compound's concentration in the water. Tissue residues of the pesticide quickly declined after the fish were placed in pesticide-free water (37). Therefore, it is unlikely that the compound would pose a threat to endangered aquatic organisms or to other organisms that consumed the fish.

Effects on other organisms (non-target species)

Fenoxycarb is practically nontoxic to bees (10).

Environmental fate

Breakdown in soil and groundwater

Fenoxycarb is of low persistence in the soil environment, with a reported field half-life of 1 day (13). It is readily broken down in soil by the chemical action of water (hydrolysis) and by microbial action.

The compound also has a low potential for leaching from the soil and has a moderate to strong tendency for soil binding (36). These characteristics of fenoxycarb in soil indicate that it is unlikely to contaminate groundwater.

Breakdown in water

The compound is stable to hydrolysis in acidic water. It breaks down very rapidly in the presence of sunlight (photodegrades) in water. Half of the initial amount of the compound is broken down by this means within 5 hours (36). It readily attaches onto organic matter, which may limit its persistence in water. Residues in the water could be detected for only 2 days following an aerial treatment of ponds for the control of mosquitoes (37).

Breakdown in vegetation

Fenoxycarb is expected to break down relatively quickly in plants (10).

Physical properties

Chemical name: ethyl 2-(4-phenoxyphenoxy)ethylcarbamate (10)
CAS #: 79127-80-3
Molecular weight: 301.3 (10)
Solubility in water: 6 mg/L @ 20°C (10)
Solubility in other solvents: s. hexane; v.s. in acetone, chloroform, diethyl ether, and methanol (10)
Melting point: 53–54°C (10)
Vapor pressure: 0.0078 mPa @ 20°C (10)
Partition coefficient (octanol/water): 20,000 (10)
Adsorption coefficient: 1500 (13)

Exposure guidelines

ADI: Not available
HA: Not available
RfD: Not available
PEL: Not available

Basic manufacturer

Ciba-Geigy Corp.
P.O. Box 18300
Greensboro, NC 27419–8300
Telephone: 800-334-9481
Emergency: 800-888-8372

3.2.7 *Methomyl*

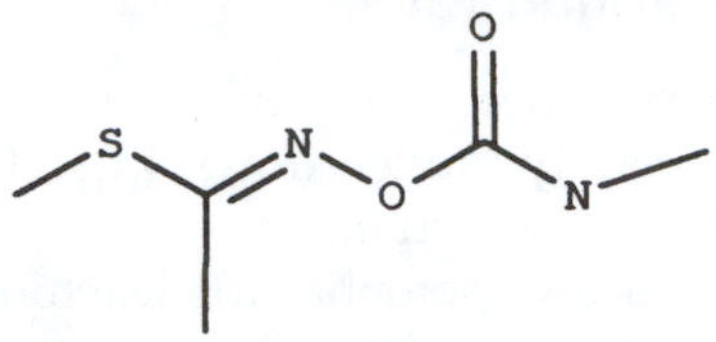

Figure 3.8 Methomyl.

Trade or other names

Common names include metomil and mesomile. Trade names include Acinate, Agrinate, DuPont 1179, Flytek, Kipsin, Lannate, Lanox, Memilene, Methavin, Methomex, Nudrin, NuBait, Pillarmate, and SD 14999.

Regulatory status

Methomyl is a highly toxic compound in EPA toxicity class I. It is classified as Restricted Use Pesticide (RUP) by EPA because of its high acute toxicity to humans. The Signal Words for products containing methomyl depend upon the formulation of the product. RUPs may be purchased and used only by certified applicators. Reentry periods for farm workers of 1 to 7 days are required, depending on the crop.

Introduction

Methomyl was introduced in 1966 as a broad-spectrum insecticide. It is also used as an acaricide to control ticks and spiders. It is used for foliar treatment of vegetable, fruit, and field crops, cotton, commercial ornamentals, and in and around poultry houses and dairies. It is also used as a fly bait. Methomyl is effective in two ways: (1) as a "contact insecticide," because it kills target insects upon direct contact, and (2) as a "systemic insecticide" because of its capability to cause overall "systemic" poisoning in target insects, after it is absorbed and transported throughout the pests that feed on treated plants. It is capable of being absorbed by plants without being "phytotoxic" or harmful, to the plant.

Toxicological effects

Acute toxicity

Methomyl is highly toxic via the oral route, with reported oral LD_{50} values of 17 to 24 mg/kg in rats (10), 10 mg/kg in mice, and 15 mg/kg in guinea pigs (5). Symptoms of methomyl exposure are similar to those caused by other carbamates and cholinesterase inhibitors (5). These may include weakness, blurred vision, headache, nausea, abdominal cramps, chest discomfort, constriction of pupils, sweating, muscle tremors, and decreased pulse. If there is severe poisoning, symptoms of twitching, giddiness, confusion, muscle incoordination, slurred speech, low blood pressure, heart irregularities, and loss of reflexes may also be experienced. Death can result from discontinued breathing, paralysis of muscles of the respiratory system, intense constriction of the openings of the lung, or all three (5).

It is moderately toxic via inhalation with a reported 4-hour inhalation LC_{50} in male rats of 0.3 mg/L (4). Inhalation of dust or aerosol may cause irritation, lung and eye problems, with symptoms of chest tightness, blurred vision, tearing, wheezing and headaches appearing upon exposure. Other systemic symptoms of cholinesterase inhibition may appear within a few minutes to several hours of exposure (39).

It is slightly toxic via the dermal route, with a reported dermal LD_{50} of 5880 mg/kg in rabbits (10), and is absorbed only slowly through the skin (5). However, if sufficient amounts are absorbed through the skin, symptoms similar to those induced by ingestion or inhalation will develop (5). Within 15 minutes to 4 hours of exposure, the immediate area of contact may show localized sweating and uncoordinated muscular contractions (5).

In rabbits, application of methomyl resulted in mild eye irritation (5). Pain, shortsightedness, blurring of distant vision, tearing, and other eye disturbances may occur within a few minutes of eye contact with methomyl (11).

Chronic toxicity

Prolonged or repeated exposure to methomyl may cause symptoms similar to the pesticide's acute effects (5). Repeated exposure to small amounts of methomyl may cause an unsuspected inhibition of cholinesterase, resulting in flu-like symptoms, such as weakness, lack of appetite, and muscle aches. Cholinesterase-inhibition may persist for 2 to 6 weeks. This condition is reversible if exposure is discontinued. Since cholinesterase is increasingly inhibited with each exposure, severe cholinesterase-inhibition symptoms may be produced in a person who has had previous methomyl exposure, while a person without previous exposure may not experience any symptoms at all (5).

In a 24-month study with rats fed doses of 2.5, 5, or 20 mg/kg/day, effects were only observed at the highest dose tested, 20 mg/kg/day. At this very high dose, red blood cell counts and hemoglobin levels were significantly reduced in female rats (39). In a 2-year feeding study with dogs, 5 mg/kg/day caused no observed adverse effects (39). It is not likely that chronic effects would be seen in humans unless exposures were unexpectedly high, as with chronic misuse.

Reproductive effects

Methomyl fed to rats at dietary doses of 2.5 or 5 mg/kg for three generations caused no adverse effect on reproduction, nor was there any evidence of congenital abnormalities (8). No fetotoxicity was observed in offspring of pregnant rats given 33.9 mg/kg/day on day 6 to 21 of gestation (5). Based on these data, it appears unlikely that methomyl will have reproductive effects.

Teratogenic effects

No teratogenic effects were found in the fetuses of female rabbits that were fed approximately 15 to 30 mg/kg/day during the 8th to 16th day of gestation (8). In rats, no embryonic or teratogenic effects were observed at the highest dietary dose administered, approximately 34 mg/kg/day (5). Thus, methomyl does not appear to be teratogenic.

Mutagenic effects

In a number of assays (including Ames test, a reverse mutation assay, a recessive lethal assay, three DNA damage studies, an unscheduled DNA synthesis assay, and *in vivo* and *in vitro* cytogenetic assays), methomyl was not mutagenic (8,39). There is no evidence that methomyl is a mutagenic or genotoxic.

Carcinogenic effects

There was no evidence of carcinogenicity in either rats or dogs that ingested high doses of methomyl in 2-year feeding studies (39). Methomyl was not carcinogenic in 22- and 24-month studies with rats fed doses of up to 20 mg/kg, nor in a 2-year study with mice fed dietary doses of up to 93.4 mg/kg/day (5). The evidence suggests that methomyl is not carcinogenic.

Organ toxicity

Lungs, skin, eyes, gastrointestinal tract, kidneys, spleen, and blood-forming organs have been affected in various experiments, depending on route of entry, duration of exposure, and dosage.

Fate in humans and animals

Methomyl is quickly absorbed through the skin, lungs, and gastrointestinal tract, and is broken down in the liver. Breakdown products are readily excreted via respiration and urine (10). Although methomyl does not appear to accumulate in any particular body tissue, it may alter many other enzymes besides the cholinesterases (8).

Ecological effects

Effects on birds

Methomyl is highly toxic to birds. The acute oral LD_{50} in bobwhite quail is 24.2 mg/kg (39). The oral LD_{50} of methomyl is 28 mg/kg in hens. All deaths occurred within 10 minutes of dosing. The clinical signs of toxicity included tearing of the eyes, salivation, occasional convulsions, and respiratory disorders. In Japanese quail, the LD_{50} is 34 mg/kg (39). The LD_{50} of a 90% pure formulation is 15.9 mg/kg in 8-month-old mallards, and 15.4 mg/kg in 3- to 4-month-old male pheasants (22). The LD_{50} for starlings is 42 mg/kg, and for redwinged blackbirds is 10 mg/kg (39).

Effects on aquatic organisms

Methomyl is moderately to highly toxic to fish and highly toxic to aquatic invertebrates. The 96-hour LC_{50} in rainbow trout for a liquid formulation of methomyl is 3.4 mg/L and for bluegill sunfish is 0.8 mg/L (10). The 48-hour LC_{50} for *Daphnia magna* (a small, freshwater crustacean) is 0.0287 mg/L (14). A 28-day fish residue study indicated that methomyl did not accumulate in fish tissue (39). Methomyl is unlikely to bioconcentrate in aquatic systems (14).

Effects on other organisms (non-target species)

Methomyl is highly toxic to bees both by direct contact and through ingestion (39).

The LD_{50} for a 90% pure formulation of methomyl is 11.0 to 22.0 mg/kg in mule deer (22). Symptoms of acute poisoning in these animals included drowsiness, drooling, diarrhea, and tremors (22).

Environmental fate

Breakdown in soil and groundwater

Methomyl has low persistence in the soil environment, with a reported half-life of approximately 14 days (14). Because of its high solubility in water and low affinity for soil binding, methomyl may have potential for groundwater contamination (14).

It is very mobile in sandy loam and silty clay loam soils, but only slight leaching was observed in a silt loam and in a sandy soil. Methomyl is rapidly degraded by soil microbes (14). Methomyl residues are not expected to be found in treated soil after the growing season in which it is applied (14).

Breakdown in water

Aqueous solutions of methomyl have been reported to decompose more rapidly on aeration, in sunlight, or in alkaline media (14). The estimated aqueous half-life for the insecticide is 6 days in surface water and over 25 weeks in groundwater (14). In one experiment, the hydrolysis half-lives of methomyl in solutions at pH values of 6.0, 7.0, and 8.0 were 54, 38, and 20 weeks, respectively. In pure water, the hydrolysis half-life has been estimated to be 262 days (14).

Breakdown in vegetation

Following soil treatment, plants take up methomyl through their roots and move it throughout the plant by a process called "translocation." When methomyl is applied to plants, its residues are short-lived (40). After it is applied to leaves, it has a 3- to 5-day half-life (10). Less than 3% methomyl remained in cabbage plants 1 week after they were given foliar treatment with the insecticide (8).

Physical properties

Methomyl is a white crystalline solid with a slight sulfurous odor.

Chemical name: S-methyl N-(methylcarbamoyloxy)thioacetimidate (10)
CAS #: 16752-77-5
Molecular weight: 162.21 (10)
Water solubility: 57.9 g/L @ 25°C (10)
Solubility in other solvents: v.s. in methanol, acetone, ethanol, and isopropanol (10)
Melting point: 79°C (10)
Vapor pressure: 6.65 mPa @ 25°C (10)
Partition coefficient (octanol/water): Not available
Adsorption coefficient: 72 (13)

Exposure guidelines

ADI: 0.03 mg/kg/day (10)
HA: 0.2 mg/L (39)
RfD: 0.025 mg/kg/day (20)
PEL: 2.5 mg/m^3 (8-hour) (31)

Basic manufacturer

DuPont Agricultural Products
Walker's Mill, Barley Mill Plaza
P.O. Box 80038
Wilmington, DE 19880-0038
Telephone: 800-441-7515
Emergency: 800-441-3637

3.2.8   Oxamyl

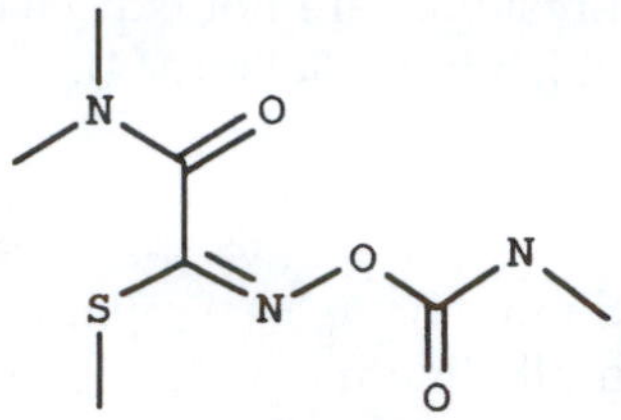

Figure 3.9 Oxamyl.

Trade or other names

Trade names include Blade, DPX 1410, Oxamil, Oxamimidic Acid, Pratt, Thioxamil, and Vydate.

Regulatory status

Oxamyl is a highly toxic compound in EPA toxicity class I. The U.S. Environmental Protection Agency (EPA) has classified most products containing oxamyl as Restricted Use Pesticides (RUPs) due to oxamyl's acute toxicity to humans and its toxicity to birds and mammals. RUPs may be purchased and used only by certified applicators. Granular formulations of oxamyl are banned in the U.S. Labels on containers of oxamyl and/or its formulated products must bear the Signal Words DANGER — POISON.

Introduction

Oxamyl is a carbamate insecticide/acaricide/nematicide that controls a broad spectrum of insects, mites, ticks, and roundworms. It may work both through systemic distribution in the target pest and on contact. Oxamyl is used on field crops, vegetables, fruits, and ornamental plants and may be applied directly onto plants or the soil surface. It is available in both liquid and granular form, but as is stated above, the granular form is banned in the U.S.

Toxicological effects

Acute toxicity

Oxamyl is highly toxic via the oral route, with a reported oral LD_{50} of 5.4 mg/kg in rats (10). Exposure to oxamyl will cause similar effects to the other carbamate compounds, and as with other carbamates, the cholinesterase-inhibiting effects are short-term and reversible (5). These effects include headache, nausea, sweating, tearing, tremors, and blurred vision.

It is slightly toxic via the dermal route, with a reported dermal LD_{50} for technical oxamyl of 2960 mg/kg in rabbits (10). Skin and eye exposure may cause poisoning, although absorption through the skin and eye is slower than through the gastrointestinal tract (5). Exposure of rabbit eyes to technical oxamyl caused mild irritation (5).

Oxamyl is highly toxic via the inhalation route as well with a reported 4-hour LC_{50} of 0.12 to 0.17 mg/L in rats (10).

Chronic toxicity

Prolonged or repeated exposure to oxamyl may cause symptoms similar to those caused by acute exposure. In a 2-year mouse feeding study, no effects were observed at a dose of 1.25 mg/kg/day, although at the very high doses of 2.5 and 3.75 mg/kg/day, there was decreased body weight and changed nutritional performance (41). Liver impairment was suggested by a slight biochemical change seen in dogs that were fed 3.75 mg/kg/day as a part of a 2-year feeding study (5).

Male rats fed the very high dose of 7.5 mg/kg/day oxamyl for 2 years had decreased organ weight of the heart, testes, and adrenals. In females, there was an increase in the relative weights of the brain, heart, lungs, and adrenals at these doses (42).

Reproductive effects

When pregnant rats were fed oxamyl, no effects were observed (relative to unexposed animals) on the number of implantation sites, resorptions, and live fetuses in rats fed doses of 2.5, 5, 7.5, or 15 mg/kg/day on days 6 through 15 of pregnancy, There were, however, dose-related decreases in maternal body weights and in food consumption rates (42). Litter size, viability, lactation and body weights of offspring were decreased in a three-generation (two litters per generation) reproduction study in rats fed very high doses, approximately 5 and 7.5 mg/kg/day technical oxamyl (42). Based on these studies, oxamyl is unlikely to cause reproductive effects in humans at expected exposure levels.

Teratogenic effects

No teratogenic effects were observed in the offspring of rabbits fed 2 and 4 mg/kg/day during days 6 to 19 of gestation (42). Oxamyl was not teratogenic or embryotoxic in the offspring of pregnant rats fed up to 15 mg/kg/day of the insecticide (3,42). No teratogenic effects were observed in rabbits given oxamyl levels at doses of approximately 4 mg/kg/day in the diet (5). Thus, it appears that oxamyl does not cause teratogenic effects.

Mutagenic effects

Oxamyl was not mutagenic in several test systems (5,42).

Carcinogenic effects

In a study of mice fed up to 9 mg/kg/day, no carcinogenic effects were seen (5). In addition, no evidence of carcinogenic potential was seen following long-term dietary exposure of rats (42). This indicates that oxamyl is not carcinogenic.

Organ toxicity

Effects noted in animal studies include changes in liver function, and changes in organ weights of the brain, heart, lungs, testes, and adrenals (42).

Fate in humans and animals

When oxamyl was administered to rats, most of the dose was rapidly eliminated in the urine and feces as breakdown by-products, or metabolites (5). Carbamates generally are excreted rapidly and do not accumulate in mammalian tissue (5).

Ecological effects

Effects on birds

Oxamyl is very highly toxic to birds (10). The acute oral LD_{50} for oxamyl in quail is 4.18 mg/kg (10). Hens given single oral doses of oxamyl at 20 to 40 mg/kg body weight followed by intramuscular injections of 0.5 mg atropine (an antidote) exhibited early symptoms of cholinesterase inhibition with full recovery after 12 hours. No signs of delayed neurotoxicity were observed in these same hens (42). The oral LD_{50} for oxamyl is 2.6 mg/kg in mallard ducks, and 9.4 mg/kg in bobwhite quail (17).

Effects on aquatic organisms

Oxamyl is moderately to slightly toxic to fish (10). The reported 96-hour LC_{50} values for technical oxamyl are 5.6 mg/L in bluegill sunfish, 27.5 mg/L in goldfish, 4.2 mg/L in rainbow trout, and 17.5 mg/L in channel catfish (10). Concentrations of 0.5 to 5.0 mg/L may have an effect on *Daphnia magna*, an aquatic invertebrate (43).

Effects on other organisms (non-target species)

Oxamyl is highly toxic to bees (10).

Environmental fate

Breakdown in soil and groundwater

Oxamyl is of low persistence in soil with reported field half-lives of 4 to 20 days (13). Loss is due to decomposition by aerobic and anaerobic bacteria (44). Oxamyl is hydrolyzed rapidly in neutral and alkaline soils and more slowly in acid soils (42). It does not readily bind, or "adsorb," to soil or sediments and it has been shown to leach in soil (13,42). Its adsorption is strongest in soils of high organic matter, and on sandy loam is fairly weak. An increase in temperature causes a decrease in adsorption (44).

Since oxamyl degrades relatively quickly in the presence of bacteria, it is more likely to be found in groundwater than in surface water. It has been found in very small amounts in the states of New York (1 to 60 µg/L) and Rhode Island (1 µg/L) (45). Wherever conditions favor very rapid movement of leachate, oxamyl may reach the groundwater.

Breakdown in water

In a river water study, oxamyl had a half-life of 1 to 2 days (42).

Breakdown in vegetation

Oxamyl has a residual period in plants of approximately 1 to 2 weeks. It is considered nontoxic to plants (10). Plants take up oxamyl through both leaves and roots; it is translocated in treated plants (10). Oxamyl is metabolized rapidly by plants (42).

Physical properties

Oxamyl is a colorless or white crystalline solid with a garlic-like odor.

Chemical name: N,N-dimethyl-2-methylcarbamoyloxyimino-2-(methylthio)-acetamide (10)

CAS #: 23135-22-0

Molecular weight: 219.36 (10)

Water solubility: 280 g/L @ 25°C (10)

Solubility in other solvents: v.s. in acetone, ethanol, 2-propanol, and methanol; s. in toluene (10)

Melting point: 100–102°C (10), at which point it changes to a different crystalline structure that melts at 108–110°C (10)

Vapor pressure: 31 mPa @ 25°C (10)

Partition coefficient (octanol/water): Not available

Absorption coefficient: 25 (13)

Exposure guidelines

ADI: 0.03 mg/kg/day (10)

MCL: 0.2 mg/L (19)

RfD: 0.025 mg/kg/day (20)

PEL: Not available

Basic manufacturer

DuPont Agricultural Products

Walker's Mill, Barley Mill Plaza

P.O. Box 80038

Wilmington, DE 19880-0038

Telephone: 800-441-7515

Emergency: 800-441-3637

3.2.9   Propoxur

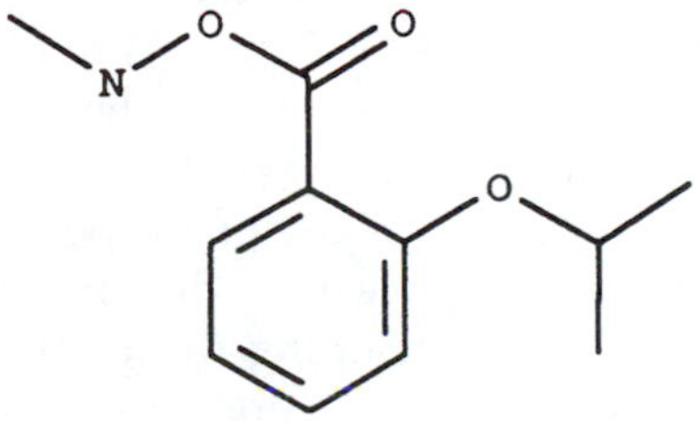

Figure 3.10 Propoxur.

Trade or other names

Trade and other names for propoxur include Arprocarb, Bay 9010, Baygon, Bayer 39007, Bifex, Blattanex, Brifur, Bolfo, BO Q 5812315, ENT 25671, Invisi-Gard, OMS

33, PHC, Pillargon, Prentox Carbamate, Propogon, Proprotox, Propyon, Rhoden, Sendran, Suncide, Tendex, Tugen, Unden, and Undene.

Regulatory status

Propoxur is a highly toxic compound; various formulations are in different toxicity classes. It is a General Use Pesticide (GUP), although some formulations may be for professional use only. Labels for pesticide products containing propoxur must bear the Signal Word, DANGER, WARNING, or CAUTION, depending on the formulation.

Introduction

Propoxur is a nonsystemic insecticide introduced in 1959. It is compatible with most insecticides and fungicides except alkalines, and may be found in combination with azinphosmethyl, chlorpyrifos, cyfluthrin, dichlorvos, disulfoton, or methiocarb. It is used on a variety of insect pests such as chewing and sucking insects, ants, cockroaches, crickets, flies, and mosquitoes, and may be used for control of these in agricultural or (as Baygon) in non-agricultural (e.g., private or public facilities and grounds) applications. Agricultural applications include cane, cocoa, fruit, grapes, maize, rice, sugar, vegetables, cotton, lucerne, forestry, and ornamentals. It has contact and stomach action that is long-acting when it is in direct contact with the target pest. Propoxur is available in several types of formulations and products, including emulsifiable concentrates, wettable powders, baits, aerosols, fumigants, granules, and oilsprays.

Toxicological effects

Acute toxicity

Propoxur is highly toxic via the oral route, with reported LD_{50} values of approximately 100 mg/kg in rats and mice (10). An oral LD_{50} of 40 mg/kg is reported for guinea pigs (5). Propoxur is only slightly toxic via the dermal route, with reported dermal LD_{50} values of 1000 mg/kg to greater than 2400 mg/kg in rats, and greater than 500 mg/kg in rabbits (5). Tests show that propoxur does not cause skin or eye irritation in rabbits (10). Via the inhalation route, it is also slightly toxic, with a reported 1-hour inhalation LC_{50} of greater than 1.44 mg/L (10).

Like other carbamates, propoxur can inhibit the action of cholinesterase and disrupt nervous system function. Depending on the severity of exposure, this effect may be short-term and reversible (5). Signs of propoxur intoxication can include nausea, vomiting, abdominal cramps, sweating, diarrhea, excessive salivation, weakness, imbalance, blurring of vision, breathing difficulty, increased blood pressure, incontinence, or death (5). In rats, propoxur poisoning resulted in brain pattern and learning ability changes at lower concentrations than those which caused cholinesterase-inhibition and/or organ weight changes (8).

During wide-scale spraying of propoxur in malarial control activities conducted by the World Health Organization (WHO), only mild cases of poisoning were noted. Applicators who used propoxur regularly showed a pronounced daily fall in whole blood cholinesterase activity and a distinct recovery after exposure stopped. No adverse cumulative effects on cholinesterase activity were demonstrated (8). Human adults have ingested single doses of 50 mg propoxur without apparent symptoms (5).

Chronic toxicity

Prolonged or repeated exposure to propoxur may cause symptoms similar to acute effects. Propoxur is very efficiently detoxified (transformed into less toxic or practically nontoxic forms), thus making it possible for rats to tolerate daily doses approximately equal to the LD_{50} of the insecticide for long periods, provided that the dose is spread out over the entire day rather than ingested all at once (5).

Reproductive effects

In female rats given high dietary doses of approximately 18 mg/kg/day of propoxur as a part of a three-generation reproduction study, reduced parental food consumption, growth, lactation, and litter size were observed (5). At 25 mg/kg/day administered to pregnant rats, there was a decrease in the number of offspring (5). Dietary doses of approximately 2.25 mg/kg/day did not affect fertility, litter size, or lactation, but parental food intake and growth were depressed in the exposed group (5). This evidence suggests that reproductive effects in humans are unlikely at expected exposure levels.

Teratogenic effects

Offspring of female rats fed 5 mg/kg/day propoxur during gestation and weaning exhibited reduced birth weight, retarded development of some reflexes, and evidence of central nervous system impairment (5). In another rat study, growth reduction was observed in the offspring of pregnant rats given doses of 3, 9, and 30 mg/kg/day, but no other physiological or anatomical abnormalities were observed. The evidence suggests that teratogenic effects will only occur at high doses.

Mutagenic effects

Propoxur did not cause mutations in six different types of bacteria (46). The evidence indicates that propoxur is not mutagenic.

Carcinogenic effects

No carcinogenic effects have been reported for propoxur (5).

Organ toxicity

As determined in animal tests and data from human autopsies in poisoned individuals, the nervous system and liver are the organs principally affected by propoxur (5,46).

Fate in humans and animals

Propoxur is broken down and excreted rapidly in urine (5). In humans given a single oral dose of 92.2 mg Baygon, 38% of the dose was excreted in urine over the first 24 hours, with most of it excreted in the first 8 to 10 hours (5). Carbamates generally are excreted rapidly and do not accumulate in mammalian tissue (5).

Ecological effects

Effects on birds

Propoxur is very highly to highly toxic to many bird species, but its toxicity varies by species. The reported LD_{50} is 25.9 mg/kg in bobwhite quail (10). Other reported oral LD_{50} values (for a 97% to 98% technical grade propoxur product) are

4 mg/kg in mourning doves and house finches, 6 mg/kg in Canada geese, 10.5 mg/kg in mallards, 20 mg/kg in pheasants, 26 and 28 mg/kg in California and Japanese quail, respectively, and 120 mg/kg in grouse (17). The 5-day dietary LC_{50} for Japanese quail is greater than 5000 ppm (10).

Acute symptoms of propoxur poisoning in birds include eye tearing, salivation, muscle incoordination, diarrhea, and trembling (47). Depending on the type of bird, poisoning signs can appear within 5 minutes of exposure, with deaths occurring between 5 and 45 minutes, or overnight. Symptoms in survivors disappeared from 90 minutes to several days after treatment (47).

Effects on aquatic organisms

Propoxur is moderately to slightly toxic to fish and other aquatic species. The reported 96-hour LC_{50} values are 3.7 mg/L in rainbow trout, and 6.6 mg/L in bluegill sunfish (10). The oral LD_{50} for propoxur in bullfrogs is 595 mg/kg (17). The compound is not expected to accumulate significantly in aquatic organisms. The calculated accumulation factor for propoxur is nine times the ambient water concentration.

Effects on other organisms (non-target species)

Propoxur is highly toxic to honeybees (10). The oral LD_{50} for propoxur in mule deer is 100 to 350 mg/kg (17).

Environmental fate

Breakdown in soil and groundwater

Propoxur is of moderate to low persistence in the soil environment, with reported field half-lives of 14 to 50 days (13). It has a low affinity for soil binding, and so may be mobile in many soils (13). Because it is highly soluble in water, is moderately persistent, and does not adsorb strongly to soil particles, propoxur has a high potential for groundwater penetration (14,48). In one study, there was practically no loss of propoxur from a silt-loam soil to which it was applied during a 6-month period, but 25% of applied Baygon was lost from sand in 100 days (48). In another study, propoxur was very mobile in sandy loam, silt loam, and silty clay soils. The rate of biodegradation in soil increases in soils that have been previously exposed to propoxur or other methylcarbamate pesticides (48).

Breakdown in water

Propoxur hydrolyzes, or breaks down in water, at a rate of 1.5% per day in a 1% aqueous solution at pH 7.0 (48).

Breakdown in vegetation

Propoxur can enter the roots of a plant and travel to the leaves, where it can then poison insects that feed on the leaves, and can have residual activity of up to 1 month when applied to plant surfaces (8).

Physical properties

Technical propoxur is a white to cream-colored crystalline solid.

Chemical name: 2-isopropoxyphenyl methylcarbamate (10)

CAS #: 114-26-1

Molecular weight: 209.25 (10)

Water solubility: 2000 mg/L @ 20°C (10)

Solubility in other solvents: Soluble in most polar organic solvents, e.g., acetone, methanol, acetone, cyclohexanone, chloroform, and toluene (10)

Melting point: 84–87°C (10)

Vapor pressure: 300 mPa @ 120°C (10)

Partition coefficient (octanol/water) (log): 0.14 (48)

Adsorption coefficient: 30 (13)

Exposure guidelines

ADI: 0.02 mg/kg/day (10)

HA: 0.003 mg/L (48)

RfD: Not available

PEL: 0.5 mg/m^3 (8-hour) (31)

Basic manufacturer

Atomergic Chemetals Corp.

222 Sherwood Avenue

Farmingdale, NY 11735-1718

Telephone: 516-694-9000

Emergency: 800-424-9300

References

(1) Fukuto, R. T. Organophosphates and carbamate esters: the anticholinesterase insecticides. In *Fate of Pesticides in the Environment, Publication Number 3320.* Biggar, J. W. and Silber, J. N., Eds. University of California Agricultural Experiment Station Publications, Davis, CA, 1987.

(2) Ware, G. W. *Fundamentals of Pesticides, A Self-Instruction Guide.* Thompson Publications, Fresno, CA, 1986.

(3) American Crop Protection Association. Pesticides production: Total gained in 1993, despite fungicides decline. *Chem. Eng. News.* 73(66), 44, 1995.

(4) Stevens, J. T. and Sumner, D. D. Herbicides. In *Handbook of Pesticide Toxicology.* Hayes, W. J., Jr. and Laws, E. R., Jr., Eds. Academic Press, New York, 1991.

(5) Baron, R. L. Carbamate insecticides. In *Handbook of Pesticide Toxicology.* Hayes, W. J., Jr. and Laws, E. R., Jr., Eds. Academic Press, New York, 1991.

(6) Kaloyanova, F. P. and El Batawi, M. A. *Human Toxicology of Pesticides.* CRC Press, Boca Raton, FL, 1991.

(7) U.S. Environmental Protection Agency. *Health Advisory Summary: Carbaryl.* Office of Drinking Water, Washington, D.C., 1987.

(8) U.S. National Library of Medicine. *Hazardous Substances Data Bank.* Bethesda, MD, 1995.

(9) U.S. Environmental Protection Agency. *Health Advisory: Oxamyl.* Office of Drinking Water, Washington, D.C., 1987.

(10) Kidd, H. and James, D. R., Eds. *The Agrochemicals Handbook,* 3rd ed. Royal Society of Chemistry Information Services, Cambridge, U.K., 1991 (as updated).

(11) E. I. DuPont de Nemours. *Technical Data Sheet for Methomyl.* Agricultural Products Division, Wilmington, DE, 1989.

(12) E. I. DuPont de Nemours. *Technical Bulletin for Lannate Insecticide.* Agricultural Products Division, Wilmington, DE, 1991.

(13) Wauchope, R. D., Buttler, T. M., Hornsby, A. G., Augustijn-Beckers, P. W. M., and Burt, J. P. SCS/ARS/CES pesticides properties database for environmental decisionmaking. *Rev. Environ. Contam. Toxicol.* 123, 1–157, 1992.

(14) Howard, P. H., Ed. *Handbook of Environmental Fate and Exposure Data for Organic Chemicals: Pesticides.* Lewis, Boca Raton, FL, 1991.

(15) U.S. Environmental Protection Agency. *Health Advisories for 50 Pesticides.* Office of Drinking Water, Washington, D.C., 1987.

(16) Chambon, C., Declune, C., and Derach, R. Effects of the insecticidal carbamate derivitives (carbofuran, primicarb, and aldicarb) on the activity of acetylcholinesterase in tissue from pregnant rats and fetuses. *Toxicol. Appl. Pharmacol.* 49, 203–208, 1979.

(17) Smith, G. J. *Toxicology and Pesticide Use in Relation to Wildlife: Organophosphorus and Carbamate Compounds.* C. K. Smoley, Boca Raton, FL, 1992.

(18) Johnson, W. W. and Finley, M. T. *Handbook of Acute Toxicity of Chemicals to Fish and Aquatic Invertebrates.* U.S. Department of the Interior, Washington, D.C., 1980.

(19) U.S. Environmental Protection Agency. *National Primary Drinking Water Standards, 810-F-94-001A.* Washington, D.C., 1994.

(20) U.S. Environmental Protection Agency. *Integrated Risk Information System,* Washington, D.C., 1994.

(21) Townsend L. and Potter T. *Earthworms: Thatch-Busters.* College of Agriculture, University of Kentucky, Lexington, KY, 1995.

(22) Tucker, R. *Handbook of Toxicity of Pesticides to Wildlife.* U.S. Department of Interior, Fish & Wildlife Service, Washington, D.C., 1970.

(23) Worthing, C. R. *The Pesticide Manual: A World Compendium, Eighth Edition.* The British Crop Protection Council, Croydon, U.K., 1987.

(24) U.S. Environmental Protection Agency. *Health Advisory Draft Report: Carbaryl.* Office of Drinking Water, Washington, D.C., 1987.

(25) Cranmer, M. F. Carbaryl: a toxicological review and risk assessment. *Neurotoxicology.* 7(1), 247–332, 1986.

(26) Siebert, D. and Eisenbrand, G. Induction of mitotic gene conversion in *Sachromyces cerevisiae* by N-nitrosated pesticides. *Mutat. Res.* 22, 121–132, 1974.

(27) Elespuru, R., Lijinski, W., and Setlow, J. K. Nitrosocarbaryl as a potent mutagen of environmental significance. *Nature (London).* 247, 386–387, 1974.

(28) *Evaluation of Carcinogenic, Teratogenic, and Mutagenic Activities of Selected Pesticides and Industrial Chemicals. Vol. 1: Carcinogenic Study.* National Technical Information Service, Washington, D.C., 1968.

(29) Regan, J. D., Setlow, R. B., Francis, A. A. and Lijinsky, W. Nitrosocarbaryl: its effect on human DNA. *Mutat. Res.* 38, 293–301, 1976.

(30) Carpenter, C. P. Mammalian toxicity of 1-napthyl-*n*-methylcarbamate (Sevin insecticide). *J. Agric. Food Chem.* 9(1), 30–39, 1961.

(31) U.S. Environmental Protection Agency. *Health Advisory Summary: Carbofuran.* Office of Drinking Water, Washington, D.C., 1989.

(32) Gosselin, R. E., Smith, R. P., and Hodge, H. C. *Clinical Toxicology of Commercial Products,* 5th ed. Williams & Wilkins, Baltimore, MD, 1984.

(33) U.S. Environmental Protection Agency. *Environmental Fact Sheet for Carbofuran.* Office of Pesticides and Toxic Substances, Washington, D.C., 1991.

(34) Hill, E. F. and Camardese, M. B. Lethal Dietary Toxicities of Environmental Contaminants and Pesticides to Coturnix. U.S. Department of Interior, Fish and Wildlife Service, Washington, D.C., 1986.

(35) Weed Science Society of America. *Herbicide Handbook,* 6th ed. Champaign, IL, 1989.

(36) U.S. Environmental Protection Agency. *Pesticide Fact Sheet Number 78: Fenoxycarb.* Office of Pesticides and Toxic Substances, Washington, D.C., 1986.

(37) Schaefer, C. H., Wilder, W. H., Mulligan, F. S. and Dupras, E. E. Efficacy of fenoxycarb against mosquitoes (*Diptera culicidae*) and its persistence in the laboratory and field. *Econ. Entomol.* 80(1), 126–130, 1987.

(38) Miura, T., Schaefer, C. H., and Stewart, R. J. Impact of carbamate insect growth regulators on the selected aquatic organisms: a preliminary study. *Proc. Annu. Conf. California Mosquito Vector Control Assoc.* 54, 36–38, 1986.

(39) U.S. Environmental Protection Agency. *Health Advisory Summary: Methomyl.* Office of Drinking Water, Washington, D.C., 1987.

(40) McEwen, F. L. and Stephenson, G. R. *The Use and Significance of Pesticides in the Environment.* John Wiley & Sons, New York, 1979.

(41) E. I. DuPont de Nemours. *Technical Data Sheet: Oxamyl.* Agricultural Products Division, Wilmington, DE, 1984.

(42) U.S. Environmental Protection Agency. *Health Advisory Summary: Oxamyl.* Office of Drinking Water, Washington, D.C., 1987.

(43) Mayer, F. L. and Ellersieck, M. R. *Manual of Acute Toxicity: Interpretation and Data Base for 410 Chemicals and 66 Species of Freshwater Animals. Resource Publication 160.* U.S. Department of Interior, Fish and Wildlife Service, Washington, D.C., 1986.

(44) Wagenet, L. P. *A Review of Physical-Chemical Parameters Related to the Soil and Groundwater Fate of Selected Pesticides in New York State, Report No. 30.* Cornell University Agricultural Experiment Station, Ithaca, NY, 1985.

(45) Cohen, S. Z. *Monitoring Groundwater for Pesticides.* American Chemical Society, Washington, D.C., 1986.

(46) Hallenbeck, W. H. and Cunningham-Burns, K. M. *Pesticides and Human Health.* Springer-Verlag, New York, 1985.

(47) Hudson, R. H., Tucker, R. K., and Haegele, M. A. *Handbook of Toxicity of Pesticides to Wildlife, Resource Publication 153.* U.S. Department of the Interior, Fish and Wildlife Service, Washington, D.C., 1984.

(48) U.S. Environmental Protection Agency. *Health Advisory Summary: (Baygon) Propoxur.* Office of Drinking Water, Washington, D.C., 1988.

chapter four

Thio- and dithiocarbamates

4.1. Class overview and general description

Background

Thio- and dithiocarbamates are each a special subclass of carbamates. As the class names imply, thiocarbamates have one sulfur atom substituted for an oxygen atom; dithiocarbamates, have two oxygen atoms replaced by sulfur. The general structures of the thiocarbamates and dithiocarbamates are illustrated in Figure 4.1; Table 4.1 lists the various thiocarbamate and dithiocarbamate compounds. As is true with the carbamates, the R groups may refer to alkyl, aryl, alkoxy, amide, or metallic substituents (1). The ethylene(bis)dithiocarbamates (EBDCs) commonly contain a metal in a complex or in a polymeric form (1).

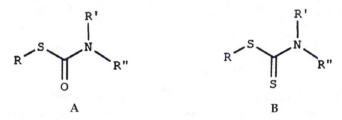

A   B

Figure 4.1 Generic structures for thiocarbamates (A) and dithiocarbamates (B).

In the past, the EBDCs were the focus of considerable media and public attention because of concern about the long-term effects of exposure to the EBDCs and their breakdown product, ethylene thiourea (ETU) (2). These issues are discussed in greater detail in the section regarding the toxicological effects.

Thiocarbamate and dithiocarbamate use

The thio- and dithiocarbamates are widely used, mainly as fungicides on crops in the field or to protect against fungal diseases or rot during harvesting, transport, and storage. Some of them, such as EPTC and molinate, may be used for other purposes, such as the control of non-crop plants. They may be used in several forms, including granular formulations and emulsifiable concentrates. Most, if not all, of the thiocarbamates and dithiocarbamates are registered as General Use Pesticides (GUPs) (3,4).

Mode of action and toxicology

Thiocarbamates will have similar action to the carbamates, whose primary effects on target and non-target species are through the inhibition of a enzyme known as

Table 4.1.a Thiocarbamates

Butylate*	Molinate*
Cartap	Orbencarb
Cycloate	Pebulate
Diallate	Prosulfocarb
Dimepiperate	Pyributicarb
EPTC*	Thiobencarb
Esprocarb	Thiocarbazil
Fenothiocarb	Triallate*
Methasulfocarb	Vernolate

Note: * indicates that a profile for this compound is included in this chapter.

Table 4.1.b Dithiocarbamates

Methyldithiocarbamates	
Metham-sodium	
Dimethyldithiocarbamates	
Dimethyldithiocarb (DDC)	Thiram*
Ferbam	Ziram*
Diethyldithiocarbamates	
Sulfallate	
Ethylene(bis)dithiocarbamates	
Anobam	Maneb*
Cufraneb	Metiram*
Mancozeb*	Zineb*

Note: * indicates that a profile for this compound is included in this chapter.

acetylcholinesterase (AChE) (1). Acetylcholine (ACh) is a substance that transmits a nerve impulse from a nerve cell to a specific receptor such as another nerve cell or a muscle cell (5). ACh, in essence, acts much like a chemical switch. When it is released to a nerve cell at the synapse, it turns the receiving nerve cell "on" and results in transmission of a nerve impulse. The transmission of nervous energy continues until the ACh is broken down (by cleavage of the ester bond) into choline and acetic acid by AChE. Thiocarbamate inhibition of ACh is a reversible process. Estimates of the recovery time in humans range from immediate up to 4 days, depending on the dose, the specific pesticide, and the method of exposure (i.e., inhalation or ingestion) (6). The effects on nerve cells may result in incoordination, muscular weakness, disruption of concentration or reasoning abilities, disruption in regulation of heartbeat and breathing, and in extreme cases, convulsions (6).

The dithiocarbamates may also result in nervous system effects, but not through the same mechanism as the thiocarbamates (1). The dithiocarbamates do not readily interact with AChE, but instead affect the nervous system through their main metabolite, carbon disulfide (1). This compound affects the ability of the nerve cell to effectively conduct nervous impulses by altering the permeability of the nerve cell membrane and myelin sheath (1).

A major toxicological concern with respect to the EBDCs is the metabolite, ethylenethiourea (ETU), which has been shown to cause thyroid and carcinogenic effects in test animals. The precise mechanism by which this may occur is not well understood (1,2).

Acute toxicity

Most members of the thiocarbamate and dithiocarbamate classes are slightly to practically nontoxic via ingestion, dermal and inhalation routes. Most cause skin and/or eye irritation and may cause skin sensitization (allergic contact reaction). Via the oral route, the reported acute LD_{50} values range from 300 to 400 mg/kg to greater than 5000 mg/kg in rats and other test animals, indicating they are practically nontoxic (1,4,6,7). Slight toxicity is also observed via the dermal route (2,3). The reported acute dermal LD_{50} values for the thio- and dithiocarbamate pesticides in rats are almost all greater than 2000 mg/kg (3,6,7). For many, mild to moderate skin sensitization and/or skin and eye irritation has been observed in test animals. Although precise acute inhalation LC_{50} values were not available for all of the thio- and dithiocarbamates, most are of moderate to slight toxicity by this route (1,4,6,7).

Effects due to exposure to some of the thio- and/or dithiocarbamates (notably EPTC) may include those similar to exposure to carbamates, i.e., cholinesterase inhibition (1,4,6,7). These include blurred vision, fatigue, headache, dizziness, abdominal cramps, and diarrhea. Severe inhibition of cholinesterase may cause excessive sweating, tearing, slowed heartbeat, giddiness, slurred speech, confusion, excessive fluid in the lungs, convulsions, and coma. Dithiocarbamates are partially metabolized to carbon disulfide, a neurotoxin capable of interfering with nerve transmission (6–8).

Chronic toxicity

Thiocarbamates

Some thiocarbamates have been associated with cholinesterase inhibition, degeneration of nervous tissue (of the spinal cord, muscles, and heart), and increased liver and thyroid weights in long-term animal studies. The doses required to produce these effects ranged from 2 mg/kg/day (in rats over two generations) to greater than 15 mg/kg/day (in dogs over 2 years) (1,6,8). High variability in some toxic responses was seen across species. Repeated application of some thiocarbamates caused skin irritation (6,8).

Dithiocarbamates

Symptoms of chronic exposure to dithiocarbamates may include neurological and/or behavioral effects (e.g., drowsiness, incoordination, weakness, and paralysis) in addition to those due to acute exposure. Doses on the order of 40 to 50 mg/kg/day were required to produce these effects in rats; doses of about 5 mg/kg/day (e.g., thiram and ziram) reportedly caused no observed effects in dogs over 1 year. Repeated or prolonged exposure to dithiocarbamates can also cause skin sensitization (1,6).

Ethylene(bis)dithiocarbamates (EBDCs)

As mentioned above, a major toxicological concern in cases of chronic exposure to the EBDCs (e.g., mancozeb, maneb, metiram, and ziram) is ethylenethiourea (ETU), which may be produced during metabolism, and also may be introduced as a contaminant during manufacture (1). ETU may also be generated in small amounts when EBDC residues are present on produce for long periods, or during cooking

(1). In test animals, ETU has caused thyroid enlargement (also known as goiter) and impaired thyroid function, birth defects, and cancers (1). Indeed, it may be that ETU (formed during metabolism) is responsible for the thyroid effects seen in test animals after long-term exposure to EBDCs (1).

The EBDC dose levels required to produce observable effects in long-term (2-year) animal dietary studies range from about 2 to 5 mg/kg/day in dogs and rats (for mancozeb) (1,9) and greater than 12.5 mg/kg/day in rats (for maneb) (1,10). Effects observed were thyroid enlargement and impairment. At higher daily doses, some EBDCs caused gastrointestinal and nervous system disturbances (muscular weakness and tremor) in these animals. For metiram, no observed adverse effects were seen at doses of up to 45 mg/kg/day following a course of 90 days of exposure in dogs (1,11). Zineb failed to produce effects on survival, growth, and blood chemistry in dogs fed doses of 250 mg/kg/day over a 1-year period (1). Field studies of some EBDCs have shown that they may cause skin or eye irritation and/or contact dermatitis (1).

Reproductive effects

For the majority of the thio- and dithiocarbamates, no reproductive effects were observed in test animals receiving doses of about 20 to 30 mg/kg/day during pregnancy (1,6,8). At higher doses (e.g., 50 mg/kg/day for mancozeb and 100 mg/kg/day for EPTC), effects such as decreased offspring weight gain and lower fertility rates were seen. The EBDCs have produced reproductive effects in some animal systems, but only at extremely high levels (1,2,9–11). It does not seem likely that these classes of pesticides will produce reproductive effects in humans under normal circumstances.

Teratogenic effects

With the majority of members of the thio- and dithiocarbamate class, teratogenic effects were observed in various single- and multi-generational rat and rabbit studies at dietary doses ranging from 50 to as high as 1000 mg/kg/day (1,6,12). With respect to the EBDCs, developmental toxicity was observed at lower doses such as 5 to 10 mg/kg/day (2,9–11). Teratogenic effects were also observed with some EBDCs (e.g., mancozeb) following single massive oral doses during pregnancy. The EDBCs have been shown to be teratogenic in rats and hamsters, but not in mice (1,12). Teratogenic effects in humans are unlikely at normal levels of exposure.

Mutagenic effects

Most of the thio- and dithiocarbamates appear to be nonmutagenic or only very weakly mutagenic as indicated by a variety of mutagenicity assays (1,4,6).

A notable exception to this general finding is ziram (an EBDC), which has been observed to increase chromosomal aberrations in tissues from occupationally exposed workers and test animals (1,6,8).

Carcinogenic effects

No carcinogenic activity is reported for the majority of the thio- and dithiocarbamates studied. There is, however, evidence for the carcinogenicity of many of the EBDCs at high doses. This may be due to ETU, a major metabolite of EBDCs, which has been demonstrated to be carcinogenic in test animals (1).

Organ toxicity

The primary target organs for the thio- and dithiocarbamates include the nervous system, thyroid, and liver. Kidney injury was observed as a result of exposure to some pesticides in these classes (1,6,8).

Fate in humans and animals

In general, thiocarbamates and dithiocarbamates are rapidly absorbed into the bloodstream from the gastrointestinal tract, and to a lesser extent into the lung. Thiocarbamates are readily broken down and excreted by treated animals (1,6). Some dithiocarbamates may be less well absorbed (e.g., ziram) and others may accumulate to some degree at sites where toxicity may occur (e.g., the thyroid, liver, nervous system, etc.) (6). In the body, carbon disulfide results from metabolism of the thio- and dithiocarbamates, and may contribute to their toxic effects (6). Members of the EBDC chemical family are generally well absorbed through all routes. Breakdown products in animal systems include carbon disulfide and ETU (1,2).

Ecological effects

Effects on birds

Thio- and dithiocarbamates are generally of slight toxicity to birds. In most cases, the reported acute oral LD_{50} values are greater than 5000 mg/kg. The 8-day dietary LC_{50} values for the thio- and dithiocarbamates in birds, greater than 2000 ppm, indicate they are practically nontoxic to avian species (3,7,12–15).

Effects on aquatic organisms

The toxicity of the thio- and dithiocarbamates to aquatic organisms is variable, but the majority are moderately to highly toxic, with reported 96-hour LC_{50} values of about 1 to 10 mg/L. Others may be only slightly toxic (e.g., EPTC and metiram), with reported 96-hour LC_{50} values of about 20 mg/L (3,7,16–18). Reported bioconcentration factors and residence times in various types of fish generally indicate that members of these chemical families will not significantly accumulate in these organisms (16–19).

Effects on other organisms (non-target species)

Most thio- and dithiocarbamates are nontoxic to bees, both through contact and ingestion (3,4).

Environmental fate

Fate in soil and groundwater

Most of the thio- and dithiocarbamates are of low to moderate persistence, with reported field half-lives of a few days to several weeks (19–21). They are also poorly bound to soils, reasonably soluble in water, and therefore somewhat mobile. As a result, they may present a risk to groundwater, especially in highly porous soils with very little soil organic matter. Most are subject to microbial breakdown and volatilization.

Members of the EBDC group are of low persistence, with reported field half-lives of 1 to several days (19–21). They all rapidly and spontaneously degrade to

ETU in the presence of water and oxygen (19–21). Because the EDBCs are strongly bound to soils, practically insoluble in water, and show a very short field residence time under normal circumstances, they generally do not present a risk to ground-water (19–21).

ETU, though, is less strongly bound, more water soluble, and may persist for several weeks to a few months (19–21). ETU therefore does have the potential to be mobile and contaminate groundwater supplies. However, ETU, the primary break-down product of the EDBCs in soils, has been detected (at 16 ppb) in only 1 out of 1295 drinking water wells tested (22).

Fate in water

Many of the thio- and dithiocarbamates undergo hydrolysis, and breakdown by microbial action may be slow. The low water solubility of the compounds means that they will either be broken down or sorbed to sediment, where they may persist for several weeks or months. Breakdown of the EBDCs to ETU is very rapid, mainly by hydrolysis, and to a lesser degree by photodegradation (19–21). None of these compounds are expected to be persistent in the surface water environment.

Fate in vegetation

Most thio- and dithiocarbamates are readily taken up and translocated within plants, and rapidly processed into carbon dioxide and fatty acids (6,7). Most are not persistent within plants and do not leave significant residues. In most plant systems, the EBDCs are not readily taken up and translocated (4). If they are, they are degraded to ETU and then rapidly metabolized further to less-toxic breakdown products.

4.2   Individual Profiles

4.2.1   Butylate

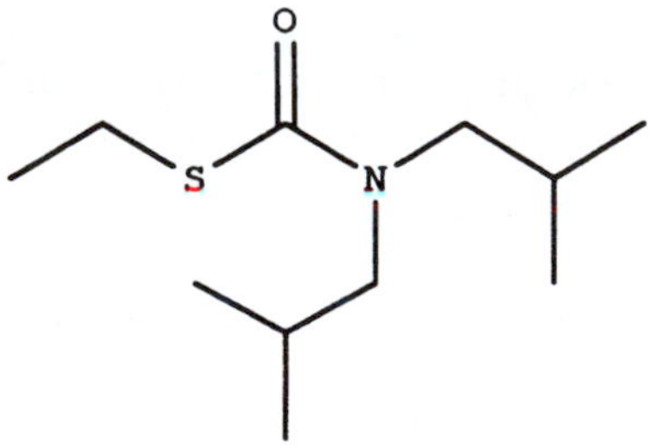

Figure 4.2 Butylate.

Trade or other names

Trade names include Anelirox, Anelda Plus, Aneldazin, Butilate, Carbamic Acid, Ethyl N, Genate, Genate Plus, N-Diisobutylthiocarbamate, R1910, Stauffer R-1, Sutan, and Sutan 6E.

Regulatory status

Butylate is classified by the U.S. Environmental Protection Agency as a General Use Pesticide (GUP), with applications limited to corn fields. It is categorized toxicity class III — slightly toxic. Products containing butylate bear the Signal Word CAUTION.

Introduction

Butylate is a herbicide and a member of the thiocarbamate class of chemicals. It is registered only for use in corn to control grassy weeds such as nutgrass and millet grass, as well as some broadleaf weeds. It is applied to soil immediately before corn is planted, often in combination with atrazine and/or cyanazine. Butylate acts selectively on seeds of weeds that are in the germination stage of development. It is absorbed from the soil by shoots of grass seedlings before they emerge, causing shoot growth to be slowed and leaves to become twisted.

Toxicological effects

Acute toxicity

The major routes of exposure to butylate are through the skin and by inhalation. Butylate is a thiocarbamate, a class of chemicals known for their tendency to irritate the skin and the mucous membranes of the respiratory tract. It may cause symptoms of scratchy throat, sneezing, and coughing when large amounts of dusts or spray are inhaled (4,7). Slight eye irritation can be caused by butylate, potentially leading to permanent eye damage (7,22).

Skin irritation was observed in rabbits topically exposed to 2000 mg technical butylate (85.71% pure) for 24 hours. The acute dermal LD_{50} for butylate is greater than 4640 mg/kg in rabbits (7). Butylate causes irritation to the eyes of rabbits (23).

The oral LD_{50} for butylate ranges from 1659 mg/kg in male guinea pigs to 5431 mg/kg in female rats. Butylate's inhalation LC_{50} (2-hour) is 19 mg/L (3,23).

Chronic toxicity

Application of 21 doses of 20 and 40 mg/kg/day to the skin of rabbits caused no effects other than local skin irritation (23). Liver changes were produced by doses of 180 mg/kg/day in a 56-week rat study with butylate. Blood clotting was affected by 10 mg/kg/day in the same experiment (24). Several studies have shown that long-term exposure to high doses of butylate causes increases in liver weights in test animals (23).

When butylate was fed to rats at doses of 50, 100, 200, or 400 mg/kg/day for 2 years, body weights were decreased and liver-to-body weight ratios increased at all but the lowest dose tested. In rats fed 20, 80, or 120 mg/kg/day for 2 years, no effects were observed at the 20 mg/kg dose, but kidney and liver lesions formed at the two higher doses. Butylate fed to rats at 10, 30, and 90 mg/kg/day for 56 weeks affected blood clotting at all doses. At the two higher doses, body weight and testes:body weight ratios decreased, liver:body weight ratios increased, and lesions formed on the testes. In a study of dogs fed 5, 25, or 100 mg/kg/day for 12 months, decreased body weights, increased liver weights, and increased incidence of liver lesions were observed at the highest dose (23).

Reproductive effects

No reproductive effects were observed in test animals receiving doses of up to 24 mg/kg/day of butylate (24). Long-term consumption of water containing butylate at very high doses caused damage to testes in rats (4). Butylate is unlikely to cause reproductive effects in humans at expected exposure levels.

Teratogenic effects

No teratogenic effects were observed in offspring of mice ingesting 4 to 24 mg/kg/day of Sutan during days 6 through 18 of pregnancy. No teratogenic effects were observed in the offspring of rats given up to 1000 mg/kg/day on days 6 through 20 of pregnancy or in the offspring of rabbits given doses of up to 500 mg/kg/day on days 6 through 18 of gestation (23,24).

However, in a study of two generations of offspring from rats fed for 63 days before mating, decreased brain weights were observed in the first generation of offspring at the 50-mg/kg/day dose level. At 200 mg/kg/day, adverse effects on the eyes and kidneys of the first generation were observed. This evidence suggests that butylate is unlikely to cause teratogenic effects in humans under normal circumstances.

Mutagenic effects

Mutations were seen in mice given very high oral doses of 1000 mg/kg/day of the herbicide (25). It was not mutagenic in the Ames test performed on Salmonella bacteria (4,24). Butylate thus is nonmutagenic or very weakly mutagenic.

Carcinogenic effects

There was no tumor formation related to doses of up to 320 mg/kg/day herbicide in a 24-month study of rats. Thus, butylate does not appear to be carcinogenic (24).

Organ toxicity

Animal studies have shown the liver and male reproductive system as the target organs.

Fate in humans and animals

Butylate is rapidly metabolized and excreted in animals (24). Within 48 hours after administration of butylate to rats by gavage, 27.3 to 31.5% of the material is eliminated through the urine, 60.9 to 64% is expired as carbon dioxide, and 3.3 to 4.7% is excreted in the feces. Only 2.2 to 2.4% of the compound is retained in the body, with most of this located in the blood, kidneys, and liver (23).

Ecological effects

Effects on birds

Given its low toxicity, butylate is considered a minimal hazard to birds (24). Technical butylate has an acute oral LD_{50} greater than 4640 mg/kg in mallard ducks. Its 8-day dietary LC_{50} in bobwhite quail is estimated at 40,000 ppm (22).

Effects on aquatic organisms

Butylate is moderately toxic to fish (3). It has a low to moderate potential for bioaccumulation in fish (23). The LC_{50} for a 96-hour exposure to technical Sutan ranges from 4.2 mg/L in rainbow trout to 6.9 mg/L in bluegill sunfish (24).

Effects on other organisms (non-target species)

Butylate is not harmful to bees if it is used appropriately (3). It appears to pose few, if any, acute toxicological hazards to non-target wildlife (24).

Environmental fate

Breakdown in soil and groundwater

Butylate has a low to moderate persistence in soil. The soil half-life is 3 to 10 weeks in moist soils under aerobic conditions. Under anaerobic conditions, butylate has a half-life of 13 weeks (23). In loamy soil, at 70 to 80°F, its half-life is 3 weeks (7). Soil half-lives of 12 days, and $1\frac{1}{2}$ to 3 weeks have also been reported (7,20).

Butylate is one of the pesticide compounds that the EPA considers to have the greatest potential for leaching into groundwater although it is only slightly soluble in water (23). Butylate does not strongly adsorb to soil particles and is slightly to highly mobile in soils, depending on the soil type (20,23). Leaching is more likely to occur in sandy, dry soils, and is less likely to occur in soil with higher amounts of organic matter and clay. An EPA study found butylate in 2 out of 152 groundwater samples analyzed (23).

Butylate degrades to sulfoxide in soil (8). Butylate has a residual activity in soil of approximately 4 months, when it is applied at 5 to 6 mg/hectare (3). When applied to dry soil surfaces, very little butylate is lost through vaporization. However, it can be lost by vaporization when applied to the surface of wet soils without being sufficiently incorporated (7).

Breakdown in water

Very low concentrations of butylate (maximum of 0.0047 mg/L) were found in 91 of 836 surface water samples analyzed (23).

Breakdown in vegetation

Butylate is readily adsorbed by plant leaves, but does not usually come in contact with foliage. It is rapidly taken up by the roots of corn plants and moved upward throughout the entire plant (7). Butylate is rapidly broken down in corn roots and leaves, to carbon dioxide, fatty acids, and certain natural plant constituents (7,22).

It is not thought to persist in plants since it disappeared from the stems and leaves of corn plants 7 to 14 days after treatment. The injury that it causes is not limited to that part of the plant to which it is applied (26).

Physical properties

Technical butylate is a clear amber to yellow liquid with an aromatic odor (3).

Chemical name: S-ethyl-di-isobutylthiocarbamate (3)
CAS #: 2008-41-5
Molecular weight: 217.38 (3)
Water solubility: 45 mg/L in water @ 22°C (3)

Solubility in other solvents: kerosene v.s.; xylene v.s.; acetone v.s.; ethyl alcohol
v.s. (3)
Vapor pressure: 170 mPa @ 25°C (3)
Partition coefficient (octanol/water): 14,000 (3)
Adsorption coefficient: 400 (20)

Exposure guidelines

ADI: Not available
HA: 0.35 mg/L (4)
RfD: 0.05 mg/kg/day (27)
PEL: Not available

Basic manufacturer

Zeneca Ag Products
1800 Concord Pike
Wilmington, DE 19897
Telephone: 800-759-4500
Emergency: 800-759-2500

4.2.2 EPTC

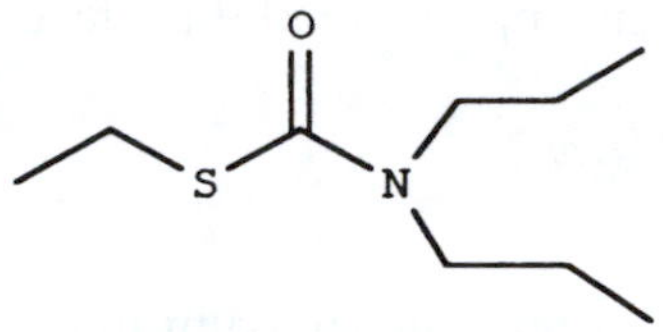

Figure 4.3 EPTC.

Trade or other names

Trade names include Alirox, Eptam, Eradicane, Eradicane Extra, Genep, Genep
Plus, and Shortstop.

Regulatory status

EPTC is a slightly toxic compound in EPA toxicity class III. It is a General Use
Pesticide (GUP); labels for products containing EPTC must bear the Signal Word
CAUTION.

Introduction

EPTC is a selective thiocarbamate herbicide used for control of annual grassy
weeds, perennial weeds, and some broadleaf weeds in beans, forage legumes, potatoes, corn, and sweet potatoes. It is usually applied pre-emergence (i.e., before weed
seeds germinate) and is usually incorporated into the soil immediately after application either mechanically or by overhead irrigation. EPTC is available as emulsifiable concentrates and granular formulations.

Toxicological effects

Acute toxicity

EPTC is slightly toxic via ingestion, with reported oral LD_{50} values of 1632 mg/kg in rats, 3160 mg/kg in mice, 112 mg/kg in cats, and 2460 mg/kg in rabbits (4,7). It is slightly toxic via the dermal route as well, with reported dermal LD_{50} values of 5000 mg/kg in rabbits and 3200 mg/kg in rats (7). The reported 1-hour inhalation LC_{50} in rats of 31.56 mg/L indicates slight toxicity by this route (3). It is a mild to moderate skin irritant in rabbits, a weak skin sensitizer in guinea pigs, and a mild eye irritant in rabbits (7).

EPTC is a cholinesterase inhibitor. Early symptoms of cholinesterase inhibition are blurred vision, fatigue, headache, vertigo, nausea, pupil contraction, abdominal cramps, and diarrhea. Severe inhibition of cholinesterase may cause excessive sweating, tearing, slowed heartbeat, giddiness, slurred speech, confusion, excessive fluid in the lungs, convulsions, and coma.

Workers subjected to inhalation exposure to EPTC experienced headaches, nausea, general malaise, and impaired working capacity. Animals poisoned in experimental tests displayed excitement, salivation, tearing, spasmodic winking, and depression (28,29).

Chronic toxicity

In a 16-week study of dogs fed 45 mg/kg/day, effects of brain cholinesterase inhibition and gastric mucosal changes were reported (28). In a 54-week feeding study of EPTC, no effects were observed at 20 mg/kg/day (28). In a two-generation study with rats fed 10 or 40 mg/kg/day, technical EPTC caused degeneration of tissues of the spinal cord, nerves, muscle, and heart tissue. No evidence of these effects was seen in a survey of workers who produced and formulated technical EPTC (29).

Reproductive effects

In a study where oral doses of 30, 100, or 300 mg/kg/day were administered on days 6 to 15 of pregnancy, maternal mortality and decreased weight gain and food consumption occurred at the highest dose. Decreased fetal body weight and increased loss of fetuses occurred at 100 and 300 mg/kg/day (27). It is not likely that EPTC will cause reproductive effects in humans under normal circumstances.

Teratogenic effects

No effects were observed in a teratogenic study in which rats were given 300 mg/kg/day (7,27). The available evidence suggests that EPTC is not teratogenic.

Mutagenic effects

EPTC was not mutagenic when tested in a series of assays with microbial and human cell culture lines (7,29).

Carcinogenic effects

In a 2-year feeding and oncogenicity study of EPTC in mice, no excess tumors were seen at doses up to 20 mg/kg/day (7). The available evidence suggests that EPTC is not carcinogenic.

Organ toxicity

In lifetime studies with animals, the target organs of technical EPTC toxicity were nerves, muscle, and heart tissue (30).

Fate in humans and animals

In rats, low oral amounts of EPTC (approximately 0.6 mg) were mainly eliminated via expired air, and much smaller amounts were eliminated via the urine and feces. When the amount was increased to 100 mg, the relative proportion excreted via urine and feces was increased (31).

Ecological effects

Effects on birds

EPTC is slightly toxic to relatively nontoxic to birds. The oral LC_{50} for technical EPTC in bobwhite quail is 20,000 ppm for a 7-day feed treatment (31).

Effects on aquatic organisms

EPTC is slightly toxic to fish and aquatic organisms. The reported 96-hour LC_{50} values for EPTC are 19 mg/L in rainbow trout, 27 mg/L in bluegill sunfish, 17 mg/L in mosquito fish, 17 mg/L in cutthroat trout, and 16 mg/L in lake trout (3,7,16). The 24-hour LC_{50} in the blue crab is greater than 20 mg/L (7). Bioconcentration values for fish range from 37 to 190 times the ambient water concentration, indicating that the compound will not significantly accumulate in these organisms (30).

Effects on other organisms (non-target species)

EPTC is practically nontoxic to bees, with a reported LD_{50} of 0.011 mg per bee (3).

Environmental fate

Breakdown in soil and groundwater

EPTC is of low persistence in the soil environment, with reported field half-lives of 6 to 32 days; a representative field half-life for most soil regimes is 6 days (20). It is not strongly bound to soils, especially those lower in organic matter and clay contents (7,20). Microbial breakdown and volatilization are the main mechanisms by which EPTC is lost from soils (20). Due to its short half-life, it is not a threat to groundwater.

Breakdown in water

There is little chance that it will enter surface waters, due to its short half-life.

Breakdown in vegetation

EPTC is readily absorbed by the roots of plants and translocated upward to the leaves and stems. EPTC is rapidly metabolized by plants to carbon dioxide and naturally occurring plant constituents (7,30).

Physical properties

EPTC is a colorless to pale yellow liquid with an aromatic odor (3).

Chemical name: s-ethyl dipropylthiocarbamate (3)
CAS #: 759-94-4
Molecular weight: 189.32 (3)
Water solubility: 375 mg/L @ 25°C (3)
Solubility in other solvents: v.s. in acetone, ethyl alcohol, kerosene, methyl isobutyl ketone, and xylene (3)
Melting point: Not available
Vapor pressure: 4700 mPa @ 25°C (3)
Partition coefficient (octanol/water): 1600 (3)
Adsorption coefficient: 200 (20)

Exposure guidelines

ADI: Not available
HA: Not available
RfD: Not available
PEL: Not available

Basic manufacturer

Zeneca Ag Products
1800 Concord Pike
Wilmington, DE 19897
Telephone: 800-759-4500
Emergency: 800-759-2500

4.2.3   Mancozeb

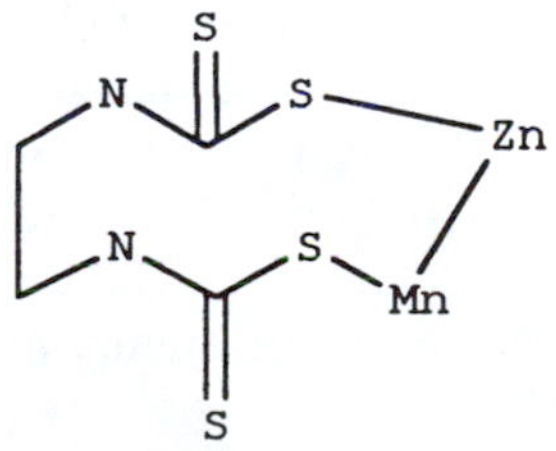

Figure 4.4 Mancozeb.

Trade or other names

Trade names include Dithane, Dithane-Ultra, Fore, Green-Daisen M, Karamate, Mancofol, Mancozeb, Mancozin, Manzate 200, Manzeb, Manzin Nemispor, Nemispot, Policar, Riozeb, and Zimaneb.

Regulatory status

Mancozeb is a practically nontoxic EDBC in EPA toxicity class IV. It is registered as a General Use Pesticide (GUP). Labels for products containing mancozeb must bear the Signal Word CAUTION.

Introduction

Mancozeb is used to protect many fruit, vegetable, nut, and field crops against a wide spectrum of fungal diseases, including potato blight, leaf spot, scab (on apples and pears), and rust (on roses). It is also used for seed treatment of cotton, potatoes, corn, safflower, sorghum, peanuts, tomatoes, flax, and cereal grains. Mancozeb is available as dusts, liquids, water-dispersible granules, wettable powders, and ready-to-use formulations. It is commonly found in combination with zineb and maneb.

Toxicological effects

Acute toxicity

Mancozeb is practially nontoxic via the oral route, with reported oral LD_{50} values of greater than 5000 mg/kg to greater than 11,200 mg/kg in rats (1,3). Via the dermal route, it is practically nontoxic as well, with reported dermal LD_{50} values of greater than 10,000 mg/kg in rats, and greater than 5000 mg/kg in rabbits (4). It is a mild skin irritant and sensitizer, and a mild to moderate eye irritant in rabbits (4,32). Workers with occupational exposure to mancozeb have developed sensitization rashes (1).

Mancozeb is a cholinesterase inhibitor. Early symptoms of cholinesterase inhibition are blurred vision, fatigue, headache, vertigo, nausea, pupil contraction, abdominal cramps, and diarrhea. Severe inhibition of cholinesterase may cause excessive sweating, tearing, slowed heartbeat, giddiness, slurred speech, confusion, excessive fluid in the lungs, convulsions, and coma.

Chronic toxicity

No toxicological effects were apparent in rats fed dietary doses of 5 mg/kg/day in a long-term study (1). Impaired thyroid function was observed as lower iodine uptake after 24 months in dogs fed doses of 2.5 and 25 mg/kg/day mancozeb, but not in those dogs fed 0.625 mg/kg/day (1).

A major toxicological concern in situations of chronic exposure is the generation of ethylene thiourea (ETU) in the course of mancozeb metabolism, and as a contaminant in mancozeb production (1,33). ETU may also be produced when EBDCs are used on stored produce or during cooking (9). In addition to having the potential to cause goiter, a condition in which the thyroid gland is enlarged, this metabolite has produced birth defects and cancer in experimental animals (9).

Reproductive effects

In a three-generation rat study with mancozeb at a dietary level of 50 mg/kg/day, there was reduced fertility but no indication of embryotoxic effects (1,9). In another study in which pregnant rats were exposed to mancozeb by inhalation, toxic effects on the pups were observed only at exposure levels (55 mg/m³) that were also toxic to the dams (1). It is unlikely that mancozeb will produce reproductive effects in humans under normal circumstances.

Teratogenic effects

No teratogenic effects were observed in a three-generation rat study with mancozeb at a dietary level of 50 mg/kg/day (1). Developmental abnormalities of the body wall, central nervous system, eye, ear, and musculoskeletal system were observed in experimental rats given a very high dose of 1320 mg/kg of mancozeb on the 11th day of pregnancy (25). Mancozeb was not teratogenic to rats when it was inhaled by pregnant females at airborne concentrations of 0.017 mg/L (32). In pregnant rats fed 5 mg/kg/day (the lowest dose tested), developmental toxicity was observed in the form of delayed hardening of the bones of the skull in offspring (9). In view of the conflicting evidence, the teratogenicity of mancozeb is not known.

Mutagenic effects

Mancozeb was found to be mutagenic in one set of tests, while in another it did not cause mutations (9). Mancozeb is thought to be similar to maneb, which was not mutagenic in the Ames test (32). Data regarding the mutagenicity are inconclusive but suggest that mancozeb is either not mutagenic or weakly mutagenic.

Carcinogenic effects

No data are available regarding the carcinogenic effects of mancozeb. While studies of other EBDCs indicate they are not carcinogenic, ETU (a mancozeb metabolite), has caused cancer in experimental animals at high doses (9,10). Thus, the carcinogenic potential of mancozeb is not currently known.

Organ toxicity

The main target organ of mancozeb is the thyroid gland; the effects may be due to the metabolite ETU (9,10).

Fate in humans and animals

Mancozeb is rapidly absorbed into the body from the gastrointestinal tract, distributed to various target organs, and almost completely excreted in 96 hours. ETU is the major mancozeb metabolite of toxicologic significance, with carbon disulfide as a minor metabolite (10).

Ecological effects

Effects on birds

Mancozeb is slightly toxic to birds, with reported 5-day dietary LC_{50} values in bobwhite quail and mallard ducklings of greater than 10,000 ppm (32). The 10-day dietary LC_{50} values of 6400 and 3200 ppm were reported for mallard ducks and Japanese quail, respectively (4).

Effects on aquatic organisms

Mancozeb is moderately to highly toxic to fish and aquatic organisms. Reported 48-hour LC_{50} values are 9 mg/L in goldfish, 2.2 mg/L in rainbow trout, 5.2 mg/L in catfish, and 4.0 mg/L in carp (4). The reported 72-hour LC_{50} for mancozeb in crayfish is greater than 40 mg/L; the 48-hour LC_{50} is 3.5 mg/L in tadpoles (32).

Effects on other organisms (non-target species)

Mancozeb is not toxic to honeybees (4).

Environmental fate

Breakdown in soil and groundwater

Mancozeb is of low soil persistence, with a reported field half-life of 1 to 7 days (20). Mancozeb rapidly and spontaneously degrades to ETU in the presence of water and oxygen (10). ETU may persist for longer, on the order of 5 to 10 weeks (20). Because mancozeb is practically insoluble in water, it is unlikely to infiltrate ground-water (3). Studies do indicate that ETU, a metabolite of mancozeb, has the potential to be mobile in soils (9). However, ETU has been detected (at 0.016 mg/L) in only 1 out of 1295 drinking water wells tested (10).

Breakdown in water

Mancozeb degrades in water with a half-life of 1 to 2 days in slightly acidic to slightly alkaline conditions (32).

Breakdown in vegetation

When used as directed, mancozeb is not poisonous to plants (4).

Physical properties

Mancozeb is a grayish-yellow powder (3).

Chemical name: manganese ethylenebis(dithiocarbamate) (polymeric) (3)
CAS #: 8018-01-7
Molecular weight: 266.31 (4)
Water solubility: 6 mg/L (3)
Solubility in other solvents: Practically insoluble in most organic solvents (3)
Melting point: Decomposes without melting @ 192°C (3)
Vapor pressure: Negligible @ 20°C (3)
Partition coefficient (octanol/water): Not available
Adsorption coefficient: >2000 (20)

Exposure guidelines

ADI: 0.03 mg/kg/day (33)
HA: Not available
RfD: 0.003 mg/kg/day (27)
PEL: Not available

Basic manufacturer

DuPont Agricultural Products
Walker's Mill, Barley Mill Plaza
P.O. Box 80038
Wilmington, DE 19880-0038
Telephone: 800-441-7515
Emergency: 800-441-3637

4.2.4   Maneb

Trade or other names

Trade names include Farmaneb, Manesan, Manex, Manzate, Nereb, and Newspor.

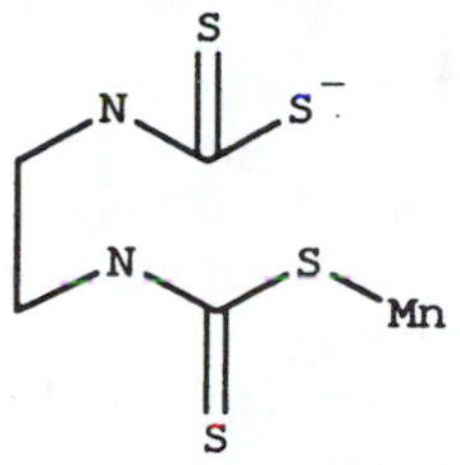

Figure 4.5 Maneb.

Regulatory status

Maneb is a practically nontoxic ethylene(bis)dithiocarbamate in EPA toxicity class IV. It is registered as a General Use Pesticide (GUP). Labels for products containing mancozeb must bear the Signal Word CAUTION.

Introduction

Maneb is an ethylene(bis)dithiocarbamate fungicide used in the control of early and late blights on potatoes and tomatoes and many other diseases of fruits, vegetables, field crops, and ornamentals. Maneb controls a wider range of diseases than other fungicides. It is available as granular, wettable powder, flowable concentrate, and ready-to-use formulations.

Toxicological effects

Acute toxicity

Maneb is practically nontoxic by ingestion, with oral LD_{50} values of greater than 5000 to 8000 mg/kg in rats, and 8000 mg/kg in mice (1,3). Via the dermal route, it is slightly toxic, with a dermal LD_{50} in rats of greater than 5000 mg/kg (3). Inflammation or irritation of the skin, eyes, and respiratory tract have resulted from contact with maneb (1,8). The 4-hour inhalation LC_{50} is greater than 3.8 mg/L, indicating slight toxicity.

Acute exposure to maneb may result in effects such as hyperactivity and inco-ordination, loss of muscular tone, nausea, vomiting, diarrhea, loss of appetite, weight loss, headache, confusion, drowsiness, coma, slowed reflexes, respiratory paralysis, and death (1,8).

Chronic toxicity

Rat feeding trials over 2 years showed no evidence of adverse health effects at dietary doses of about 12.5 mg/kg/day (1). Goiter (increased thyroid weight) and reduced growth rate were seen in rats fed daily doses of 62.5 mg/kg/day after 97 days (10). Dogs that received maneb orally at doses of 200 mg/kg/day for 3 or more months developed tremors, lack of energy, gastrointestinal disturbances, and inco-ordination. In addition, they experienced damage to the spinal cord, but not the thyroid gland (1). Rats that received 1500 mg/kg/day for 10 days showed weight loss, weakness of the hind legs, and increased mortality (1).

Reproductive effects

Female rats that were given 50 mg/kg/day on every other day during gestation showed increased rates of embryo death and stillbirth, and decreased newborn survival (4). In rats given a single dose of 770 mg/kg maneb (the lowest dose tested) on the 11th day of gestation, early fetal deaths occurred (4). In mice, the lowest single oral toxic dose administered during gestation that caused toxicity to the fetus was 1420 mg/kg (4). It appears that a very high level of exposure is necessary to cause reproductive effects in humans, and this level of exposure is not likely under normal circumstances.

Teratogenic effects

Fetal abnormalities of the eye, ear, body wall, central nervous system, and musculoskeletal system were seen in rats given single doses of 770 mg/kg (4). Maneb is metabolized to ethylene thiourea (ETU), a compound that has been shown to cause birth defects in laboratory animals such as rats, mice, and hamsters (12). From these data and information about other EBDCs, it is likely that maneb will not be teratogenic in humans under normal circumstances.

Mutagenic effects

Several tests have shown that maneb is not mutagenic (10).

Carcinogenic effects

In one study, maneb did not display significant carcinogenicity in laboratory tests with experimental animals (8). In another study, malignant tumors were observed in rats given scrotal injections of 12.5 mg/kg body weight of 82.6% pure maneb (4). Based on these data and evidence from other EBDCs, maneb is unlikely to cause cancer in humans (1,10).

Organ toxicity

Target organs affected by maneb include the thyroid, kidneys, and heart.

Fate in humans and animals

Animal studies show that maneb is readily absorbed through the gastrointestinal tract, and is rapidly eliminated. In rats, 55% of an administered dose of over 300 mg/kg was eliminated within 5 days (1). A study in mice showed that elimination was mainly through the feces. The metabolites of maneb include ethylenediamine, ethylene(bis)thiurammonosulfide, and ethylenethiourea (1). Maneb was not found to accumulate in the tissues of rats given 125 mg/kg/day over 2 years, nor in dogs given 75 mg/kg/day for 1 year (1).

Ecological effects

Effects on birds

Maneb is practically nontoxic to birds; the 5-day dietary LC_{50} for maneb in bobwhite quail and mallard ducklings is greater than 10,000 ppm (34).

Effects on aquatic organisms

Maneb is highly toxic to fish and aquatic species. The 96-hour LC_{50} for maneb is 1 mg/L in bluegill sunfish (34). The reported 48-hour LC_{50} is 1.9 mg/L in rainbow trout, and 1.8 mg/L in carp (34). The 72-hour LC_{50} is more than 40 mg/L in crayfish, and the 48-hour LC_{50} is 40 mg/L in tadpoles (34).

Effects on other organisms (non-target species)

Maneb-treated crop foliage may be toxic to livestock (26). The fungicide is not thought to be toxic to bees (3).

Environmental fate

Breakdown in soil and groundwater

Maneb is similar in its environmental fate to mancozeb (20). Like mancozeb, maneb is of low persistence (with a reported field half-life of 12 to 36 days), but it is readily transformed into ETU, which is more persistent (20). Since it is strongly bound by most soils and is not highly soluble in water (20), it should not be very mobile. It therefore does not represent a significant threat to groundwater. Its breakdown product, ETU, may however be more highly mobile. Maneb breaks down under both aerobic and anaerobic soil conditions (4). In one study, residues of maneb did not leach below the top 5 inches of soil (4).

Breakdown in water

Maneb degraded completely within 1 hour under anaerobic aquatic conditions.

Breakdown in vegetation

The main metabolite of maneb in plants is ethylene thiourea (ETU); this is then rapidly metabolized further. Significant amounts of ETU were formed in cooking vegetables that had been experimentally treated with maneb.

Physical properties

Maneb is a yellow powder with a faint odor (3). It is a polymer of ethylene-(bis)dithiocarbamate units linked with manganese.

Chemical name: manganese ethylenebis(dithiocarbamate) (polymeric) (3)
CAS #: 12427-38-2
Molecular weight: 265.29 (single manganese-TBDC unit) (3)
Water solubility: 6 mg/L (estimated) (3)
Solubility in other solvents: Practically insoluble in common organic solvents (3)
Melting point: Decomposes before melting @ approximately 192°C (3)
Vapor pressure: Negligible @ 20°C (3)
Partition coefficient (octanol/water): Not available
Adsorption coefficient: <2000 (estimated) (20)

Exposure guidelines

ADI: 0.03 mg/kg/day (33)
HA: Not available
RfD: 0.05 mg/kg/day (27)
PEL: Not available

Basic manufacturer

ELF Atochem North America, Inc.
2000 Market Street
Philadelphia, PA 19103–3222
Telephone: 215-419-7219
Emergency: 800-523-0900

4.2.5   Metiram

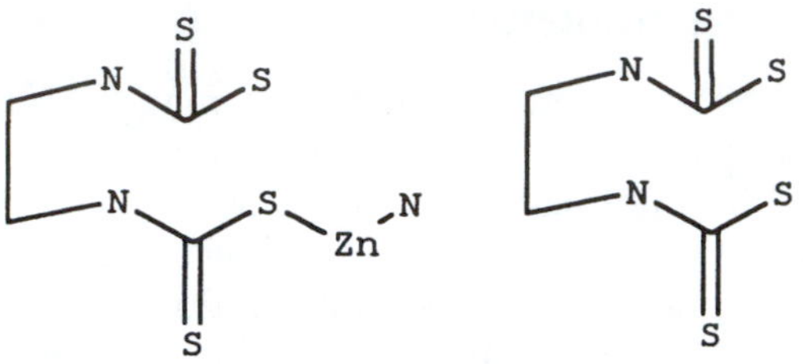

Figure 4.6 Metiram.

Trade or other names

Trade or other names for metiram include arbatene, NIA 9102, Polyram, Polyram-Combi, and Zinc metiram.

Regulatory status

Metiram is a practically nontoxic compound in EPA toxicity class IV. Labels for products containing it must bear the Signal Word CAUTION. It is a General Use Pesticide (GUP).

Introduction

Metiram may be used to prevent crop damage in the field, during storage, or transport. Metiram is effective against a broad spectrum of fungi and is used to protect fruits, vegetables, field crops, and ornamentals from foliar diseases and damping off.

Toxicological effects

Acute toxicity

Metiram is practically nontoxic when ingested, with reported oral LD_{50} values of greater than 6180 mg/kg to greater than 10,000 mg/kg in rats; greater than 5400 mg/kg in mice; and 2400 to 4800 mg/kg in guinea pigs (1,4). The dermal LD_{50} is greater than 2000 mg/kg in rats, indicating slight toxicity (3). It is reported to be a mild skin and eye irritant (4). Via the inhalation route, it is slightly toxic, with a reported 4-hour inhalation LC_{50} of greater than 5.7 mg/L (3).

Metiram is a cholinesterase inhibitor. Early symptoms of cholinesterase inhibition are blurred vision, fatigue, headache, vertigo, nausea, pupil contraction, abdominal cramps, and diarrhea. Severe inhibition of cholinesterase may cause excessive sweating, tearing, slowed heartbeat, giddiness, slurred speech, confusion, excessive fluid in the lungs, convulsions, and coma.

Chronic toxicity

When rats were fed metiram at dietary doses of 50 mg/kg/day, 5 days a week for 2 weeks, no symptoms of illness were produced. Adverse effects did occur at 500 mg/kg/day (1). No ill effect was observed in dogs that received 45 mg/kg daily of the fungicide for 90 days, or 7.5 mg/kg daily for almost 2 years (1). When metiram was fed to rats at dietary doses of 0.25, 1, 4, or 16 mg/kg/day, the only effect observed was muscle atrophy in rats receiving 16 mg/kg/day (11).

The major toxicological concern in situations of chronic exposure to metiram, however, is ethylenethiourea (ETU), a contaminant and a breakdown product of metiram that has been shown to cause birth defects and cancer in experimental animals.

Reproductive effects

Pregnant rats fed 80 and 160 mg/kg/day exhibited reduced rates of body weight gain. Litter size was reduced for rats fed 0.25, 2, or 16 mg/kg/day metiram. Rats receiving the 16-mg/kg/day dose also exhibited decreases in parental body weight and in food consumption (1). The evidence suggests that reproductive effects are unlikely in humans under normal circumstances.

Teratogenic effects

No teratogenic effects were found in female rats fed 40, 80, and 160 mg/kg/day of metiram (1,11).

Mutagenic effects

The majority of mutagenicity studies on metiram have been negative. Out of six tests performed, only one highly sensitive test indicated that metiram may be mutagenic (10,11). These data indicate that metiram is either nonmutagenic or weakly mutagenic.

Carcinogenic effects

All of the EBDC pesticides can be degraded or metabolized into ethylenethiourea (ETU), which has been shown to produce cancer in mice and rats (31). However, other EBDCs do not appear to be carcinogenic. There were no data available regarding the carcinogenic properties of metiram itself.

Organ toxicity

No data were available regarding the target organs of metiram. It is likely that its target organs will be similar to those affected by closely related compounds such as maneb and mancozeb. Those compounds exert their principal effects on the thyroid (1).

Fate in humans and animals

Metiram is not well absorbed through the skin; less than 1% of a 240-mg/kg dose, applied topically, was absorbed through the skin of rats after 8 hours (31). Metabolic fate studies in rats indicate that ingested metiram is readily absorbed by the body and eliminated through the urine and feces. Residues remaining in the body were highest in the kidneys, thyroid, and gastrointestinal tract and were higher in females than in males (31). In mammalian tissues the ethylene(bis)dithiocarbamates break down into ETU (2).

Ecological effects

Effects on birds

Metiram is slightly toxic to birds, with 5- to 8-day LC_{50} values in both mallard ducks and bobwhite quail of greater than 3712 ppm (3,4).

Effects on aquatic organisms

Metiram is slightly to moderately toxic to fish; reported 96-hour LC_{50} values are 85 mg/L in carp and 1.1 mg/L in rainbow trout (1,11). The 48-hour LC_{50} for metiram in harlequin fish is 17 mg/L (1,11).

Effects on other organisms (non-target species)

Metiram is practically nontoxic to bees; the reported oral LD_{50} is greater than 40 µg per bee, and the contact LD_{50} is reported to be greater than 16 µg per bee (3).

Environmental fate

Breakdown in soil and groundwater

Metiram is probably similar in its environmental fate to closely related compounds such as maneb and mancozeb. They are of low persistence and are strongly bound to most soils (20). This property, and their low water solubilities, indicate that they probably do not pose a significant risk to groundwater. They are unstable in the presence of atmospheric moisture and oxygen and are rapidly degraded in biological systems to ETU and other metabolites (31). These products are of moderate persistence and more mobile, and therefore may pose a slight risk to groundwater. ETU, the primary metabolite of metiram in water, has been detected (at 0.016 mg/L) in only 1 out of 1295 drinking water wells tested (2).

Breakdown in water

Breakdown of metiram to ETU is very rapid, mainly by hydrolysis, and to a lesser degree by photodegradation (31).

Breakdown in vegetation

Metiram is not taken up by plants to a significant degree (3).

Physical properties

Metiram is a yellow powder at room temperature (3).

Chemical name: zinc ammoniate ethylenebis(dithiocarbamate)-poly(ethylene
   thiuram disulfide) (3)
CAS #: 9006-42-2
Molecular weight: 1088.7 (3)
Water solubility: <1 mg/L (3)
Solubility in other solvents: Practically insoluble in most organic solvents (3)
Melting point: Decomposes at 140°C (3)
Vapor pressure: 0.01 mPa @ 20°C (3)

Partition coefficient (octanol/water): 2 (3)
Adsorption coefficient: 500,000 (estimated) (20)

Exposure guidelines

ADI: 0.03 (33)
HA: Not available
RfD: Not available
PEL: Not available

Basic manufacturer

BASF Corp.
Agricultural Products Group
P.O. Box 13528
Research Triangle Park, NC 27709-3528
Telephone: 800-669-2273
Emergency: 800-832-4357

4.2.6   Molinate

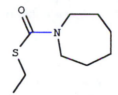

Figure 4.7 Molinate.

Trade or other names

Trade names include Hydram, Molinate, Ordram, and Yalan.

Regulatory status

Molinate is a slightly to moderately toxic compound in EPA toxicity class III, and is a registered as a General Use Pesticide (GUP). Products containing molinate must bear the Signal Word WARNING or CAUTION.

Introduction

Molinate is a selective thiocarbamate herbicide used to control broad-leaved and grassy plants in rice and other crops. Molinate is available in granular and emulsifiable liquid formulations.

Toxicological effects

Acute toxicity

Molinate is moderately toxic by ingestion with reported oral LD_{50} values of 369 to 720 mg/kg in rats, and 530 to 795 mg/kg in mice. Dermal LD_{50} values are 4000 to 4800 mg/kg in rats (3,7). It is mildly irritating to rabbit skin and moderately irritating to rabbit eyes, and is not a skin sensitizer (7). The 4-hour inhalation LC_{50} of 1.36 mg/L indicates moderate toxicity by this route as well (4). Some formulations show a lower degree of acute toxicity (6,7). Symptoms of exposure to molinate include nausea, diarrhea, abdominal pain, fever, weakness, and conjunctivitis (7,13).

Chronic toxicity

Chronic dietary exposure of dogs to 22.5 mg/kg/day over 13 weeks did not cause adverse health effects, but doses of 45 mg/kg/day caused increased thyroid weight over the same period (27). Increased organ weights have been reported in rats at doses of 2 mg/kg/day over 2 years, although not at 8 mg/kg/day over 13 weeks in rats (27).

The only reported human exposure to molinate resulting in adverse health effects comes from a report of well contamination in Japan. After field application of approximately 60 kg active ingredient to a 2-hectare rice paddy, several people noticed an odor emitted from a nearby well, and fell ill as a result of repeated consumption of water from that well (6). Their symptoms, which were apparently quite rapid in onset, included abdominal and gastrointestinal disorders, fever, weakness, and conjunctivitis (6). These symptoms disappeared following the use of an alternative water source, and there were no reports of long-term complications or lingering effects due to this exposure (6). The concentration of the well water sampled 15 days following the first reported symptoms was 6 µg/L; it is not known what the initial concentration was (6).

Reproductive effects

Administration of molinate to young male rats at a dose of 3.6 mg/kg/day for 2 months caused changes in spermatozoa but did not decrease sperm fertility (6). When these rats were mated to normal females, many of the embryos were resorbed and postnatal mortality was increased (6). It is unlikely that such effects will occur in humans at expected exposure levels.

Teratogenic effects

Reports on the teratogenicity of molinate are conflicting, with one suggestion that it is teratogenic (35) and another that it is not (7). Thus, its teratogenicity is unknown.

Mutagenic effects

No data were located regarding the potential mutagenic effects of molinate although it has been reported to be nongenotoxic (4,7).

Carcinogenic effects

In a 2-year assay in rats, no carcinogenic activity was reported at doses up to 2 mg/kg/day (6).

Organ toxicity

The primary target organ affected by molinate is the thyroid.

Fate in humans and animals

Molinate is only fairly well absorbed through oral, dermal, and inhalation exposure (4). It is metabolized in the rat liver, and rapid excretion occurs primarily through the urine (88% of the applied dose) with a small amount lost in the feces (11% of the applied dose). Excretion by rats was practically complete within 48 hours (6).

Ecological effects

Effects on birds

Molinate appears to be practically nontoxic to birds. The reported 5-day dietary LC_{50} in Japanese quail is greater than 5000 ppm, and that in mallards is greater than 13,000 ppm (4,13).

Effects on aquatic organisms

The reported toxicity to fish varies greatly, from slightly to highly toxic. One source reports the 96-hour LC_{50} values at 0.21 mg/L in rainbow trout and 0.32 mg/L in bluegill sunfish (16), while another reports them as 1.3 and 29 mg/L, respectively, (3). A 96-hour LC_{50} value of 30 mg/L in goldfish has also been reported (4).

Fish kills of carp due to molinate were observed in Japan. The pesticide caused an anemia-like condition in these fish (4). Reported 96-hour LC_{50} values in aquatic invertebrates such as *Daphnia* and stoneflies are about 0.3 to 0.6 mg/L, indicating that molinate is highly toxic to these invertebrates (4,16).

Effects on other organisms (non-target species)

No data are currently available.

Environmental fate

Breakdown in soil and groundwater

Molinate is of low persistence in the soil environment, with a field half-life of 5 to 21 days (20). It is poorly bound to soils, soluble in water, and thus may be mobile (20) and present a risk of groundwater contamination. Soil microorganisms are responsible for most molinate breakdown (4). Molinate may rapidly volatilize if not plowed into the soil, and may undergo breakdown by sunlight (4).

Breakdown in water

Molinate may be degraded by hydrolysis (reaction with water).

Breakdown in vegetation

Molinate is rapidly taken up by plant roots and transported to the leaves. In the leaves, molinate inhibits leaf growth and development. It is rapidly metabolized to carbon dioxide and other naturally occurring plant products such as amino acids and organic acids in nonsusceptible plants.

Physical properties

Molinate is a noncorrosive, clear liquid with an aromatic or spicy odor (3).

Chemical name: S-ethyl hexhydro-1 H-azepine-1-carbothioate (3)
CAS #: 2212-67-1
Molecular weight: 187.30 (3)

Solubility in water: 880 mg/L (3)
Solubility in other solvents: v.s. in acetone, xylene, ethanol, kerosene, and 4-
    methylpentan-2-one (3)
Vapor pressure: 746 mPa @ 25°C (3)
Melting point: Not available
Partition coefficient (octanol/water): 760 (3)
Adsorption coefficient: 190 (20)

Exposure guidelines

ADI: Not available
HA: Not available
RfD: 0.002 mg/kg/day (27)
PEL: Not available

Basic manufacturer

Zeneca Ag Products
1800 Concord Pike
Wilmington, DE 19897
Telephone: 800-759-4500
Emergency: 800-759-2500

4.2.7   *Thiram*

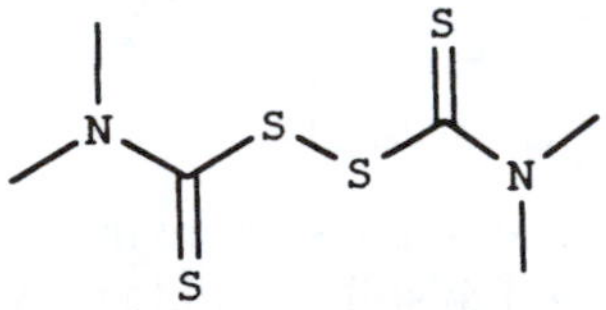

Figure 4.8 Thiram.

Trade or other names

Common names include thiram (U.S.), thiuram (Japan), and TMTD (former
U.S.S.R.), TMT, and TMTDS. Trade names include AAtack, Arasan, Aules, Fermide
850, Fernasan, FMC 2070, Hexathir, Mercuram, Micropearls, Nomersan, Pomarsol,
Puralin, Rezifilm, Rhodiasan Express, Spotrete, Tersan, Thiosan, Thiotex, Thiramad,
Thirame, Thiuramin, Thirasan, Tirampa, Tiuramyl, TMTC, TMTD 50 Borches,
Trametan, Tuads, and Tulisan.

Regulatory status

Thiram is registered as a General Use Pesticide (GUP) by the U.S. Environmental
Protection Agency (EPA). It is classified as toxicity class III — slightly toxic. Pesticide
products containing thiram bear the Signal Word CAUTION on the product label.

Introduction

Thiram is a dimethyl dithiocarbamate compound used as a fungicide to prevent
crop damage in the field and to protect harvested crops from deterioration in storage

or transport. Thiram is also used as a seed protectant and to protect fruit, vegetable, ornamental, and turf crops from a variety of fungal diseases. In addition, it is used as an animal repellent to protect fruit trees and ornamentals from damage by rabbits, rodents, and deer. Thiram is available as dust, flowable, wettable powder, water dispersible granules, and water suspension formulations, and in mixtures with other fungicides.

Thiram has been used in the treatment of human scabies, as a sunscreen, and as a bactericide applied directly to the skin or incorporated into soap.

Toxicological effects

Acute toxicity

Thiram is slightly toxic by ingestion and inhalation, but it is moderately toxic by dermal absorption. Acute exposure in humans may cause headaches, dizziness, fatigue, nausea, diarrhea, and other gastrointestinal complaints. In rats and mice, large doses of thiram produced muscle incoordination, hyperactivity followed by inactivity, loss of muscular tone, labored breathing, and convulsions. Most animals died within 2 to 7 days (4).

Thiram is irritating to the eyes, skin, and respiratory tract. It is a skin sensitizer. Symptoms of acute inhalation exposure to thiram include itching, scratchy throat, hoarseness, sneezing, coughing, inflammation of the nose or throat, bronchitis, dizziness, headaches, fatigue, nausea, diarrhea, and other gastrointestinal complaints. Persons with chronic respiratory or skin disease are at increased risk from exposure to thiram (4).

Ingestion of thiram and alcohol together may cause stomach pains, nausea, vomiting, headache, slight fever, and possible dermatitis. Workers exposed to thiram during application or mixing operations within 24 hours of moderate alcohol consumption have been hospitalized with symptoms.

The 4-hour inhalation LC_{50} for thiram is greater than 500 mg/L in rats. Reported oral LD_{50} values for thiram are 620 to over 1900 mg/kg in rats; 1500 to 2000 mg/kg in mice; and 210 mg/kg in rabbits (1,3). The dermal LD_{50} is greater than 1000 mg/kg in rabbits (4) and in rats (1,3).

Chronic toxicity

Symptoms of chronic exposure to thiram in humans include drowsiness, confusion, loss of sex drive, incoordination, slurred speech, and weakness, in addition to those due to acute exposure. Repeated or prolonged exposure to thiram can also cause allergic reactions such as dermatitis, watery eyes, sensitivity to light, and conjunctivitis (1).

Except for the occurrence of allergic reactions, harmful chronic effects from thiram have been observed in test animals only at very high doses. In one study, a dietary dose of 125 mg/kg/day thiram was fatal to all rats within 17 weeks. Oral doses of about 49 mg/kg/day to rats for 2 years produced weakness, muscle incoordination, and paralysis of the hind legs. Rats fed 52 to 67 mg/kg/day for 80 weeks exhibited hair loss, and paralysis and atrophy of the hind legs. Symptoms of muscle incoordination and paralysis from thiram poisoning have been shown to be associated with degeneration of nerves in the lower lumbar and pelvic regions. Day-old white leghorn chicks fed 30 and 60 ppm for 6 weeks exhibited bone malformations (1).

At doses of about 10% of the LD_{50} for 15 days, thiram reduced blood platelet and white blood cell counts, suppressed blood formation, and slowed blood coagulation in rabbits (1).

Reproductive effects

Very high oral doses of approximately 1200 mg/kg/day thiram to mice on days 6 to 17 of pregnancy caused resorption of embryos and retarded fetal development. In another study, doses of 132 mg/kg/day for 13 weeks produced infertility in male mice, while doses of 96 mg/kg/day for 14 days delayed the estrous cycle in females (1). The feeding of 50 mg/kg/day thiram from day 16 of pregnancy to 21 days after birth caused reduced growth and survival of the pups. Pups that were transferred to untreated dams at birth remained healthy, while pups transferred from untreated to treated dams showed toxic effects (1). These data suggest that reproductive effects occur at high doses not likely to be experienced by humans.

Teratogenic effects

Cleft palate, wavy ribs and curved long leg bones were observed in the offspring of mice that ingested very high thiram doses of 1200 mg/kg/day on days 6 to 17 of pregnancy. Maternal doses of 125 mg/kg/day thiram were teratogenic in hamsters, causing incomplete formation of the skull and spine, fused ribs, abnormalities of the legs, heart, great vessels and kidneys (1). Developmental toxicity was observed in a three-generation study of rats fed 5.0 mg/kg/day (1,4). These data suggest that high doses are required to cause teratogenic effects.

Mutagenic effects

Thiram has been found to be mutagenic in some test organisms but not in others (1). Thus, the evidence is inconclusive.

Carcinogenic effects

When administered to mice at the highest dose possible, thiram was not carcinogenic. Dietary levels as high as 125 mg/kg/day for two years did not cause tumors in rats (1). These data indicate that thiram is not carcinogenic.

Organ toxicity

Studies have shown evidence of damage to the liver by thiram in the form of decreased liver enzyme activity and increased liver weight (1). Thiram may also cause damage to the nervous system, blood, and kidneys (4).

Fate in humans and animals

In the body, carbon disulfide is formed from the breakdown of thiram and does contribute to the toxicity of thiram to the liver (1,3).

Thiram is not a member of the ethylene(bis)dithiocarbamate (EBDC) chemical family, and thus it should not generate ethylene thiourea (ETU) (1).

Ecological effects

Effects on birds

Thiram is practically nontoxic to birds. The reported dietary LC_{50} of thiram in Japanese quail is greater than 5000 ppm (36). Reported dietary LC_{50} values in pheasants and mallard ducks are 2800 ppm and 673 ppm, respectively (14). The LD_{50} for the compound in red-winged blackbirds is greater than 100 mg/kg (3).

Effects on aquatic organisms

Thiram is highly toxic to fish (4). The LC_{50} for the compound is 0.23 mg/L in bluegill sunfish, 0.13 mg/L in trout, and 4 mg/L in carp (17). Thiram is not expected to bioconcentrate in aquatic organisms (19).

Effects on other organisms (non-target species)

Thiram is nontoxic to bees (3).

Environmental fate

Breakdown in soil and groundwater

Thiram is of low to moderate persistence. It is nearly immobile in clay soils or in soils high in organic matter. Because it is only slightly soluble in water (30 mg/L) and has a strong tendency to adsorb to soil particles, thiram is not expected to contaminate groundwater. The soil half-life for thiram is reported as 15 days (20).

Thiram degrades more rapidly in acidic soils and in soils high in organic matter. In a humus sandy soil, at pH 3.5, thiram decomposed after 4 to 5 weeks, while at pH 7.0, thiram decomposed after 14 to 15 weeks. Thiram persisted for over 2 months in sandy soils, but disappeared within 1 week from a compost soil. The major metabolites of thiram in the soil are copper dimethyldithiocarbamate, dithiocarbamate, dimethylamine, and carbon disulfide (19).

In soil, thiram will be degraded by microbial action or by hydrolysis under acidic conditions. Thiram will not volatilize from wet or dry soil surfaces (19).

Breakdown in water

In water, thiram is rapidly broken down by hydrolysis and photodegradation, especially under acidic conditions. Thiram may adsorb to suspended particles or to sediment (19).

Breakdown in vegetation

No data are currently available.

Physical properties

Thiram is a white to yellow crystalline powder with a characteristic odor (3).

Chemical name: tetramethylthiuram disulfide (3)
CAS #: 137-26-8
Molecular weight: 240.44 (3)
Water solubility: 30 mg/L at 25°C (3)
Solubility in other solvents: s.s. in ethanol; s. in acetone and chloroform (3)
Melting point: 146°C (3)
Vapor pressure: Negligible at room temperature (3)

Partition coefficient (octanol/water): Not available
Adsorption coefficient: 670 (11)

Exposure guidelines

ADI: 0.01 mg/kg/day (33)
HA: Not available
RfD: 0.005 mg/kg/day (27)
PEL: 5 mg/m^3 (8-hour) (28)

Basic manufacturer

ELF Atochem North America, Inc.
2000 Market Street
Philadelphia, PA 19103-3222
Telephone: 215-419-7219
Emergency: 800-523-0900

4.2.8   Triallate

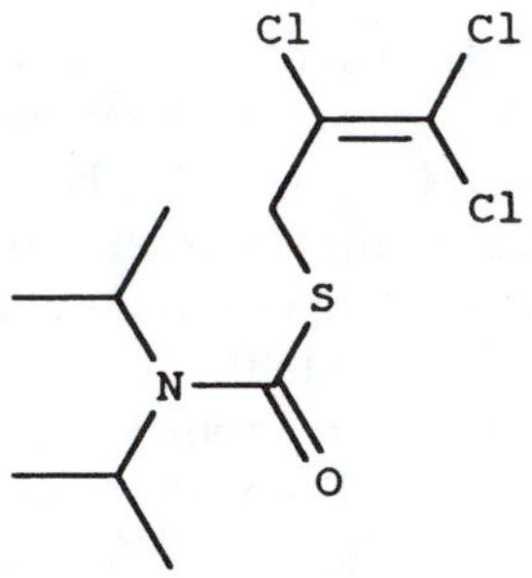

Figure 4.9 Triallate.

Trade or other names

Trade names for triallate include Avadex BW, Buckle, Carbamothoic acid, CP 23426, Dipthal, Far-Go, Showdown, and TDTC Technical.

Regulatory status

Triallate is a General Use Pesticide (GUP). It is classified as toxicity class III — slightly toxic. Formulations of triallate bear the Signal Word CAUTION.

Introduction

Triallate belongs to the thiocarbamate chemical class. It is a pre-emergence selective herbicide used to control grass weeds in field and pulse crops. It is used selectively to control wild oats, black grass, and annual meadow grass in barley, wheat, peas, lentils, rye, maize, beets, brassicas, carrots, and onions. Depending on the crop that is treated, the herbicide is incorporated in the soil before or after planting. Triallate is available as emulsifiable concentrate and granular formulations.

Toxicological effects

Acute toxicity

Technical triallate is slightly toxic by ingestion and practically nontoxic via dermal exposure or inhalation (4). The oral LD_{50} for technical triallate in rats is 800 to 2165 mg/kg, and in mice is 930 mg/kg (3,4). The oral LD_{50} in rats for emulsifiable concentrate formulations is 2700 mg/kg, and for granular formulations is greater than 12,000 mg/kg (7). The dermal LD_{50} for technical triallate is 8200 mg/kg in rabbits, and 3500 mg/kg in rats. The inhalation 4-hour LC_{L0} in cats is 0.4 mg/L (4).

In rats fed triallate at doses of 50 to 2000 mg/kg, abnormal behavior was observed at doses of 100 mg/kg and above. No changes in nerve tissue occurred. At doses of 600 mg/kg and above, death and reduced body weight occurred (37). Sheep may be poisoned by 300 mg/kg of triallate, with symptoms of depression, lack of appetite, mouth watering, weakness, and convulsions (38). Inhalation exposure to large amounts of thiocarbamates may cause itching, scratchy throat, sneezing, and coughing (7). Triallate is moderately irritating to the skin and is a mild eye irritant. Tests on guinea pigs indicate that technical triallate does not cause allergic skin reactions (38).

Although triallate is a carbamate, it does not inhibit cholinesterase activity. No symptoms occurred, and cholinesterase activity was not affected in rats fed single doses of 1500 and 3000 mg/kg (4).

Chronic toxicity

Prolonged or repeated exposure to triallate may cause symptoms similar to those caused by acute exposure.

Oral doses of 100 mg/kg/day triallate to hamsters for 22 months resulted in decreased body weight gain, changes in blood chemistry, slight anemia, increased liver weights, and decreased spleen weights. Mice fed 3 and 12.5 mg/kg/day triallate for 2 years exhibited increased liver and heart weights, changes in the liver and spleen, and mineralization in the brain and cornea. No adverse effects were observed in dogs fed 1.5, 5, and 15 mg/kg/day triallate for 2 years (37,38).

At high-dose levels in subchronic exposure studies, neurological effects have been observed in rats. Rat deaths at these high levels were probably due to a variety of systemic effects such as liver and stomach pathological changes, decreased food consumption, and loss of body weight. Neurological effects were not observed in rats at doses of 50 mg/kg/day or below (37,38).

Reproductive effects

Reduced body and pup weights, reduced pregnancy rate and length, reduced pup survival, and effects on other reproductive parameters occurred when rats were fed 30 mg/kg/day triallate during mating, pregnancy, and nursing for two successive generations (37,38). This suggests that triallate can cause reproductive effects at high doses.

Teratogenic effects

No birth defects were observed in the offspring of rabbits given triallate doses of 5, 15, and 45 mg/kg/day on days 6 to 28 of pregnancy. No birth defects were observed in the offspring of rats given doses of 10, 30, 90 mg/kg/day on days 6 to 20 of pregnancy. In both of these studies, the highest dose administered caused poisoning symptoms in both the mothers and their offspring (38). These data indicate that triallate is not teratogenic.

Mutagenic effects

No genetic changes occurred in tests using live animals (fruit flies, hamsters, and mice). In tests using bacterial and animal cell cultures, both positive and negative results have been reported (37,38). This suggests that triallate is either nonmutagenic or weakly mutagenic.

Carcinogenic effects

When fed dietary doses of about 2.5, 7.5, and 30 mg/kg/day technical triallate over a long time, the incidence of liver tumors increased in a strain of mice normally prone to spontaneous production of liver tumors. Several other long-term feeding studies showed no incidence of tumors (37). Triallate did not produce tumors in rats fed up to 12.5 mg/kg/day for 2 years (38). No tumors appeared when hamsters were fed dietary doses up to 100 mg/kg triallate for 22 months (37). These data indicate that triallate is not carcinogenic (39).

Organ toxicity

Changes in the cellular processes of the brain, liver, and spleen were observed in pigs given triallate (4). Studies on other species have indicated that the thymus, kidneys, and reproductive organs are potential targets as well.

Fate in humans and animals

In general, thiocarbamates, the chemical class in which triallate is included, are rapidly absorbed into the bloodstream from the gastrointestinal tract, readily broken down into metabolites, and then excreted by treated animals. It is rarely possible to detect thiocarbamates in the blood (40).

A single oral dose of 500 mg/kg triallate was rapidly absorbed from the gastrointestinal tract of rabbits. It was then found to be present in all organs tested within 15 to 20 minutes after dosing. The largest amount of the herbicide accumulated in the liver, lungs, kidneys, and spleen. All traces were gone by the 7th day. Triallate was reported to be completely eliminated from the body of rabbits within 7 to 10 days (4). The meat of sheep poisoned by approximately 300 mg/kg triallate had detectable traces 84 days later in cold storage. No traces of the herbicide were detected in the eggs, meat, or internal organs of hens fed 0 to 4% of an LD_{50} dose (4,38).

Ecological effects

Effects on birds

Triallate is slightly toxic to relatively nontoxic to birds. The acute oral LD_{50} for triallate in bobwhite quail is 2251 mg/kg. The 8-day dietary LD_{50} is greater than 5000 ppm in both mallards and bobwhite quail (7,38).

Effects on aquatic organisms

Triallate is highly toxic to fish and other aquatic organisms. The 48-hour LC_{50} in *Dapnia magna*, a small freshwater crustacean, is 0.06 to 0.10 mg/L for the 95% technical material, and the LC_{50} is 0.05 to 0.07 mg/L for the 46% emulsifiable concentrate (17). The 96-hour LC_{50} in algae is 0.12 mg/L (38). The 96-hour LC_{50} for technical material has been reported as 0.62 mg/L in rainbow trout (1.0 mg/L for the emulsifiable concentrate), and 1.7 mg/L in channel catfish (1.1 mg/L for the emulsifiable concentrate) (17). A 96-hour LC_{50} of 1.3 mg/L is reported in bluegill (38).

When technical triallate concentrations were measured in bluegill sunfish over a 7-week period, marked bioaccumulation occurred. The concentration in the fish was 1600 times the ambient water concentration. However, after 2 weeks in water without triallate, the compound was nearly completely eliminated by the fish (38).

Effects on other organisms (non-target species)

Triallate is nontoxic to bees (3).

Environmental fate

Breakdown in soil and groundwater

Triallate has a moderate persistence in the soil environment. It adsorbs strongly to loam and clay soils and is not readily dissolved in water. This indicates that triallate is not likely to move through the soil, even though it has an average soil half-life of 82 days (7,20). However, if there is significant moisture and/or a low level of organic matter in the soil, leaching and groundwater contamination may be possible. The EPA suggests that triallate does not pose a threat to the environment due to leaching because it is generally used where the water table is relatively low (39).

Triallate is reported to be degraded in soil primarily by soil microbes (7). Plants also degrade triallate, lessening its potential to accumulate in the soil (39). If applied to the soil surface at high temperatures, without incorporation into the soil, triallate can be lost to the atmosphere through volatilization. Its volatility increases with soil water content (39). Triallate must be incorporated into the soil after application to prevent its loss from soil at high temperatures. Triallate can persist into the next growing season, especially in colder climates in which it is less likely to be broken down (39).

Photodecomposition, or breakdown in the presence of sunlight, is considered an insignificant method of degradation for triallate (4).

Breakdown in water

Triallate is stable to ultraviolet degradation and will probably be found adsorbed to suspended sediment in the water column or in hydrosoils due to its slight water solubility and its ability to bind to particulates (7). Typical breakdown times in hydrosoils may be longer than in terrestrial systems due to lower oxygen availability for microbial degradation.

Breakdown in vegetation

Studies indicate that triallate does not bioaccumulate in plants. Triallate is absorbed and metabolized by plants (39).

Physical properties

Triallate is an amber, oily liquid (3).

Chemical name: S-(2,3,3-trichloro-2-propenyl)bis(1-methylethyl) carbamothioate (3)
CAS #: 2303-17-5

Molecular weight: 304.66 (3)
Water solubility: 4 mg/L @ 25°C (3)
Solubility in other solvents: s. in acetone, ether, ethyl alcohol, heptane, benzene,
    ethyl acetate, and most organic solvents (3)
Melting point: 29–30°C (3)
Vapor pressure: 16 mPa @ 25°C (3)
Partition coefficient (octanol/water) not available
Adsorption coefficient: 2400 (20)

Exposure guidelines

ADI: Not available
HA: Not available
RfD: 0.013 mg/kg/day (27)
PEL: Not available

Basic manufacturer

Monsanto Company
800 N. Lindbergh Blvd.
St. Louis, MO 63167
Telephone: 314-694-6640
Emergency: 314-694-4000

4.2.9   *Zineb*

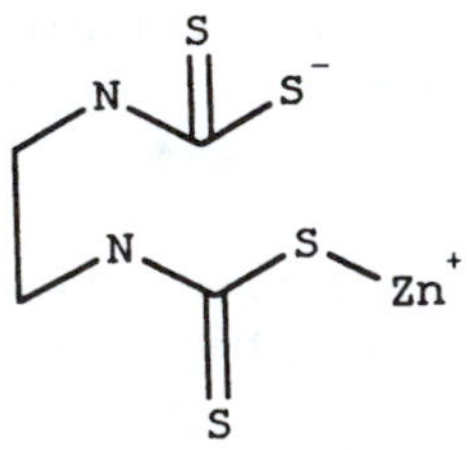

Figure 4.10 Zineb.

Trade or other names

Trade names include Aspor, Chem Zineb, Devizeb, Dipher, Discon-Z, ethylene-
(bis)dithiocarbamate (EBDC), Hexathane, Kypzin, Lodaco, Lonacol, Mancozan,
Parazate, Parzate, Tiezene, Zebtox, Ziden, and Zinosan. The compound may also be
found in formulations with other pesticides.

Regulatory status

Zineb was formerly registered in the U.S. as a General Use Pesticide (GUP) and
was rated as a pesticide of low toxicity — EPA toxicity class IV. Products containing
zineb were required to carry the Signal Word CAUTION on the label.

Following an EPA Special Review of all the ethylene(bis)dithiocarbamate pesticides (EDBCs), including zineb, all registrations for zineb were voluntarily canceled
by the manufacturer. All tolerances for zineb in agricultural commodities in the U.S.

(except grapes used in winemaking) were revoked, effective 12/31/94. The tolerance for grapes used in winemaking was revoked, effective 12/31/97.

Introduction

The EBDCs are dithiocarbamate fungicides used to prevent crop damage in the field and to protect harvested crops from deterioration during storage or transport. Zineb was used to protect fruit and vegetable crops from a wide range of foliar and other diseases. It was available in the U.S. as wettable powder and dust formulations. Zineb can be formed by combining nabam and zinc sulfate in the spray tank.

Toxicological effects

Acute toxicity

The oral LD_{50} for zineb in rats is 1850 to 8900 mg/kg; in mice is 7600 to 8900 mg/kg; and in rabbits is 4450 mg/kg. The LC_{lo} (inhalation) in rats is 0.8 mg/L (4-hour). The dermal LD_{50} in rats is over 2500 mg/kg, the highest dose possible to administer (3,4).

Zineb is slightly toxic when ingested. Following a single large dose of zineb, rats and mice exhibited incoordination, hyperactivity followed by inactivity, loss of muscle tone, and loss of hair (1,4). Sheep died within 3 weeks of being given oral doses of 500 mg/kg zineb.

In spray or dust forms, zineb is moderately irritating to the skin, eyes, and respiratory mucous membranes. It may also be a dermal sensitizer, with possible cross-sensitization to maneb and mancozeb (4). This irritation may result in itching, scratchy throat, sneezing, coughing, inflammation of the nose or throat, and bronchitis (40). Early symptoms from exposure of humans to zineb include tiredness, dizziness, and weakness. More severe symptoms include headache, nausea, fatigue, slurred speech, convulsions, and unconsciousness (1,4). These effects may be exacerbated with concurrent exposure to alcohol. Acute neurotoxic effects are probably due to carbon disulfide, a metabolite of zineb (1).

Animal studies indicate that changes in thyroid may occur following a single, large dose (1), but that these may be reversible (1). Ethylene thiourea (ETU), a potentially toxic metabolite of zineb, may be involved in thyroid effects (1).

Chronic toxicity

The survival, growth, and blood chemistry of dogs were not affected by dietary levels up to 250 mg/kg/day for 1 year. However, thyroid size and weight increases were observed at this dose level. No effects on the thyroid were detected at 100 mg/kg/day (1). In a 2-year study of rats fed 500 mg/kg/day, diminished growth, kidney pathology, and increased thyroid weight and size were observed (1). Sheep showed no adverse effects from dosages of 100 and 250 mg/kg/day for 19 weeks (1).

Occupational inhalation of zineb can lead to changes in liver enzymes, moderate anemia and other blood changes, increased incidence of poisoning symptoms during pregnancy, and chromosomal changes in the lymphocytes (4). Liver functioning was affected in workers exposed to zineb. Moderate anemia and other blood changes were also reported in 150 workers exposed to zineb in a chemical plant (1). A 5-month study of zineb showed that concentrations of 20 and 200 mg/L caused decreases in the activity of cholinesterase, an essential enzyme of the nervous system. Inhalation exposure to zineb may decrease the size of the bronchial passages (1).

Repeated or prolonged dermal exposure may cause dermatitis or conjunctivitis (4). Farm workers who were repeatedly exposed to zineb, in fields sprayed with 0.5% suspension of the fungicide, reported severe and extensive contact dermatitis (31).

Ethylene thiourea formation during metabolism of zineb or other EBDC pesticides may potentially result in goiter, a condition in which the thyroid gland is enlarged (2).

Reproductive effects

It is advisable that pregnant women avoid exposure to zineb, as it can damage the fetus, as well as cause adverse reproductive system effects (1,35). A single intraperitoneal injection of approximately 160 mg/kg zineb in mice, during the second half of pregnancy, resulted in abortions and weak offspring. Oral doses of zineb, at a rate of 100 mg/kg/day for 2 months or more, produced sterility, resorption of fetuses, and abnormal tails in offspring (1,4).

After oral ingestion of zineb, similar concentrations of ETU, a metabolite of this pesticide, were found in both maternal and fetal tissues of rats.

Teratogenic effects

Offspring of rats given a near lethal oral dose of 2000 mg/kg/day zineb on days 11 or 13 of pregnancy, showed a high incidence of skeletal malformations, as well as defects in the closing of the neural tube, an embryonic tube that eventually develops into the brain and spinal cord (12). A very high single oral dose of 8000 mg/kg to rats on day 11 of gestation produced numerous deformities in the offspring (1). Teratogenic effects were also seen when pregnant mice were given intraperitoneal injections of 150 mg/kg (41). In pregnant rats fed 5.0 mg/kg/day, the lowest dose tested, developmental toxicity was observed in the form of delayed hardening of the skull bones in offspring.

Zineb's metabolite, ethylene thiourea (ETU), may cause abnormal fetal development (35). It has been shown to be teratogenic in hamsters, but not in mice (2).

Mutagenic effects

Results of mutagenicity assays of zineb and its metabolite, ETU, are inconclusive (4). The data suggest that zineb may be a weak mutagen.

Carcinogenic effects

Available data clearly show that low doses of zineb are not carcinogenic. Very high doses have caused tumors in some test animals. In two strains of mice, the maximum tolerated lifetime dose of zineb did not cause tumors. Oral doses of 3500 mg/kg/week for 6 weeks caused one of two mice strains tested to develop benign lung tumors after 3 weeks (1). Two rat feeding studies showed no evidence of tumor formation (4). Overall, evidence of carcinogenicity in chronic oral feeding studies in rats and mice is inconclusive, although they suggest it is not likely to be carcinogenic in humans (1).

Organ toxicity

Zineb appears to be harmful to the thyroid, liver and, muscles. Liver and kidney injury were observed in autopsies done on sheep that died after a 3-week exposure to oral doses of 500 mg/kg/day zineb (1).

Studies of the effects of zineb on test animals have shown rapid reduction in the uptake of iodine and swelling of the thyroid (i.e., goiter) (1).

Fate in humans and animals

Approximately 68 to 74% of ingested zineb was recovered unchanged in the feces after administration at various dietary levels. It is estimated that only 11 to 17% of an oral dose of zineb was absorbed into the body from the gastrointestinal tract of the rat (1). In general, zineb is rapidly excreted from the body following ingestion. Zineb is metabolized in mammalian tissues into ETU and carbon disulfide (1).

Ecological effects

Effects on birds

Zineb is practically nontoxic to birds. The oral LD_{50} for zineb in mallards and young pheasants is greater than 2000 mg/kg (42).

Effects on aquatic organisms

Zineb is moderately toxic to fish. The 96-hour LC_{50} in perch is 2 mg/L (3,4).

Effects on other organisms (non-target species)

Zineb is not toxic to bees (3,4). Little or no reduction in the numbers of beneficial predatory and parasitic arthropods was seen when zineb was used in Nova Scotian orchards at recommended dosages. Mites appear to be sensitive to zineb (15).

Environmental fate

Breakdown in soil and groundwater

Zineb is subject to chemical breakdown (hydrolysis) and is of low persistence in soil. It adsorbs strongly to soil particles and usually does not move below the upper layer of soil (21). For this reason, zineb is unlikely to contaminate groundwater. Its bioactive half-life in the field is 16 days. Within 4 months after a field planted with alfalfa was sprayed, 99.7% of the applied zineb was lost (19).

ETU, a metabolite of zineb, has been detected (at 0.016 mg/L) in only 1 out of 1295 drinking water wells tested (2).

Breakdown in water

Zineb is practically insoluble in water (3). It is unstable in water and hydrolyzes rapidly, producing ETU and other compounds (19).

Breakdown in vegetation

Zineb is generally not poisonous to plants, except in zinc-sensitive varieties such as tobacco and cucurbits. Pears have been slightly injured by this fungicide in a few cases (4). ETU is the major zineb metabolite in plants (3).

Physical properties

Zineb is a light-colored powder or crystal (3). Zineb is a polymer of ethylene-(bis)thiocarbamate units linked with zinc.

Chemical name: zinc ethylenebis(dithiocarbamate) (3)
CAS #: 12122-67-7
Molecular weight: 275.74 (EBDC-zinc unit) (3)
Water solubility: 10 mg/L @ 25°C (3)
Solubility in other solvents: s. in carbon disulfide; s.s. in pyridine; practically i.s.
    in common organic solvents (3)
Melting point: Thermal decomposition @ 157°C (3)
Vapor pressure: <0.01 mPa @ 20°C (3)
Partition coefficient (octanol/water): <20 at 20°C (3)
Adsorption coefficent: 1000 (estimated) (21)

Exposure guidelines

ADI: 0.03 mg/kg/day (33)
HA: Not available
RfD: 0.05 mg/kg/day (27)
PEL: Not available

Basic manufacturers

ELF Atochem North American
2000 Market Street
Philadelphia, PA 19103-3222
Telephone: 215-419-7219
Emergency: 800-523-0900

4.2.10 Ziram

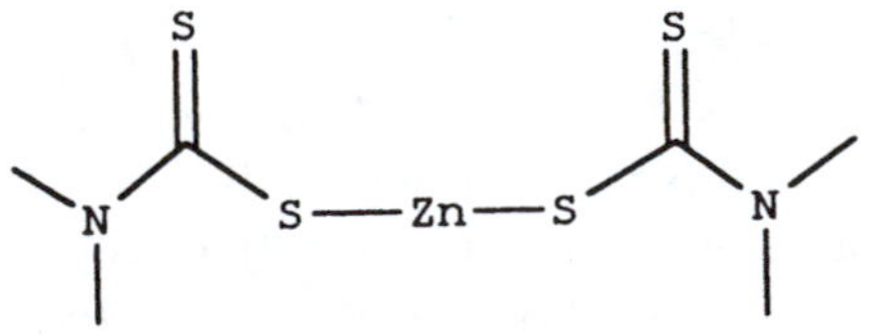

Figure 4.11 Ziram.

Trade or other names

Trade names for products containing ziram include AAprotect, AAvolex, Antene, Attivar, Carbazinc, Corozate, Cuman, Drupine, Fuklasin, Fungostop, Mezene, Milbam, Pomarsol Z Forte, Prodaram, Tricarbamix, Triscabol, Z-C Spray, Zerlate, Zincmate, Zinkcarbamate, Ziram, Zirasan, Zirbeck, and Zirex. The compound may be found in formulations with other fungicides such as bitertanol, dodine, myclobutanil, thiram, and zineb.

Regulatory status

Ziram is a General Use Pesticide (GUP) in the U.S. It is a slightly to moderately toxic compound, EPA toxicity class III. Ziram carries the Signal Word DANGER on its label due to eye irritation hazard.

Introduction

Ziram is an agricultural dithiocarbamate fungicide used on a wide variety of plant fungi and diseases. It may be applied to the foliage of plants, but it is also used as a soil and/or seed treatment. Ziram is used primarily on almonds and stone fruits. It is also used as an accelerator in rubber manufacturing, packaging materials, adhesives, and textiles. Another use of the compound is as a bird and rodent repellent.

Ziram is often marketed as a wettable powder or as granules. Granules or grains are sifted into water and agitated prior to application.

Toxicological effects

Acute toxicity

Acute exposure among industrial and farm workers in the former U.S.S.R. caused irritation of the skin, nose, eyes, and throat (1).

The oral LD_{50} for ziram is 1400 mg/kg in rats, and 480 and 400 mg/kg in mice and rabbits, respectively. Ziram has an LD_{50} of 100 to 150 mg/kg in guinea pigs (4). The acute dermal LD_{50} in rats is greater than 6000 mg/kg. Ziram is corrosive to eyes and may cause irreversible eye damage (43).

Chronic toxicity

Female rats administered relatively small doses of ziram in their diets (2.5 mg/kg/day) for 9 months showed decreased antibody formation. Rats fed doses of 1 to 2 mg/kg/day ziram for an unknown time period exhibited poor growth and development (44). In a 1-year feeding study with rats, no effects were seen at the low dose of 5 mg/kg/day, nor were any effects seen in weanlings receiving 5 mg/kg/day in their diet for 30 days (4). A study with dogs fed ziram in their diets showed no harmful effects after 12 months at 5 mg/kg/day (44).

Reproductive effects

When female and male rats were given moderate doses of ziram (50 mg/kg/day) for nearly 2 months prior to pregnancy, the rats had marked reductions in fertility and litter size. The rats in this study became largely sterile. A lower dose of 10 mg/kg had no effect on reproduction (1,44).

Female mice fed moderate doses (50 mg/kg/day) of ziram for 15 days exhibited reduced fertility, but no effects on fertility appeared in male mice (4). Wasting away of the testes has been noted as a toxic effect of ziram (45).

Based on these data, reproductive effects in humans are unlikely at normal levels of exposure.

Teratogenic effects

Pregnant rats administered ziram at doses of 12.5 to 100 mg/kg/day during the organ-forming period of pregnancy showed embryotoxic effects at doses of 25 mg/kg/day and greater. The compound also had a slight growth-inhibiting effect on the embryos at 100 mg/kg. Maternal toxicity was observed at all test levels (4). No teratogenic effects were observed.

Mutagenic effects

Numerous tests have established that ziram is mutagenic. For example, there was an increase in the number of chromosome changes in bone marrow cells in mice treated with oral doses of 100 mg/kg/day (4).

Chromosomal changes have also been observed in workers exposed to the compound in industrial settings for 3 to 5 years (1). The concentration in the air averaged 1.95 mg/L but reached as high as 71.3 mg/L in some of these cases. Thus, there is a risk to humans chronically exposed to ziram at moderate to high concentrations.

Carcinogenic effects

A carcinogenicity study was performed on rats and mice exposed to ziram for a 103-week period. Under the conditions of the study, ziram was carcinogenic to male rats, causing an increase in thyroid cancer. There was no increase in carcinogenicity in female rats or in male mice. Female mice showed an increase in lung tumors, but this was complicated by a virus infection, thus making interpretation difficult (45). Ziram's carcinogenicity is not determinable from current evidence.

Organ toxicity

The primary target organ is the thyroid, as shown in a study of workers who experienced thyroid enlargement after ziram exposure (45).

Fate in humans and animals

Ziram is poorly absorbed in the absence of oils. However, it may be readily absorbed into the body in the presence of oil, including through the skin.

Rats that had been fed low doses (30 mg/kg/day) of the compound for 2 years had very low levels in their livers (0.03 mg). However, the zinc component of the parent compound is stored in the body to a slightly higher degree. The amount of zinc in bone was related to the dose over a 2-year experiment. Female rats had some water-soluble residues in blood, kidneys, liver, ovaries, spleen, and thyroid 24 hours following a single oral dose (4). Ziram that remained unchanged in the rat was excreted in the feces (4). This indicates that, though ziram has only a slight potential to persist and concentrate in living tissue, the compound may be selectively localized in the body, as are other dithiocarbamates, at sites where toxicity may occur.

The highest concentrations of zinc after ziram exposure are found in the male reproductive system and specifically in the prostate. High concentrations are also found in bone, liver, kidney, pancreas, and endocrine glands.

Rats that were fed low doses of ziram followed by ethyl alcohol had higher alcohol levels in their bloodstream over a 4-hour period (4).

Ecological effects

Effects on birds

Toxicity of ziram to birds will vary from essentially nontoxic to moderately toxic. Its LD_{50} is 100 mg/kg in European starlings and red-wing blackbirds. In a 2-year study, the dietary LC_{50} in quail was 3346 ppm (13). In chickens, doses of 56 mg/kg were toxic (44). Ziram has an antifertility action in laying hens. When given to chickens under unspecified conditions, there were adverse effects on body weight and retarded testicular development (4).

Effects on aquatic organisms

Based on data from only one species, the goldfish, the compound appears to be moderately toxic to fish. The 5-hour LC_{50} for ziram in goldfish was between 5 and 10 mg/L (3). Based on its low solubility in water, ziram should have a low bioconcentration potential (13).

Effects on other organisms (non-target species)

No data are currently available.

Environmental fate

Breakdown in soil and groundwater

Ziram has not been detected in groundwater (19). In soils with medium to high content of soil organic matter, ziram will be moderately bound. A field half-life of 30 days has been estimated for ziram (21), indicating a low to moderate persistence.

Breakdown in water

Of the metallic dithiocarbamate fungicides, ziram is the most stable. Because the compound is toxic to bacteria, biodegradation in sediment may be rather slow, or occur only at very low concentrations. If ziram gets to the bottom of bodies of water, it may persist for months (19).

Breakdown in vegetation

On plants, persistent breakdown products were formed. A significant amount of carbon disulfide was released during the breakdown process. The leaf surface was slightly acidic probably due to dissolved carbon dioxide (4).

Physical properties

Ziram is an odorless powder at room temperature (3).

Chemical name: zinc bis(dimethyldithiocarbamate) (3)
CAS #: 137-30-4
Molecular weight: 305.83 (3)
Solubility in water: 65 mg/L (3)
Solubility in other solvents: s. in alcohol, acetone, benzene, and carbon tetrachloride (3)
Melting point: 240–244°C (3)
Vapor pressure: Negligible at room temperature (3)
Partition coefficient (octanol/water): Not available
Adsorption coefficient: 400 (estimated) (21)

Exposure guidelines

ADI: 0.02 mg/kg/day (33)
HA: Not available
RfD: Not available
PEL: Not available

Basic manufacturer

FMC Corporation
Agricultural Chemicals Group
1735 Market Street
Philadelphia, PA 19103
Telephone: 215-299-6565
Emergency: 800-331-3148

References

(1) Edwards, I. R., Ferry, D. G. and Temple, W. A. Fungicides and related compounds, In *Handbook of Pesticide Toxicology.* Hayes, W. J. and Laws, E. R., Eds. Academic Press, New York, 1991.

(2) U.S. Environmental Protection Agency. Ethylene bisdithiocarbamates (EBDCs); Notice of intent to cancel and conclusion of Special Review. *Fed. Reg.* 57, 7434–7530, 1992.

(3) Kidd, H. and James, D. R., Eds. *The Agrochemicals Handbook,* 3rd ed. Royal Society of Chemistry Information Services, Cambridge, U.K., 1991 (as updated).

(4) U.S. National Library of Medicine. *Hazardous Substances Data Bank.* Bethesda, MD, 1995.

(5) Guyton, A. C. *Textbook of Medical Physiology,* 8th ed. W. B. Saunders, Philadelphia, PA, 1991.

(6) Stevens, J. T. and Sumner, D. D. Herbicides. In *Handbook of Pesticide Toxicology.* Hayes, W. J. and Laws, E. R., Eds. Academic Press, New York, 1991.

(7) Weed Science Society of America. *Herbicide Handbook,* 7th ed. Champaign, IL, 1994.

(8) Gosselin, R. E., Smith, R. P., and Hodge, H. C. *Clinical Toxicology of Commercial Products,* 5th ed. Williams and Wilkins, Baltimore, MD, 1984.

(9) U.S. Environmental Protection Agency. *Pesticide Fact Sheet Number 125: Mancozeb.* Office of Pesticides and Toxic Substances, Washington, D.C., 1987.

(10) U.S. Environmental Protection Agency. *Guidance for the Registration of Pesticide Products Containing Maneb as the Active Ingredient.* Washington, D.C., 1988.

(11) U.S. Environmental Protection Agency. *Pesticide Fact Sheet Number 181: Metiram.* Office of Pesticides and Toxic Substances, Washington, D.C., 1988.

(12) Shepard, T. H. *Catalog of Teratogenic Agents,* 5th ed. Johns Hopkins University Press, Baltimore, MD, 1986.

(13) Smith, G.J. *Pesticide Use and Toxicology in Relation to Wildlife: Organophosphorus and Carbamate Compounds.* C. K. Smoley, Boca Raton, FL, 1992.

(14) Hudson, R. H., Tucker, R. K., and Haegele, M. A. *Handbook of Toxicity of Pesticides to Wildlife. Resource Publication 153.* U.S. Department of Interior, Fish and Wildlife Service, Washington, D.C., 1984.

(15) Pimentel, D. *Ecological Effects of Pesticides on Nontarget Species.* Executive Office of the President, Office of Science and Technology, U.S. Government Printing Office, Washington, D.C., 1971.

(16) Johnson, W. W. and Finley, M. T. *Handbook of Acute Toxicity of Chemicals to Fish and Aquatic Invertebrates. Resource Publication 137.* U.S. Department of the Interior, Fish and Wildlife Service, Washington, D.C., 1980.

(17) Mayer, F. L. and Ellersieck, M. R. *Manual of Acute Toxicity: Interpretation and Data Base for 410 Chemicals and 66 Species of Freshwater Animals. Resource Publication 160.* U.S. Department of Interior, Fish and Wildlife Service, Washington, D.C., 1986.

(18) Menzie, C. M. *Metabolism of Pesticides. Update III. (Report. No. 232).* U.S. Department of the Interior, Fish and Wildlife Service, Washington, D.C., 1980.

(19) Howard, P. H., Ed. *Handbook of Environmental Fate and Exposure Data for Organic Chemicals: Pesticides.* Lewis, Boca Raton, FL, 1989.

(20) Wauchope, R. D., Buttler, T. M., Hornsby A. G., Augustijn-Beckers, P. W. M., and Burt, J. P. SCS/ARS/CES pesticide properties database for environmental decisionmaking. *Rev. Environ. Contam. Toxicol.* 123, 1–157, 1992.

(21) Augustijn-Beckers, P. W. M., Hornsby, A. G., and Wauchope, R. D. SCS/ARS/CES Pesticide properties database for environmental decisionmaking. II. Additional compounds. *Rev. Environ. Contam. Toxicol.* 137, 1–82, 1994.

(22) U.S Environmental Protection Agency. *Guidance for the Reregistration of Pesticide Products Containing Butylate as the Active Ingredient.* Washington, D.C., 1983.

(23) U.S. Environmental Protection Agency. *Health Advisory: Butylate.* Office of Drinking Water, Washington, D.C., 1989.

(24) U.S. Environmental Protection Agency. *Pesticide Fact Sheet Number 7: Butylate.* Office of Pesticides and Toxic Substances, Washington, D.C., 1984.

(25) National Institute for Occupational Safety and Health. *Registry of Toxic Effects of Chemical Substances.* Cincinnati, OH, 1995.

(26) Thomson, W. T. Fungicides. In *Agricultural Chemicals: Book IV.* Thomson Publications, Fresno, CA, 1985.

(27) U.S. Environmental Protection Agency. *Integrated Risk Information System.* Washington, D.C., 1995.

(28) U.S. Occupational Safety and Health Administration. *Permissible Exposure Limits for Air Contaminants. (29 CFR 1910. 1000, Subpart Z).* U.S. Department of Labor, Washington, D.C., 1994.

(29) ICI Americas Inc. *Material Safety Data Sheet: Eptam Technical.* Wilmington, DE, 1992.

(30) U.S. Environmental Protection Agency. *Pesticide Fact Sheet Number 6: EPTC.* Office of Pesticides and Toxic Substances, Washington, D.C., 1983.

(31) Wagner, S. L. *Clinical Toxicology of Agricultural Chemicals.* Oregon State University Environmental Health Sciences Center, Corvallis, OR, 1981.

(32) E. I. DuPont de Nemours. *Technical Data Sheet for Mancozeb.* Biochemicals Department, Wilmington, DE, 1983.

(33) Lu, F. C. A review of the acceptable daily intakes of pesticides assessed by the World Health Organization. *Regul. Toxicol. Pharmacol.* 21, 351–364, 1995.

(34) E. I. DuPont de Nemours. *Technical Data Sheet for Maneb.* Agricultural Chemicals Department, Wilmington, DE, 1983.

(35) Hallenbeck, W. H. and Cunningham-Burns, K. M. *Pesticides and Human Health.* Springer-Verlag, New York, 1985.

(36) Hill, E. F. and Camardese, M. B. *Lethal Dietary Toxicities of Environmental Contaminants to Coturnix, Technical Report Number 2.* U.S. Department of Interior, Fish and Wildlife Service, Washington, D.C., 1986.

(37) Hammond, B. Letter of December 23, 1991. Monsanto Agricultural Company, St. Louis, MO, 1991.

(38) Monsanto Company. *Toxicology Information Summary for Triallate.* St. Louis, MO, 1989.

(39) U.S. Environmental Protection Agency. *Triallate Decision Document.* Washington, D.C., 1980.

(40) Morgan, D. P. *Recognition and Management of Pesticide Poisonings,* 3rd ed. U.S. Environmental Protection Agency. Washington, D.C., 1982.

(41) Cornell University. *1988 New York State pesticide recommendations. 49th Annu. Pest Control Conf.* Ithaca, NY, 1987.

(42) Tucker, R. and Crabtree, D. G. *Handbook of Toxicity of Pesticides to Wildlife.* U.S. Department of Interior, Fish and Wildlife Service, Washington, D.C., 1970.

(43) *Material Safety Data Sheet for Ziram.* FMC Corporation, Philadelphia, PA, 1991.

(44) National Research Council. *Drinking Water and Health.* National Academy of Sciences, Washington, D.C., 1977.

(45) National Toxicology Program. *Carcinogenesis Bioassay of Ziram (CAS No. 137-30-4)in F344/N Rats and B6CF1 Mice (Feed Study), (Technical Report No. 238).* National Institutes of Health, Bethesida, MD, 1983.

chapter five

Organophosphates

5.1   Class overview and general description

Background

Organophosphates are the most commonly used insecticides. They are also employed as herbicides and fungicides. Although developed in the early 19th century, it was not until 1932 that the effects of these compounds on insects were discovered (1). Organophosphates (OPs) are characterized by a central phosphorus atom and numerous side chains. Most of the OPs used as insecticides are dimethoxy and diethoxy compounds. These broad groups of pesticides contain the well-known insecticides malathion and diazinon. The generalized structure of the organophosphate compounds is shown in Figure 5.1 (2).

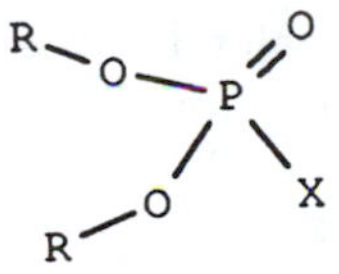

Figure 5.1 Generic organophosphate structure.

One feature of the OPs that has led to their wide usage in agriculture and in the home is that they are much less persistent in the environment than the organochlorines, such as DDT. The compounds in the latter group were the pesticides of choice before the development of the OPs. The OPs are considerably more acutely toxic to vertebrates than the organochlorines (3). The OPs are being replaced in some applications by the carbamate insecticides, which have lower toxicities to humans and wildlife. The commonly used OPs are listed in Table 5.1.

Organophosphate usage

Organophosphate insecticides can be effective whether they are ingested by the pest or are absorbed through the cuticle (skin) of the insect. However, some of the OPs are specifically formulated as stomach poisons or as contact poisons. The OPs are used against a wide array of insects and mites.

OPs are used extensively on cotton, corn, wheat, and a variety of other agricultural crops. They are also used for domestic pest control.

Some of the OPs, such as dichlorvos (DDVP), are administered orally to livestock to control internal parasites like the bot larva. Several others are used externally on livestock to control parasites on the animal's skin.

Table 5.1 Organophosphates

Acephate	Formothion
Azinphos-methyl*	Isofenphos*
Bensulide*	Malathion*
Carbophenothion	Methidathion*
Chlorpyrifos*	Methyl-parathion*
Coumaphos*	Mevinphos*
Demeton-S-methyl	Monocrotophos
Diazinon*	Naled*
Dichlorvos/DDVP*	Parathion
Dicrotophos	Phorate*
Dimethoate*	Phosalone
Disulfoton*	Phosmet*
Endothion	Phosphamidon
Ethion*	Phoxim
Fenamiphos*	Propetamphos*
Fenitrothion	Temephos*
Fenthion*	Terbufos*
Fonofos*	Trichlorfon*

Note: * indicates that a profile for this compound is included
in this chapter.

In 1982, OPs accounted for 67% of all insecticides used in the U.S. (4). This represented nearly 50 million pounds applied annually in the U.S. Also in that year, OP production in the U.S. was greater than 143 million pounds. Exports to the international community accounted for over 50% of the total U.S. organophosphate production.

In 1990, OPs accounted for 33% of all pesticidal poisoning reports in the U.S. Diazinon and chlorpyrifos led the list, accounting for over 50% of the reports (5).

Mechanism of action and toxicology

The toxic mechanism of action of OP compounds is the same for insects and mammals. The OP compounds cause the enzyme acetylcholinesterase (AChE) to become inactivated. This enzyme speeds the breakdown of acetylcholine (ACh) which is produced in the nerve cells. ACh allows the transfer of a nerve impulse from one nerve cell to a receptor cell, such as from a muscle cell or another nerve cell. The nerve impulse continues until AChE breaks down ACh by chemical inactivation. Without this regulation of ACh by the enzyme, the nerve transmission continues indefinitely, causing a wide variety of symptoms in mammals such as weakness or paralysis of the muscles (2).

In humans, OP insecticides can be absorbed through the skin, can be inhaled, or can enter the body through direct ingestion. Skin absorption is a slow process, so significant absorption occurs only after prolonged contact with the pesticide (5). Absorption is considerably faster when the skin is inflamed; thus, dermatitis could lead to much more serious poisoning than would normally occur.

Acute toxicity

Members of this group of pesticides are cholinesterase inhibitors and among the most acutely toxic of all the pesticides in current use (1). They are highly toxic by all routes of exposure. When inhaled, the first effects are usually respiratory and may include bloody or runny nose, coughing, chest discomfort, difficulty in breathing, and wheezing due to constriction or excess fluid in the bronchial tubes. Skin contact with organophosphates may cause localized sweating and involuntary muscle contractions. Eye contact will cause pain, bleeding, tears, pupil constriction, and blurred vision. Following exposure by any route, other systemic effects may begin within a few minutes or be delayed for up to 12 hours. These may include pallor, nausea, vomiting, diarrhea, abdominal cramps, headache, dizziness, eye pain, blurred vision, contraction or dilation of the pupils, tears, salivation, sweating, and confusion.

Severe poisoning will affect the central nervous system and the peripheral nervous system (6). The central nervous system includes the spinal cord and the brain, while the peripheral nervous system includes all of the nerves and fibers that are not associated with the central nervous system. Typically, the signs of acute exposure become noticeable when the normal activity of AChE is reduced by about one half.

Symptoms may include some of the following: incoordination, slurred speech, loss of reflexes, weakness, fatigue, involuntary muscle contractions, twitching, tremors of the tongue or eyelids, and eventually paralysis of the body extremities and the respiratory muscles. In severe cases, there may also be involuntary defecation or urination, psychosis, irregular heart beats, unconsciousness, convulsions, and coma. Death may occur when enzyme activity falls to between 10 and 20% of normal functioning levels (7) and be caused by respiratory failure or cardiac arrest (8). Acetylcholine is also found in red blood cells and in blood plasma and, thus, levels of cholinesterase in the bloodstream provide a good indicator of organophosphate and carbamate exposure (6).

All of these symptoms may vary with the dose and the specific type of nerve cells that are affected. Generally, the toxic effects can be broken down into three broad categories: 1) effects on smooth muscles, including the heart and endocrine glands (muscarinic receptors); 2) effects on motor nerve endings in skeletal muscles and autonomic nervous system (nicotinic signs); and 3) central nervous system effects (7).

In addition, it has been suggested that the organophosphates may have synergistic effects with pyrethroid insecticides (3); i.e., the combined toxicity is greater than the sum of the individual toxicities.

Chronic toxicity

Effects of repeated low-dose exposure to organophosphates have been shown in pesticide workers and applicators. Repeated or prolonged exposure to OPs may result in the same effects as acute exposure including the delayed symptoms. Other effects reported in workers repeatedly exposed include impaired memory and concentration, disorientation, severe depression, irritability, confusion, headache, speech difficulties, delayed reaction times, nightmares, sleepwalking, drowsiness, and insomnia. An influenza-like condition with headache, nausea, weakness, loss of appetite, and malaise has also been reported (2,8).

Reproductive effects

When coumaphos or malathion was administered at high doses, it caused a decrease in the number of pregnancies, litter size, and surviving offspring and also depressed cholinesterase activity of the fetus (2,8). However, other OP compounds at lower doses (such as azinphos-methyl at 0.25 mg/kg/day and DDVP at 5 mg/kg/day) have not produced any reproductive effects (2).

It appears that OP compounds will be unlikely to cause reproductive effects in humans at expected exposure levels.

Teratogenic effects

Some experiments have shown that some OPs have crossed the placental barrier. However, in rat and rabbit experiments with various OP compounds, no teratogenic effects were detected (2,8,9).

Therefore, organophosphate compounds appear unlikely to cause teratogenic effects.

Mutagenic effects

The overwhelming majority of OP compounds are not mutagenic (2,8,9). However, there may be some exceptions. It has been suggested that diazinon has some potential to cause mutagenic effects, though there is no conclusive evidence. Malathion has produced detectable mutations in three different types of human cells, but mutagenic risks to humans are unlikely at expected exposure levels (2).

Carcinogenic effects

When OP compounds were fed to animals in laboratory experiments, there were no noticeable tumor growths (2,8). However, there is one exception. Dichlorvos has been classified as a possible human carcinogen by the U.S. EPA because in an experiment with female rats, there was an increase in benign tumors of the mammary glands (10,11).

The overwhelming majority of the evidence suggests that OP compounds will be unlikely to cause carcinogenic effects in humans.

Organ toxicity

As previously mentioned in the acute toxicity section, OP compounds are cholinesterase inhibitors and their effects occur throughout the body in many organs such as the brain, nervous system, adrenal glands, and liver.

Fate in humans and animals

Organophosphate compounds are metabolized and excreted rapidly in animals. The half-life for diazinon is approximately 12 hours (2). The metabolites are eliminated rapidly in the urine and feces and there is no evidence of bioaccumulation in body tissues (2,12).

Ecological effects

Effects on birds

The avian toxicity of organophosphate compounds (OPs) varies from slightly toxic to highly toxic. However, a majority of OPs such as coumaphos, dichlorvos, fonofos, methidathion, and parathion are highly toxic to wild birds, mallard ducks, and pheasants (8,13–15).

Effects on aquatic organisms

OPs are moderately to highly toxic to fish. For example, the 96-hour LC_{50} of azinphos-methyl in rainbow trout is 0.003 mg/L; the LC_{50} of dichlorvos is 0.9 mg/L in bluegills; and the LC_{50} of bensulide is 0.7 mg/L in rainbow trout (8,13,16).

Studies also show that OPs are highly toxic to aquatic invertebrates. For example, the LC_{50} for azinphos-methyl ranged from 0.13 to 56 mg/L in these species (17), and for coumaphos was 0.015 µg/L in amphipods (18).

Effects on other organisms (non-target species)

Organophosphate compounds are moderately toxic to highly toxic to bees (8,13).

Environmental fate

Breakdown in soil and groundwater

The behavior and fate of organophosphates in the soil environment is largely governed by soil moisture, soil organic matter, acidity, temperature, and the mineral content of the soil (19,20). Generally, though, OPs (along with the carbamates) are much less persistent in soils than most other pesticides. They are categorized as low to moderately persistent compounds that persist in soil at the application site from a few hours through several weeks to months (21).

Although generalizations are difficult to make with such a large and diverse group of compounds, the OPs are less mobile in soils with high organic content and a high inorganic metal concentration. One notable exception is an increase in degradation of diazinon and chlorpyrifos when in contact with inorganic copper (21). Generally, the pesticides are more stable under acidic conditions than under alkaline conditions.

Due to the relatively short half-lives under many field conditions, the OPs do not represent a great threat to surface or groundwater over the long term. However, lakes and streams may be susceptible to pesticide runoff if application occurs prior to rainfall (21).

Breakdown in water

The temperature effects on degradation of OPs in surface water are striking. For example, methyl parathion has a half-life of 8 days during the summer and 38 days in winter (12). A similar response is shown for a variety of other organophosphate compounds (12,21).

Also, pH affects the rate of degradation. The breakdown rates increase substantially with increasing alkalinity (21).

Breakdown in vegetation

The effects of organophosphate compounds in plants depend on several factors such as the rate and frequency of application, the nature of the plant surface, and the weather conditions. Plants absorb OPs mainly through the roots and translocate them to other parts of the plant, although leafy vegetables will also absorb OPs through the foliage (8).

OPs do not usually bioaccumulate. For example, residues of azinphos-methyl, chlorpyrifos, and diazinon remain in plants only between 1 and 3 weeks (13,22).

5.2   Individual profiles

5.2.1   Azinphos-methyl

Trade or other names

Common names include azinphos-methyl and metiltriazotion. Trade names include Azimil, Bay 9027, Bay 17147, Carfene, Cotnion-methyl, Gusathion, Gusathion-M, Guthion, and Methyl-Guthion.

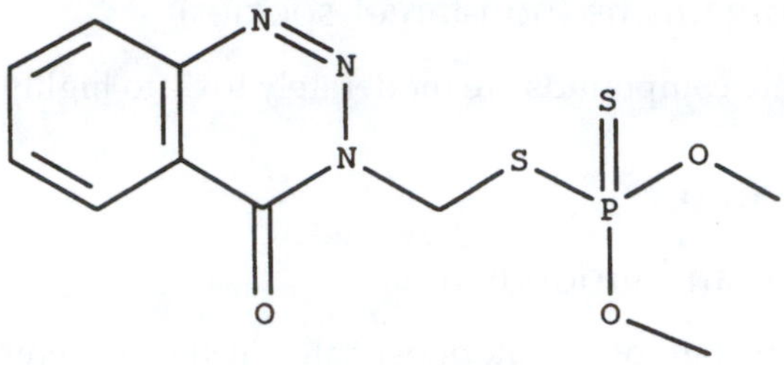

Figure 5.2 Azinphos-methyl.

Regulatory status

All azinphos-methyl liquids with a concentration greater than 13.5% are classi-fied as Restricted Use Pesticides (RUPs) by the U.S. Environmental Protection Agency (EPA) because of the inhalation hazard and acute toxicity they present, as well as their potential adverse effects on mammalian species, birds, and aquatic organisms. RUPs may be purchased and used only by certified applicators. The EPA has imposed a 24-hour reentry interval for this material. It is toxicity class I — highly toxic. Products containing azinphos-methyl bear the Signal Words DANGER — POISON.

Introduction

Azinphos-methyl is a highly persistent, broad-spectrum insecticide. It is also toxic to mites and ticks, and poisonous to snails and slugs. It is a member of the organophosphate class of chemicals. It is nonsystemic, meaning that it is not trans-ported from one plant part to another. It is used primarily as a foliar application against leaf-feeding insects. It works as both a contact insecticide and a stomach poison.

Azinphos-methyl is registered for use in the control of many insect pests on a wide variety of fruit, vegetable, nut, and field crops, as well as on ornamentals, tobacco, and forest and shade trees. Outside the U.S., azinphos-methyl is used in lowland rice production. Azinphos-methyl is available in emulsifiable liquid, liquid flowable, ULV liquid, and wettable powder formulations.

Toxicological effects

Acute toxicity

Azinphos-methyl is one of the most toxic of the OP insecticides (2,23). It is highly toxic by inhalation, dermal absorption, ingestion, and eye contact (2). Like all orga-nophosphate chemicals, azinphos-methyl is a cholinesterase inhibitor. It damages normal functioning of cholinesterase, an enzyme essential to proper nervous system function. Individuals with a history of reduced lung function, convulsive disorders, or recent exposure to other cholinesterase inhibitors are at increased risk from expo-sure to azinphos-methyl (2,8).

There is wide variation in the recorded LD_{50} values for azinphos-methyl, depend-ing on the route of exposure and the test animal. The oral LD_{50} for azinphos-methyl is 4.4 to 16 mg/kg in rats, 80 mg/kg in guinea pigs, and 8 to 20 mg/kg in mice (2,8,13). The dermal LD_{50} is 88 to 220 mg/kg in rats, and 65 mg/kg in mice (2,8,13). The 1-hour inhalation LC_{50} for azinphos-methyl in rats is 0.4 mg/L (13).

For humans, ingestion of azinphos-methyl in amounts above 1.5 mg/day can cause severe poisoning with symptoms such as dimness of vision, salivation, excessive sweating, stomach pain, vomiting, diarrhea, unconsciousness, and death (2,23). Inhalation of the dust or aerosol preparation of azinphos-methyl may cause wheezing, tightness in the chest, blurred vision, and tearing of the eyes. Complete symptomatic recovery may occur within 1 week after sublethal poisoning; i.e., poisoning from an exposure that is just below the amount necessary to be fatal (23).

Pure azinphos-methyl is easily absorbed by the skin, and lethal amounts can build up in the body after dermal exposure. Symptoms of illness caused by this type of exposure include nausea, vomiting, blurred vision, and muscle cramps (2,23).

Eye contact with concentrated solutions of azinphos-methyl can be life-threatening. Within a few minutes of eye exposure, azinphos-methyl may cause pain, blurring of distant vision, tearing, and other problems. Symptoms of cholinesterase inhibition may also occur, such as respiratory difficulties, gastrointestinal problems, and central nervous system disturbances (23).

Some organophosphates may cause delayed symptoms beginning 1 to 4 weeks after an acute exposure that may or may not have produced immediate symptoms. In such cases, numbness, tingling, weakness, and cramping may appear in the lower limbs and progress to incoordination and paralysis. Improvement may occur over months or years, and in some cases residual impairment will remain (2,23).

Chronic toxicity

Long-term exposure to azinphos-methyl, above the average 8-hour standard set by the Occupational Safety and Health Administration (OSHA), can impair concentration and memory, and cause headache, irritability, nausea, vomiting, muscle cramps, and dizziness (23).

Cholinesterase inhibition from exposure to azinphos-methyl may persist for 2 to 6 weeks (1). Repeated exposure to small amounts may result in an unexpected inhibition of cholinesterase, causing symptoms that resemble other flu-like illnesses, including general discomfort, weakness, and lack of appetite (1,2). The effects of azinphos-methyl exposure may be greater in a previously exposed person than in an individual with no previous exposure. Rats tolerated dietary doses of 0.25 mg/kg/day for 60 days without cholinesterase inhibition, 1 mg/kg/day resulted in questionable growth effects and a slight inhibition of brain and red blood cell cholinesterase.

In chronic oral toxicity studies, rats and dogs were fed doses of 0.125, 0.5, 1, or 2.5 mg/kg/day. The 2.5-mg/kg/day dose was increased to 5 mg/kg/day after 47 weeks. At 0.125 mg/kg/day, cholinesterase was not affected in rats and dogs. At 1 mg/kg/day, the plasma and red blood cell cholinesterases in the rat were initially inhibited, but returned to normal after 65 weeks. The 5-mg/kg/day doses produced convulsions in some animals. In dogs, 0.5 mg/kg/day produced a slight, irregular decrease in red blood cell cholinesterase (24,25).

Rats fed about 5 to 10 mg/kg/day azinphos-methyl for 2 years had depressed red blood cell counts and brain cholinesterase activity. Dietary levels of about 0.5 mg/kg/day or less had no negative effects (2).

Reproductive effects

In a two-generation reproduction study, there were no observed reproductive or maternal effects in rats at 0.25 mg/kg/day (25). However, at oral doses of 20 mg/kg to 8-day pregnant mice, Guthion was toxic to the fetus (25). These data indicate that reproductive effects in humans are unlikely at expected exposure levels.

Teratogenic effects

In a teratology study, no maternal or developmental effects were observed in rats at doses of 2 mg/kg/day (25). A 16-mg/kg oral dose to 8-day pregnant rats caused specific development abnormalities in the muscles and bones. It appears that teratogenic effects are not likely in humans under expected exposure conditions.

Mutagenic effects

No mutagenic effects were observed in the Ames test on bacteria and a test on human cell cultures (25). These data suggest that azinphos-methyl is not mutagenic.

Carcinogenic effects

Although one carcinogenicity study on rats suggested that tumors of the pancreas and selected thyroid cells may have been associated with azinphos-methyl (2), two other studies at doses up to 10 mg/kg/day did not show an increase in the incidence of tumors in mice from azinphos-methyl (2,17). The carcinogenicity of azinphos-methyl is not clear from current evidence.

Organ toxicity

Toxicity from azinphos-methyl is primarily manifested in cholinesterase inhibition, which affects the nervous system. Dogs fed 9 mg/kg/day showed tremors, weakness, abnormal quietness, and some weight loss (2).

Fate in humans and animals

One study suggests that Guthion is rapidly broken down into nonpoisonous forms in the body (24). Azinphos-methyl is eliminated in the feces and urine of mammals within 2 days of administration.

Ecological effects

Effects on birds

Azinphos-methyl is slightly to moderately toxic to birds. Acute symptoms of azinphos-methyl poisoning include regurgitation, wing drop, wing spasms, diarrhea, and lack of movement (26). Chickens fed azinphos-methyl at doses of 40 mg/kg developed leg weakness.

The oral LD_{50} for azinphos-methyl is 136 mg/kg in young mallards, 74.9 mg/kg in young pheasants, 84.2 mg/kg in young chukar partridges, 262.0 mg/kg in chickens, and 32.2 mg/kg in bobwhite quail (13,17,27). The dietary LC_{50} for azinphos-methyl is 639 ppm in Japanese quail, 1821 ppm in ring-necked pheasant, and 1940 ppm in mallard duck (13,17).

Effects on aquatic organisms

Azinphos-methyl is moderately to very highly toxic to freshwater fish. For most species, the LC_{50} values are less than 1 mg/L. The 96-hour LC_{50} for azinphos-methyl in rainbow trout is 0.003 mg/L (8,13).

Guthion-poisoned fish exhibit central nervous system impairment, including erratic swimming accompanied by uncontrolled convulsions. Rapid gill movements, paralysis, and death follow in rapid succession (8).

Azinphos-methyl is highly toxic to aquatic invertebrates, shellfish, frogs, and toads (8). The LC_{50} values are below 1 µg/L for many of the species (8,17).

Effects on other organisms (non-target species)

Several studies have indicated that azinphos-methyl causes adverse effects in wildlife. Wild mammals and aquatic organisms appear to be more vulnerable than birds to hazards created by this material (29). The EPA requires endangered species labeling for certain azinphos-methyl uses (17).

Azinphos-methyl is toxic to honeybees and other beneficial insects (8,24). It will cause severe bee losses if used when bees are present at treatment time or within a day thereafter (30). A 90% mortality rate is seen in pollinating leaf-cutting bees after a 9-day exposure to greenhouse alfalfa treated with azinphos-methyl (27).

Environmental fate

Breakdown in soil and groundwater

Persistence of azinphos-methyl in soil is quite variable, but is generally low under field conditions (19,31). The half-life in sandy loam soil is 5 days. Its half-life in nonsterile soil is 21 days when oxygen is present, or 68 days under oxygen-free conditions. In sterile soil, the half-life is reported to be 355 days.

Azinphos-methyl is fairly immobile in soil because it adsorbs strongly to soil particles and has low water solubility. It has low leaching potential and is unlikely to contaminate groundwater (19,31). It was not detected in 54 groundwater samples collected in New York state (32). Azinphos-methyl is one of 118 synthetic organic chemicals that the state of Florida has designated for groundwater monitoring (33). It was detected in only 5 out of 1628 wells sampled in ten states from 1983 to 1991 (34).

The disappearance of azinphos-methyl from soil is more rapid in the surface layers (0 to 2.5 cm deep) than it is in the next deeper layer (2.4 to 7.5 cm). Biodegradation and evaporation are the primary routes of disappearance for azinphos-methyl. Azinphos-methyl is also subject to degradation by ultraviolet (UV) light from the sun and hydrolytic decomposition. Photodecomposition is particularly rapid at high levels of soil moisture and in the presence of UV light (31). Rapid degradation of Gusathion was observed at temperatures higher than 37°C (29).

Breakdown in water

In general, organophosphates such as azinphos-methyl are dissipated rapidly in water (35). In pond water, it is subject to degradation by sunlight and microorganisms, with a half-life of up to 2 days. Volatilization from water is unlikely. Chemical hydrolysis is important in alkaline waters (12). Azinphos-methyl is very stable in water below pH 10.0. Above pH 11.0, it is rapidly hydrolyzed to anthranilic acid, benzamide, and other metabolites. Azinphos-methyl has a low to medium tendency to adsorb to sediments or suspended solids (12).

Breakdown in vegetation

Residue levels of azinphos-methyl in crops are dependent on the rate and frequency of application, nature of the plant surface, and weather conditions such as rainfall, temperature, sunlight, humidity, and wind (24). The half-life on vegetable and forage crops is 3 to 5 days under field conditions (24). It gives effective protection for 2 or more weeks (36). On treated apple trees, the half-life of this pesticide was about 2.6 to 6.3 days (28). Hawthorn and American Linden trees have been injured by this material. It has also caused russeting on certain varieties of fruit (24).

Physical properties

Pure azinphos-methyl is a white crystalline solid. Technical azinphos-methyl is a brown waxy solid (13,37).

> Chemical name: S-(3,4-dihydro-4-oxobenzo[d]-[1,2,3]-triazin-3-ylmethyl) O,O-dimethyl phosphorodithioate (13)
> CAS #: 86-50-0
> Molecular weight: 317.33 (13)
> Water solubility: 30 mg/L @ 25°C (13)
> Solubility in other solvents: dichloromethane v.s.; toluene v.s. (13)
> Melting point: 65–68°C (technical) (13); 73–74°C (pure form) (13)
> Vapor pressure: <1 mPa @ 20°C (13)
> Partition coefficient (octanol/water): Not available
> Adsorption coefficient: 1000 (19)

Exposure guidelines

> ADI: 0.005 mg/kg/day (38)
> HA: Not available
> RfD: Not available
> PEL: 0.2 mg/m^3 (8-hour) (skin) (39)

Basic manufacturer

> Miles, Inc.
> P.O. Box 4913
> 8400 Hawthorn Road
> Kansas City, MO 64120
> Telephone: 816-242-2429
> Emergency: 816-242-2582

5.2.2   *Bensulide*

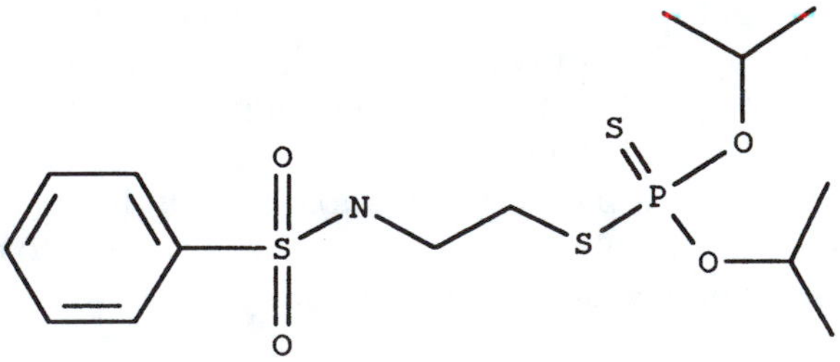

Figure 5.3 Bensulide.

Trade or other names

Trade names for bensulide include Betamec, Betasan, Bensumec, Benzulfide, Disan, Exporsan, Prefar, Pre-San, and R-4461. It is used in combination with other pesticides such as thiobencarb and molinate.

Regulatory status

Bensulide is classified as a General Use Pesticide (GUP) by the U.S. Environmental Protection Agency. It is classified toxicity class III — slightly toxic. Products with bensulide bear the Signal Word CAUTION.

Introduction

Bensulide is a selective organophosphate herbicide. It is one of a few OPs used as a herbicide. Most of the others are used as insecticides. It is used on vegetable crops such as carrots, cucumbers, peppers, and melons, and on cotton and turfgrass to control annual grasses such as bluegrass and crabgrass and broadleaf weeds. It is often applied before the weed seeds germinate (pre-emergence) in order to prevent them from germinating. It is available as granules or an emulsifiable concentrate.

Estimates place the total U.S. use of bensulide at about 632,000 pounds annually. Application rates may be relatively heavy (up to 22.6 kg/ha) when it is used.

Toxicological effects

Acute toxicity

Although its toxicity is not high, bensulide can cause convulsions in humans when large amounts are ingested. Other symptoms of acute poisoning range from nausea and vomiting at mild exposure levels to abdominal cramps, loss of muscle coordination, slurring of speech, coma, and death at higher levels of acute exposure (8).

The oral LD_{50} in rats ranges from 271 to 770 mg/kg (13,40). The dermal LD_{50} is 3950 mg/kg in rats and 2000 mg/kg in rabbits. Thus, bensulide's dermal acute toxicity is low. In tests with rodents, Betasan (a bensulide-containing product) did not cause eye irritation. Rabbits exposed to bensulide suffered minor eye irritation (8).

Chronic toxicity

Bensulide inhibits cholinesterase, a chemical that is critical to the proper functioning of the nervous system. Symptoms of human chronic exposure are fairly typical of other organophosphate pesticides and may include chest tightness, nausea, abdominal cramps, diarrhea, headache, dizziness, weakness, blurring, tearing, loss of muscle coordination, and face muscle twitches (8). The lowest dose that resulted in no adverse effects for a 90-day feeding study with rats was 25 mg/kg/day (8).

Reproductive effects
No data are currently available.

Teratogenic effects
No data are currently available.

Mutagenic effects
Bensulide was not mutagenic in the one bacterial assay that was performed (8). No other data are currently available.

Carcinogenic effects

Bensulide does not appear to be carcinogenic. In a 90-day feeding trial, rats and dogs tolerated daily doses close to the lethal dose without any noticeable tumor growth (8).

Organ toxicity

Bensulide can inhibit the enzyme cholinesterase and affect brain, nerve, and some blood cells (41). It may cause mild eye irritation.

Fate in humans and animals
No data are currently available.

Ecological effects

Effects on birds

Bensulide is only slightly toxic to birds. The bensulide herbicide, Betasan, was fed to adult Japanese quail for 3 weeks, and egg hatchability was significantly reduced at the highest dose (about 50 mg/kg/day), but fertility was not affected. Blood cholinesterase was inhibited at lower doses, but recovered within 2 weeks after the treatments stopped (8). The oral LD_{50} in bobwhite quail is 1386 mg/kg (13).

Effects on aquatic organisms

Bensulide is moderately to highly toxic to aquatic organisms, including rainbow trout and bluegill (8). The LC_{50} for bensulide is 1.1 mg/L in rainbow trout, 1.4 mg/L in bluegill, and 1 to 2 mg/L in goldfish. The compound is moderately toxic to aquatic invertebrates like the amphipod *Gammarus lacustrus* (6). The calculated bioconcentration is low and it is not expected to bioaccumulate (6).

Effects on other organisms (non-target species)

Bensulide is very highly toxic to bees (13). The LD_{50} of bensulide is 0.0016 mg per bee (13).

Environmental fate

Breakdown in soil and groundwater

Bensulide is highly persistent in both plants and soil (8). Because it strongly binds to the top 0 to 2 inches of soil, bensulide does not evaporate easily but can be carried off-site with sediment or dust. The rate of application, temperature, soil organic matter, and soil acidity can all affect its breakdown. Bensulide leaches very little in sand, clay, or organic soils.

Bensulide is slowly broken down by soil microorganisms. The rate of degradation increases with increasing soil temperature and organic matter, but decreases with increasing basicity (8). At 70 to 80°F, the half-life of bensulide is 4 months in a moist loam soil, and 6 months in a moist, loamy sand (13). As of 1988, it had not been found in groundwater or in well water (42).

Breakdown in water

In flooded rice fields, the half-life of bensulide averages 4 to 6 days (13). Some decomposition by sunlight occurs over several days (8).

Breakdown in vegetation

Bensulide is rapidly absorbed by roots and foliage and is translocated to the active growing portions of the plant (root or stem tips) where it works to stop cell division and plant growth (13). When applied to roots, bensulide is not translocated to leaves except as metabolites (8).

Physical properties

Bensulide is a viscous, colorless liquid or a white crystalline solid (13).

Chemical name: O,O-diisopropyl S-2-phenylsulfonylaminoethyl phospho-
    rodithioate (13)
CAS #: 741-58-2
Molecular weight: 397.54 (13)
Solubility in water: 25 mg/L @ 20°C (13)
Solubility in other solvents: kerosene s.; acetone v.s, ethanol v.s., xylene v.s (13)
Melting point: 34.4°C (13)
Vapor pressure: 0.133 mPa @ 25°C (13)
Partition coefficient (octanol/water): 16,500 (13)
Adsorption coeffient: 1000 (estimated) (19)

Exposure guidelines

ADI: Not available
MCL: Not available
RfD: Not available
PEL: Not available

Basic manufacturer

ICI Americas, Inc.
Agricultural Products
New Murphey Road
Wilmington, DE 19897
Telephone: 302-866-1000 or 800-323-8633

5.2.3   *Chlorpyrifos*

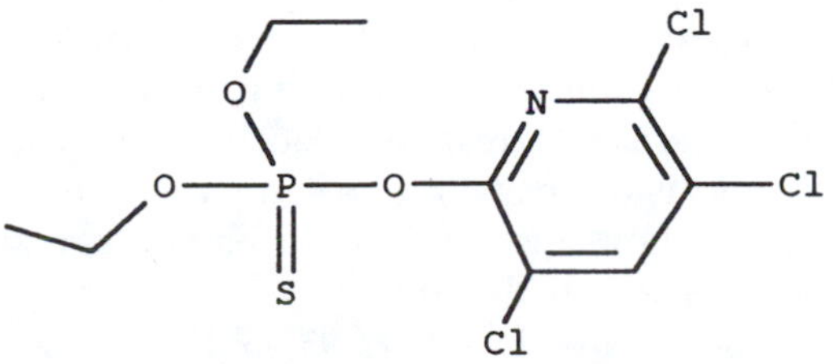

Figure 5.4 Chlorpyrifos.

Trade or other names

Trade names for chlorpyrifos include Brodan, Detmol UA, Dowco 179, Dursban, Empire, Eradex, Lorsban, Paqeant, Piridane, Scout, and Stipend.

Regulatory status

The EPA has established a 24-hour reentry interval for crop areas treated with emulsifiable concentrate or wettable powder formulations of chlorpyrifos unless workers wear protective clothing. Chlorpyrifos is toxicity class II — moderately toxic. Products containing chlorpyrifos bear the Signal Word WARNING or CAUTION, depending on the toxicity of the formulation. It is classified as a General Use Pesticide (GUP).

Introduction

Chlorpyrifos is a broad-spectrum organophosphate insecticide. While originally used primarily to kill mosquitoes, it is no longer registered for this use. Chlorpyrifos is effective in controlling cutworms, corn rootworms, cockroaches, grubs, flea beetles, flies, termites, fire ants, and lice. It is used as an insecticide on grain, cotton, field, fruit, nut and vegetable crops, as well as on lawns and ornamental plants. It is also registered for direct use on sheep and turkeys, for horse site treatment, dog kennels, domestic dwellings, farm buildings, storage bins, and commercial establishments.

Chlorpyrifos acts on pests primarily as a contact poison, with some action as a stomach poison. It is available as granules, wettable powder, dustable powder, and emulsifiable concentrate.

Toxicological effects

Acute toxicity

Chlorpyrifos is moderately toxic to humans (43). Poisoning from chlorpyrifos may affect the central nervous system, the cardiovascular system, and the respiratory system. It is also a skin and eye irritant (2). While some organophosphates are readily absorbed through the skin, studies in humans suggest that skin absorption of chlorpyrifos is limited (2).

Symptoms of acute exposure to organophosphate or cholinesterase-inhibiting compounds may include the following: numbness, tingling sensations, incoordination, headache, dizziness, tremor, nausea, abdominal cramps, sweating, blurred vision, difficulty breathing or respiratory depression, and slow heartbeat. Very high doses may result in unconsciousness, incontinence, and convulsions or fatality.

Persons with respiratory ailments, recent exposure to cholinesterase inhibitors, cholinesterase impairment, or liver malfunction are at increased risk from exposure to chlorpyrifos. Some organophosphates may cause delayed symptoms beginning 1 to 4 weeks after an acute exposure that may or may not have produced immediate symptoms (2). In such cases, numbness, tingling, weakness, and cramping may appear in the lower limbs and progress to incoordination and paralysis. Improvement may occur over months or years and, in some cases, residual impairment will remain (2). Plasma cholinesterase levels activity have been shown to be inhibited when chlorpyrifos particles are inhaled (8).

The oral LD_{50} for chlorpyrifos in rats is 95 to 270 mg/kg (2,13). The LD_{50} for chlorpyrifos is 60 mg/kg in mice, 1000 mg/kg in rabbits, 32 mg/kg in chickens, 500 to 504 mg/kg in guinea pigs, and 800 mg/kg in sheep (2,13,44). The dermal LD_{50} is greater than 2000 mg/kg in rats, and 1000 to 2000 mg/kg in rabbits (2,13,45). The 4-hour inhalation LC_{50} for chlorpyrifos in rats is greater than 0.2 mg/L (46).

Chronic toxicity

Repeated or prolonged exposure to organophosphates may result in the same effects as acute exposure, including the delayed symptoms. Other effects reported in workers repeatedly exposed include impaired memory and concentration, disorientation, severe depression, irritability, confusion, headache, speech difficulties, delayed reaction times, nightmares, sleepwalking, and drowsiness or insomnia. An influenza-like condition with headache, nausea, weakness, loss of appetite, and malaise has also been reported (8).

When technical chlorpyrifos was fed to dogs for 2 years, increased liver weight occurred at 3.0 mg/kg/day. Signs of cholinesterase inhibition occurred at 1 mg/kg/day. Rats and mice given technical chlorpyrifos in the diet for 104 weeks showed no adverse effects other than cholinesterase inhibition (43). Two-year feeding studies using doses of 1 and 3 mg/kg/day chlorpyrifos in rats showed moderate depression of cholinesterase. Cholinesterase levels recovered when the experimental feeding was discontinued (2). Identical results occurred in a 2-year feeding study with dogs. No long-term health effects were seen in either the dog or rat study (2,47).

A measurable change in plasma and red blood cell cholinesterase levels was seen in workers exposed to chlorpyrifos spray. Human volunteers who ingested 0.1 mg/kg/day chlorpyrifos for 4 weeks showed significant plasma cholinesterase inhibition (47).

Reproductive effects

Current evidence indicates that chlorpyrifos does not adversely affect reproduction. In two studies, no effects were seen in animals tested at dose levels up to 1.2 mg/kg/day (8). No effects on reproduction occurred in a three-generation study with rats fed dietary doses as high as 1 mg/kg/day (43,47). In another study in which rats were fed 1.0 mg/kg/day for two generations, the only effect observed was a slight increase in the number of deaths of newborn offspring (2).

Teratogenic effects

Available evidence suggests that chorpyrifos is not teratogenic. No teratogenic effects in offspring were found when pregnant rats were fed doses as high as 15 mg/kg/day for 10 days. When pregnant mice were given doses of 25 mg/kg/day for 10 days, minor skeletal variations and a decrease in fetal length occurred (43,45). No birth defects were seen in the offspring of male and female rats fed 1.0 mg/kg/day during a three-generation reproduction and fertility study (2,47).

Mutagenic effects

There is no evidence that chlorpyrifos is mutagenic. No evidence of mutagenicity was found in any of four tests performed (43).

Carcinogenic effects

There is no evidence that chlorpyrifos is carcinogenic. There was no increase in the incidence of tumors when rats were fed 10 mg/kg/day for 104 weeks, nor when mice were fed 2.25 mg/kg/day for 105 weeks (43).

Organ toxicity

Chlorpyrifos primarily affects the nervous system through inhibition of cholinesterase, an enzyme required for proper nerve functioning.

Fate in humans and animals

Chlorpyrifos is readily absorbed into the bloodstream through the gastrointestinal tract if it is ingested, through the lungs if it is inhaled, or through the skin if there is dermal exposure (8). In humans, chlorpyrifos and its principal metabolites are eliminated rapidly (2). After a single oral dose, the half-life of chlorpyrifos in the blood appears to be about 1 day (41).

Chlorpyrifos is eliminated primarily through the kidneys (8). Following oral intake of chlorpyrifos by rats, 90% is removed in the urine and 10% is excreted in the feces (13). It is detoxified quickly in rats, dogs, and other animals (8). The major metabolite found in rat urine after a single oral dose is trichloropyridinol (TCP). TCP does not inhibit cholinesterase and it is not mutagenic (8).

Chlorpyrifos does not have a significant bioaccumulation potential (8). Following intake, a portion is stored in fat tissues but it is eliminated in humans, with a half-life of about 62 hours (2). When chlorpyrifos (Dursban) was fed to cows, unchanged pesticide was found in the feces, but not in the urine or milk (48). However, it was detected in the milk of cows for 4 days following spray dipping with a 0.15% emulsion. The maximum concentration in the milk was 0.304 ppm (2). In a rat study, chlorpyrifos did not accumulate in any tissue except fat (49).

Ecological effects

Effects on birds

Chlorpyrifos is moderately to very highly toxic to birds (43). Its oral LD_{50} is 8.41 mg/kg in pheasants, 112 mg/kg in mallard ducks, 21.0 mg/kg in house sparrows, and 32 mg/kg in chickens (8,13,43). The LD_{50} for a granular product (15G) in bobwhite quail is 108 mg/kg (13,43).

At 125 ppm, mallards laid significantly fewer eggs (43). There was no evidence of changes in weight gain, or in the number, weight, or quality of eggs produced by hens fed dietary levels of 50 ppm chlorpyrifos (8).

Effects on aquatic organisms

Chlorpyrifos is very highly toxic to freshwater fish, aquatic invertebrates, and estuarine and marine organisms (43). Cholinesterase inhibition was observed in acute toxicity tests of fish exposed to very low concentrations of this insecticide. Application of concentrations as low as 0.01 lb active ingredient per acre may cause fish and aquatic invertebrate deaths (43).

Chlorpyrifos toxicity to fish may be related to water temperature. The 96-hour LC_{50} for chlorpyrifos is 0.009 mg/L in mature rainbow trout, 0.098 mg/L in lake trout, 0.806 mg/L in goldfish, 0.01 mg/L in bluegill, and 0.331 mg/L in fathead minnow (50).

When fathead minnows were exposed to Dursban for a 200-day period during which they reproduced, the first generation of offspring had decreased survival and growth, as well as a significant number of deformities. This occurred at approximately 0.002-mg/L exposure for a 30 day-period (8).

Chlorpyrifos accumulates in the tissues of aquatic organisms. Studies involving continuous exposure of fish during the embryonic through fry stages have shown bioconcentration values of 58 to 5100 (51).

Due to its high acute toxicity and its persistence in sediments, chlorpyrifos may represent a hazard to sea bottom dwellers (52). Smaller organisms appear to be more sensitive than larger ones (50).

Effects on other organisms (non-target species)

Aquatic and general agricultural uses of chlorpyrifos pose a serious hazard to wildlife and honeybees (13,48).

Environmental fate

Breakdown in soil and groundwater

Chlorpyrifos is moderately persistent in soils. The half-life of chlorpyrifos in soil is usually between 60 and 120 days, but can range from 2 weeks to over 1 year, depending on the soil type, climate, and other conditions (12,19). The soil half-life of chlorpyrifos was from 11 to 141 days in seven soils ranging in texture from loamy sand to clay and with soil pH values from 5.4 to 7.4. Chlorpyrifos was less persistent in the soils with a higher pH (51). Soil half-life was not affected by soil texture or organic matter content. In anaerobic soils, the half-life was 15 days in loam and 58 days in clay soil (43). Adsorbed chlorpyrifos is subject to degradation by UV light, chemical hydrolysis, and soil microbes. When applied to moist soils, the volatility half-life of chlorpyrifos was 45 to 163 hours, with 62 to 89% of the applied chlorpyrifos remaining on the soil after 36 hours (51). In another study, 2.6 and 9.3% of the chlorpyrifos applied to sand or silt loam soil remained after 30 days (51). Chlorpyrifos adsorbs strongly to soil particles and it is not readily soluble in water (19,51). It is therefore immobile in soils and unlikely to leach or to contaminate groundwater (51). TCP, the principal metabolite of chlorpyrifos, adsorbs weakly to soil particles and appears to be moderately mobile and persistent in soils (43).

Breakdown in water

The concentration and persistence of chlorpyrifos in water will vary depending on the type of formulation. For example, a large increase in chlorpyrifos concentrations occurs when emulsifiable concentrations and wettable powders are released into water. As the pesticide adheres to sediments and suspended organic matter, concentrations rapidly decline. The increase in the concentration of insecticide is not as rapid for granules and controlled-release formulations in the water, but the resulting concentration persists longer (50).

Volatilization is probably the primary route of loss of chlorpyrifos from water. Volatility half-lives of 3.5 and 20 days have been estimated for pond water (51). The photolysis half-life of chlorpyrifos is 3 to 4 weeks during midsummer in the U.S. Its change into other natural forms is slow (52). Research suggests that this insecticide is unstable in water, and the rate at which it is hydrolyzed increases with temperature, decreasing by 2.5- to 3-fold with each 10°C drop in temperature. The rate of hydrolysis is constant in acidic to neutral waters, but increases in alkaline waters. In water at pH 7.0 and 25°C, it had a half-life of 35 to 78 days (12).

Breakdown in vegetation

Chlorpyrifos may be toxic to some plants, such as lettuce (36). Residues remain on plant surfaces for approximately 10 to 14 days. Data indicate that this insecticide and its soil metabolites can accumulate in certain crops (8).

Physical properties

Technical chlorpyrifos is an amber to white crystalline solid with a mild sulfur odor (13).

Chemical name: O,O-diethyl O-3,5,6-trichloro-2-pyridyl phosphorothioate (13)
CAS #: 2921-88-2
Molecular weight: 350.62 (13)
Water solubility: 2 mg/L @ 25°C (13)
Solubility in other solvents: benzene s.; acetone s.; chloroform s.; carbon disul-
    fide s.; diethyl ether s.; xylene s.; methylene chloride s.; methanol s. (13)
Melting point: 41.5–44°C (13)
Vapor pressure: 2.5 mPa @ 25°C (13)
Partition coefficient (octanol/water): 50,000 (13)
Adsorption coefficient: 6070 (19)

Exposure guidelines

ADI: 0.01 mg/kg/day (38)
HA: 0.02 mg/L (lifetime) (53)
RfD: 0.003 mg/kg/day (53)
PEL: 0.2 mg/m^3 (8-hour) (skin) (47)

Basic manufacturer

DowElanco
9330 Zionsville Rd.
Indianapolis, IN 46268-1054
Telephone: 317-337-7344
Emergency: 800-258-3033

5.2.4   Coumaphos

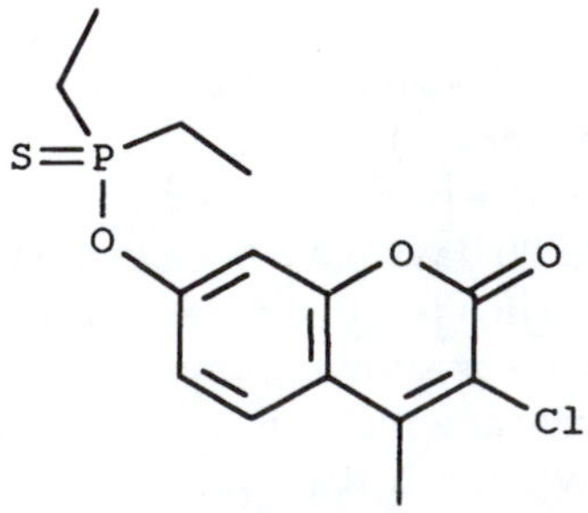

Figure 5.5 Coumaphos.

Trade or other names

Trade names for coumaphos include Agridip, Asunthol, Bay 21, Baymix, Co-Ral, Dilice, Meldame, Muscatox, Negashunt, Resistox, Suntol, and Umbethion.

Regulatory status

The U.S. Environmental Protection Agency (EPA) classifies most formulations of coumaphos as General Use Pesticides (GUPs). The formulations 11.6% EC and 42% flowable concentrate end-use products have been classified as Restricted Use Pesticides (RUPs) because they pose a hazard of acute poisoning from ingestion. RUPs may be purchased and used only by certified applicators. Coumaphos is classified as toxicity class II — moderately toxic. Products containing coumaphos bear the Signal Word WARNING.

Introduction

Coumaphos is an organophosphate insecticide used for control of a wide variety of livestock insects, including cattle grubs, screw-worms, lice, scabies, flies, and ticks. It is used against ectoparasites, which are insects that live on the outside of host animals such as sheep, goats, horses, pigs, and poultry. It is added to cattle and poultry feed to control the development of fly larvae that breed in manure. It is also used as a dust, dip, or spray to control mange, horn flies, and face flies of cattle. Because of its low toxicity to fish, it is also used in water as an agent to control mosquito larvae. Coumaphos is considered a selective insecticide because it kills specific insect species while sparing other non-target organisms.

Toxicological effects

Acute toxicity

Coumaphos is highly toxic by ingestion, and moderately toxic by inhalation and dermal absorption (8). As with all organophosphates, coumaphos is readily absorbed through the skin. Skin and eye contact with this insecticide may cause mild irritation, as well as cholinesterase inhibition. Coumaphos does not cause skin sensitization allergies (18). Toxic symptoms in humans are largely caused by the inhibition of cholinesterase. Individuals with respiratory ailments, impaired cholinesterase production, or with liver malfunction may be at increased risk from exposure to coumaphos. High ambient temperatures or exposure to UV light may increase the toxicity of coumaphos (8). Signs of poisoning include diarrhea, drooling, difficulty in breathing, and leg and neck stiffness (41). Some of the symptoms of acute inhalation of coumaphos include headaches, dizziness, and incoordination. Moderate poisoning is characterized by muscle twitching and vomiting. Severe poisoning is indicated by diarrhea, fever, toxic psychosis, fluid retention (edema) of the lungs, and high blood pressure. Symptoms of sublethal poisoning may continue for 2 to 6 weeks (8).

Some organophosphates may cause delayed symptoms beginning 1 to 4 weeks after an acute exposure that may or may not have produced immediate symptoms. In such cases, numbness, tingling, weakness, and cramping may appear in the lower limbs and progress to incoordination and paralysis. Improvement may occur over months or years, and in some cases residual impairment will remain (8).

The oral LD_{50} for coumaphos is 13 to 41 mg/kg in rats, 28 to 55 mg/kg in mice, 58 mg/kg in guinea pigs, and 80 mg/kg in rabbits (2,8). The dermal LD_{50} is 860 mg/kg in rats, and 500 to 2400 mg/kg in rabbits (18). The 1-hour inhalation LC_{50} for coumaphos is 0.34 mg/L in female rats and 1.1 mg/L in male rats (18).

Chronic toxicity

Repeated or prolonged exposure to organophosphates may result in the same effects as acute exposure, including the delayed symptoms. Other effects reported in workers repeatedly exposed include impaired memory and concentration, disorientation, severe depressions, irritability, confusion, headache, speech difficulties, delayed reaction times, nightmares, sleepwalking, and drowsiness or insomnia. An influenza-like condition with headache, nausea, weakness, loss of appetite, and malaise has also been reported (8).

Reproductive effects

Once in the bloodstream, coumaphos may cross the placenta. Mice fed coumaphos at a dietary level of 100 mg/kg/day exhibited a decrease in the number of pregnancies, litter size, and surviving offspring. No reproductive effects were observed in three generations of mice fed dietary doses of 1.25 mg/kg/day (8). Coumaphos is unlikely to cause reproductive effects in humans at expected exposure levels.

Teratogenic effects

Based on studies with rats and rabbits, coumaphos does not appear to be teratogenic (18). No developmental effects occurred in rat offspring at maternal doses up to 25 mg/kg/day. Similar results were observed when pregnant rabbits were given doses up to 18 mg/kg/day on days 7 to 19 of pregnancy (54). No increase in embryonic deaths or teratogenesis was observed in heifers given dermal applications of coumaphos during various stages of gestation (8).

Mutagenic effects

Gene mutation and DNA damage studies performed on bacterial cultures showed no evidence of mutagenicity (18,54).

Carcinogenic effects

Coumaphos was not found to be carcinogenic in tests done on mice and rats. There was no increase in the number of tumors reported in rats given doses of 1.25 or 5 mg/kg/day coumaphos in a 2-year chronic feeding study (55).

Organ toxicity

Coumaphos primarily affects the nervous system through cholinesterase inhibition, the blockage of an enzyme required for proper nerve functioning. No organ effects were seen in acute or chronic studies of coumaphos.

Fate in humans and animals

Following oral administration to mammals, coumaphos is rapidly broken down into nontoxic products that are eliminated in urine and feces with no evidence of bioaccumulation (18). Some 70% of an oral dose given to rats was eliminated in 7 days. With dermal doses, 5% was eliminated. Single oral doses produced no changes in metabolism and no evidence of bioaccumulation in rats (54). Coumaphos was found in the milk of dermally treated cows (8).

Unchanged coumaphos and other breakdown products were found in the excreta of hens that were dusted with the insecticide. Similar results were found after oral treatment of hens with coumaphos (8).

Ecological effects

Effects on birds

Coumaphos is highly toxic to birds (18). The symptoms of acute toxicity in mallards given a dietary concentration of 29.8 mg/kg include spraddle-legged walking, wing twitching, wing drop, tearing of the eyes, and spread wings. These symptoms persisted in some survivors for up to 13 days, accompanied by weight loss. Death usually occurred between 2 and 12 hours after treatment. Severe acute toxicity, and eventual death, was caused in hens after they were given daily oral doses of 10 mg/kg/day for 1 to 8 days. Hens given single oral doses of 50 mg/kg recovered from the initial effects of cholinesterase inhibition and developed signs of delayed nerve poisoning (8). The oral LD_{50} for coumaphos is 3 mg/kg in wild birds, 29.4 mg/kg in mallard ducks, 7.94 mg/kg in pheasants, and 14 mg/kg in chickens (8,13).

Effects on aquatic organisms

Coumaphos is moderately toxic to fish and highly toxic to aquatic invertebrates (8). The LC_{50} (96-hour) in channel catfish is 0.8 mg/L, in largemouth bass is 1.1 mg/L, and in walleye is 0.8 mg/L (13,27). The LC_{50} (96-hour) in rainbow trout is 5.9 mg/L, in bluegill sunfish is 5 mg/L, and in freshwater invertebrates (amphipods) is 0.00015 mg/L (18).

Coumaphos tends to accumulate slightly in fish. For example, bluegill sunfish showed a bioconcentration factor of 331 times the ambient water concentration; however, mortality was high among the fish at the concentrations tested (0.1 mg/L).

Effects on other organisms (non-target species)

Coumaphos poses a moderate hazard to honeybees and a slight hazard to other beneficial insects (8).

Environmental fate

Breakdown in soil and groundwater

Based on the general characteristics of organophosphates, coumaphos is expected to have low to moderate persistence in soil. Coumaphos was relatively immobile in a sandy loam soil and is unlikely to contaminate groundwater. A general characteristic of organophosphates such as coumaphos is that they bind fairly well to soil particles. Therefore, they do not readily move (leach) with water percolating through the soil (8).

Breakdown in water

Coumaphos is resistant to breakdown in water (hydrolysis). It is nearly insoluble in water, and is stable over a wide pH range (8).

Breakdown in vegetation

No data are currently available.

Physical properties

Technical coumaphos is a tan crystalline solid with a slight sulfur odor (13).

Chemical name: 3-chloro-7-diethoxyphosphinothioyloxy-4-methylcoumarin (13)
CAS #: 56-72-4
Molecular weight: 362.5 (13)
Water solubility: i.s. in water (11); 1.5 mg/L at 20°C (13)
Solubility in other solvents: acetone s.s.; chloroform s.s.; ethanol s.s. (13)
Melting point: 90–92°C (technical) (13)
Vapor pressure: 0.013 mPa @ 20°C (13)
Partition coefficient (octanol/water): Not available
Adsorption coefficient: Not available

Exposure guidelines

ADI: Not available
MCL: Not available
RfD: Not available
PEL: Not available

Basic manufacturer

Bayer Corporation
Animal Health
Box 390
Shawnee Mission, KS 66201
Telephone: 913-631-4800

5.2.5   Diazinon

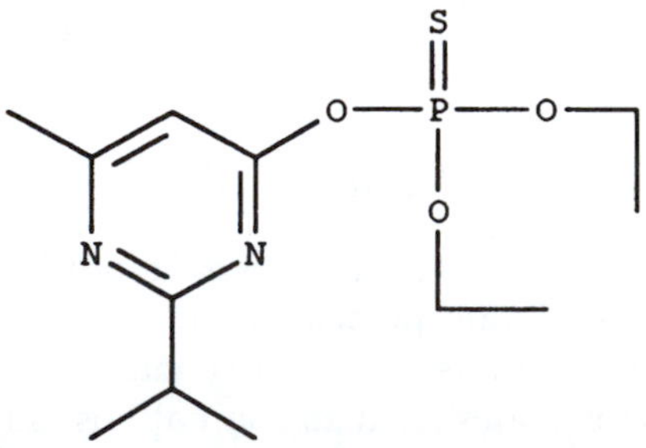

Figure 5.6 Diazinon.

Trade or other names

Trade names of this product include Basudin, Dazzel, Gardentox, Kayazol, Knox Out, Nucidol, and Spectracide. Diazinon may be found in formulations with a variety of other pesticides, such as pyrethrins, lindane, and disulfoton.

Regulatory status

Diazinon is classified as a Restricted Use Pesticide (RUP) and is for professional pest control operator use only. In 1988, the EPA canceled registration of diazinon for use on golf courses and sod farms because of die-offs of birds that often congregated in these areas. It is classified toxicity class II — moderately toxic, or toxicity class III — slightly toxic, depending on the formulation. Products containing diazinon bear the Signal Word WARNING or CAUTION.

Introduction

Diazinon is a nonsystemic organophosphate insecticide used to control cockroaches, silverfish, ants, and fleas in residential, non-food buildings. Bait is used to control scavenger yellow-jackets in the western U.S. It is used on home gardens and farms to control a wide variety of sucking and leaf-eating insects. It is used on rice, fruit trees, sugarcane, corn, tobacco, potatoes, and horticultural plants. It is also an ingredient in pest strips. Diazinon has veterinary uses against fleas and ticks. It is available in dust, granules, seed dressings, wettable powder, and emulsifiable solution formulations.

Toxicological effects

Acute toxicity

Toxic effects of diazinon are due to the inhibition of acetylcholinesterase, an enzyme needed for proper nervous system function. The range of doses that results in toxic effects varies widely with formulation and with the individual species being exposed. The toxicity of encapsulated formulations is relatively low because diazinon is not released readily while in the digestive tract. Some formulations of the compound can be degraded to more toxic forms. This transformation may occur in air, particularly in the presence of moisture, and by ultraviolet radiation. Most modern diazinon formulations in the U.S. are stable and do not degrade easily (8).

The symptoms associated with diazinon poisoning in humans include weakness, headaches, tightness in the chest, blurred vision, nonreactive pinpoint pupils, salivation, sweating, nausea, vomiting, diarrhea, abdominal cramps, and slurred speech. Death has occurred in some instances from both dermal and oral exposures at very high levels (2,8).

The LD_{50} is 300 to 400 mg/kg for technical grade diazinon in rats (2,13). The inhalation LC_{50} (4-hour) in rats is 3.5 mg/L (13). In rabbits, the dermal LD_{50} is 3600 mg/kg (13).

Chronic toxicity

Chronic effects have been observed at doses ranging from 10 mg/kg/day for swine to 1000 mg/kg/day for rats. Inhibition of red blood cell cholinesterase and enzyme response occurred at lower doses in the rats. Enzyme inhibition has been documented in red blood cells, in blood plasma, and in brain cells at varying doses and with different species (2).

Reproductive effects

No data are currently available.

Teratogenic effects

The data on teratogenic effects due to chronic exposure are inconclusive. One study has shown that injection of diazinon into chicken eggs resulted in skeletal and spinal deformities in the chicks. Bobwhite quail born from eggs treated in a similar manner showed skeletal deformities but no spinal abnormalities. Acetylcholine was significantly affected in this latter study (56). Tests with hamsters and rabbits at low doses (0.125 to 0.25 mg/kg/day) showed no developmental effects, while tests with dogs and pigs at higher levels (1.0 to 10.0 mg/kg/day) revealed gross abnormalities (57).

Mutagenic effects

While some tests have suggested that diazinon is mutagenic, current evidence is inconclusive (2).

Carcinogenic effects

Diazinon is not considered carcinogenic. Test on rats over a 2-year period at moderate doses (about 45 mg/kg) did not cause tumor development in the test animals (2).

Organ toxicity

Diazinon itself is not a potent cholinesterase inhibitor. However, in animals, it is converted to diazoxon, a compound that is a strong enzyme inhibitor (2).

Fate in humans and animals

Metabolism and excretion rates for diazinon are rapid. The half-life of diazinon in animals is about 12 hours. The product is passed out of the body through urine and in the feces. The metabolites account for about 70% of the total amount excreted. Cattle exposed to diazinon may store the compound in their fat over the short term (8). One study showed that the compound cleared the cows within 2 weeks after spraying stopped. Application of diazinon to the skin of cows resulted in trace amounts in milk 24 hours after the application (8).

Ecological effects

Effects on birds

Birds are quite susceptible to diazinon poisoning. In 1988, the EPA concluded that the use of diazinon in open areas poses a "widespread and continuous hazard" to birds. Bird kills associated with diazinon use have been reported in every area of the country and at all times of the year. Canadian geese and mallard ducks may be exposed to LC_{50} concentrations in very short periods of time after application (from 15 to 80 minutes, depending on the application rate of the pesticide). Birds are significantly more susceptible to diazinon than other wildlife. LD_{50} values for birds range from 2.75 to 40.8 mg/kg (8).

Effects on aquatic organisms

Diazinon is highly toxic to fish. In rainbow trout, the diazinon LC_{50} is 2.6 to 3.2 mg/L (13). In hard water, lake trout and cutthroat trout are somewhat more resistant. Warmwater fish such as fathead minnows and goldfish are even more resistant, with diazinon LC_{50} values ranging up to 15 mg/L (8). There is some evidence that salt-water fish are more susceptible than freshwater fish. Bioconcentration ratios range from 200 in minnows to 17.5 for guppies. These studies show that diazinon does not bioconcentrate significantly in fish (12).

Effects on other organisms (non-target species)

Diazinon is highly toxic to bees (13).

Environmental fate

Breakdown in soil and groundwater

Diazinon has a low persistence in soil. The half-life is 2 to 4 weeks (19). Bacterial enzymes can speed the breakdown of diazinon and have been used in treating emergency situations such as spills (12).

Diazinon seldom migrates below the top half inch in soil, but in some instances it may contaminate groundwater. The pesticide was detected in 54 wells in California and in tap water in Ottawa, Canada, and in Japan (12).

Breakdown in water

The breakdown rate is dependent on the acidity of water. At highly acidic levels, one half the compound disappeared within 12 hours while in a neutral solution, the pesticide took 6 months to degrade to one half the original concentration (12).

Breakdown in vegetation

In plants, a low temperature and a high oil content tend to increase the persistence of diazinon (58). Generally, the half-life is rapid in leafy vegetables, forage crops, and grass. The range is from 2 to 14 days. In treated rice plants, only 10% of the residue was present after 9 days (58). Diazinon is absorbed by plant roots when applied to the soil and translocated to other parts of the plant (13).

Physical properties

Diazinon is a colorless to dark brown liquid. It has a flashpoint of 180°F (13).

Chemical name: O,O-diethyl 0-2-isopropyl-6-methyl(pyrimidine-4-yl) phosphor-
   othioate (13)
CAS #: 333-41-5
Molecular weight: 304.35 (13)
Solubility in water: 40 mg/L @ 20°C (13)
Solubility in other solvents: petroleum ether v.s.; alcohol v.s.; benzene v.s. (13)
Melting point: Decomposes @ >120°C (8)
Vapor pressure: 0.097 mPa @ 20°C (13)
Partition coefficient (octanol/water): not available (8)
Adsorption coefficient: 1000 (estimated) (19)

Exposure guidelines

ADI: 0.002 mg/kg/day (38)
HA: 6×10^{-4} mg/L (lifetime) (8)
RfD: 9×10^{-5} mg/kg/day (53)
TLV: 0.1 mg/m^3 (8-hour) (47)

Basic manufacturer

Ciba-Geigy Corp.
P.O. Box 18300
Greensboro, NC 27419-8300
Telephone: 800-334-9481
Emergency: 800-888-8372

5.2.6   *Dichlorvos (DDVP)*

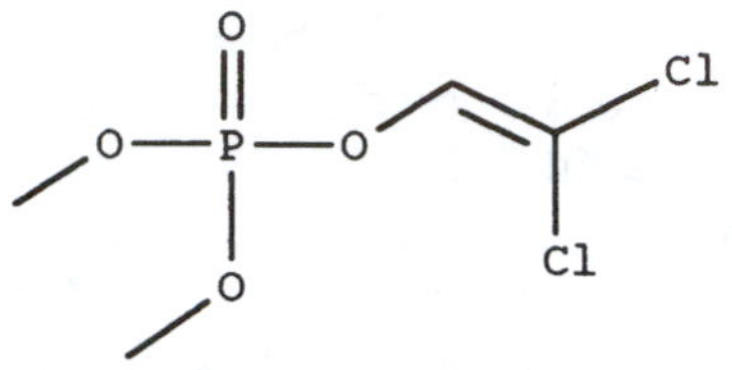

Figure 5.7 Dichlorvos(DDVP).

Trade or other names

Dichlorvos is also called DDVP. Trade names include Apavap, Benfos, Cekusan, Cypona, Derriban, Derribante, Devikol, Didivane, Duo-Kill, Duravos, Elastrel, Fly-Bate, Fly-Die, Fly-Fighter, Herkol, Marvex, No-Pest, Prentox, Vaponite, Vapona, Verdican, Verdipor, and Verdisol. Trade names used outside the U.S. include Doom, Nogos, and Nuvan.

Regulatory status

The EPA has classified it as toxicity class I — highly toxic, because it may cause cancer and there is only a small margin of safety for other effects. Products containing dichlorvos must bear the Signal Words DANGER — POISON. Dichlorvos is a Restricted Use Pesticide (RUP) and may be purchased and used only by certified applicators.

Introduction

Dichlorvos is an organophosphate compound used to control household, public health, and stored product insects. It is effective against mushroom flies, aphids, spider mites, caterpillars, thrips, and white flies in greenhouse, outdoor fruit, and vegetable crops. Dichlorvos is used to treat a variety of parasitic worm infections in dogs, livestock, and humans. Dichlorvos can be fed to livestock to control botfly larvae in the manure. It acts against insects as both a contact and a stomach poison. It is used as a fumigant and has been used to make pet collars and pest strips. It is available as an aerosol and soluble concentrate.

Toxicological effects

Acute toxicity

Dichlorvos is highly toxic by inhalation, dermal absorption, and ingestion (2,8). Because dichlorvos is volatile, inhalation is the most common route of exposure. As with all organophosphates, dichlorvos is readily absorbed through the skin.

Acute illness from dichlorvos is limited to the effects of cholinesterase inhibition. Compared to poisoning by other organophosphates, dichlorvos causes a more rapid onset of symptoms, which is often followed by a similarly rapid recovery (2,8). This occurs because dichlorvos is rapidly metabolized and eliminated from the body. Persons with reduced lung function, convulsive disorders, liver disorders, or recent exposure to cholinesterase inhibitors will be at increased risk from exposure to dichlorvos. Alcoholic beverages may enhance the toxic effects of dichlorvos. High environmental temperatures or exposure of dichlorvos to light may enhance its toxicity (2,8).

Dichlorvos is mildly irritating to skin (8). Concentrates of dichlorvos may cause burning sensations or actual burns (2).

Application of 1.67 mg/kg dichlorvos in rabbits' eyes produced mild redness and swelling, but no injury to the cornea (8).

Symptoms of acute exposure to organophosphate or cholinesterase-inhibiting compounds may include the following: numbness, tingling sensations, incoordination, headache, dizziness, tremor, nausea, abdominal cramps, sweating, blurred vision, difficulty in breathing or respiratory depression, and slow heartbeat. Very high doses may result in unconsciousness, incontinence, and convulsions or fatality.

Some organophosphates may cause delayed symptoms beginning 1 to 4 weeks after an acute exposure that may or may not have produced immediate symptoms. In such cases, numbness, tingling, weakness, and cramping may appear in the lower limbs and progress to incoordination and paralysis. Improvement may occur over months or years, but some residual impairment may remain (8).

The oral LD_{50} for dichlorvos is 61 to 175 mg/kg in mice, 100 to 1090 mg/kg in dogs, 15 mg/kg in chickens, 25 to 80 mg/kg in rats, 157 mg/kg in pigs, and 11 to 12.5 mg/kg in rabbits (2,8,13). The dermal LD_{50} for dichlorvos is 70.4 to 250 mg/kg in rats, 206 mg/kg in mice, and 107 mg/kg in rabbits (2,8,13). The 4-hour LC_{50} for dichlorvos is greater than 0.2 mg/L in rats (8).

Chronic toxicity

Repeated or prolonged exposure to organophosphates may result in the same effects as acute exposure, including the delayed symptoms. Other effects reported in workers repeatedly exposed include impaired memory and concentration, disorientation, severe depression, irritability, confusion, headache, speech difficulties, delayed reaction times, nightmares, sleepwalking, and drowsiness or insomnia. An influenza-like condition with headache, nausea, weakness, loss of appetite, and malaise has also been reported (8).

Repeated, small doses generally have no effect on treated animals. Doses up to 4 mg/kg of a slow-release formulation, given to cows to reduce flies in their feces, had no visibly adverse effects on the cows; but blood tests of these cows indicated cholinesterase inhibition (2).

Feeding studies indicate that a dosage of dichlorvos very much larger than doses that inhibit cholinesterase are needed to produce illness. Rats tolerated dietary doses as high as 62.5 mg/kg/day for 90 days with no visible signs of illness, while a dietary level of 0.25 mg/kg/day for only 4 days produced a reduction in cholinesterase levels (2).

Rats exposed to air concentrations of 0.5 mg/L dichlorvos over a 5-week period exhibited significantly decreased cholinesterase activity in the plasma, red blood cells, and brain. Dogs fed dietary doses of 1.6 or 12.5 mg/kg/day for 2 years showed decreased red blood cell cholinesterase activity, increased liver weights, and increased liver cell size (10). Chronic exposure to dichlorvos will cause fluid to build up in the lungs (pulmonary edema).

Liver enlargement has occurred in pigs maintained for long periods of time on high doses (2). Dichlorvos caused adverse liver effects, and lung hemorrhages may occur at high doses in dogs (8). In male rats, repeated high doses caused abnormalities in the tissues of the lungs, heart, thyroid, liver, and kidneys (8).

Reproductive effects

There is no evidence that dichlorvos affects reproduction. When male and female rats were given a diet containing 5 mg/kg/day dichlorvos just before mating, and through pregnancy and lactation for females, there were no effects on reproduction or on the survival or growth of the offspring, even though severe cholinesterase inhibition occurred in the mothers and significant inhibition occurred in the offspring. The same results were observed in a three-generation study with rats fed dietary levels up to 25 mg/kg/day (2). Once in the bloodstream, dichlorvos may cross the placenta (8).

Teratogenic effects

There is no evidence that dichlorvos is teratogenic. A dose of 12 mg/kg/day was not teratogenic in rabbits and did not interfere with reproduction in any way. There was no evidence of teratogenicity when rats and rabbits were exposed to air concentrations up to 6.25 mg/L throughout pregnancy. Dichlorvos was not teratogenic when given orally to rats (2).

Mutagenic effects

Dichlorvos can bind to molecules such as DNA. For this reason, there has been extensive testing of dichlorvos for mutagenicity. Several studies have shown dichlorvos to be a mutagen (10); for example, dichlorvos is reported positive in the Ames mutagenicity assay and in other tests involving bacterial or animal cell cultures. However, no evidence of mutagenicity has been found in tests performed on live animals. Its lack of mutagenicity in live animals may be due to rapid metabolism and excretion (2).

Carcinogenic effects

Dichlorvos has been classified as a possible human carcinogen because it caused tumors in rats and mice in some studies but not others (11). When dichlorvos was administered by gavage (stomach tube) to mice for 5 days per week for 103 weeks at doses of 20 mg/kg/day in males and 40 mg/kg/day in females, there was an increased incidence of benign tumors in the lining of the stomach in both sexes. When rats were given doses of 4 or 8 mg/kg/day for 5 days per week for 103 weeks, there was an increased incidence of benign tumors of the pancreas and of leukemia in male rats at both doses. At the highest dose, there was also an increased incidence of benign lung tumors in males. In female rats, there was an increase in the incidence of benign tumors of the mammary gland (10). However, no tumors caused by dichlorvos were found in rats fed up to 25 mg/kg/day for 2 years, or in dogs fed up to 11 mg/kg/day for 2 years. No evidence of carcinogenicity was found when rats were exposed to air containing up to 5 mg/L for 23 hours/day for 2 years (11). A few tumors were found in the esophagus of mice given dichlorvos orally, even though tumors of this kind are normally rare (8). In sum, current evidence about the carcinogenicity of dichlorvos is inconclusive.

Organ toxicity

Dichlorvos primarily affects the nervous system through cholinesterase inhibition, the blockage of an enzyme required for proper nerve functioning.

Fate in humans and animals

Among the organophosphates, dichlorvos is remarkable for its rapid metabolism and excretion by mammals. Exposure of rats to 11 mg/L (250 times the normal exposure) for 4 hours was required before dichlorvos was detectable in rats (2). Even then, it was detected only in the kidneys. Following exposure to 50 mg/L, the half-life for dichlorvos in the rat kidney was 13.5 minutes (2). The reason for this rapid disappearance of dichlorvos is the presence of degrading enzymes in both tissues and blood plasma. When dichlorvos is absorbed after ingestion, it is moved rapidly to the liver where it is rapidly detoxified. Thus, poisoning by nonlethal doses of dichlorvos is usually followed by rapid detoxification in the liver and recovery (2). Rats given oral or dermal doses at the LD_{50} level either died within 1 hour of dosing or recovered completely (2).

Dichlorvos does not accumulate in body tissues and has not been detected in the milk of cows or rats, even when the animals were given doses high enough to produce symptoms of severe poisoning (2).

Ecological effects

Effects on birds

Dichlorvos is highly toxic to birds, including ducks and pheasants (13); the LD_{50} in wild birds fed dichlorvos is 12 mg/kg.

Effects on aquatic organisms

UV light makes dichlorvos 5 to 150 times more toxic to aquatic life (8). Grass shrimp are more sensitive to dichlorvos than the sand shrimp, hermit crab, or mummichog. The LC_{50} (96-hour) for dichlorvos is 11.6 mg/L in fathead minnow, 0.9 mg/L in bluegill, 5.3 mg/L in mosquito fish, 0.004 mg/L in sand shrimp, 3.7 mg/L in mummichog, and 1.8 mg/L in American eel. The LC_{50} (24-hour) for dichlorvos in bluegill sunfish is 1.0 mg/L (10). Dichlorvos does not significantly bioaccumulate in fish (12).

Effects on other organisms (non-target species)

Dichlorvos is toxic to bees (13).

Environmental fate

Breakdown in soil and groundwater

Dichlorvos has low persistence in soil. Half-lives of 7 days were measured on clay, sandy clay, and loose sandy soil (12,20). In soil, dichlorvos is subject to hydrolysis and biodegradation. Volatilization from moist soils is expected to be slow. The pH of the media determines the rate of breakdown (12). Breakdown is rapid in alkaline soils and water, but it is slow in acidic media. For instance, at pH 9.1, the half-life of dichlorvos is about 4.5 hours. At pH 1 (very acidic), the half-life is 50 hours (12).

Dichlorvos does not adsorb to soil particles and is likely to contaminate groundwater (12,20). When spilled on soil, dichlorvos leached into the ground, with 18 to 20% penetrating to a depth of 12 inches within 5 days (12).

Breakdown in water

In water, dichlorvos remains in solution and does not adsorb to sediments. It degrades primarily by hydrolysis, with a half-life of approximately 4 days in lakes and rivers. This half-life will vary from 20 to 80 hours between pH 4 and pH 9. Hydrolysis is slow at pH 4 and rapid at pH 9 (8,12). Biodegradation may occur under acidic conditions, which slow hydrolysis, or where populations of acclimated microorganisms exist, as in polluted waters. Volatilization from water is slow; it has been estimated at 57 days from river water and over 400 days from ponds (12).

Breakdown in vegetation

Except for cucumbers, roses, and some chrysanthemums, plants tolerate dichlorvos very well (8).

Physical properties

Dichlorvos is a colorless to amber liquid with a mild chemical odor (13).

Chemical name: 2,2-dichlorovinyl dimethyl phosphate (13)
CAS #: 62-73-7
Molecular weight: 220.98 (13)
Solubility in water: 10,000 mg/L (estimated) (13)
Solubility in other solvents: dichloromethane v.s.; 2-propanol v.s.; toluene v.s.;
    ethanol s.; chloroform s.; acetone s.; kerosene s. (13)
Vapor pressure: 290 mPa @ 20°C (13)
Partition coefficient (octanol water): Not available
Adsorption coefficient: 30 (estimated) (20)

Exposure guidelines

ADI: 0.004 mg/kg/day (38)
MCL: Not available
RfD: 0.0005 mg/kg/day (53)
PEL: 1.0 mg/m³ (8-hour) (skin) (39)

Basic manufacturer

Amvac Chemical Corp.
4100 E. Washington Blvd.
Los Angeles, CA 90023
Telephone: 213-264-3910
Emergency: 800-228-5635, ext. 169

5.2.7 Dimethoate

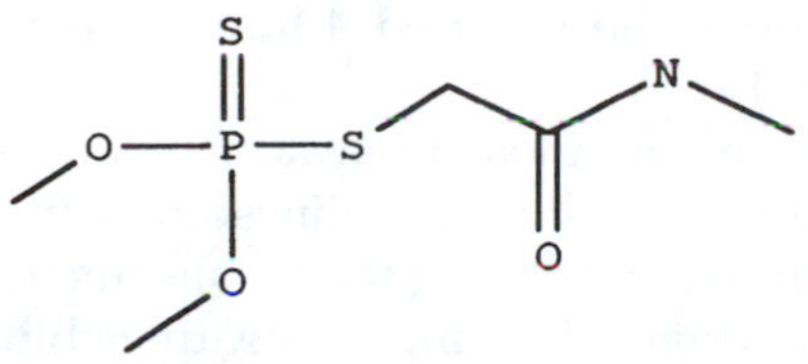

Figure 5.8 Dimethoate.

Trade or other names

Trade names for dimethoate include Cekuthoate, Chimigor 40, Cygon 400, Daphene, De-Fend, Demos NF, Devigon, Dicap, Dimate 267, Dimet, Dimethoat Tech 95%, Dimethopgen, Ferkethion, Fostion MM, Perfekthion, Rogodan, Rogodial, Rogor, Roxion, Sevigor, and Trimetion.

Regulatory status

Dimethoate is a moderately toxic compound in EPA toxicity class II. Labels for products containing dimethoate must bear the Signal Word WARNING. Dimethoate is a General Use Pesticide (GUP).

Introduction

Dimethoate is an organophosphate insecticide used to kill mites and insects systemically and on contact. It is used against a wide range of insects, including aphids, thrips, planthoppers, and whiteflies on ornamental plants, alfalfa, apples, corn, cotton, grapefruit, grapes, lemons, melons, oranges, pears, pecans, safflower, sorghum, soybeans, tangerines, tobacco, tomatoes, watermelons, wheat, and other vegetables. It is also used as a residual wall spray in farm buildings for house flies. Dimethoate has been administered to livestock for control of botflies. Dimethoate is available in aerosol spray, dust, emulsifiable concentrate, and ULV concentrate for-mulations. *Unless otherwise specified, the data summarized in this profile refer to the technical product.*

Toxicological effects

Acute toxicity

Dimethoate is moderately toxic by ingestion, inhalation, and dermal absorption. The reported acute oral LD_{50} values for the technical product range from 180 to 330 mg/kg in the rat; although an oral LD_{50} of as low as 28 to 30 mg/kg has been reported, it is regarded by some as less reflective of the toxicity of current products (2,13). Reported oral LD_{50} values in other species are 160 mg/kg in mice, and 400 to 500 mg/kg in rabbits (2,13). In guinea pigs, the oral toxicity is reported as 550 to 600 mg/kg for the pure and laboratory grade of the compound, but for the technical grade is only 350 to 400 mg/kg (2). It is not clear whether the increased toxicity results from impurities present initially in the technical product or whether these may be formed from degradation over time (2).

Reported dermal LD_{50} values for dimethoate are 100 to 600 mg/kg in rats, again with a much lower value for an earlier product (2,13). Dimethoate is reportedly not irritating to the skin and eyes of lab animals (8,13). Severe eye irritation has occurred in workers manufacturing dimethoate, although this may be due to impurities (2).

Via the inhalation route, the reported 4-hour LC_{50} is greater than 2.0 mg/L, indicating slight toxicity (13).

Effects of acute exposure are those typical of organophosphates. Symptoms of acute exposure to organophosphate or cholinesterase-inhibiting compounds may include the following: numbness, tingling sensations, incoordination, headache, dizziness, tremor, nausea, abdominal cramps, sweating, blurred vision, difficulty in breathing or respiratory depression, and slow heartbeat. Very high doses may result in unconsciousness, incontinence, and convulsions or fatality. Persons with respiratory ailments, recent exposure to cholinesterase inhibitors, impaired cholinesterase production, or liver malfunction may be at increased risk from exposure to dimethoate. High environmental temperatures or exposure of dimethoate to visible or UV light may enhance its toxicity (2).

Chronic toxicity

There was no cholinesterase inhibition in an adult human who ingested 18 mg (about 0.26 mg/kg/day) dimethoate per day for 21 days. No toxic effects and no cholinesterase inhibition were observed in individuals who ingested 2.5 mg/day (about 0.04 mg/kg/day) for 4 weeks. In another study with humans given oral doses of 5, 15, 30, 45, or 60 mg/day for 57 days, cholinesterase inhibition was observed only in the 30-mg/day and higher dosage groups (2).

Repeated or prolonged exposure to organophosphates may result in the same effects as acute exposure, including the delayed symptoms. Other effects reported in workers repeatedly exposed include impaired memory and concentration, disorientation, severe depression, irritability, confusion, headache, speech difficulties, delayed reaction times, nightmares, sleepwalking, and drowsiness or insomnia. An influenza-like condition with headache, nausea, weakness, loss of appetite, and malaise has also been reported (2).

Reproductive effects

When mice were given 9.5 to 10.5 mg/kg/day dimethoate in their drinking water, there was decreased reproduction, pup survival, and growth rates of surviving pups. Adults in this study exhibited reduced weight gain, but their survival was not affected. In a three-generation study with mice, 2.5 mg/kg/day did not decrease reproductive performance or pup survival (2). Once in the bloodstream, dimethoate may cross the placenta (2). Impaired reproductive function in humans is not likely under normal conditions.

Teratogenic effects

Dimethoate is teratogenic in cats and rats (2,8). A dosage of 12 mg/kg/day given to pregnant cats increased the incidence of extra toes on kittens (2,8). The same dosage given to pregnant rats produced birth defects related to bone formation, runting, and malfunction of the bladder. Dosages of 3 or 6 mg/kg/day were not teratogenic in cats or rats (2). No effects were observed in cats and rats at doses of 2.8 mg/kg/day. There were no teratogenic effects seen in the offspring of mice given 9.5 to 10.5 mg/kg/day dimethoate in their drinking water (2). It is not likely that teratogenic effects will be seen in humans under normal circumstances.

Mutagenic effects

Mutagenic effects due to dimethoate exposure were seen in mice. They were more prominent in male mice given a single high dose of dimethoate than in male mice given one twelfth the same dose daily for 30 days (2). Mutagenic effects are unlikely in humans under normal circumstances.

Carcinogenic effects

An increase in malignant tumors was reported in rats given oral doses of 5, 15, or 30 mg/kg/day dimethoate for over a year. The increases were not, however, dose dependent (2). That is, higher doses did not necessarily result in higher tumor rates. Thus, the evidence of carcinogenicity, even with high-dose, long-term exposure, is inconclusive. This suggests carcinogenic effects in humans are unlikely.

Organ toxicity

Target organs as determined through animal tests include the testicles, kidneys, liver, and spleen (2).

Fate in humans and animals

Dimethoate is rapidly metabolized by mammals. Rats excreted about 50 to 60% of administered doses in urine, expired air, and feces within 24 hours (2). Human volunteers excreted 76 to 100% of administered dimethoate within 24 hours (2). The rate of metabolism and elimination varied in several species tested. Among several mammalian species tested, dimethoate appears to be less toxic to those animals with higher liver-to-body weight ratios and to those with the highest rate of dimethoate metabolism (2).

Following application of dimethoate to the backs of cows at 30 mg/kg, the concentration of dimethoate reached a maximum level of 0.02 ppm in blood and milk in about 3 hours, and decreased to 0.01 ppm within 9 hours (2).

Ecological effects

Effects on birds

Dimethoate is moderately to very highly toxic to birds. In Japanese quail, a 5-day dietary LC_{50} of 341 ppm is reported (14). It may be very highly toxic to other birds; reported acute oral LD_{50} values are 41.7 to 63.5 mg/kg in mallards and 20.0 mg/kg in pheasants (6). Birds are not able to metabolize dimethoate as rapidly as mammals do, which may account for its relatively higher toxicity in these species (26).

Effects on aquatic organisms

Dimethoate is moderately toxic to fish, with reported LC_{50} values of 6.2 mg/L in rainbow trout, and 6.0 mg/L in bluegill sunfish (16). It is more toxic to aquatic invertebrate species such as stoneflies and scuds (16).

Effects on other organisms (non-target species)

Dimethoate is highly toxic to honeybees. The 24-hour topical LD_{50} for dimethoate in bees is 0.12 µg per bee (13).

Environmental fate

Breakdown in soil and groundwater

Dimethoate is of low persistence in the soil environment. Soil half-lives of 4 to 16 days, or as high as 122 days, have been reported, but a representative value may be on the order of 20 days (12,19). Because it is rapidly broken down by soil micro-organisms, it will be broken down faster in moist soils.

Dimethoate is highly soluble in water, and it adsorbs only very weakly to soil particles so it may be subject to considerable leaching (12,19). However, it is degraded by hydrolysis, especially in alkaline soils and evaporates from dry soil surfaces. Losses due to evaporation of 23 to 40% of applied dimethoate have been reported (12). Biodegradation may be significant, with a 77% loss reported in a nonsterile clay loam soil after 2 weeks (12).

Breakdown in water

In water, dimethoate is not expected to adsorb to sediments or suspended particles, nor to bioaccumulate in aquatic organisms (12). It is subject to significant hydrolysis, especially in alkaline waters. The half-life for dimethoate in raw river water was 8 days, with disappearance possibly due to microbial action or chemical degradation (12). Photolysis and evaporation from open waters are not expected to be significant (12).

Breakdown in vegetation

Dimethoate is not toxic to plants (13).

Physical properties

Dimethoate is a grey-white crystalline solid at room temperature (13).

Chemical name: O,O-dimethyl S-methylcarbamoylmethyl phosphorodithioate (13)
CAS #: 60-51-5
Molecular weight: 229.28 (13)
Solubility in water: 25 g/L @ 21°C (13)
Solubility in other solvents: s. in methanol and cyclohexane; s.s. in aliphatic hydrocarbons, aromatic hydrocarbons, diethyl ether, carbon tetrachloride, hexane, and xylene; v.s. in chloroform, benzene, toluene, alcohols, esters, ketones, methylene chloride, acetone, and ethanol; i.s. in petroleum ether (13)
Melting point: 43–45°C (technical) (13)
Vapor pressure: 1.1. mPa @ 25°C (13)
Partition coefficient: (octanol/water): 5 (13)
Adsorption coefficient: 20 (19)

Exposure guidelines:

ADI: 0.01 mg/kg/day (38)
HA: Not available
RfD: 0.0002 mg/kg/day (53)
PEL: Not available

Basic manufacturer

BASF Corp.
Agricultural Products Group
P.O. Box 13528
Research Triangle Park, NC 27709-3528
Telephone: 800-669-2273
Emergency: 800-832-4357

5.2.8 Disulfoton

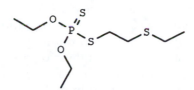

Figure 5.9 Disulfoton.

Trade or other names

Trade names for disulfoton include Bay S276, Disyston, Disystox, Dithiodemeton, Dithiosystox, Frumin AL, Solvigram, and Solvirex.

Regulatory status

All products formulated at greater than 2% disulfoton are classified as Restricted Use Pesticides (RUPs). RUPs may be purchased and used only by certified applicators. Disulfoton is classified as toxicity class I — highly toxic. Products containing disulfoton bear the Signal Word DANGER.

Introduction

Disulfoton is a selective, systemic organophosphate insecticide and acaricide that is especially effective against sucking insects. It is used to control aphids, leafhoppers, thrips, beet flies, spider mites, and coffeeleaf miners. Disulfoton products are used on cotton, tobacco, sugar beets, cole crops, corn, peanuts, wheat, ornamentals, cereal grains, and potatoes.

Toxicological effects

Acute toxicity

Disulfoton is very highly toxic to all mammals by all routes of exposure. Whether absorbed through the skin, ingested, or inhaled, early symptoms in humans may include blurred vision, fatigue, headache, dizziness, sweating, tearing, and salivation. It inhibits cholinesterase and affects nervous system function (2,8). Symptoms occurring at high doses include defecation, urination, fluid accumulation in the lungs, convulsions, or coma. Death can occur if high doses stop respiratory muscles or constrict the windpipe (2,8).

Ingestion of high doses can lead to rapid onset of effects on the stomach. Symptoms resulting from skin exposure may be delayed for up to 12 hours. Complete recovery from acute effects takes at least 1 week, but complete restoration of the blood to normal cholinesterase enzyme levels may take up to 3 months (8).

The oral LD_{50} ranges from 6.2 to 12.5 mg/kg in male rats, and from 1.9 to 2.5 mg/kg in female rats (1,59). For weanling male rats, the oral LD_{50} is 5.4 mg/kg (60). The dermal LD_{50} is 3.6 mg/kg in female rats, and 15.9 mg/kg in male rats (60). The inhalation LC_{50} for 1 hour is 0.3 mg/L in male rats (13).

Chronic toxicity

Rats have survived daily doses of 0.5 mg/kg/day for 90 days. Some studies have shown that rats can acquire a tolerance for the chemical, so they are able to adjust to the lower cholinesterase levels resulting from chronic low-level exposures (8).

In a 2-year rat study, males fed disulfoton (95.5% pure) daily at levels below the LD_{50} had increased spleen, liver, kidney, and pituitary weights, while females with similar treatment had decreased weights in these organs. Also, at all dietary levels, male brains decreased in weight while female brain weights increased. At the highest doses, cholinesterase activity was inhibited in both sexes in the brain, plasma, and red blood cells (59).

In a 23-month mouse study, kidney weights increased in females fed high daily doses. At that level, cholinesterase activity was decreased in both sexes (59).

Reproductive effects

In a long-term reproduction study, 98.5% pure disulfoton was fed at a dose of 0.5 mg/kg/day to both male and female albino rats. The number of animals per litter was reduced by 21% in the first and third generations, and a 10 to 25% lower pregnancy rate was noted. Some third-generation litters whose parents were exposed to this dose developed fatty deposits and swelling in their livers. Exposed adults and litters had a 60 to 70% inhibition of red blood cell cholinesterase (59). This suggests that disulfoton is unlikely to cause reproductive effects in humans at expected exposure levels.

Teratogenic effects

In one study, pregnant rats were given disulfoton (98.2% pure) at doses ranging from 0.1 to 1.0 mg/kg/day through a stomach tube during the sensitive period of gestation. Cholinesterase activity was decreased. In the fetuses, no developmental defects were seen except at high doses, where incomplete bone development was noted (59,60). In another study, rabbits were given disulfoton (97.3% pure) during the sensitive period. At the higher doses (1.5 and 2.0 mg/kg/day), fetal growth was not affected (59). These studies indicate that disulfoton is unlikely to cause birth defects.

Mutagenic effects

Disulfoton has been shown to be mutagenic in studies on bacteria (59).

Carcinogenic effects

There is no evidence that disulfoton is carcinogenic. Studies of rats and mice fed high doses for 2 years did not show significant tumor growth (59,60).

Organ toxicity

Continual daily absorption may cause flu-like symptoms, loss of appetite, weakness, and uneasiness. While repeated exposure to disulfoton may inhibit the cholinesterase enzyme and thus interfere with the nervous system, 30-day human exposures have not resulted in significant enzyme inhibition (1). Workers chronically exposed to organophosphates like disulfoton have developed irritability, delayed reaction times, anxiety, slowness of thinking, and memory defects (1). Chronic exposure of workers may also lead to cataracts.

Fate in humans and animals

Disulfoton is rapidly absorbed by the gastrointestinal tract, metabolized, and excreted via urine. In one study, in which both male and female rats received single doses, females excreted the chemical at a slower rate than did the males. Males excreted 50% of the dose in the urine within 4 to 6 hours after dosing, while it took females 30 to 32 hours to excrete 50% through the urine. Within 10 days after dosing, both male and female rats lost, on average, 81.6% of the initial dose via the urine, 7.0% in the feces, and 9.2% in expired air (59).

Ecological effects

Effects on birds

Disulfoton is moderately toxic to birds. The 5-day acute dietary LC_{50} for disulfoton is 692 ppm in mallard ducks, and 544 ppm in quail (13).

Effects on aquatic organisms

Disulfoton-containing products are highly toxic to cold- and warmwater fish, crab, and shrimp (8). The LC_{50} values for the compound are 0.038 mg/L in bluegill sunfish, 0.25 mg/L in guppies, 1.85 mg/L in rainbow trout, and 6.5 mg/L in goldfish (13).

The bioconcentration factor of 460 indicates that there is a low to moderate potential for this compound to concentrate in living organisms (60).

Effects on other organisms (non-target species)

Use of disulfoton on certain crops may pose a risk to some aquatic and terrestrial endangered species (60). Disulfoton is toxic to bees.

Environmental fate

Breakdown in soil and groundwater

Disulfoton has a low to moderate persistence in soils. Disulfoton is not strongly bound to soil (8,19). Some metabolites are more mobile than the parent disulfoton in sandy loam, clay loam, and silty clay loam soils. Mobility decreases as organic matter content of the soil increases. In addition, these metabolites can persist longer than disulfoton. In a study on sandy loam soils, disulfoton had a half-life of 1 week, and 90% loss in 5 weeks. One metabolite had a half-life of 8 to 10 weeks, and another was fairly stable for 42 weeks (59).

Disulfoton has been found in groundwater in Virginia and Wisconsin at levels up to 1 mg/L (8).

Breakdown in water

Like other organophosphorous insecticides, disulfoton will break down in water under alkaline conditions, and is most stable at normal surface water acidities. The degradation of the compound is temperature dependent (8).

Breakdown in vegetation

When applied to the soil, disulfoton is actively taken up by plant roots and translocated to all parts of the plant (2). Such systemic distribution is especially effective against sucking insects, while predators and pollinating insects are not destroyed. Control may persist for 6 to 8 weeks (60,61). Breakdown in plants follows the same chemical pathway as in soil.

Physical properties

Disulfoton, a member of the organophosphate chemical family, is a yellowish oil (13).

> Chemical name: O,O-diethyl S-[2-(ethylthio)ethyl]phosphorodithioate (13)
> CAS #: 298-04-4
> Molecular weight: 274.4 (13)
> Solubility in water: 25 mg/L @ 23°C (13)
> Solubility in other solvents: Soluble in most organic solvents and fatty oils (13)
> Melting point: 25°C (13)
> Vapor pressure: 24 mPa @ 20°C (13)
> Partition coefficient (octanol/water): Not available
> Adsorption coeffient: 600 (estimated) (19)

Exposure guidelines

> ADI: 3×10^{-3} mg/kg/day (38)
> HA: 3×10^{-4} mg/L (8)
> RfD: 4×10^{-5} mg/kg/day (53)
> TLV: 0.1 mg/m^3 (8-hour) (skin) (47)

Basic manufacturer

> Sanex, Inc.
> 5300 Harvester Road
> Burlington, Ontario
> L7L 5N5 Canada
> Telephone: 905-639-7535

5.2.9   Ethion

Trade or other names

Trade names for ethion include Acithion, Aqua Ethion, Ethanox, Ethiol, Hylmox, Nialate, Rhodiacide, Rhodocide, RP-Thion, Tafethion, and Vegfru Fosmite.

Regulatory status

Ethion is a highly to moderately toxic compound in EPA toxicity class II — moderately toxic. Products containing this compound are General Use Pesticides (GUPs), and labels for these products must bear the Signal Word WARNING.

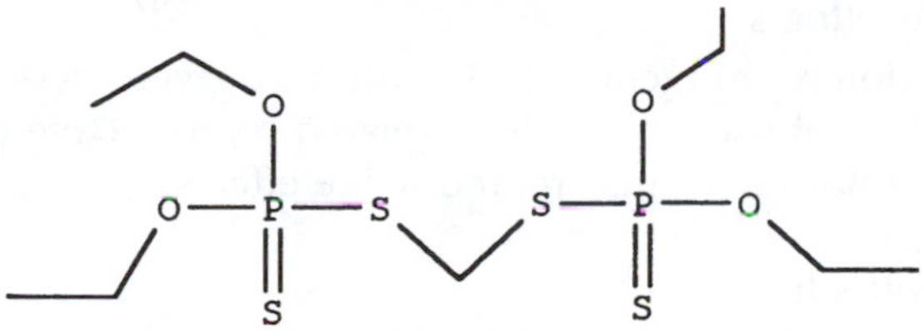

Figure 5.10 Ethion.

Introduction

Ethion is an organophosphate pesticide used to kill aphids, mites, scales, thrips, leafhoppers, maggots, and foliar-feeding larvae. It may be used on a wide variety of food, fiber, and ornamental crops, including greenhouse crops, lawns, and turf. Ethion is often used on citrus and apples. It is mixed with oil and sprayed on dormant trees to kill eggs and scales. Ethion may also be used on cattle. It is available in dust, emulsifiable concentrate, emulsifiable solution, granular, and wettable powder formulations.

Toxicological effects

Acute toxicity

Ethion is highly to moderately toxic by the oral route, with reported oral LD_{50} values for pure ethion in rats of 208 mg/kg, and for technical ethion of 21 to 191 mg/kg (8,13). Other reported oral LD_{50} values (for the technical product) are 40 mg/kg in mice and guinea pigs (13). Ethion is moderately toxic via inhalation, with a reported 4-hour LC_{50} in rats of 0.864 mg/L (8). It is highly to moderately toxic via the dermal route as well, with a reported dermal LD_{50} of 62 mg/kg in rats (13), 915 mg/kg in guinea pigs, and 890 mg/kg in rabbits (62).

Acute effects are typical of organophosphate exposure and will vary according to the degree of exposure. Effects could include nausea, cramps, diarrhea, excessive salivation, blurred vision, headache, fatigue, tightness in chest, abnormal heart beat and breathing, loss of coordination, convulsions, coma, and death. Skin exposure may cause contact burns. Persons with respiratory ailments, recent exposure to cholinesterase inhibitors, impaired cholinesterase production, or liver malfunction may be at increased risk from exposure to ethion. High environmental temperatures or exposure of ethion to visible or UV light may enhance its toxicity (2,8).

Chronic toxicity

In a chronic toxicity study with rats fed 0.1, 0.2, or 2 mg/kg/day for 18 months, decreased cholinesterase levels occurred in the high-dose group. No other toxic effects were observed (62).

Repeated or prolonged exposure to organophosphates may result in the same effects as acute exposure, including the delayed symptoms. Other effects reported in workers repeatedly exposed include impaired memory and concentration, disorientation, severe depression, irritability, confusion, headache, speech difficulties, delayed reaction times, nightmares, sleepwalking, and drowsiness or insomnia. An influenza-like condition with headache, nausea, weakness, loss of appetite, and malaise has also been reported (2,8).

Reproductive effects

A three-generation reproduction study with rats given dietary doses as high as 1.25 mg/kg/day did not show any ethion-related reproductive effects (62,63). This suggests that ethion does not cause reproductive effects.

Teratogenic effects

When rats were given doses of 0.2, 0.6, or 2.5 mg/kg/day on days 6 to 15 of pregnancy, developmental effects were seen only in the highest dose tested. In fetuses of the high-dose group, there was an increased incidence of delayed ossification of the pubic bones. When rabbits were given doses of 0.6, 2.4, or 9.6 mg/kg/day on days 6 to 18 of pregnancy, fetuses from the highest dose group exhibited an increased incidence of fused sternal bones (62). Teratogenic effects observed in lab animals are not likely in humans under normal circumstances.

Mutagenic effects

Assays on gene mutation, structural chromosomal aberration, and unscheduled DNA synthesis indicate that ethion is not mutagenic (62,63).

Carcinogenic effects

Ethion was not carcinogenic in rats and mice (7). There was no increase in the incidence of tumors in rats fed dietary doses as high as 2 mg/kg/day for 18 months. No evidence of carcinogenicity was observed in mice fed dietary doses up to 1.2 mg/kg/day for 2 years (62).

Organ toxicity

Ethion primarily affects the nervous system through cholinesterase inhibition, the blockage of an enzyme required for proper nerve functioning.

Fate in humans and animals

Based on its similarity to other organophosphates, ethion is probably degraded in the same general way as other members of this class. These chemicals are rapidly metabolized and excreted via urine, with a biological half-life of 1 or 2 days.

Ecological effects

Effects on birds

Ethion is highly toxic to practically nontoxic to birds, depending on the species. Ethion is highly toxic to songbirds; for example, the LD_{50} in red-winged blackbirds is 45 mg/kg. It is moderately toxic to medium-sized birds such as bobwhite quail (LD_{50} is 128.8 mg/kg) and starlings (LD_{50} is greater than 304 mg/kg). Ethion is practically nontoxic to larger upland game birds (ring-necked pheasant) and water-fowl (mallard duck). Ethion is persistent in the environment and may be stored in plant and animal tissues (62).

Effects on aquatic organisms

Ethion is very highly toxic to freshwater and marine fish and to freshwater invertebrates (8,63). The 96-hour LC_{50} for ethion in rainbow trout is 0.5 mg/L (8). The acute LC_{50} is 0.049 mg/L in Atlantic silversides, 0.210 mg/L in bluegill sunfish, and 0.72 mg/L in cutthroat trout and flathead minnows. The LD_{50} for freshwater invertebrates is 0.056 µg/L to 0.0077 mg/L, depending on the species, and 0.05 to 0.049 mg/L for marine and estuarine invertebrates. Ethion accumulates in the tissues of fish (16,62,63).

Effects on other organisms (non-target species)

Ethion is practically nontoxic to honeybees. Its LD_{50} is 20.55 µg per bee (62,63).

Environmental fate

Breakdown in soil and groundwater

Ethion is moderately to highly persistent. Under laboratory conditions, the soil half-life of ethion was 1.3 to 8 weeks; but in a greenhouse with an organic soil, it was more persistent, with half-lives of 16 to 49 weeks, depending on the degree of watering (12,19). When used repeatedly, ethion residues in soil will increase from one year to the next (13).

Ethion adsorbs strongly to soil particles and is nearly insoluble in water (12,19). It is therefore unlikely to leach or contaminate groundwater (12). In soil, ethion is subject to microbial degradation. It is resistant to hydrolysis, except in alkaline conditions (pH 9 or above) (12).

Breakdown in water

Ethion is almost insoluble in water (13). In open waters, it is likely to adsorb to suspended particles and bottom sediments. The persistence half-life of ethion varied from 4 to 22 weeks when tested in three different natural waters under laboratory conditions. It breaks down slowly in irrigation canal water (half-life = 26 days) (12). Its hydrolysis half-lives at 25°C are 63, 58, 25, and 8.4 weeks at pH 5, 6, 7, and 8, respectively (12). The half-life was 1 day at pH 10 and 30°C. Microbial degradation of ethion may be insignificant in open waters. Volatilization may be important only in shallow, rapidly moving streams. Photooxidation may occur in sunlight (12). Bioconcentration of ethion may be significant (8).

Breakdown in vegetation

No data are currently available.

Physical properties

Technical ethion is an odorless amber liquid (13).

Chemical name: O,O,O′,O′-tetraethyl S,S′-methylene bis(phosphorodithioate) (13)
CAS #: 563-12-2
Molecular weight: 384.48 (13)
Solubility in water: 1.10 mg/L (13)
Solubility in other solvents: Soluble in most organic solvents (13)
Melting point: −15 to −12°C (13)
Vapor pressure: 0.2 mPa @ 25°C (13)
Partition coefficient (octanol/water) (log): 5.073 (8)
Adsorption coefficient: 10,000 (19)

Exposure guidelines

ADI: 0.002 mg/kg/day (38)
HA: Not available
RfD: 0.0005 mg/kg/day (53)
TLV: 0.4 mg/m³ (8-hour) (skin) (47)

Basic manufacturer

FMC Corporation
Agricultural Chemicals Group
1735 Market Street
Philadelphia, PA 19103
Telephone: 215-299-6661
Emergency: 800-331-3148

5.2.10 *Fenamiphos*

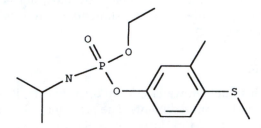

Figure 5.11 Fenamiphos.

Trade or other names

Trade names for products containing fenamiphos include Bay 68138, Nemacur, and Phenamiphos. It may also be found in formulations with other pesticides such as isofenphos, carbofuran, and disulfoton.

Regulatory status

Fenamiphos is an highly toxic compound in EPA toxicity class I. It is a Restricted Use Pesticide (RUP). RUPs may be purchased and applied only by a certified applicator. Labels for products containing fenamiphos must carry the Signal Word DANGER.

Introduction

Fenamiphos is an organophosphate nematicide used to control a wide variety of nematode (roundworm) pests. Nematodes can live as parasites on the outside or inside of a plant. They may be free living or associated with cyst and root-knot formations in plants. Fenamiphos is used on a variety of plants, including tobacco, turf, bananas, pineapples, citrus and other fruit vines, and some vegetables and grains.

The compound is absorbed by roots and then translocated throughout the plant. Fenamiphos, as is typical of other organophosphates, blocks the enzyme acetylcholinesterase in the target pest. The pesticide also has secondary activity against other invertebrates such as sucking insects and spider mites. It is available in emulsifiable concentrate, granular, or emulsion formulations.

Toxicological effects

Acute toxicity

Fenamiphos is highly toxic via the oral route, with reported LD_{50} values of 2 to 19 mg/kg in the rat, and 56 to 100 mg/kg in the guinea pig. It is also highly toxic to dogs and rabbits (8,13). The acute dermal toxicity of the compound is also high, with reported dermal LD_{50} values of 72 to 154 mg/kg in rats (8,13). The inhalation toxicity of the compound is also high, with reported inhalation LC_{50} values in rats of 0.11 to 0.17 mg/L (8,13). Longer exposures at moderately lower concentrations also caused rat mortality (8). The compound has the potential to cause significant eye damage at acute exposure levels. It is nonirritating to the skin (13).

Symptoms of acute toxic exposure to the nematicide are consistent with those of other organophosphate compounds and include difficulty in breathing, diarrhea, urination, and slowness of the heart. Other symptoms include muscle twitching and tremors (8,13).

Chronic toxicity

A number of long-term feeding studies have been conducted with this compound on several different species of animals. In dogs, dietary doses of 0.0125 to 0.25 mg/kg/day over 2 years produced depressions in cholinesterase activity at middle doses and above. No effects were noted in the liver or in blood chemistry, even at the highest dose (8).

Rats exposed to 1.5 mg/kg/day over 2 years experienced increases in thyroid gland and lung weights in females, and increased heart weight in males. There were no organ weight changes noted in the rats at doses below 0.5 mg/kg/day (64). Brain weights have also been affected by exposure to moderate amounts of the compound (8).

Two studies have been conducted on the potential risk to pesticide workers (loaders and applicators) from the use of Nemacur. One study concluded that occupational exposure levels were more than 100 times lower than the level that causes cholinesterase inhibition in animals, and thus the use of the compound did not pose a significant risk to the users (8). Another study concluded that the main threat to applicators was through the skin on the hands. However, the levels of exposure on the hands were significantly below the level that had caused chronic toxicity in mice. It was concluded that the pesticide could be used safely (65).

Reproductive effects

Both male and female rats fed moderate to high doses of fenamiphos (0.15 to 1.5 mg/kg/day) over three generations showed no compound-related reproductive effects at the middle doses tested (0.5 mg/kg/day). At the higher doses, the second generation of pups showed a decrease in body weight gain. This effect was not seen in the third generation (64). It is unlikely that this compound would cause reproductive effects in humans.

Teratogenic effects

A single study of pregnant rats fed fenamiphos during gestation over a range of doses (up to 1 mg/kg/day) showed a decrease in the maternal weight at doses of 0.3 mg/kg/day and above. At the highest dose, a higher number of the pups from the exposed group had died relative to the unexposed controls, and the pups that survived had decreased weights (64).

In tests with pregnant rabbits fed up to 0.4 mg/kg, no birth defects were noted (8). However, another reference stated that teratogenic studies were positive in rabbits, though the effects in the offspring were induced at doses much higher than those that cause maternal toxicity (8).

The results from these studies suggest that teratogenic effects in humans are unlikely.

Mutagenic effects

A number of studies evaluating the mutagenic potential of fenamiphos have all shown the compound to be nonmutagenic. The test subjects included bacterial cells and male mice (8,64).

Carcinogenic effects

Two studies, one conducted with mice and the other with rats, indicated that fenamiphos is not carcinogenic (8,64). One study was conducted for $1\frac{1}{2}$ years at very high levels (up to 7.5 mg/kg/day in mice), and the other study was conducted over 2 years (up to 1.5 mg/kg/day in rats) (64).

Organ toxicity

Target organs identified in studies of test animals and exposed workers are the central nervous system, heart, lungs, and thyroid.

Fate in humans and animals

Fenamiphos is readily absorbed through the digestive tract and lungs. One study placed the amount absorbed near 95% of the ingested dose. The compound is rapidly broken down within the organism, and the by-products are excreted in the urine. The majority of a dose was recovered in urine within 15 hours after treatment (65).

Ecological effects

Effects on birds

Fenamiphos is very highly toxic to birds, with a reported acute oral LD_{50} for the most sensitive species tested, the ring-necked pheasant, of 0.5 mg/kg (6). LD_{50} values for other species range from 1.0 to 2.4 mg/kg, all of which indicate that this is a very highly toxic compound (6). In a controlled experiment, fenamiphos was determined to be the most toxic of thirteen different cholinesterase inhibitors (66). In tests with wild songbirds (red-winged blackbirds and house sparrows), an unspecified dose of Nemacur was highly toxic to these species, with death of the birds occurring within an hour of eating the granules (66).

Effects on aquatic organisms

The toxicity of fenamiphos to aquatic species varies from moderate to high. Bluegill sunfish are extremely sensitive to the presence of the compound. The LC_{50} for fenamiphos is 9.6 mg/L in this species. Other species tested include the rainbow trout (LC_{50} is 0.11 mg/L) and the goldfish (LC_{50} is 3.2 mg/L) (8). The compound is not expected to bioaccumulate appreciably in aquatic organisms (6).

Effects on other organisms (non-target species)

Fenamiphos is practically nontoxic to honeybees (8).

Environmental fate

Breakdown in soil and groundwater

Fenamiphos is of moderate persistence in the soil environment, with a reported soil half-life of about 50 days (19). The compound appears to have no effect on the activity of soil bacteria (8). Aerobic processes are most important for breakdown of the compound. Fenamiphos is not strongly adsorbed to soils (19), but neither it nor its breakdown products have been found in over 1200 wells tested in six states.

Breakdown in water

Fenamiphos disappears quickly from water in acidic and alkaline water, but it is stable in neutral water when held in the dark. The compound, when in the presence of artificial light, disappears very rapidly. In neutral solution, half of the initial amount of the compound degraded within 4 hours.

Breakdown in vegetation

In plants, the compound is absorbed through the roots and translocated to the leaves. It is broken down within the plant. The products of its breakdown are relatively persistent and can also inhibit cholinesterase (6).

Physical properties

Fenamiphos is a colorless crystal in its pure form, or a tan waxy solid in the technical form (13).

Chemical name: ethyl 4-methylthio-*m*-tolyl isopropylphosphoramidate (13)
CAS #: 22224-92-6
Molecular weight: 303.40 (13)
Solubility in water: 700 mg/L @ 20°C (13)
Solubility in other solvents: s. in dichloromethane, isopropanol, and toluene (13)
Melting point: 46°C (technical) (13)
Vapor pressure: 0.12 mPa @ 20°C (13)
Partition coefficient (octanol/water): Not available
Adsorption coefficient: 100 (19)

Exposure guidelines

ADI: 0.005 mg/kg/day (38)
HA: 0.002 mg/L (lifetime) (64)
RfD: 0.00025 mg/kg/day (53)
TLV: 0.1 mg/m^3 (8-hour) (skin) (47)

Basic manufacturer

Miles, Inc.
P.O. Box 4913
Kansas City, MO 64120
Telephone: 816-242-2429
Emergency: 816-242-2582

5.2.11 Fenthion

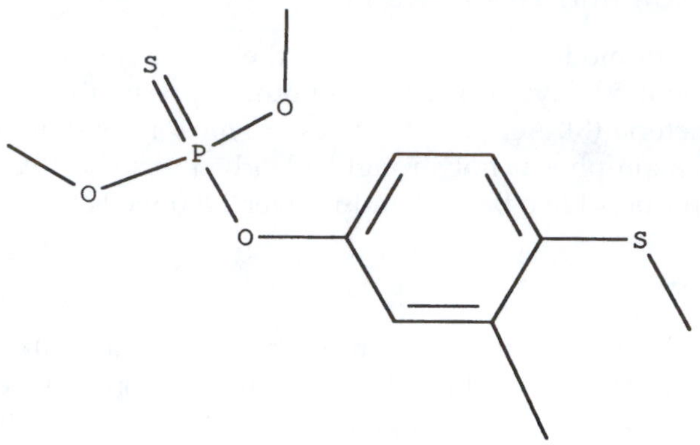

Figure 5.12 Fenthion.

Trade or other names

Fenthion was formerly called DMTP. Trade names for fenthion include Bay 29493, Baycid, Baytex, Dalf, DMTP, Entex, Lebaycid, Mercaptophos, Prentox Fenthion 4E, Queletox, S 1752, Spotton, Talodex, and Tiguvon.

Regulatory status

Fenthion is a moderately toxic compound in EPA toxicity class II. It is classified by the U.S. Environmental Protection Agency (EPA) as a Restricted Use Pesticide (RUP) due to the special handling warranted by its toxicity. All bird control products, as well as nondomestic, nongranular formulations of 70% and greater are RUPs. RUPs may be purchased and used only by trained, certified applicators. Fenthion may not be used on food crops. Labels for products containing fenthion must bear the Signal Word WARNING.

Introduction

Fenthion is a contact and stomach organophosphate insecticide used against many sucking, biting pests, especially fruit flies, stem borers, mosquitoes, and Eurygaster cereal bugs. In mosquitoes, it is toxic to both the adult and immature forms (larvae). Once used extensively in the U.S. for controlling intestinal worms, fenthion no longer has FDA approval due to poisoning deaths. Fenthion is available in dust, emulsifiable concentrate, granular, liquid concentrate, spray concentrate, ULV, and wettable powder formulations.

While it is effective as an insecticide, it is also moderately toxic to mammals and highly toxic to birds. Based on its high toxicity to birds, fenthion is used in various parts of the world for weaver bird control. Pest control operators have used it to control pigeons around public buildings as well. For bird control, use is made of fenthion's contact action and its ready absorption through the skin. It is applied as a paste to roosting areas when utilized for such purposes.

Toxicological effects

Acute toxicity

Fenthion is moderately toxic via the oral route, with reported oral LD_{50} values of 180 to 298 mg/kg in rats, 150 mg/kg in rabbits, and 88 to 145 mg/kg in mice (8,13). It is moderately toxic via the dermal route as well, with reported dermal LD_{50} values of 330 to 1000 mg/kg in rats, and 500 mg/kg in mice (8,13). It is slightly toxic via inhalation, with a reported 1-hour airborne LC_{50} for fenthion in rats of 2.4 to 3.0 mg/L (13). Acute effects of fenthion are similar to those caused by other organophosphates, but may take somewhat longer to develop (2). Fenthion is of sufficiently low toxicity that it has been investigated as an agent against insect parasites in animals (e.g., dogs) (2).

Symptoms of acute exposure to organophosphate or cholinesterase-inhibiting compounds may include the following: numbness, tingling sensations, incoordination, headache, dizziness, tremor, nausea, abdominal cramps, sweating, blurred vision, difficulty in breathing or respiratory depression, and slow heartbeat. Very high doses may result in unconsciousness, incontinence, and convulsions or fatality.

Chronic toxicity

In rats, 12.5 mg/kg/day caused weight loss and 85% inhibition of normal brain cholinesterase activity within 4 weeks (2). Much less severe, but still detectable decreases were noticeable at doses of 2.5 mg/kg/day (2). Repeated or prolonged exposure to organophosphates may result in the same effects as acute exposure, including the delayed symptoms. There was no evidence of weight loss or decreased food consumption in dogs given dietary doses of 1.25 mg/kg/day for 1 year (13).

In Nigerian sprayers, those not wearing protective clothing while spraying showed decreased whole-blood cholinesterase activity (2). Veterinary clinic workers who did not use skin protection when applying a 20% topical application to dogs experienced symptoms ranging from tingling and numbness of the hands and feet to generalized weakness and shooting pains (2). Other possible effects are similar to those caused by the other organophosphates.

Reproductive effects

Single injections of 40 or 80 mg/kg fenthion into the abdominal cavities of pregnant female mice caused poisoning in the developing fetuses, particularly when administered on days 10 to 12 of gestation (2). Fetuses were injured primarily by dosages that caused toxicity in the maternal mouse (8). No influence was seen on reproduction in other three-generation studies of mice (8). These data indicate that reproductive effects are unlikely in humans.

Teratogenic effects

Some reduction in fetal weight occurred, but no defects were found in mice given intraperitoneal doses up to 80 mg/kg fenthion in single-day or 3-day periods during the period of gestation in which organs are formed (8,9). No teratogenic effects were seen in five generations of mice that drank water containing 60 mg/L fenthion (9). Other tests on mice and rats did not show teratogenic effects from fenthion (8). These data indicate that fenthion is not teratogenic.

Mutagenic effects

Tests on mice did not show mutagenic effects from fenthion (8). However, available data are insufficient to draw a conclusion regarding the mutagenicity of the compound.

Carcinogenic effects

One carcinogenicity test on fenthion indicated that this insecticide may be a carcinogen in male mice (9). However, no carcinogenic effects were observed in other 2-year feeding studies of rats and mice (8). Available data are insufficient to draw conclusions regarding the carcinogenicity of fenthion.

Organ toxicity

As identified through animal tests and human-use experience, target organs affected by fenthion exposure include the central and peripheral nervous systems, as well as the heart.

Fate in humans and animals

In animals, fenthion is quickly absorbed into the bloodstream through the digestive tract, lungs, and skin, and is systemically distributed (2). It is eliminated through the urine and the feces (22). A single dose of the insecticide has prolonged action, suggesting that much of it is stored in body fat and later released for metabolism (8).

Fenthion and its metabolites were found in the fat of steers slaughtered 3 days after dermal application of fenthion (2). When cows were given a dermal application of 9 mg fenthion per kilogram, 45 to 55% of the dose was excreted in the urine, 2.0 to 2.5% in the feces, and 1.5 to 2.0% was recovered in the milk (2).

Ecological effects

Effects on birds

Fenthion is very highly to highly toxic to birds, with reported LD_{50} values for various species ranging from less than 4 mg/kg in bobwhite quail to 26 mg/kg in ducks (27). Birds that showed acute LD_{50} values within this interval were California quail, Japanese quail, Canada geese, finches, starlings, sparrows, mallards, mourning doves, and chukars (27). Acute symptoms of fenthion poisoning in birds include tearing of the eyes, foamy salivation, lack of movement, tremors, congestion of the windpipe, lack of coordination in walking, and an abnormally rapid rate of breathing or difficult breathing (27). Chickens developed leg weakness when fed 25-mg/kg doses of fenthion (26). The acute oral LD_{50} in poultry is 15 to 30 mg/kg (22).

The approximate dietary LC_{50} values (over an unspecified number of days) for fenthion are: Japanese quail, 130 ppm; pheasant, 200 ppm; bobwhite, 30 ppm; and mallard, 230 ppm (27). After administration of 0.5 mg/kg/day for 30 days, the eggs laid by surviving mallards had markedly reduced fertility (26).

Effects on aquatic organisms

Fenthion is moderately toxic to fish, with reported 96-hour LC_{50} values of 9.3 mg/L in rainbow trout, 1.33 mg/L in brown trout, 1.32 mg/L in coho salmon, 1.16 mg/L in carp, 1.54 mg/L in largemouth bass, 1.38 mg/L in bluegill, 1.65 mg/L in yellow perch, 2.40 mg/L in fathead minnow, and 3.40 mg/L in goldfish (13,16). Brown bullheads were not affected by the insecticide when it was applied to a

California refuge at 0.01 pounds per acre (26). It may be very highly toxic to some freshwater aquatic invertebrates (16).

Effects on other organisms (non-target species)

Fenthion is toxic to bees (13).

Environmental fate

Breakdown in soil

Fenthion is of moderate persistence in soil, with an average field half-life of 34 days under most conditions (19). In soil, residues of fenthion may persist for approximately 4 to 6 weeks (67). It adsorbs fairly strongly to soil particles, and so is not likely to move (or leach) through the soil (19,67).

Breakdown in water

In one study of its persistence in water, 50% of applied fenthion remained in river water 2 weeks later, while 10% remained after 4 weeks (68). It is more rapidly degraded under alkaline conditions (8).

Breakdown in vegetation

Fenthion is phytotoxic (or harmful to plants) to American linden, Hawthorn, and sugar maple trees, and to certain rose varieties (8). It is not considered phytotoxic when used at recommended rates, although injury has occurred in certain varieties of apples and cotton. Plant foliage should not be sprayed when temperatures exceed 90°F (22). Only about 10% of applied fenthion remained on rice plants after 6 hours. Almost half of the activity was found in the rice bran, 6.5% was found in the husk, and 14.7% was found in polished rice. Water-soluble metabolites were found 14 days after fenthion application to rice plants (8).

Physical properties

Pure fenthion is a colorless liquid. Technical fenthion is a yellow or brown oily liquid with a weak garlic odor (13).

Chemical name: O,O-dimethy O-4-methylthio-*m*-tolyl phosphorothioate (13)
CAS #: 55-38-9
Molecular weight: 278.33 (13)
Water solubility: 2 mg/L @ 20°C (13)
Solubility in other solvents: s. in most organic solvents, including alcohols, ethers, esters, halogenated aromatics, petroleum ethers, etc. (13)
Melting point: 7.5°C (13)
Vapor pressure: 4 mPa @ 20°C; 10 mPa @ 30°C (13)
Partition coefficient (octanol/water) (log): 4.091 (8)
Adsorption coefficient: 1500 (19)

Exposure guidelines

ADI: 0.001 mg/kg/day (38)
HA: Not available

RfD: Not available
PEL: 0.2 mg/m³ (8-hour) (skin) (39)

Basic manufacturer

Miles, Inc.
P.O. Box 4913
8400 Hawthorn Road
Kansas City, MO 64120
Telephone: 816-242-2659
Emergency: 816-242-2582

5.2.12 Fonofos

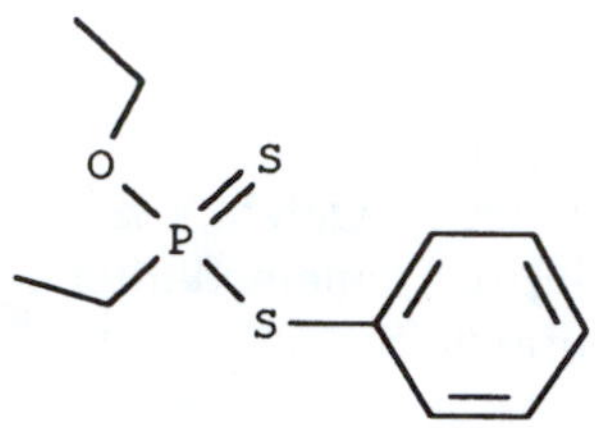

Figure 5.13 Fonofos.

Trade or other names

Trade names for fonofos include Capfos, Cudgel, Difonate, Dyfonate, Dypho-nate, and Stauffer N 2790.

Regulatory status

Fonofos is a highly toxic organophosphate insecticide in EPA toxicity class I. Labels for products containing it must bear the Signal Words DANGER — POISON. Some or all formulations of fonofos are classified as Restricted Use Pesticides (RUPs) by the EPA. RUPs may be purchased and used only by certified applicators.

Introduction

Fonofos is a soil organophosphate insecticide primarily used on corn. It is also used on sugarcane, peanuts, tobacco, turf, and some vegetable crops. It controls aphids, corn borers, corn rootworms, corn wireworms, cutworms, white grubs, and some maggots. It is available in granular, microgranular, emusifiable concentrate, suspension concentrate, microcapsule suspension, and seed treatment formulations.

Toxicological effects

Acute toxicity

Fonofos is highly toxic via the oral route, with reported oral LD_{50} values in male rats ranging from 6.8 to 18.5 mg/kg (2,28), and from 3.2 to 7.9 mg/kg in female rats (69). It is also highly toxic via the dermal route, with reported dermal LD_{50} values of 25 mg/kg in female rabbits, 147 mg/kg in rats, and 278 mg/kg in guinea pigs (2). The inhalation toxicity is moderate, with a reported 4-hour airborne LC_{50} of 0.9 mg/L (13).

Symptoms of fonofos exposure may be delayed from a few minutes to up to 12 hours after exposure. Early symptoms include blurred vision, headache, and dizziness. Skin contact often brings about sweating and muscle twitching. Eye contact causes tearing, pain, and blurring. Ingestion may cause nausea, abdominal cramps, and diarrhea (47). Deaths resulting from high exposures are often due to respiratory arrest. While these effects are similar to those caused by other organophosphates, fonofos may cause them at lower doses than others.

A number of human poisonings by fonofos have been recorded. One woman exposed orally to a large amount of fonofos, developed nausea, sweating, and respiratory arrest in addition to muscle twitching, low blood pressure and pulse rate, and pinpoint pupils. A pancreatic cyst (attributed to fonofos exposure) was located and drained externally in the course of her treatment. She recovered after 2 months of hospitalization (2).

Chronic toxicity

Chronic effects may be similar to those observed after acute exposure, with some delay (2,8). In hens, administration of 90 oral doses of 8 mg/kg (at unspecified intervals) didn't produce delayed neurotoxicity (2). Dietary feeding of fonofos at low levels to dogs for 14 weeks produced no effects at or below doses of 0.20 mg/kg (47).

Reproductive effects

A long-term reproduction test in rats showed no effects on female reproductive ability at doses of 2.0 mg/kg/day fonofos (69,70). This evidence suggests that fonofos does not cause reproductive effects.

Teratogenic effects

Pregnant mice were fed very high doses of fonofos during the sensitive period of gestation. At these levels, some abnormal bone development and brain changes were observed in the fetuses (69). Teratogenic effects in humans are not likely under normal conditions.

Mutagenic effects

Fonofos was not mutagenic in five microbial assays nor in a DNA synthesis test using human cells (69), suggesting that it is not mutagenic.

Carcinogenic effects

Male and female rats that ingested very high daily doses (relative to the LD_{50}) of fonofos for 2 years showed no cancerous effects (69).

Organ toxicity

Fonofos affects the eyes, respiratory system, and central nervous system.

Fate in Animals and Humans

Fonofos is readily absorbed through the skin, gastrointestinal tract, and respiratory tract (2). Once absorbed, it is systemically distributed. Fonofos is quickly excreted in animals; 96 hours after rats received a single, high oral dose of fonofos, three quarters of the dose was found in the urine, a third in the feces, a small amount in expired air, and only a trace in the tissues (71). Virtually complete elimination occurred within 2 to 16 days after exposure (69). In another study, rats given nearly pure fonofos excreted almost all of it within 4 days (69).

A fonofos product, Dyfonate, is broken down in rat livers into several metabolites, one of which is a potent cholinesterase inhibitor known as an oxon (69). This also is rapidly eliminated. Other than the fonofos oxon (an anticholinesterase metabolite), fonofos' breakdown products are less toxic than the parent compound (2).

Ecological effects

Effects on birds

Fonofos is highly toxic to birds, with a reported acute LD_{50} of 128 mg/kg in mallards (13). Other reported LD_{50} values are 16.9 mg/kg in mallards, 12 to 14 mg/kg in Northern bobwhite, and 10 mg/kg in red-winged blackbirds (6). It has a reported 5-day dietary LC_{50} of 284 ppm in Japanese quail (14). Despite its high acute toxicity and widespread use, wildlife die-offs due to fonofos use have not been reported (14).

Effects on aquatic organisms

Fonofos is highly toxic to freshwater fish and saltwater organisms. Its acute freshwater 96-hour LC_{50} ranges from 0.028 mg/L (13) to 44 mg/L (72) in bluegill sunfish, and 0.05 mg/L (13) to 44 mg/L (72) in rainbow trout, perhaps reflecting a difference in formulation. Organophosphates, such as fonofos, do not bioaccumulate in the environment or in animals (1).

Effects on other organisms (non-target species)

Fonofos is toxic to bees (13).

Environmental fate

Breakdown in soil and groundwater

Fonfos is of moderate persistence in soils, with a field half-life of about 40 days (19). Field residence times depend on variables such as soil type, organic matter content, rainfall, and sunlight. On a silty clay loam, the half-life was 82 days after a high level of application, while at lower application rates the half-life was 46 days (73).

Fonofos is moderately well-bound to soils, depending on organic matter content (19). It can be transported in runoff. It is immobile in sandy loam and silt loam soils, but is mobile in quartz sand. Soil microbes, such as fungi, rapidly degrade fonofos (73). Fonofos resists soil leaching, and has been detected only rarely in groundwater and then at very low levels. Fonofos has been detected in California groundwater at 0.01 to 0.03 µg/L, and in Iowa groundwater at 0.1 µg/L (73).

Breakdown in water

Although it is practically insoluble in water, it is quickly broken down by hydrolysis (8,13).

Breakdown in vegetation

Fonofos is rapidly metabolized in plant tissues to nontoxic compounds (8). The compound is not readily absorbed by plant foliage and it is not translocated throughout plants (8).

Physical properties

Fonofos is a clear to yellow liquid at room temperature (13).

Chemical name: O-ethyl S-phenyl (RS)-ethylphosphonodithioate (13)
CAS #: 944-22-9
Molecular weight: 246.32 (13)
Solubility in water: 13 mg/L (13)
Solubility in solvents: v.s. in organic solvents like kerosene, xylene, and isobutyl methyl ketone (13)
Melting point: Not available
Vapor pressure: 28 mPa @ 25°C (13)
Partition coefficient (octanol/water): 8000 (13)
Adsorption coefficient: 870 (19)

Exposure guidelines

ADI: Not available
HA: 0.01 mg/L (69)
RfD: 0.002 mg/kg/day (53)
PEL: 0.1 mg/m^3 (8-hour) (skin) (39)

Basic manufacturer

Zeneca Ag Products
1800 Concord Pike
Wilmington, DE 19897
Telephone: 800-759-4500
Emergency: 800-759-2500

5.2.13 Isofenphos

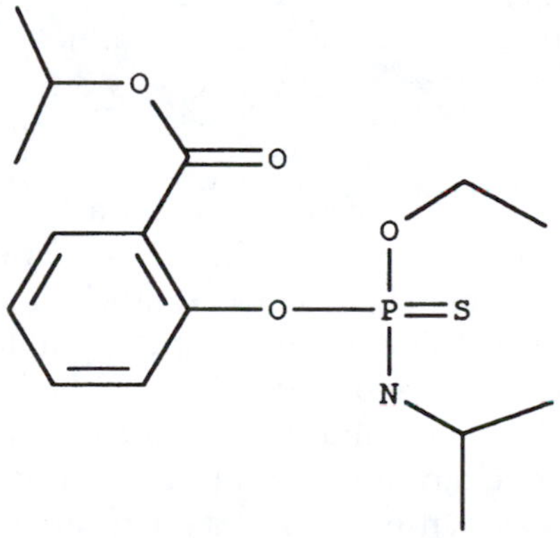

Figure 5.14 Isofenphos.

Trade or other names

Trade names for products containing isofenphos include Amaze, Oftanol, and Pryfon.

Regulatory status

Isofenphos is a highly toxic insecticide in EPA toxicity class I. Products containing it must bear the Signal Words DANGER — POISON on their labels. Isofenphos is a Restricted Use Pesticide (RUP), which may be purchased and applied only by certified applicators.

Introduction

Isofenphos is an organophosphate insecticide used to control soil-dwelling insects such as white grubs, cabbage root flies, corn roundworms, and wireworms. The product is used on vegetables, including maize and carrots, on soil insects in fruit crops like bananas, and in soils with turfgrass. It is a selective contact and stomach poison in insects. It is applied as a preplant or pre-emergence soil treatment. It is also used to control termites in and around structures. The insecticide may also be found in formulations with the fungicide thiram. It is available in emulsifiable concentrate, dry seed treatment, granular, and wettable powder formulations.

Toxicological effects

Acute toxicity

Isofenphos is highly toxic via the oral route, with a reported acute oral LD_{50} of 28 to 38 mg/kg in rats, and 91.3 to 127 mg/kg in mice (8,13). It is moderately toxic via the dermal route, with a reported dermal LD_{50} of 188 mg/kg in rats (13,41). The acute dermal LD_{50} in rabbits ranges from 162 to 315 mg/kg (8,13). The compound causes no damage to the skin or mucous membranes in rats. It is highly toxic via the inhalation route, with reported airborne LC_{50} values of 1.3 and 0.144 mg/L in rats (8,13). The hamster 4-hour inhalation LC_{50} is 0.23 mg/L (8).

Typical of other organophosphate insecticides, this compound is a cholinesterase inhibitor. Effects that may occur from exposure include increased secretions, difficulty in breathing, diarrhea, urination, pupil contraction, and slowness of the heart (74). At very high doses, convulsions and coma may ensue. Isofenphos may be more toxic when it is combined with the insecticide malathion (74).

Chronic toxicity

The primary chronic effect of isofenphos in animals and in humans is suppression of cholinesterase activity. Doses in the diets of rats and mice below 0.05 mg/kg/day had no effect on bloodstream (plasma) cholinesterase activity. Rats and dogs fed 1 mg/kg/day isofenphos for 3 months exhibited no compound-related effects. Rats fed low doses of isofenphos for 2 years showed no effects on cholinesterase activity related to the compound below the 1-mg/kg/day dose (8). Repeated doses at 2 mg/kg/day for 90 days did not affect the nerves of the hens. Other tests on rats, however, showed evidence of nerve damage after repeated low-level exposures to some organophosphates (75).

Reproductive effects

Female rats fed isofenphos at doses of 0.05 to 5.0 mg/kg/day through three successive litters exhibited no adverse effects on reproduction at the lowest dose, but maternal body weight reductions occurred at doses above 0.05 mg/kg/day. At 0.5 mg/kg/day, isofenphos produced decreases in reproductive success in some studies, but not others (8). However, some evidence of abnormalities associated with toxicity to the developing embryos were reported at 0.15 mg/kg/day (8). The available data on reproductive effects due to isofenphos are inconclusive, but suggest that such effects are unlikely in humans at expected exposure levels.

Teratogenic effects

Pregnant rats fed low to moderate amounts (0.3 to 3 mg/kg/day) of isofenphos during gestation had no compound-induced malformations in their offspring (8). Pregnant rabbits fed isofenphos over a range of doses from 1 to 5 mg/kg/day during gestation had offspring with no teratogenic effects (8). These data indicate that isofenphos is not teratogenic.

Mutagenic effects

A number of tests on the mutagenic potential of isofenphos were all negative (8,74). Additional tests using concentrations as high as 3.15 ml per plate confirmed the nonmutagenicity of isofenphos (8,74).

Carcinogenic effects

Mice fed isofenphos at doses up to 5 mg/kg/day for 2 years in their food had no dose-related increases in tumors. There was, however, high mortality in all of the groups in the study, including the controls (74). Rats fed low to high doses of isofenphos (0.05 to 5 mg/kg/day) for 2 years had no treatment-related alterations in thirty different tissues examined (8,74). The evidence suggests that isofenphos is not carcinogenic.

Organ toxicity

Target organs affected by isofenphos exposure include the central and peripheral nervous systems and the blood.

Fate in humans and animals

Rats, pigs, and cows all eliminate isofenphos rapidly. Rats excreted nearly all of the compound in urine and the remaining small amount in feces within 3 days of the initial dosing. Pigs eliminated 80% in urine and most of the remaining amount in feces within a day. Cows had a very similar pattern of excretion over a 2-day interval. Less than 1% of the initial dose was detected in the milk (8).

When domestic hens were fed moderate doses of isofenphos (4 mg/kg/day) for 3 days, a significant amount of the compound was eliminated within 2 days. Residues of the compound in the tissues and eggs were less than 3% of the administered dose. The major residues found in excreta, tissues, and eggs were isofenphos and isopropyl salicylate (8).

Though the compound can be temporarily stored in tissues, the concentrations drop after exposure to the compound ceases. Rats fed isofenphos at high doses (15 mg/kg/day) for 6 days had small amounts of the compound in muscle, liver tissue, fat, and kidneys. Within 5 days after dosing, the concentration in these tissues had fallen to very low levels (15). Cows had a peak blood plasma level 2 hours after the dose. Levels fell sharply within a day.

Ecological effects

Effects on birds

Isofenphos is highly toxic to birds. The reported 5-day dietary LC_{50} in Japanese quail is 299 ppm (6,14). In Northern bobwhite quail, the reported acute LD_{50} is 13 to 19 mg/kg (6). For hens, the LD_{50} ranges between 6 and 20 mg/kg. Red-winged blackbirds are also susceptible to isofenphos (66). Laboratory studies on blackbirds, and the death of over 100 birds linked to the application of the product Oftanol, support its toxicity to this species.

Effects on aquatic organisms

Isofenphos is moderately toxic to fish. The LC_{50} values of the compound for various species of fish are relatively consistent. The reported 96-hour LD_{50} values are 2 mg/L in goldfish, 2 to 4 mg/L in carp, 1 to 2 mg/L in orfe, and 1 mg/L in rudd (13). Studies indicate possible adverse effects on aquatic protozoa at concentrations exceeding 20 mg/L (76).

The potential of the compound to significantly bioaccumulate is relatively low. When channel catfish were exposed to 0.01 mg/L for a month, residues accumulated to 75 times the water concentration within the first 7 days and only slowly declined during the treatment period. After the end of treatment, fish tissue residues decreased by 87% within the first day and by 96% within 10 days. All of the extractable tissue residue was isofenphos (76).

Effects on other organisms (non-target species)

No data are currently available.

Environmental fate

Breakdown in soil and groundwater

Isofenphos is moderately to highly persistent in the soil environment. At a normal field-use rate, a single application of isofenphos had a half-life of 30 to 300 days, with a typical time of 150 days (19). Isofenphos is poorly bound to soils, but has a low water solubility (19). Isofenphos slowly leaches in all soils and the breakdown products leach more rapidly.

The first step in the breakdown of the compound requires the presence of oxygen (oxidation). The products of this process are isopropyl salicylate and cyclic isofenphos. Degradation was not greatly influenced by the addition of organic matter to the soil (77). The evaporation of breakdown products accounts for a substantial portion of the loss of these residues from the soil.

Breakdown in water

The maximum concentrations expected as the result of runoff following a field application of the compound would be very low in water (0.07 mg/L) and in sediment (0.04 mg/L) (76). The rate of its chemical breakdown in the presence of water (hydrolysis) is markedly increased by acidic or basic conditions and by increases in temperature. Its breakdown in the presence of sunlight is very slow.

Breakdown in vegetation

Plants, such as corn and onions, absorb isofenphos from the soil and, in the case of corn, translocate it to plant leaves and stems. Plants generally transform the compound to its oxygen analog. The oxygen analog is then moved within the plant to the upper plant parts, where further changes occur. Cereal grains, leafy and root vegetables, and edible oil crops grown in soil treated 9 months earlier can take up the residues of the insecticide even when it is present at very low concentrations (0.005 mg/kg). The principal residues found in plants are isofenphos and an isofenphos-oxygen analog (8).

Physical properties

Isofenphos is a colorless oil at room temperature (13).

Chemical name: O-ethyl O-2-isopropoxycarbonylphenyl isopropylphosphoramidothioate (13)
CAS #: 25311-71-1
Molecular weight: 345.40 (13)
Solubility in water: 24 mg/L @ 20°C (13)
Solubility in other solvents: s. in cyclohexone, toluene, acetone, and diethyl ether (13)
Melting point: Less than −12°C (13)
Vapor pressure: 0.53 mPa @ 20°C (13)
Partition coefficient (octanol/water): not available
Adsorption coefficient: 600 (19)

Exposure guidelines

ADI: 0.0001 mg/kg/day (38)
HA: Not available
RfD: Not available
PEL: Not available

Basic manufacturer

Miles Inc.
8400 Hawthorn Rd.
P.O. Box 4913
Kansas City, MO 64120
Telephone: 816-242-2429
Emergency: 816-242-2582

5.2.14 Malathion

Trade or other names

Malathion is also known as carbophos, maldison, and mercaptothion. Trade names for products containing malathion include Celthion, Cythion, Dielathion, El 4049, Emmaton, Exathios, Fyfanon, Hilthion, Karbofos, and Maltox.

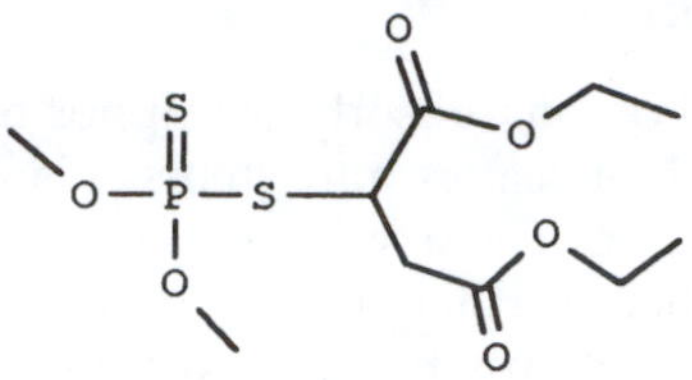

Figure 5.15 Malathion.

Regulatory status

Malathion is a slightly toxic compound in EPA toxicity class III. Labels for products containing it must carry the Signal Word CAUTION. Malathion is a General Use Pesticide (GUP). It is available in emulsifiable concentrate, wettable powder, dustable powder, and ULV liquid formulations.

Introduction

Malathion is a nonsystemic, wide-spectrum organophosphate insecticide. It was one of the earliest organophosphate insecticides developed (introduced in 1950). Malathion is suited for the control of sucking and chewing insects on fruits and vegetables, and is also used to control mosquitoes, flies, household insects, animal parasites (ectoparasites), and head and body lice. Malathion may also be found in formulations with many other pesticides.

Toxicological effects

Acute toxicity

Malathion is slightly toxic via the oral route, with reported oral LD_{50} values of 1000 mg/kg to greater than 10,000 mg/kg in the rat, and 400 mg/kg to greater than 4000 mg/kg in the mouse (2,13). It is also slightly toxic via the dermal route, with reported dermal LD_{50} values of greater than 4000 mg/kg in rats (2,13). Effects of malathion are similar to those observed with other organophosphates, except that larger doses are required to produce them (2,8). It has been reported that single doses of malathion may affect immune system response (2).

Symptoms of acute exposure to organophosphate or cholinesterase-inhibiting compounds may include the following: numbness, tingling sensations, incoordination, headache, dizziness, tremor, nausea, abdominal cramps, sweating, blurred vision, difficulty in breathing or respiratory depression, and slow heartbeat. Very high doses may result in unconsciousness, incontinence, and convulsions or fatality.

The acute effects of malathion depend on product purity and the route of exposure (33). Other factors that may influence the observed toxicity of malathion include the amount of protein in the diet and gender. As protein intake decreased, malathion was increasingly toxic to the rats (78). Malathion has been shown to have different toxicities in male and female rats and humans due to metabolism, storage, and excretion differences between the sexes, with females being much more susceptible than males (79).

Numerous malathion poisoning incidents have occurred among pesticide workers and small children through accidental exposure. In one reported case of malathion poisoning, an infant exhibited severe signs of cholinesterase inhibition after exposure to an aerosol bomb containing 0.5% malathion (44).

Chronic toxicity

Human volunteers fed very low doses of malathion for $1\frac{1}{2}$ months showed no significant effects on blood cholinesterase activity. Rats fed dietary doses of 5 to 25 mg/kg/day over 2 years showed no symptoms apart from depressed cholinesterase activity. When small amounts of the compound were administered for 8 weeks, rats showed no adverse effects on whole-blood cholinesterase activity (2). Weanling male rats were twice as susceptible to malathion as adults.

Reproductive effects

Several studies have documented developmental and reproductive effects due to high doses of malathion in test animals (2). Rats fed high doses of 240 mg/kg/day during pregnancy showed an increased rate of newborn mortality. However, malathion fed to rats at low dosages caused no reproductive effects (8). It is not likely that malathion will cause reproductive effects in humans under normal circumstances.

Teratogenic effects

Rats fed high doses (240 mg/kg/day) showed no teratogenic effects. Malathion and its metabolites can cross the placenta of the goat and depress cholinesterase activity of the fetus (8). Chickens fed diets at low doses for 2 years showed no adverse effects on egg hatching (8). Current evidence indicates that malathion is not teratogenic.

Mutagenic effects

Malathion produced detectable mutations in three different types of cultured human cells, including white blood cells and lymph cells (2,8). It is not clear what the implications of these results are for humans.

Carcinogenic effects

Female rats on dietary doses of approximately 500 mg/kg/day malathion for 2 years did not develop tumors (2). Adrenal tumors developed in the males at low doses, but not at the high doses (80), suggesting that malathion was not the cause. Three of five studies that investigated the carcinogenicity of malathion found that the compound did not produce tumors in the test animals. The two other studies were determined to be unacceptable studies and the results were discounted (2,8,80). Available evidence suggests that malathion is not carcinogenic, but the data are not conclusive.

Organ toxicity

The pesticide has been shown in animal testing and from use experience to affect the central nervous system, immune system, adrenal glands, liver, and blood.

Fate in humans and animals

Malathion is rapidly and effectively absorbed by practically all routes, including the gastrointestinal tract, skin, mucous membranes, and lungs. Malathion undergoes similar detoxification mechanisms to other organophosphates, but it can also be rendered nontoxic via another simple mechanism, splitting of either of the carboxy ester linkages. Animal studies indicate it is very rapidly eliminated though urine, feces, and expired air, with a reported half-life of approximately 8 hours in rats and approximately 2 days in cows (2).

Autopsy samples from one individual who had ingested large amounts of malathion showed a substantial portion in the stomach and intestines, a small amount in fat tissue, and no detectable levels in the liver. Malathion requires conversion to malaoxon in order to become an active anticholinesterase agent. Most of the occupational evidence indicates a low chronic toxicity for malathion. One important exception to this was traced to impurities in the formulation of the pesticide (2).

Ecological effects

Effects on birds

Malathion is moderately toxic to birds. The reported acute oral LD_{50} values are: in mallards, 1485 mg/kg; in pheasants, 167 mg/kg; in blackbirds and starlings, over 100 mg/kg; and in chickens, 525 mg/kg (2,6). The reported 5- to 8-day dietary LC_{50} is over 3000 ppm in Japanese quail, mallard, and Northern bobwhite, and is 2639 ppm in ring-neck pheasants (6). Furthermore, 90% of the dose to birds was metabolized and excreted in 24 hours via urine (79).

Effects on aquatic organisms

Malathion has a wide range of toxicities in fish, extending from very highly toxic in the walleye (96-hour LC_{50} of 0.06 mg/L) to highly toxic in brown trout (0.1 mg/L) and cutthroat trout (0.28 mg/L), moderately toxic in fathead minnows (8.6 mg/L), and slightly toxic in goldfish (10.7 mg/L) (8,13,16). Various aquatic invertebrates are extremely sensitive, with EC50 values ranging from 1 μg/L to 1 mg/L (28).

Malathion is highly toxic to aquatic invertebrates and to the aquatic stages of amphibians. Because of its very short half-life, malathion is not expected to bioconcentrate in aquatic organisms. However, brown shrimp showed an average concentration of 869 and 959 times the ambient water concentration in two separate samples (12).

Effects on other organisms (non-target species)

The compound is highly toxic to honeybees (13).

Environmental fate

Breakdown in soil and groundwater

Malathion is of low persistence in soil, with reported field half-lives of 1 to 25 days (19). Degradation in soil is rapid and related to the degree of soil binding (12). Breakdown occurs by a combination of biological degradation and nonbiological reaction with water (12). If released to the atmosphere, malathion will break down rapidly in sunlight, with a reported half-life in air of about 1.5 days (12).

It is moderately bound to soils and is soluble in water, so it may pose a risk of groundwater or surface water contamination in situations that may be less conducive to breakdown. The compound was detected in 12 of 3252 different groundwater sources in two different states, and in small concentrations in several wells in California, with a highest concentration of 6.17 μg/L (33).

Breakdown in water

In raw river water, the half-life is less than 1 week, whereas malathion remained stable in distilled water for 3 weeks (12). Applied at 1 to 6 lb/acre in log ponds for mosquito control, it was effective for 2.5 to 6 weeks (12). In sterile seawater, the degradation increases with increasing salinity. The breakdown products in water are mono- and dicarboxylic acids (12).

Breakdown in vegetation

Residues were found mainly associated with areas of high lipid content in the plant. Increased moisture content increased degradation (33).

Physical properties

Technical malathion is a clear, amber liquid at room temperature (13).

Chemical name: diethyl (dimethoxy thiophosphorylthio) succinate (13)
CAS #: 121-75-5
Molecular weight: 330.36 (13)
Solubility in water: 130 mg/L (13)
Solubility in other solvents: v.s in most organic solvents (13)
Melting point: 2.85°C (13)
Vapor pressure: 5.3 mPa @ 30°C (13)
Partition coefficient (octanol/water): 560 (13)
Adsorption coefficient: 1800 (19)

Exposure guidelines

ADI: 0.02 mg/kg/day (38)
HA: 0.2 mg/L (lifetime) (53)
RfD: 0.02 mg/kg/day (53)
PEL: 15 mg/m³ (8-hour) (dust) (39)

Basic manufacturer

Drexel Chemical Company
1700 Channel Avenue
Memphis, TN 38113
Telephone: 901-774-4370

5.2.15 Methidathion

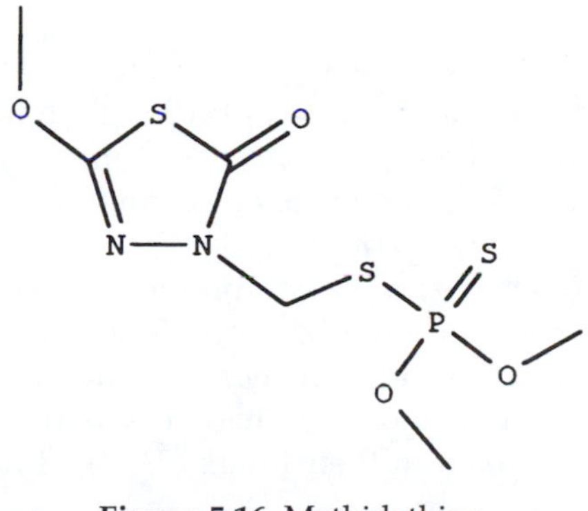

Figure 5.16 Methidathion.

Trade or other names

Trade names for products containing methidathion include Somonic, Somonil, Supracide, Suprathion, and Ultracide. The compound may be found in formulations with many other pesticides.

Regulatory status

Methidathion is a highly toxic compound in EPA toxicity class I. Labels for products containing it must bear the Signal Word DANGER. Methidathion is a Restricted Use Pesticide (RUP), except for use in nurseries, and on safflower and sunflowers.

Introduction

Methidathion is a nonsystemic organophosphorous insecticide and acaricide with stomach and contact action. The compound is used to control a variety of insects and mites in many crops such as fruits, vegetables, tobacco, alfalfa, and sunflowers, and also in greenhouses and on rose cultures. It is especially useful against scale insects. It works by inhibiting certain enzyme actions in the target pests. It is available in emulsifiable concentrate, wettable powder, and ultra-low volume (ULV) liquid formulations.

Toxicological effects

Acute toxicity

Methidathion is highly toxic via the oral route, with reported acute oral LD_{50} values of 25 to 54 mg/kg in the rat, and 18 to 25 mg/kg in the mouse (2,13). Other reported oral LD_{50} values include 25 mg/kg in guinea pigs, 80 mg/kg in rabbits, and 200 mg/kg in dogs (2). It is highly toxic via the dermal route as well, with reported dermal LD_{50} values of 85 to 94 mg/kg in the rat (2). Methidathion is only a mild skin irritant and is nonirritating to the eyes (in rabbits). Via the inhalation route, it may be slightly toxic, with a reported 4-hour inhalation LC_{50} of 3.6 mg/L in rats (13).

Effects due to acute methidathion exposures are similar to those caused by other organophosphate pesticides, and may include nausea, vomiting, cramps, diarrhea, salivation, headache, dizziness, muscle twitching, difficulty in breathing, blurred vision, and tightness in the chest (2). High acute exposure may cause intense breathing problems, including paralysis of the respiratory muscles.

Chronic toxicity

Beagle dogs fed small doses of the compound for 2 years experienced no compound-related effects at or below the dose of 0.10 mg/kg/day (2,8). At doses of 0.4 mg/kg/day and above, the dogs experienced enzymatic changes and liver alterations. Inhibition of red blood cell cholinesterase, an enzyme, was observed only at the highest dose tested (1.6 mg/kg/day) (2).

Rats also have a low tolerance for the compound. Compound-related effects were first noted in the rats at doses of 2 mg/kg and above, and included cholinesterase inhibition in the blood and brain and some nerve-related effects. At the highest dose of 5 mg/kg, the rats ate more food but had less body weight gain. They also developed skin lesions and foam in their lungs (2,8,81). Rhesus monkeys fed small amounts of the compound developed changes in blood cholinesterase activity at doses of 1 mg/kg/day and above.

Humans ingesting very small amounts of the compound at doses of 0.11 mg/kg/day for 6 weeks had no noticeable clinical effects (8). A study of exposure levels of mixer/loaders of methidathion (Supracide applications) in California showed that the greatest exposure potential to the compound was through the skin (dermal) (82).

Reproductive effects

Moderate amounts of methidathion caused a number of adverse reproductive effects. When male and female rats were fed moderate amounts of methidathion over two successive litters, the parents experienced tremors, decreased food consumption, and decreased ovary weights at 1.25 mg/kg/day (2,8). The low dose of 0.25 mg/kg/day disrupted mating behavior and also affected nursing offspring. At 2.5 mg/kg/day (the highest dose tested), stillbirths and decreased pup survival were observed (2). Reproductive effects in humans as a result of methidathion exposure are unlikely under normal circumstances.

Teratogenic effects

Small to moderate amounts of methidathion administered to pregnant rats and rabbits produced no birth defects in the offspring. The pregnant females experienced several compound-related effects, most of which were typical of cholinesterase inhibition (2). The compound is unlikely to pose a developmental risk to humans.

Mutagenic effects

Methidathion did not induce any genetic changes in a number of tests for gene mutation, chromosomal aberrations, and DNA damage. The various gene mutation studies were conducted on hamster bone marrow cells, in mammalian cells, and on several species of bacteria (2). These data indicate that methidathion is not mutagenic.

Carcinogenic effects

Methidathion caused malignant and benign liver tumors (adenomas) in male mice fed 2.5 mg/kg/day for 2 years. Additional tumors (carcinomas) were found in the male mice fed 5 mg/kg/day over the same time period. This higher feeding level also produced numerous other signs of toxicity (2,8). Since these results apply to only one sex in one species, the carcinogenic potential of methidathion is unclear.

Organ toxicity

Target organs in animal studies include the nervous system, liver, gall bladder, and ovaries.

Fate in animals and humans

Methidathion is rapidly absorbed, broken down, and eliminated in animals (8). Following absorption of the compound, the majority is lost as a breakdown product through the lungs (2,6). Between 30 and 50% of the ingested amount is eliminated (as breakdown products) in urine (2,6). Half of the initial amount of the compound is removed from mammals within 6 hours (2,6).

The breakdown products of the parent compound are not of toxicological concern (8). Only very small amounts of various metabolic products of methidathion have been detected in milk from cows (6) and in chicken eggs (2).

Ecological effects

Effects on birds

Methidathion is highly toxic to birds following acute exposure. The reported oral LD_{50} values for the compound are 23 to 33 mg/kg in mallards, 8.41 mg/kg in Canadian geese, 33.2 mg/kg in ring-necked pheasants, and 225 mg/kg in chukar partridges (8,6).

Effects on aquatic organisms

The compound is very highly acutely toxic to aquatic organisms (both vertebrate and invertebrate); the reported LC_{50} values of the compound are 10 to 14 µg/L in rainbow trout, and 2 to 9 µg/L in bluegill sunfish (13,72). Tests on lobsters indicated that the combination of methidathion and another organophosphate insecticide, phosphamidon, was more toxic than either compound separately or than would be expected if the toxicities were added together (13).

Studies with bluegill sunfish indicate that there is only a slight potential that the compound would accumulate in fish tissues (6). Maximum levels of the residues of the pesticide after 1 month of exposure to very low concentrations in the water (0.05 µg/L) were 1.0 µg/kg in the edible tissue, 3.9 µg/kg in nonedible tissue, and 2.4 µg/kg in whole fish. These concentrations indicate a low bioconcentration factor of 46 for whole fish. After 2 weeks in water without methidathion, the concentration in whole fish fell by nearly 80% (2).

Effects on other organisms (non-target species)

Methidathion is slightly toxic to bees (13).

Environmental fate

Breakdown in soil and groundwater

Methidathion is of low persistence in the soil environment; reported field half-lives are 5 to 23 days, with a representative value of about 7 days (19). Breakdown of the compound in soil occurs through the action of soil microorganisms (83). Under alkaline conditions, methidathion is rapidly degraded by chemical action (8).

Methidathion and its breakdown products are poorly bound by soils, and thus may be mobile (19,81). However, they have not been detected in any groundwater sources. This is probably due to the short half-life of the compound and its degradates.

Breakdown in water

No data are currently available.

Breakdown in vegetation

In plants, methidathion is rapidly metabolized (13). Oranges sprayed with Supracide at a rate of nearly 2 pounds per acre had residues of the compound of about 0.1 µg/ml (81). Within 2 days, over 60% of the compound was removed from the outside of the fruit, and within 1 week, less than 1% of the compound remained (81).

Physical properties

Methidathion is a colorless crystalline compound at room temperature (13).

Chemical name: S-2,3-dihydro-5-methoxy-2-oxo-1,3,4-thiadiazol-3-ylmethyl
O,O-dimethylphosphorodithioate (13)
CAS #: 950-37-8
Molecular weight: 302.33 (13)
Melting point: 39–40°C (13)
Solubility in water: 240 mg/L @ 20°C (13)
Solubility in other solvents: v.s. in octanol, ethanol, xylene, acetone, and cyclo-
hexane (13)
Melting point: 39.5°C (13)
Vapor pressure: 186 mPa @ 20°C (13)
Partition coefficient (octanol/water): 53,000 (8)
Adsorption coefficient: 400 (estimated) (19)

Exposure guidelines

ADI: 0.001 mg/kg/day (38)
HA: Not available
RfD: 0.001 mg/kg/day (53)
PEL: Not available

Basic manufacturer

Ciba-Giegy Corporation
P.O. Box 18300
Greensboro, NC 27419-8300
Telephone: 800-334-9481
Emergency: 800-888-8372

5.2.16 Methyl parathion

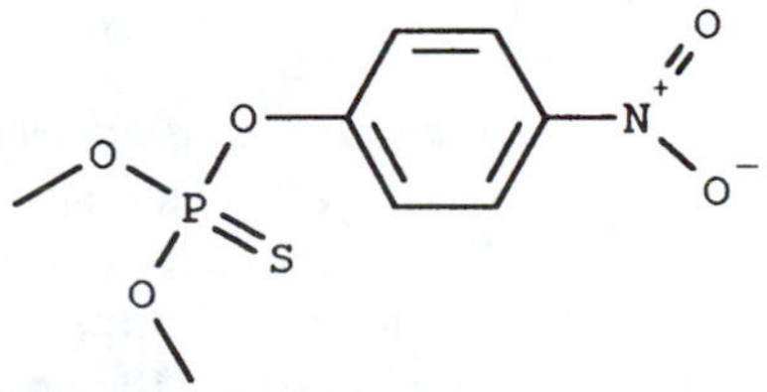

Figure 5.17 Methyl parathion.

Trade or other names

Alternate common names are parathion-methyl and metafos. Trade names include Bladan M, Cekumethion, Dalf, Dimethyl Parathion, Devithion, E 601, Folidol-M, Fosferno M50, Gearphos, Kilex Parathion, Metacide, Metaphos, Metron, Nitrox 80, Partron M, Penncap-M, and Tekwaisa.

Regulatory status

Methyl parathion is a highly toxic insecticide in EPA toxicity class I. Some or all formulations of methyl parathion may be classified as Restricted Use Pesticides (RUPs). RUPs may be purchased and used only by certified applicators. Labels for products containing methyl parathion must bear the Signal Word DANGER.

Introduction

Methyl parathion is an organophosphate insecticide and acaricide used to control boll weevils and many biting or sucking insect pests of agricultural crops, primarily on cotton. It kills insects by contact, stomach, and respiratory action. Methyl parathion is available in dust, emulsifiable concentrate, ULV liquid, and wettable powder formulations.

Toxicological effects

Acute toxicity

Methyl parathion is highly toxic via the oral route, with reported oral LD_{50} values of 6 to 50 mg/kg in rats, 14.5 to 19.5 mg/kg in mice, 420 mg/kg in rabbits, 1270 mg/kg in guinea pigs and 90 mg/kg in dogs (2,13). It is highly toxic via the dermal route as well, with reported dermal LD_{50} values of 67 mg/kg in rats, 1200 mg/kg in mice, and 300 mg/kg in rabbits (2,13). The 1-hour inhalation LC_{50} for methyl parathion in rats is 0.24 mg/L (2).

Effects associated with acute exposure to methyl parathion are similar to those associated with exposure to other organophosphate pesticides (8). Symptoms of acute exposure to organophosphate or cholinesterase-inhibiting compounds may include the following: numbness, tingling sensations, incoordination, headache, dizziness, tremor, nausea, abdominal cramps, sweating, blurred vision, difficulty in breathing or respiratory depression, and slow heartbeat. Very high doses may result in unconsciousness, incontinence, and convulsions or fatality. Persons with respiratory ailments, recent exposure to cholinesterase inhibitors, cholinesterase impairment, or liver malfunction are at increased risk from exposure to methyl parathion.

Chronic toxicity

Studies with human volunteers have found that doses of 1 to 22 mg/person/day have no effect on cholinesterase activity. In a 4-week study of volunteers given 22, 24, 26, 28, or 30 mg/person/day, mild cholinesterase inhibition appeared in some individuals in the 24-, 26-, and 28-mg dosage groups. In the 30-mg/person/day (about 0.43 mg/kg/day) group, red blood cholinesterase activity was depressed by 37%. When methyl parathion was fed to dogs for 12 weeks, a dietary level of 1.25 mg/kg caused a significant depression of red blood cell and plasma cholinesterase. A dietary level of 0.125 mg/kg produced no effects (2).

Reproductive effects

In a three-generation study with rats fed dietary levels of 0.5 or 1.5 mg/kg/day, reduced weanling survival and birthweight occurred at both doses, as well as an increase in the number of stillbirths at 1.5 mg/kg/day (2,8). Single injections of 15 mg/kg in rats on day 12 of pregnancy, and single injections of 60 mg/kg on day 10 of pregnancy, in mice caused suppression of fetal growth and bone formation in the offspring that survived. In another study, there were no adverse effects observed in the offspring of rats given oral doses of 4 or 6 mg/kg on day 9 or 15 of pregnancy (2). Reproductive effects in humans are not likely under normal circumstances.

Teratogenic effects

In a three-generation study with rats fed dietary levels of 0.5 or 1.5 mg/kg/day, there were no compound-related teratogenic effects (2). Single injections of 5 to 10 mg/kg in rats on day 12 of pregnancy, and single injections of 20 mg/kg on day 10 of pregnancy, in mice caused no statistically significant changes in the offspring (2,8). Oral administration of 4 to 6 mg/kg on day 9 or 15 of pregnancy in rats resulted in no fetal anomalies. Available evidence indicates that methyl parathion does not cause teratogenic effects.

Mutagenic effects

No signs of mutagenicity were seen in mice given dosages of 5 to 100 mg/kg, nor in mice fed methyl parathion for 7 weeks (2). Available evidence suggests that methyl parathion is nonmutagenic.

Carcinogenic effects

Available evidence suggests that methyl parathion is not carcinogenic (84).

Organ toxicity

Methyl parathion primarily affects the nervous system.

Fate in humans and animals

Methyl parathion is rapidly absorbed into the bloodstream through all normal routes of exposure. Following administration of a single oral dose, the highest concentration of methyl parathion in body tissues occurred within 1 to 2 hours (2). Methyl parathion does not accumulate in the body, and is almost completely excreted by the kidneys (urine) within 24 hours as phenolic metabolites (2,13).

Ecological effects

Effects on birds

Methyl parathion is very highly to highly toxic to birds. Reported acute oral LD_{50} values are 3 mg/kg in American kestrels, 7.5 mg/kg in European starlings, 6 to 10 mg/kg in mallards, 8 mg/kg in Northern bobwhites, 10 to 24 mg/kg in red-wing blackbirds, and 8 mg/kg in ring-neck pheasants (6). The 5- to 8-day dietary LC_{50} values reported for methyl parathion include 69 ppm in Japanese quail, 330 to 680 ppm in mallard, 90 ppm in Northern bobwhite, and 91 ppm in ring-neck pheasant (6).

Effects on aquatic organisms

Methyl parathion is moderately toxic to fish and to animals that eat fish (13,8). Reported 96-hour LC_{50} values are from 1.9 to 8.9 mg/L in the following fish species: coho salmon, cutthroat trout, rainbow trout, brown trout, lake trout, goldfish, carp, fathead minnow, black bullhead, channel catfish, green sunfish, bluegill, largemouth bass, and yellow perch. Reported 96-hour LC_{50} values indicate very high toxicity for aquatic invertebrates such as *Daphnia* spp., scuds, and sideswimmers (8).

Effects on other organisms (non-target species)

Methyl parathion is toxic to bees (13).

Environmental fate

Breakdown in soil and groundwater

Methyl parathion is of low persistence in the soil environment, with reported field half-lives of 1 to 30 days (19). A representative value is estimated to be 5 days (19). The rate of degradation increases with temperature and with exposure to sunlight. Methyl parathion is moderately adsorbed by most soils, and is slightly soluble in water (19). Due to its low residence time and soil binding affinity, it is not expected to be significantly mobile. 4-Nitrophenol, a breakdown product of methyl parathion, does not adsorb well to soil particles and may contaminate groundwater. When large concentrations of methyl parathion reach the soil, as in an accidental spill, degradation will occur only after many years, with photolysis being the dominant route (12). Some volatilization of applied methyl parathion may occur.

Breakdown in water

Methyl parathion degrades rapidly in seawater and lake and river waters, with 100% degradation occurring within 2 weeks to 1 month or more. Degradation is faster in the presence of sediments, and is faster in fresh water than in salt water. Complete breakdown occurs at a rate of 5 to 11% in 4 days in rivers, and more slowly in marine waters. In water, methyl parathion is subject to photolysis, with a half-life of 8 days during the summer and 38 days in winter (12).

Breakdown in vegetation

Uptake and metabolism of methyl parathion in plants is fairly rapid. Within 4 days after applying methyl parathion to the leaves of corn, it was almost completely metabolized (12).

Physical properties

Pure methyl parathion is a colorless crystalline solid. The technical product is light to dark tan, with about 80% purity (13).

Chemical name: O,O-dimethyl O-4-nitrophenyl phosphorothioate (13)
CAS #: 298-00-0
Molecular weight: 263.21 (13)
Water solubility: 55–60 mg/L @ 25°C (13)
Solubility in other solvents: s. in dichloromethane, 2-propanol, toluene, and most
    organic solvents; i.s. in *n*-hexane (13)
Melting point: 35–36°C (13)
Vapor pressure: 1.3 mPa @ 20°C (13)

Partition coefficient (octanol/water): 3300–6900 (8)
Adsorption coefficient: 5100 (19)

Exposure guidelines

ADI: 0.02 mg/kg/day (38)
HA: 0.002 mg/L (8)
RfD: 0.00025 mg/kg/day (53)
PEL: 0.2 mg/m^3 (8-hour) (skin) (39)

Basic manufacturer

Drexel Chemical Company
1700 Channel Avenue
Memphis, TN 38113
Telephone: 901-774-4370

5.2.17 *Mevinphos*

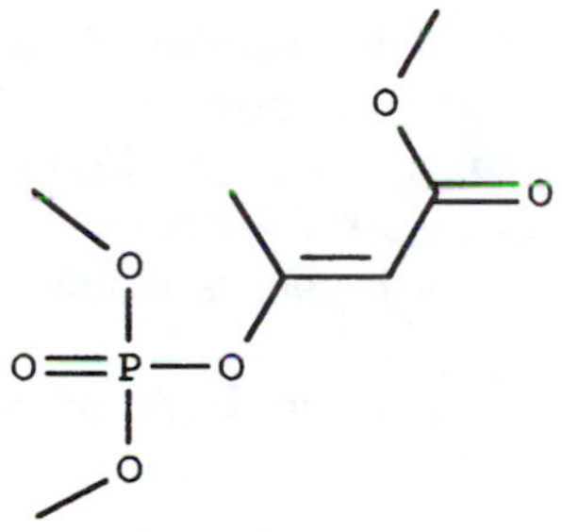

Figure 5.18 Mevinphos.

Trade or other names

Trade names for mevinphos include Apavinphos, CMDP, ENT 22374, Fosdrin, Gesfid, Meniphos, Menite, Mevinox, Mevinphos, OS-2046, PD5, Phosdrin, and Phosfene.

Regulatory status

Mevinphos is a highly toxic compound in EPA toxicity class I. All emulsifiable and liquid concentrates of mevinphos are classified as Restricted Use Pesticides (RUPs) by the U.S. Environmental Protection Agency (EPA), due to their acute oral and dermal toxicity and residue effects on mammalian, aquatic, and bird species. RUPs may be purchased and used only by certified applicators. Products containing mevinphos must bear the Signal Words DANGER — POISON.

Introduction

Mevinphos is an organophosphate insecticide used to control a broad spectrum of insects, including aphids, grasshoppers, leafhoppers, cutworms, caterpillars, and many other insects on a wide range of field, forage, vegetable, and fruit crops. It is also an acaricide that kills or controls mites and ticks. It acts quickly both as a contact insecticide, acting through direct contact with target pests, and as a systemic insecticide that becomes absorbed by plants on which insects feed. It is available in concentrate and liquid formulations.

Toxicological effects

Acute toxicity

Mevinphos is highly toxic via the oral route, with reported oral LD_{50} values of 3 to 12 mg/kg in rats, and 4 to 18 mg/kg in mice (2,13). Via the dermal route, it is highly toxic as well, with reported dermal LD_{50} values for mevinphos of 4.2 mg/kg in rats, and 40 mg/kg in mice (2,13). The 1-hour LC_{50} for mevinphos in rats is 0.125 mg/L, indicating high toxicity by inhalation as well (2). Acute pulmonary edema (or the filling up of lungs with fluid) and changes in the structure or function of salivary glands were seen in rats exposed to this concentration for 1 hour. Effects of acute exposure to mevinphos are similar to those due to exposure to other organophosphates, except that these may occur at lower doses than with other organophosphates.

Symptoms of acute exposure to organophosphate or cholinesterase-inhibiting compounds may include the following: numbness, tingling sensations, incoordination, headache, dizziness, tremor, nausea, abdominal cramps, sweating, blurred vision, difficulty in breathing or respiratory depression, and slow heartbeat. Very high doses may result in unconsciousness, incontinence, and convulsions or fatality.

In humans, symptoms of poisoning have appeared within as little as 15 minutes to 2 hours after exposure to mevinphos, but onset of symptoms has been delayed for as long as 2 days (2). Several children were made ill by unknowingly wearing clothing that had been contaminated with mevinphos. Cholinesterase inhibition resulting from acute exposure is reversible, but it can persist for 2 to 6 weeks.

Chronic toxicity

The lowest oral dose of mevinphos that produced toxic effects (peripheral nervous system effects) in humans was 690 µg/kg when it was given intermittently over 28 days (2,8). Repeated low-level exposure to mevinphos may cause similar effects to those observed with acute exposures (2,8). Severe symptoms of cholinesterase inhibition may be produced in a previously exposed person, whereas symptoms of cholinesterase inhibition may not occur in someone who has not been previously exposed to mevinphos (2,8).

In a 2-year feeding study with rats, no observable effects occurred at 0.025 mg/kg/day. A similar result was found in a 2-year feeding study with dogs (8). No effects on general health and no poisoning symptoms were seen in a 2-year study with rats fed 4 mg/kg/day, nor in dogs given dietary doses of 5 mg/kg/day (8). However, rats were killed by doses of 20 mg/kg/day mevinphos in their diets for 13 weeks. Dietary doses of 10 mg/kg/day for 14 weeks were lethal for dogs (8). Rats given dietary doses of 10 or 20 mg/kg/day for 13 weeks exhibited degeneration of the liver, kidneys, and cells lining the salivary, tear, and other glands (13).

Reproductive effects

In a three-generation reproduction study with rats, there were no reproductive effects at 1.2 mg/kg/day (85), indicating that mevinphos is unlikely to cause reproductive toxicity.

Teratogenic effects

No teratogenic effects were seen in rabbits given doses of 1.0 mg/kg/day (85), suggesting that this pesticide is not teratogenic.

Mutagenic effects

No data are currently available.

Carcinogenic effects

A 2-year study with rats fed dietary doses up to 0.75 mg/kg/day mevinphos produced no evidence of tumor formation (85). This evidence suggests that mevinphos is not carcinogenic.

Organ toxicity

Target organs affected by mevinphos include the nervous system, liver, kidneys, and cells lining the salivary, tear, and other glands.

Fate in humans and animals

Mevinphos is rapidly degraded by the liver. Samples taken by autopsy from a man who had died within 45 minutes of drinking mevinphos showed the following bodily concentrations of the compound: 3400 ppm in the stomach wall, 360 ppm in blood, 240 ppm in the liver, 86 ppm in skeletal muscle, 20 ppm in the kidneys, 8 ppm in urine, and 3 ppm in the brain (2,8).

In a case of moderate poisoning associated with occupational exposure to mevinphos, the concentration of metabolites of mevinphos in the urine collected during the first 12 hours after onset of symptoms was 0.4 ppm. The concentration declined rapidly until 36 hours after onset, and excretion was nearly complete within 50 hours (2).

Ecological effects

Effects on birds

Mevinphos is highly toxic to birds. The acute symptoms of mevinphos poisoning in birds include lack of muscle coordination, curled toes, salivation, diarrhea, trembling, and wing beating. Pheasant deaths all occurred 8 to 18 minutes after treatment; deaths of mallards and grouse occurred 5 to 40 minutes post-treatment (26). The oral LD_{50} for mevinphos is 7.52 mg/kg in chickens, 3 mg/kg in wild bird species, 4.63 mg/kg in female mallards, 1.37 mg/kg in male pheasants, and 0.75 to 1.50 mg/kg in male grouse. Its dermal LD_{50} in ducks is 11 mg/kg (41).

Effects on aquatic organisms

Mevinphos is very highly toxic to fish. The 96-hour LC_{50} for technical mevinphos is 0.012 mg/L in rainbow trout, and 0.022 mg/L in bluegill sunfish (13). The 48-hour LC_{50} for mevinphos is 0.017 mg/L in rainbow trout; the 24-hour LC_{50} is 0.034 mg/L in rainbow trout, 0.041 mg/L in bluegills, and 0.8 mg/L in mosquito fish (27).

Effects on other organisms (non-target species)

Mevinphos is highly toxic to bees, especially when they are exposed to direct treatment or residues on crops (30).

Environmental fate

Breakdown in soil and groundwater

Mevinphos is of low persistence in the soil environment, with reported half-lives of 2 to 3 days (19). One study indicated that this material lost its insecticidal capability in 2 to 4 weeks (85). It is poorly adsorbed to soil particles, and thus may be mobile (19). Its capacity to contaminate groundwater may be limited by its short half-life. No harmful effects to soil microorganisms have been observed from applications of mevinphos formulations (86).

Breakdown in water

Mevinphos dissolves and is readily broken down by water (hydrolyzed), losing its insecticidal activity within 2 to 4 weeks (31). In aqueous solution, mevinphos is hydrolyzed with half-lives of 1.4 hours at pH 11, 3 days at pH 9, 35 days at pH 7, and 120 days at pH 6 (8).

Breakdown in vegetation

When mevinphos is used as directed, it is not phytotoxic (toxic to plants). Plants rapidly degrade it to less-toxic products (13). However, some crops may be sensitive to solvents in which the active ingredient is formulated, as well as to excessive dosages (86).

Physical properties

Pure mevinphos is a colorless liquid; technical grade mevinphos is a pale yellow liquid that has a very mild odor or is odorless (13).

Chemical name: 2-methoxycarbonyl-1-methylvinyl dimethyl phosphate (13)
CAS #: 7786-34-7
Molecular weight: 224.15 (13)
Water solubility: >10 g/L (13); completely miscible with H_2O
Solubility in other solvents: Technical mevinphos is v.s. in alcohols, ketones, chlorinated hydrocarbons, aromatic hydrocarbons and most organic solvents; s.s. in aliphatic hydrocarbons and petroleum ether (13)
Melting point: 21°C (E-isomer); 6.9°C (Z-isomer) (13)
Vapor pressure: 17 mPa @ 20°C (13)
Partition coefficient (octanol/water): not available
Adsorption coefficient: 44 (19)

Exposure guidelines

ADI: 0.0015 mg/kg/day (38)
HA: Not available
RfD: Not available
TLV: 0.092 mg/m³ (8-hour) (47)

Basic manufacturer

Amvac Chemical Corp.
4100 E. Washington Blvd.
Los Angeles, CA 90023
Telephone: 213-264-3910
Emergency: 800-228-5635, ext. 169

5.2.18 Naled

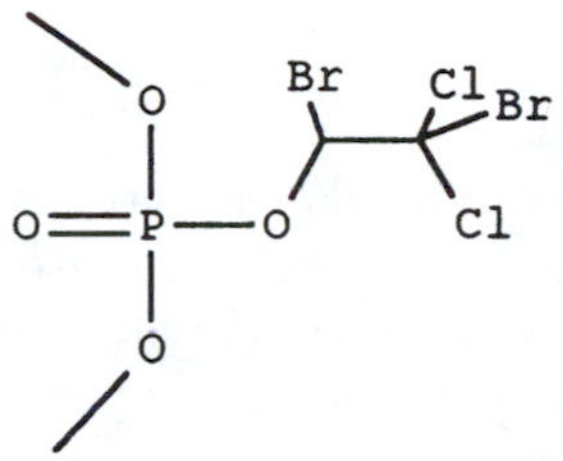

Figure 5.19 Naled.

Trade or other names

Trade names for naled include Bromex, Dibrom, Fly Killer-D, Lucanal, and RE 4355.

Regulatory status

Naled is a moderately toxic compound in EPA toxicity class I. Products containing naled must bear the Signal Words DANGER — POISION because it is corrosive to the eyes. Naled is a General Use Pesticide (GUP).

Introduction

Naled is a fast-acting, nonsystemic contact and stomach organophosphate insecticide used to control aphids, mites, mosquitoes, and flies on crops and in greenhouses, mushroom houses, animal and poultry houses, kennels, food-processing plants, and aquaria, and in outdoor mosquito control. Liquid formulations can be applied to greenhouse heating pipes to kill insects by vapor action. It has been used by veterinarians to kill parasitic worms (other than tapeworms) in dogs. Naled is available in dust, emulsion concentrate, liquid, and ULV formulations. *Unless otherwise specified this profile refers to the technical product of naled.*

Toxicological effects

Acute toxicity

Naled is highly to moderately toxic via the oral route, with reported oral LD_{50} values of 91 to 430 mg/kg in rats, and 330 to 375 mg/kg in mice (2,13). It is moderately toxic through skin exposure; reported dermal LD_{50} values are 1100 mg/kg in rabbits and 800 mg/kg in rats (2,13). Naled may cause dermatitis (skin rashes) and skin sensitization (allergies) (2,8), and may be corrosive to the skin and eyes. Mice exposed to 1.5 mg/L in air for 6 hours showed no adverse effects (13).

Naled is used to combat parasitic infestations (such as worms) in dogs at recommended doses of 16.7 mg/kg (2).

Effects due to naled exposure will be similar to those caused by other organophosphate pesticides, including inhibition of cholinesterase and neurological and neuromuscular effects (2). Symptoms of acute exposure to organophosphate or cholinesterase-inhibiting compounds may include the following: numbness, tingling sensations, incoordination, headache, dizziness, tremor, nausea, abdominal cramps, sweating, blurred vision, difficulty in breathing or respiratory depression, and slow heartbeat. Very high doses may result in unconsciousness, incontinence, and convulsions or fatality.

Chronic toxicity

Chronic exposure to organophosphates may also cause the neurological and neuromuscular effects associated with cholinesterase inhibition (2). Rats have tolerated a dosage of 28 mg/kg/day for 9 weeks with no visible signs of poisoning and with only moderate inhibition of cholinesterase (2).

Reproductive effects
No data are currently available.

Teratogenic effects
No data are currently available.

Mutagenic effects
Naled did not affect the ability of one bacterial species (*Proteus mirabilis*) to repair DNA damage, but did increase the frequency of mutations in another bacterial species (*Salmonella typhimurium*) (8). These data are insufficient to determine its potential for mutagenicity.

Carcinogenic effects
No data are currently available.

Organ toxicity
Naled primarily affects the nervous system through cholinesterase inhibition.

Fate in humans and animals
Naled is readily absorbed into the bloodstream through the skin and lung and intestinal tissue. Rat studies suggest that accumulation may occur in bone (8).

Ecological effects

Effects on birds

Naled is highly to moderately toxic to birds. The reported acute oral LD_{50} for naled is 52 mg/kg in mallard ducks, 65 mg/kg in sharp-tailed grouse, 36 to 50 mg/kg in Canadian geese, 120 mg/kg in ring-neck pheasants, and 59 mg/kg in chickens (6,13). Reported 5- to 8-day dietary LC_{50} values indicate slight toxicity in species studied. These were 1328 ppm in Japanese quail, 2724 ppm in mallard duck, 2117 ppm in Northern bobwhite, and 2538 ppm in ring-neck pheasant (6,13,14).

Effects on aquatic organisms

Naled is highly to moderately toxic to fish (16). Reported 96-hour LC_{50} values range from 0.127 mg/L in cutthroat trout, 0.195 mg/L in rainbow trout, and 0.087 mg/L in lake trout, to higher values of 3.3 mg/L in fathead minnow, 2.2 mg/L in bluegill sunfish, and 1.9 mg/L in largemouth bass (16). The reported LC_{50} for goldfish is 2 to 4 mg/L (13). Naled may be very highly toxic to aquatic invertebrate species, with reported 96-hour LC_{50} values of 0.4 µg/L in *Daphnia*, 8 µg/L in stoneflies, and 18 µg/L in scuds and sideswimmers (16).

Effects on other organisms (non-target species)

Naled is toxic to bees (13). The reported acute oral LD_{50} in mule deer is 200 mg/kg (6).

Environmental fate

Breakdown in soil and groundwater

Naled is practically nonpersistent in the environment, with reported field half-lives of less than 1 day (19). It rapidly degrades in the presence of sunlight to dichlorvos (2,13). For more information on the environmental fate of dichlorvos, see the pesticide profile for dichlorvos. Naled is not strongly bound to soils, and is not highly soluble in water (19). It is rapidly broken down if wet, and it is moderately volatile (8). Soil microorganisms break down most of the naled in the soil. It therefore should not present a hazard to groundwater.

Breakdown in water

Naled is rapidly broken down in water, with a reported half-life of about 2 days (8). Naled is moderately volatile.

Breakdown in vegetation

Plants reductively eliminate bromine from naled to form dichlorvos (DDVP), which may evaporate or be further metabolized (13).

Physical properties

Technical naled is a colorless liquid with a slightly pungent odor (13).

Chemical name: 1,2-dibromo-2,2-dichloroethyl dimethyl phosphate (13)
CAS #: 300-76-5
Molecular weight: 380.84 (13)
Solubility in water: <1 mg/L @ 20°C (13)
Solubility in other sovents: v.s. in alcohols, aromatic solvents; s. in aliphatic hydrocarbons, aromatic hydrocarbons, chlorinated hydrocarbons, and ketones; s.s. in mineral oils and petroleum solvents (13)
Melting point: 26–27.5°C (13)
Vapor pressure: 260 mPa @ 20°C (13)
Partition coefficient (octanol/water): Not available
Adsorption coefficient: 180 (19)

Exposure guidelines

ADI: Not available
HA: Not available
RfD: 0.002 mg/kg/day (53)
PEL: 3 mg/m^3 (8-hour) (39)

Basic manufacturer

Amvac Chemical Corp.
4100 E. Washington Blvd.
Los Angeles, CA 90023
Telephone: 213-264-3910
Emergency: 800-228-5635, ext. 169

5.2.19 *Phorate*

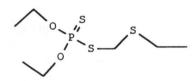

Figure 5.20 Phorate.

Trade or other names

Trade names for phorate include AC 8911, Agrimet, Geomet, Granutox, Phorate 10G, Rampart, Terrathion, Thimenox, Thimet, Timet, Vegfru, and Vegfru Foratox.

Regulatory status

Phorate is a highly toxic compound in EPA toxicity class I. Labels for products containing it must bear the Signal Words DANGER — POISON. It is a Restricted Use Pesticide (RUP). RUPs may be purchased and used only by certified applicators.

Introduction

Phorate is an organophosphorus insecticide and acaricide used to control sucking and chewing insects, leafhoppers, leafminers, mites, and some nematodes and root-worms (1,2). Phorate is used in pine forests and on root and field crops, including corn, cotton, coffee, and some ornamental and herbaceous plants and bulbs. It is available in granular and emulsifiable concentrate formulations.

Toxicological effects

Acute toxicity

Phorate is highly toxic via the oral route, with reported oral LD$_{50}$ values of 1.1 to 3.7 mg/kg in rats (2,13), and 2.25 to 6.59 mg/kg in mice (2,87). It is highly toxic via the dermal route as well, with reported dermal LD$_{50}$ values of 2.5 to 6.2 mg/kg in rats (2,13), and 5.2 mg/kg in rabbits (2,87). Guinea pigs reportedly have a dermal

LD_{50} of 20 to 30 mg/kg during a 24-hour exposure (13,87). The acute 1-hour inhalation LC_{50} for rats is reported as 0.06 mg/L (13).

Symptoms of acute exposure to phorate are similar to those caused by exposure to other organophosphate pesticides, except that they may occur at lower doses. Symptoms of acute exposure to organophosphate or cholinesterase-inhibiting compounds may include the following: numbness, tingling sensations, incoordination, headache, dizziness, tremor, nausea, abdominal cramps, sweating, blurred vision, difficulty in breathing or respiratory depression, and slow heartbeat. Very high doses may result in unconsciousness, incontinence, and convulsions or fatality. Toxicity appears to vary with age, with the young being more susceptible (8). Several poisoning cases involved workers from 16 to 18 years old, wearing inadequate protection while applying phorate to crops, or working around machines used to apply phorate (2,8). Studies indicate that direct eye exposure may cause blurring, tearing, and ocular pain (2).

Chronic toxicity

Repeated low-level exposures may result in cholinesterase inhibition and the associated neurological and neuromuscular effects (2). A survey of workers exposed to phorate revealed toxic effects in 60% of the males tested (after a 2-week exposure). Symptoms included a lowering of the heart rate. Effects on cholinesterase in the blood of the workers were also noted in this study (2,8).

In a study on dogs, moderate to high doses of phorate, 6 days each week for 13 to 15 weeks, lowered cholinesterase activity, but produced no tissue damage (87).

Reproductive effects

Long-term studies of mice fed high doses of 98.7% pure phorate showed no effects on fertility, gestation, or viability (87). Maternal and embryo toxicity occurred at dietary doses of 0.5 mg/kg/day fed to rats (88). Available data suggest that phorate is unlikely to cause reproductive effects.

Teratogenic effects

No birth defects were found in two studies on the rat (87,88). Available data suggest that phorate does not cause birth defects.

Mutagenic effects

Studies of phorate in both bacterial systems and in mice indicate that it is nonmutagenic (2).

Carcinogenic effects

Studies in both rats and mice produced no evidence of carcinogenicity (88).

Organ toxicity

Phorate's main target organ, as determined by animal testing and human use experience, is the nervous system.

Fate in animals and humans

Phorate is readily absorbed by the skin and the gastrointestinal tract. The major breakdown products of phorate in mammals are more toxic and have greater anticholinesterase activity than phorate (2). Phorate may have a long residence time in mammalian systems; for example, rats given a high oral dose excreted less than 40% in 6 days. The liver, kidney, lung, brain, and glandular tissue held the remaining residues (89).

Ecological effects

Effects on birds

Phorate is very highly toxic to highly toxic to birds. The reported acute oral LD_{50} values are 12.8 mg/kg in chukar, 7.5 mg/kg in starlings, 0.6 to 2.5 mg/kg in mallards, 7 to 21 mg/kg in Northern bobwhite quail, 1 mg/kg in red-winged blackbirds, and 7 mg/kg in ring-neck pheasants (6). The 5- to 8-day dietary LC_{50} values are reported as 370 to 580 ppm in Japanese quail, mallard, Northern bobwhite quail, and ring-neck pheasant (6).

Effects on aquatic organisms

Phorate is very highly toxic to fish. Reported 96-hour LC_{50} values range from 2 to 13 µg/L in cutthroat trout, bluegill sunfish, and largemouth bass. Other 96-hour LC_{50} values are 110 µg/L in Northern pike and 280 µg/L in channel catfish (16). Reported 96-hour LC_{50} values for the compound in freshwater invertebrates such as stoneflies and scuds are 4 µg/L, also indicating very high toxicity. Other LC_{50} values are 0.006 µg/L for amphipods, and 0.11 to 1.9 µg/L in other freshwater invertebrates (88). The acute oral LD_{50} of phorate is 85 mg/kg in bullfrogs (6).

Effects on other organisms (non-target species)

Phorate is toxic to bees, with a reported topical application LD_{50} of 10 µg per bee (13).

Environmental fate

Breakdown in soil and groundwater

Phorate is of moderate persistence in the soil environment, with reported field half-lives of 2 to 173 days. A representative value may be approximately 60 days (19). Actual residence times may be influenced by soil clay and organic matter contents, rainfall, and soil pH (8). Soil treatments often leave more residues in plants than foliar treatments because the compound persists in the soil and is readily taken up by plant roots (8,13).

Phorate binds moderately well to most soils and is slightly soluble in water (19). It should therefore not be highly mobile in most soils, and should mainly be transported with runoff via sediment and water. Phorate has minimal potential to leach through the soil and contaminate groundwater. This is most likely where soils are sandy and aquifers are shallow. Field studies indicate that leaching is very low in soils high in clay and organic matter contents, and lower in sandy soils (8).

Breakdown in water

The half-life of phorate in acidic water solutions is between a few days and a few weeks, depending on temperature; the half-life in alkaline (basic) water may be much shorter (8,88). Phorate is degraded by waterborne microorganisms and hydrolysis (8,88). As it breaks down in water, nontoxic, water-soluble products are formed.

Breakdown in vegetation

Phorate itself is not persistent in plants, but plants metabolize phorate to very potent anticholinesterase agents such as the sulfoxide and sulfone derivatives of the compound (2). This activity will usually peak several days following application before decreasing (8). Phorate and its soil metabolites are absorbed from the soil by plant roots and are translocated to above-ground portions of the plant.

Following treatment with a 10% granular formulation at 1 pound a.i. per acre, phorate residues persisted at very low levels for 28 days in the kernels, cobs, or husks. After 83 days, there were no detectable residues of phorate or breakdown products (8).

Physical properties

Technical phorate is a clear liquid at room temperature (13).

Chemical name: O,O-diethyl S-ethylthiomethyl phosphorodithioate (13)
CAS #: 298-02-2
Molecular weight: 260.38 (13)
Solubility in water: 50 mg/L @ 25°C (13)
Solubility in other solvents: v.s. in xylene, carbon tetrachloride, dioxane, methyl cellosolve, dibutylphthalate vegetable oils, ethanol, ether, and aliphatic hydrocarbons (13)
Melting point: Less than −15°C (13)
Vapor pressure: 110 mPa @ 20°C (13)
Partition coefficient (octanol/water): 8410 (13)
Adsorption coefficient: 1000 (estimated) (19)

Exposure guidelines

ADI: 0.0002 mg/kg/day (38)
HA: Not available
RfD: Not available
TLV: 0.05 mg/m^3 (8-hour) (skin) (47)

Basic manufacturer

American Cyanamid Co.
One Cyanamid Plaza
Wayne, NJ 07470-8426
Telephone: 201-831-2000
Emergency: 201-835-3100

5.2.20 Phosmet

Trade or other names

Trade names for products containing phosmet include Appa, Decemthion, Fesdan, Imidan, Kemolate, Prolate, PMC, and Safidon. It is also found in combination with other insecticides such as carbophenothion.

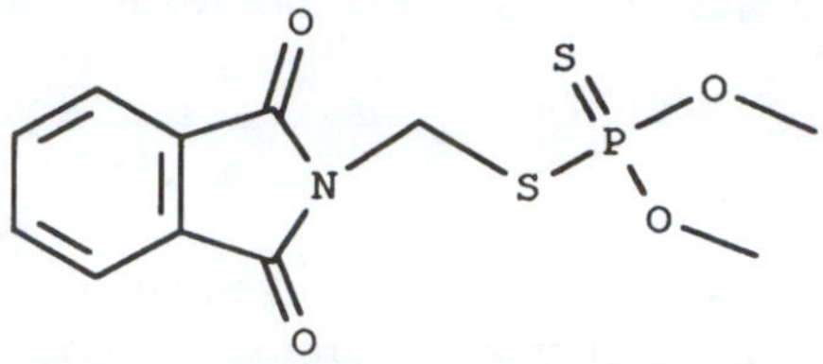

Figure 5.21 Phosmet.

Regulatory status

Phosmet is a moderately toxic compound, falling in EPA toxicity class II. It is a General Use Pesticide (GUP). Products containing phosmet must bear the Signal Word WARNING on the label. Some tolerances for phosmet in processed foods were revoked by the EPA in 1994.

Introduction

Phosmet is a nonsystemic, organophosphate insecticide used on both plants and animals. It is mainly used on apple trees for control of coddling moth, though it is also used on a wide range of fruit crops, ornamentals, and vines for the control of aphids, suckers, mites, and fruit flies. The compound is also an active ingredient in some dog collars. Phosmet is used on approximately 1.1 million acres in the U.S. each year.

Toxicological effects

Acute toxicity

Phosmet is moderately toxic via the oral route, with a reported acute oral LD_{50} of 113 to 160 mg/kg in rats (6,13). It is slightly toxic via the dermal route, with a dermal LD_{50} of 3160 to greater than 4640 mg/kg in rabbits (6,13). Phosmet is a mild irritant to the eyes, and only mildly irritating to the skin of rabbits (13). Via the inhalation route it is slightly toxic as well, with a reported 1-hour inhalation LC_{50} value of 2.76 mg/L (8,13). Signs of acute poisoning are rapid, generally occurring within 30 minutes after exposure.

The principal effects due to acute exposure to phosmet are similar to those caused by acute effects of other organophosphate pesticides (e.g., inhibition of cholinesterase) (90). Symptoms of acute exposure to organophosphate or cholinesterase-inhibiting compounds may include the following: numbness, tingling sensations, incoordination, headache, dizziness, tremor, nausea, abdominal cramps, sweating, blurred vision, difficulty in breathing or respiratory depression, and slow heartbeat. Very high doses may result in unconsciousness, incontinence, and convulsions or fatality.

Chronic toxicity

Rats fed phosmet for 16 weeks at 22.5 to 300 mg/kg/day suffered some mortality and exhibited a number of toxic effects (90). Over a 6-month period, doses of phosmet of 1 mg/kg/day in the diets of rats produced no observable chronic effects (8,90). Doses of 1 mg/kg/day in a 2-year feeding study also produced no effects (8,90). Rabbits that had phosmet applied to their skin for 5 days a week for 3 weeks suffered high mortality rates at doses of 300 to 600 mg/kg/day. At 50 mg/kg/day, there was significant brain enzyme (cholinesterase) depression (8). The signs and symptoms of chronic toxicity are generally consistent with those for the class of organophosphate insecticides.

Reproductive effects

A three-generation study with rats indicated that there were no reproductive effects when the animals were fed phosmet at a dose of 2.0 mg/kg/day during gestation for the first generation, and a dose of 4 mg/kg/day for the second and third generations (8). Female rabbits given phosmet both dermally and orally for 3 weeks prior to mating and for 18 consecutive days of gestation showed no effects on reproductive parameters. The doses tested ranged from 10 to 60 mg/kg/day for both routes of exposure (91). These data indicate that phosmet does not cause reproductive toxicity.

Teratogenic effects

No birth defects were noted in studies with pregnant rabbits fed 10 to 60 mg/kg/day for 3 weeks during pregnancy, or in monkeys given 8 to 12 mg/kg/day on days 22 to 32 of gestation. Offspring from rats fed 10 to 30 mg/kg/day on days 6 through 15 of gestation showed no abnormalities. In another study, however, doses of 30 mg/kg/day administered to rats between day 9 and 13 of gestation produced a dose-dependent increase in brain damage (hydrocephaly) in 33 of the 55 embryos examined (91). These data suggest that phosmet is not likely to cause teratogenic effects.

Mutagenic effects

Almost all tests of phosmet on bacteria indicate that it does not cause any mutations (90). However, there is a suggestion that workers producing the compound Safidon show some changes in their chromosomes (8). A definite conclusion cannot be drawn from current evidence.

Carcinogenic effects

Rats fed diets containing 1 to 20 mg/kg/day phosmet for 2 years showed no increase in cancer. However, a 2-year mouse study showed that phosmet is associated with a significant increase in liver tumors in male mice (90). In female mice, there was a positive dose-related trend for liver tumors and carcinomas (90). The available data are not sufficient to draw a firm conclusion about the carcinogenicity of phosmet.

Organ toxicity

The primary target organ for phosmet is the nervous system.

Fate in humans and animals

Phosmet is rapidly absorbed, distributed, and eliminated in mammals; rats given single doses of 23 to 35 mg/kg phosmet excreted greater than 75% of the dose in urine and about 15% in the feces. Less than 3% was found in body tissues after 2 days (91). Other studies show nearly 80% eliminated in the urine and 20% eliminated in the feces after 3 days (90). Rat studies indicate that phosmet crosses the placenta (8). Phosmet's breakdown products in steers include phthalamic and phthalic acids (8). The situation is similar in other animal systems as well; the major metabolite is phthalamic acid with phthalic acid produced in smaller quantities (13).

Ecological effects

Effects on birds

Phosmet is variable in its toxicity to birds. In red-wing blackbirds, it is highly toxic, with a reported acute oral LD_{50} of 18 mg/kg. In starlings and pheasants, it is moderately toxic, with reported acute oral LD_{50} values of greater than 100 and 237 mg/kg, respectively. In mallards, it is only slightly toxic, with a reported acute oral LD_{50} of 1830 mg/kg. The 5- to 8-day dietary LC_{50} values (greater than 2000 ppm) indicate slight toxicity in Japanese quail, mallard duck, and ring-neck pheasant (6). Phosmet may be moderately toxic to Northern bobwhite quail, with a reported dietary LC_{50} of 501 ppm (6). Phosmet can cause reproductive difficulties in birds whose feed contains phosmet residues (92). No delayed neurotoxic effects were noted in chickens fed diets containing moderate levels of phosmet for 6 weeks (92).

Effects on aquatic organisms

Phosmet's toxicity to aquatic organisms is species-specific, varying from highly to very highly toxic. The reported 96-hour LC_{50} values for phosmet are less than 1 mg/L in bluegill sunfish, small- and largemouth bass, rainbow trout, and chinook salmon, and are less than 10 mg/L in fathead minnow and channel catfish (16). The reported 96-hour LC_{50} values in aquatic invertebrates and crustaceans such as *Daphnia* spp., scuds, and sideswimmers indicate very high toxicity (16). Phosmet has a reported bioconcentration factor of 6 to 37 in fish (90), indicating it has very little potential for accumulation in aquatic organisms.

Effects on other organisms (non-target species)

Phosmet is very toxic to honeybees; the contact LD_{50} for the compound is 0.0001 mg per bee (13).

Environmental fate

Breakdown in soil and groundwater

Phosmet is of low persistence in the soil environment, with reported field half-lives of 4 to 20 days. A representative half-life may be about 19 days (19). Phosmet is rapidly broken down in soil to nontoxic products (6,13). Degradation is by hydrolysis (the chemical action of water) and microbial action, and may be faster under wet, basic conditions (6). It is moderately bound by soils, and not highly soluble in water (19), and thus is not expected to be highly mobile or a significant risk to groundwater. There is little leaching or runoff associated with the compound, even after repeated applications (8).

Breakdown in water

In water, phosmet is rapidly broken down by the chemical action of the water (hydrolysis) and by sunlight (photolysis) (90). Under moderately acidic conditions (pH 5), half the compound degrades within 9 days; but in neutral solution (pH 7), its half-life is 18 hours. Under alkaline conditions (pH 9), the half-life is reported as 16 hours (8,90).

Breakdown in vegetation

Phosmet breaks down rapidly in plant systems, primarily due to passive oxidation and hydrolysis due to air contact (90). The breakdown products are much less toxic than the parent compound (13). On apricots and nectarines treated at unknown rates, there were residues of less than 5 mg/kg at 7 days after treatment, and less than 1 mg/kg at 21 days after treatment (8). Maize used for silage showed a rapid decline in residues before being made into silage, but the half-life of the silage residue was about 92 days (8). Washing and blanching of fruits and vegetables can reduce residue levels by 50 to 80% and thus reduce the potential human exposure to the pesticide (91).

Physical properties

Technical phosmet is an off-white to pink solid at room temperature (13).

Chemical name: O,O-dimethyl S-phthalimidomethyl phosphorodithioate (13)
CAS #: 732-11-6
Molecular weight: 317.33 (13)
Solubility in water: 25 mg/L @ 25°C (13)
Solubility in other solvents: v.s. in acetone, xylene, methanol, benzene, toluene, and methyl isobutyl ketone; s. in kerosene (13)
Melting point: 66–69°C (technical) (13)
Vapor pressure: 133 mPa @ 50°C (13)
Partition coefficient (octanol/water): 1100 (13)
Adsorption coefficient: 820 (19)

Exposure guidelines

ADI: 0.02 mg/kg/day (38)
HA: Not available
RfD: 0.02 mg/kg/day (53)
PEL: Not available

Basic manufacturer

Zeneca Ag Products
1800 Concord Pike
Wilmington, DE 19897
Telephone: 800-759-4500
Emergency: 800-759-2500

5.2.21 Propetamphos

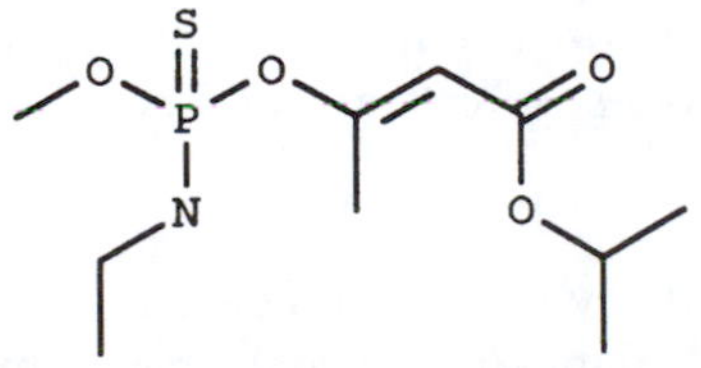

Figure 5.22 Propetamphos.

Trade or other names

Trade names for products containing propetamphos include Blotic, Safrotin, and Seraphos.

Regulatory status

Propetamphos is a moderately toxic compound in EPA toxicity class II. Labels for products containing propetamphos must carry the Signal Word WARNING. Most formulations of propetamphos are General Use Pesticides (GUPs), but some may be Restricted Use Pesticides (RUPs). RUPs may be purchased and applied only by certified applicators.

Introduction

Propetamphos is an organophosphate insecticide designed to control cockroaches, flies, ants, ticks, moths, fleas, and mosquitoes in households and where vector eradication is necessary to protect public health. It is also used in veterinary applications to combat parasites such as ticks, lice, and mites in livestock. Commercial products include aerosols, emulsified concentrates, liquids, and powders.

Toxicological effects

Acute toxicity

Propetamphos is moderately toxic via the oral route, with reported oral LD_{50} values for propetamphos of 75 to 119 mg/kg in rats (6,13). Via the dermal route it is slightly to practically nontoxic, with reported acute dermal LD_{50} values of 2300 to greater than 3100 mg/kg in rats, and greater than 10,000 mg/kg in rabbits (6,13). The 4-hour inhalation LC_{50} is greater than 2.04 mg/L in rabbits, indicating slight toxicity via this route (13).

Effects due to acute exposure to propetamphos include those that occur with exposure to other orghanophosphate pesticides, including neurological and neuromuscular effects due to cholinesterase inhibition (8). Symptoms of acute exposure to organophosphate or cholinesterase-inhibiting compounds may include the following: numbness, tingling sensations, incoordination, headache, dizziness, tremor, nausea, abdominal cramps, sweating, blurred vision, difficulty in breathing or respiratory depression, and slow heartbeat. Very high doses may result in unconsciousness, incontinence, and convulsions or fatality.

Chronic toxicity

Rats fed propetamphos for 13 weeks exhibited no effects at a low dose of 0.2 mg/kg/day. Over a 77-week study, the rats exhibited no adverse effects at or below the very low dose of 0.05 mg/kg/day. In 2-year feeding studies with rats, there were no effects noted at or below a dose of 6 mg/kg in their diets. Dogs fed the compound for 6 months showed no adverse effects at the dose of 0.05 mg/kg/day (93).

Reproductive effects

A three-generation rat study showed no significant effects in litters at doses of 1 mg/kg/day (93). Available data suggest that propetamphos does not cause reproductive toxicity.

Teratogenic effects

A teratology study in rabbits was negative (93). Available data indicate that propetamphos is not teratogenic.

Mutagenic effects

In studies with the fruit fly *Drosophila*, propetamphos did not cause chromosome damage (94). However, in mouse tissue, high levels of the compound caused some mild chromosome damage (93,94). These data suggest that the compound is non-mutagenic or weakly mutagenic.

Carcinogenic effects

A 2-year carcinogenicity test on rats and a lifetime carcinogenesis study on mice were both negative. The highest dose administered to the rats was 6 mg/kg/day, and the maximum dose administered to the mice was 21 mg/kg/day (8). This evidence suggests that propetamphos does not cause cancer.

Organ toxicity

The primary target organ affected by propetamphos is the nervous system.

Fate in humans and animals

Cultured preparations of house fly, cockroach, and mouse liver cells have all shown the ability to break down the compound (95).

Ecological effects

Effects on birds

Propetamphos is moderately toxic to birds. The acute oral LD_{50} for propetamphos in the mallard ranges from 45 to nearly 200 mg/kg (8,13). The dietary LC_{50} for the compound in the mallard ranges from about 700 ppm to greater than 1780 ppm (8). The LC_{50} for propetamphos in the quail ranges from 138 to 250 ppm (8). Propetamphos is not generally used in outdoor settings and thus poses little risk to wildlife (6).

Effects on aquatic organisms

The compound is highly toxic to fish such as bluegill sunfish and rainbow trout. The LC_{50} values range from 0.13 mg/L in bluegill and 0.36 mg/L in rainbow trout, to 3.7 to 8.8 mg/L in carp (moderately toxic range for carp) (13,8). Propetamphos may be very highly toxic to aquatic invertebrates, with reported LC_{50} values ranging between 0.68 and 14.5 µg/L in *Daphnia magna* (8).

Effects on other organisms (non-target species)

No data are currently available.

Environmental fate

Breakdown in soil and groundwater

No data are currently available.

Breakdown in water

In water, propetamphos is rapidly degraded only under extreme pH conditions (acidic or basic), or in the presence of sunlight (13,96).

Breakdown in vegetation

No data are currently available.

Physical properties

Technical propetamphos is a yellowish, oily liquid at room temperature (13).

Chemical name: (E)-O-2-isopropoxycarbonyl-1-methylvinyl O-methyl eth-
   ylphosphoramidothioate (13)
CAS #: 31218-83-4
Molecular weight: 281.3 (13)
Solubility in water: 110 mg/L @ 24°C (13)
Solubility in other solvents: v.s in acetone, chloroform, ethanol, and hexane (13)
Melting point: Not available
Vapor pressure: 1.9 mPa @ 20°C (13)
Partition coefficient (octanol/water): Not available
Adsorption coefficient: Not available

Exposure guidelines

ADI: Not available
HA: Not available
RfD: 0.02 mg/kg/day (53)
PEL: Not available

Basic manufacturer

Sandoz Agro, Inc.
1300 E. Touhy Ave.
Des Plaines, IL 60018
Telephone: 708-699-1616
Emergency: 708-699-1616

5.2.22 *Temephos*

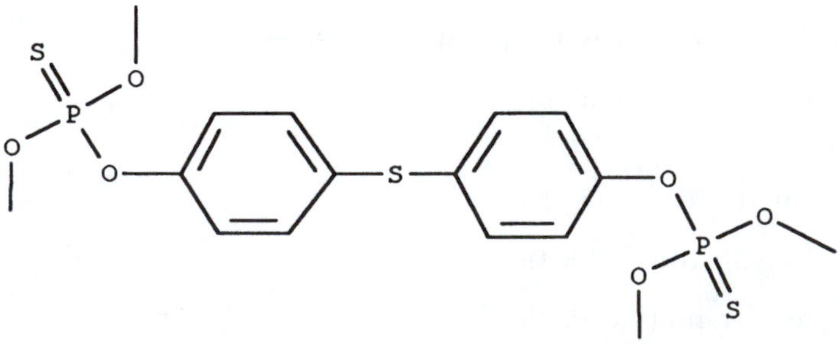

Figure 5.23 Temephos.

Trade or other names

Trade names for products containing temephos include Abat, Abate, Abathion, Acibate, Biothion, Bithion, Difennthos, Ecopro, Nimitox, and Swebate. The compound may also be found in mixed formulations with other insecticides, including trichlorfon.

Regulatory status

Temephos is a General Use Pesticide (GUP). Temephos-containing products are slightly toxic compounds (EPA toxicity class III) that carry the Signal Word WARNING on their labels despite the relatively low toxicity of the technical compound.

Introduction

Temephos is a nonsystemic organophosphorus insecticide used to control mosquito, midge, and black fly larvae. It is used in lakes, ponds, and wetlands. It may also be used to control fleas on dogs and cats and to control lice on humans. Temephos is available in up to 50% emulsifiable concentrates, 50% wettable powder, and up to 5% granular forms.

Toxicological effects

Acute toxicity

Typical of other organophosphate insecticides, temephos inhibits the action of the group of enzymes called cholinesterases. These enzymes are most important in the nervous system, the brain, and the musculoskeletal systems in controlling nerve signal transmission.

Symptoms of acute exposure are similar to other organophosphates and may include nausea, salivation, headache, loss of muscle coordination, and difficulty in breathing (8). Temephos produces signs and symptoms typical of cholinesterase inhibition at moderate levels of exposure, but mortality does not occur unless very large doses of the compound are administered (2,8).

Reported oral LD_{50} values of temephos range from 1226 to 13,000 mg/kg in rats (2,13), and 460 to 4700 mg/kg in mice. The LD_{50} for a 2% powder formulation of temephos in dogs and cats is greater than 5000 mg/kg for both species.

Temephos may potentiate (greatly increase) the observed toxicity of malathion when used in combination with it at very high doses (2).

Chronic toxicity

Rats, rabbits, guinea pigs, and chickens fed temephos at doses of approximately 20 mg/kg/day for extended periods showed no clinical effects (2). Dogs tolerated 3 to 4 mg/kg/day for an extended period, although there was a slight decrease in cholinesterase activity in the blood and the brain (2). Severe effects were seen in dogs given 14 mg/kg/day for an extended period, and 15.3 mg/kg/day produced leg weakness in chickens over a 30-day period (2).

As noted under carcinogenicity, a reduction in liver weights was noted in a study on rats fed small doses of temephos over a 2-year period. In another study of rabbits, findings of minor pathological changes in the liver at doses of 10 mg/kg/day were noted, but were not found at a dose of 1 mg/kg/day (2). No other effects on organs have been reported.

Thus, while the LD_{50} values for acute toxicity indicate that the compound is relatively nontoxic or only slightly toxic, the compound has the potential to cause significant toxic effects (depression of the activity of the enzyme cholinesterase in the blood and the brain) in mammals exposed over long periods of time.

Temephos was used in cisterns and other potable water sources in some locations in the U.S. and in the West Indies for the control of mosquito larvae. Subsequent tests on the residents that had used the water sources showed no observable effects in the exposed individuals (2). Humans ingested 256 mg/kg/day for 5 days, and 64 mg/kg/day for 4 weeks, without any symptoms or detectable effects on blood cholinesterase activity (2).

Reproductive effects

Neither of two studies of rats fed small amounts of temephos showed any reproductive difficulties in the test animals. The maximum dose (25 mg/kg/day) had no effect on the number of litters, litter size, or viability in the young, and produced no congenital defects in the offspring. The concentration of temephos in the diet of the test animals was, however, sufficient to produce cholinesterase inhibition and some toxic symptoms (2).

Low oral doses of temephos of up to 2.5 mg/kg administered in feed over $1\frac{1}{2}$ years caused no reproductive effects in sheep or in their offspring (2). These data indicate that temephos does not cause reproductive toxicity.

Teratogenic effects

There were no birth defects noted in the offspring of pregnant rabbits fed temephos in two separate studies utilizing different formulations of temephos, a 2% formulation and a 90% formulation. In both studies, maternal toxicity and depression of cholinesterase activity occurred during the study (97). These data suggest that temephos poses little teratogenic risk.

Mutagenic effects

The potential of a commercial product containing temephos (Abate) to cause mutations was tested on several strains of bacteria. Though the conclusion of the study was that the compound was not mutagenic, weakly mutagenic effects were noted in one of the strains. Additional tests on rabbits and on other strains of bacteria have shown the compound to be nonmutagenic (8,97).

Carcinogenic effects

Only one study of the carcinogenic potential of temephos has been conducted with rats. The rats were fed doses of the compound over a 2-year interval. No tumors or cancer-related changes were noted in the test animals at 15 mg/kg/day, the highest dose used (8,97). During the study, the rats experienced a reduction in liver weight at the lowest dose of 0.5 mg/kg/day (8,97). These data suggest that temephos is not carcinogenic.

Organ toxicity

Animal studies indicate that target organs include the nervous system and liver.

Fate in humans and animals

In general, organophosphate insecticides are readily absorbed through the lungs, skin, and digestive tract (8). A single oral dose of temephos reached peak concentration in the bloodstream of rats between 5 and 10 hours after it was administered (2), and was eliminated with a half-life of 10 hours. Some of the compound was also found in the digestive tract and some in fat in mammals. Most of the compound is eliminated unchanged through the feces and urine, though some breakdown products have been detected (2).

Ecological effects

Effects on birds

Tests with various wildlife species indicate that the compound is highly toxic to some bird species and moderately toxic to others. The LD_{50} of temephos ranges from 18.9 mg/kg in the California quail to 240 mg/kg in the chukar partridge (15). The LD_{50} values in other bird species studied (Japanese quail, pheasant, and rock dove) were between 35 and 85 mg/kg (15). Mallards fed diets containing moderate amounts of temephos showed no changes in reproduction, except in the frequency of egg laying (98).

Effects on aquatic organisms

Temephos shows a wide range of toxicity to aquatic organisms, depending on the formulation. Generally, the technical grade compound (tech) is moderately toxic and the emulsifiable concentrate (ec) and wettable powder (wp) formulations are highly to very highly toxic. The most sensitive species of fish is the rainbow trout, with a temephos LD_{50} ranging from 0.16 mg/L (ec) to 3.49 mg/L (tech) (16). Other 96-hour LD_{50} values are reported as: coho salmon, 0.35 mg/L (ec); largemouth bass, 1.44 mg/L (ec); channel catfish, 3.23 mg/L (ec) to >10 mg/L (tech); bluegill sunfish, 1.14 mg/L (ec) to 21.8 mg/L (tech); and Atlantic salmon, 6.7 mg/L (ec) to 21 mg/L (tech) (6,8,13,16).

Freshwater aquatic invertebrates such as amphipods are very highly susceptible to temephos, as are some marine invertebrates such as mysids. The 96-hour LD_{50} of temephos in *Gammarus lacustris* is 0.08 mg/kg, and in stoneflies is 0.01 to 0.03 mg/kg (6,8,16). Because the compound is an insecticide and is used effectively to control the aquatic larval stages of mosquitoes, black flies, and midges, its high toxicity to these organisms is not surprising. The product Abate 4E (46% emulsifiable concentrate) is very highly toxic to saltwater species such as the pink shrimp ($LC_{50} = 0.005$ mg/L) and the Eastern oyster ($LC_{50} = 0.019$ mg/L) (8). The compound is nearly nontoxic to the bull frog, with an LD_{50} of greater than 2000 mg/kg (8).

Temephos has the potential to accumulate in aquatic organisms. The bluegill sunfish accumulated 2300 times the concentration present in the water. Nearly 75% of the compound was eliminated from the fish after exposure ended (8).

Effects on other organisms (non-target species)

The compound is highly toxic to bees, with a direct contact LC_{50} of 1.55 μg per bee (13).

Environmental fate

Breakdown in soil and groundwater

There is little information available about the fate and behavior of temephos in the environment. Based on its very low solubility in water, it would probably have a high affinity for soil. Based on this, a half-life of 30 days has been estimated (19), indicating a low to moderate persistence.

Breakdown in water

Weekly application of temephos at twice the normal application rates on pond water resulted in the rapid disappearance of the compound from the water and from the sediments (6). At even higher application rates to pond water, there were still only traces of the compound detected 1 week after application. Temephos will be photolyzed in water (8).

Temephos was sprayed over an intertidal mangrove community in Florida. Between 15 and 70% of the sprayed amount reaching the leaf surface entered the water below the trees. Additional amounts were washed into the water during rainfall. Pesticide residues were detected in the water 2 hours, but not 4 hours, after application, indicating a very short persistence in the water. However, in simulated tide pools, the compound persisted for up to 4 days. It also persisted in oysters for 2 days after application (99).

Temephos has low persistence in water.

Breakdown in vegetation

Breakdown in plants is very slow.

Physical properties

Temephos is a solid at room temperature and is composed of colorless crystals. As a liquid, it is brown and viscous (13).

Chemical name: O,O'-(thiodi-4,1-phenylene)bis(O,O-dimethyl phosphorothio-ate) (13)
CAS #: 3383-96-8
Molecular weight: 466.46 (13)
Solubility in water: 0.001 mg/L (13)
Solubility in other solvents: s. in common organic solvents; i.s. in hexane and methylcyclohexane (13)
Melting point: 30–30.5°C (13)
Vapor pressure: Not available
Partition coefficient (octanol/water): 89,900 (13)
Adsorption coefficient: 100,000 (estimated) (19)

Exposure guidelines

ADI: Not available
HA: Not available
RfD: 0.02 mg/kg/day (53)
TLV: 10 mg/m^3 total dust; 5 mg/m^3 respirable fraction (8-hour) (47)

Basic manufacturer

American Cyanamid
One Cyanamid Plaza
Wayne, NJ 07470-8426
Telephone: 210-831-2000
Emergency: 210-835-3100

5.2.23 *Terbufos*

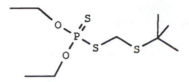

Figure 5.24 Terbufos.

Trade or other names

Trade names for terbufos include AC 92100, Aragran, Contraven, Counter, and Plydox.

Regulatory status

Products containing 15% or more terbufos are classified as Restricted Use Pesticides (RUPs). RUPs may be purchased and used only by certified applicators. Terbufos is classified as toxicity class I — highly toxic. Terbufos products are labeled with the Signal Word DANGER.

Introduction

Terbufos is an organophosphate insecticide and nematicide used on corn, sugar beets, and grain sorghum. Primarily formulated as granules, it is applied at planting in a band or directly to the seed furrow. Terbufos controls wireworms, seedcorn maggots, white grubs, corn rootworm larvae, and other pests.

Toxicological effects

Acute toxicity

The oral LD_{50} of terbufos is from 1.3 to 1.57 mg/kg in female rats, and from 1.6 to 1.74 mg/kg in male rats (100). The oral LD_{50} for technical terbufos is 3.5 mg/kg in male mice, 9.2 mg/kg in female mice, 4.5 mg/kg in male dogs, and 6.3 mg/kg in female dogs (101). Rabbits given a single dose of 0.1 mg to the eyes died within 2 to 24 hours after dosing (102). The dermal LD_{50} in rabbits is 1.1 mg/kg for 24 hours (8). Terbufos is highly toxic by both dermal and oral routes of exposure.

Symptoms of acute toxicity often include nausea, abdominal cramps, vomiting, salivation, excessive sweating, and diarrhea within 45 minutes of ingestion. Absorption into the bloodstream may cause inhibition of cholinesterase, an enzyme essential for normal functioning of the nervous system. This, in turn, can lead to chest tightness, wheezing, blurred vision, fatigue, headache, slurred speech, and confusion.

Symptoms from skin absorption, such as localized sweating, may be delayed up to 12 hours. At high enough doses, death may result from respiratory arrest, respiratory muscle paralysis, and/or constriction of the lungs.

No neurotoxic effects were observed in chickens given a single dose of 40 mg/kg, the highest dose tested (100).

Chronic toxicity

Slow thinking, memory loss, irritability, delayed reaction times, and anxiety have been noted in workers chronically exposed to organophosphates like terbufos.

When rats were fed terbufos for 90 days, no effects were observed at doses up to 0.02 mg/kg/day. Cholinesterase inhibition occurred at higher doses (101). Similar results were obtained in a 1-year study with rats. No effects were observed at doses up to 0.025 mg/kg/day, with cholinesterase inhibition occurring at 0.05 mg/kg/day, the highest dose tested (101).

Reproductive effects

In a long-term study in rats, no chronic reproductive effects were observed after daily exposure to low doses of terbufos (102). In another study, when rats were fed 0.0125 or 0.05 mg/kg/day for 6 months, there was an increase in the number of litters with dead offspring at the highest dose tested (101).

Rabbits were given doses up to 0.4 mg/kg/day on days 7 to 19 of pregnancy. Cesarean sections were performed on day 29. No adverse effects on the offspring were seen at any dose. Toxic effects on the mothers occurred at the highest dose tested (101).

These data suggest that terbufos is unlikely to cause reproductive effects in humans.

Teratogenic effects

Available evidence indicates that terbufos does not cause birth defects. There were no birth defects in the offspring of rats given up to 0.2 mg/kg/day on days 6 to 15 of pregnancy (101). In a similar study on rabbits, no birth defects were observed in the offspring of rabbits given 0.25 mg/kg/day, the highest dose tested. The mothers exhibited reduced body weight gain at this dose (100).

Mutagenic effects

Several tests have shown that terbufos is not mutagenic. These include a dominant lethal study in rats, an Ames test, a DNA repair chromosomal aberration test, and a test for DNA repair in rat liver cells (100,101).

Carcinogenic effects

No tumors were found in mice given 1.8 mg/kg/day, the highest dose tested, for 18 months. The same results occurred in a 2-year study with rats given up to 0.40 mg/kg/day, the highest dose tested (100,101). These data indiate that terbufos is not carcinogenic.

Organ toxicity

This pesticide can affect the eyes, lungs, skin, and central nervous system, depending on the route of exposure and the concentration.

Fate in humans and animals

In rats given a single oral dose of terbufos, 10% remained in the liver 6 hours after dosing. Breakdown products were found in the kidney 12 hours after dosing. Of the original dose administered, 83% was excreted in the urine within 7 days after dosing, and 3.5% was found in the feces. Terbufos and its metabolites did not accumulate in tissues (101,102).

No detectable residues have been found in the eggs, milk, or body tissues of animals (hens and cows) fed very high dietary doses of terbufos and its cholinesterase-inhibiting metabolites (102).

Ecological effects

Effects on birds

Terbufos is extremely toxic to birds. Its acute LD_{50} is 28.6 mg/kg in bobwhite quail. Its dietary LC_{50} is 143 to 157 ppm in bobwhites (102). Reported dietary LC_{50} values in Japanese quail are 194 to 265 ppm in 5-day studies (14).

There were no effects on bird reproduction from chronic exposure to terbufos (101).

Effects on aquatic organisms

Terbufos is extremely toxic to fish and aquatic invertebrates (102). The reported 96-hour LC_{50} for terbufos in *Daphnia magna*, a small freshwater invertebrate, is 0.00031 mg/L (102); in *Gammarus psendogomnaens*, also a freshwater invertebrate, it is 0.0002 mg/L (72). Reported 96-hour LC_{50} values for the technical material (88%) are 0.008 to 0.0013 mg/L in rainbow trout, 0.390 mg/L in the fathead minnow, and 0.0017 to 0.0024 mg/L in the bluegill (72).

The compound has a moderate potential to accumulate in living tissue in aquatic organisms. The bluegill sunfish accumulated 680 times the ambient water concentration (in whole fish); 14 days after the fish were placed in terbufos-free water, between 84 and 93% of the compound was eliminated (8).

Effects on other organisms (non-target species)

Terbufos is expected to be extremely toxic to mammals and reptiles (102). It is nontoxic to bees.

Environmental fate

Breakdown in soil and groundwater

Terbufos is low to moderately persistent in soil. It is rapidly converted to its metabolites, which tend to persist in the soil and may be detected at harvest time (103). Terbufos and its metabolites quickly degrade during the first 15 to 30 days after application, then gradually stabilize. Only 3% of the original application stayed in field-study soils after 1 month, with 1.5% of the chemical present after 60 days (104).

In a study on silty clay loam soil in South Dakota, the half-life of terbufos was about 2 weeks. The half-life for the metabolite, terboxon sulfone, was two to three times longer (103). Other reported field half-lives are 5 to 30 days (19). Terbufos dissipation is generally faster in soils with very low organic carbon, while binding increases with increasing organic carbon content (104).

Soil moisture does not appear to affect the degradation of terbufos. This chemical will break down at about the same rate in soils regardless of the level of wetness (104). As temperature increases, terbufos degrades more quickly.

Terbufos is generally immobile and is therefore unlikely to leach into or contaminate groundwater (102). Much of the chemical can be recovered near the site of application. In one study, over 90% of the applied terbufos was recovered in the top 4 inches of a soil profile despite heavy rainfall, and thorough incorporation down to 2½ inches (104).

Being of low water solubility, terbufos is not often found in groundwater. Terbufos has been found in a few groundwater samples collected from locations across the U.S. at concentrations of approximately 0.01 mg/L (42).

Breakdown in water

Terbufos hydrolyzes rapidly. At a concentration of 4.6 mg/L, its hydrolysis half-lives were 4.5, 5.5, and 8.5 days at pH 5, 7, and 9, respectively (101,102). In another study, terbufos hydrolyzed with a half-life of 2.2 weeks at pH 5, 7, and 9. Formaldehyde was the major degradate detected (101).

Breakdown in vegetation

Terbufos moves from the soil into plants, where it is broken down rapidly. Little of the parent compound is found in plants. At 57 days after seeding and application, the total residues in broccoli were very low, while the marketable heads of broccoli harvested 90 days after seeding held only traces (less than 0.01 ppm, fresh weight) of residues. Under the same conditions, marketable cabbage and cauliflower had trace to nondetectable levels of total residues (105). Field corn banded with 1.12 kg/ha had no detectable residues 60 days after treatment. Sweet corn and popcorn grain harvested at maturity also showed no residue, even though the surrounding soil contained 10 to 14 ppm (105).

Physical properties

Terbufos is a clear, slightly brownish-yellow liquid usually formulated into granules for agricultural applications (13).

Chemical name: S-*tert*-butylthiomethyl O,O-diethyl phosphorodithioate (13)
CAS #: 13071-79-9
Molecular weight: 288.43 (13)
Solubility in water: 5 mg/L (13)
Solubility in other solvents: s. in acetone, aromatic hydrocarbons, chlorinated hydrocarbons, and alcohols (13)
Melting point: −29.2°C (13)
Vapor pressure: 34.6 mPa @ 25°C (13)
Partition coefficient (octanol/water): 33,000 (13)
Adsorption coefficient: 500 (19)

Exposure guidelines

ADI: 0.002 mg/kg/day (38)
HA: 0.0009 mg/L (lifetime) (101)

RfD: 0.0001 mg/kg/day (53)
PEL: Not available

Basic manufacturer

American Cyanamid Co.
One Cyanamid Plaza
Wayne, NJ 07470-8426
Telephone: 201-831-2000
Emergency: 201-835-3100

5.2.24 Trichlorfon

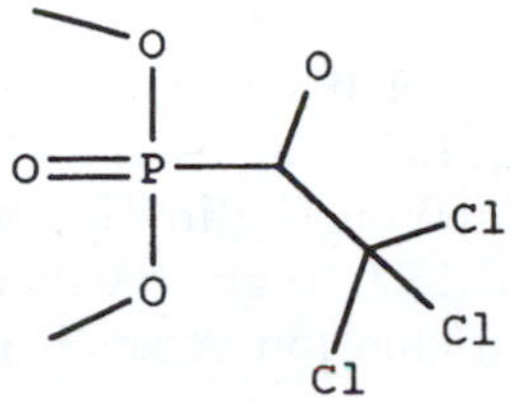

Figure 5.25 Trichlorfon.

Trade or other names

Trade names for trichlorfon include Anthon, Bovinos, Briten, Chlorophos, Ciclo-som, Dipterex, Ditrifon, Dylox, Dyrex, Equino-Aid, Foschlor, Leivasom, Neguvon, Masoten, Pronto, Phoschlor, Proxol, Totalene, Trichlorophene, Trichlorophon, Trinex, Tugon, and Vermicide Bayer 2349. The common name used in Great Britain is trichlorphon, in Turkey is dipterex, and in the former U.S.S.R. is chlorofos. When this material is used as a drug, it is called metrifonate or metriphonate.

Regulatory status

Trichlorfon is classified by the U.S. Environmental Protection Agency (EPA) as a General Use Pesticide (GUP). It is in toxicity class II — moderately toxic. Products containing trichlorfon bear the Signal Word WARNING.

Introduction

Trichlorfon is an organophosphate insecticide used to control cockroaches, crickets, silverfish, bedbugs, fleas, cattle grubs, flies, ticks, leafminers, and leafhoppers. It is applied to vegetable, fruit, and field crops; livestock; ornamental and forestry plant-ings; in agricultural premises and domestic settings; in greenhouses; and for control of parasites of fish in designated aquatic environments. It is also used for treating domestic animals for control of internal parasites (2,6). Trichlorfon is available in dust, emulsifiable concentrate, granular, fly bait, and soluble powder formulations with percent active ingredient ranging from 40 (soluble powder) to 98% (technical).

Trichlorfon is a selective insecticide, meaning that it kills selected insects, but spares many or most other organisms. Trichlorfon is toxic to target insects through direct application and via ingestion.

Toxicological effects

Acute toxicity

Trichlorfon is moderately toxic by ingestion or dermal absorption. As with all organophosphates, trichlorfon is readily absorbed through the skin. Skin sensitivity (allergies) can result from dermal exposure (106). Trichlorfon decreases activity of the cholinesterase enzyme, which is necessary for normal nervous system function.

Symptoms of acute exposure include headache, giddiness, nervousness, blurred vision, weakness, nausea, cramps, loss of muscle control or reflexes, convulsion, or coma (2). It has been suggested that impurities or additives may be associated with some cases of delayed polyneuropathy (damage to nerve cells) attrributed to ingestion of large amounts of trichlorfon (2). These delayed symptoms may occur following recovery from the initial acute effects. Pure trichlorfon is reported to be less toxic than the technical material (2,8).

The oral LD_{50} for trichlorfon is 450 to 650 mg/kg in rats, and 300 to 860 mg/kg in mice (2). Other reported oral LD_{50} values are 94 mg/kg in cats, 400 mg/kg in dogs, 420 mg/kg in dogs, and 160 mg/kg in rabbits (41). The dermal LD_{50} is 2000 to 5000 mg/kg in rats (2,8), and 1500 to greater than 2100 mg/kg in rabbits (2,41). The 4-hour inhalation LC_{50} for trichlorfon in rats is greater than 0.5 mg/L (13).

Chronic toxicity

Repeated or prolonged exposure to organophosphates may result in the same effects as acute exposure. Other effects reported in workers repeatedly exposed include impaired memory and concentration, disorientation, severe depression, irritability, confusion, headache, speech difficulties, delayed reaction times, nightmares, sleepwalking, and drowsiness or insomnia. An influenza-like condition with headache, nausea, weakness, loss of appetite, and malaise has also been reported (41).

When 45 mg/kg/day was administered to dogs for 3 months, serum cholinesterase was reduced to 60% of normal (2,8). A dietary level of about 10.5 mg/kg/day for 12 weeks produced a similar effect (2,8). During a 60-day testing period with repeated doses of trichlorfon at 100 mg/kg/day, the cholinesterase activity of rats was reduced to less than half of normal levels. Doses of 50 mg/kg/day reduced the activity to 50 to 75% of normal levels (2,8). Trichlorfon produced no pathological changes in rats fed 500 mg/kg/day of the insecticide for 1 year (8).

Edema (excessive accumulation of fluid) of the brain, congestion of organs, degeneration of various parts of the liver, inflammation of the lungs, and heart muscle changes were observed in rats given very high daily oral doses of 300 mg/kg/day technical trichlorfon for 5 days. Brain disturbances and changes in the liver, kidneys, spleen, lungs, and testicles were seen in bulls that were given oral doses of 1, 2, or 5 mg/kg formulated trichlorfon (Chlorophos) daily, or 5 mg/kg at weekly intervals for 6 months (108).

Reproductive effects

Trichlorfon is suspected of having negative reproductive effects (8). An increased number of embryonic deaths, a decreased number of live fetuses, and an increased number of fetal abnormalities were observed in rats given a single oral dose of 80 mg/kg body weight, by stomach tube, on the day 13 of pregnancy (8). During a three-generation study of the effect of trichlorfon on rat reproduction, a dose of about

150 mg/kg/day resulted in a marked decrease in the rate of pregnancy and under-developed rat pups at birth, none of which survived to weaning. A dose of 50 mg/kg/day reduced the number of pups per litter, as well as the weight of individual pups. Doses of 15 mg/kg/day had no detectable effect on reproduction (2,8). Once in the bloodstream, trichlorfon may cross the placenta (41). It is unlikely that reproductive effects will occur in humans at expected exposure levels.

Teratogenic effects

Trichlorfon caused inability to walk and tremors in pig offspring if administered at day 55 at a dose of 55 mg/kg (2,8). Dipterex was teratogenic when given to pregnant rats through a stomach tube, at a dose level of 480 mg/kg/day, on days 6 through 15 of pregnancy, but not when administered only on days 8 or 10 of pregnancy (8). Teratogenic effects were also seen in hamsters given 400 mg/kg/day on days 7 through 11 of pregnancy (8). There was no evidence of teratogenesis in a three-generation study with rats fed dietary doses as high as 150 mg/kg/day (2) or in a metabolite study of rabbits at doses of 50 to 75 mg/kg/day (2,8). Thus, the evidence suggests that reproductive effects occur only at high doses and are unlikely in humans at expected exposure levels.

Mutagenic effects

Studies indicate that trichlorfon, or its degradation products, can be mutagenic in bacterial and mammalian cells (107). The insecticide produced mutations in mice when it was given in the highest tolerable single dose and in smaller, repeated doses (2,8).

Carcinogenic effects

One study suggests that oral doses of 37.5 to 75 mg/kg/day trichlorfon contribute to the production of tumors in rats (107). Carcinogenic effects were also seen in rats given oral doses of 186 mg/kg or intramuscular doses of 183 mg/kg/day for 6 weeks (8). Benign tumors called "papillomas" developed in the lining of the forward portion of the stomach when trichlorfon was administered orally or subcutaneously to rats. Rats that survived for 6 months had varying degrees of liver damage (2,8). However, no evidence of carcinogenicity was found in rats given the insecticide orally or intraperitoneally for 90 weeks (2,8). In addition, no evidence of carcinogenicity was observed when trichlorfon was administered orally, intraperitoneally, or dermally to mice (2,8). Thus, the carcinogenic data are inconclusive.

Organ toxicity

Trichlorfon primarily affects the nervous system through inhibition of cholinesterase, an enzyme required for proper nerve functioning. Other target organs include the liver, lungs, and bone marrow (blood-forming tissue).

Fate in humans and animals

The absorption, distribution, and excretion of trichlorfon is rapid. About 70 to 80% of a dose administered orally to mice was excreted during the first 12 hours following treatment (108). Similar rapid elimination was seen in pigs following intraperitoneal injection (2,8). A hypothesized metabolite of trichlorfon, dichlorvos (DDVP), was found in some body tissues of exposed cows. Trichlorfon was found in cows' milk following "pour-on" applications of the insecticide (2,8).

Ecological effects

Effects on birds

Trichlorfon is moderately to highly toxic to birds. Signs of intoxication in birds include regurgitation, imbalance, trembling, slowness, lack of movement, and wing-beat convulsions. Signs of poisoning appear as soon as 10 minutes after exposure, and death usually occurs within 30 minutes to 3 hours of treatment (15).

The dietary LC_{50} for trichlorfon is 700 to 800 ppm in bobwhites. The estimated dietary LC_{50} is about 1800 ppm in 2-week-old Japanese quail that were fed treated feed for 5 days, followed by untreated feed for 3 days (14). Some 77% of exposed hen embryos were killed when 100 ppm trichlorfon (in acetone) was injected into their eggs (2,8). The acute oral LD_{50} for trichlorfon is 36.8 mg/kg in mallards, 22.4 mg/kg in bobwhite quail, 59.3 mg/kg in California quail, 95.9 mg/kg in male pheasant, and 23 mg/kg in rock dove (15).

Effects on aquatic organisms

Trichlorfon, in both technical and formulated forms, is very highly toxic to many aquatic species such as *Daphnia,* stoneflies, crayfish, and several freshwater fish species (14,15). Reported LC_{50} (96-hour) values are 0.18 mg/L (48-hour) in *Daphnia,* 0.01 mg/L in stoneflies, 7.8 mg/L in crayfish, 1.4 mg/L in rainbow trout, 2.5 mg/L in brook trout, 0.88 mg/L in channel catfish, and 0.26 mg/L in bluegill (14). Toxicity in the field can be affected by many factors, including temperature, pH, and water hardness, which may have different effects across species (14,15). In some species, temperature differences of 10°C could result in differences of 7- to 60-fold in observed 96-hour LC_{50} values (14). Effects of changing pH from 6.5 to 8.5 resulted in changes of 13- to 20-fold in several species (15). Generally, toxicity increased (i.e., observed LC_{50} was lower) with increasing temperature and pH.

Studies did not show a potential for trichlorfon to accumulate in fish (107).

Effects on other organisms (non-target species)

Trichlorfon has moderate to high acute toxicity toward certain beneficial or non-target insects (109). This pesticide may be toxic to other wildlife (107). Data indicate that trichlorfon has a low toxicity to bees; it can be used around bees with minimum injury (110).

Environmental fate

Breakdown in soil and groundwater

Trichlorfon breaks down, or degrades, rapidly in aerobic soils, with a half-life of between 3 and 27 days. An average half-life of 10 days has been reported (19). Its major breakdown product is dichlorvos (DDVP) (8). Trichlorfon is of low persistence in soil environments.

Trichlorfon does not adsorb strongly to soil particles, is readily soluble in water, and is very mobile in soils of varying textures and organic contents. It is therefore likely to contaminate groundwater (8). Soil organic matter content does not appear to influence trichlorfon's movement in soil (110).

Breakdown in water

Trichlorfon degrades rapidly in alkaline pond water (pH 8.5). Approximately 99% of applied trichlorfon was broken down within 2 hours. It was stable in the same pond water kept under acidic (pH 5.0) conditions for 2 hours. The major breakdown product of trichlorfon in water is dichlorvos (DDVP) (110). This insecticide was shown to persist at detectable levels for 526 days in water at 20°C (27).

Breakdown in vegetation

Studies on the dissipation of trichlorfon in forest environments indicate that it does not persist in leaves or leaf litter (107).

The approximate residual period is 7 to 10 days on plants. Injury has been reported on the foliage of apples, and on carnations and zinnias (109).

Physical properties

Trichlorfon is a pale clear, white, or yellow crystalline solid with an ethyl ether odor. Trichlorfon is a solid at room temperature (13).

Chemical name: dimethyl 2,2,2-trichloro-1-hydroxyethylphosphonate (13)
CAS #: 52-68-6
Molecular weight: 257.44 (13)
Water solubility: 120,000 mg/L @ 20°C (13)
Solubility in other solvents: s. in alcohols, ketones, dichloromethane, 2-propanol, methylene chloride, and toluene; s.s. in aromatic solvents; i.s. in *n*-hexane (13)
Melting point: 75–79°C (13)
Vapor pressure: 0.21 mPa @ 20°C (13)
Partition coefficient (octanol/water) (log): 5.75 (8)
Adsorption coefficient: 10 (19)

Exposure guidelines

ADI: 0.01 mg/kg/day (38)
HA: Not available
RfD: Not available
PEL: Not available

Basic manufacturer

Miles, Inc.
8400 Hawthorn Road
P.O. Box 4913
Kansas City, MO 64120
Telephone: 816-242-2429
Emergency: 816-242-2582

References

(1) Wagner, S. L. The acute health hazards of pesticides. In *Chemistry, Biochemistry, and Toxicology of Pesticides*. Witt, J. M., Ed. Oregon State University Cooperative Extension Service, Corvallis, OR, 1989.

(2) Gallo, M. A. and Lawryk, N. J. Organic phosphorus pesticides. In *Handbook of Pesticide Toxicology*. Hayes, W. J., Jr. and Laws, E. R., Jr., Eds. Academic Press, New York, 1991.

(3) Ware, G. W. *Fundamentals of Pesticides: A Self-Instruction Guide*. Thompson Publications, Fresno, CA, 1986.

(4) U.S. Department of Agriculture. Emergency Preparedness Branch. *The Pesticide Review*. Washington, D.C., 1991.

(5) U.S. Environmental Protection Agency. Memorandum from the Office of Pesticides and Toxic Substances to Office of Pesticide Programs Division Director, Washington, D.C., 1991.

(6) Smith, G. J. *Toxicology and Pesticide Use in Relation to Wildlife: Organophosphorus and Carbamate Compounds*. C. K. Smoley, Boca Raton, FL, 1993.

(7) Timbrell, J. A. *Principles of Biochemical Toxicology*. Taylor and Francis, Washington, D.C., 1991.

(8) National Library of Medicine. *Hazardous Substance Data Bank*. Washington, D.C., 1995.

(9) Shepard, T. H. *Catalog of Teratogenic Agents*, 5th ed. Johns Hopkins University Press, Baltimore, MD, 1986.

(10) U.S. Environmental Protection Agency. Dichlorvos: initiation of special review. *Fed. Reg.* 53, 5542–5549, 1988.

(11) U.S. Environmental Protection Agency. Dichlorvos: revocation of tolerance and food additive regulation. *Fed. Reg.* 56, 5788–5789, 1991.

(12) Howard, P. H., Ed. *Handbook of Environmental Fate and Exposure Data for Organic Chemicals*. Vol. 3, *Pesticides*. Lewis, Boca Raton, FL, 1991.

(13) Kidd, H. and James, D. R., Eds. *The Agrochemicals Handbook*, 3rd ed. Royal Society of Chemistry Information Services, Cambridge, U.K., 1991 (as updated).

(14) Hill, E. F. and Camardese, M. B. *Lethal Dietary Toxicities of Environmental Contaminants to Coturnix, Technical Report Number 2*. U.S. Department of Interior, Fish and Wildlife Service, Washington, D.C., 1986.

(15) Hudson, R. H., Tucker, R. K., and Haegele, M. A. *Handbook of Toxicity of Pesticides to Wildlife. Resource Publication 153*. U.S. Department of Interior, Fish and Wildlife Service, Washington, D.C., 1984.

(16) Johnson, W. W. and Finley, M. T. *Handbook of Acute Toxicity of Chemicals to Fish and Aquatic Invertebrates. Resource Publication 137*. U.S. Department of Interior, Fish and Wildlife Service, Washington, D.C., 1980.

(17) U.S. Environmental Protection Agency. *Pesticide Fact Sheet Number 100: Azinphos-methyl*. Office of Pesticides and Toxic Substances, Washington, D.C., 1986.

(18) U.S. Environmental Protection Agency. *Pesticide Fact Sheet Number 207: Coumaphos*. Office of Pesticides and Toxic Substances, Washington, D.C., 1989.

(19) Wauchope, R. D., Buttler, T. M., Hornsby A. G., Augustijn-Beckers, P. W. M., and Burt, J. P. SCS/ARS/CES pesticide properties database for environmental decisionmaking. *Rev. Environ. Contam. Toxicol.* 123, 1–157, 1992.

(20) Augustijn-Beckers, P. W. M., Hornsby, A. G., and Wauchope, R. D. SCS/ARS/CES pesticide properties database for environmental decisionmaking. II. Additional Compounds. *Rev. Environ. Contam. Toxicol.* 137, 1–82, 1994.

(21) Menzer, R. E. Water and soil pollutants. In *Casarett and Doull's Toxicology*, 4th ed. Amdur, M. O., Doull, J., and Klaassen, C. D., Eds. Pergamon Press, New York, 1991.

(22) Thomson, W. T. *Insecticides, Acaricides, and Ovicides. Agricultural Chemicals. Book I*. Thomson Publications, Fresno, CA, 1982.

(23) New York State Department of Health. *Chemical Fact Sheet: Guthion*. Bureau of Toxic Substances Management, Albany, NY, 1984.

(24) Anderson, C. A., Cavagnol, J. C., and Cohen, C. J. Guthion (azinphosmethyl): organophosphoros insecticide. *Residue Rev.* 51, 123–180, 1974.

(25) U.S. Environmental Protection Agency. Pesticide tolerance for O,O-dimethyl S-[(4-oxo-1,2,3-benzotriazin-3(4H)-yl)methyl] phosphorodithioate (azinphos-methyl). *Fed. Reg.* 54, 46082–46084, 1989.

(26) Tucker, R. and Crabtree, D. G. *Handbook of Toxicity of Pesticides to Wildlife.* U.S. Department of Agriculture, Fish and Wildlife Service, Washington, D.C., 1970.

(27) Pimentel, D. *Ecological Effects of Pesticides on Nontarget Species.* Executive Office of the President's Office of Science and Technology. U.S. Government Printing Office, Washington, D.C., 1971.

(28) Menzie, C. M. *Metabolism of Pesticides. Update III. Special Scientific Report, Wildlife No. 232.* U.S. Department of Interior, Fish and Wildlife Service, Washington, D.C., 1980.

(29) U.S. Environmental Protection Agency. *Guidance for the Reregistration of Pesticide Products Containing Azinphos Methyl as the Active Ingredient. Case Number 235.* Office of Pesticide Programs. U.S. Government Printing Office, Washington, D.C., 1986.

(30) Morse, R. A. Bee poisoning. In *1988 New York State Pesticide Recommendations. 49th Annu. Pest Control Conf.* Cornell University, Ithaca, NY, 1987.

(31) Wagenet, L. P. *A Review of Physical-chemical Parameters Related to the Soil and Groundwater Fate of Selected Pesticides in New York State, Report No. 30.* Cornell University Agricultural Experiment Station, Ithaca, NY, 1985.

(32) Holden, P. W. *Pesticides and Groundwater Quality: Issues and Problems in Four States.* National Academy Press, Washington, D.C., 1986.

(33) National Research Council. *Drinking Water and Health.* National Academy of Sciences. Washington, D.C., 1977.

(34) U.S. Environmental Protection Agency. *Pesticides in Groundwater Database.* Washington, D.C., 1992.

(35) Gillett, J. W. The biological impact of pesticides in the environment. *Environmental Health Sciences Series No. 1.* Oregon State University, Corvallis, OR, 1970.

(36) McEwen, F. L. and Stephenson, G. R. *The Use and Significance of Pesticides in the Environment.* John Wiley & Sons, New York, 1979.

(37) U.S. Environmental Protection Agency. *Chemical Profile: Azinphos-methyl.* Washington, D.C., 1985.

(38) Lu, F. C. A review of the acceptable daily intakes of pesticides assessed by the World Health Organization. *Regul. Toxicol. Pharmacol.* 21, 351–364, 1995.

(39) U.S. Occupational Safety and Health Administration. *Permissible Exposure Limits for Air Contaminants.* (29 CFR 1910. 1000, Subpart Z). U.S. Department of Labor, Washington, D.C., 1994.

(40) Buchel, K. H., Ed. and Holmwood, G. M., translator. *Chemistry of Pesticides.* John Wiley & Sons, New York, 1983.

(41) National Institute for Occupational Safety and Health. *Registry of Toxic Effects of Chemical Substances.* Cincinnati, OH, 1981–1986.

(42) U.S. Environmental Protection Agency. *Pesticides in Ground Water Data Base. 1988 Interim Report.* Washington, D.C., 1988.

(43) U.S. Environmental Protection Agency. *Registration Standard (Second Round Review) for the Reregistration of Pesticide Products Containing Chlorpyrifos.* Washington, D.C., 1989.

(44) Gosselin, R. E., Smith, R. P., and Hodge, H. C. *Clinical Toxicology of Commercial Products,* 5th ed. Williams and Wilkins, Baltimore, MD, 1984.

(45) Dow Chemical Company. *Summary of Acute Dermal Toxicity Study on Chlorpyrifos in Fischer 344 rats.* Indianapolis, IN, 1986.

(46) Dow Elanco Company. *Material Safety Data Sheet: Dursban Insecticidal Chemical-Unflaked.* Agricultural Products Division, Indianapolis, IN, 1992.

(47) American Conference of Governmental Industrial Hygienists, Inc. *Documentation of the Threshold Limit Values and Biological Exposure Indices, Fifth Edition.* Cincinnati, OH, 1986.

(48) U.S. Environmental Protection Agency. *Pesticide Fact Sheet Number 37: Chlorpyrifos.* Office of Pesticide and Toxic Substances, Washington, D.C., 1984.

(49) National Academy of Sciences. *Possible Long-term Health Effects of Short-term Exposure to Chemical Agents.* Vol. 1. *Anti-cholinesterases and Anticholinergics.* National Academy Press, Washington, D.C., 1982.

(50) U.S. Environmental Protection Agency. *Ambient Water Quality Criteria for Chlorpyrifos — 1986*. Washington, D.C., 1986.

(51) Racke, K. D. The environmental fate of chlorpyrifos. *Rev. Environ. Contam. Toxicol.* 131, 1–151, 1992.

(52) Schimmel, S. C. et al. Acute toxicity, bioconcentration, and persistence of AC 222, 705, benthiocarb, chlorpyrifos, fenvalerate, methyl parathion, and permethrin in the estuarine environment. *J. Agric. Food Chem.* 31(2), 399–407, 1983.

(53) U.S. Environmental Protection Agency. *Integrated Risk Information System Database*. Washington, D.C., 1994.

(54) U.S. Environmental Protection Agency. *Registration Standard for Pesticide Products Containing Coumaphos as the Active Ingredient*. Office of Pesticide Programs, Washington, D.C., 1989.

(55) National Cancer Institute. *Bioassay of Coumaphos for Possible Carcinogenicity. (Technical Report No. 96. NCI-CG-TR-96.)* National Institutes of Health, Bethesda, MD, 1979.

(56) Eisler, R. *Diazinon Hazards to Fish, Wildlife and Invertebrates: A Synoptic Review (Contaminant Hazard Review No. 9)*. U.S. Department of the Interior, Fish and Wildlife Service, Washington, D.C., 1986.

(57) Vettorazzi, G. Carbamate and organophosphorous pesticides used in agriculture and public health. *Residue Rev.* 63, 1–44, 1976.

(58) Bartsch, E. Diazinon. II. Residue in plants, soil and water. *Residue Rev.* 51, 37–68, 1974.

(59) U.S. Environmental Protection Agency. *Health Advisory Draft: Disulfoton*. Office of Drinking Water, Washington, D.C., 1987.

(60) U.S. Environmental Protection Agency. *Pesticide Fact Sheet Number 43: Disulfoton*. Office of Pesticides and Toxic Substances, Washington, D.C., 1984.

(61) Mobay Chemical Corporation. *Di-syston Insecticide*. Technical information. Chemagro Agricultural Division, Kansas City, MO, 1976.

(62) U.S. Environmental Protection Agency. *Registration Standards for Pesticide Products Containing Ethion as the Active Ingredient*. Office of Pesticides and Toxic Substances, Washington, D.C., 1989.

(63) U.S. Environmental Protection Agency. *Pesticide Fact Sheet Number 209: Ethion*. Office of Pesticides and Toxic Substances, Washington, D.C., 1989.

(64) U.S. Environmental Protection Agency. *Health Advisories for 50 Pesticides: Fenamiphos*. Washington, D.C., 1988.

(65) Knaak, J. B., Jacobs, K. C., and Wang, G. M. Estimating the hazard to humans applying nemacur 3EC with rat dermal-dose ChE response data. *Bull. Environ. Contam. Toxicol.* 37(2), 159–163, 1986.

(66) Balcomb, R., Stevens, R., and Bowen, C. Toxicity of 16 granular insecticides to wild caught songbirds. *Bull. Environ. Contam. Toxicol.* 33, 302–307, 1984.

(67) Harding, W. C. *Pesticide Profiles: Insecticides and Miticides. Bulletin 267*. University of Maryland Cooperative Extension Service, College Park, MD, 1979.

(68) Khan, M. A. Q., Ed. *Pesticides in Aquatic Environments*. Plenum Press, New York, 1977.

(69) U.S. Environmental Protection Agency. *Draft Health Advisory Summary: Fonofos*. Office of Drinking Water, Washington, D.C., 1987.

(70) U.S. Environmental Protection Agency. *Pesticide Fact Sheet Number 36: Fonofos*. Office of Pesticides and Toxic Substances, Washington, D.C., 1984.

(71) Hoffman, L. J., Ford, I. M., and Menn, J. J. Dyfonate metabolism studies. I. Absorption, distribution, and excretion of dyfonate in rats. *Pest. Biochem. Physiol.* 1, 349–355, 1971.

(72) Mayer, F. L. and Ellersieck, M. R. *Manual of Acute Toxicity: Interpretation and Data Base for 410 Chemicals and 66 Species of Freshwater Animals. Resource Publication 160*. U.S. Department of Interior, Fish and Wildlife Service, Washington, D.C., 1986.

(73) Ahmad, N., Walgenbach, D. D., and Sutter, G. R. Comparative disappearance of fonofos, phorate and terbufos soil residues under similar South Dakota field conditions. *Bull. Environ. Contam. Toxicol.* 23, 423–429, 1979.

(74) Broadberg, R. K. *Estimation of Exposure of Persons in California to Pesticide Products Containing Isofenphos*. California Department of Food and Agriculture. Division of Pest Management, Sacramento, CA, 1990.

(75) Wilson, B. W., Hooper, M., Chow, E., Higgins, R. J., and Knaack, J. B. Antidotes and neuropathic potential of isofenphos. *Bull. Environ. Contam. Toxicol.* 33, 386–394, 1984.

(76) U.S. Environmental Protection Agency. *Pesticide Environmental Fate One-Line Summary: Isofenphos.* Environmental Fate and Effects Division, Washington, D.C., 1990.

(77) Somasundaram, L., Racke, K. D., and Coats, J. R. Effects of manuring on the persistence and degradation of soil insecticides. *Bull. Environ. Contam. Toxicol.* 39, 579–586, 1987.

(78) Carlson, G. P. Factors modifying toxicity. In *Toxic Substances and Human Risk: Principles of Data Interpretation.* Tardiff, R. G. and Rodricks, J. V., Eds. Plenum Press, New York, 1987.

(79) Menzer, R. E. Selection of animal models for data interpretation. In *Toxic Substances and Human Risk: Principles of Data Interpretation.* Robert, G. T. and Rodricks, J. V., Eds. Plenum Press, New York, 1987.

(80) National Cancer Institute. *Bioassay of Malathion for Possible Carcinogenicity. Technical Reports 192.* National Institutes of Health, Bethesda, MD, 1979.

(81) U.S. Environmental Protection Agency. *Guidance for the Reregistration of Pesticide Products Containing Methidathion as the Active Ingredient.* Washington, D.C., 1988.

(82) Maddy, K. T., Gibbons, D., Richmond, D. M., and Fredrickson, S. A. *Potential Exposure of Leader/Applicators to Methidathion (SUPRACIDE) During Applications to Citrus In Riverside County, California in 1982.* California Department of Food and Agriculture, Division of Pest Management, Sacramento, CA, 1983.

(83) Gauthier, M. J., Berge, J. B., Cuany, A., Breittmayer, V., and Fournier, D. Microbial degradation of methidathion in natural environments and metabolism of this pesticide by *Bacillus coagulans. Pestic. Biochem. Physiol.* 31, 61–66, 1988.

(84) National Cancer Institute. *Bioassy of Methyl Parathion for Possible Carcinogenicity. DHEW Pub. No. (NIH) 79-1713. National Institute of Health,* Bethesda, MD, 1979.

(85) U.S. Environmental Protection Agency. Methyl 3-[(dimethoxyphosphinyl)oxy] butenoate, alpha and beta isomers: proposed tolerance. *Fed. Reg.* 49, 11854–11855, 1984.

(86) Shell Chemical Company. *Summary of Basic Data for Mevinphos Insecticide.* Technical data bulletin. San Ramon, CA, 1972.

(87) American Cyanamid. *Toxicological Summary for Thimet.* Wayne, NJ, 1992.

(88) U.S. Environmental Protection Agency. *Pesticide Fact Sheet Number 34.1: Phorate.* Office of Pesticides and Toxic Substances, Washington, D.C., 1985.

(89) Vettorazzi, G. Phorate. In *International Regulatory Aspects for Pesticide Chemicals.* Vol. 1. CRC Press, Boca Raton, FL, 1979.

(90) U.S. Environmental Protection Agency. *Chemical Information Fact Sheet. Technical Sheet 766C.* Washington, D.C., 1983–1985.

(91) Food and Drug Administration. *The FDA Surveillance Index.* National Technical Information Service, Springfield, VA, 1986.

(92) Blewett, C. T. and Krieger, R. I. *Estimation of Exposure of Persons in California to Pesticide Products that Contain Phosmet and Estimation of Effectiveness of Exposure Reduction Measures.* California Department of Food and Agriculture, Division of Pest Management, Sacramento, CA, 1988.

(93) U.S. Environmental Protection Agency. Tolerance for pesticides in food: propetamphos. *Fed. Reg.* 46, 43964–43965, 1981.

(94) Kumari, J. and Krishnamurthy, N. B. Mutagenicity studies with safrotin in *Drosophila melanogaster* and mice. *Environ. Res.* 41, 44–52, 1986.

(95) Wells, D. S., Afifi, L. M., Motoyama, N., and Dauterman, W. C. *In vitro* metabolism of propetamphos by house fly, cockroach, and mouse liver preparations. *J. Agric. Food Chem.* 34, 79–86, 1986.

(96) Rettich, F. Residual toxicity of wall-sprayed organophosphates, carbamates, and pyrethroids to mosquito. *J. Hyg. Epidemiol. Microbiol. Immunol.* 24, 110–117, 1980.

(97) U.S. Environmental Protection Agency. *Toxicology One-Line Summary: Temephos.* Environmental Fate and Effects Division, Washington, D.C., 1985.

(98) Franson, J. C. and Spann, J. W. Effects of dietary ABATE on reproductive success, duckling survival, behavior, and clinical pathology in game-farm mallards: temephos. *Arch. Environ. Contam. Toxicol.* 12, 529–534, 1983.

(99) Pierce, R. H., Brown, R. B., Hardman, K. R., Henry, M. S., Palmer, C. L., Miller, T. W., and Witcherman, G. Fate and toxicity of temephos applied to an intertidal mangrove community. *J. Am. Mosq. Control Assoc.* 4, 569–578, 1989.

(100) U.S. Environmental Protection Agency. Pesticide tolerance for terbufos. *Fed. Reg.* 54, 35896–35897, 1989.

(101) U.S. Environmental Protection Agency. *Health Advisory: Terbufos.* Office of Drinking Water, Washington, D.C., 1988.

(102) U.S. Environmental Protection Agency. *Pesticide Fact Sheet Number 5.2: Terbufos.* Office of Pesticide and Toxic Substances, Washington, D.C., 1988.

(103) Cobb, G. P., Hol, E. T., Allen, P. W., Gagne, J. A., and Kudall, R. J. Uptake, metabolism and toxicity of terbufos in the earthworm (*Lumbirens terrestris*) exposed to counter-15G in artificial soils. *Environ. Toxicol. Chem.* 14(2), 279–285, 1995.

(104) Felsot, A., Wei, L., and Wilson, J. Environmental chemodynamic studies with terbufos ("Counter") insecticide in soil under laboratory and field conditions. *J. Environ. Sci. Health.* B17(6), 649–673, 1982.

(105) Szeto, S. Y., Brown, M. J., Mackenzie, J. R., and Vernon, R. S. Degradation of terbufos in soil and its translocation into cole crops. *J. Agric. Food Chem.* 34, 876–879, 1986.

(106) U.S. Environmental Protection Agency. *Chemical Profile: Trichlorophon.* Washington, D.C., 1985.

(107) U.S. Environmental Protection Agency. *Guidance for Registration of Pesticide Products Containing Trichlorfon as the Active Ingredient.* Washington, D.C., 1984.

(108) Hallenbeck, W. H. and Cunningham-Burns, K. M. *Pesticides and Human Health.* Springer-Verlag, New York, 1985.

(109) Lambert, W. P. *Dylox: A Profile of Its Behavior in the Environment.* Roy F. Weston, Inc., West Chester, PA, not dated.

(110) U.S. Environmental Protection Agency. *Pesticide Fact Sheet Number 30: Trichlorfon.* Office of Pesticides and Toxic Substances, Washington, D.C., 1984.

chapter six

Chlorinated hydrocarbons

6.1 Class overview and general description

Background

Chlorinated hydrocarbons, also known as organochlorines, were used widely from the 1940s to the 1960s for agricultural pest control and for malarial control programs. Since the 1960s their use in the U.S. has been curtailed greatly because of their persistence in the environment, in wildlife, and in humans. The pesticide most responsible for this reduction was dichlorodiphenyltrichloroethane (DDT). DDT use has been eliminated in the U.S. though it is still applied in many regions throughout the world.

The organochlorines can be divided into three groups: 1) dichlorodiphenyl-ethanes (DDT and related compounds) (Figure 6.1A), 2) cyclodiene compounds (Figure 6.1B), and 3) other related compounds. In addition, particular organochlorines may consist of a number of related compounds. For example, toxaphene is made up of more than 177 related compounds.

Although there is no structure common to all organochlorines, they are all characterized by one or more chlorine atoms positioned around one or more hydrocarbon rings. Members of each group of organochlorines share similar or identical compositions although they may have very different three-dimensional structures and shapes. These isomers may differ significantly in their toxicities and other characteristics. The generic structures of dichlorophenyl ethanes and cyclodienes are shown in Figures 6.1A and 6.1B, respectively. The latter is a member of the cyclodiene group. Dichlorophenylethanes, cyclodienes, and other chlorinated hydrocarbons are listed in Table 6.1.

Chlorinated hydrocarbon usage

Organochlorines are powerful pesticides, and members of this group can be produced at relatively low cost. At one time, DDT sold to the World Health Organization (WHO) cost less than $0.22 per pound. DDT use reached a peak in 1961 when 160 million pounds were manufactured; 80% of that volume was used for agriculture. The other organochlorines also saw a great upsurge in use following World War II. Many of the commercially viable products, especially the cyclodienes such as aldrin, dieldrin, and heptachlor, were developed in the 1950s.

Lindane, also known as BHC, is an expensive compound to produce and is thus reserved for nonagricultural uses such as louse and mite control lotions.

When chlorinated hydrocarbon usage diminished in the 1960s and 1970s, they were replaced by the organophosphates (OPs) despite the higher mammalian acute toxicities of the OPs (1). Organochlorines still in use in the U.S. are utilized to protect a variety of crops and ornamental flowers, as well as to control house pests.

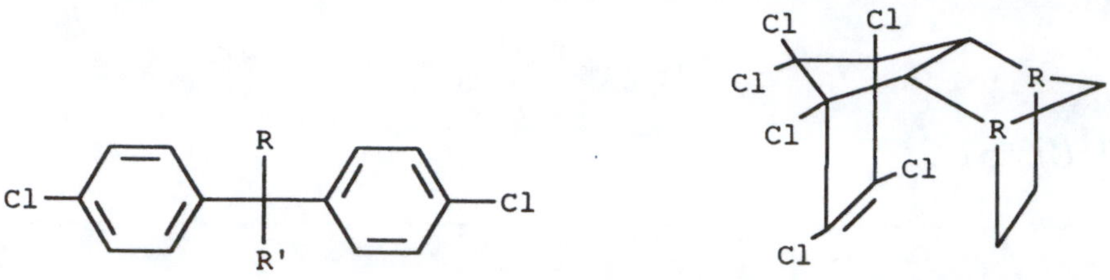

Figure 6.1 Structures of generic cyclodienes (A) and dichlorophenylethanes (B).

Table 6.1 Chlorinated Hydrocarbons

Dichlorophenylethanes
  Chlorobenzilate*
  DDT
  Dicofol*
  Methoxychlor*
Cyclodienes and related compounds
  Aldrin
  Chlordane*
  Dieldrin
  Endosulfan*
  Endrin
  Heptachlor*
  Toxaphene
Other chlorinated hydrocarbons
  Chlorothalonil*
  Dalapon
  Dienochlor
  Hexachlorobenzene (HCB)*
  Lindane*
  Mirex
  PCNB (Quintozene)*
  Pentachlorophenol*

Note: * indicates that a profile for this compound is included in this chapter.

Mechanism of action and toxicology

Mechanism of action

The chlorinated hydrocarbons are stimulants of the nervous system. Their mode of action is similar in insects and humans. They affect nerve fibers, along the length of the fiber, by disturbing the transmission of the nerve impulse. More specifically, the members of this group of pesticides disrupt the sodium/potassium balance that surrounds the nerve fiber. The result of this imbalance is a nerve that sends transmissions continuously rather than in response to stimuli.

Despite the similarity of many of the compounds within each of the three subgroups, the individual toxicities vary greatly (2). The compounds also vary greatly in their ability to be stored in tissue. For example, the structure of methoxychlor is very similar to DDT, but its toxicity is far lower, as is its tendency to accumulate in fatty tissue. Storage in fatty tissue is a strategy that the body uses to remove toxic materials from active circulation. Fatty storage prevents the toxic agent from reaching the target organ until it is remobilized in an organism, generally through metabolism of fat.

The toxicity of organochlorines, DDT in particular, is directly related to their concentration in nerve tissue. Acute and chronic effects are rapidly reversible when the concentration falls below some threshold level. The threshold levels vary with each compound. The abatement of symptoms, however, does not necessarily mean that the pesticide has been removed from the body, but rather that the compound has been removed from active circulation in the body (2).

Acute toxicity

Although each of the three subgroups of the chlorinated hydrocarbon compounds have rather distinctive sets of symptoms, they, as a class, mainly affect the central nervous system, and the symptoms of poisoning are muscular and behavioral effects. The most common symptom across the entire range of organochlorines is nervousness and hyperexcitement leading to tremors (3). The tremors may progress gradually to the point of convulsions. Some organochlorines, however, cause convulsions immediately after exposure (2).

These pesticides may be responsible for the onset of fever, although the specific reasons for the fever are currently unclear. It may be due to the direct poisoning of the temperature-control center in the brain, or the body's inability to rapidly get rid of heat generated by a convulsion, or other causes. Other symptoms of organochlorine poisoning include vomiting, nausea, confusion, and uncoordinated movement (2).

Chronic toxicity

Reproductive effects

Organochlorine compounds may adversely affect fertility and reproduction at high doses. In a 3-week dietary mouse study of chlordane, fertility was reduced by about 50% at a dose of 22 mg/kg/day (4).

In another study, rat offspring only experienced adverse effects when the doses, 6.25 and 12.5 mg/kg/day dicofol, were high enough to cause maternal toxicity (5).

At doses up to 100 mg/kg/day of another organochlorine, chlorobenzilate, there were no adverse reproductive effects in rats (6).

It is unlikely that organochlorine compounds will cause reproductive effects in humans at expected exposure levels.

Teratogenic effects

Most of the animal studies with organochlorine compounds have shown that there were no teratogenic effects (2). However, two of the organochlorine compounds, hexachlorobenzene (HCB) and dieldrin, have been shown to cause birth defects at high doses. In a rat study with HCB, some offspring had an extra rib and cleft palates (7). In a dietary study of dieldrin, mice experienced delayed bone formation and an increase in rib bones (2).

Based on all of the evidence, organochlorine compounds are unlikely to produce teratogenic effects in humans.

Mutagenic effects

In studies of nearly all of the commonly used organochlorine compounds, no mutagenic effects were found. The only exception was endosulfan, which was found to be mutagenic to bacterial and yeast cells (2).

Carcinogenic effects

In several chronic, high-dose exposure rat studies with organochlorine compounds such as chlordane, heptachlor, and pentachlorophenol, there were increased incidences of liver tumors. Because the above compounds have caused liver tumors in rats, they have been classified by the U.S. EPA as probable human carcinogens (8).

Ecological effects

Effect on birds

Organochlorine compounds are only slightly acutely toxic to birds. For example, the LD_{50} dose of lindane in bobwhite quail is 120 to 130 mg/kg (9). The LC_{50} value for DDT is 611 ppm in bobwhite quail, 311 ppm in pheasant, and 1869 ppm in mallard duck (10).

The evidence of bioaccumulation is most notable at the top of the food chain in the terrestrial community. Predatory birds contain the highest body burdens and thus suffer the most effects, generally reproductive failure. DDT and the other organochlorines can cause reproductive failure by disrupting the bird's ability to mobilize calcium, thus resulting in thin, brittle eggshells that may be crushed by the parents during incubation or attacked by bacteria (10).

Effects on aquatic organisms

The acute toxicity of organochlorine compounds to aquatic life varies but may be very high. For example, the LC_{50} value for toxaphene is <0.001 mg/L in freshwater fish. However, the LC_{50} value for lindane is 0.1 mg/L in freshwater fish (11).

The evidence of bioaccumulation is most notable at the top of the food chain in the aquatic community. Predatory fish contain the highest body burdens and thus suffer the most from reproductive failure. Fish reproduction can be affected when organochlorines, such as DDT, concentrate in the egg sac. At a DDT residue level of 2.4 mg/kg, eggs of the winter flounder contained abnormal embryos in the laboratory (10).

Effects on other organisms (non-target species)

Organochlorine compounds range from highly toxic to nontoxic to bees. Compounds such as chlordane and lindane are highly toxic, while dicofol and HCB are nontoxic to bees (9).

Environmental fate

Breakdown in soil and groundwater

Organochlorines are not mobile in soil because they are tightly bound to soil particles and do not dissolve in water. Some localized or regional movement of chlorinated hydrocarbon compounds may occur while attached to soil particles, either through the blowing of dust and soil or through soil erosion. Because organochlorine compounds bind tightly to soil, they resist leaching into the groundwater (12).

Of particular significance is the ability of organochlorines to persist for long periods in the environment in biologically active forms and to accumulate in living systems.

Most notable within this group of long-lasting insecticides are DDT and dieldrin. The average time it takes for half of a chlorinated hydrocarbon compound to disappear after it is applied to soil is between 2 and 10 years (13–15). For a compound with a half-life of 10 years, over 12% of the compound would remain after a 30-year period. The compound's resistance to biochemical degradation, coupled with its solubility in fats (lipids), leads to bioaccumulation in living organisms (12).

Breakdown in water

Most organochlorine compounds are insoluble in water (9) or dissolve very slowly in water. Methoxychlor has been detected at the Niagara River in New York at a very low concentration of 0.001 µg/L (12). Therefore, it is more likely that organochlorines will be found in the sediment.

Breakdown in vegetation

Organochlorines may accumulate in fruits and vegetables. For example, chlorobenzilate residues have been found in the peels of citrus fruits (16). When chlorobenzilate was sprayed on treated crops, it caused the browning of the edges and veins of leaves (17).

Worldwide dispersion

Recent evidence points to organochlorine movement throughout the world. Organochlorine compounds like DDT and toxaphene, while banned for use in the U.S., are still being used in other parts of the world. These compounds slowly evaporate and are translocated throughout the world by wind and rain. For example, toxaphene, prior to its ban in 1982, was used in the southern U.S. on a variety of crops. Even though it was not used in the northern U.S., it has been found as a widespread contaminant throughout the Great Lakes region and in marine fish (18).

Also, cyclodiene insecticides, such as chlordane, have been found in rainwater and organisms in Scandinavia though they have never been used in that area (19). Earlier notions about these pesticides remaining on or very near their application site have been revised as the result of recent studies. The physical and chemical properties of the organochlorines have led to their worldwide dispersion in the environment.

6.2   Individual profiles

6.2.1   Chlordane

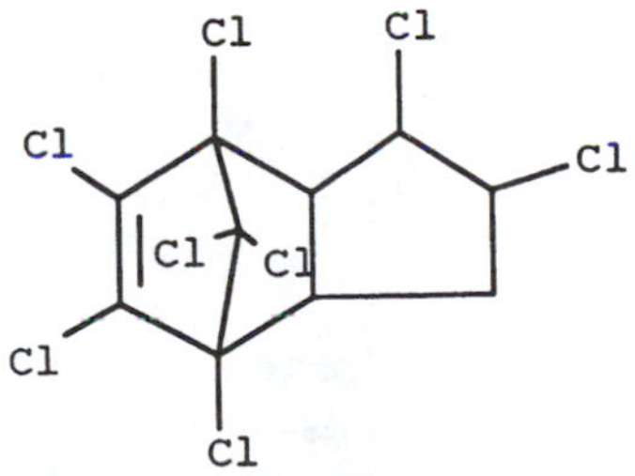

Figure 6.2 Chlordane.

Trade or other names

In addition to chlordane, common names have included chlordan and clordano. Trade names include Belt, Chlor Kil, Chlortox, Corodane, Gold Crest C-100, Kilex Lindane, Kypchlor, Niran, Octachlor, Synklor, Termex, Topiclor 20, Toxichlor, and Velsicol 1068.

Regulatory status

Because of concern about the risk of cancer, use of chlordane was canceled in April 1988. Between 1983 and 1988, the only permitted use for chlordane was for control of subterranean termites. Chlordane is no longer distributed in the U.S. The only commercial use still permitted is for fire ant control in power transformers. It was classified toxicity class II — moderately toxic. Products containing chlordane bear the Signal Word WARNING.

Introduction

Chlordane is a persistent organochlorine insecticide. It kills insects when ingested and on contact. Formulations include dusts, emulsifiable concentrates, granules, oil solutions, and wettable powders.

Toxicological effects

Acute toxicity

Chlordane is moderately to highly toxic through all routes of exposure. Symptoms usually start within 45 minutes to several hours after exposure to a toxic dose. Convulsions may be the first sign of poisoning or they may be preceded by nausea, vomiting, and gut pain. Initially, poisoning victims may appear agitated or excited, but later they may become depressed, uncoordinated, tired, or confused. Other symptoms reported in cases of chlordane poisoning include headaches, dizziness, vision problems, irritability, weakness, or muscle twitching. In severe cases, respiratory failure and death may occur. Complete recovery from a toxic exposure to chlordane is possible if proper medical treatment is administered (2,20). Chlordane is very irritating to the skin and eyes (21,22).

Chlordane affects liver function; thus, many interactions between medicines and this pesticide may occur. Among these are decreased effectiveness of anticoagulants, phenylbutazone, chlorpromazine, steroids, birth control pills, and diphenhydramine. Increased activity of thyroid hormone may also occur (23).

The oral LD_{50} for chlordane in rats is 200 to 700 mg/kg, in mice is 145 to 430 mg/kg, in rabbits is 20 to 300 mg/kg, and in hamsters is 1720 mg/kg (2,9). The dermal LD_{50} in rabbits is 780 mg/kg, and in rats is 530 to 690 mg/kg (9,17). The 4-hour inhalation LD_{50} in cats is 100 mg/L (17,24).

Chronic toxicity

Liver lesions and changes in blood serum occurred in rats exposed to 1.0 mg/L chlordane in air. Increased kidney weights occurred in rats exposed to 10 mg/L. For monkeys, increased liver weight occurred at 10 mg/L (20).

Animal studies have shown that consumption of chlordane caused damage to the liver and the central nervous system (20,21). In a 2-year feeding study with rats, a near-lethal dose of 300 mg/kg/day produced eye and nose hemorrhaging, severe changes in the tissues of the liver, kidney, heart, lungs, adrenal gland, and spleen. In this same study, no adverse effects were observed in rats fed 5 mg/kg/day. In a long-term feeding study with mice, body weight loss, increased liver weight, and death occurred at doses of 22 to 63.8 mg/kg/day. Dogs fed doses of 15 and 30 mg/kg/day exhibited increased liver weights (2,20).

Reproductive effects

Chlordane has been shown to affect reproduction in test animals. Fertility was reduced by about 50% in mice injected with chlordane at 22 mg/kg once a week for 3 weeks (25). The data suggest that reproductive effects in humans are unlikely at expected exposure levels.

Teratogenic effects

No teratogenic effects were observed in rats born to dams fed chlordane at 5 to 300 mg/kg/day for 2 years (20). It is unlikely that chlordane will cause teratogenic effects in humans.

Mutagenic effects

Chlorinated hydrocarbon insecticides (such as chlordane) are generally not mutagenic (2). It was reported that 15 of 17 mutagenicity tests performed with chlordane showed no mutagenic effects (25). Thus, chlordane is weakly or nonmutagenic.

Carcinogenic effects

The EPA has classified chlordane as a probable human carcinogen. Chlordane has caused liver cancer in mice given doses of 30 to 64 mg/kg/day for 80 weeks (24). However, a study was done on workers at a manufacturing plant who had been exposed to chlorinated hydrocarbons for 34 years, including chlordane. No increase in any type of cancer was found (24,25).

Organ toxicity

In clinical studies of acute or chronic exposure to chlordane, the effects most frequently observed were central nervous system effects, liver effects, and blood disorders (25). Chronic exposure to chlordane may cause jaundice in humans. Chlordane may also cause blood diseases, including aplastic anemia and acute leukemia in rats (20).

Fate in humans and animals

Chlordane is absorbed into the body through the lungs, stomach, and skin. It is stored in fatty tissues as well as in the kidneys, muscles, liver, and brain (2,20). Chlordane has been found in human fat samples at concentrations of 0.03 to 0.4 mg/kg in U.S. residents (20). Chlorinated hydrocarbons stored in fatty tissues can be released into circulation if these fatty tissues are metabolized, as in starvation or intense activity (2). Chlordane that is not stored in the body is excreted through the urine and feces. Chlordane has been found in human breast milk (25).

Rats that breathed chlordane vapor for 30 minutes retained 77% of the total amount inhaled. Rabbits that received four doses of chlordane stored it in fatty tissues, the brain, kidneys, liver, and muscles (2).

Excretion of orally administered chlordane is slow and can take days to weeks. The biological half-life of chlordane in the blood serum of a 4-year-old child who drank an emulsifiable concentrate of chlordane was 88 days. In another accidental poisoning of a 20-month-old child, the half-life was 21 days (20,25).

Ecological effects

Effects on birds

Chlordane is moderately to slightly toxic to birds. The LD_{50} in bobwhite quail is 83 mg/kg. The 8-day dietary LC_{50} for chlordane is 858 ppm in mallard duck, 331 ppm in bobwhite quail, and 430 ppm in pheasant (9,26).

Effects on aquatic organisms

Chlordane is very highly toxic to freshwater invertebrates and fish. The LC_{50} (96-hour) for chlordane in bluegill is 0.057 to 0.075 mg/L, and 0.042 to 0.090 mg/L in rainbow trout (9,17,26).

Chlordane bioaccumulates in bacteria and in marine and freshwater fish species (17). Expected bioaccumulation factors for chlordane are in excess of 3000 times background water concentrations, indicating that bioconcentration is significant for this compound.

Effects on other organisms (non-target species)

Chlordane is highly toxic to bees and earthworms (26). Studies done in the late 1970s showed that the fatty tissues of land and water wildlife contained large amounts of cyclodiene insecticides, including chlordane (20).

Environmental fate

Breakdown in soil and groundwater

Chlordane is highly persistent in soils, with a half-life of about 4 years. Several studies have found chlordane residues in excess of 10% of the initially applied amount 10 years or more after application (20). Sunlight may break down a small amount of the chlordane exposed to light (9). Evaporation is the major route of removal from soils (20). Chlordane does not chemically degrade and is not subject to biodegradation in soils. Despite its persistence, chlordane has a low potential for groundwater contamination because it is both insoluble in water and rapidly binds to soil particles, making it highly immobile within the soil (14). Chlordane molecules usually remain adsorbed to clay particles or to soil organic matter in the top soil layers and slowly volatilize into the atmosphere (14,20). However, very low levels of chlordane (0.01 to 0.001 µg/L) have been detected in both ground and surface waters in areas where chlordane was heavily used (21,25). Sandy soils allow the passage of chlordane to groundwater.

Breakdown in water

Chlordane does not degrade rapidly in water. It can exit aquatic systems by adsorbing to sediments or by volatilization. The volatilization half-life for chlordane in lakes and ponds is estimated to be less than 10 days (20).

Chlordane has been detected in surface water, groundwater, suspended solids, sediments, bottom detritus, drinking water, sewage sludge, and urban runoff, but not in rain water. Concentrations detected in surface water have been very low, while those found in suspended solids and sediments are always higher (<0.03 to 580 µg/L). The presence of chlordane in drinking water has almost always been associated with an accident rather than with normal use (20).

Breakdown in vegetation

No data are currently available.

Physical properties

Technical chlordane is actually a mixture of at least 23 different components, including chlordane isomers, other chlorinated hydrocarbons, and by-products. It is a viscous, colorless or amber-colored liquid with a chlorine-like odor (9).

Chemical name: 1,2,4,5,6,7,8,8-octachloro-2,3,3a,4,7,7a-hexahydro-4,7-metha-noindene (9)
CAS #: 57-74-9
Molecular weight: 409.83 (9)
Water solubility: 0.1 mg/L @ 25°C (9)
Solubility in other solvents: s. in most organic solvents, including petroleum oils (9)
Melting point: 104–107°C (9)
Vapor pressure: 1.3 mPa @ 25°C (9)
Partition coefficient (octanol/water) (log): 2.78 (17)
Adsorption coefficient: 20,000 (14)

Exposure guidelines

ADI: 0.0005 mg/kg/day (27)
MCL: 0.002 mg/L (8)
RfD: 0.00006 mg/kg/day (8)
PEL: 0.5 mg/m^3 (8-hour) (8)

Basic manufacturer

Velsicol Chemical Corporation
10400 W. Higgins Rd.
Rosemont, IL 60018–5119
Telephone: 708-298-9000

6.2.2 Chlorobenzilate

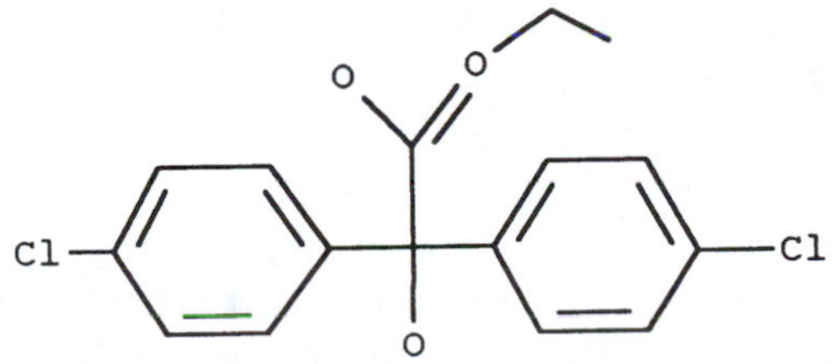

Figure 6.3 Chlorobenzilate.

Trade or other names

Trade names for chlorobenzilate include Acaraben, Akar 338, Benzilan, Benz-o-chlor, ECB, Folbex, Geigy 338, and Kop-mite.

Regulatory status

The U.S. Environmental Protection Agency (EPA) has classified all formulations containing chlorobenzilate as Restricted Use Pesticides (RUPs). RUPs may be purchased and used only by certified applicators. It is classified as an RUP based on its ability to cause tumors in mice and its effects on the testes of rats. Aerial and ground foliar sprays are restricted to citrus use in the states of Arizona, California, Florida, and Texas for the control of mites. Considered toxicity class III — slightly toxic, products containing chlorobenzilate bear the Signal Word CAUTION.

Introduction

Chlorobenzilate is a chlorinated hydrocarbon compound. It is used for mite control on citrus crops and in beehives. It has narrow insecticidal action, killing only ticks and mites. Products are available as emulsifiable concentrate or wettable powder formulations.

Toxicological effects

Acute toxicity

Chlorobenzilate is slightly toxic to humans. Symptoms of acute poisoning from ingestion of chlorobenzilate include incoordination, nausea, vomiting, fever, apprehension, confusion, muscle weakness or pain, dizziness, wheezing, and coma. Symptoms may occur within several hours after exposure. Death may result from discontinued breathing or irregular heartbeats (2,17). Chlorobenzilate is a severe eye irritant. It is mildly irritating to skin (2,17).

The oral LD_{50} is 2784 to 3880 mg/kg for chlorobenzilate in rats. The dermal LD_{50} is greater than 10,000 mg/kg in rats and rabbits (2,9).

Chronic toxicity

Prolonged or repeated exposure to chlorobenzilate may cause the same effects as acute exposure (2,17). After continuous exposure to chlorobenzilate, 16 out of 73 workmen tested had abnormal electrical activity of the brain. The most severe brain activity changes were seen in those persons exposed to the herbicide for 1 to 2 years (2,17). Chronic skin exposure to chlorobenzilate may cause inflamed skin or rashes. Chronic eye exposure may cause conjunctivitis (2,17).

Autopsies revealed intestinal irritation and bleeding in the lungs of rats poisoned by dietary doses of 25 mg/kg/day chlorobenzilate (2,17). Liver damage may be caused by repeated or prolonged contact (2,17).

Reproductive effects

A three-generation rat reproduction study resulted in reduced testicular weights, but did not affect reproduction. The results of another study indicate that chlorobenzilate does not adversely affect reproductive performance at dosage levels up to 100 mg/kg/day (2,29). Atrophy of testes was observed in a 2-year study of rats (2,17). It is unlikely that chlorobenzilate will cause reproductive toxicity in humans at expected exposure levels.

Teratogenic effects
No data are currently available.

Mutagenic effects
No data are currently available.

Carcinogenic effects
Chlorobenzilate is a suspected carcinogen in animals and a possible human carcinogen. It has produced liver tumors in mice, but the evidence for carcinogenicity in rats is uncertain (2).

Organ toxicity
Exposure to chlorobenzilate may affect the central nervous system, the kidneys, and the liver (2,17).

Fate in humans and animals
Chlorobenzilate is rapidly excreted by humans, usually within 3 to 4 days (2,17). After doses of 12.8 mg/kg/day to dogs, for 5 days a week, for 35 weeks, about 40% of the dose was excreted unchanged or as urinary metabolites. No significant storage in fat of dogs or rats was reported (2,17).

Detectable traces of chlorobenzilate were found in urine collected from Texas and Florida citrus-grove growers and workers. The results showed low levels in harvest-season pickers exposed to little or no chlorobenzilate, and higher levels among permanent or semipermanent workers employed during the spraying season. Among all workers, urinary values ranged from 0 to 63.6 ppm (30). This acaricide has not been found in human milk in the U.S. (17).

Ecological effects

Effects on birds

Chlorobenzilate is slightly toxic to practically nontoxic to birds. The 7-day dietary LC_{50} for chlorobenzilate is 3375 ppm in bobwhite quail. Its 5-day dietary LC_{50} in mallard ducks is greater than 8000 ppm (31).

Effects on aquatic organisms

An LC_{50} (96 hour) of 0.7 mg/L in rainbow trout and 1.8 mg/L in the bluegill indicate that chlorobenzilate is moderately to highly toxic to different species of fish (9,17). Chlorobenzilate is not expected to bioconcentrate in aquatic organisms (12).

Effects on other organisms (non-target species)

Chlorobenzilate is nontoxic to beneficial insects, including honeybees (9).

Environmental fate

Breakdown in soil and groundwater

Chlorobenzilate has a low persistence in soils (12,14). Its half-life in fine sandy soils was 10 to 35 days after application of 0.5 to 1.0 ppm chlorobenzilate. The removal is probably due to microbial degradation (12). Because chlorobenzilate is practically insoluble in water and it adsorbs strongly to soil particles in the upper soil layers, it is expected to exhibit low mobility in soils, and therefore be unlikely to leach to groundwater (12). Following a 5-day application of chlorobenzilate to several different citrus groves employing various tillage treatments, chlorobenzilate was not found in subsurface drainage waters, nor in surface runoff waters (32). Due to its strong adsorption to soil particles and low vapor pressure, chlorobenzilate is not expected to volatilize from soil surfaces (12,32).

Breakdown in water

Chlorobenzilate adsorbs to sediment and suspended particulate material in water. It is practically insoluble in water (17). It is not expected to volatilize but may be subject to biodegradation (12).

Breakdown in vegetation

Chlorobenzilate is fairly persistent on plant foliage and may be phytotoxic (or poisonous) to some plants (33). It is not absorbed or transported throughout a plant.

Chlorobenzilate residues have been found in the peel of citrus fruit. Its half-life in lemon and orange peels was from 60 to over 160 days (17). Spraying 200, 1000, and 5000 ppm chlorobenzilate in emulsions or suspensions caused leaf-browning on most treated crops (17). When chlorobenzilate was applied to the surface of soybean leaves, the miticide was quite stable and very little was absorbed and moved (or translocated) from one part of the plant to another (34).

Physical properties

Technical chlorobenzilate, a brownish liquid, contains approximately 90% active compound (17). Pure chlorobenzilate is a yellow solid (9).

Chemical name: ethyl 4,4'-dichlorobenzilate (9)
CAS #: 510-15-6
Molecular weight: 325.21 (9)
Water solubility: 10 mg/L @ 20°C (9)
Solubility in other solvents: benzene v.s.; acetone v.s.; methyl alcohol v.s.; toluene
    v.s.; hexane and alcohol v.s. (9)
Melting point: 37.5°C (9)
Vapor pressure: 0.12 mPa @ at 20°C (9)
Partition coefficient (octanol/water): Not available
Adsorption coefficient: 2000 (estimated) (14)

Exposure guidelines

ADI: 0.02 mg/kg/day (27)
MCL: Not available
RfD: 0.02 mg/kg/day (8)
PEL: Not available

Basic manufacturer

Ciba-Geigy Corp.
P.O. Box 18300
Greensboro, NC 27419–8300
Telephone: 800-334-9481
Emergency: 800-888-8372

6.2.3   Chlorothalonil

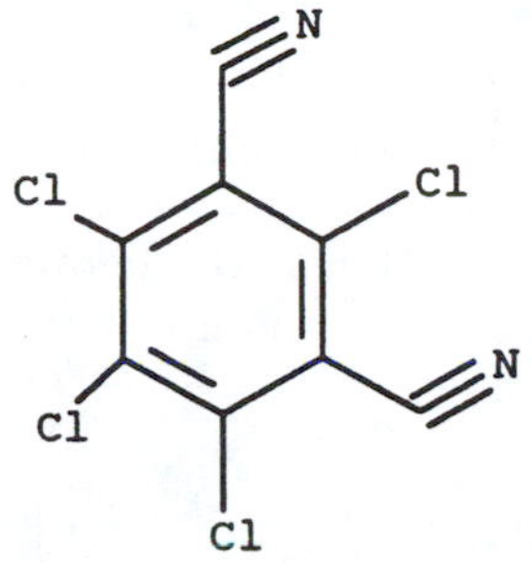

Figure 6.4 Chlorothalonil.

Trade or other names

Trade names for chlorothalonil include Bravo, Chlorothalonil, Daconil 2787, Echo, Exotherm Termil, Forturf, Mold-Ex, Nopcocide N-96, Ole, Pillarich, Repulse, and Tuffcide. The compound can be found in formulations with many other pesticide compounds.

Regulatory status

Chlorothalonil is classified as a General Use Pesticide (GUP) by the U.S. Environmental Protection Agency. It is classified as toxicity class II — moderately toxic, due to its potential for eye irritation. Chlorothalonil-containing products have a range of Signal Words, including WARNING (Bravo 720, 500), CAUTION (Exotherm Termil), and DANGER (Bravo W-75, Daconil W-75). Each of these products has a different formulation and product concentration and thus requires a different Signal Word.

Introduction

Chlorothalonil is a broad-spectrum organochlorine fungicide used to control fungi that threaten vegetables, trees, small fruits, turf, ornamentals, and other agricultural crops. It also controls fruit rots in cranberry bogs.

Toxicological effects

Acute toxicity

Chlorothalonil is slightly toxic to mammals, but it can cause severe eye and skin irritation in certain formulations (2). Very high doses may cause a loss of muscle coordination, rapid breathing, nose bleeding, vomiting, hyperactivity, and death.

Dermatitis, vaginal bleeding, bright yellow and/or bloody urine, and kidney tumors may also occur (17).

The oral LD_{50} is greater than 10,000 mg/kg in rats, and 6000 mg/kg in mice (9,17). The acute dermal LD_{50} in both albino rabbits and albino rats is 10,000 mg/kg (9,17). In albino rabbits, 3 mg chlorothalonil applied to the eyes caused mild irritation that subsided within 7 days of exposure (35).

Chronic toxicity

In a number of tests of varying lengths of time, rats fed a range of doses of chlorothalonil generally showed no effects on physical appearance, behavior, or survival (35). Skin contact with chlorothalonil may result in dermatitis or light sensitivity (35). Human eye and skin irritation is linked to chlorothalonil exposure; 14 of 20 workers exposed to 0.5% chlorothalonil in a wood preservative developed dermatitis. All workers showed swelling and inflammation of the upper eyelids (35). Allergic skin responses have also been noted in farm workers (7).

Reproductive effects

Administration of high doses of chlorothalonil to pregnant rabbits through the stomach during the sensitive period of gestation was required to induce abortion in four of the nine mothers. This and other studies suggest that chlorothalonil will not affect human reproduction at expected exposure levels (35).

Teratogenic effects

Long-term studies indicate that high doses fed to rats caused reduced weight gains for males and females in each generation studied (35). Female rats given high doses of chlorothalonil through the stomach during the sensitive period of gestation had normal fetuses, even though that dose was toxic to the mothers (35). A study of birth defects in rabbits showed no effects (36). Chlorothalonil is not expected to produce birth defects in humans.

Mutagenic effects

Mutagenicity studies on various animals, bacteria, and plants indicate that chlorothalonil does not cause any genetic changes (17,35,36). The compound is not expected to pose mutagenic risks to humans.

Carcinogenic effects

Based on evidence from animal studies, chlorothanolil's carcinogenic potential is unclear. Male and female rats fed chlorothalonil daily over a lifetime developed carcinogenic and benign kidney tumors at the higher doses (35). In another study, where mice were fed high daily doses of chlorothalonil for 2 years, females developed tumors in the fore-stomach area (attributed to irritation by the compound) and males developed carcinogenic and benign kidney tumors (35).

Organ toxicity

Chronic studies of rats and dogs fed high dietary levels show that chlorothalonil is toxic to the kidney. In addition to less urine output, changes in the kidney included enlargement, greenish-brown color, and development of small grains (37).

Fate in humans and animals

Chlorothalonil is rapidly excreted, primarily unchanged, from the body. It is not stored in animal tissues. Rats and dogs fed very high doses for 2 years eliminated almost all of the chemical in urine, feces, and expired air (17,38). At lower concentrations, chlorothalonil leaves the body within 24 hours. Residues have not been found in the tissues or milk of dairy cows fed chlorothalonil (17).

Ecological effects

Effects on birds

Chlorothalonil is practically nontoxic to birds. The LD_{50} in mallard ducks is 5000 mg/kg (9). Most avian wildlife are not significantly affected by this compound (17).

Effects on aquatic organisms

Chlorothalonil and its metabolites are highly toxic to fish, aquatic invertebrates, and marine organisms. Fish, such as rainbow trout, bluegills, and channel catfish, are noticeably affected even when chlorothalonil levels are low (less than 1 mg/L). The LC_{50} is 0.25 mg/L in rainbow trout, 0.3 mg/L in bluegills, and 0.43 mg/L in channel catfish (9).

Chlorothalonil does not store in fatty tissues and is rapidly excreted from the body. Its bioaccumulation factor is quite low (17).

Effects on other organisms (non-target species)

The compound is nontoxic to bees (9).

Environmental fate

Breakdown in soil and groundwater

Chlorothalonil is moderately persistent. In aerobic soils, the half-life is from 1 to 3 months. Increased soil moisture or temperature increases chlorothalonil degradation. It is not degraded by sunlight on the soil surface (17).

Chlorothalonil has high binding and low mobility in silty loam and silty clay loam soils, and has low binding and moderate mobility in sand (35).

Chlorothalonil was not found in any of 560 groundwater samples collected from 556 U.S. sites (35).

Breakdown in water

In very basic water (pH 9.0), about 65% of the chlorothalonil was degraded into two major metabolites after 10 weeks. Chlorothalonil was found in one surface water location in Michigan at 6.5 mg/L (35).

Breakdown in vegetation

Chlorothalonil's residues may remain on above-ground crops at harvest, but will dissipate over time. Chlorothalonil is a fairly persistent fungicide on plants, depending on the rate of application. Small amounts of one metabolite may be found in harvested crops (37).

Physical properties

Chlorothalonil is an aromatic halogen compound, a member of the chloronitrile chemical family. It is a grayish to colorless crystalline solid that is odorless to slightly pungent (9).

Chemical name: tetrachloroisophthalonitrile (9)
CAS #: 1897-45-6
Molecular weight: 265.92 (9)
Solubility in water: 0.6 mg/L @ 25°C (9)
Solubility in solvents: acetone s.s.; dimethyl sulfoxide s.s.; cyclohexanone s.s.;
    kerosene i.s.; xylene s.s. (9)
Melting point: 250–251°C (9)
Vapor pressure: 1.3 Pa @ 40°C (9)
Partition coefficient (octanol/water) (log) 437 (calc.): 20.9 (17)
Adsorption coefficient: 1380 (14)

Exposure guidelines

ADI: 0.03 mg/kg/day (27)
HA: 0.5 mg/L (longer-term) (35)
RfD: 0.015 mg/kg/day (8)
PEL: Not available

Basic manufacturer

Crystal Chemical Inter-America
10303 N.W. Freeway, Suite 512
Houston, TX 77083
Telephone: 713-956-6196

6.2.4   Dalapon

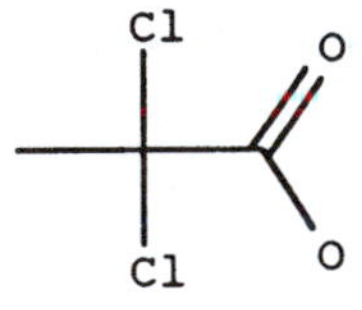

Figure 6.5 Dalapon.

Trade or other names

Trade names for dalapon include Alatex, Basinex P, Dalacide, Dalapon-Na (Dalapon-Sodium), Devipon, Ded-Weed, Dowpon, DPA, Gramevin, Kenapon, Liropon, Radapon, Revenge, and Unipon. Dalapon is also called sodium dalapon or magnesium dalapon.

Regulatory status

Dalapon is classified by the U.S. Environmental Protection Agency (EPA) as a General Use Pesticide (GUP). Dalapon is in toxicity class II — moderately toxic. Products containing the herbicide bear labels with the Signal Word WARNING.

Introduction

Dalapon is an organochlorine herbicide and plant growth regulator used to control specific annual and perennial grasses, such as quackgrass, Bermuda grass, Johnson grass, as well as cattails and rushes. The major food crop use of dalapon is on sugarcane and sugar beets. It is also used on various fruits, potatoes, carrots, asparagus, alfalfa, and flax, as well as in forestry, home gardening, and in or near water to control reed and sedge growth. Dalapon is applied both before the target plant comes up and after the plant emerges. Commercial products consist of the sodium salt or mixed sodium and magnesium salts of dalapon.

Toxicological effects

Acute toxicity

Dalapon is moderately toxic to humans. Skin and inhalation exposure could be of significance to dalapon production workers, pesticide applicators, and some agricultural workers (39). Symptoms of high acute exposure include loss of appetite, slowed heartbeat, skin irritation, eye irritation such as conjunctivitis or corneal damage, gastrointestinal disturbances such as vomiting or diarrhea, tiredness, pain, and irritation of the respiratory tract (40). Dalapon is an acid that may cause corrosive injury to body tissues (17). Eye exposure to this material can cause permanent eye damage. Skin burns may occur from dermal exposure to dalapon, especially when skin is moist.

Oral LD_{50} values range from 9330 mg/kg in male rats to 7570 mg/kg in female rats (4). The oral LD_{50} is 3860 mg/kg in female rabbits, and greater than 4600 mg/kg in female guinea pigs. Dalapon is moderately irritating to skin and eyes (17,41). The application of the sodium salt of dalapon (in a dry powder formulation) to rabbit eyes produced pain and irritation, followed by severe conjunctivitis and corneal injury, which healed after several days (17).

Chronic toxicity

Long-term dalapon feeding studies in dogs and rats did show increased kidney weights in animals fed very high daily doses (17,41). Rats fed 50 mg/kg/day for 2 years showed a slight average increase in kidney weight. No adverse effects were seen in this study in rats fed 15 mg/kg/day. In a 1-year feeding study with dogs fed 100 mg/kg/day, there was a slight average increase in kidney weight. No adverse effects were seen at 50 mg/kg/day (17,41). These mild effects on the kidneys are consistent with data that show that ingested dalapon is rapidly excreted in the urine.

Reproductive effects

Tests indicate that dalapon does not produce adverse effects on fertility or reproduction, except at extremely high doses (41).

Teratogenic effects

Sodium dalapon was not teratogenic in the rat at doses as high as 2000 mg/kg/day (42).

Mutagenic effects

Dalapon was not mutagenic when tested in several organisms (42).

Carcinogenic effects

No carcinogenic effects were seen in rats fed the sodium salt of dalapon at 5, 15, or 50 mg/kg/day for 2 years (41,42).

Organ toxicity

Dalapon dust and vapor may be irritating to the respiratory tract (17). Repeated or prolonged exposure to dalapon may cause irritation to the mucous membrane linings of the mouth, nose, throat, and lungs, and to the eyes (17). Chronic skin contact can lead to moderate irritation or even mild burns, although occasional contact is not likely to produce irritation. Dalapon is not absorbed through the skin in toxic amounts (41).

Fate in humans and animals

The half-life of dalapon in human blood is 1.5 to 3 days (39). Dalapon's half-life in the blood system of dogs is about 12 hours (39,41).

Dalapon and all of its known breakdown products dissolve easily in water. They are readily washed from cells and tissues. Because dalapon is insoluble in organic solvents and lipids, it does not build up in animal tissues. A nonmetabolized form of dalapon was excreted in the urine of animals fed the herbicide. Less than 1% of the ingested dose appeared as residues in the milk of dairy cows that were fed dalapon (17,39).

Ecological effects

Effects on birds

Dalapon is practically nontoxic to birds. When dalapon was fed to 2-week old birds for 5 days, followed by untreated feed for 3 days, the LC_{50} of dalapon was more than 5000 ppm in mallards, ring-necked pheasants, and Japanese quail (17,43). The acute oral LD_{50} of dalapon is 5660 mg/kg for chickens. While dalapon is practically nontoxic to birds, reproduction rates of birds are decreased at very high doses (17). Reproduction was depressed in mallard ducks fed one fourth the dose of dalapon that caused death (43).

Effects on aquatic organisms

Dalapon is practically nontoxic to fish (43). While there were no deaths reported in goldfish after a 24-hour exposure to 100 mg/L dalapon, all fish died after a similar exposure to 500 mg/L or above (17). The 1-to-21-day LC_{50} values for dalapon in fish are all on the order of 100 mg/L for several species tested (9,17). The LC_{50} for dalapon in bluegill is 105 mg/L (9,17).

Its toxicity to aquatic invertebrates varies, depending on the species. Values can be as low as the 48-hour effective concentration (EC) of 1 mg/L in brown shrimp, or as high as the 96-hour LC_{50} of 4800 mg/L in other crustaceans. Aquatic crustaceans and insects are the most dalapon sensitive of the aquatic invertebrates. Dalapon is only slightly toxic to mollusks (9,17).

Effects on other organisms (non-target species)

Dalapon is relatively nontoxic to honeybees and other insects and has low toxicity to soil microorganisms (9,17).

Environmental fate

Breakdown in soil and groundwater

Dalapon has a low to moderate persistence in soil. It remains in the soil for 2 to 8 weeks (14). Dalapon has residual activity in soil for 3 to 4 months when applied at high rates (22 kg/ha) (9).

Dalapon does not readily bind to soil particles. In clay and clay loam soils, there may be no adsorption. Since it does not adsorb to soil particles, dalapon has a high degree of mobility in all soil types and leaching does occur. However, dalapon movement in soil is usually limited by rapid and complete breakdown of the herbicide into naturally occurring compounds by soil microorganisms (12,14). Dalapon is not found below the first 6-inch soil layer. Higher temperatures and increased soil moisture speed up degradation. At higher temperatures, dalapon can also be degraded by UV light from the sun (39). In a national groundwater survey, dalapon was not found in groundwater (17).

Breakdown in water

In ponds and streams, dalapon disappears via microbial degradation, hydrolysis, and photolysis (12). Microbial degradation tends to be the most active form of its breakdown in water. In the absence of microbial degradation, the half-life of dalapon, by chemical hydrolysis, is several months at temperatures less than 25°C. Hydrolysis is accelerated with increasing temperature and pH (39,42).

Breakdown in vegetation

Dalapon is absorbed by plant roots and leaves and moved (or translocated) within plants (9). It tends to build up in the areas of greatest plant metabolic activity, such as developing seeds and in the tips of roots, shoots, and leaves. At high rates of application, dalapon precipitates out of solution as an acid, and has immediate and local acute effects on foliage (17). It is easily washed off foliage. In addition to herbicidal activity, dalapon is a plant-growth inhibitor. Conditions of increased light and high temperature may cause nutrient solutions or soil applications of dalapon to build up in the tops of plants, via transpiration (17).

Physical properties

Dalapon is a type of acid that is usually formulated with sodium and magnesium salts (44). The acid itself is not used directly. Commercial products usually contain 85% sodium salt or mixed sodium and magnesium salts of dalapon (17). In its pure acid form, dalapon is a colorless liquid with an acrid odor. As sodium-magnesium salts, it is a white to off-white powder (9,39).

Chemical name: 2,2-dichloropropionic acid (9)
CAS #: 127-20-8 (sodium salt); 75-99-0 (acid)
Molecular weight: 164.95 (sodium salt) (9)
Water solubility: 900,000 mg/L @ 25°C (sodium salt) (9)
Solubility in other solvents: alkali solvents v.s.; ethanol v.s.; acetone, benzene, and methanol s. (9)
Melting point: (with decomposition) 166.5°C (sodium salt) (9)

Vapor pressure: 0.01 mPa @ 20°C (sodium salt) (9)
Partition coefficient (octanol/water) (log): 0.778 (17)
Adsorption coefficient: 1 (sodium salt) (14)

Exposure guidelines

ADI: Not available
MCL: 0.2 mg/L (8)
RfD: 0.03 mg/kg/day (8)
PEL: Not available

Basic manufacturer

BASF Corp.
Agricultural Products Group
P.O. Box 13528
Research Triangle Park, NC 27709–3528
Telephone: 800-669-2273
Emergency: 800-832-4357

6.2.5   Dicofol

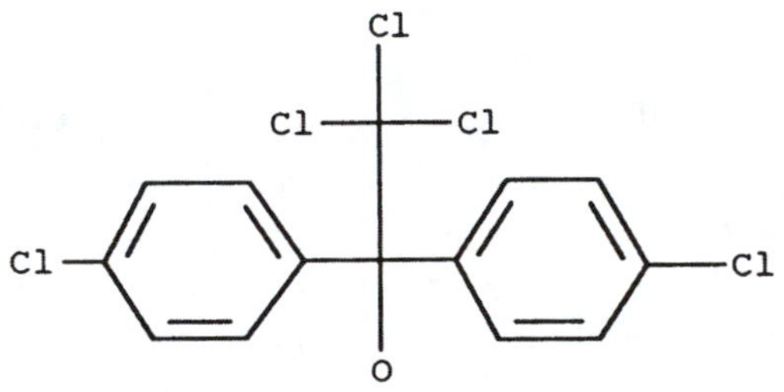

Figure 6.6 Dicofol.

Trade or other names

Trade names for dicofol include Acarin, Cekudifol, Decofol, Dicaron, Dicomite, Difol, Hilfol, Kelthane, and Mitigan.

Regulatory status

The EPA has classified dicofol as toxicity class II — moderately toxic, and toxicity class III — slightly toxic, depending on the formulation. Products containing dicofol bear the Signal Word WARNING or CAUTION, depending on the formulation. Products containing dicofol are designated General Use Pesticides (GUPs).

Introduction

Dicofol is an organochlorine miticide used on a wide variety of fruit, vegetable, ornamental, and field crops.

Dicofol is manufactured from DDT. In 1986, use of dicofol was temporarily canceled by the EPA because of concerns raised by high levels of DDT contamination.

However, it was reinstated when it was shown that modern manufacturing processes can produce technical grade dicofol that contains less than 0.1% DDT.

Toxicological effects

Acute toxicity

Dicofol is moderately toxic to practically nontoxic and may be absorbed through ingestion, inhalation, or skin contact. Symptoms of exposure include nausea, dizziness, weakness and vomiting from ingestion or respiratory exposure, skin irritation or rash from dermal exposure, and conjunctivitis from eye contact. Poisoning may affect the liver, kidneys, or the central nervous system. Overexposure by any route may cause nervousness and hyperactivity, headache, nausea, vomiting, unusual sensations, and fatigue. Very severe cases may result in convulsions, coma, or death from respiratory failure (44,45).

Dicofol is a moderate skin and eye irritant (17,45). Since dicofol is stored in fatty tissues, intense activity or starvation may mobilize the pesticide, resulting in the reappearance of toxic symptoms long after actual exposure (17).

The oral LD_{50} for dicofol in rats is 575 to 960 mg/kg, in rabbits and guinea pigs is 1810 mg/kg, and in mice is 420 to 675 mg/kg. The dermal LD_{50} in rats is 1000 to 5000 mg/kg, and in rabbits is between 2000 and 5000 mg/kg. The inhalation LC_{50} (4-hour) in rats is greater than 5 mg/L (9,7,45).

Chronic toxicity

In a 2-year dietary study with rats, liver growth, enzyme induction, and other changes in the liver, adrenal gland, and urinary bladder were observed at doses of 2.5 mg/kg/day and above. Effects on the liver, kidney, and adrenals, and reduced body weights were observed at doses of 6.25 mg/kg/day and above in a 3-month dietary study with mice (45).

When dicofol was fed to rats for 3 months, fewer than half of the animals survived a 75-mg/kg/day dose. Liver enzyme induction was observed at 75 mg/kg/day and above. Decreased body weights, decreased cortisone levels, and toxic changes in the liver, adrenal glands, and kidneys were noted at 25 mg/kg/day. Similar results were observed in a 3-month feeding study with mice (44).

When dogs were fed dicofol for 3 months, only 2 dogs out of 12 survived at 25 mg/kg/day. Poisoning symptoms and effects on the liver, heart, and testes were observed at the 7.5-mg/kg/day dose (44). When dicofol was fed to dogs, 4.5 mg/kg/day for 1 year caused toxic effects on the liver. Long-term dermal exposure of rats to dicofol as an emulsifiable concentrate formulation also produced toxic effects on the liver (44).

Reproductive effects

Reproductive effects in rat offspring have been observed only at doses high enough to also cause toxic effects on the livers, ovaries, and feeding behavior of the parents. Rats fed diets containing dicofol through two generations exhibited adverse effects on the survival and/or growth of newborns at 6.25 and 12.5 mg/kg/day (44).

Teratogenic effects

No teratogenic effects were observed when rats were given up to 25 mg/kg/day on days 6 through 15 of pregnancy (44).

Mutagenic effects

Five separate laboratory tests have shown that dicofol is not mutagenic (44,45).

Carcinogenic effects

No evidence of carcinogenicity was observed in rats fed up to 47 mg/kg/day for 78 weeks. A 2-year oncogenicity study in mice showed an increased incidence of liver tumors in male mice at dietary concentration levels of 13.2 and 26.4 mg/kg/day (45). It is unlikely that dicofol poses a carcinogenic risk to humans.

Organ toxicity

Chronic exposure to dicofol can cause damage to the kidney, liver, and heart.

Prolonged or repeated exposure to dicofol can cause the same effects and symptoms as acute exposure (17). Prolonged or repeated skin contact can cause moderate skin irritation and/or sensitization of the skin (45).

Fate in humans and animals

Dicofol is converted in rats to the metabolites 4,4'-dichlorobenzophenone and 4,4'-dichlorodicofol (2,46). Studies of the metabolism of dicofol in rats, mice, and rabbits have shown that ingested dicofol is rapidly absorbed, distributed primarily to fat, and readily eliminated in feces. When mice were given a single oral dose of 25 mg/kg dicofol, approximately 60% of the dose was eliminated within 96 hours, 20% in the urine, and 40% in the feces. Concentrations in body tissues peaked between 24 and 48 hours following dosing, with 10% of the dose found in fat, followed by the liver and other tissues. Levels in tissues other than fat declined sharply after the peak. When rats were given a single oral dose of 50 mg/kg dicofol, all but 2% of the dose was eliminated within 192 hours, with peak concentrations in body tissues occurring between 24 and 48 hours after dosing (44).

Ecological effects

Effects on birds

Dicofol is slightly toxic to birds. The 8-day dietary LC_{50} is 3010 ppm in bobwhite quail, 1418 ppm in Japanese quail, and 2126 ppm in ring-necked pheasant. Eggshell thinning and reduced offspring survival were noted in the mallard duck, American kestrel, ring dove, and screech owl (45).

Effects on aquatic organisms

Dicofol is highly toxic to fish, aquatic invertebrates, and algae. The LC_{50} is 0.12 mg/L in rainbow trout, 0.37 mg/L in sheepshead minnow, 0.06 mg/L in mysid shrimp, 0.015 mg/L in shell oysters, and 0.075 mg/L in algae (45).

Effects on other organisms (non-target species)

Dicofol is not toxic to bees (9).

Environmental fate

Breakdown in soil and groundwater

Dicofol is moderately persistent in soil, with a half-life of 60 days (14,46). Dicofol is susceptible to chemical breakdown in moist soils (12). It is also subject to degradation by UV light. In a silty loam soil, its photodegradation half-life was 30 days. Under anaerobic soil conditions, the half-life for dicofol was 15.9 days (46).

Dicofol is practically insoluble in water and adsorbs very strongly to soil particles. It is therefore nearly immobile in soils and unlikely to infiltrate groundwater. Even in sandy soil, dicofol was not detected below the top 3 inches in standard soil column tests. It is possible for dicofol to enter surface waters when soil erosion occurs (46,14).

Breakdown in water

Dicofol degrades in water or when exposed to UV light at pH levels above 7. Its half-life in solution at pH 5 is 47 to 85 days. Because of its very high absorption coefficient (K_{oc}), dicofol is expected to adsorb to sediment when released into open waters (12).

Breakdown in vegetation

In a number of studies, dicofol residues on treated plant tissues have been shown to remain unchanged for up to 2 years (46).

Physical properties

Pure dicofol is a white crystalline solid. Technical dicofol is a red-brown or amber viscous liquid with an odor like fresh-cut hay (9,45).

Chemical name: 2,2,2-trichloro-1,1-bis(y-chlorophenyl)ethanol (9)
CAS #: 115-32-2
Molecular weight: 370.51 (9)
Water solubility: 0.8 mg/L @ 25°C (9)
Solubility in other solvents: s. in most organic solvents (9)
Melting point: 78.5–79.5°C for pure dicofol (1,5); 50°C for technical dicofol (9,45)
Vapor pressure: Negligible at room temperature (9,45)
Partition coefficent (octanol/water): 19,000 (9,45)
Adsorption coefficient: 5000 (estimated) (14)

Exposure guidelines

ADI: 0.002 mg/kg/day (27)
MCL: Not available
RfD: Not available
PEL: Not available

Basic manufacturer

Rohm and Haas Co.
Agricultural Chemicals
100 Independence Mall West
Philadelphia, PA 19106
Telephone: 215-592-3000

6.2.6 Dienochlor

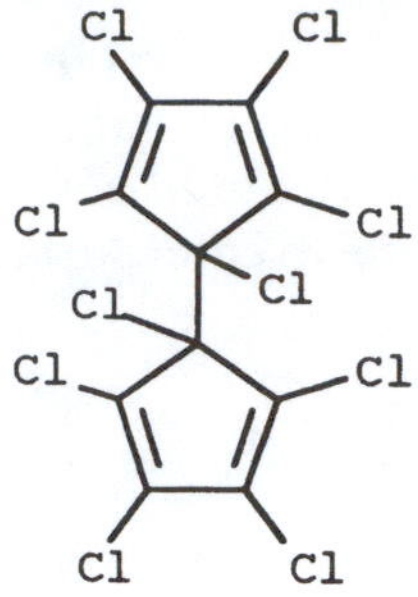

Figure 6.7 Dienochlor.

Trade or other names

Trade names for products containing dienochlor include Pentac WP and Pentac Aquaflow. The compound may be found in formulations with a wide variety of other common pesticides.

Regulatory status

Dienochlor is a General Use Pesticide (GUP). The U.S. EPA has classified it as toxicity class III — slightly toxic. Products containing dienochlor bear the Signal Word WARNING because they are moderately toxic when inhaled.

Introduction

Dienochlor is an organochlorine insecticide with contact action. It is used for the control of plant-damaging mites on a variety of ornamental shrubs and trees outdoors and in greenhouses. The compound may also be used on non-food ornamental crops. Dienochlor disrupts the egg-laying ability (oviposition) of female mites.

Toxicological effects

Acute toxicity

Symptoms of acute dienochlor exposure are similar to those of other organochlorine compounds and may include stimulation of the central nervous system (tremors, convulsions, agitation, and nervousness), slowing of breathing, nausea, vomiting, and diarrhea (47).

The oral LD_{50} for technical dienochlor is 3160 mg/kg in male rats, indicating that the compound is only slightly toxic by this route of exposure (47). Dienochlor is only slightly toxic by exposure through the skin. The dermal LC_{50} for the compound is greater than 3160 mg/kg in rabbits (47). Acute inhalation studies with the product Pentac 50 WP indicate that dienochlor is moderately toxic by this route of exposure. The LC_{50} value ranged between 1.4 and 2.4 mg/L in rats (47).

Dienochlor is not a primary skin irritant or a skin sensitizer, and is only a mild eye irritant. Rabbits exposed to a single dose of the technical product (dose undisclosed) experienced corneal opacity and irritation. The condition abated completely within 7 days (47).

Chronic toxicity

Two subchronic feeding studies were conducted over 3-month periods. Above 6.3 mg/kg/day, rats experienced a reduction in body weight gain. At 64 mg/kg/day, mice experienced increased mortality, inactivity, hunchbacked-walk, decreased body weight gain, changes in blood and urine chemistry, and altered organ weights. The spleen and thymus also showed atrophy (17,47).

Rats fed dienochlor in their diets over 2 weeks had no effects at or below 5 mg/kg/day (17,47).

Reproductive effects

No data are currently available.

Teratogenic effects

No birth defects appeared in the offspring of pregnant rats fed up to 50 mg/kg/day dienochlor in their food (17).

Mutagenic effects

Tests evaluating the mutagenicity of dienochlor have produced mixed results, but suggest that the compound is nonmutagenic or weakly mutagenic (47). This indicates that the mutagenic risk to humans is unlikely.

Carcinogenic effects

No data are currently available.

Organ toxicity

Animal tests have shown the liver, kidneys, spleen, and thymus to be affected by dienochlor exposure.

Fate in humans and animals

Female rats fed a single, low dose (1 mg/kg) of dienochlor excreted nearly 90% of the breakdown products of the compound in the feces and only 2% in the urine (48). Nearly all of the dienochlor was broken down in the rats within 1 day. At this dose after 4 days, only 2% of the initial dose remained in the rat in the liver, kidneys, stomach, and intestines. Dienochlor is poorly absorbed through the stomach and intestines. This may account for its low oral toxicity (high oral LD_{50}) (48).

When the compound was administered on the skin of the rats, only a very small amount passed through the skin to the bloodstream (2%) (48). It is expected that even less would penetrate the skin of humans. Only 1% of the applied dose was detected in the urine, and less than 0.2% in the tissues.

Ecological effects

Effects on birds

Dienochlor is practically nontoxic to bobwhite quail and mallard ducks. The oral LD_{50} for the compound in the quail is 705 mg/kg, and the 8-day dietary LC_{50} for dienochlor in mallards is nearly 4000 ppm (9,17).

Effects on aquatic organisms

Tests with several species of fish indicate that the compound is highly to very highly toxic to this group of organisms. The LC_{50} for dienochlor is 0.6 mg/L in bluegill sunfish and 0.05 mg/L in rainbow trout (9,49). Dienochlor is only moderately toxic to the freshwater invertebrate *Daphnia magna*. There are no data available on the potential of the compound to accumulate in aquatic organisms.

Effects on other organisms (non-target species)

Dienochlor is practically nontoxic to bees (9).

Environmental fate

There is very little information about the fate of the compound in the environment. Few studies have been conducted in this area. One study indicated that the compound is nonpersistent (50). A nonpersistent compound only lasts in the environment from a few hours to up to 12 weeks (14).

Dienochlor is readily broken down by the action of sunlight (50).

Physical properties

Pure dienochlor is a colorless crystalline solid. The technical product is a light yellow powder (9).

Chemical name: perchloro-1,1'-bicyclopenta-2,4-dienyl (9)
CAS #: 2227-17-0
Molecular weight: 474.64 (9)
Solubility in water: 25 mg/L @ 20–25°C (9)
Solubility in other solvents: s.s. in hot ethanol, acetone, and cyclohexanone; m.s. in benzene, xylene, and other aromatic hydrocarbons (9)
Melting point: 122–123°C (9)
Vapor pressure: 1.3 mPa @ 25°C (9)
Partition coefficient (octanol/water): 1411–2011 (9)
Adsorption coefficient: 1000 (estimated) (14)

Exposure guidelines

ADI: Not available
HA: Not available
RfD: Not available
TLV: Not available

Basic manufacturer

Sandoz Agro, Inc.
1300 E. Touhy Ave.
Des Plaines, IL 60018
Telephone: 708-699-1616
Emergency: 708-699-1616

6.2.7   Endosulfan

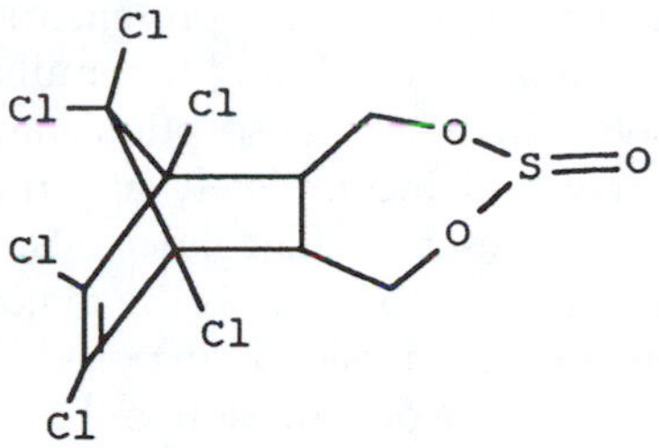

Figure 6.8 Endosulfan.

Trade or other names

Trade or other names for endosulfan products include Afidan, Beosit, Cyclodan, Devisulfan, Endocel, Endocide, Endosol, FMC 5462, Hexasulfan, Hildan, Hoe 2671, Insectophene, Malix, Phaser, Thiodan, Thimul, Thifor, and Thionex.

Regulatory status

Endosulfan is a highly toxic pesticide in EPA toxicity class I. It is a Restricted Use Pesticide (RUP). Labels for products containing endosulfan must bear the Signal Words DANGER — POISON, depending on formulation.

Introduction

Endosulfan is a chlorinated hydrocarbon insecticide and acaricide of the cyclo-diene subgroup, which acts as a poison to a wide variety of insects and mites on contact. Although it may also be used as a wood preservative, it is used primarily on a wide variety of food crops, including tea, coffee, fruits, and vegetables, as well as on rice, cereals, maize, sorghum, or other grains. Formulations of endosulfan include emsulsifiable concentrate, wettable powder, ultra-low volume (ULV) liquid, and smoke tablets. It is compatible with many other pesticides and may be found in formulations with dimethoate, malathion, methomyl, monocrotophos, pirimicarb, triazophos, fenoprop, parathion, petroleum oils, and oxine-copper. It is not compat-ible with alkaline materials.

Technical endosulfan is made up of a mixture of two molecular forms (isomers) of endosulfan, the alpha- and beta-isomers. *Information presented in this profile refers to this technical product unless otherwise stated.*

Toxicological effects

Acute toxicity

Endosulfan is highly toxic via the oral route, with reported oral LD_{50} values ranging from 18 to 160 mg/kg in rats, 7.36 mg/kg in mice, and 77 mg/kg in dogs (2,9). It is also highly toxic via the dermal route, with reported dermal LD_{50} values in rats ranging from 78 to 359 mg/kg (2,9). Endosulfan may be only slightly toxic via inhalation, with a reported inhalation LC_{50} of 21 mg/L for 1 hour, and 8.0 mg/L for 4 hours (2). It is reported not to cause skin or eye irritation in animals (2).

The alpha-isomer is considered to be more toxic than the beta-isomer (2). Animal data indicate that toxicity may also be influenced by species and by the level of protein in the diet; rats that have been been deprived of protein are nearly twice as susceptible to the toxic effects of endosulfan (2). Solvents and/or emulsifiers used with endosulfan in formulated products may influence its absorption into the system via all routes; technical endosulfan is slowly and incompletely absorbed into the body, whereas absorption is more rapid in the presence of alcohols, oils, and emulsifiers (2).

Stimulation of the central nervous system is the major characteristic of endosulfan poisoning (51). Symptoms noted in acutely exposed humans include those common to the other cyclodienes, e.g., incoordination, imbalance, difficulty in breathing, gagging, vomiting, diarrhea, agitation, convulsions, and loss of consciousness (2). Reversible blindness has been documented for cows that grazed in a field sprayed with the compound. The animals completely recovered after a month following the exposure (2). In an accidental exposure, sheep and pigs grazing on a sprayed field suffered a lack of muscle coordination and blindness (2).

Chronic toxicity

In rats, oral doses of 10 mg/kg/day caused high rates of mortality within 15 days, but doses of 5 mg/kg/day caused liver enlargement and some other effects over the same period (2). This dose level also caused seizures commencing 25 to 30 minutes following dose adiministration that persisted for approximately 60 minutes (2). There is evidence that administration of this dose over 2 years in rats also caused reduced growth and survival, changes in kidney structure, and changes in blood chemistry (2,51).

Reproductive effects

Rats fed doses of endosulfan of 2.5 mg/kg/day for three generations showed no observable reproductive effects, but 5.0 mg/kg/day caused increased dam mortality and resorption (2,51). Female mice fed the compound for 78 weeks at 0.1 mg/kg/day had damage to their reproductive organs (52). Oral dosage for 15 days at 10 mg/kg/day in male rats caused damage to the semeniferous tubules and lowered testes weights (2,5). It is unlikely that endosulfan will cause reproductive effects in humans at expected exposure levels.

Teratogenic effects

An oral dose of 2.5 mg/kg/day resulted in normal reproduction in rats in a three-generational study, but 5 and 10 mg/kg/day resulted in abnormalities in bone development in the offspring (2,51). Teratogenic effects in humans are unlikely at expected exposure levels.

Mutagenic effects

Endosulfan is mutagenic to bacterial and yeast cells (51). The metabolites of endosulfan have also shown the ability to cause cellular changes (2,51). This compound has also caused mutagenic effects in two different mammalian species (51). Thus, evidence suggests that exposure to endosulfan may cause mutagenic effects in humans if exposure is great enough.

Carcinogenic effects

In a long-term study done with both mice and rats, the males of both groups experienced such a high mortality rate that no conclusions could be drawn (52).

However, the females of both species failed to develop any carcinogenic conditions 78 weeks after being fed diets containing up to about 23 mg/kg/day. The highest tolerated dose of endosulfan did not cause increased incidence of tumors in mice over 18 months, and a later study also showed no evidence of carcinogenic activity in mice or rats (2,52). It appears that endosulfan is not carcinogenic.

Organ toxicity
Data from animal studies reveal the organs most likely to be affected include kidneys, liver, blood, and the parathyroid gland (51).

Fate in humans and animals
Endosulfan is rapidly degraded into mainly water-soluble compounds and eliminated in mammals with very little absorption in the gastrointestinal tract (2). In rabbits, the beta-isomer is cleared from blood plasma more quickly than the alpha-isomer, with reported blood half-lives of approximately 6 hours and 10 days, respectively (2), which may account in part for the observed differences in toxicity. The metabolites are dependent on the mixture of isomers and the route of exposure (2). Most of the endosulfan seems to leave the body within a few days to a few weeks.

Ecological effects

Effects on birds

Endosulfan is highly to moderately toxic to bird species, with reported oral LD_{50} values in mallards ranging from 31 to 243 mg/kg (9,53), and in pheasants ranging from 80 to greater than 320 mg/kg (53). The reported 5-day dietary LC_{50} is 2906 ppm in Japanese quail (54). Male mallards from 3 to 4 months old exhibited wings crossed high over their back, tremors, falling, and other symptoms as soon as 10 minutes after an acute oral dose. The symptoms persisted for up to a month in a few animals (53).

Effects on aquatic organisms

Endosulfan is very highly toxic to four fish species and two aquatic invertebrates studied; in fish species, the reported 96-hour LC_{50} values were (in $\mu g/L$): rainbow trout, 1.5; fathead minnow, 1.4; channel catfish, 1.5; and bluegill sunfish, 1.2. In two aquatic invertebrates, scuds (*G. lacustris*) and stoneflies (*Pteronarcys*), the reported 96-hour LC_{50} values were, respectively, 5.8 and 3.3 $\mu g/L$ (55). The bioaccumulation for the compound may be significant; in the mussel (*Mytelus edulis*), the compound accumulated to 600 times the ambient water concentration (17).

Effects on other organisms (non-target species)

It is moderately toxic to bees and is relatively nontoxic to beneficial insects such as parasitic wasps, lady bird beetles, and some mites (9,17).

Environmental fate

Breakdown in soil and groundwater

Endosulfan is moderately persistent in the soil environment, with a reported average field half-life of 50 days (14). The two isomers have different degradation times in soil. The half-life for the alpha-isomer is 35 days, and is 150 days for the beta-isomer under neutral conditions. These two isomers will persist longer under more acidic conditions. The compound is broken down in soil by fungi and bacteria (9).

Endosulfan does not easily dissolve in water, and has a very low solubility (9,14). It has a moderate capacity to adhere or adsorb to soils (14). Transport of this pesticide is most likely to occur if endosulfan is adsorbed to soil particles in surface runoff. It is not likely to be very mobile or to pose a threat to groundwater. It has, however, been detected in California well water (12).

Breakdown in water

In raw river water at room temperature and exposed to light, both isomers disappeared in 4 weeks (12). A breakdown product first appeared within the first week. The breakdown in water is faster (5 weeks) under neutral conditions than at more acidic conditions or basic conditions (5 months) (12). Under strongly alkaline conditions, the half-life of the compound is 1 day.

Large amounts of endosulfan can be found in surface water near areas of application (51). It has also been found in surface water throughout the country at very low concentrations (12).

Breakdown in vegetation

In plants, endosulfan is rapidly broken down to the corresponding sulfate (9). On most fruits and vegetables, 50% of the parent residue is lost within 3 to 7 days (9). Endosulfan and its breakdown products have been detected in vegetables (0.0005 to 0.013 ppm), in tobacco, in various seafoods (0.2 ppt to 1.7 ppb), and in milk (12).

Physical properties

Pure endosulfan is a colorless crystal. Technical grade is a yellow-brown color (9).

Chemical name: 6,7,8,9,10,10-hexachloro-1,5,5a,6,9,9a-hexahydro-6,9-methano-2,4,3-benzadioxathiepin 3-oxide (9)
CAS #: 115-29-7 (alpha-isomer, 959-98-8; beta-isomer, 33213-65-9)
Molecular weight: 406.96 (9)
Solubility in water: 0.32 mg/L @ 22°C (9)
Solubility in other solvents: s. in toluene and hexane (9)
Melting point: Technical material, 70–100°C (9).
Vapor pressure: 1200 mPa @ 80°C (9)
Partition coefficient (octanol/water): Not available
Adsorption coefficient: 12,400 (14)

Exposure guidelines

ADI: 0.006 mg/kg/day (27)
HA: Not available
RfD: 0.00005 mg/kg/day (8)
TLV: 0.1 mg/m³ (8-hour) (56)

Basic manufacturer

FMC Corporation
Agricultural Chemicals Group
1735 Market Street
Philadelphia, PA 19103
Telephone: 215-299-6661
Emergency: 800-331-3148

6.2.8 Heptachlor

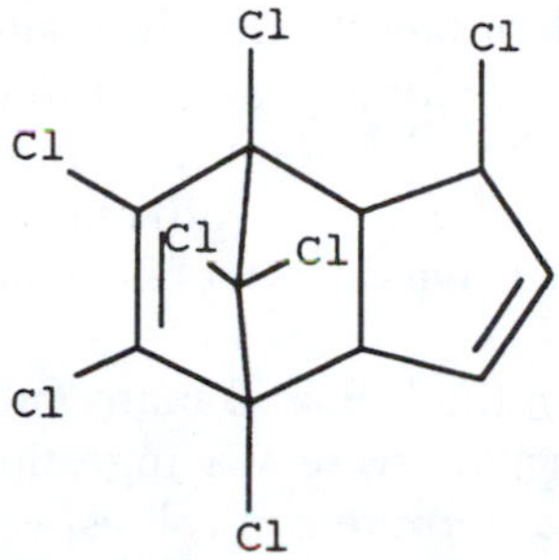

Figure 6.9 Heptachlor.

Trade or other names

Trade names for heptachlor include Biarbinex, Cupincida, Drinox, E 3314, Fennotox, Heptagran, Heptamul, Heptox, Termide, and Velsicol 104.

Regulatory status

Heptachlor is a moderately toxic compound in EPA toxicity class II. In 1988, the EPA canceled all uses of heptachlor in the U.S. The phase-out of heptachlor use began in 1978. The only commercial use still permitted is for fire ant control in power transformers. Heptachlor is still available outside the U.S.

Introduction

Heptachlor is an organochlorine cyclodiene insecticide, first isolated from technical chlordane in 1946. During the 1960s and 1970s, it was used primarily by farmers to kill termites, ants, and soil insects in seed grains and on crops, as well as by exterminators and homeowners to kill termites. Before heptachlor was banned, formulations available included dusts, wettable powders, emulsifiable concentrates, and oil solutions. It acts as a nonsystemic stomach and contact insecticide. An important metabolite of heptachlor is heptachlor epoxide, which is an oxidation product formed from heptachlor by many plant and animal species.

Toxicological effects

Acute toxicity

Heptachlor is highly to moderately toxic compound via the oral route, with reported oral LD_{50} values of 100 to 220 mg/kg in rats, 30 to 68 mg/kg in mice, 116 mg/kg in guinea pigs, 100 mg/kg in hamsters, and 62 mg/kg in chickens (2,57). It is moderately toxic via the dermal route as well, with reported dermal LD_{50} values of 119 to 320 mg/kg in rats, and greater than 2000 mg/kg in rabbits (9). It is reported not to be a skin or eye irritant (9). Heptachlor, like many organochlorines, may interfere with nerve transmission and may also cause an increase in activity of the enzymes involved in the breakdown of foreign chemicals (57). This may lead to serious toxicities from drugs taken for medical reasons. The acute toxicity of heptachlor epoxide, the main and most persistent of heptachlor's metabolites (see below), may be greater. Effects due to heptachlor exposure may include hyperexcitation of the central nervous system, liver damage, lethargy, incoordination, tremors, convulsions, stomach cramps or pain, and coma (2,57).

In humans exposed to chlordane, a closely related organochlorine insecticide that usually contains 10% heptachlor, signs of neurotoxicity such as irritability, salivation, lethargy, dizziness, labored respiration, muscle tremors, and convulsions have been observed (2,57). In severe cases, death may occur due to respiratory failure. Persons with underlying convulsive disorders or liver damage are at increased risk from exposure (2,57).

Prior to it being banned in the U.S. and many other countries, the main routes of human exposure to heptachlor were via ingestion of residues in food or via inhalation in homes treated for termite control, especially where applications were done improperly (58).

Chronic toxicity

Chronic exposure to heptachlor may cause the same effects as acute exposure. No effects were observed in rats fed dietary doses of 0.25 mg/kg/day over 2 years (57). Increased mortality was observed in mice fed heptachlor/heptachlor epoxide for 2 years at dose levels of 1.5 mg/kg/day (2,57). When fed to dogs for 60 days, 0.1 mg/kg/day resulted in no observed adverse effects. Liver damage produced by doses of 0.35 mg/kg/day in rats over a 50-week period were reversible (57). Changes that occurred in rat liver tissues after dosages of 0.35 mg/kg/day for 50 weeks returned to normal after 30 additional weeks without dosing (2). The photoisomer of heptachlor (photoheptachlor) and the major metabolite of heptachlor (heptachlor epoxide) may both be more toxic than the parent compound (2).

Reproductive effects

There is evidence that heptachlor and heptachlor epoxide are associated with infertility and improper development of offspring. Animal studies have shown that females were less likely to become pregnant when both males and females were fed heptachlor. Decreased postnatal survival was reported in the progeny of rats that were fed 0.25 mg/kg/day heptachlor for 60 days and during pregnancy (57). Dosage of 6.9 mg/kg/day for 3 days significantly reduced fertility in rats and reduced survival by one-third of young during the first weeks. A dose of 1 mg/kg/day had no adverse effects on reproduction. No increase in fetal mortality or malformations occurred when pregnant rats were given up to 20 mg/kg/day on days 7 to 17 of gestation (57). Because the available data are inconclusive, it is not possible to make conclusions about the possible reproductive effects of heptachlor in humans.

Teratogenic effects

No teratogenic effects were observed in rats, rabbits, chickens, and beagle dogs (58). No increase in malformations occurred when pregnant rats were given up to

20 mg/kg/day on days 7 to 17 (2,57). In studies where reproductive effects were noted (e.g., increased mortality), decreased viablility of offspring was observed (57). In one study, rats born to mothers fed relatively low doses of heptachlor showed a tendency to develop cataracts shortly after their eyes opened, but these results have not been reproduced elsewhere (2,57). In another study, doses of 5 mg/kg/day over 3 days produced developmental abnormalities in rats (2,57). Overall, these data suggest that teratogenic effects in humans are unlikely at expected exposures.

Mutagenic effects

Laboratory tests indicate that neither heptachlor nor heptachlor epoxide are mutagenic (57,58).

Carcinogenic effects

It is reported that "a few large doses" of heptachlor given to suckling rats did not result in observable tumor incidence over an observation period of 106 to 110 weeks (2,57). In another study, doses of approximately 1.2 mg/kg/day of either heptachlor or heptachlor epoxide increased the incidence of liver carcinomas in rats (2,57). There is evidence that heptachlor promotes development of tumors in rats after initiation with a known tumor initiator (2,57). Available evidence is not sufficient to assess the potential of heptachlor to cause cancer in humans.

Organ toxicity

Results of animal tests show that chronic exposure to heptachlor or heptachlor epoxide adversely affect the liver, kidney, and red blood cells.

Fate in humans and animals

Heptachlor is readily taken up through the skin, lungs, and gastrointestinal tract (57). Once absorbed, it is systemically distibuted and moves into body fat (2,57). In mammals, heptachlor is readily converted to its most persistent and toxic metabolite, heptachlor epoxide, in the liver (2,57). Heptachlor epoxide is stored mainly in fatty tissue, but also in liver, kidney, and muscle tissues (57).

Rats fed diets containing 30 ppm had the highest concentration of heptachlor in their fatty tissues after 2 to 4 weeks. At 12 weeks after cessation of exposure, heptachlor disappeared completely from fatty tissues, while heptachlor epoxide was found in the rats' fatty, liver, kidney, and muscle tissues (57). Heptachlor is able to cross the placenta and has been found in human milk. Intense activity or starvation may mobilize the pesticide as body fat is burned, resulting in the reappearance of toxic symptoms long after uptake from the environment (58). It is excreted in the urine and feces (2,58).

Heptachlor is generally not detectable in the human population, but heptachlor epoxide has been found in human fat, blood, organs, and milk (58). In localities where heptachlor was used regularly, it was found at higher concentrations in human milk than in dairy milk (58). Rats retained 77% of the heptachlor that they inhaled during a 30-minute period (9).

Ecological effects

Effects on birds

Heptachlor is moderately to highly toxic to bird species; the reported acute oral LD_{50} in mallard ducks is 2080 mg/kg (53). The reported 5-day dietary LC_{50} in

Japanese quail is 99 ppm (54). Other reported 8-day dietary LC_{50} values for heptachlor are 450 to 700 ppm in bobwhite quail, and 250 to 275 ppm in pheasants (9). It is also reported to decrease the survivability of chicken eggs (17). Heptachlor and its more potent metabolite, heptachlor epoxide, have been found in the fat of fish and birds. They have also been found in the liver, brain, muscle, and eggs of birds (58).

Effects on aquatic organisms

Both heptachlor and its epoxide are very highly toxic to most fish species tested. The reported 96-hour LC_{50} values are: 5.3 to 13 μg/L in bluegill sunfish, 7.4 to 20 μg/L in rainbow trout, 6.2 μg/L in northern pike, 23 μg/L in fathead minnow, and 10 μg/L in largemouth bass (55). Heptachlor is also very highly toxic to freshwater aquatic invertebrates (like snails, worms, crayfish, etc.) (17,55). Heptachlor is also toxic to marine aquatic life, but its toxicity varies greatly from species to species; crustaceans and younger life stages of fish and invertebrates are most sensitive (58).

Both heptachlor and heptachlor epoxide have been shown to bioconcentrate in aquatic organisms such as fish, mollusks, insects, plankton, and algae (57). It has been found in several fish, mollusks and other aquatic species at concentrations of 200 to 37,000 times the concentration of heptachlor in the surrounding waters (57,58).

Effects on other organisms (non-target species)

Heptachlor is highly toxic to bees.

Environmental fate

Breakdown in soil and groundwater

Heptachlor and heptachlor epoxide are highly persistent in soils, with a reported representative field half-life of 250 days (15). Data collected in Mississippi, New Jersey, and Maryland showed a soil half-life for heptachlor of 0.4 to 0.8 years. The mean disappearance rates of heptachlor from soil ranged from 5.25 to 79.5% per year, depending upon the soil type and mode of application. The highest rates of degradation were observed in sandy soils following application of a granular formulation. Soil incorporation also led to rapid disappearance rates in all soil types. Without incorporation, volatilization from soil surfaces, especially wet ones, is the major route of loss of heptachlor (58). This compound has sometimes been detected in soil in trace amounts 14 to 16 years after application (58).

Heptachlor and its epoxide are moderately bound to soils (15) and should not be highly mobile (57,58). Over their long residence times, even low mobility may result in appreciable movement, and thus heptachlor and its metabolite (heptachlor epoxide) may be considered to pose a risk of groundwater contamination over time (57,58). Very low levels of heptachlor have been found in well water (58).

Heptachlor epoxide is not very susceptible to biodegradation, photolysis, oxidation, or hydrolysis in the environment (57).

Breakdown in water

Heptachlor is almost insoluble in water, and will enter surface waters primarily through drift and surface runoff. In water, heptachlor readily undergoes hydrolysis to a compound which is then readily processed (preferentially under anaerobic conditions) by microoorganisms into heptachlor epoxide (2,3,6). After hydrolysis, volatilization, adsorption to sediments, and photodegradation may be significant routes for disappearance of heptachlor from aquatic environments (57).

Breakdown in vegetation

In plants, the major breakdown product of heptachlor is the epoxide (9). Heptachlor is nonphytotoxic when used as directed (9).

Physical properties

Pure heptachlor is a white or light tan, crystalline solid with a mild camphor or cedar-like odor; the technical heptachlor is a soft wax (9).

Chemical name: 1,4,5,6,7,8,8-heptachloro-3a,4,7,7a-tetrahydro-4,7-methanoindene (9)
CAS #: 76-44-8
Molecular weight: 373.34 (9)
Water solubility: 0.056 mg/L (9)
Solubility in other solvents: v.s. in acetone, alcohol, benzene, carbon tetrachloride, cyclohexanone, kerosene, and xylene (9)
Melting point: 95–96°C (pure); 46–74°C (technical) (9)
Vapor pressure: 53 mPa @ 25°C (9)
Partition coefficient (octanol/water) (log): 5.44 (57)
Adsorption coefficient: 24,000 (15)

Exposure guidelines

ADI: 0.0001 mg/kg/day (27)
MCL: 0.0004 mg/L (8)
RfD: 0.005 mg/kg/day (8)
PEL: 0.5 mg/m^3 (8-hour) (skin) (28)

Basic manufacturer

Velsicol Chemical Corporation
10400 W. Higgins Road
Rosemont, IL 60018–5119
Telephone: 708-298-9000

6.2.9   *Hexachlorobenzene*

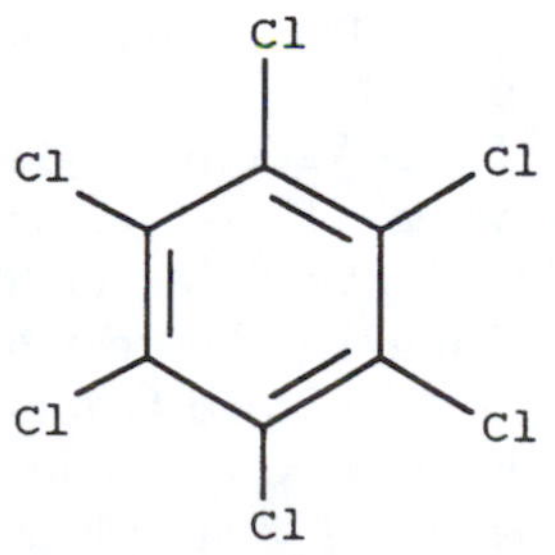

Figure 6.10 Hexachlorobenzene.

Trade or other names

Trade names for hexachlorobenzene (HCB) products include Anticarie, Bent-cure, Bent-No-more, Ceku C.B., Granero, No Bunt, Perchlorobenzene, and Res-Q.

NOTE: This compound should not be confused with hexachlorocyclohexane (HCH), which has historically been referred to as benzene hexachloride and is also commonly known as lindane.

Regulatory status

Hexachlorobenzene is a practically nontoxic compound in EPA toxicity class IV. It has been banned from use in the U.S.

Introduction

Hexachlorobenzene is a chlorinated hydrocarbon, selective fungicide used as a seed protectant treatment, especially on wheat to control common and dwarf bunt. It may be used with or without other seed treatments, fungicides, and/or insecticides. It has fumigant action on fungal spores and is available as a dry seed treatment or slurry for seed treatment. *Unless otherwise stated, the data presented in this profile are for technical hexachlorobenzene.*

Toxicological effects

Acute toxicity

Hexachlorobenzene is slightly to practically nontoxic via the oral route of exposure. The reported acute oral LD_{50} values are 3500 mg/kg in the rat, 4000 mg/kg in the mouse, 2600 mg/kg in the rabbit, and 1700 mg/kg in the cat (7). Its toxicity via the dermal route has not been determined (7,9,59). It is reported to be a possible skin irritant (7). Hexachlorobenzene is slightly to moderately toxic via inhalation, with reported inhalation LC_{50} values of 1.6 mg/L for the cat, 3.6 mg/L for the rat, and 4 mg/L for the mouse (17).

Single doses of HCB are relatively nontoxic, though repeated doses, even at small amounts, are toxic (60). Unlike humans, rodents exhibit neurological symptoms, including tremors, paralysis, muscle incoordination, weakness, and convulsions at high single doses (7). Rats have been observed to withstand an acute subcutaneous (injected just below the skin) dose of 500 mg/kg without detectable injury (including liver injury) (7).

Chronic toxicity

Dietary doses to rats of approximately 50 mg/kg/day killed 95% of the females and 30% of the males within 4 months (7). Almost all survived when the dose rate was reduced to approximately 5 mg/kg/day (7). Rats receiving doses of 25 or 50 mg/kg/day showed nervous system effects such as tremor, hyperexcitablity, and lethargy; skin eruptions; as well as increases in weights of liver, kidneys, spleen, and lungs (7). Those receiving only 5 mg/kg/day remained clinically well (i.e., had no outward manifestations of illness), but blood factors such as total hemoglobin and blood enzymes were decreased in females, and males showed increased liver weight (7,59). Findings of increased weights of liver, spleen, and kidney were reproduced in a second study in rats with a dose regime of 50 mg/kg/day on alternating days over a period of 53 weeks (7,59). Increases in weights of the adrenal gland due to HCB exposure were also observed. These increases were reversible within a 38-week period after HCB administration was stopped (59).

A syndrome called "porphyria" is associated with HCB exposure as well (7,59). HCB is one of the most effective compounds at inducing porphyria in humans and in animals. Approximately 15 mg/kg/day produced porphyria in rabbits (leading to death), and 50 mg/kg/day led to similar effects and fatalities in pigs over 90 days (7,59). A dose of 5 mg/kg/day did not produce any symptoms (7). In a Turkish population accidentally exposed to HCB in the 1950s (from eating treated seed grain), estimated total dietary exposure of 50 to 200 mg/person (a different dose to each individual according to body weight) resulted in several thousand cases of porphyria (7). In this population, recovery usually followed termination of exposure, although many people remain seriously disfigured and at least 10% of those people affected died (7).

Symptoms of HCB-induced porphyria consist of blistering/scarring of the skin, light sensitivity, susceptibility to skin infection, and possibly osteoporosis (decreased bone calcium content) (7). Porphyria is a general disruption in the normal metabolism of porphyrin compounds (often an overproduction), of which hemoglobin and its chemical building blocks are members (7). It is recognizable by the measurable increases in porphyrin compounds in the body and bodily fluids (e.g., blood, urine, feces, etc.) and increased activity of liver enzymes (7,59).

Reproductive effects

Doses of 4 mg/kg/day blocked ovulation in one female monkey of four and reduced estrogen levels in all four over an unspecified period (7). Doses of 10 mg/kg/day did not cause increased fetal mortality or miscarriage in rabbits (7). In a four-generation reproduction study of rats at doses of approximately 8 mg/kg/day, decreased fertility was observed (7). It does not appear that exposure to HCB at normal levels will cause reproductive effects in human populations.

Teratogenic effects

Doses of 80 mg/kg/day for 4 or more days of pregnancy reduced fetal birth weights and caused a slight increase in 14th ribs in rats (7). Mice showed cleft palate and kidney malformations at maternal doses of 100 mg/kg/day on days 7 to 16 of pregnancy (7). In a four-generation reproduction study of rats at doses of approximately 8 mg/kg/day, high pup mortality was observed (7). Survival was affected down to doses of approximately 2 mg/kg/day (7). No gross deformities were noted at any dose level in this study. The potential for the compound to cause birth defects in human populations is likely to be small at common levels of exposure.

Mutagenic effects

Dominant lethal assays in rats were negative for mutagenicity (7). HCB was shown to be nonmutagenic in several tests with bacteria and was mutagenic with yeast cells (17). Available data suggest that it is unlikely that HCB is mutagenic.

Carcinogenic effects

Hexachlorobenzene caused increased numbers of tumors per animal in a long-term study where the lowest dose tested was 4 mg/kg/day (7). Increases in tumors of the lung, thyroid, liver, and spleen were noted (7,59). Mice fed for 2 years had increases in liver tumors at doses of approximately 12 and 24 mg/kg/day (7,59). The potential for hexachlorobenzene to cause carcinogenic effects in humans at normal levels of exposure is not known.

Organ toxicity

Available data from animal tests indicate that the nervous system, liver, kidneys, spleen, and lungs are the target organs affected by exposure to HCB (59).

Fate in humans and animals

Animal studies show that absorption of HCB from the gastrointestinal tract following oral administration is variable and depends on the solvent vehicle used (7,59). Following absorption, HCB may be processed into pentachlorophenol and other more water-soluble compounds (e.g., tetrachlorohydroquinone or pentachlorothiophenol) in the liver and then excreted in the urine, or in some instances it may be excreted intact into the intestine with the bile (59). Absorption may also occur via lymphatic system uptake and direct deposition into body fat, with no processing in the liver (59). Movement into body fat may be increased by lack of protein in the diet (7). In rats, half a single dose of the product is lost within 3 to 4 months; but in monkeys, it takes $2\frac{1}{2}$ to 3 years (59).

Ecological effects

Effects on birds

Hexachlorobenzene is slightly to moderately toxic to bird species. In Japanese quail it has a 5-day dietary LC_{50} of 568 ppm (54). The reported acute oral LD_{50} values were 575 mg/kg in bobwhite quail, and 1450 mg/kg in mallard duck (17).

Effects on aquatic organisms

Hexachlorobenzene is slightly toxic to fish species, with reported 96-hour LD_{50} values of 11 to 16 mg/L in channel catfish, greater than 50 mg/L in coho salmon, 22 mg/L in fathead minnow, and 12 mg/L in bluegill and largemouth bass (55). Rainbow trout have been shown to accumulate residues of 3800 to 8900 times the exposure level of 0.1 to 2.0 µg/L (ppb) within 28 days (55). Likewise, *Daphnia* accumulated residues nearly 900 times the exposure levels of 0.05 to 0.15 µg/L within 48 hours, and neither trout nor *Daphnia* significantly degraded hexachlorobenzene (55). The bioaccumulation ratio in algae is 570, and the major proportion of compounds detected is the parent compound, indicating little degradation (61).

These data indicate a significant potential for bioaccumulation.

Effects on other organisms (non-target species)

The compound is nontoxic to bees (9).

Environmental fate

Breakdown in soil and groundwater

HCB is a highly persistent compound, with reported field half-lives in the soil environment ranging from 2.7 to 7.5 years (15). Evaporation is rapid while it is on soil surfaces, but considerably less so when it is mixed into the soil (62). HCB is moderately to strongly bound by most soils (15). Data from testing on hydrosoils indicate that it may be degraded both aerobically and anaerobically (55).

It has a low water solubility and thus is likely to show low mobility in the soil environment. Due to its lengthy persistence, however, even low mobility may result in appreciable travel; therefore, HCB may pose some risk of groundwater contamination. Hexachlorobenzene has been found in well water in several states at low concentrations ranging from 1 to 5.6 ppb and only in a very small percentage of all the wells tested (63).

Breakdown in water

HCB is of low water solubility, so it would most likely reach surface waters via surface runoff by attachment to soil particles. Once in the aquatic environment, it is likely to be short-lived; HCB underwent very rapid, almost complete (i.e., less than 5 days) degradation to pentachlorophenol and related compounds in innoculated hydrosoil samples under both aerobic and anaerobic conditions (15).

Breakdown in vegetation

Breakdown in vegetation appears rapid, with residue levels in grass at approximately 1% of the initial amount after 15 days, and at approximately 0.01% after 19 months (62).

Physical properties

HCB is a colorless crystalline solid at room temperature (9).

Chemical name: hexachlorobenzene (9)
CAS #: 118-74-1
Molecular weight: 284.81 (9)
Solubility in water: 0.005 mg/L (9)
Solubility in other solvents: i.s. in ethanol; s. in hot benzene, chloroform, and ether (9)
Melting point: 226°C (9)
Vapor pressure: 1.45 mPa @ 20°C (9)
Partition coefficient (octanol/water): Not available
Adsorption coefficient: 50,000 (estimated) (15)

Exposure guidelines

ADI: Not available
MCL: 0.001 mg/L (8)
RfD: 0.0008 mg/kg/day (8)
PEL: 0.025 mg/m^3 (8-hour) (28)

Basic manufacturer

Atomergic Chemetals Corp.
222 Sherwood Avenue
Farmingdale, NY 11735-1718
Telephone: 516-694-9000
Emergency: 800-424-9300

6.2.10 *Lindane*

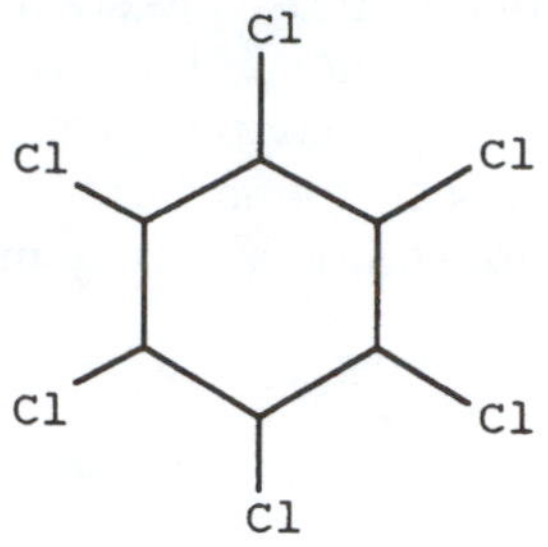

Figure 6.11 Lindane.

Trade or other names

Trade or other names for lindane include Agrocide, Ambrocide, Aparasin, Aphitiria, Benesan, Benexane, benhexachlor, benzene hexachloride, BHC, BoreKil, Borer-Tox, Exagama, Gallogama, Gamaphex, gamma-BHC, Gamma-Col, gamma-HCH, Gammex, Gammexane, Gamasan, Gexane, hexachlorocyclohexane, HCH, Isotox, Jacutin, Kwell, Lindafor, Lindagronox, Lindaterra, Lindatox, Lintox, Lorexane, New Kotol, Noviagam, Quellada, Steward, Streunex, and Tri-6.

NOTE: Lindane (or hexachlorocyclohexane, HCH) has historically and widely been inappropriately referred to as "benzene hexachloride" or "BHC." This compound should not be confused with hexachlorobenzene, or HCB.

Regulatory status

Lindane is a moderately toxic compound in EPA toxicity class II. Labels for products containing it must bear the Signal Word WARNING. Some formulations of lindane are classified as Restricted Use Pesticides (RUPs), and as such may only be purchased and used by certified pesticide applicators. Lindane is no longer manufactured in the U.S., and most agricultural and dairy uses have been canceled by the EPA because of concerns about the compound's potential to cause cancer.

Introduction

Lindane is an organochlorine insecticide and fumigant that has been used on a wide range of soil-dwelling and plant-eating (phytophagous) insects. It is commonly used on a wide variety of crops, in warehouses, in public health to control insect-borne diseases, and (with fungicides) as a seed treatment. Lindane is also presently used in lotions, creams, and shampoos for the control of lice and mites (scabies) in humans.

Technical lindane is comprised of the gamma-isomer of hexachlorocyclohexane, HCH. Five other isomers (molecules with a unique structural arrangement, but identical chemical formulas) of HCH are commonly found in technical lindane, but the gamma-isomer is the predominant one, comprising at least 99% of the mixture of isomers. *Data presented in this profile are for the technical product unless otherwise stated; lindane, HCH, or BHC refer to technical lindane, i.e., gamma-hexachlorocyclohexane.* Gamma-HCH has been shown to be the insecticidally effective isomer.

Lindane may also be found in formulations with a host of fungicides and insecticides. It is available as a suspension, emulsifiable concentrate, fumigant, seed treatment, wettable and dustable powder, and ultra-low volume (ULV) liquid.

Toxicological effects

Acute toxicity

Lindane is a moderately toxic compound via oral exposure, with a reported oral LD_{50} of 88 to 190 mg/kg in rats (2). Other reported oral LD_{50} values are 59 to 562 mg/kg in mice, 100 to 127 mg/kg in guinea pigs, and 200 mg/kg in rabbits (2,9). Gamma-HCH is generally considered to be the most acutely toxic of the isomers following single administration (2).

It is moderately toxic via the dermal route as well, with reported dermal LD_{50} values of 500 to 1000 mg/kg in rats, 300 mg/kg in mice, 400 mg/kg in guinea pigs, and 300 mg/kg in rabbits (2,9). Notably, a 1% solution of lindane in vanishing creme resulted in a sixfold increase in acute toxicity via the dermal route in rabbits, with a reported dermal LD_{50} of 50 mg/kg (2). It is reported to be a skin and eye irritant (9). Younger animals may be more susceptible to lindane's toxic effects (2). Calves are especially susceptible to dermal application (2).

Effects of high acute exposure to lindane may include central nervous system stimulation (usually developing within 1 hour), mental/motor impairment, excitation, clonic (intermittent) and tonic (continuous) convulsions, increased respiratory rate and/or failure, pulmonary edema, and dermatitis (2). Other symptoms in humans are more behavioral in nature, such as a loss of balance, grinding of the teeth, and hyperirritability (2).

Most acute effects in humans have been due to accidental or intentional ingestion, although inhalation toxicity occurred (especially among children) when it was used in vaporizers. Workers may be exposed to the product through skin absorption and through inhalation if handled incorrectly. Lotions (10%) applied for scabies have resulted in severe intoxication in some children and infants (17).

It is reported that single administrations of 120 mg/kg inhibited the ability of white blood cells to attack and kill foreign bacteria in the blood of rats, and 60 mg/kg inhibited antibody formation to human serum albumin (3). It is not clear whether these effects were temporary, or for how long they may have lasted (2).

Chronic toxicity

Doses of 1.25 mg/kg/day in mice, rats, and dogs produced no observable effects over periods of up to 2 years (2,3). Doses of 40 to 80 mg/kg/day were rapidly fatal to dogs in a study over 2 years, and doses of 2.6 to 5.0 mg/kg/day resulted in convulsions in some test animals (2). This same dose level caused liver lesions in rats (2). In one study, 6 to 10 mg/kg/day was reported to have no observable effects on mice; but in another study, that dose caused apparent metabolic changes in the liver (2). Other studies in mice have demonstrated liver damage at higher doses (2). In a 2-year rat study, significant liver changes were attributed to the dietary intake of approximately 5 mg/kg/day (2). Sufficiently high repeated administration of lindane has caused kidney, pancreas, testes, and nasal mucous membrane damage in test animals (2).

There have been reported links of lindane to immune system effects; however, these have not been amply demonstrated in test animals or in humans in a long-term study (2).

Long-term toxicity of the gamma-isomer may be less than that of the alpha- and beta-isomers due to its more rapid transformation and elimination and lesser storage in the body (2).

Sixty male workers in a lindane-producing factory had no signs of neurological impairment or perturbation after 1 to 30 years exposure (17). Another study of chronically exposed workers showed subtle differences between their electrocardiographs (graphs of the heartbeat impulses) and those of unexposed workers (17).

Reproductive effects

In rats, doses of 10 mg/kg/day for 138 days resulted in marked reductions in fecundity and litter size (2), and half that dose (5 mg/kg/day) reportedly had no effect (2). In another study in rats, doses as low as 0.5 mg/kg/day over 4 months caused observable disturbances in the rat estrus cycle, lengthened gestation time, decreased fecundity, and increased fetal mortality (2).

Lindane was found to be slightly estrogenic to female rats and mice, and also caused the testes of male rats to become atrophied. Semeniferous tubules and Leydig cells (important for production of sperm) were completely degenerated at doses of 8 mg/kg/day over a 10-day period (2). Reversible decreases in sperm cell production were noted in male mice fed approximately 60 mg/kg/day for 8 months (2). It is unlikely that lindane will cause effects at the low levels of exposure expected in human populations.

Teratogenic effects

In rats, doses as low as 0.5 mg/kg/day over 4 months caused decreased growth in offspring (2). Beagles given 7.5 or 15 mg/kg/day from day 5 throughout gestation did not produce pups with any noticeable birth defects. Pregnant rats given small amounts of lindane in their food had offspring unaffected by the pesticide (17). It appears that lindane is unlikely to cause developmental effects at levels of exposure expected in human populations.

Mutagenic effects

Most tests on mice and on microorganisms have shown no mutagenicity due to lindane exposure (2). However, lindane has been shown to induce some changes in the chromosomes of cultured human lymphocytes at 5 and 10 ppm in the culture medium. Some chromosomal damage was also noted at a concentration of 1 ppm in this study as well. An *in vivo* (in live animals) study of the effects of lindane on rat leukocytes (white blood cells) did not find chromosomal abnormalities after a single administration of 75 mg/kg (2). It is unlikely that lindane would pose a mutagenic risk in humans at normal exposure levels.

Carcinogenic effects

No tumors were found in groups of 20 mice fed the beta-, gamma-, and delta-isomers of HCH at about 64 mg/kg/day, but tumors occurred in 100% of the mice fed the alpha-isomer only (2). In rats, similar findings were noted at doses of about 49 mg/kg/day (2). Other studies suggest that rodents may suffer from liver tumors from high doses of the gamma-isomer (lindane) (2). HCH was not found to promote tumors initiated by benz[a]anthracene (2). The available evidence is contradictory, and does not allow assessment of the potential for carcinogenic effects in humans from lindane exposure.

Organ toxicity

Data from animal tests indicate that lindane may affect the central nervous system, liver, kidney, pancreas, testes, and nasal mucous membranes.

Fate in humans and animals

Animal studies show that lindane is readily absorbed through the gastrointestinal tract, skin, and lungs (2). Studies show that systemic distribution may be similarly rapid (2). The metabolism of the different isomers of HCH is complex and occurs via many different pathways. The metabolism of lindane, while complex, is nonetheless fairly rapid (2,9). The main pathways include stepwise elimination of chlorines to form tri- and tetrachlorophenols and conjugation with sulfates or glucuronides and subsequent elimination (2). Other pathways involve the ultimate formation of mercapturates (2). These water-soluble end-products are eliminated via the urine (2). Less is known of the metabolism of the other isomers (2).

While all isomers of HCH are stored in fat, the gamma-isomer is stored at very much larger rates than the other isomers, which are more readily metabolized and eliminated (2). Storage equilibrium, at low levels, of lindane in rats is reached after 2 to 7 days, while that for beta-HCH or other isomers may take longer (2). Of a single dose of 40 mg/kg to rats, 80% was excreted in urine and 20% in feces (2). Half of the administered lindane is excreted in 3 or 4 days (2).

Ecological effects

Effects on birds

Lindane is moderately to practically nontoxic to bird species, with a reported LD_{50} of more than 2000 mg/kg in the mallard duck. The 5-day dietary LC_{50} of lindane in Japanese quail is 490 ppm (54). The LC_{50} values of lindane in pheasant and bobwhite quail are 561 and 882 ppm, respectively (64). Egg-shell thinning and reduced egg production has occurred in birds exposed to lindane (64). Lindane can be stored in the fat of birds; birds of prey in the Netherlands contained up to 89 ppm in their fat (64). Residues can also find their way into egg yolks at measurable concentrations for 32 days after dosing (64).

Effects on aquatic organisms

Lindane is highly to very highly toxic to fish and aquatic invertebrate species. Reported 96-hour LC_{50} values range from 1.7 to 90 µg/L in trout (rainbow, brown, and lake), coho salmon, carp, fathead minnow, bluegill, largemouth bass, and yellow perch (55). Water hardness did not seem to alter the toxicity to fish, but increased temperature caused increased toxicity for some species and decreased toxicity for others. Reported 96-hour LC_{50} values in aquatic invertebrates were: in *Daphnia*, 460 µg/L; in scuds, 10 to 88 µg/L; and in *Pteronarcys* (stone flies), 4.5 µg/L (55). The bioconcentration factor for the compound is 1400 times ambient water concentrations, indicating significant bioaccumulation (64).

Effects on other organisms (non-target species)

Lindane is highly toxic to bees (9).

Environmental fate

Breakdown in soil and groundwater

Lindane is highly persistent in most soils, with a field half-life of approximately 15 months (14). When sprayed on the surface, the half-life was typically much shorter than when incorporated into the soil (64). It shows a low affinity for soil binding (14), and may be mobile in soils with especially low organic matter content or subject to high rainfall. It may pose a risk of groundwater contamination.

The pesticide has been found in a significant number of groundwater samples in New Jersey, California, Mississippi, South Carolina, and in Italy at concentrations of less than 1 µg/L (ppb) (12). Lindane is a contaminant in water in the Great Lakes at very low concentrations as well (17).

Breakdown in water

Lindane is very stable in both fresh- and saltwater environments, and is resistant to photodegradation (9). It will disappear from the water by secondary mechanisms such as adsorption on sediment, biological breakdown by microflora and fauna, and adsorption by fish through gills, skin, and food (64).

Breakdown in vegetation

Plants may pick up residues from not only direct application, but through water and vapor phases. Persistence is seen when plants are rich in lipid content, and crops like cauliflower and spinach will build up less residue than crops like carrots (64). The metabolism in plants is not well understood, but carrots are estimated to metabolize lindane with a half-life of just over 10 weeks (based on plant uptake), whereas it may have a half-life in lettuce of only 3 to 4 days (64).

Physical properties

Lindane is a colorless crystal compound.

Chemical name: gamma-1,2,3,4,5,6-hexachlorocyclohexane (9)
CAS #: 58-89-9
Molecular weight: 290.85 (9)
Solubility in water: 7.3 mg/L @ 25°C (9)
Solubility in other solvents: v.s. acetone, benzene, and ethanol (9)
Melting point: Approximately 113°C (9)
Vapor pressure: 5.6 mPa @ 20°C (9)
Partition coefficient (octanol/water): Not available
Adsorption coefficient: 1100 (14)

Exposure guidelines:

ADI: 0.008 mg/kg/day (27)
MCL: 0.0002 mg/L (8)
RfD: 0.0003 mg/kg/day (18)
PEL: 0.5 mg/m³ (8-hour) (28)

Basic manufacturer

Drexel Chemical Company
1700 Channel Avenue
Memphis, TN 38113
Telephone: 901-774-4370

6.2.11 Methoxychlor

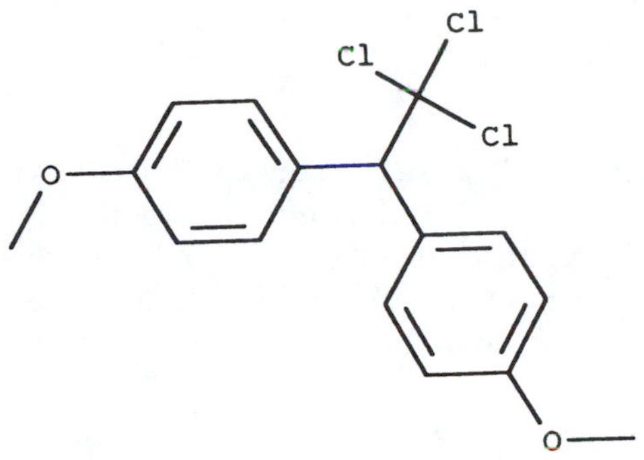

Figure 6.12 Methoxychlor.

Trade or other names

Trade names for methoxychlor include Chemform, Dimethoxy-DT, DMDT, ENT 1716, Higalmetox, Methoxychlore, Marlate, Methoxy-DDT, OMS 466, and Prentox.

Regulatory status

Methoxychlor is a practically nontoxic compound in EPA toxicity class IV. It is a General Use Pesticide (GUP), and labels for products containing it must bear the Signal Word CAUTION.

Introduction

Methoxychlor is an organochlorine insecticide effective against a wide range of pests encountered in agriculture, households, and ornamental plantings. It is registered for use on fruits, vegetables, forage crops, and in forestry. Methoxychlor is also registered for veterinary use to kill parasites on dairy and beef cattle.

Methoxychlor is one of a few organochlorine pesticides that have seen an increase in use since the ban on DDT in 1972. It is quite similar in structure to DDT, but has relatively low toxicity and relatively short persistence in biological systems. It is available in wettable and dustable powders, emulsifiable conentrates, granules, and as an aerosol. It may be found in formulations with malathion, parathion, piperonyl butoxide, and pyrethrins.

Toxicological effects

Acute toxicity

Methoxychlor is practically nontoxic via the oral route, with reported oral LD_{50} values of 5000 to 6000 mg/kg in rats (2,9), 1850 mg/kg in mice, and 2000 mg/kg in hamsters (2). The lowest oral dose that can cause lethal effects in humans is estimated to be 6400 mg/kg, and the lowest dose through the skin that produces toxic effects in humans is 2400 mg/kg based on behavioral symptoms (17). It is reportedly slightly to practically nontoxic dermally, with a reported dermal LD_{50} in rabbits of greater than 2000 mg/kg (9).

Symptoms of high acute exposure include central nervous system depression, progressive weakness, and diarrhea (2). Extremely high doses can cause death within 36 to 48 hours.

Chronic toxicity

Rats fed methoxychlor at doses of 500 mg/kg/day for 2 years showed practically no weight gain, but this was attributed to refusal of food rather than any toxic effects of the compound (2,17). At doses of 1500 mg/kg/day in rats, severe reductions in weight appeared, and most animals died within 45 days (2).

Rabbits were more susceptible than rats; doses of 200 mg/kg/day were fatal in most cases within 15 days (2).

Data from experiments in dogs are contradictory; dogs experienced weight loss at approximately 25 mg/kg/day over 6 months, and doses of 50 mg/kg/day caused convulsions and subsequent death in some animals within 9 weeks (2). In other studies, dogs fed up to 300 mg/kg/day in the diet for 1 year, and about 63 mg/kg/day by stomach tube for 5 months showed no signs of injury or observable effects (2).

Massive doses in swine, rats, and monkeys produced pathological changes in liver, kidney, mammary glands, and uteri. Other data suggest that the liver effects in rats may be temporary, and one study showed no effects on the liver in rats (2).

Human volunteers taking oral doses up to 2.0 mg/kg/day for 8 weeks showed no detectable effects in overall health, blood, and enzyme biochemistry, and no observable changes in bone marrow, liver, small intestine, or testes (2).

Reproductive effects

Available evidence suggests that high doses of technical methoxychlor (88 to 90% pure) or its metabolites may have estrogenic or reproductive effects (2). In rats, dietary doses of about 125 mg/kg/day reduced mating, and many did not produce litters (2). Rats fed doses of about 50 mg/kg/day had normal fertility and fecundity, but their offspring had abnormal reproductive functioning (2). Male and female weanling rats fed methoxychlor through puberty and mating had normal fertility overall, but female rats had reduced fertility when paired with untreated males (2).

In mice, 200 mg/kg/day administered on days 6 to 15 of pregnancy decreased fertility and birthweight (2). Testicular atrophy was observed in rats at levels of approximately 500 mg/kg/day over an unspecified period, but not in dogs at doses of 100 mg/kg/day (2). Wistar rats given 100 mg/kg/day for 14 (females) to 70 (males) days showed pathological changes in reproductive tissues (2).

It is unlikely that methoxychlor will cause reproductive effects in humans at expected exposure levels.

Teratogenic effects

In mice, 200 mg/kg/day administered on days 6 to 15 of pregnancy resulted in delayed ossification and wavy ribs in the offspring (2). When a methoxychlor formulation containing 50% active ingredient and 50% unknown compounds was administered to pregnant female rats, adverse effects on the fetus occurred only at doses large enough to be toxic to the dams (65). At 400 mg/kg/day, the pesticide killed rat embryos (65). These suggest that teratogenic effects in humans are unlikely under normal conditions.

Mutagenic effects

Most mutation assays have proven negative (65). There is no convincing evidence that methoxychlor is toxic to genetic material.

Carcinogenic effects

Tumor incidence was statistically similar in unexposed rats and those given as much as 80 mg/kg/day over 2 years (2). Dogs given about 250 mg/kg/day over an unspecified period did not show evidence of tumors (17). Two strains of mice fed diets containing up to approximately 90 mg/kg/day methoxychlor for 2 years showed no significant incidence of liver tumors, but one strain did have increased testicular tumors (17). In rats, about 25 mg/kg/day produced slight increases in liver cancers. The data suggest that methoxychlor is unlikely to show carcinogenic activity in humans.

Organ toxicity

Central nervous system depression occurs with acute exposure; data from animal studies indicate that target organs for methoxychlor include the kidneys, liver, mammary glands, and uterus.

Fate in humans and animals

Available evidence suggests that methoxychlor does not accumulate to any significant degree in fat or other tissues of mammals. At high dietary doses in rats, storage was minimal over the 18-week course of the study, and nondetectable within 2 weeks after the study (2). Mice excreted 98.3% of a single oral 50-mg/kg dose in urine and feces within 24 hours (2). The major metabolites in mouse feces and urine were the monophenol and bisphenol (2,9). Other metabolites (e.g., dihydroxybenzophenone) were also present, but are not the primary metabolites (2,9). These compounds are typically eliminated in a conjugated form (i.e., bound to an innocuous molecule such as glutathione, sulfate, etc.) (2). It is thought that these metabolites may form reactive intermediates if not successfully conjugated and eliminated (2).

Lactating cows treated twice in 14 days with sprays of 0.25 to 0.5% methoxychlor (2 quarts per animal) had residues of 2 to 3 ppm in milk. After 14 days, levels were at the limit of detection (0.005 ppm) (17).

Ecological effects

Effects on birds

Methoxychlor is slightly toxic to bird species, with reported acute oral LD_{50} values of greater than 2000 mg/kg in the mallard duck, sharp-tailed grouse, and California quail (53). The reported 5-day dietary LC_{50} in Japanese quail is greater than 5000 ppm. Reported 8-day LC_{50} values are greater than 5000 ppm in bobwhite quail and ring-necked pheasants (9). Dietary levels of as high as about 145 mg/kg/day had no effects on reproductive function of male and female chickens over 8 to 16 weeks (2).

Effects on aquatic organisms

Methoxychlor is very highly toxic to fish and aquatic invertebrates. Reported 96-hour LC_{50} values (for the technical grade material, ca. 90% pure) are less than 20 µg/L for cutthroat trout, Atlantic salmon, brook trout, lake trout, Northern pike, and largemouth bass (55). Reported LC_{50} values are between 20 and 65 µg/L in rainbow trout, goldfish, fathead minnow, channel catfish, bluegill, and yellow perch (55). Aquatic invertebrates with 96- or 48-hour LC_{50} values of less than 0.1 mg/L include *Daphnia*, scuds, sideswimmers, and stoneflies (55).

Predicted bioconcentration factors were the highest in the mussel (12,000) and in the snail (8570) (66). This indicates that methoxychlor would accumulate in aquatic organisms that do not rapidly metabolize the compound. Practically no metabolism was seen in *Daphnia* or mayflies (55). Fish reportedly break down methoxychlor fairly rapidly and thus tend not to accumulate it appreciably (12), but this may vary according to species and/or life stage. No magnification of residues was observed in largemouth bass fingerlings fed contaminated *Daphnia*, and no evidence of metabolism was seen in rainbow trout (55).

Effects on other organisms (non-target species)

The compound is nontoxic to bees (9).

Environmental fate

Breakdown in soil and groundwater

Methoxychlor is very persistent in soil, with a reported representative half-life of approximately 120 days (14). However, rates may be as fast as 1 week in some instances. Methoxychlor degrades much more rapidly in soil that has a supply of oxygen (aerobic) than in soil without oxygen (anaerobic).

Methoxychlor is tightly bound to soil and is insoluble in water, so it is not expected to very mobile in moist soils (9,14). Actual mobility will depend on site-specific factors (e.g., soil organic matter and rainfall). The risk to groundwater should be slight, but may be greater if application rates are very high, or the water table is very shallow (17).

Movement of the pesticide is more likely via adsorption to suspended soil particles in runoff. In the EPA pilot groundwater survey, methoxychlor was found in a number of wells in New Jersey (not quantified) and at extremely low concentrations (from 0.1 to 1.0 ng/L, or ppt) in water from the Niagara River, the James River, and a Lake Michigan tributary (65).

Breakdown in water

Methoxychlor is practically insoluble in water, and thus will most likely reach surface waters via runoff as described above. In hydrosoils (sediments in an aquatic environment), degradation of methoxychlor to methoxychlor olefin (MDE) occurred only under aerobic conditions (55).

In open water, the major products of breakdown in a neutral solution are anisoin, anisil, and *p,p*-dimethoxydichloroethene (DMDE). The half-life in distilled water is 37 to 46 days; but in some river waters, the half-life may be as rapid as 2 to 5 hours (9,17). Methoxychlor evaporates very slowly, but the evaporation may contribute to the cycling of the product in the environment (12).

Breakdown in vegetation

On mature soybean foliage, the washoff rate was 8% per centimeter rainfall, with a total of 33.5% washoff for a season (17). Dislodgeable residues account for less than 1% of the amount applied.

Physical properties

Pure methoxychlor is a colorless crystalline solid; technical methoxychlor (88 to 90% pure) is a grey powder (9).

Chemical name: 1,1,1-trichloro-2,2-bis(4-methoxyphenyl)ethane (9)
CAS #: 72-43-5
Molecular weight: 345.65 (9)
Solubility in water: 0.1 mg/L @ 25°C (9)
Solubility in other solvents: v.s. in most organic solvents (9)
Melting point: 77°C (technical product) (9)
Vapor pressure: Very low (9)
Partition coefficient (octanol/water): Not available
Adsorption coefficient: 80,000 (14)

Exposure guidelines

ADI: 0.1 mg/kg/day (27)
MCL: 0.04 mg/L (8)
RfD: 0.005 mg/kg/day (8)
TLV: 10.0 mg/m³ (8-hour) (56)

Basic manufacturer

Drexel Chemical Company
1700 Channel Avenues
Memphis, TN 38113
Telephone: 901-774-4370

6.2.12 PCNB (Quintozene)

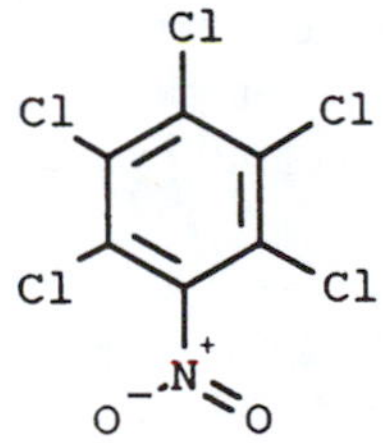

Figure 6.13 PCNB (Quintozene).

Trade or other names

The common name for PCNB (pentachloronitrobenzene) is quintozene. Trade or other names for PCNB or products containing it include Avicol, Botrilex, Brassicol, Earthcide, Folosan, Kobu, Kobutol, pentachloronitrobenzene, Pentagen, Saniclor, Terraclor, Terrazan, Tilcarex, Tri-PCNB, Triquintam, Tritisan, Tubergran, and Turfcide.

Regulatory status

Pentachloronitrobenzene is a slightly toxic compound in EPA toxicity class III. Labels for products containing PCNB must bear the Signal Word CAUTION. Most products containing PCNB have been canceled for use in the U.S.

Introduction

PCNB is an organochlorine fungicide used as a seed dressing or soil treatment to control a wide range of fungal species in such crops as potatoes, wheat, onions, lettuce, tomatoes, tulips, garlic, and others. Depending on the producer and the manufacturing procedure, PCNB impurities can include hexachlorobenzene, pentachlorobenzene, and tetrachloronitrobenzene. The fungicide is often used in combination with insecticides and fungicides, including carbaryl, imazalil, tridimenol, etridiazole, and fuberidazole. It is available as a dustable or wettable powder, in granular form, emulsifiable concentrate, and seed treatment.

Toxicological effects

Acute toxicity

PCNB is a slightly toxic compound via oral administration. The reported oral LD_{50} values for test animals are 1710 mg/kg in male rats, 1650 mg/kg in female rats, and 800 mg/kg in rabbits (7,9). A single oral administration of over 2500 mg/kg in dogs reportedly caused no fatal response (7). It is slightly toxic via inhalation as well, with a reported 4-hour inhalation LC_{50} of 6.49 mg/L (9).

PCNB is slightly toxic via the dermal route, with reported dermal LD_{50} values of 2000 to 4000 mg/kg in rats (17). A dermal dose of 4000 mg/kg in rabbits did not result in observable signs of toxicity or skin irritation (7). Humans exhibited no skin irritation after a single, 48-hour contact; but a second exposure (2 weeks later) elicited a sensitivity reaction in 13 of 50 test subjects, indicating a potential for skin sensitization (7). No reports of skin sensitization have been reported from nonexperimental exposures (7). It has a potential for eye irritation, and has caused a reversible case of conjunctivitis with some corneal injury (7,17).

Administration of 1600 mg/kg to cats caused disruption of normal transport of oxygen by hemoglobin (methemoglobin), as well as changes in red blood cells (7). If these effects occur in humans as a result of very high acute exposure, the likely effects could include fatigue, possible dizziness, and lethargy. It seems unlikely — based on the reported levels of methemoglobin (a maximum of about 11%) — that loss of consciousness or brain anoxia could occur (7,17).

Chronic toxicity

Rats fed diets of about 275 mg/kg/day PCNB over 3 months experienced reduced body weights and growth rates (7). These effects were seen in males at only half that dose (7). At doses of about 35 mg/kg/day, both sexes of rats showed increased relative liver weights, and males had elevated relative liver weights at doses as low as 3.5 mg/kg/day (7).

In a rat study over the course of 2 years, the sex differences in reduced weight gain were reversed, and the females showed reduced weight gain at doses at which males did not (7). These decreases in either sex were not markedly different at low and high doses, and did not show a strong relationship to PCNB dose (7). Dogs showed no growth effects at doses of 20 mg/kg/day, but there was some enlargement in liver cells (7). In another study in dogs, impairment of liver bile production and fat metabolism was observed at doses of slightly more than 3.5 mg/kg/day, but no adverse effects were seen at about 0.6 mg/kg/day (7).

Reproductive effects

A three-generation study of rats fed PCNB doses of 25 mg/kg/day (and higher for weanlings) showed no effects from the treatment (7). Between the treatment and control groups of rats, there were no differences in numbers and positions of implantations, dead/resorbed fetuses, litter size, or fetal sex ratios at maternal doses as high as 125 mg/kg/day on days 6 to 15 of pregnancy (7). Similar findings (lack of reproductive effects) were observed in another rat study at doses up to 200 mg/kg/day on unspecified days (7). PCNB does not appear to cause reproductive effects.

Teratogenic effects

There were no increases in visceral or skeletal abnormalities or birth weights at maternal doses as high as 125 mg/kg/day on days 6 to 15 of pregnancy (7). Findings of missing kidneys and cleft palates in mice at maternal doses of 500 mg/kg/day on days 7 to 11 of gestation may have been attributable to the presence of hexachlorobenzene in the test product (at a level of 11%, or approximately 55 mg/kg/day) (7,17). The data indicate that PCNB is not teratogenic.

Mutagenic effects

No significant increases in mutation rates were observed in mice following injection of PCNB just underneath the skin (7). The dominant lethal assay for mutation in mice was also negative. It has been observed to increase frequency of mutation in *Escherichia coli*, but it is not clear whether the underlying mutation rate changed or PCNB interfered with DNA repair processes in the cells (7). It appears that PCNB is either nonmutagenic or weakly mutagenic, and it is unlikely that mutagenic effects would be seen in humans under normal circumstances.

Carcinogenic effects

The maximum tolerated dosage of about 464 mg/kg/day from day 7 to 28 and about 150 mg/kg/day for 78 weeks (when they were sacrificed) resulted in increased production of liver tumors in mice (7). Under slightly different conditions, the same results were obtained, also in mice (7). It is possible that these studies were performed with PCNB samples that may have contained hexachlorobenzene, which may have contibuted to the observed increases. There is insufficient information to determine the carcinogenic potential of PCNB.

Organ toxicity

The main target organ, following chronic exposure to PCNB, is the liver. Acute effects may occur in the red blood cells and hemoblobin in the circulatory system.

Fate in humans and animals

PCNB is very rapidly absorbed from the gastrointestinal tract of monkeys (7). The speed and thoroughness of absorption varies among species. In mammals, PCNB is eliminated unchanged in the feces or as metabolites in the urine (7). While PCNB metabolism may be fairly rapid, the metabolism of the common contaminants of PCNB (hexa- and pentachlorobenzene) may be much slower (7). The major metabolites are pentachloroaniline, pentachlorophenol, and pentachloroanisole, which may be stored in body fat (9). Storage of PCNB in the tissues did not occur in rats, dogs, or cows. Eggs from chickens fed low doses of PCNB in their diets were considered safe for human consumption. Only trace amounts were found in the milk of cows fed 25 mg/day (17).

Ecological effects

Effects on birds

PCNB is practically nontoxic to bird species; the reported acute oral LD_{50} values were greater than 2000 mg/kg in mallards and pheasants, and 170 mg/kg in bobwhite quail (17,53). No tissue build-up was observed, and no lesions were found on internal organs (17).

Effects on aquatic organisms

PCNB is highly toxic to fish, with reported LC_{50} values of 0.55 mg/L in rainbow trout and 0.1 mg/L in bluegill sunfish (17). However, other data suggest that PCNB may be less toxic; a water concentration of 1.2 mg/L (its solubility limit in water) reportedly caused no fatalities in rainbow trout and golden orfe (9). PCNB has been shown to accumulate in aquatic animals and in aquatic plants (68).

Effects on other organisms (non-target species)

The compound is nontoxic to bees when used as directed (9).

Environmental fate

Breakdown in soil and groundwater

PCNB is of varying persistence in the environment; various half-lives in soil have been reported, ranging from less than 3 weeks to over a year (68). The soil type is the main source of variation, with more rapid breakdown occurring in sandier soils (68). Metabolites in the soil include pentachloroaniline, pentachlorobenzene, hexachlorobenzene, and pentachlorothioanisole (12). These will persist for 2 or 3 years. PCNB is lost mainly through either evaporation or biotransformation (68). Under oxygenated conditions, many soil bacteria aid in the breakdown of this product (68).

PCNB shows an appreciable capacity for soil binding, and is not expected to be extensively mobile in most soils (12,14). Even very low mobility of a highly persistent compound, though, can result in appreciable movement over time. PCNB has been found in very low amounts in well water in Ohio, and in groundwater in California, Missouri, and Texas, and Ontario, Canada (12).

Breakdown in water

PCNB is unchanged by sunlight, and is stable in acidic and neutral solutions (9,12). It is rapidly adsorbed to sediments and suspended solids, absorbed by biota, and volatilized out of the water column, with an estimated half-life of 1.8 to 5 days (12).

Breakdown in vegetation

Plants take up PCNB from both soil and water and it may be translocated throughout the plant (17). Major breakdown products in plants may include pentachloroaniline and methylthiopentachlorobenzene (9). One week after young cotton plants were treated with 300 ppm PCNB, they contained 91 ppm parent compound and 2.0 ppm pentachlorobenzene, 2.77 ppm hexachlorobenzene, 0.65 ppm aniline, 1.36 ppm sulfide, and traces of other metabolites (17).

Physical properties

PCNB is a pale yellow crystalline solid at room temperature (9).

Chemical name: pentachloronitrobenzene (9)
CAS #: 82-68-8
Molecular weight: 295.34 (9)
Solubility in water: 0.44 mg/L @ 20°C (9)
Solubility in other solvents: v.s. in alcohols, benzene, and chloroform (9)
Melting point: 144°C (9)
Vapor pressure: 6.6 mPa @ 20°C (9)
Partition coefficient (octanol/water) (log): 4.46 (17)
Adsorption coefficient: 5000 (estimated) (14)

Exposure guidelines

ADI: 0.007 mg/kg/day (8)
HA: Not available
RfD: 0.003 mg/kg/day (8)
TLV: 0.5 mg/m^3 (8-hour) (56)

Basic manufacturer

Uniroyal Chemical Co., Inc.
Benson Road
Middlebury, CT 06749
Telephone: 203-573-2000
Emergency: 203-723-3670

6.2.13 *Pentachlorophenol*

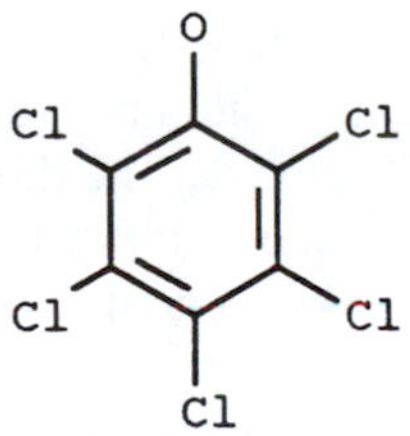

Figure 6.14 Pentachlorophenol.

Trade or other names

Trade names for pentachlorophenol include Dowicide, PCP, Penchlorol, Penta, Penta Plus, Pentachloral, Pentacon, Penwar, Priltox, Santobrite, Santophen, Sinituho, and Weedone.

Regulatory status

Pentachlorophenol is a moderately toxic compound in EPA toxicity class II. It is a Restricted Use Pesticide (RUP) in its formulations as a wood preservative, but a

General Use Pesticide (GUP) for other purposes; labels for products containing it must bear the Signal Word DANGER.

Introduction

Pentachlorophenol (PCP) is a chlorinated hydrocarbon insecticide and fungicide. It is primarily used to protect timber from fungal rot and wood-boring insects, but may also be used as a preharvest defoliant in cotton, a general pre-emergence herbicide, and as a biocide in industrial water systems. It is available in blocks, flakes, granules, liquid concentrates, wettable powders, or ready-to-use petroleum solutions.

Data presented in this profile are for technical grade pentachlorophenol unless otherwise stated. Technical grade PCP has historically contained dioxins (e.g., tetra-, hexa-, and octochlorodibenzo-*p*-dioxin) and hexachlorobenzene as manufacturing by-products. Technical grade PCP is typically about 86% pure. The discovery of these compounds in technical grade PCP may be one reason for its being phased out of use. Pentachlorophenol is also a major product of the metabolism of hexachlorobenzene in mammals.

Toxicological effects

Acute toxicity

Pentachlorophenol is moderately toxic via the oral route, with reported oral LD_{50} values for various formulations ranging from 27 to 211 mg/kg in rats (69,70). In mice, the oral LD_{50} is 74 to 130 mg/kg, and in rabbits is 70 to 300 mg/kg (69,70). It is moderately toxic via inhalation as well, with a reported inhalation LC_{50} of 0.2 to 2.1 mg/L in rats (17). The time frame for this LC_{50} (e.g., 4-hour, 1-hour, etc.) was not given. Inhalation LD_{50} values (i.e., the median lethal doses, not concentrations, via the inhalation route) of 225 mg/kg in rats and 355 mg/kg in mice are reported, also without a time frame (70). Another calculated LD_{50} in rats via the inhalation route is 11.7 mg/kg for 28 to 44 minutes of exposure and assuming a breathing rate of 80 mL/minute (69).

Pentachlorophenol causes irritation to the mucous membranes, skin, and eyes of test animals (9). Via the dermal route, it is moderately toxic, with reported dermal LD_{50} values ranging from 96 to 330 mg/kg in the rat, and 40 to greater than 1000 mg/kg in the rabbit (depending on formulation) (69). Skin penetration may be the most dangerous route of exposure, being responsible for about 50 known cases of PCP poisoning, 30 of which have resulted in death. Immersion of a human hand in a 0.4% PCP solution for 10 minutes caused pain and inflammation. Technical PCP resulted in chloracne on the ears of rabbits, and edema in chicks, but pure PCP did not (70).

High acute exposure to PCP can cause elevated temperature, profuse sweating, dehydration, loss of appetite, decreased body weight, nausea, and neurological effects such as tremors, uncoordinated movement, leg pain, muscle twitching, and coma (69,70). Some of the symptoms may be due to the impurities in the formulation rather than the pentachlorophenol itself (69).

Chronic toxicity

Much research on PCP has been performed with poorly characterized technical material, and the chronic toxicity observed may depend in large measure on the proportion of chlorodibenzo-*p*-dioxins present in the mixture (69). In a 90-day feeding trial in rats, 30 mg/kg/day produced depressed red blood cell and hemoglobin levels as well as liver degeneration, and even lower doses resulted in irregular blood chemistry and enzyme levels, along with increased liver and kidney weights (69,71). Pure PCP, and also technical PCP without dioxin contamination, produced only slight enlargement of livers and kidneys (69). Purified PCP also did not produce toxic effects such as liver damage and immune system alterations, which had previously been reported for the technical product (69,71).

In humans, the most common exposure to PCP is inhalation in the workplace. Abdominal pain, nausea, fever, and respiratory irritation, as well as eye, skin, and throat irritation, may result from such exposure (70), while very high levels may cause obstruction of the circulatory system in the lungs and cause heart failure (70). Survivors of toxic exposures may suffer permanent visual and central nervous system damage (70). Persons regularly exposed to PCP tend to tolerate higher levels of PCP vapors than persons having little contact with these vapors (70,71).

Reproductive effects

Rats fed PCP at doses of 30 mg/kg/day for 62 days before mating and during lactation showed weight loss, but no decreases in fecundity and fertility (69). Sperm of male mice given technical or purified PCP for 5 days at 50 mg/kg/day showed no abnormalities within 35 days of treatment (69). The evidence indicates that PCP does not cause reproductive effects.

Teratogenic effects

Offspring of rats fed PCP at doses of 30 mg/kg/day for 62 days before mating and during lactation showed lowered survival and growth rates (69); 3 mg/kg/day did not have any effects (69). Maternal doses of 5 mg/kg/day of technical PCP in rats produced toxicity to the fetus or embryo; and 50 mg/kg/day on days 6 to 15, 8 to 11, or 12 to 15 of gestation produced increases resorptions, swelling, dilated ureters, and skeletal anomalies (69). It is unlikely that PCP has teratogenic effects in humans at normal exposure levels.

Mutagenic effects

PCP is not mutagenic in bacteria or houseflies, but is weakly mutagenic in mice and may be mutagenic in yeast (71). One study of chromosomal aberrations in occupationally exposed workers showed no increased incidence of sister-chromatid exchanges, while another did find increases (71). Weak mutagenic effects were seen in human lymphocyte cultures exposed to PCP (71). The evidence suggests that PCP is nonmutagenic or weakly mutagenic.

Carcinogenic effects

Studies of two formulated PCP products (Dowcide and Penta) showed increases in cancers of the spleen, liver, and adrenal gland in test mice or rats at doses of about 17 to 18 mg/kg/day (71). These findings were not replicated for Dowcide in mice in a second study (71).

There have been reports of a possible association between occupational exposures to technical PCP and Hodgkin's disease, acute leukemia, and soft-tissue sarcoma, but confounding factors such as concurrent exposure to other substances makes interpretation of these data problematic (71). No convincing evidence of PCP's caricinogenic effects in humans is available (71).

Current evidence is not sufficient to assess the potential of PCP cause carcinogenic effects in humans.

Organ toxicity

Data from animal studies indicate that the major target organs for PCP are the liver, kidneys, and central nervous system.

Fate in humans and animals

PCP is rapidly absorbed through the gastrointestinal tract following ingestion (71). Accumulation is not common, but if it does occur, the major sites are the liver, kidneys, plasma protein, brain, spleen, and fat (69,71). Unless kidney and liver functions are impaired, PCP is rapidly eliminated from blood and tissues, and is excreted, mainly unchanged or in conjugated form, via the urine (71). Single doses of PCP have half-lives in blood of 15 hours in rats, 78 hours in monkeys, and 30 to 50 hours in humans (69).

Ecological effects

Effects on birds

The compound is slightly toxic to practically nontoxic to bird species. The reported 5-day dietary LC_{50} value in Japanese quail is greater than 5139 ppm (54). Reported acute oral LD_{50} values for PCP are 380 mg/kg in mallard duck and 504 mg/kg in pheasant (55).

Effects on aquatic organisms

PCP may be highly to very highly toxic to many species of fish; reported 96-hour LC_{50} values are 68 µg/L in chinook salmon, 52 µg/L in rainbow trout, 205 µg/L in fathead minnow, 68 µg/L in channel catfish, and 32 µg/L in bluegill sunfish (55).

Several species of fish, invertebrates, and algae have had levels of PCP that were significantly higher (up to 10,000 times) than the concentration in the surrounding waters (71). Once absorbed by fish, pure PCP is rapidly excreted, as is its metabolite, with a biological half-life of only 10 hours (71). Biomagnification (i.e., is the progressively higher concentration of a compound as it passes up the food chain) is not thought to be significant because of PCP's rapid breakdown in living organisms (71).

Effects on other organisms (non-target species)

Cattle and other farm animals have ingested PCP by chewing and licking outdoor wood structures, or from being housed in wooden pens that were treated with PCP solutions. This has caused sickness and death in some of these animals (17).

Environmental fate

Breakdown in soil and groundwater

PCP is moderately persistent in the soil environment, with a reported field half-life of 45 days (15). PCP degrades most rapidly in flooded or anaerobic (airless) soils, at higher temperatures, and in the presence of organic matter in the soil (12,15). Breakdown is mainly by anaerobic biodegradation; breakdown by sunlight and hydrolysis do not appear to be significant processes (15). It is poorly sorbed at neutral and alkaline conditions, and may be mobile in many soils (12,15). Sorption will be slightly greater (and mobility slightly lesser) in soils with higher proportions of soil organic matter (12). The compound has been found in groundwater in California,

Oregon, and Minnesota at very low concentrations, ranging from 0.06 ppt to 0.64 ppb (15).

Breakdown in water

In the water environment, PCP is mainly bound to sediments and suspended particles in water (12). PCP will dissociate by releasing a hydrogen ion and may then be more readily degraded by sunlight or microorganisms (12). In water, biodegradation occurs, mainly at the surface, with a half-life ranging from hours to days (12). It does not evaporate to a significant degree. PCP has been detected at very low levels in rivers and streams (0.01 to 16 µg/L), surface water systems (1.3 to 12 µg/L), and seawater (0.02 to 11 µg/L) (12).

Breakdown in vegetation

PCP may be taken up by plants; lettuce grown on soil containing PCP contained low levels of PCP residues (12). Uptake and accumulation vary according to plant species. PCP is strongly toxic to plants (9).

Physical properties

At room temperature, pentachlorophenol is a colorless crystalline solid with a phenolic odor (9). Color may vary from white to dark grayish brown, depending on the purity of the compound (9).

Chemical name: pentachlorophenol (9)
CAS #: 87-86-5
Molecular weight: 266.34 (9)
Solubility in water: 80 mg/L @ 20°C (9)
Solubility in solvents: v.s. in acetone, alcohols, ether, and benzene; s. in petroleum ether, carbon tetrachloride, and paraffins (9)
Melting point: 191°C (9)
Vapor pressure: 16,000 mPa @ 20°C (9)
Partition coefficient (octanol/water) (log): 5.12 (17)
Adsorption coefficient: 30 (at pH 7) (estimated) (15)

Exposure guidelines

ADI: Not available
MCL: 0.001 mg/L (67)
RfD: 0.03 mg/kg/day (8)
PEL: 0.5 mg/m^3 (8-hour) (28)

Basic manufacturer

ISK Biosciences
5966 Heisley Road
P.O. Box 8000
Mentor, OH 44061–8000
Telephone: 216-357-4100
Emergency: 216-357-7070

References

(1) Ware, G. W. *Fundamentals of Pesticides: A Self-Instruction Guide.* Thompson Publications, Fresno, CA, 1986.

(2) Smith, A. G. Chlorinated hydrocarbon insecticides. In *Handbook of Pesticide Toxicology.* Hayes, W. J., Jr.and Laws, E. R., Jr., Eds. Academic Press, New York, 1991.

(3) Matsumura, F. *Toxicology of Insecticides,* 2nd ed. Plenum Press, New York, 1985.

(4) U.S. Environmental Protection Agency. *Health Advisory: Chlordane.* Office of Drinking Water, Washington, D.C., 1987.

(5) Hurt, S. S. *Dicofol: Toxicological Evaluation of Dicofol Prepared for the WHO Expert Group on Pesticide Residues (Report No. 91 R-1017).* Toxicology Department, Rohm & Haas, Spring House, PA, 1991.

(6) U.S. Environmental Protection Agency. *Guidance for the Reregistration of Pesticide Products Containing Chlorobenzilate as the Active Ingredient.* Washington, D.C., 1983.

(7) Edwards, I. R., Ferry, D. G., and Temple, W. A. Fungicides and related compounds. In *Handbook of Pesticide Toxicology.* Hayes, W. J., Jr. and Laws, E. R., Jr., Eds. Academic Press, New York, 1991.

(8) U.S. Environmental Protection Agency. *Integrated Risk Information System,* Washington, D.C., 1995.

(9) Kidd, H. and James, D. R., Eds. *The Agrochemicals Handbook,* 3rd ed. Royal Society of Chemistry Information Services, Cambridge, U.K., 1991 (as updated).

(10) World Health Organization. *DDT and its Derivatives: Environmental Aspects. Environmental Health Criteria 83.* WHO, Geneva, Switzerland, 1989.

(11) Murty, A.S. *Toxicity of Pesticides to Fish.* Vol. II. CRC Press, Boca Raton, FL, 1986.

(12) Howard, P. H., Ed. *Handbook of Environmental Fate and Exposure Data for Organic Chemicals. Pesticides.* Lewis, Boca Raton, FL, 1991.

(13) Buhler, D. R. Transport, accumulation, and disappearance of pesticides. In *Chemistry, Biochemistry, and Toxicology of Pesticides.* Pesticide Education Program. Witt, J. M., Ed. Oregon State University Extension Service, Corvallis, OR, 1989.

(14) Wauchope, R. D., Buttler, T. M., Hornsby, A. G., Augustijn-Beckers, P. W. M., and Burt, J. P. SCS/ARS/CES pesticide properties database for environmental decision making. *Rev. Environ. Contam. Toxicol.* 123, 1–157, 1992.

(15) Augustijn-Beckers, P. W. M., Hornsby, A. G., and Wauchope, R. D. SCS/ARS/CES pesticide properties database for environmental decisionmaking. II. Additional Compounds. *Rev. Environ. Contam. Toxicol.* 137, 1–82, 1994.

(16) Paasivirta, J. *Chemical Ecotoxicology.* Lewis, Boca Raton, FL, 1991.

(17) U.S. National Library of Medicine. *Hazardous Substances Data Bank.* Bethesda, MD, 1995.

(18) Bidleman, T. F., Zaranski, M. T., and Walla, M. D. Toxaphene: usage, aerial transport and deposition. In *Toxic Contamination in Large Lakes. Vol. I. Chronic Effects of Toxic Contaminants in Large Lakes.* Schmidtke, N. W., Ed. Lewis, Boca Raton, FL, 1988.

(19) World Health Organization. *Environmental Health Criteria 38: Heptachlor.* Geneva, Switzerland. 1984.

(20) U.S. Agency for Toxic Substances and Disease Registry. *Toxicological Profile for Chlordane (ATSDR/TP-89/06).* Atlanta, GA, 1989.

(21) U.S. Environmental Protection Agency. *Health Advisory Summary: Chlordane.* Office of Drinking Water, Washington, D.C., 1989.

(22) Aldrich, F. D. and Holmes, J. H. Acute chlordane intoxication in a child: case report with toxicological data. *Arch. Environ. Health.* 19, 129–132, 1969.

(23) Martin, E. W. *Hazards of Medication: A Manual on Drug Interactions, Incompatibilities, Contraindications, and Adverse Effects.* Lippincott, Philadelphia, PA, 1971.

(24) National Institute for Occupational Safety and Health. *Registry of Toxic Effects of Chemical Substances.* Cincinnati, OH, 1981–1986.

(25) U.S. Environmental Protection Agency. *Health Advisory Summary: Chlordane.* Office of Drinking Water, Washington, D.C., 1987.

(26) U.S. Environmental Protection Agency. *Pesticide Fact Sheet Number 109: Chlordane.* Office of Pesticides and Toxic Substances, Washington, D.C., 1986.

(27) Lu, F. C. A review of the acceptable daily intakes of pesticides assessed by the World Health Organization. *Regul. Toxicol. Pharmacol.* 21, 351–364, 1995.

(28) U.S. Occupational Safety and Health Administration. *Permissible Exposure Limits for Air Contaminants* (29 CFR 1910. 1000, Subpart Z). U.S. Department of Labor, Washington, D.C., 1994.

(29) U.S. Environmental Protection Agency. *Guidance for the Reregistration of Pesticide Products Containing Chlorobenzilate as the Active Ingredient.* Washington, D.C., 1983.

(30) Griffeth, J. and Duncan, R. C. Urinary chlorobenzilate residues in citrus fieldworkers. *Bull. Environ. Contam. Toxicol.* 35, 496–499, 1985.

(31) U.S. Environmental Protection Agency. *Pesticide Fact Sheet Number 15: Chlorobenzilate.* Office of Pesticides and Toxic Substances, Washington, D.C., 1984.

(32) Lyman, W. J. *Handbook of Chemical Property Estimation Methods. Environmental Behavior of Organic Compounds.* McGraw-Hill, New York, 1983.

(33) McEwen, F. L. and Stephenson, G. R. *The Use and Significance of Pesticides in the Environment.* John Wiley & Sons, New York, 1979.

(34) Menzie, C. M. *Metabolism of Pesticides. Special Scientific Report: Wildlife.* U.S. Department of the Interior, Fish and Wildlife Service, U.S. Government Printing Office, Washington, D.C., 1974.

(35) U.S. Environmental Protection Agency. *Chlorothalonil Health Advisory. Draft Report.* Office of Drinking Water, Washington, D.C., 1987.

(36) U.S. Environmental Protection Agency. Pesticide tolerance for chlorothalonil. *Fed. Reg.* 50, 26592–26593, 1985.

(37) Vettorazzi, G. *International Regulatory Aspects for Pesticide Chemicals.* CRC Press, Boca Raton, FL, 1979.

(38) Chin, B. H., Heilman, R. D., Bachand, R. T., Chernenko, G., and Barrowman, J. Absorption and biliary excretion of chlorothalonil and its metabolites in the rat. *Toxicol. Lett.* 5(1), 150, 1980.

(39) Doyle, R. *Dalapon Information Sheet.* Food and Drug Administration, Washington, D.C., 1984.

(40) Hallenbeck, W. H. and Cunningham-Burns, K. M. *Pesticides and Human Health.* Springer-Verlag, New York, 1985.

(41) Weed Science Society of America. *Herbicide Handbook,* 6th ed. Champaign, IL, 1989.

(42) U.S Environmental Protection Agency. *Health Advisory Summary: Dalapon.* Office of Drinking Water, Washington, D.C., 1988.

(43) Pimentel, D. *Ecological Effects of Pesticides on Nontarget Species.* President's Office of Science and Technology, Washington, D.C., 1971.

(44) Hurt, S. S. *Dicofol: Toxicological Evaluation of Dicofol Prepared for the WHO Expert Group on Pesticide Residues (Report No. 91R-1017).* Toxicology Department, Rohm and Haas Company, Spring House, PA, 1991.

(45) Rohm and Haas Company. *Material Safety Data Sheet for Kelthane Technical B Miticide.* Philadelphia, PA, 1991.

(46) Tillman, A. *Residues, Environmental Fate and Metabolism Evaluation of Dicofol Prepared for the FAO Expert Group on Pesticide Residues. (Report No. AMT 92–76).* Rohm and Haas Company, Philadelphia, PA, 1992.

(47) U.S. Environmental Protection Agency. *Toxicology One-Line Summary: Dienolchlor.* Environmental Fate and Effects Division, Washington, D.C., 1990.

(48) Quistad, G. B., Mulholland, K. M., and Skinner, W. S. The fate of dienochlor administered orally and dermally to rats. *Toxicol. Appl. Pharmacol.* 85(2), 215–220, 1986.

(49) U.S. Environmental Protection Agency. *Dienochlor.* Washington, D.C., 1981.

(50) Quistad, G. B., and Mulholland, K. M. Photodegradation of dienochlor. *J. Agric. Food Chem.* 31(3), 621–624, 1986.

(51) U.S. Agency for Toxic Substances and Disease Registry. *Toxicological Profile for Endosulfan. Draft Report.* Atlanta, GA, 1990.

(52) National Cancer Institute. *Bioassay of Endosulfan for Possible Carcinogenicity, (Technical Report Series No. 62).* National Institutes of Health, Bethesda, MD, 1978.

(53) Hudson, R. H., Tucker, R. K., and Haegele, M. A. *Handbook of Acute Toxicity of Pesticides to Wildlife, Resource Publication 153.* U.S. Department of Interior, Fish and Wildlife Service, Washington, D.C., 1984.

(54) Hill, E. F. and Camardese, M. B. *Lethal Dietary Toxicities of Environmental Contaminants to Coturnix, Technical Report Number 2.* U.S. Department of Interior, Fish and Wildlife Service, Washington, D.C., 1986.

(55) Johnson, W. W. and Finley, M. T. *Handbook of Toxicity of Chemicals to Fish and Aquatic Invertebrates, Resource Publication 137.* U.S. Department of Interior, Fish and Wildlife Service, Washington, D.C., 1980.

(56) American Conference of Governmental Industrial Hygienists, Inc. *Documentation of the Threshold Limit Values and Biological Exposure Indices,* 5th ed. Publications Office, Cincinnati, OH, 1986.

(57) Agency for Toxic Substances and Disease Registry. *Toxicological Profile for Heptachlor/Heptachlor Epoxide, ATSDR/TP-88/16.* Atlanta, GA, 1989.

(58) World Health Organization. *Environmental Health Criteria 38: Heptachlor.* Geneva, Switzerland, 1984.

(59) U.S. Agency for Toxic Substance and Diseases Registry. *Toxicological Profile for Hexachlorobenzene (Update) Draft for Public Comment.* Atlanta, GA, 1994.

(60) Ecobichon, D. J. Toxic effects of pesticides. In *Casarett and Doull's Toxicology,* 4th ed. Amdur, M. O., Doull, J., and Klaassen, C. D., Eds. Pergamon Press, New York, 1991.

(61) Metcalf, R. L., Kapoor, I. P., Lu, P., Schuth, C. K., and Sherman, P. Model ecosystem studies of environmental fate of six organochlorine pesticides. *Environ. Health Perspect.* 4, 35–44, 1973.

(62) Beall, M. L., Jr. Persistence of aerially applied hexachlorobenzene on grass and soil. *J. Environ. Qual.* 5(4), 367–369, 1976.

(63) Williams, W. M., Holden, P. W., Parsons, D. W., and Lorber, M. N. *Pesticides in Ground Water Data Base 1988. Interim Report.* U.S. Environmental Protection Agency, Office of Pesticide Programs. Washington, D.C., 1988.

(64) Ulman, E. *Lindane, Monograph of an Insecticide.* Schillinger Verlag, Federal Republic of Germany, 1972.

(65) U.S. Environmental Protection Agency. *Health Advisory Summaries: Methoxychlor.* Office of Drinking Water, Washington, D.C., 1989.

(66) Trabalka, J. R. and Garten, C. T., Jr. *Development of Predictive Models for Xenobiotic Bioaccumulation in Terrestrial Ecosystems. Environmental Sciences Division Publication No. 2037.* Oak Ridge National Laboratory, Oak Ridge, TN.

(67) U.S. Environmental Protection Agency. *National Primary Drinking Water Standards (EPA 810-F94-001-A).* Washington, D.C., 1994.

(68) U.S. Environmental Protection Agency. 1968–81. *Pesticide Abstracts, 75-0098, 78–2944, 79–1635, 80-0246, 81–1983.* Washington, DC.

(69) Gasiewicz, T. A. Nitro compounds and related phenolic pesticides. In *Handbook of Pesticide Toxicology.* Hayes, W. J., Jr. and Laws, E. R., Jr., Eds. Academic Press, New York, 1991.

(70) Wagner, S. L. *Clinical Toxicology of Agricultural Chemicals.* Oregon State University Environmental Health Sciences Center, Corvallis, OR, 1981.

(71) U.S. Agency for Toxic Substance and Disease Registry. *Toxicological Profile for Pentachlorophenol. Draft Report.* Atlanta, GA, 1992.

chapter seven

Phenoxy and benzoic acid herbicides

7.1   Class overview and general description

Background

The phenoxy and benzoic acid herbicides are treated together because of their similar structures and mechanisms of action. The prototypical phenoxy herbicide is 2,4-D. This was the first compound in this group of herbicides to be produced and is currently used in more than 1500 different commercial products and in many different herbicide formulations (1).

The phenoxy herbicides are all based on the ring-like structure shown in Figure 7.1.A. The phenoxy herbicides have at least one atom of chlorine attached to the ring at various positions. The benzoic acids have a similar structure though they also have an additional CO_2H group attached to the ring, as shown in Figure 7.1.B. The respective compounds in these two categories are listed in Table 7.1.

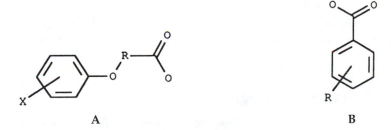

Figure 7.1 Generic structures for phenoxy herbicides (A) and benzoic acid herbicides (B).

One of the most controversial phenoxy compounds is 2,4,5-T. This compound was used extensively in the formulations of Agent Orange and other defoliants applied throughout this country and in Southeast Asia during the Vietnam War. The manufacture of 2,4,5-T produced small but significant amounts of a by-product, dioxin, a powerful cancer-causing compound in test animals. All uses of 2,4,5-T have been banned in the U.S.

Phenoxy herbicide usage

In 1987, the total use of phenoxy herbicides in the U.S. exceeded 37.4 million pounds (1). Nearly 60% is attributed to the use of 2,4-D. During the same year, phenoxy compounds accounted for slightly less than 5% of the total herbicide use in the U.S. (1).

Table 7.1 Phenoxy and
Benzoic Acid Herbicides

Phenoxy Acids
2,4-D*
2,4-DB*
2,4,5-T
Aclonifen
Clonifen
Cloprop
Dichlorprop
Fenoxaprop
Fluazifop
Fluazifop-p-butyl*
Haloxyfop
Isoxapyrifop
MCPA*
MCPB
Mecoprop/MCPP
Quizalofop
Quizalofop-p-ethyl*
Benzoic Acids
Chloramben*
DCPA*
Dicamba*

Note: * indicates that a profile for
this compound is included
in this chapter.

Phenoxy herbicides are used against a wide variety of broad-leaved weeds to protect crops such as sorghum, maize, cereal grains, fruit trees, and some vegetables. Phenoxy compounds have varying degrees of selectivity for certain types of broad-leaved weeds and crops. These compounds have very little effect on grasses (2).

Benzoic acid herbicide usage

In 1987, benzoic acid herbicide compounds accounted for less than 2% of the total herbicide use in the U.S. (1).

Mechanisms of action and toxicology

The phenoxy herbicides, such as 2,4-D, are taken up by the roots or absorbed through the leaves of a plant and then distributed throughout the plant. They accumulate in the active growth regions at the tips of the stems and roots where they disrupt normal cell growth. Both the phenoxy herbicides and the benzoic acids act like plant growth hormones (auxins). They stimulate the growth of old cells and the rapid expansion of new cells. This rapid growth in cell size, without normal cell division, effectively crushes the plant's water and nutrient transport system, the phloem, in the active growing regions (2).

In mammals, the phenoxy herbicide compounds affect the peripheral nervous system. The mechanism of toxic action is poorly understood, although in several cases mammalian toxicity has been significant (1). Although the compound acts as a growth hormone in plants, no hormone activity is triggered by these compounds in mammals (1,3).

Acute toxicity

Phenoxy herbicides

Symptoms of acute exposure to phenoxy herbicides include involuntary twitching of the muscles, a "pins and needles" type of sensation, or a loss of sensation in parts of the body. Other signs and symptoms of poisoning include headache, dizziness, nausea, vomiting, abdominal pains, diarrhea, aching and tender muscles, weakness, and fatigue (1,3). Most phenoxy herbicides are moderately toxic to animals. For example, the oral LD_{50} for 2,4-D is 370 mg/kg in mice, though for MCPA it is 700 to 1330 mg/kg in rats (1). The phenoxy compounds also irritate the skin and may irritate the eyes, nose, and mouth. Contact dermatitis may occur and is characterized by reddening and burning of the skin (4).

Benzoic acids

Benzoic acid herbicides are slightly toxic and can cause dermatitis through dermal exposure. Most of their effects are similar to those from acute exposure to phenoxy compounds. The LD_{50} of chloramben in rats is 3500 mg/kg (5), and of dicamba in rats is 757 to 1701 mg/kg (1,6,7).

Chronic toxicity

Phenoxy herbicides

Doses of 100 mg/kg/day haloxyfop caused kidney damage in adult rats (8). In another study, oral doses of 9 mg/kg/day mecoprop given to female rats and oral doses of 27 mg/kg/day given to male rats caused kidney damage (9).

Benzoic acids

In long-term tests on rats, mice, and dogs, no chloramben-related effects on body weight, body function, and organs were observed even at very high concentrations (7,10). Chronic studies in rats with dicamba have shown no obervable effects on survival, body weight, food consumption, organ weight, blood chemistry, or tissue structure at low doses (1,5,11). It has caused enlargement of liver cells in mice at very high doses in long-term studies (1).

Reproductive effects

Phenoxy herbicides

Phenoxy herbicides have caused some reproductive effects at high doses. For example, female rats exposed to 75 mg/kg/day 2,4-DB showed a decrease in ovarian weights. Also, fewer offspring were born and there were lower overall body weights (1,6,7). In another study, dogs receiving 8 to 16 mg/kg/day MCPA for 13 weeks had various adverse sperm and testes changes (12).

Reproductive effects in humans are unlikely at expected levels of exposure (7).

Benzoic acids

In long-term, three-generation studies of rats exposed to benzoic acids such as chloramben and dicamba, there were no effects on the reproductive capacity (1,10).

However, in another study, rabbits given doses up to 20 mg/kg/day dicamba from days 6 through 18 of pregnancy experienced toxic effects in the mothers, slightly reduced fetal body weights, and an increased loss of fetuses at the 10-mg/kg/day dose level (11).

Therefore, benzoic acids are unlikely to cause reproductive effects in humans at expected levels of exposure.

Teratogenic effects

Phenoxy herbicides

Phenoxy herbicides have caused various birth defects in laboratory animals at high doses. For example, oral doses of 50 mg/kg/day haloxyfop in rats between days 6 and 16 of pregnancy caused developmental abnormalities in the offspring's urogenital system and also death to the fetus (13).

In another study, oral doses of 125 mg/kg/day mecoprop (MCPP) in pregnant rats from day 6 to day 15 of pregnancy caused increased intrauterine deaths, decreased body lengths, and an increased incidence of delayed or absent bone formation in the offspring (9).

Teratogenic effects in humans are not likely at expected exposure levels.

Benzoic acids

One of the representative benzoic acids caused developmental effects in rats at very high doses. Pregnant rats fed lower doses of chloramben during the sensitive period of gestation were not affected, but fetal deaths increased and fetal skeletal development was not complete (10). These data suggest that teratogenic effects in humans are unlikely under anticipated exposure conditions.

Mutagenic effects

Phenoxy herbicides

There is limited information regarding the mutagenicity of phenoxy herbicides. Some phenoxys are weakly mutagenic or only mutagenic at high doses. For example, MCPA is weakly mutagenic to hamster bone marrow and ovarian cells (7), and mecoprop causes an increase in sister-chromatid exchanges only after very high oral doses of 470 and 3800 mg/kg in Chinese hamsters (9). These data indicate that the phenoxy herbicides are weakly or nonmutagenic.

Benzoic acids

Benzoic acids do not appear to be mutagenic (6,11).

Carcinogenic effects

Phenoxy herbicides

The majority of phenoxy herbicides do not produce carcinogenic effects in laboratory animals, though rats fed moderate doses of 2,4-D (45 mg/kg/day) for 2 years showed an increase in brain tumors (1).

A study of people employed in the manufacture of phenoxy herbicides showed an association between these herbicides and cancer of soft tissues and non-Hodgkins lymphoma (14). However, other data do not support this conclusion (15). Thus, available evidence is inconclusive.

Benzoic acids

Data from laboratory studies are inadequate to determine whether benzoic acids are carcinogenic. However, rats fed dicamba at 25 mg/kg/day for 2 years showed no increased incidence of tumors (15).

Organ toxicity

Phenoxy herbicides

Some of the phenoxy herbicides have caused adverse effects in the liver and kidneys.

Benzoic acids

One of the benzoic acids, chloramben, irritates the throat, lungs, skin, and eyes (7). One long-term study in mice found effects on the liver (1).

Fate in humans and animals

Phenoxy herbicides

Mammals eliminate phenoxy herbicides through the urine and feces rather quickly. For example, rats eliminated through the urine nearly all of a single oral dose of MCPA within 24 hours (1). In another study, quizalofop-P-ethyl was rapidly broken down in mammals; more than 90% of a single dose was eliminated in urine within 3 days (6).

Benzoic acids

Benzoic acids are rapidly absorbed into the bloodstream from the gastrointestinal tract, but are then rapidly excreted in the urine and feces and no residues remain in the tissues.

For instance, when lactating cows were fed chloramben, 88% of the compound left the body via urine and 5% left the body via feces within 4 days (7).

In a study of the metabolism of dicamba in cows, 73% of the pesticide was unchanged and 20% was eliminated as a gluconiride conjugate. In another study on rats, 96% of the dicamba was excreted in the urine after 24 hours (16).

Ecological effects

Effects on birds

Phenoxy herbicides

Phenoxy herbicides can range from practically nontoxic to moderately toxic to birds. For example, fluazifop-p-butyl, haloxyfop, mecoprop and quizalofop-p-ethyl are nontoxic to birds. The haloxyfop LD_{50} in bobwhite quail is 5620 mg/kg (6,7).

However, MCPA is moderately toxic to birds, with a LD_{50} value of 377 mg/kg in bobwhite quail (6). This variability makes generalization about the group difficult. Most of the effects on wildlife occur through the changes to habitat following the application of the herbicides rather than through direct acute toxicological effects (17).

Benzoic acids

Dicamba, a representative benzoic acid compound, is moderately toxic to mallard ducks and bobwhite quail, with an LD_{50} of 2009 mg/kg in the ducks (5).

Effects on aquatic organisms

Phenoxy herbicides

Phenoxy herbicides vary in their toxicity to fish. Haloxyfop and mecoprop are virtually nontoxic to fish (7,8). MCPA is slightly toxic to freshwater fish, with an LC_{50} value of 232 mg/L in rainbow trout (6). 2,4-D, depending on the formulation, may be highly toxic to fish. The LC_{50} of 2,4-D in cutthroat trout ranges from 1.0 to 100 mg/L (6).

Benzoic acids

Dicamba, a representative benzoic acid compound, is of slight toxicity to fish and aquatic invertebrates. The 96-hour LC_{50} for dicamba in rainbow trout is 135 mg/L, and about 100 mg/L in grass shrimp (5,6).

Effects on other organisms (non-target species)

Phenoxy herbicides

Most phenoxy herbicides are nontoxic to bees (7). An exception is quizalofop-p-ethyl, which has an LD_{50} of 0.1 mg/kg in bees (6).

Benzoic acids

Chloramben and dicamba, representative benzoic acid compounds, are not toxic to bees (7,16).

Environmental fate

Breakdown in soil and groundwater

Phenoxy herbicides

Most of the phenoxy compounds are of low persistence in soil, lasting up to about 2 weeks (18). In one study after the application of 2,4-D and MCPA to soil, half of the initial amounts degraded after about 7 to 11 days (6).

Other phenoxy compounds like haloxyfop and 2,4,5-T are moderately persistent in soil, with half-lives ranging between 55 and 100 days and 5 months (6,19).

Despite the relatively rapid breakdown of many of these herbicides, they tend to be mobile in soil and thus have the ability to move from soil into surface water or groundwater (20–22). 2,4-D, MCPA, 2,4,5-T, dichlorprop, and fluazifop-p-butyl have all been detected in well water. Of all the phenoxy compounds, 2,4-D has the highest number of detections in well water. It was found in 2.3% of 6142 wells tested (23).

Benzoic acids

Benzoic acid herbicides do not bind to soil particles and are highly soluble in water. The rate of biodegradation increases with temperature and increasing soil moisture (20).

The benzoic acids show a pattern similar to the phenoxy compounds in terms of persistence and mobility in soil. Breakdown is through the action of microbes, and both dicamba and chloramben have been detected in well water (23).

Breakdown in water

Phenoxy herbicides

Phenoxy herbicides are degraded by sunlight and microorganisms. Under aerobic conditions and increasing pH, the breakdown is more rapid.

In one study, the half-life of 2,4-D under aerobic conditions was 1 to several weeks (20). In another study, the half-life of haloxyfop ranged from 33 days at pH 5, to 5 days at pH 7, to a few hours at pH 9 (7).

Benzoic acids

Benzoic acid herbicides are highly soluble in water and are not expected to volatilize significantly from surface water (20).

Breakdown in vegetation

Phenoxy herbicides

There is limited information regarding the fate of phenoxy herbicides in vegetation. For example, 2,4-D interferes with normal plant processes though it is generally nonpersistent. In one study, when 2,4-D was applied to grass, there was 80 ppm at day zero, 45 ppm after 14 days, and 6 ppm after 56 days (24).

Benzoic acids

Benzoic acids are rapidly taken up by the leaves and roots of plants. They easily translocate throughout the plant. Dicamba may accumulate in the tips of mature leaves of broadleaf plants, such as fruit trees, and cause harm during their growth and development stages (5).

7.2   *Individual profiles*

7.2.1   *2,4-D*

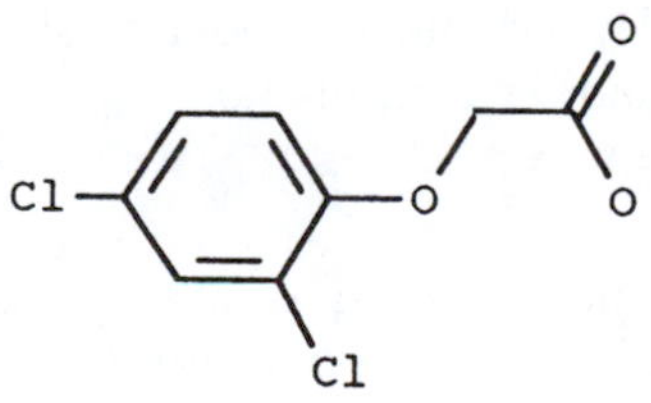

Figure 7.2 2,4-D.

Trade or other names

2,4-D is used in many commercial products. Commercial names for products containing 2,4-D include Aqua-Kleen, Barrage, Lawn-Keep, Malerbane, Planotox, Plantgard, Savage, Salvo, Weedone, and Weedtrine-II.

Regulatory status

2,4-D is a General Use Pesticide (GUP) in the U.S. The diethylamine salt is toxicity class III — slightly toxic orally, but toxicity class I — highly toxic by eye exposure. It bears the Signal Word DANGER — POISON because 2,4-D has produced serious eye and skin irritation among agricultural workers.

Introduction

There are many forms or derivatives of 2,4-D including, esters, amines, and salts. *Unless otherwise specified, this document will refer to the acid form of 2,4-D.* 2,4-D, a chlorinated phenoxy compound, functions as a systemic herbicide and is used to control many types of broadleaf weeds. It is used in cultivated agriculture, in pasture and rangeland applications, forest management, in the home, and garden, and to control aquatic vegetation. It may be found in emulsion form, in aqueous solutions (salts), and as a dry compound.

The product Agent Orange, used extensively throughout Vietnam, was about 50% 2,4-D. However, the controversies associated with the use of Agent Orange were associated with a contaminant (dioxin) in the 2,4,5-T component of the defoliant.

Toxicological effects

Acute toxicity

The acid form is of slight to moderate toxicity. The oral LD_{50} of 2,4-D ranges from 375 to 666 mg/kg in the rat, 370 mg/kg in mice, and from less than 320 to 1000 mg/kg in guinea pigs. The dermal LD_{50} values are 1500 mg/kg in rats and 1400 mg/kg in rabbits (1,5,7).

In humans, prolonged breathing of 2,4-D causes coughing, burning, dizziness, and temporary loss of muscle coordination (1). Other symptoms of poisoning can be fatigue and weakness, with possible nausea. On rare occasions following high levels of exposure, there can be inflammation of the nerve endings with muscular effects (25).

Chronic toxicity

Rats given high amounts, 50 mg/kg/day, of 2,4-D in the diet for 2 years showed no adverse effects. Dogs fed lower amounts in their food for 2 years died, probably because dogs do not excrete organic acids efficiently. A human given a total of 16.3 g in 32 days therapeutically lapsed into a stupor and showed signs of incoordination, weak reflexes, and loss of bladder control (1,5,7).

Reproductive effects

High levels of 2,4-D (about 50 mg/kg/day) administered orally to pregnant rats did not cause any adverse effects on birth weights or litter size. Higher doses (188 mg/kg/day) resulted in fetuses with abdominal cavity bleeding and increased mortality (1,5,7). DNA synthesis in the testes was significantly inhibited when mice were fed large amounts (200 mg/kg/day) of 2,4-D (7). The evidence suggests that if 2,4-D causes reproductive effects in animals, this only occurs at very high doses. Thus, reproductive problems associated with 2,4-D are unlikely in humans under normal circumstances.

Teratogenic effects

2,4-D may cause birth defects at high doses. Rats fed 150 mg/kg/day on days 6 to 15 of pregnancy had offspring with increased skeletal abnormalities such as delayed bone development and wavy ribs (7). This suggests that 2,4-D exposure is unlikely to be teratogenic in humans at expected exposure levels.

Mutagenic effects

2,4-D has been very extensively tested and was found to be nonmutagenic in most systems. 2,4-D did not damage DNA in human lung cells. However, in one study, significant effects occurred in chromosomes in cultured human cells at low exposure levels (26). The data suggest that 2,4-D is not mutagenic or has low mutagenic potential.

Carcinogenic effects

2,4-D fed to rats for 2 years caused an increase in malignant tumors (7). Female mice given a single injection of 2,4-D developed cancer (reticulum cell sarcomas) (7). Another study in rodents shows a low incidence of brain tumors at moderate exposure levels (45 mg/kg/day) over a lifetime (1,7). However, a number of questions were raised about the validity of this evidence and thus about the carcinogenic potential of 2,4-D.

In humans, a variety of studies give conflicting results. Several studies suggest an association between 2,4-D exposure and cancer. An increased occurrence of non-Hodgkin's lymphoma was found among a Kansas and Nebraska farm population associated with the spraying of 2,4-D (25,27). Other studies done in New Zealand, Washington, New York, Australia, and on Vietnam veterans from the U.S. were all negative. There remains considerable controversy about the methods used in the various studies and their results (28). Thus, the carcinogenic status of 2,4-D is not clear.

Organ toxicity

Most symptoms of 2,4-D exposure disappear within a few days, but there is a report of liver dysfunction from long-term exposure (1,25).

Fate in humans and animals

The absorption of 2,4-D is almost complete in mammals after ingestion, and nearly all of the dose is excreted in the urine. The compound is readily absorbed through the skin and lungs. Men given 5 mg/kg excreted about 82% of the dose as unchanged 2,4-D.

The half-life is between 10 and 20 hours in living organisms. There is no evidence that 2,4-D accumulates to significant levels in mammals or in other organisms (20). Between 6 and 8 hours after doses of 1 mg/kg, peak concentrations of 2,4-D were found in the blood, liver, kidney, lungs, and spleen of rats. There were lower levels in muscle and brain. After 24 hours, there were no detectable tissue residues.

Only traces of the compound have been found in the milk of lactating animals for 6 days following exposure. 2,4-D passes through the placenta in pigs and rats. In rats, about 20% was detected in the uterus, placenta, fetus, and amniotic fluid (27). Chickens given moderate amounts of 2,4-D in drinking water from birth to maturity had very low levels of the compound in eggs (7).

Ecological effects

Effects on birds

2,4-D is slightly toxic to wildfowl, and slightly to moderately toxic to birds. The LD$_{50}$ is 1000 mg/kg in mallards, 272 mg/kg in pheasants, and 668 mg/kg in quail and pigeons (5–7).

Effects on aquatic organisms

Some formulations of 2,4-D are highly toxic to fish while others are less so. For example, the LC_{50} ranges between 1.0 and 100 mg/L in cutthroat trout, depending on the formulation used. Channel catfish had less than 10% mortality when exposed to 10 mg/L for 48 hours (1,9). Green sunfish, when exposed to 110 mg/L for 41 hours, showed no effect on swimming response. Limited studies indicate a half-life of less than 2 days in fish and oysters (24).

Concentrations of 10 mg/L for 85 days did not adversely affect the survival of adult dungeness crabs. For immature crabs, the 96-hour LC_{50} is greater than 10 mg/L, indicating that 2,4-D is only slightly toxic. Brown shrimp showed a small increase in mortality at exposures of 2 mg/L for 48 hours (7,20).

Effects on other organisms (non-target species)

Moderate doses of 2,4-D severely impaired honeybees' brood production. At lower levels of exposure, exposed bees lived significantly longer than the controls. The honeybee LD_{50} is 0.0115 mg per bee (6,7).

Environmental fate

Breakdown in soil and groundwater

2,4-D has low soil persistence. The half-life in soil is less than 7 days (21). Soil microbes are primarily responsible for its disappearance (20).

Despite its short half-life in soil and in aquatic environments, the compound has been detected in groundwater supplies in at least five states and in Canada (20). Very low concentrations have also been detected in surface waters throughout the U.S. (23).

Breakdown in water

In aquatic environments, microorganisms readily degrade 2,4-D. Rates of breakdown increase with increased nutrients, sediment load, and dissolved organic carbon. Under oxygenated conditions, the half-life is 1 week to several weeks (20).

Breakdown in vegetation

2,4-D interferes with normal plant growth processes. Uptake of the compound is through leaves, stems, and roots. Breakdown in plants is by a variety of biological and chemical pathways (10). 2,4-D is toxic to most broadleaf crops, especially cotton, tomatoes, beets, and fruit trees (7).

Physical properties

2,4-D is a white powder (6).

Chemical name: (2,4-dichlorophenoxy)acetic acid (6)
CAS #: 94–75–7
Molecular weight: 221.04 (6)
Solubility in water: 900 mg/L @ 25°C (acid) (5)
Solubility in other solvents: ethanol v.s.; diethylether v.s.; toluene s.; xylene s. (6)
Melting point: 140.5°C (6)

Vapor pressure: 0.02 mPa @ 25°C (acid) (5)
Partition coefficient (octanol/water) (log): 2.81 (20)
Adsorption coefficient: 20 (acid) (21)

Exposure guidelines

ADI: 0.3 mg/kg/day (29)
MCL: 0.07 mg/L (30)
RfD: 0.01 mg/kg/day (31)
PEL: 10 mg/m³ (8-hour) (32)

Basic manufacturer

Rhone-Poulenc Ag. Co.
P.O. Box 12014
2 T.W. Alexander Dr.
Research Triangle Park, NC 27709
Telephone: 919-549-2000
Emergency: 800-334-7577

7.2.2   2,4-DB

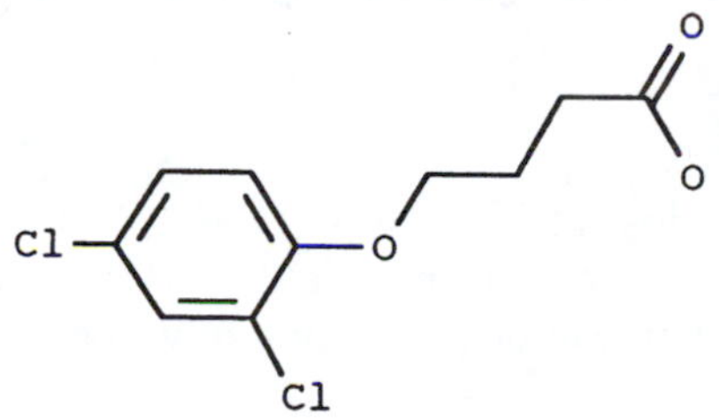

Figure 7.3 2,4-DB.

Trade or other names

2,4-DB is 4-(2,4-dichlorophenoxy)butyric acid. Trade names for products containing 2,4-DB include Butoxone, Butyrac, Butirex, Embutone, Embutox, and Venceweed. 2,4-DB may also be found in formulations with other herbicides such as cyanazine, MCPA, benazolin, linuron, and mecoprop.

Regulatory status

2,4-DB is a General Use Pesticide (GUP). The U.S. Environmental Protection Agency has classified it as toxicity class III — slightly toxic. Products containing it bear the Signal Word CAUTION.

Introduction

2,4-DB is a selective systemic herbicide in the phenoxy family. It is used for the control of many annual and perennial broad-leaved weeds in alfalfa, peanuts, soybeans, and other crops. In the plant, the compound changes to 2,4-D and inhibits growth at the tips of stems and roots.

This compound is not to be confused with another phenoxy compound, 2,4-D, and its derivatives, or with the derivatives of 2,4-DB such as the sodium salt, the isooctyl ester, or the butyl ester. Each of these are slightly different compounds and thus have different toxicities and environmental characteristics. It is found in formulations as a soluble concentrate and as an emulsified concentrate. *All of the material in this profile refers to the acid form of 2,4-DB unless specifically stated otherwise.* Well over 1 million pounds of 2,4-DB were used in the U.S. in 1990.

Toxicological effects

Acute toxicity

2,4-DB is a slightly toxic compound through oral exposure. It has an oral LD_{50} ranging between 370 and 700 mg/kg in rats. The sodium salt has a substantially lower toxicity, with an LD_{50} of 1500 mg/kg in rats and 400 mg/kg in mice (6,7). In rabbits, 2,4-DB is only slightly toxic when exposure is through the skin. The acute dermal LD_{50} of 2,4-DB is greater than 2000 mg/kg (6).

Chronic toxicity

At 25 and 80 mg/kg/day, dogs experienced changes in body weight and tissue effects (33). In a longer-term chronic feeding study with rats, 30-mg/kg effects noted included decreases in body weight, a decrease in heart weight, and changes in blood chemistry (33).

Reproductive effects
Female rats fed high doses of 75 mg/kg/day 2,4-DB, experienced a number of chronic effects, including lower ovarian weights and fewer offspring born (33). These data suggest that reproductive effects are not likely in humans at expected exposure levels.

Teratogenic effects
No data are currently available.

Mutagenic effects
Chromosome changes occurred due to exposure to 2,4-DB in Chinese hamster cells; however, the exposure was relatively high (just below a threshold of significant cell toxicity) and for relatively prolonged periods of time (17 or more hours). Tests over 2 hours produced no chromosome changes (33). The Ames test produced no chromosomal abnormalities in bacterial cells. The data, though minimal, indicate that 2,4-DB is weakly or nonmutagenic.

Carcinogenic effects
Preliminary evidence suggests that 2,4-DB can cause liver cancer in mice (hepatocellular carcinomas) (33). However, in another study with rats fed lower doses (to 30 mg/kg/day), no cancer-related changes occurred (33). Thus the carcinogenic status of 2,4-DB is not clear.

Organ toxicity
Target organs identified in long-term animal studies include the heart and liver.

Fate in Animals and Humans
No data are currently available.

Ecological effects

Effects on birds

Several dietary feeding tests indicate that 2,4-DB is practically nontoxic (LC_{50} values in excess of 5000 ppm) to waterfowl and upland game birds (33).

Effects on aquatic organisms

2,4-DB is slightly to moderately toxic to fish. The LC_{50} value for the compound is 18 mg/L in the fathead minnow, 7.5 to 17 mg/L in the bluegill sunfish, and 2 to 14 mg/L in the rainbow trout (33,34). No studies were available on the toxicity of the compound to freshwater or marine invertebrates.

Effects on other organisms (non-target species)

2,4-DB is nontoxic to bees (7).

Environmental fate

Breakdown in soil and groundwater

In soil, 2,4-DB is broken down by the action of soil microbes to the product 2,4-D. The half-life for 2,4-DB is about 7 days (21). Because of the risk of groundwater contamination from the whole family of 2,4-D related compounds, special advisories are required on the labels of end-use products related to 2,4-D including 2,4-DB. 2,4-DB has been detected in only one well in Texas (out of 928 tested) (23). It had a concentration of 0.002 mg/L of 2,4-DB.

Breakdown in water

No data are currently available.

Breakdown in vegetation

The compound is taken up by the roots and moved throughout the plant. In plants, the compound 2,4-DB is degraded to 2,4-D, which is then broken down further to less-toxic materials. In plants tolerant to the herbicide, the breakdown from 2,4-DB to 2,4-D is very slow (6).

Physical properties

2,4-DB is a colorless to white crystal with a slightly phenolic odor. It is slightly corrosive to iron (6).

Chemical name: 4-(2,4-dichlorophenoxy)butyric acid (6)
CAS #: 94–82–6
Molecular weight: 249.10 (6)

Solubility in water: 46 mg/L @ 25°C (6)
Solubility in other solvents: acetone v.s.; ethanol v.s.; diethyl ether v.s.; benzene
     s.s.; toluene s.s.; and kerosene s.s. (6)
Melting point: 117–119°C (6)
Vapor pressure: Negligible (acid and salts) (6)
Partition coefficient (octanol/water): Not available
Adsorption coefficient: 20 (estimated) (salt) (21)

Exposure guidelines

ADI: Not available
HA: Not available
RfD: 8×10^{-3} mg/kg/day (31)
PEL: Not available

Basic manufacturer

Rhone Poulenc Ag. Co.
P.O. Box 12014
2 T.W. Alexander Drive
Research Triangle Park, NC 27709
Telephone: 919-549-2000
Emergency: 800-334-7577

7.2.3   *Chloramben*

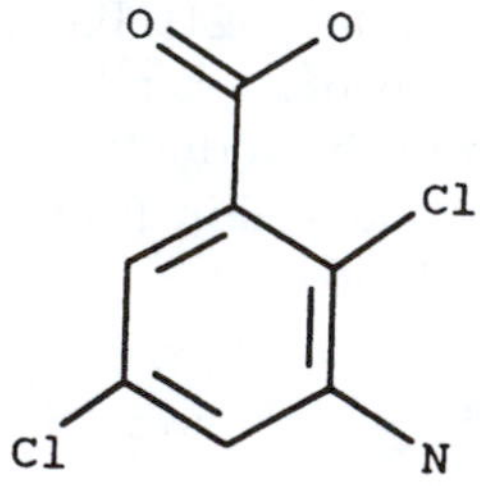

Figure 7.4 Chloramben.

Trade or other names

The trade names and synonyms for chloramben are Ambiben, Amiben, Amiben DS, Chlorambene, Ornamental Weeder, and Vegiben.

Regulatory status

Chloramben is classified as a General Use Pesticide (GUP) by the U.S. Environmental Protection Agency. A toxicity class IV pesticide, chloramben is practically nontoxic. Chloramben products bear the Signal Word CAUTION. It is no longer produced or sold in the U.S.

Introduction

Chloramben is a selective, pre-emergence benzoic acid herbicide that is primarily soil-applied to control annual grass and broad-leaved weed seedlings. It was mostly used for soybeans, but has also been used for dry beans, peanuts, sunflowers, peppers, cotton, sweet potatoes, squash, hardwood trees, shrubs, and some conifers. Chloramben inhibits seedling root development and causes plants to bend and die as they emerge from the soil.

Toxicological effects

Acute toxicity

The oral LD_{50} for chloramben in rats is 3500 mg/kg (5,7) and is 3725 mg/kg in mice.

Chloramben irritates the throat, lungs, skin, and the eyes. However, chloramben is not known to move through the skin. On rats, a low-dose skin application caused mild irritation that subsided within 24 hours (7).

The dermal LD_{50} for rabbits is 3136 mg/kg (6,7); for albino rats, it is greater than 3160 mg/kg. This indicates that chloramben has about the same toxicity when applied to the skin as when ingested.

The major health concern for humans is dermatitis resulting from skin exposure to chloramben.

Chronic toxicity

In long-term tests on rats, mice, and dogs, no chloramben-related effects on body weight, body function, and organs were observed, even at very high concentrations (7,10).

Reproductive effects

No effects on fertility, survival of the fetus, or on nursing ability were found in a long-term study of rats fed large daily doses of chloramben (9). These data suggest that chloramben does not cause reproductive toxicity.

Teratogenic effects

When pregnant rabbits were given high doses of chloramben during the sensitive period of gestation, no developmental defects were seen in the fetuses. Pregnant rats fed lower doses of chloramben during the sensitive period of gestation were not affected, but fetal deaths increased and fetal skeletal development was not complete. At even lower doses, no increase in fetal deaths was seen, but the bones were not fully developed. No changes were seen at the lowest doses (9). This evidence indicates that chloramben is unlikely to cause teratogenic effects at expected exposure levels.

Mutagenic effects

Most bacterial and mammalian cell assays of chloramben indicate that it is not toxic to genetic material. However, mutagenicity was seen in one test using Chinese hamster ovary cells (9). These data suggest chloramben shows slight or no mutagenic potential.

Carcinogenic effects

Rats given high daily doses of chloramben for 80 weeks did not show carcinogenicity. However, mice given similar doses (up to 300 mg/kg) each day for 80 weeks developed cancerous tumors (7). Thus, the carcinogenic status of chloramben is not clear.

Organ toxicity

Long-term animal studies did not show evidence of chronic injury to the organs examined.

Fate in humans and animals

Several animal studies have shown that once ingested, chloramben is rapidly excreted in the urine and feces, with no residues found in tissues. Very small amounts of chloramben fed daily to lactating cows left the body via urine (88%) and the feces (5%) within 4 days. No residues were found in the milk (7). Similarly, dogs fed very small amounts showed rapid chloramben excretion, with no residues found in the tissues (7).

Chloramben is quickly absorbed from the gastrointestinal tract of female rats. About 97% of an oral dose is absorbed and then excreted through the urine and expired air. In rats fed a single dose, 70% of that recovered 24 hours later from the urine was the original compound. Only 0.6% of the dose remained in the body after 3 to 4 days (9).

Ecological effects

Effects on birds

The LC_{50} of chloramben is 4640 mg/L in mallard ducks, indicating that the compound is practically nontoxic to this species (6).

Effects on aquatic organisms

Chloramben is not toxic to fish (5).

Effects on other organisms (non-target species)

Chloramben is not toxic to bees (6,7).

Environmental fate

Breakdown in soil and groundwater

Chloramben is active in the soil. It is moderately persistent. The wetter the soil, the greater the toxicity of this compound to target plants. Rainfall or irrigation within 10 to 14 days of application is needed for chloramben to be effective against weeds.

In a soil solution, chloramben formulations readily break down into forms that are quite mobile in the soil (34). Thus, much of it is lost from the soil due to leaching, although some loss is due to breakdown by soil microorganisms. Leaching is greatest in sandy soil (5). As the organic matter content of the soil increases, binding of chloramben also increases and leaching decreases. Like most pre-emergence herbicides, the activity is nearly the same on muck soils as on mineral soils. In general, chloramben activity lasts for about 6 to 8 weeks in soil (6,21). The compound does not evaporate appreciably.

Of 188 groundwater samples tested, chloramben was found in only one, at a concentration of 0.0017 mg/L (9).

Breakdown in water

Chloramben readily degrades in water exposed to sunlight (9).

Breakdown in vegetation

Chloramben reacts with plant constituents to form a stable and nontoxic product in both susceptible and tolerant plants. The ease of formation of this metabolite seems to be the reason the herbicide is not totally degraded within the plant (5).

When soil-applied, plant roots are the primary site of absorption and chloramben action. Roots of resistant species, like soybeans, can absorb large amounts of the compound, but translocate very little of it to the above-ground portions of the plant. In contrast, the susceptible barley plant translocates a significant amount into its leaves, and thus dies (5).

Physical properties

Chloramben is a colorless and odorless crystalline solid (6).

Chemical name: 3-amino-2,5-dichlorobenzoic acid (6)
CAS #: 133–90–4
Molecular weight: 206.03 (6)
Solubility in water: 700 mg/L @ 25°C (6)
Solubility in other solvents: alcohol v.s.; ether s.; acetone v.s.; ethanol v.s. (6)
Melting point: 200–201°C (6)
Vapor pressure: 930 mPa @ 100°C (6)
Partition coefficient (octanol/water): Not available
Adsorption coefficient: 15 (estimated) (salts at pH < 5) (21)

Exposure guidelines

ADI: Not available
HA: 0.1 mg/L (lifetime) (9)
RfD: 0.015 mg/kg/day (31)
PEL: Not available

Basic manufacturer

Chloramben is no longer produced or sold in the U.S.

7.2.4   DCPA

Trade or other names

DCPA is also called chlorthal or chlorthal-dimethyl. Trade names for products containing DCPA include DAC 893, Dacthal, and Dacthalor. It may be formulated with other herbicides such as methazole and propachlor.

Regulatory status

DCPA is classified as a General Use Pesticide (GUP). The U.S. EPA classifies it as toxicity class IV — practically nontoxic. Products containing DCPA bear the Signal

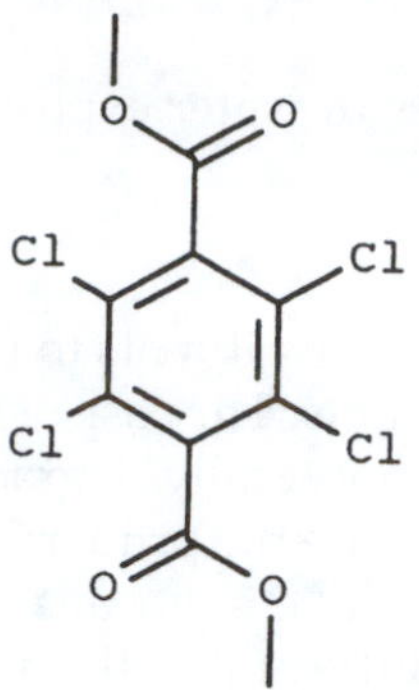

Figure 7.5 DCPA.

Word CAUTION. Its only restrictions apply in the state of Washington, which requires supplemental labeling for Dacthal W-75.

Introduction

DCPA is a phthalate pre-emergent herbicide used on annual grasses and annual broadleaf weed species in a wide range of vegetable crops. About 20% of the use of this compound in the U.S. is for homes and gardens. Products containing DCPA may be formulated as wettable powders, granules, or as suspension concentrates.

While there is no toxicological concern for DCPA per se, the technical product DCPA may contain very small amounts of dioxin (2,3,7,8-TCDD) and hexachloroben-zene (HCB) as impurities, which may be of toxicological concern.

Toxicological effects

Acute toxicity

The compound has a very low toxicity to mammals. The LD_{50} values for DCPA in rats range from greater than 3000 to 12,500 mg/kg. DCPA in rabbits and beagle dogs has an LD_{50} of greater than 10,000 mg/kg. The dermal LD_{50} in rabbits is greater than 2000 mg/kg. DCPA is not a skin sensitizer. It is a mild eye irritant. The inhalation LC_{50} (4-hour) is greater than 5.7 mg/L for rats (5,7).

Chronic toxicity

A 3-mg dose in a rabbit eye produced mild irritation that disappeared in 24 hours. Dogs given high doses of 800 mg/kg/day for a month showed some adverse effects in the liver. In longer-term studies with rats (90 days), similar doses (about 750 mg/kg/day) caused no adverse effects (35). In a 2-year study with rats, a dose of around 50 mg/kg/day was responsible for changes in the adrenal weights of the females and in the kidney weights of the males (35).

Reproductive effects

Rats fed high doses of DCPA (500 mg/kg/day) showed no changes in fertility, gestation, live births, or lactation (35). The study was conducted over one full generation. These data suggest that the compound does not cause reproductive effects.

Teratogenic effects

Available data indicate that DCPA is not teratogenic. Pregnant rabbits fed moderate doses (up to 300 mg/kg) of DCPA on days 8 to 16 of gestation showed no skeletal or organ abnormalities in the offspring (35).

Mutagenic effects

No mutagenicity was seen in a number of tests, including mutation frequency and activity, cytogenetic tests, DNA repair, and dominant lethal tests (35). This evidence indicates that DCPA is not mutagenic.

Carcinogenic effects

No carcinogenic effects were noted in rats in a 2-year study where diets contained up to 500 mg/kg/day DCPA (35). Thus, DCPA does not appear to be carcinogenic.

Organ toxicity

Long-term studies in test animals have indicated the liver and adrenal glands as target organs.

Fate in humans and animals

Much of the compound that is ingested is not absorbed. Cows excreted nearly all of a small dose of DCPA within 5 days, and dogs absorbed only small amounts (3%) of the compound. The remaining amount was eliminated within 4 days.

When dairy cows were fed diets with up to 200 ppm DCPA for 24 days, 0.26 ppm of the compound or its metabolites were found in milk, while 30 to 90 ppm for 9 or 23 days resulted in residues of 0.036 and 0.066 ppm in milk, respectively. Residues in other tissues were generally less than 1 ppm (8).

Ecological effects

Effects on birds

DCPA appears to be moderately toxic to some young wildfowl, and practically nontoxic to the young of other species and to adult birds. The LD_{50} in young bobwhite quail is 5500 mg/kg (7). Young mallards and young quail were more sensitive to the herbicide than adult birds. Diets containing about 250 mg/kg caused heavier mortality to young ducks in the first 5 days. Older birds had a higher survival rate (35).

Effects on aquatic organisms

DCPA is slightly toxic to practically nontoxic to fish, depending on the species. It is practically nontoxic to bluegill sunfish and slightly toxic to rainbow trout. The compound is practically nontoxic to estuarine and marine organisms (invertebrates and some fish). The available data suggest that DCPA poses no hazard to endangered aquatic species (35).

Effects on other organisms (non-target species)

At high doses of DCPA, there was only 3% bee mortality. Thus, DCPA is only slightly toxic to bees (7).

Environmental fate

Breakdown in soil and groundwater

DCPA is moderately persistent. The half-life is from 14 to 100 days in most soils (21). However, moisture is essential for degradation. In one study, there was no apparent build-up of pesticide residues in soil even after repeated application. The DCPA concentration declined slowly to 75 or 80% in 28 days. Later sampling showed a continued decline of DCPA and its breakdown products (7,34).

The DCPA metabolite, tetrachloroterephthalic acid (TTA or diacid), is much more water soluble than the parent compound and is subject to leaching in some soils. This metabolite has been detected in groundwater in the onion-growing areas of eastern Oregon (23). It has been detected in several other states in the U.S. as well (7).

Breakdown in water

There is virtually no degradation of DCPA in water ranging from moderately acidic to moderately alkaline (pH 5.0 to 9.0). Breakdown is due to the action of sunlight and the half-life is greater than 1 week (34).

Breakdown in vegetation

Plants may metabolize DCPA to the same two breakdown products that are seen in soils, with the proportion varying in different species. DCPA affects the seed and pre-emergence stage, but has little effect on crops or weeds after they have emerged. Limited information suggests that plants may remove the chlorine molecules of DCPA (8). In one study, pine trees took up nearly 1% of the soil-applied chemical. The majority of the compound taken up by the trees remained in the root system where it was rapidly diluted.

Physical properties

DCPA consists of colorless crystals (6).

Chemical name: dimethyl-2,3,5,6-tetrachlorobenzene-1,4-dicarboxylic acid (6)
CAS #: 1861–32–1
Molecular weight: 303.9 (6)
Solubility in water: 0.5 mg/L @ 25°C (6)
Solubility in other solvents: benzene v.s.; toluene v.s.; acetone v.s.; carbon tetra-
    chloride s. (6)
Melting point: 155–156°C (6)
Vapor pressure: 0.33 mPa @ 25°C (5)
Partition coefficient (octanol/water): Not available
Adsorption coefficient: 5000 (chlorthal-dimethyl) (21)

Exposure guidelines

ADI: Not available
HA: 4.0 mg/L (lifetime) (35)
RfD: 0.5 mg/kg/day (31)
PEL: Not available

Basic manufacturer

ISK Biosciences
5966 Heisley Road
P.O. Box 8000
Mentor, OH 44061-8000
Telephone: 216-357-4100
Emergency: 216-357-7070

7.2.5 Dicamba

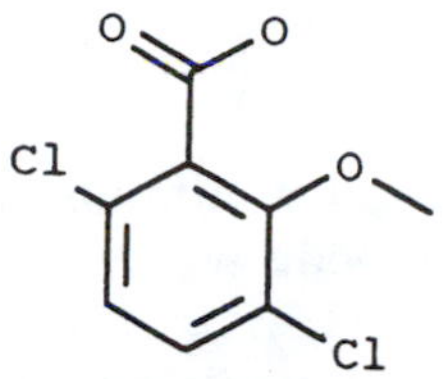

Figure 7.6 Dicamba.

Trade or other names

Trade names for products containing dicamba include Banfel, Banvel, Banvel CST, Banvel D, Banvel XG, Dianat, Dicazin, Fallowmaster, Mediben, Metambane, Tracker, and Trooper.

Regulatory status

The EPA has classified this General Use Pesticide (GUP) as toxicity class III — slightly toxic. Products containing dicamba bear the Signal Word WARNING. This is because of its irritating, corrosive effect on skin and eyes.

Introduction

Dicamba is a benzoic acid herbicide. It can be applied to the leaves or to the soil. Dicamba controls annual and perennial broadleaf weeds in grain crops and grass-lands, and it is used to control brush and bracken in pastures. It will kill broadleaf weeds before and after they sprout. Legumes will be killed by dicamba. In combi-nation with a phenoxyalkanoic acid or other herbicide, dicamba is used in pastures, range land, and non-crop areas such as fence rows and roadways to control weeds.

Toxicological effects

Acute toxicity

Dicamba is slightly toxic by ingestion and slightly toxic by inhalation or dermal exposure (7). The oral LD_{50} for dicamba is 757 to 1707 mg/kg in rats, 1190 mg/kg in mice, 2000 mg/kg in rabbits, and 566 to 3000 mg/kg in guinea pigs (1,6). The dermal LD_{50} in rabbits is greater than 2000 mg/kg (7). The inhalation LC_{50} for dicamba in rats is greater than 200 mg/L (7).

Symptoms of poisoning with dicamba include loss of appetite (anorexia), vom-iting, muscle weakness, slowed heart rate, shortness of breath, central nervous system effects (victim may become excited or depressed), benzoic acid in the urine, incontinence, cyanosis (bluing of the skin and gums), and exhaustion following repeated muscle spasms (1,5). In addition to these symptoms, inhalation can cause irritation of the linings of the nasal passages and the lungs, and loss of voice (7). Most individuals who have survived severe poisoning from dicamba have recovered within 2 to 3 days with no permanent effects (7).

Dicamba is very irritating and corrosive and can cause severe and permanent damage to the eyes (7). The eyelids may swell and the cornea may be cloudy for a week after dicamba is splashed in the eyes (1).

In some individuals, dicamba is a skin sensitizer and may cause skin burns (7). There is no evidence that dicamba is absorbed into the body through the skin (1).

Chronic toxicity

Doses of 25 mg/kg/day in the diet administered to rats for 2 years produced no observable effects on survival, body weight, food consumption, organ weight, blood chemistry, or tissue structure (1,5,11).

Consumption of dicamba at high levels over a long period of time was shown to cause changes in the liver and a decrease in body weight in rats (1,5,11). In mice, some enlargement of liver cells has occurred (1).

Reproductive effects

In a three-generation study, dicamba did not affect the reproductive capacity of rats (1). When rabbits were given doses of 0.5, 1, 3, 10, or 20 mg/kg/day technical dicamba from days 6 through 18 of pregnancy, toxic effects on the mothers, slightly reduced fetal body weights, and increased loss of fetuses occurred at the 10-mg/kg dose (5,10). These data suggest that dicamba is unlikely to cause reproductive effects in humans at expected exposure levels.

Teratogenic effects

No teratogenic effects have been shown in lab animals such as rabbits and rats exposed to dicamba (36).

Mutagenic effects

Dicamba has not been shown to be a mutagen (5,10).

Carcinogenic effects

Rats fed dicamba up to 25 mg/kg/day for 2 years showed no increased incidence of tumors (5,10). This evidence suggests that dicamba is not carcinogenic.

Organ toxicity

Chronic exposure can lead to the development of the same symptoms as described for acute exposure.

Fate in humans and animals

Dicamba was excreted rapidly by rats, mainly in the urine, when administered orally or subcutaneously; 1 to 4% was excreted in the feces (1). Mice, rats, rabbits, and dogs excreted 85% of an oral dose as unmetabolized dicamba in the urine within 48 hours of dosing. Eventually, between 90 and 99% of the dose was excreted unmetabolized in the urine. This indicates that dicamba is rapidly absorbed into the bloodstream from the gastrointestinal tract (10).

When dicamba was ingested daily in the feed, the concentrations in different organs reached a steady state within 2 weeks. When daily intake stopped, storage in the organs declined rapidly (1). It is therefore concluded that dicamba does not bioaccumulate in mammalian tissues.

Ecological effects

Effects on birds

Dicamba is practically nontoxic to birds. The LD_{50} for technical dicamba in mallard ducks is 2009 mg/kg. The 8-day dietary LC_{50} in mallards and in bobwhite quail is greater than 10,000 ppm (5,7).

Effects on aquatic organisms

Dicamba is of low toxicity to fish (5,7). The LC_{50} (96-hour) for technical dicamba is 135 mg/L in rainbow trout and bluegill sunfish, greater than 100 mg/L in grass shrimp, and greater than 180 mg/L in fiddler crab and sheepshead minnow (5). The LC_{50} (48-hour) for dicamba is 35 mg/L in rainbow trout, 40 mg/L in bluegill, 465 mg/L in carp, and 110 mg/L in *Daphnia magna,* a small freshwater crustacean (5,7).

Effects on other organisms (non-target species)

Dicamba poses little threat to wildlife. Dicamba is not toxic to bees (7).

Environmental fate

Breakdown in soil and groundwater

Dicamba is moderately persistent in soil. The half-life of dicamba in soil is typically 1 to 4 weeks (21). Under conditions suitable for rapid metabolism, the half-life is less than 2 weeks (20).

Metabolism by soil microorganisms is the major pathway of loss under most soil conditions. The rate of biodegradation increases with temperature and increasing soil moisture, and tends to be faster when soil is slightly acidic. When soil moisture increases above 50%, the rate of biodegradation declines (5).

Dicamba slowly breaks down in sunlight (5). Volatilization from soil surfaces is probably not significant, but some volatilization may occur from plant surfaces (20). It is stable to water and other chemicals in the soil (6). Dicamba does not bind to soil particles and is highly soluble in water. It is therefore highly mobile in the soil and may contaminate groundwater (21). In humid areas, dicamba will be leached from the soil in 3 to 12 weeks (20).

Breakdown in water

In water, microbial degradation is the main route of dicamba disappearance. Photolysis may also occur. Aquatic hydrolysis, volatilization, adsorption to sediments, and bioconcentration are not expected to be significant (20).

Breakdown in vegetation

Dicamba is rapidly taken up by the leaves and roots of plants and it is readily translocated to other plant parts. In some plant species, dicamba accumulates in the tips of mature leaves (5). Desirable broadleaf plants such as fruit trees, and tomatoes may be harmed during their growth and developmental stages (5).

Residues of dicamba on treated plants can disappear through exudation from the roots into the surrounding soil, metabolism within the plant, or by loss from leaf surfaces (5).

Physical properties

Pure dicamba is an odorless, white crystalline solid. The technical acid is a pale buff crystalline solid (6).

Chemical name: 3,6-dichloro-o-anisic acid (6)
CAS #: 1918-00–9
Molecular weight: 221.04 (6)
Solubility in water: 6500 mg/L @ 25°C (6)
Solubility in solvents: acetone s.; dichloromethane s.; dioxane v.s.; ethanol s.; toluene s.; xylene s. (6)
Melting point: 114–116°C (6)
Vapor pressure: 4.5 mPa @ 25°C (6)
Partition coefficient (octanol/water): 0.29 (5)
Adsorption coefficient: 2 (salt) (21)

Exposure guidelines

ADI: Not available
HA: 0.2 mg/L (10)
RfD: 0.03 mg/kg/day (31)
PEL: Not available

Basic manufacturer

Sanex, Inc.
5300 Harvester Road
Burlington, Ontario
L7L 5N5 Canada
Telephone: 905-639-7535

7.2.6   *Fluazifop-p-butyl*

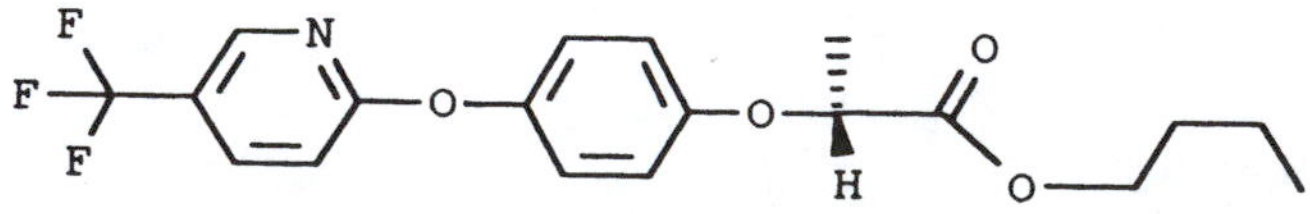

Figure 7.7 Fluazifop-p-butyl.

Trade or other names

Trade names for products containing fluazifop-p-butyl include Fluazifop-p-butyl ester, Fusilade 2000, Fusilade DX, Fusilade Five, Fusilade Super, Fusion, Horizon, Ornamec, PP005, and Tornado.

Regulatory status

Fluazifop-p-butyl is a slightly to practically nontoxic compound in EPA toxicity class IV (1,2). It is a General Use Pesticide (GUP), and labels for products containing it must bear the Signal Word CAUTION.

NOTE: This profile is for fluazifop-p-butyl herbicide, which is a different compound than either fluazifop-p or fluazifop-butyl. Fluazifop-p-butyl is the n-butyl ester of the acid fluazifop-p. Fluazifop-p is the R-enantiomer ("dextro" or "plus") of the acid, while fluazifop typically contains equal proportions of both the R- and S-enantiomers (2,3). Fluazifop-p (the R-enantiomer) is the herbicidally active portion of the mixture.

Introduction

Fluazifop-p-butyl is a selective post-emergence phenoxy herbicide used for control of most annual and perennial grass weeds in cotton, soybeans, stone fruits, asparagus, coffee, and other crops. It may often be used with an oil adjuvant or nonionic surfactant to increase efficiency. It has essentially no activity on broadleaf species. It is compatible with a wide variety of other herbicides and may also be found in formulations with other products such as fenoxaprop ethyl ester (in Horizon and Fusion) and fomesafen (in Tornado). It is available as an emulsifiable concentrate. Fluazifop-p-butyl formulations may contain some fluazifop-butyl.

Toxicological effects

Acute toxicity

Fluazifop-p-butyl is slightly to practically nontoxic via the oral route. The reported acute oral LD_{50} values for technical fluazifop-p-butyl are 3680 (6) to 4096 mg/kg (5) in male rats, and 2451 to 2721 mg/kg (5) in female rats. The reported acute oral LD_{50} values for one formulated product (Fusilade DX) are higher, indicating practically no toxicity via the oral route (5).

A single, large oral dose of a formulated compound (Fusilade 2000) can cause severe stomach and intestinal disturbances (39). Ingestion of large quantities may also cause problems in the central nervous system such as drowsiness, dizziness, loss of coordination, and fatigue (37).

Fluazifop-p-butyl is slightly toxic via the dermal route as well. The reported dermal LD_{50} for the compound is greater than 2400 mg/kg in rabbits (5,6). It is reported to cause only slight skin and mild eye irritation in rabbits, but no skin sensitization in guinea pigs (5,6). The formulation Fusilade DX is reported to have similar acute toxicity via the dermal route and does not cause skin sensitization in guinea pigs, but may cause moderate skin and mild eye irritation in rabbits (5).

The formulation Fusilade DX is reported to have a 4-hour inhalation LC_{50} of greater than 0.54 mg/L in male rats and 0.77 mg/L in female rats, indicating moderate toxicity via the inhalation route (5). Breathing small amounts of the product Fusilade 2000 may cause vomiting and severe lung congestion; larger amounts may ultimately lead to labored breathing, coma, and death (37).

Chronic toxicity

Rats fed small amounts of fluazifop-p-butyl for 90 days developed no compound-induced effects at doses at or below 10 mg/kg/day (6).

Reproductive effects
No data are currently available.

Teratogenic effects
No data are currently available.

Mutagenic effects
Numerous tests have shown the compound to be nonmutagenic (5).

Carcinogenic effects
No data are currently available.

Organ toxicity
Organ toxicity has not been seen in experimental animals (6).

Fate in humans and animals
No data are currently available.

Ecological effects

Effects on birds

Fluazifop-p-butyl is practically nontoxic to bird species; the reported acute oral LD_{50} for the technical product in mallards is greater than 3528 mg/kg (5,6). The reported 5-day dietary LC_{50} in mallard duck is greater than 4321 ppm, and in bobwhite quail is greater than 4659 ppm (5).

Effects on aquatic organisms

Fluazifop-p-butyl may be highly to moderately toxic to fish, but only slightly toxic to other aquatic species such as invertebrates. The reported 96-hour LC_{50} values for the technical product in fish species are 0.53 mg/L in bluegill sunfish and 1.37 mg/L in rainbow trout (5), indicating very high to high toxicity. The 48-hour LC_{50} in *Daphnia magna* (an aquatic invertebrate) is reported as greater than 10 mg/L (5), indicating only slight toxicity.

Effects on other organisms (non-target species)

The compound is of low toxicity to bees. Oral and contact LD_{50} values for bees are greater than 0.20 mg per bee (6).

Environmental fate

Breakdown in soil and groundwater

Fluazifop-p-butyl is of low persistence in moist soil environments, with a reported half-life in these conditions of less than 1 week (5,6). Fluazifop-p-butyl breaks down rapidly in moist soils to the fluazifop acid, which is also of low persistence. Fluazifop-p-butyl and fluazifop-p are both reported to be of low mobility in soils and not to present appreciable risks for groundwater contamination (5). The reported soil adsorption coefficient for fluazifop-p indicates a moderate to low affinity for soil (21).

Breakdown in water

Fluazifop-p-butyl is rapidly hydrolyzed (cleaved apart by water) under most conditions to the fluazifop acid (2,3). It is relatively stable to breakdown by UV or sunlight, and nonvolatile (5,6).

Breakdown in vegetation

After uptake by the leaves of plants, fluazifop-p-butyl is rapidly broken down in the presence of water to fluazifop-p, which is translocated throughout the plant (5,6). The compound accumulates in the actively growing regions of the plant (meristems of roots and shoots, root rhizomes, and stolons of grass) where it interferes with energy (ATP) production and cell metabolism in susceptible species (5,6).

Physical properties

The compound is a pale, straw-colored liquid (6).

Chemical name: (R)-2-[4-(5-trifluoromethyl-2-pyridyloxy)phenoxy]propionic acid (6)
CAS #: 79241-46-6
Molecular weight: 383.4 (6)
Solubility in water: 1.1 mg/L @ 25°C (5)
Solubility in other solvents: v.s. in most organic solvents (6)
Melting point: ca. 5°C (6)
Vapor pressure: 0.054 mPa @ 20°C (6)
Partition coefficient (octanol/water): 31,620 (6)
Adsorption coefficient: 5700 (21)

Exposure guidelines

ADI: Not available
HA: Not available
RfD: Not available
PEL: Not available

Basic manufacturer

Zeneca Ag Products
1800 Concord Pike
Wilmington, DE 19897
Telephone: 800-759-4500
Emergency: 800-759-2500

7.2.7   MCPA

Trade or other names

Trade or other names for MCPA or products containing it include: Agritox, Agroxone, Agrozone, Agsco MXL, Banlene, Blesal MC, Bordermaster, Cambilene, Cheyenne, Chimac Oxy, Chiptox, Class MCPA, Cornox Plus, Dakota, Ded-Weed, Empal, Envoy, Gordon's Amine, Kilsem, Legumex, Malerbane, Mayclene, MCP,

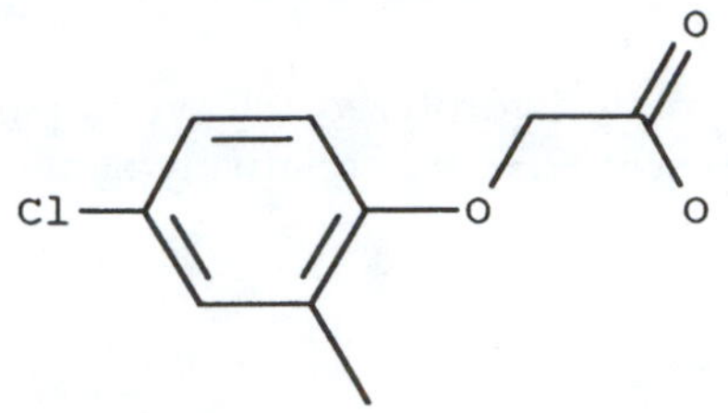

Figure 7.8 MCPA.

Mephanac, Midox, Phenoxylene, Rhomene, Rhonox, Sanaphen-M, Shamrox, Selectyl, Tiller, U 46 M-Fluid, Vacate, Weed-Rhap, and Zhelan.

Regulatory status

MCPA is a slightly toxic compound in EPA toxicity class III, and is a General Use Pesticide (GUP). Labels for products containing MCPA must carry the Signal Word DANGER due to its potential to cause severe eye irritation.

Introduction

MCPA is a systemic post-emergence phenoxy herbicide used to control annual and perennial weeds (including thistle and dock) in cereals, flax, rice, vines, peas, potatoes, grasslands, forestry applications, and on rights-of-way. This herbicide is very compatible with many other compounds and may be used in formulation with many other products, including bentazone, bromoxynil, 2,4-D, dicamba, fenoxaprop, MCPB, mecoprop, thifensulfuron, and tribenuron.

NOTE: As with some of the other phenoxy herbicides, MCPA is an acid, but is often formulated as a salt (e.g., dimethylamine salt) or an ester (e.g., isooctyl ester). *Unless otherwise indicated, this document will refer to the acid form.*

Toxicological effects

Acute toxicity

MCPA acid is slightly toxic via ingestion, with reported oral LD_{50} values for the technical product in rats ranging from 700 to 1160 mg/kg (5,6), and ranging in mice from 550 to 800 mg/kg (5,6). It is slightly toxic via the dermal route as well, with reported dermal LD_{50} values ranging from greater than 1000 mg/kg in rats to greater than 4000 mg/kg in rabbits (5,6). Symptoms in humans from very high acute exposure could include slurred speech, twitching, jerking and spasms, drooling, low blood pressure, and unconsciousness (1).

Chronic toxicity

Dietary levels of approximately 50 and 125 mg/kg/day over 7 months caused reduced feeding rates and retarded growth rates in rats (1). White blood cell counts and ratios were not affected, but some reductions in red blood cell counts and hemoglobin did appear to be associated with exposure to MCPA at oral dose levels of approximately 20 mg/kg/day.

In the same study, oral doses of approximately 5 mg/kg/day caused increased relative kidney weights, and oral doses of approximately 20 mg/kg/day caused increased relative liver weights (1). Another study in rats showed no effects on kidney or liver weights over an unspecified time period at oral doses of 60 mg/kg/day; but oral doses of 150 mg/kg/day did cause reversible increases in these weights over the course of 3 months (1).

Very high dermal doses of 500 mg/kg/day caused reduced body weight, and even higher dermal doses of 1000 and 2000 mg/kg/day resulted in increased mortality and observable changes in liver, kidney, spleen, and thymus tissue (1).

Reproductive effects

A two-generation rat study at doses up to 15 mg/kg/day affected reproductive function. Even smaller amounts of the compound were toxic to the fetuses. Dogs receiving relatively small amounts of MCPA (8 and 16 mg/kg) for 13 weeks showed adverse sperm and testes changes (8). It is unlikely that humans will experience these effects under normal exposure conditions.

Teratogenic effects

Offspring of pregnant rats fed low to moderate doses of MCPA (20 to 125 mg/kg) on days 6 to 15 of gestation had no birth defects. However, when the ethyl ester form of MCPA was fed to pregnant rats (2 to 100 mg/kg/day on days 8 to 15 of gestation), cleft palate, heart defect, and kidney anomalies were observed in the offspring (7). Mice fed 5 to 100 mg/kg/day MCPA on days 6 to 15 showed significantly reduced fetal weight and delayed bone development at the highest dose (24).

Teratogenic effects in humans are unlikely at expected exposure levels.

Mutagenic effects

MCPA is reportedly weakly mutagenic to bone marrow and ovarian cells of hamsters, but negative results were reported for other mutagenic tests (38). It was negative in a bacterial test system (both with and without metabolic activation), negative in spot tests, and negative in host-mediated tests (1). It produced no detectable increase in chromosomal aberrations in houseflies (4). Some irregularities occurred in gene transfer during cell division in brewer's yeast, although at levels that caused massive cell death (1). It appears that the compound poses little or no mutagenic risk.

Carcinogenic effects

All of the available evidence on MCPA indicates that the compound does not cause cancer (1). Forestry and agricultural workers occupationally exposed to MCPA in Sweden did not show increased cancer incidence (39).

Organ toxicity

Target organs identified in animal studies include the liver, kidneys, spleen, and thymus. Farmworker exposure has resulted in reversible anemia, muscular weakness, digestive problems, and slight liver damage (1).

Fate in humans and animals

MCPA is rapidly absorbed and eliminated from mammalian systems (1). Rats eliminated nearly all of a single oral dose within 24 hours, mostly through the urine with little or no metabolism (1,6). In another rat study, three quarters of the dose was eliminated within 2 days. All was eliminated by 8 days (1). Humans excreted about half of a 5-mg dose in the urine within a few days. No residues were found after day 5 (1). Cattle and sheep fed low to moderate doses of MCPA in the diet for 2 weeks showed no residues from levels less than about 18 mg/kg (1). The major metabolite of MCPA is 2-methyl-4-chlorophenol in the free and conjugated form, which is formed in the liver (38).

Ecological effects

Effects on birds

MCPA is moderately toxic to wildfowl; the LD_{50} of MCPA in bobwhite quail is 377 mg/kg (5,6).

Effects on aquatic organisms

MCPA is only slightly toxic to freshwater fish, with reported LC_{50} values ranging from 117 (5) to 232 mg/L in rainbow trout (6). MCPA is practically nontoxic to freshwater invertebrates, and estuarine and marine organisms.

Effects on other organisms (non-target species)

MCPA is nontoxic to bees, with a reported oral LD_{50} of 104 μg per bee (5,6).

Environmental fate

Breakdown in soil and groundwater

MCPA and its formulations are rapidly degraded by soil microorganisms and has low persistence, with a reported field half-life of 14 days to 1 month, depending on soil moisture and soil organic matter contents (21). Decreased soil moisture and microbial activity, as well as increased soil organic matter, will prolong the field half-life of MCPA (12). With less than 10% organic matter in soil, the compound is degraded in 1 day and, with greater than 10% levels in soil, it takes 3 to 9 days to degrade. The half-life is 5 to 6 days in slightly acidic to slightly alkaline soils (12).

MCPA readily leaches in most soils, but its mobility decreases with increasing organic matter (12). MCPA and its formulations show little affinity for soil.

Breakdown in water

It is relatively stable to light breakdown (5), but can be rapidly broken down by microorganisms. In sterilized water, it takes about 3 weeks for half of the compound to degrade due to the action of sunlight. In rice paddy water, however, MCPA is almost totally degraded by aquatic microorganisms in less than 2 weeks (12).

Breakdown in vegetation

MCPA is readily absorbed and translocated in most plants (5). It works by concentrating in the actively growing regions of a plant (meristematic tissue) where it interferes with protein synthesis, cell division, and ultimately the growth of non-resistant plants (7). It is actively broken down in plants, the major metabolite being 2-methyl-4-chlorophenol (5).

Physical properties

Pure MCPA occurs as colorless crystals (6).

Chemical name: (4-chloro-2-methylphenoxy)acetic acid (6)
CAS #: 94–74–6
Molecular weight: 200.62 (6)
Solubility in water: 825 mg/L @ 25°C (acid) (5)
Solubility in other solvents: v.s. in ether, ethanol, toluene, xylene; s. in methanol (6)
Melting point: 118–119°C (6)
Vapor pressure: 0.2 mPa @ 20°C (6)
Partition coefficient: Not available
Adsorption coefficient: MPCA acid, 100; MCPA salts, 20 (estimated); MCPA ester, 1000 (estimated) (21)

Exposure guidelines

ADI: Not available
HA: 0.01 mg/L (lifetime) (38)
RfD: 0.0005 mg/kg/day (31)
PEL: Not available

Basic manufacturer

Gilmore, Inc.
5501 Murray Road
Memphis, TN 38119-3703
Telephone: 901-761-5870

7.2.8   *Quizalofop-p-ethyl*

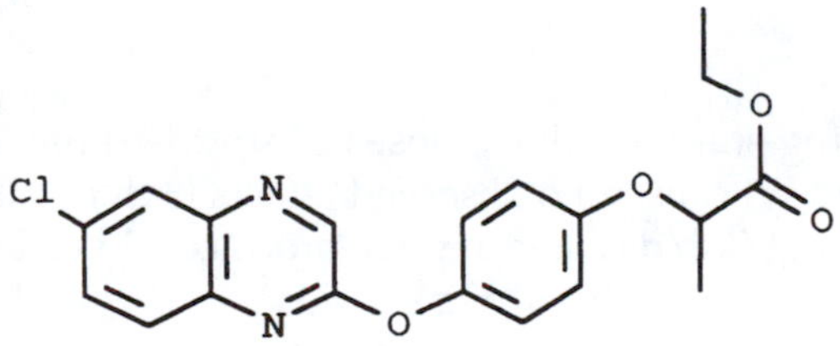

Figure 7.9 Quizalofop-p-ethyl.

Trade or other names

Trade and other names of products containing quizalofop-p-ethyl include Assure II, Copilot, Pilot Super, Sheriff, Targa D+, and Targa Super. The compound may be found in formulations with other herbicides such as benazolin and clopyralid (trade name, Benazalox).

Regulatory status

The compound is a General Use Pesticide (GUP). Quizalofop-p-ethyl is a slightly toxic compound in EPA toxicity class III. End-use products are required to have the Signal Word CAUTION on their labels.

Introduction

Quizalofop-p-ethyl is a selective, post-emergence phenoxy herbicide. It is used to control annual and perennial grass weeds in potatoes, soybeans, sugar beets, peanuts, vegetables, cotton, and flax, as well as other crops.

Quizalofop-p and quizalofop-p-ethyl should not be confused with quizalofop or quizalofop-ethyl. The latter two compounds are distinctly different from the former two. *The material in this profile refers only to the technical compound quizalofop-p-ethyl unless otherwise stated.*

Toxicological effects

Acute toxicity

Quizalofop-p-ethyl is slightly toxic by oral exposure. The reported oral LD_{50} values of the compound are 1210 to 1670 mg/kg in male rats, and 1182 to 1480 mg/kg in female rats (5,6). Mice are only slightly less susceptible to the compound. Quizalofop-p-ethyl has reported LD_{50} values of 1753 to 2350 mg/kg in male mice, and 1805 to 2360 mg/kg in female mice (5,6). For the formulated product Assure, the reported oral LD_{50} values are 6600 mg/kg in male rats and 5700 in female rats (5).

Exposure of the skin of rabbits to the compound indicated that the compound is only slightly toxic by this route as well. The acute percutaneous (absorbed through the skin) LD_{50} for quizalofop-p-ethyl in mice, rats, and rabbits is greater than 2000 mg/kg (6). For the formulated product Assure, the reported dermal LD_{50} in rabbits is greater than 5000 mg/kg (5).

Quizalofop-p-ethyl is slightly to practically nontoxic via inhalation, both in technical form and formulation. Reported 4-hour inhalation LC_{50} values are 5.8 mg/L for technical quizalofop-p-ethyl and 75 mg/L for Assure in rats (5).

Quizalofop-p-ethyl is nonirritating to the skin and only slightly irritating to the eyes in rabbits (5). It is nonsensitizing to the skin of guinea pigs. The Assure formulation, however, is severely irritating to rabbit eyes (5).

Chronic toxicity

In a 1-year feeding study on dogs, doses of up to 10 mg/kg/day (the highest dose tested in that study) caused no observed effects (31). In a 90-day feeding study in rats, doses of 6.4 mg/kg/day and higher produced liver lesions and increased liver weight (31). In a 2-year study of rats, doses of 5 mg/kg/day produced no observed effects (31).

Reproductive effects

Data from reproductive studies indicated only decreased body weight gains, and did not report findings of impaired reproductive function in test animals (31). A 6-month study in dogs found atrophy of the semeniferous tubles at doses of 2.5 mg/kg/day, but was unclear whether this was extensive enough to result in impaired reproductive function (31). These data are insufficient to draw conclusions regarding the likely reproductive effects of quizalofop-p-ethyl in animals, but do suggest that effects on human reproduction are unlikely under normal circumstances.

Teratogenic effects

In a two-generational study in rats, doses of 2.5 mg/kg/day and higher produced increased liver weights in offspring (31). No teratogenic effects were observed in another study in rats at doses up to 300 mg/kg/day (the highest doses tested) over an unspecified period, although maternal decreases in body weight, food consumption, and corpora lutea were observed at doses of 100 mg/kg/day (31). These data suggest that teratogenic or developmental effects are unlikely in humans.

Mutagenic effects

The results of many assays for mutagenicity and genotoxicity of quizalofop-p-ethyl show no mutagenic or genotoxic activity. Quizalofop-p-ethyl was not found to be mutagenic in the Ames assay, either with or without metabolic activation, nor was mutagenic activity seen in Chinese hamster ovary cell culture tests (6). Assays for chromosome structural aberrations and alterations in DNA damage repair capacity were also negative (5).

Carcinogenic effects

In an 18-month carcinogenicity study on mice, increased liver weights, changes in blood chemistry, and some changes in liver tissue structure were detected, but no carcinogenic or tumor-causing activity was reported (8). This study suggests that this compound is not carcinogenic.

Organ toxicity

Available data show that the target organ in test animals has consistently been the liver in rats and dogs (8). It is possible that testes may be a target organ in some species; e.g., in dogs (7).

Fate in animals and humans

Quizalofop-p appears to be rapidly broken down in mammals. More than 90% of a single oral dose is eliminated in urine within 3 days (6).

Ecological effects

Effects on birds

Quizalofop-p-ethyl is practically nontoxic to birds. The reported 8-day feeding (dietary) LC_{50} is greater than 5000 ppm in bobwhite quail and mallard ducks. The reported LD_{50} for quizalofop-p-ethyl is greater than 2000 mg/kg in mallard ducks (5).

Effects on aquatic organisms

Quizalofop-p-ethyl is highly to very highly toxic to fish. Reported 96-hour LC_{50} values are 10.7 mg/L in rainbow trout, and 0.46 to 2.8 mg/L in bluegill sunfish (5).

Effects on other organisms (non-target species)

Quizalofop-p-ethyl is practically nontoxic to bees, with a 48-hour contact LD_{50} of greater than 100 mg per bee (7).

Environmental fate

Breakdown in soil and groundwater

Quizalofop-p-ethyl is moderately persistent in soils, with a reported half-life of 60 days (5,6). It may be more rapidly broken down in soil with high microbial activity

(5). It is moderately to strongly sorbed to soils (5), and studies indicate very low soil mobility (5). It should not leach significantly into groundwater.

Breakdown in water

No data are currently available.

Breakdown in vegetation

No data are available regarding the breakdown of the compound; however, it is absorbed from the leaf surface and translocated throughout the plant. It accumulates in the active growing regions of stems and roots (5).

Physical properties

The compound is a pale brown crystal (6).

Chemical name: (R)-2-[4-(6-chloroquinoxalin-2-yloxy)phenoxy]propionic acid (6)
CAS #: 100646–51–3
Molecular weight: 372.8 (6)
Solubility in water: 0.4 mg/L @ 20°C (6)
Solubility in other solvents: s. in hexane; v.s. in acetone, ethanol, and xylene (6)
Melting point: 76–77°C (6)
Vapor pressure: 0.000011 mPa @ 20°C (6)
Partition coefficient (octanol/water): Not available
Adsorption coefficient: 510 (5)

Exposure guidelines

ADI: Not available
HA: Not available
RfD: Not available
PEL: Not available

Basic manufacturer

Zeneca Agricultural Products
1800 Concord Pike
Wilmington, DE 19897
Telephone: 800-759-4500
Emergency: 800-759-2500

References

(1) Stevens, J. T. and Sumner, D. D. Herbicides. In *Handbook of Pesticide Toxicology*. Hayes, W. J., Jr. and Laws, E. R., Jr., Eds. Academic Press, New York, 1991.
(2) Jordan, L. S. and Cudney, D. W. Herbicides. In *Fate of Pesticides in the Environment*. Biggar, J. W. and Seiber, J. N., Eds. University of California Agricultural Experiment Station, Davis, CA, 1987.
(3) Ecobichon, D. J. Toxic effects of pesticides. In *Casarett and Doull's Toxicology*, 4th ed. Amdur, M. O., Doull, J., and Klaassen, C. D., Eds. Pergamon Press, New York, 1991.

(4) Wagner, S. L. The acute health hazards of pesticides. In *Chemistry, Biochemistry and Toxicology of Pesticides*. Witt, J. M., Ed. Oregon State University Extension Service, Cornallis, OR, 1989.

(5) Weed Science Society of America. *Herbicide Handbook*, 7th ed. Champaign, IL, 1994.

(6) Kidd, H. and James, D. R., Eds. *The Agrochemicals Handbook*, 3rd ed. Royal Society of Chemistry Information Services, Cambridge, U.K., 1991 (as updated).

(7) U.S. National Library of Medicine. *Hazardous Substances Data Bank*. Bethesda, MD, 1995.

(8) National Institute for Occupational Safety and Health. *Registry of Toxic Effects of Chemical Substances*. 1993.

(9) U.S. Environmental Protection Agency. *Fact Sheet Number 192, 2-(2-Methyl-4-chlorophenoxy) Propionic Acid and Its Salts and Esters*. Office of Pesticides and Toxic Substances, Washington, D.C., 1988.

(10) U.S. Environmental Protection Agency. *Health Advisory Draft Report: Chloramben*. Office of Drinking Water, Washington, D.C., 1987.

(11) U.S. Environmental Protection Agency. *Health Advisory Summary: Dicamba*. Office of Drinking Water, Washington, D.C., 1988.

(12) Food and Drug Administration. *The FDA Surveillance Index*. National Technical Information Service, Springfield, VA, 1986.

(13) Machera, K. Developmental toxicity of haloxyfop ethoxyethyl ester in the rat. *Bull. Environ. Contam. Toxicol.* 51(4), 625–632, 1993.

(14) Lynge, E. A. Follow-up study of cancer incidence among workers in manufacture of phenoxy herbicides in Denmark. *Br. J. Cancer.* 52, 259–270, 1985.

(15) Buzik, S. C. *Toxicology of 2,4-Dichlorophenoxyacetic Acid (2,4-D): A Review*. Toxicology Research Center of University of Saskatchewan, Saskatoon, Saskatchewan, Canada, 1992.

(16) Wagner, S. L. *Clinical Toxicology of Agricultural Chemicals*. Oregon State University Environmental Health Sciences Center, Corvallis, OR, 1981.

(17) Mullison, W. R. Environmental fate of phenoxy herbicides. In *Fate of Pesticides in the Environment*. Biggar, J. W. and Seiber, J. N., Eds. University of California Agricultural Experiment Station, Davis, CA, 1987.

(18) Buhler, D. R. Transport, accumulation and disappearance of pesticides. In *Chemistry, Biochemistry and Toxicology of Pesticides*. Witt, J. M., Ed. Oregon State University Extension Service, Corvallis, OR, 1989.

(19) Norris, L. A. Behavior of chemicals in the forest environment. In *Chemistry, Biochemistry and Toxicology of Pesticides*. Witt, J. M., Ed. Oregon State University Extension Service, Corvallis, OR, 1989.

(20) Howard, P. H., Ed. *Handbook of Environmental Fate and Exposure Data for Organic Chemicals. Pesticides*. Lewis, Boca Raton, FL, 1991.

(21) Wauchope, R. D., Buttler, T. M., Hornsby A. G., Augustijn-Beckers, P. W. M., and Burt, J. P. SCS/ARS/CES pesticide properties database for environmental decisionmaking. *Rev. Environ. Contam. Toxicol.* 123, 1–157, 1992.

(22) Augustijn-Beckers, P. W. M., Hornsby, A. G., and Wauchope, R. D. SCS/ARS/CES pesticide properties database for environmental decisionmaking II. Additional compounds. *Rev. Environ. Contam. Toxicol.* 137, 1–82, 1994.

(23) U.S.Environmental Protection Agency. *Pesticides in Ground Water Database: A Compilation of Monitoring Studies, 1971–1991 National Summary*. Washington, D.C., 1992.

(24) National Research Council Canada. *Phenoxy Herbicides — Their Effects on Environmental Quality with Accompanying Scientific Criteria for 2,3,7,8-Tetrachlorodibenzo-p-dioxin (TCDD)*. Ottawa, Canada, 1978.

(25) Gosselin, R. E., Smith, R. P., and Hodge, H. C. *Clinical Toxicology of Commercial Products*. Williams and Wilkins, Baltimore, MD, 1984.

(26) Schlop, R. N., Hardy, M. H., and Goldberg, M. T. Comparison of the activity of topically applied pesticides and the herbicide 2,4-D in two short-term *in vivo* assays of genotoxicity in the mouse. *Fundam. Appl. Toxicol.* 15, 666–675, 1990.

(27) Hoar, S. K., Zahm, S., Weisenburger, D. D., and Babbitt, P.A. A case-control study of non-Hodgkin's lymphoma and the herbicide 2,4-dichlorophenoxyacetic acid (2,4-D) in eastern Nebraska. *Epidemiology* 1, 349–356, 1990.

(28) U.S. Environmental Protection Agency. Proposed rules. *Fed. Reg.* 55, 24116–24117, 1990.

(29) Lu, F. C. A review of the acceptable daily intakes of pesticides assessed by the World Health Organization. *Regul. Toxicol. Pharmacol.* 21, 351–364, 1995.

(30) U.S. Environmental Protection Agency. *National Primary Drinking Water Standards: (810-F-94-001A).* Office of Drinking Water, Washington, D.C., 1994.

(31) U.S. Environmental Protection Agency. *Integrated Risk Information System,* Washington, D.C., 1995.

(32) U.S. Occupational Safety and Health Administration. *Permissible Exposure Limits for Air Contaminants.* (29 CFR 1910. 1000, Subpart Z). U.S. Department of Labor, Washington, D.C., 1994.

(33) U.S. Environmental Protection Agency. *Pesticide Fact Sheet Number 179, 2,4-DB.* Office of Pesticides and Toxic Substances, Washington, D.C., 1988.

(34) Johnson, W. W. and Finley, M. T. *Handbook of Acute Toxicity of Chemicals to Fish and Aquatic Invertebrates, Resource Publication 137.* U.S. Department of Interior, Fish and Wildlife Service, Washington, D.C., 1980.

(35) U.S. Environmental Protection Agency. *Health Advisory for 50 Pesticides: Dacthal.* Office of Drinking Water, Washington, D.C., 1988.

(36) Hallenbeck, W. H. and Cunningham-Burns, K. M. *Pesticides and Human Health.* Springer-Verlag, New York, 1985.

(37) ICI Americas Inc. *Material Safety Data Sheet: Fusilade 2000.* Wilmington, DE, 1992.

(38) U.S. Environmental Protection Agency. *Health Advisory Summary: MCPA.* Office of Drinking Water, Washington, D.C., 1987.

(39) Wiklund, K. and Holm, L. E. Soft tissue sarcoma risk in Swedish agricultural and forestry workers. *J. Natl. Cancer Inst.* 76, 229–234, 1986.

chapter eight

Triazines and triazoles

8.1. Class overview and general description

Background

The triazine compounds are herbicides and most of the triazole compounds are fungicides. However, the structures of the pesticides in both groups are similar in that they consist of a single ring structure with three nitrogen atoms. Therefore, these two compound groups are treated together in this chapter.

Triazine compounds

The triazine herbicides were discovered in 1954. The first compound in the group, simazine, was manufactured in 1958 (1). Triazine herbicides are among the most widely used herbicides in the U.S. (2). This group of herbicides is represented by atrazine, simazine, cyanazine, and other compounds listed in Table 8.1.

The structures of all of the triazine herbicides have a six-member ring containing three nitrogen atoms and three carbon atoms as shown in Figure 8.1. Triazines can be symmetrical with alternating nitrogen and carbon atoms around the ring, or asymmetrical with adjacent nitrogen atoms. No differences in toxicity or action appear to be linked to these structural differences.

Triazole compounds

Triazoles are similar to triazines because they also contain three nitrogen atoms in their ring structures, though they have only two carbon atoms in the ring structure instead of three. Amitrole and triadimefon are commonly used triazoles. Different side chains (groups of atoms) that branch off the central ring structure determine the activity and potency of the compound.

Amitrole, a triazole, was the catalyst for the addition of the Delaney clause to the Food Drug and Cosmetic Act. This clause states that the residues of any cancer-causing agent are not allowed as additives in food (3).

Triazine and triazole usage

Triazine usage

Total 1987 U.S. crop application of the active ingredients of triazines was greater than 85,000 tons (4). Atrazine is the most widely used triazine compound in the U.S.

Table 8.1
Triazines and Triazoles

Triazines
    Ametryn
    Anilazine
    Atrazine*
    Aziprotryn
    Cyanazine*
    Cyromazine
    Desmetryn
    Dipropetryn
    Eglinazine ethyl
    Hexazinone*
    Metribuzin*
    Prometon
    Prometryn*
    Propazine*
    Simazine*
    Terbutryn
Triazoles
    Amitrole*
    Diclobutrazol
    Fenchlorazole
    Flusilazole
    Hexaconazole
    Triadimefon* (Bayleton)
    Triadimenol

Note: * indicates that a profile for
      this compound is included
      in this chapter.

Figure 8.1 Generic structures for triazines (A) and triazoles (B).

In a 1987 survey of U.S. herbicide use on crops, it was found that over 32,000 tons of atrazine were used on crops such as corn and sorghum (4).

    Triazine compounds are used against a wide variety of weed species. For example, atrazine is used to control broad-leaved weeds and grasses in maize, sorghum, asparagus, vines, fruit orchards, citrus, sugarcane, and other fruits and vegetables. Simazine has an equally broad usage. Triazines are generally used in pre-emergent applications and applied to the soil (5).

Triazole usage

    Amitrole is the only herbicide in the triazole family. The remaining triazoles are fungicides and are used to control powdery mildews and various fungi on cereals, coffee, vines and apples (6).

Mechanism of action and toxicology: triazines

Mechanism of action

The triazine compounds act by interfering with photosynthesis. Most of the compounds interrupt Photosystem II of photosynthesis, which utilizes chlorophyll to convert short-wavelength light (photons) into chemical energy for use by plants cells. Without Photosystem II the plants are unable to produce enough energy for growth. Susceptible plants, especially emergent seedlings, turn yellow and die. The compounds are applied to the soil and absorbed by the plant's roots. However, not all plants are susceptible to triazine herbicides. Plants such as maize and sugarcane contain an enzyme that quickly breaks down the triazine herbicides to virtually nontoxic forms. The effectiveness of triazines depends on their relative rates of uptake and breakdown in the plants (7).

Some evidence has linked the mode of toxicity of the triazine compounds in mammals to a disruption of the metabolism of vitamins (8).

Acute toxicity

Triazines have a low acute toxicity to mammals. Both atrazine and simazine have oral rat LD_{50} values of 3000 mg/kg/day. Most other compounds in this group also have high oral LD_{50} values. However, cyanazine is a notable exception. The oral rat LD_{50} of 182 to 334 mg/kg indicates that it is a moderately toxic compound (6). Typical symptoms of acute poisoning from triazine compounds include skin and eye irritation, nausea and vomiting, diarrhea, muscle weakness, and salivation (8). No cases of acute systemic poisoning have been reported for atrazine, simazine, propazine, or cyanazine. However, acute and subacute skin rashes have been reported in the former U.S.S.R. from contact with simazine. The symptoms include reddening and burning lasting up to 5 days (5).

Chronic toxicity

In 2-year oral rat studies conducted for atrazine, propazine, and simazine, the results showed no toxicological effects nor visible or microscopic signs of toxicity (5). Also 2-year chronic feeding studies of cyanazine in rats and dogs showed no signs of toxic effects at levels up to 0.65 mg/kg/day (5).

Reproductive effects

In three-generation rat studies in which 5 mg/kg/day was administered, toxicity varied from no adverse effects on reproductive capacity or development for simazine to a decrease in the mean of parental body weights for propazine (9,10). The triazine compounds caused a decrease in second-generation rat pup body weights; and in another study, cyanazine caused some toxic effects in fetal rabbits (11). Therefore, some of the triazine compounds may cause adverse reproductive effects.

Teratogenic effects

Administration of atrazine or simazine to rats did not produce teratogenic effects (5 mg/kg/day) (10,12). Doses of 100 mg/kg/day propazine in rats did not produce maternal toxicity nor teratogenic effects. At doses of 300 mg/kg/day or higher, there were reductions in maternal body weight, feed consumption, and fetal body weight (9). Cyanazine has caused teratogenic effects and developmental toxicity in two animal species, the rabbit and the rat (11). Cyanazine at doses of 25 to 75 mg/kg/day produced cleft palates and skeletal bone hardening, but only at levels that caused maternal toxicity (11). Therefore, triazines do not appear to produce teratogenic effects unless doses are high enough to cause toxic effects in the mother. Such levels do not occur from normal use of these products.

Mutagenic effects

Triazines have not produced any mutagenic effects in human or in rat liver cells (2,9). Simazine is the only triazine that has produced both positive and negative mutagenic results in fruit flies (4). The evidence suggests that triazine compounds are not likely to produce mutagenic effects in humans and animals.

Carcinogenic effects

In 2-year, oral, high-dose studies with atrazine and propazine, results showed an increase of mammary tumors in female rats (9–11). However, these effects did not occur with cyanazine (11).

Due to questions about studies that have been performed, the carcinogenic potential of atrazine, propazine, and simazine has not been determined (9,10,12).

Organ toxicity

High doses of various triazines in test animals have caused liver damage. Cyanazine has caused depression of the central nervous system (11).

Fate in humans and animals

Ingested triazines are readily absorbed through the gastrointestinal tract. In rat and dog studies, 75% of various triazines were excreted either in the urine or feces within 4 days. Small residues of triazines can bioaccumulate in the animal's fat and muscle (10).

Mechanism of action and toxicology: triazoles

Mechanism of action

Amitrole is a systemic herbicide, and the other triazoles are systemic fungicides and are absorbed by the plants' roots and leaves and readily translocated throughout the plant. The triazole fungicides inhibit the synthesis of a form of vitamin D. Their mode of action in mammals is thought to be through the inhibition of liver enzymes (8).

Acute toxicity

Triazole compounds show low acute toxicity. For example, acute oral LD_{50} values are 569 mg/kg triadimefon in male rats and about 1000 mg/kg in mice (6).

Chronic toxicity

Chronic exposure of laboratory animals to triazole compounds has caused adverse effects in the adrenal glands, kidneys, liver, and lungs, as well as the formation of kidney and urinary tract stones (3).

Reproductive effects

Low doses of triazoles (1.25 to 5 mg/kg/day) have not produced significant reproductive effects. At higher doses, there were reductions in litter size and reduced viability of the offspring, combined with lower birth weights for the second-generation offspring (3). Therefore, triazoles are not likely to produce reproductive effects in humans at normal usage levels.

Teratogenic effects

The teratogenic potential for triazole compounds is low. Triazoles have produced birth defects in pups of pregnant rabbits, rats, and mice, though only at dose levels high enough to produce signs of maternal toxicity (5).

Mutagenic effects

Various assays, including the Ames test, have shown no mutagenic effects. Therefore, triazoles are unlikely to be mutagenic risks to humans (2,14).

Carcinogenic effects

The study of carcinogenic effects in triazoles has not produced consistent results. Most triadimefon (Bayleton) studies indicated it has no potential to produce cancer in rats. Amitrole, on the other hand, has induced thyroid and liver tumors in rodents (2).

Organ toxicity

In rabbit ocular studies, triadimefon caused temporary loss of sight in rabbits, though, after 7 days this condition cleared up (13).

Moderate doses of amitrole to rats have induced changes in the liver and inhibited the activity of various liver enzymes. Also, rat dietary studies have shown that 3 to 6 mg/kg/day amitrole causes enlargement of the thyroid and reduced uptake of iodine (2).

Fate in humans and animals

When triazole compounds were administered to rats, nearly 95% of the chemicals was eliminated through the urine and feces within 2 to 3 days (2).

Ecological effects: triazines

Effects on birds

Triazines are slightly toxic to birds. The oral LC_{50} values in mallard ducks range from >19,650 mg/kg for atrazine to >10,000 mg/kg for propazine (6). The propazine dietary LC_{50} value of 8800 ppm is the same for bobwhite quail and pheasant (9).

Effects on aquatic organisms

The range of toxicities to fish is rather wide and depends on both the compound and the species. Most of the triazine compounds fall into the moderately or slightly toxic categories, based on the LC_{50} values in fish such as rainbow trout and aquatic invertebrates (5). However, several of the compounds, including desmetryn, are highly toxic to fish (5).

Effects on other organisms (non-target species)

Triazines are nontoxic to bees (6).

Ecological effects: triazoles

Effects on birds

Triazole compounds have a low oral toxicity to birds, with a LD_{50} value of 2000 mg/kg in mallard ducks. Canaries are slightly less tolerant to triadimefon, having an oral LD_{50} of 1000 mg/kg (6,15,16).

Effects on aquatic organisms

Triazole compounds are moderately toxic to fish. Bluegill sunfish are the most susceptible, with an LC_{50} value of 1.7 mg/L for flusilazole. The LC_{50} value in rainbow trout is about 5.0 mg/L (6,15,16).

Effects on other organisms (non-target species)

The triazole compounds are all nontoxic to bees (6).

Environmental fate: triazines

Breakdown in soil and groundwater

Triazines tend to quickly degrade in various soils and do not adsorb to soil particles. Therefore, triazines may leach through the soil, and traces of their breakdown products have been found in groundwater. Because of the large quantities applied, atrazine and simazine have been detected in rural domestic well water and in community water system wells (6).

Breakdown in water

Most triazine compounds such as atrazine, cyanazine, propazine, and simazine are relatively stable in water (i.e., not easily hydrolyzed). The solubility decreases with increasing pH. At pH 5, 60% of propazine remained unhydrolyzed; at pH 9, 100% remained unhydrolyzed (15,16).

Breakdown in vegetation

When triazines are applied to the soil, plants absorb it through the roots and translocate it upward throughout the plant (7).

Environmental fate: triazoles

Breakdown in soil and groundwater

Triazoles do not strongly adsorb to soil particles and are subject to degradation by soil microbes. In a loamy soil, the breakdown may occur within a week (16). After breakdown of the chemical, the products will readily leach into the groundwater within 2 to 4 weeks (6,15,16).

Breakdown in water

Most triazole compounds are stable in water. For example, at pH 4 to 9, 95% of triadimefon remained in surface water after 28 weeks (17). Other triazoles, such as flusilazole and hexaconazole, may persist in water up to 1 or 2 years (6).

Breakdown in vegetation

Triazole compounds are readily absorbed into plants through their roots and translocated throughout the plant. Plants readily metabolize triazoles within a month (15).

Current status and concerns

As of November 10, 1994, the U.S. Environmental Protection Agency began a Special Review of triazine pesticides, specifically atrazine, cyanazine, and simazine. Some animal studies indicate that these three triazines and others in the same class produce mammary tumors in some strains of rats. There are also recently published human epidemiological studies that offer conflicting but, in some cases, suggestive evidence that exposure to environmental toxicants such as pesticides may contribute to the still increasing rates of breast cancer in American women.

A Special Review is a structured procedure that encourages public involvement and is initiated whenever data on a pesticide lead the EPA to believe that a significant risk to public health or the environment may exist. The review weighs the risks and benefits of a pesticide's use, analyzes the effects of a shift to alternative pesticide control strategies, and ultimately determines whether the pesticide(s) should be canceled, further restricted, or continued to be used without further amendments.

The U.S. EPA has already taken other actions to reduce the potential risks of the triazine products. In 1990, use of atrazine was restricted to certified applicators (except for certain home turf products) and application rates were reduced. In 1992, additional label changes were approved, including a further reduction in application rates, the requirement of buffer zones near surface waters, and deletion of non-crop uses. Similar label changes have been approved for cyanazine, including restricted use. Most uses of simazine are restricted to certified applicators.

8.2   *Individual profiles*

8.2.1   *Amitrole*

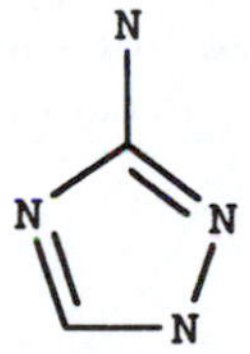

Figure 8.2 Amitrole.

Trade or other names

Amitrole is known as amino-triazole in Great Britain, France, New Zealand, and the former U.S.S.R. Trade names for amitrole include Amerol, Amino Triazole, Amitrol, Amizine, Amizol, Azolan, Azole, Cytrol, Diurol, and Weedazol.

Regulatory status

Amitrole is a Restricted Use Pesticide (RUP). RUPs may be purchased and used only by certified applicators. Amitrole is classified as toxicity class III — slightly toxic. Products containing amitrole bear the Signal Word CAUTION. All use of amitrole on food crops was canceled by the EPA in 1971 because it caused cancer in experimental animals.

Introduction

Amitrole is a nonselective systemic triazole herbicide. It has been used on non-cropland for control of annual grasses and perennial and annual broadleaf weeds, for poison ivy control, and for control of aquatic weeds in marshes and drainage ditches (1). This compound is compatible with many other herbicides. It was available as wettable powders, soluble concentrates, and water-dispersable granules.

Amitrole was involved in the Delaney clause's first enforcement. The Delaney clause prohibits any amount of any cancer-causing substance to be a food additive, prompting growers to follow pesticide label directions carefully. In 1959, amitrole was registered for post-harvest use on cranberries. Misuse left small residues on portions of cranberry crops. Just 13 days before Thanksgiving, 1587 metric tons of cranberries were seized by the Food and Drug Administration; and just 3 days before Thanksgiving, the FDA certified sufficient cranberries to meet holiday demands and end the "Cranberry Crisis."

Toxicological effects

Acute toxicity

Amitrole has a very low acute toxicity to humans and animals. Associated symptoms in humans include skin rash, vomiting, diarrhea, and nosebleeds. Poisoning by amitrole is characterized by increased intestinal peristalsis (this may lead to diarrhea), fluid in the lungs, and hemorrhages of various organs. No toxic effects were observed in a woman who ingested 20 mg/kg, but a single dose of 1200 mg/kg reduced iodine uptake by the thyroid in healthy persons (3,15). Amitrole is a mild skin and eye irritant (3,15).

The oral and dermal LD_{50} values for amitrole in rats are greater than 5000 mg/kg. Studies have reported oral LD_{50} values as high as 15,000 mg/kg in mice and 24,600 mg/kg in rats. In one study, the largest doses tested, 4080 mg/kg orally and 2500 mg/kg dermally, produced no toxic effects on rats. The dermal LD_{50} in rabbits is greater than 200 mg/kg (3,15).

Chronic toxicity

Feeding of amitrole to rats at dietary doses of 3 or 6 kg/mg/day for 2 weeks caused enlargement of the thyroid and reduced uptake of iodine. A dietary dose of 50 mg/kg/day produced significant enlargement of the thyroid after 3 days of feeding. Several studies have shown that amitrole inhibits the activity of various liver enzymes. Long-term exposure to amitrole can cause reversible goiters (16,19).

Reproductive effects

In a two-generation study in rats, dams fed 5 or 25 mg/kg/day amitrole had fewer pups per litter, and their weight at weaning was reduced. Dietary doses of 1.25 mg/kg/day had no significant effect on reproduction (3). It is unlikely that reproductive effects will occur in humans under normal circumstances.

Teratogenic effects

Birth defects have occurred in the pups of pregnant rabbits, rats, and mice exposed to amitrole, but only at doses high enough to also produce signs of toxicity in the mothers (14). Atrophy of the thymus and spleen occurred at high doses (5 or 25 mg/kg/day). Within a week after weaning, most of these pups died of a condition resembling runt disease. Similar effects have been observed in mice. Teratogenic effects in humans are unlikely under normal circumstances.

Mutagenic effects

One laboratory assay has shown amitrole to be weak mutagen. All other assays have shown no mutagenic effects (3,14). These data suggest that amitrole is weakly or nonmutagenic.

Carcinogenic effects

Amitrole has induced thyroid and liver tumors in rats and mice after lifetime high-dose exposures (3,14,18).

Organ toxicity

Animal studies have shown that amitrole's main effects are on the thyroid and liver.

Fate in humans and animals

Amitrole is rapidly and completely absorbed into the body through the gastrointestinal tract when eaten. Amitrole is excreted through the urine. The highest concentrations in all tissues generally occur within 1 hour after exposure. Concentrations begin to decline after 2 to 6 hours (3).

Most (70 to 95.5%) of the amitrole administered to rats by stomach tube was excreted in the urine during the first 24 hours. Some was detected in the rats' feces for 2 to 5 days after dosing. After 6 days, only 0.28 to 1.36% of the total dose remained, mainly in the liver (3).

Ecological effects

Effects on birds

Amitrole is practically nontoxic to upland game birds (6,18). The LD_{50} for amitrole in mallard ducks is 2000 mg/kg (18).

Effects on aquatic organisms

Amitrole is slightly toxic to various species of freshwater fish and freshwater invertebrates (18).

Effects on other organisms (non-target species)

Amitrole inhibits the growth of bacteria (16). It is nontoxic to bees (6).

Environmental fate

Breakdown in soil and groundwater

Amitrole has low soil persistence. Its half-life is 14 days (20). Microbial breakdown of amitrole takes 2 to 3 weeks in warm, moist soil (15). Some chemical degradation may also occur in soils. Loss of amitrole from soils by volatilization or photodegradation is minor (21). Amitrole has a moderate potential for groundwater contamination because it does not adsorb strongly to soil particles and is readily soluble in water.

Breakdown in water

In aquatic environments, amitrole does not break down by hydrolysis or photolysis, volatilize, nor bioaccumulate in aquatic organisms. The biodegradation half-life for amitrole in water is about 40 days. Degradation of amitrole in open waters may occur through oxidation by other chemicals (21).

Breakdown in vegetation

Amitrole is readily absorbed and rapidly translocated in the roots and leaves of higher plants (18). However, plants are able to metabolize amitrole in 1 to 4 weeks (15). Amitrole residues were not detected in crops planted into soil 1 to 50 days after treatment with amitrole (18).

Physical properties

Amitrole is a white to off-white, odorless crystalline powder with a bitter taste (6).

Chemical name: 1H-1,2,4-triazole-3-ylamine (6)
CAS #: 61–82–5
Molecular weight: 84.08 (6)
Water solubility: 280,000 mg/L @ 23°C (6)
Solubility in other solvents: chloroform, methanol, acetonitrile; acetone and non-polar solvents, i.s. (6)
Melting point: 157°C (6)
Vapor pressure: <1 mPa @ 20°C (6)
Partition coefficient (octanol/water): Not available
Adsorption coefficient: 100 (21)

Exposure guidelines

ADI: 0.0005 mg/kg/day (22)
MCL: Not available
RfD: Not available
TLV: 0.2 mg/m³ (8-hour) (16)

Basic manufacturer

Rhone Poulenc Ag Co.
P.O. Box 12014
2 T.W. Alexander Dr.
Research Triangle Park, NC 27709
Telephone: 919-549-2000
Emergency: 800-334-7577

8.2.2   *Atrazine*

Trade or other names

Trade names for atrazine include Aatrex, Aktikon, Alazine, Atred, Atranex, Atrataf, Atratol, Azinotox, Crisazina, Farmco Atrazine, G-30027, Gesaprim, Giffex 4L, Malermais, Primatol, Simazat, and Zeapos.

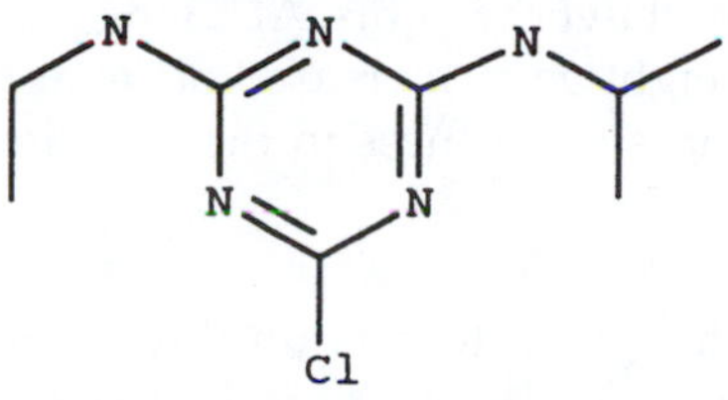

Figure 8.3 Atrazine.

Regulatory status

Atrazine has been classified as a Restricted Use Pesticide (RUP) due to its potential for groundwater contamination (2). RUPs may be purchased and used only by certified applicators. Atrazine is toxicity class III — slightly toxic. In November 1994, the EPA initiated a Special Review that could result in use restrictions or cancellation of atrazine if health data warrant such action. Products containing atrazine must bear the Signal Word CAUTION.

Introduction

Atrazine is a selective triazine herbicide used to control broadleaf and grassy weeds in corn, sorghum, sugarcane, pineapple, Christmas trees, and other crops, and in conifer reforestation plantings. It is also used as a nonselective herbicide on non-cropped industrial lands and on fallow lands. Over 64 million acres of cropland were treated with atrazine in the U.S. in 1990. It is available as dry flowable, flowable liquid, liquid, water dispersible granular, and wettable powder formulations.

Toxicological effects

Acute toxicity

Atrazine is slightly to moderately toxic to humans and other animals. It can be absorbed orally, dermally, and by inhalation. Symptoms of poisoning include abdominal pain, diarrhea and vomiting, eye irritation, irritation of mucous membranes, and skin reactions (3). At very high doses, rats show excitation followed by depression, slowed breathing, incoordination, muscle spasms, and hypothermia (3). After consuming a large oral dose, rats exhibit muscular weakness, hypoactivity, breathing difficulty, prostration, convulsions, and death (16). Atrazine is a mild skin irritant. Rashes associated with exposure have been reported.

The oral LD_{50} for atrazine is 3090 mg/kg in rats, 1750 mg/kg in mice, 750 mg/kg in rabbits, and 1000 mg/kg in hamsters. The dermal LD_{50} in rabbits is 7500 mg/kg and greater than 3000 mg/kg in rats (15,16).

The 1-hour inhalation LC_{50} is greater than 0.7 mg/L in rats. The 4-hour inhalation LC_{50} is 5.2 mg/L in rats (3,6).

Chronic toxicity

Some 40% of rats receiving oral doses of 20 mg/kg/day for 6 months died with signs of respiratory distress and paralysis of the limbs. Structural and chemical changes in the brain, heart, liver, lungs, kidney, ovaries, and endocrine organs were observed (3,16). Rats fed 5 or 25 mg/kg/day atrazine for 6 months exhibited growth retardation. In a 2-year study with dogs, 7.5 mg/kg/day caused decreased food intake and increased heart and liver weights. At 75 mg/kg/day, there were decreases in food intake and body weight gain, increased adrenal weight, lowered blood cell counts, and occasional tremors or stiffness in the rear limbs (3).

Reproductive effects

Dietary doses of atrazine given to rats on days 3, 6, and 9 of gestation up to about 50 mg/kg/day caused no adverse reproductive effects (3).

Teratogenic effects

Atrazine does not appear to be teratogenic. In mice, atrazine did not cause abnormalities in fetuses whose dams were given doses of 46.4 mg/kg/day during days 6 through 14 of gestation (3).

Mutagenic effects

The weight of evidence from more than 50 studies indicates that atrazine is not mutagenic (3).

Carcinogenic effects

Atrazine did not cause tumors when mice were given oral doses of 21.5 mg/kg/day from age 1 to 4 weeks, followed by dietary doses of 82 mg/kg for an additional 17 months. However, mammary tumors were observed in rats after lifetime administration of high doses of atrazine (3). Thus, available data regarding atrazine's carcinogenic potential are inconclusive.

Organ toxicity

Lethal doses of atrazine in test animals have caused congestion and/or hemorrhaging in the lungs, kidneys, liver, spleen, brain, and heart (3). Long-term consumption of high levels of atrazine has caused tremors, changes in organ weights, and damage to the liver and heart (3).

Fate in humans and animals

Atrazine is readily absorbed through the gastrointestinal tract. When a single dose of 0.53 mg atrazine was administered to rats by gavage, 20% of the dose was excreted in the feces within 72 hours. The other 80% was absorbed across the lining of the gastrointestinal tract into the bloodstream. After 72 hours, 65% was eliminated in the urine and 15% was retained in body tissues, mainly in the liver, kidneys, and lungs (3).

Ecological effects

Effects on birds

Atrazine is practically nontoxic to birds. The LD_{50} is greater than 2000 mg/kg in mallard ducks. At dietary doses of 5000 ppm, no effect was observed in bobwhite quail and ring-necked pheasants (15,16).

Effects on aquatic organisms

Atrazine is slightly toxic to fish and other aquatic life. Atrazine has a low level of bioaccumulation in fish. In whitefish, atrazine accumulates in the brain, gall bladder, liver, and gut (16).

Effects on other organisms (non-target species)

Atrazine is not toxic to bees (16).

Environmental fate

Breakdown in soil and groundwater

Atrazine is highly persistent in soil. Chemical hydrolysis, followed by degradation by soil microorganisms, accounts for most of the breakdown of atrazine. Hydrolysis is rapid in acidic or basic environments, but is slower at neutral pH. Addition of organic material increases the rate of hydrolysis. Atrazine can persist for longer than 1 year under dry or cold conditions (21).

Atrazine is moderately to highly mobile in soils with low clay or organic matter content. Because it does not adsorb strongly to soil particles and has a lengthy half-life (60 to >100 days), it has a high potential for groundwater contamination despite its moderate solubility in water (20).

Atrazine is the second most common pesticide found in private and community wells (16). Trace amounts have been found in drinking water and groundwater samples in a number of states (21,23). A 5-year survey of drinking water wells detected atrazine in an estimated 1.7% of community water systems and 0.7% of rural domestic wells nationwide. Levels detected in rural domestic wells sometimes exceeded the MCL (23). The recently completed National Survey of Pesticides in Drinking Water found atrazine in nearly 1% of all wells tested (23).

Breakdown in water

Atrazine is moderately soluble in water. Chemical hydrolysis, followed by biodegradation, may be the most important route of disappearance from aquatic environments. Hydrolysis is rapid under acidic or basic conditions, but is slower at neutral pH. Atrazine is not expected to strongly adsorb to sediments. Bioconcentration and volatilization of atrazine are not environmentally important (21). Atrazine has been detected in each of 146 water samples collected at eight locations from the Mississippi, Ohio, and Missouri Rivers and their tributaries. For several weeks, 27% of these samples contained atrazine concentrations above the EPA's maximum contaminent level (MCL) (24).

Breakdown in vegetation

Atrazine is absorbed by plants mainly through the roots, but also through the foliage. Once absorbed, it is translocated upward and accumulates in the growing tips and the new leaves of the plant. In susceptible plant species, atrazine inhibits photosynthesis. In tolerant plants, it is metabolized (6).

Most crops can be planted 1 year after application of atrazine. Atrazine increases the uptake of arsenic by treated plants (16).

Physical properties

Atrazine is a white crystalline solid (6).

Chemical name: 2-chloro-4-ethylamine-6-isopropylamino-S-triazine (6)
CAS #: 1912–24–9
Molecular weight: 215.69 (6)

Water solubility: 28 mg/L @ 20°C (6)
Solubility in other solvents: Chloroform v.s.; diethyl ether v.s.; dimethyl sulfoxide
    v.s. (6)
Melting point: 176°C (6)
Vapor pressure: 0.04 mPa @ 20°C (6)
Partition coefficent (octanol/water): 219 (6)
Adsorption coefficient: 100 (20)

Exposure guidelines

ADI: Not available
MCL: 0.003 mg/L (25)
RfD: 0.035 mg/kg/day (26)
TLV: 5 mg/m^3 (8-hour) (16)

Basic manufacturer

Ciba-Geigy Corp.
P.O. Box 18300
Greensboro, NC 27419–8300
Telephone: 800-334-9481
Emergency: 800-888-8372

8.2.3   Cyanazine

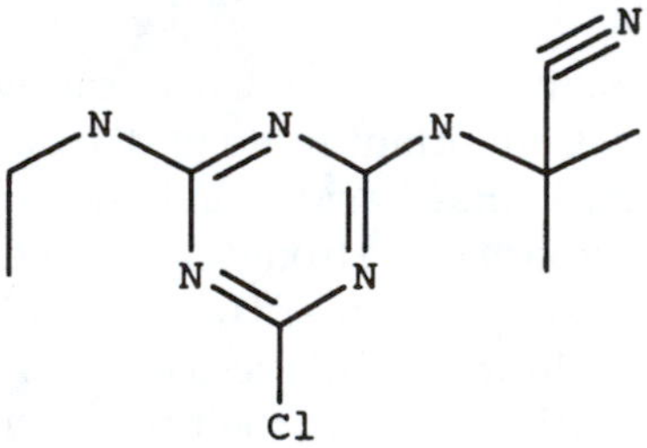

Figure 8.4 Cyanazine.

Trade or other names

Trade names for cyanazine include Bladex, DW3418, Fortrol, Match, and Payze.
Cyanazine may be used in combination with other herbicides.

Regulatory status

Cyanazine is classified by the EPA as a Restricted Use Pesticide (RUP) because
of its teratogenicity and because it has been found in groundwater. RUPs may be
purchased and used only by certified applicators. Classified as toxicity class II —
moderately toxic, products containing cyanazine bear the Signal Word WARNING.

Introduction

Cyanazine, a triazine, is used as a pre- and post-emergent herbicide to control
annual grasses and broadleaf weeds. It is used mostly on corn, some on cotton, and
less than 1% on grain sorghum and wheat fallow. Nearly 2.25 million acres of land

have been treated with cyanazine in the U.S. The compound is formulated as a wettable powder, a flowable suspension, or as granules.

Toxicological effects

Acute toxicity

Cyanazine is moderately toxic to mammals. The oral LD_{50} in rats ranges from 182 to 332 mg/kg, with females experiencing higher toxicities (i.e., lower LD_{50}). The oral LD_{50} is 380 mg/kg in mice and 141 mg/kg in rabbits (3,6,15). Poisoned animals have labored breathing and blood in their saliva (5,6,15). Cyanazine may also cause inactivity and depression in laboratory animals. The dermal LD_{50} in rabbits treated with technical cyanazine is greater than 2000 mg/kg, and the dermal LD_{50} in rats is greater than 1200 mg/kg (3,6,15). Cyanazine is only slightly toxic through inhalation. It is a mild eye irritant.

Chronic toxicity

Several long-term feeding studies in rats and mice at doses up to 225 mg/kg/day showed that cyanazine decreases body weight gain and increases liver weights (11).

Reproductive effects

Cyanazine caused decreases in maternal body weight gain in rats at doses of 30 mg/kg/day (27). It also caused maternal toxicity and decreased fetal viability in rabbits at doses of 2 mg/kg/day (27). It appears that reproductive effects are not likely in humans at expected exposure levels.

Teratogenic effects

Cyanazine can cause a variety of birth defects in animals over a wide range of doses. In a long-term study of rats fed cyanazine, moderate doses resulted in increased brain weights and decreased kidney weights in third-generation offspring (11). Toxic effects on the fetus were also observed in experiments on rabbits using comparable doses. Female rats fed cyanazine through a stomach tube during pregnancy ate less food and their fetuses had incomplete bone development. At the higher doses, fetuses showed cleft palates and the absence of, or underdeveloped eyeballs (11). Other birth defects observed in animals include abnormalities in diaphragm development and changes in the brain. Birth defects have been observed in the offspring of pregnant rats fed cyanazine during gestation at doses as low as 1 mg/kg/day.

Mutagenic effects

Cyanazine is not mutagenic (3).

Carcinogenic effects

Cyanazine does not appear to be carcinogenic. A study evaluated the carcinogenicity of the compound in mice and found, up to the maximum dose tested (50 mg/kg/day), no evidence of cancer in the animals (3,11).

Organ toxicity

Cyanazine in animals causes depression of the central nervous system.

Fate in humans and animals

Low doses of cyanazine fed to rats, dogs, and cows are rapidly absorbed from the gastrointestinal tract. In a study on rats and dogs, much of the cyanazine ingested was eliminated from animals within 4 days (11). There is some tendency for cyanazine to accumulate in the brain, liver, kidney, muscle, and fat (11).

Cows fed very low amounts of cyanazine eliminated up to 88% of the cyanazine in urine and feces within 21 days. The concentration in cows' milk was very low, at 0.022 ppm (11).

Ecological effects

Effects on birds

Cyanazine is slightly to moderately toxic in birds. The oral LD_{50} in mallards is greater than 2000 mg/kg, and in bobwhite quail is 400 mg/kg (15,16).

Effects on aquatic organisms

Cyanazine is slightly to moderately toxic to fish and aquatic invertebrates (15,16). The LC_{50} (96-hour) for cyanazine in harlequin fish is 7.5 mg/L, and 18 mg/L in sheepshead minnow (16).

Effects on other organisms (non-target species)

Cyanazine is nontoxic to bees (27).

Environmental fate

Breakdown in soil and groundwater

Cyanazine has a low to moderate persistence in soil. It quickly degrades in many soil types, mostly due to the action of microbes (28). Cyanazine has a half-life of 2 to 4 weeks in an air-dried sandy clay loam, 7 to 10 weeks in a sandy loam soil, 10 to 14 weeks in a clay soil, and 9 weeks in a fresh sandy clay soil. It undergoes slight decomposition by sunlight. The rate of evaporation of cyanazine from soil is very slow (12).

Cyanazine can be transported in runoff, sediment, and water, and it can leach through the soil to the groundwater. Cyanazine has been found in numerous groundwater samples at very low concentrations (0.001 to 0.08 mg/L) (6). A groundwater advisory statement on cyanazine product labels is required.

Breakdown in water

Cyanazine is stable to the chemical action of water (hydrolysis) and to the action of sunlight (photolysis) (28).

Breakdown in vegetation

Cyanazine is absorbed by the roots and is translocated up the plant into the leaves. It works by inhibiting photosynthesis (15).

Physical properties

Cyanazine is a member of the s-triazine chemical family. This odorless, white crystalline solid is incompatible with metals (6).

Chemical name: 2-(4-chloro-6-ethylamino-1,3,5-triazin-2-ylamino)-2-methylpro-pionitrile (6)
CAS #: 21725–46–2
Molecular weight: 240.70 (6)
Solubility in water: 171 mg/L @ 25°C (6)
Solubility in other solvents: benzene v.s.; chloroform v.s.; alcohol v.s.; hexane v.s.; xylene s.s.; ethanol s.s. (6)
Melting point: 167°C (6)
Vapor pressure: 0.0002 mPa @ 20°C (6)
Partition coefficient (octanol/water) (log): 2.22 (16)
Adsorption coefficient: 190 (20)

Exposure guidelines

ADI: Not available
HA: 0.01 mg/L (lifetime) (11)
RfD: Not available
PEL: Not available

Basic manufacturer

DuPont Agricultural Products
Walker's Mill, Barley Mill Plaza
P.O. Box 80038
Wilmington, DE 19880-0038
Telephone: 800-441-7515
Emergency: 800-441-3637

8.2.4   *Hexazinone*

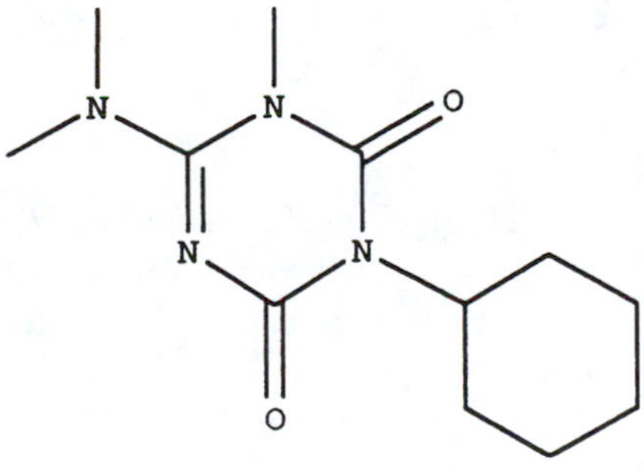

Figure 8.5 Hexazinone.

Trade or other names

Trade names for products containing hexazinone are DPX 3674, Pronone, and Velpar. It may be used in combination with other herbicides such as bromacil and diuron.

Regulatory status

Hexazinone is a slightly toxic compound in EPA toxicity class I. Labels for products containing hexazinone bear the Signal Words DANGER — POISON due its ability to cause serious and irreversible eye irritation. It is a General Use Pesticide (GUP).

Introduction

Hexazinone is a triazine herbicide used against many annual, biennial, and perennial weeds as well as some woody plants. It is mostly used on non-crop areas; however, it is used selectively for the control of weeds among sugarcane, pineapples, and lucerne. Hexazinone is a systemic herbicide that works by inhibiting photosynthesis in the target plants. Rainfall or irrigation water is needed before it becomes activated. It is available in soluble concentrate, water-soluble powder, or granular formulations.

Toxicological effects

Acute toxicity

Hexazinone is slightly toxic via the oral route, with a reported LD_{50} of 1690 mg/kg in rats and 860 mg/kg in male guinea pigs (6,15). Via the dermal route, it is practically nontoxic, with a reported dermal LD_{50} in rabbits of greater than 5278 mg/kg (6,15). Hexazinone does not cause significant skin irritation or sensitization in guinea pigs or rabbits, but it does cause severe eye irritation in rabbits (15). Hexazinone's inhalation toxicity is very low, its 1-hour inhalation LC_{50} is greater than 7.48 mg/L in rats (15).

Effects due to acute exposure may include irritatation of the eyes, nose, and throat, as well as nausea and vomiting (29).

Chronic toxicity

Over a 2-week period, male rats receiving dietary doses of 300 mg/kg/day showed no evidence of cumulative toxicity (15). Male rats receiving doses of 50 mg/kg/day over 90 days showed no effects, but higher doses caused decreased body weights. Body weight gain was seen in dogs at doses of about 35 mg/kg/day and higher over 1 year (15). Very high doses for 8 weeks did not affect hamsters and caused only increased liver weights in mice (29).

Reproductive effects

Female rats, fed moderate to high doses (up to 150 mg/kg) over two generations, showed no effects on reproduction or milk production, but only reduced offspring weight (15,29). Available evidence suggests that hexazinone is unlikely to cause reproductive effects in humans.

Teratogenic effects

Pregnant female rats receiving doses up to 100 mg/kg/day during gestation, and rabbits receiving up to 125 mg/kg/day, evidenced no fetal abnormalities (15). Teratogenic effects were observed in rats only at maternal doses greater than 400 mg/kg/day during gestation (15). It is unlikely that hexazinone would pose a teratogenic effects in humans under normal conditions.

Mutagenic effects

Hexazinone showed no mutagenic activity in the Ames assay, nor in tests using Chinese hamster ovary cell cultures (15). In living animal tests, no changes in chromosomal structure occurred. In other laboratory analyses of its capacity to induce genetic disruption, results were inconclusive (15). The evidence suggests hexazinone is either slightly or nonmutagenic.

Carcinogenic effects

Rats, mice, and dogs have been tested for 1 to 2 years on diets containing up to 500 mg/kg. Hexazinone was not carcinogenic in rats, and was only carcinogenic in mice at dietary levels of over 300 mg/kg. At these levels in mice, liver adenomas were observed (15). These studies suggest that hexazinone is unlikely to be carcinogenic to humans under normal circumstances.

Organ toxicity

Target organs affected in lab animals by chronic hexazinone exposure include the liver.

Fate in humans and animals

Hexazinone is fairly rapidly processed and excreted by animal systems. Rats typically excrete hexazinone almost completely within 3 to 6 days, the majority in urine (30). Long-term exposure does not diminish this rapid processing and elimination; rats given prior exposure for 2 weeks excreted almost all of the product within 3 days (30). Less than 1% of the parent hexazinone was detected in urine and feces. There does not appear to be any significant tissue accumulation (30).

Dairy cows given small amounts of hexazinone in their diets for 30 days had no detectable residues in milk, fat, liver, kidney, or lean muscle, but did have minute amounts of a hexazinone metabolite in their milk (30). Lactating goats given small amounts of hexazinone for 5 days also had small amounts of the compound in their milk and livers (30).

Ecological effects

Effects on birds

Hexazinone is slightly to practically nontoxic to birds. The acute oral LD_{50} of hexazinone in bobwhite quail is 2258 mg/kg (15). The 5- to 8-day dietary LC_{50} in bobwhite quail and mallard ducklings is greater than 10,000 ppm (15).

Effects on aquatic organisms

Hexazinone is slightly toxic to fish and other freshwater organisms. Some of the reported 96-hour LC_{50} values include: rainbow trout, 320 mg/L; bluegill, 370 mg/L; and fathead minnow, 274 mg/L (6,15). The 48-hour LC_{50} for hexazinone in the water flea, *Daphnia magna,* is 151 mg/L (15). The bioconcentration factor in bluegill sunfish is only seven times the ambient water concentration, indicating very low bioaccumulation in fish (30).

Effects on other organisms (non-target species)

Hexazinone is nontoxic to honeybees. The herbicide is toxic to larch trees (*Larix* spp.) and should not be used for weed control in forested areas (6).

Environmental fate

Breakdown in soil and groundwater

Hexazinone is of moderate to high persistence in the soil environment. Measured field half-lives range from less than 30 to 180 days, with a representative value of about

90 days (31). Hexazinone is broken down by soil microbes, which release carbon dioxide in the process (15). Sunlight may also break down the compound via photo-degradation (31). The rate of breakdown under natural field conditions will depend on many site-specific variables, including sunlight, rainfall, soil type, and rate of application. Hexazinone does not evaporate to any appreciable extent from soil (31).

Hexazinone is very poorly adsorbed to soil particles, very soluble in water, and slowly degraded, thus it is likely to be mobile in most soils and has the potential to contaminate groundwater.

Breakdown in water

Photodecomposition, biodegradation, and dilution are the prime mechanisms for loss of hexazinone activity in aquatic systems (15).

Breakdown in vegetation

Hexazinone is readily absorbed in the root zone and translocated throughout the plant. It is less mobile following uptake from the foliage. It is converted in nonsuscep-tible plants to less phytotoxic compounds. In susceptible plants, it is more persistent and can result in disruption of photosynthesis and chloroplast damage (15).

Physical properties

Hexazinone is a colorless, odorless crystal at room temperature (6).

Chemical name: 3-cyclohexyl-6-(dimethylamino)-1-methyl-1,3,5-triazine-2,4(1H,3H)-dione (6)
CAS #: 51235-04-2
Molecular weight: 252.32 (6)
Solubility in water: 33,000 mg/L @ 25°C (6)
Solubility in other solvents: v.s. in acetone, hexane, and methanol (6)
Melting point: 115–117°C (6)
Vapor pressure: 0.03 mPa @ 25°C (6)
Partition coefficient (octanol/water) (log): –4.40 (calculated) (16)
Adsorption coefficient: 54 (20)

Exposure guidelines

ADI: Not available
HA: 0.20 mg/L (lifetime) (29)
RfD: 0.033 mg/kg/day (26)
PEL: Not available

Basic manufacturer

DuPont Agricultural Products
Walker's Mill, Barley Mill Plaza
P.O. Box 80038
Wilmington, DE 19880-0038
Telephone: 800-441-7515
Emergency: 800-441-3637

8.2.5 Metribuzin

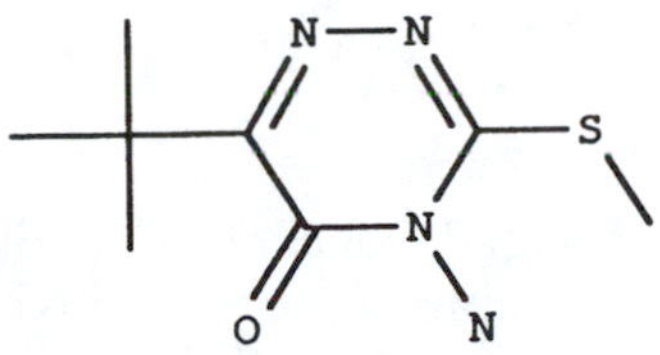

Figure 8.6 Metribuzin.

Trade or other names

Trade names include Bay 94337, Bay DIC 1468, Lexone, Sencor, Sencoral, and Sencorex.

Regulatory status

Metribuzin is a slightly toxic compound in EPA toxicity class III. It is a General Use Pesticide (GUP). Products containing metribuzin must bear the Signal Word CAUTION.

Introduction

Metribuzin is a selective triazine herbicide that inhibits photosynthesis of susceptible plant species. It is used for control of annual grasses and numerous broadleaf weeds in field and vegetable crops, in turfgrass, and on fallow lands. Metribuzin is available as liquid suspension, water-dispersible granular, and dry flowable formulations.

Toxicological effects

Acute toxicity

Metribuzin is slightly toxic via the oral route, with reported oral LD_{50} values of 1090 to 2300 mg/kg in rats, 700 mg/kg in mice, and 245 to 274 mg/kg in guinea pigs (6,15). It is practically nontoxic dermally, with a dermal LD_{50} of 20,000 mg/kg in rabbits (6,15). The 4-hour inhalation LC_{50} for metribuzin in rats is greater than 0.65 mg/L, indicating moderate toxicity via the inhalation route (15). Metribuzin has been shown not to irritate the skin or eyes of rats, rabbits, guinea pigs, or human volunteers (15,32).

Effects of high acute exposure in metribuzin-poisoned rats included narcosis (stupor) and labored breathing. Deaths occurred within 24 hours, and survivors recovered slowly without permanent effects (33).

Chronic toxicity

No ill effects were observed in dogs fed dietary doses of 12.5 mg/kg/day for 3 months (33). No effects were apparent in rats receiving 2.5 mg/kg/day over 3 months, but doses of 25 and 75 mg/kg/day caused enlarged livers and thyroid glands (33). In 2-year feeding studies with rats and dogs, results showed no observable effects at doses of 5 mg/kg/day in rats and 2.5 mg/kg/day in dogs. Reduced weight gain, an increase in the number of deaths, blood chemistry changes, and liver and kidney damage were observed in a 2-year study in which dogs were given 1500 ppm or 37.5 mg/kg/day metribuzin (34).

Reproductive effects

Doses of 15, 45, or 135 mg/kg/day technical metribuzin were administered by gavage to rabbits on days 6 through 18 of pregnancy. No effects on the mothers were observed at a dose of 45 mg/kg, but 135 mg/kg lowered maternal weight gain (35). No effects on the fetuses were observed at any of the doses tested (35). A three-generation study in rats at doses up to 15 mg/kg/day (the highest dose tested), showed no influence on reproduction (32,36). Metribuzin does not cause reproductive effects.

Teratogenic effects

In rats, reduced fetal body weights were seen at doses of 70 mg/kg/day, and developmental delays were observed at doses of 200 mg/kg/day (15). Metribuzin did not show teratogenic activity in rabbits at doses up to 85 mg/kg/day, but did decrease weight gain in offspring (15). These data suggest that metribuzin is unlikely to cause teratogenic effects in humans under normal circumstances.

Mutagenic effects

Tests on live animals and on tissue cultures have shown that metribuzin has no mutagenic activity (15,32).

Carcinogenic effects

There were no indications of carcinogenic effects in rats receiving dietary doses of up to 15 mg/kg/day for 2 years, nor in mice fed up to about 380 mg/kg/day for 2 years (32). These data suggest that metribuzin is not carcinogenic.

Organ toxicity

In single high-dose studies, metribuzin appears to depress the central nervous system. Other studies indicate that the target organs of metribuzin are the thyroid gland and the liver.

Fate in humans and animals

After metribuzin is absorbed, it is rapidly distributed in the body and excreted unchanged in the urine (6). In mammals, 90% elimination occurs within 96 hours, about equally distributed between the urine and feces (6).

Ecological effects

Effects on birds

Data indicate that metribuzin is moderately to slightly toxic to birds. The acute oral LD_{50} values are about 100 to 200 mg/kg in bobwhite quail, mallard ducks, and Japanese quail (6,15). The reported 5- to 8-day dietary LC_{50} values for these species are all greater than 4000 ppm (6,15).

Effects on aquatic organisms

Metribuzin is slightly toxic to fish. The 96-hour LC_{50} is 64 to 76 mg/L in rainbow trout, 80 mg/L in bluegill sunfish, and greater than 10 mg/L in goldfish (6,15). The reported 48-hour LC_{50} in *Daphnia magna* is 4.5 mg/L, indicating similar toxicity (15). The 96-hour LC_{50} in marine/estuarine shrimp is 48.3 mg/L (34).

Effects on other organisms (non-target species)

It is nontoxic to bees (6). Metribuzin may be phytotoxic to non-target plant species (6,15).

Environmental fate

Breakdown in soil and groundwater

Metribuzin is of moderate persistence in the soil environment (20). The half-life of metribuzin varies according to soil type and climatic conditions. Soil half-lives of 30 to 120 days have been reported; a representative value may be approximately 60 days (20). Metribuzin is poorly bound to most soils and soluble in water, giving it a potential for leaching in many soil types (20). Soil mobility is affected by many site-specific variables, including the amount of soil organic matter, particle size distribution, porosity, rainfall, and application rates. Metribuzin has been detected in Ohio rivers and Iowa wells and groundwater (34,35).

The major mechanism by which metribuzin is lost from soil is microbial degradation. Losses due to volatilization or photodegradation are not significant under field conditions (3,6).

Breakdown in water

The half-life of metribuzin in pond water is approximately 7 days (6). If present, metribuzin would most likely be found in the water column rather than the sediment due to its low binding affinity and high water solubility.

Breakdown in vegetation

Metribuzin is absorbed through the leaves when plants are given surface treatment, but the primary route for uptake is through the root system. From the roots, it is translocated upward, becoming concentrated in the roots, stems, and leaves of treated plants. In nonsusceptible plants, it is deaminized to more water-soluble conjugates; in susceptible plants, it is not metabolized and disrupts phosynthesis in the chloroplast (15).

Physical properties

Metribuzin is a white, crystalline solid with a slightly sharp, sulfurous odor (6).

Chemical name: 4-amino-6-tert-butyl-4,5-dihydro-3-methylthio-1,2,4-triazin-5-
   one (6)
CAS #: 21087–64–9
Molecular weight: 214.29 (6)

Solubility in water: 1050 mg/L (6)
Solubility in other solvents: s. in aromatic and chlorinated hydrocarbon solvents, including dimethyl formamide, cyclohexane, acetone, methanol, benzene, ethanol, xylene, and kerosene (6)
Melting point: 125–126.5°C (6)
Vapor pressure: 0.058 mPa @ 20°C (6)
Partition coefficient (octanol/water): 40 (6)
Adsorption coefficient: 60 (estimated) (20)

Exposure guidelines

ADI: Not available
HA: 0.2 mg/L (lifetime) (35)
RfD: 0.025 mg/kg/day (26)
TLV: 5 mg/m^3 (8-hour) (36)

Basic manufacturer

DuPont Agricultural Products
Walker's Mill, Barley Mill Plaza
P.O. Box 80038
Wilmington, DE 19880-0038
Telephone: 800-441-7515
Emergency: 800-441-3637

8.2.6   *Prometryn*

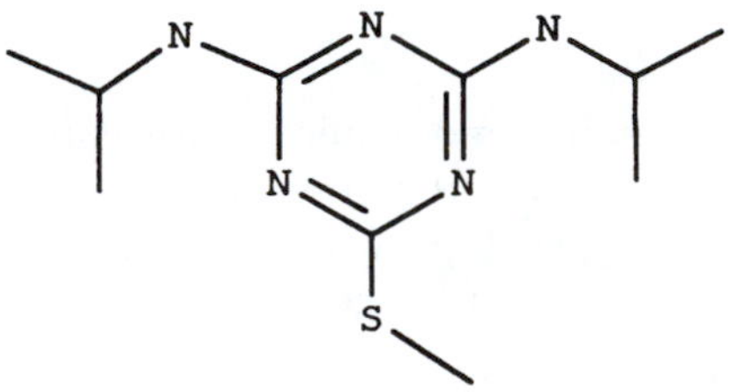

Figure 8.7 Prometryn.

Trade or other names

Trade names include Caparol, Gesagard, Mercasin, Promet, Prometrex, and Primatol Q.

Regulatory status

Prometryn is a slightly to moderately toxic compound that is classified as a member of toxicity class II or III depending on the formulation. It is classified as a General Use Pesticide (GUP). Labels for products containing prometryn must bear the Signal Words CAUTION or WARNING, depending on the formulation.

Introduction

Prometryn is a selective herbicide that controls annual grasses and broadleaf weeds in a variety of crops, including cotton and celery. It inhibits photosynthesis in susceptible species. Prometryn is available in wettable powder and liquid formulations.

Toxicological effects

Acute toxicity

Prometryn is slightly to practically nontoxic by ingestion, with reported oral LD_{50} values of 3750 to 5235 mg/kg in rats, 3750 mg/kg in mice, and greater than 2020 mg/kg in rabbits (6,15). Via the dermal route, it is slightly toxic, with reported dermal LD_{50} values of greater than 2000 mg/kg to greater than 3100 mg/kg in rabbits (6,15). Technical prometryn does not cause skin irritation in rabbits or skin sensitization in guinea pigs, but may cause slight eye irritation in rabbits (15). Some formulations (e.g., Caparol) may be mild eye irritants and/or slight skin irritants in rabbits (6,15). The 4-hour LC_{50} for prometryn in rats is 5.2 mg/L (15).

Symptoms of high acute exposure may include sedation, muscle incoordination, breathing difficulty, bulging eyes, constricted pupils, diarrhea, excessive urination, and convulsions.

Chronic toxicity

The results of long-term feeding studies do not indicate obvious, nor severe toxicity from prometryn exposure. Rats fed dietary doses of 37.5 mg/kg/day and dogs given 4 mg/kg/day over a 2-year period did not show observable gross or microscopic signs of systemic toxicity (15). Effects that occurred at higher dose rates in these animals included changes in relative weights of the kidney and liver (15).

Reproductive effects

In a three-generation study, no reproductive effects were seen in rats fed up to 5 mg/kg/day (37). In another study, reduced offspring body weights, but no other reproductive effects, were seen in rats at doses up to 75 mg/kg/day (15). From the data, it appears that prometryn is unlikely to cause reproductive effects.

Teratogenic effects

No teratogenic effects were seen in the offspring of rats fed 250 mg/kg/day, the highest dose tested (37). In another study, no teratogenic effects were seen in rats at doses of 50 mg/kg/day (15). No teratogenic or developmental effects were seen in rabbits at doses of 72 mg/kg/day (15). Prometryn does not appear to cause birth defects.

Mutagenic effects

Eleven different tests for mutagenicity involving hamsters, bacteria, or mammalian cell cultures have all produced negative results indicating that prometryn is not a mutagen (38).

Carcinogenic effects

Prometryn was not carcinogenic in a 2-year rat feeding study at doses up to 62.5 mg/kg/day (37). Carcinogenic effects were not seen in mice at doses up to 300 mg/kg/day over 18 months. The available data suggest that prometryn is not carcinogenic.

Organ toxicity

Target organs identified through animal studies include the liver, kidneys, and bone marrow.

Fate in humans and animals

The triazines are generally well-absorbed by the mammalian gut, and probably across the skin (16). While the breakdown of prometryn is not adequately understood, available data indicate that, in rats, most of the herbicide is excreted in urine and feces within 48 hours of administration (39). No detectable residues of prometryn or its metabolites were found in the muscle, fat, blood, liver, kidney, or other organs of sheep and cattle fed up to 100 ppm for 4 weeks. However, prometryn or its breakdown products were found in whole milk samples taken from cows that were fed up to 100 ppm in their diet for 21 days (39).

Ecological effects

Effects on birds

Prometryn is practically nontoxic to birds; the acute oral LD_{50} values in bobwhite quail and mallard ducks are greater than 2150 and 4640 mg/kg, respectively. The reported 5- to 7-day dietary LC_{50} values are greater than 10,000 ppm for these same species (15).

Effects on aquatic organisms

Prometryn is moderately toxic to fish, with reported 96-hour LC_{50} values of 2.5 to 2.9 mg/L in rainbow trout, 10.0 mg/L in bluegill sunfish, 3.5 mg/L in goldfish, and 8 mg/L in carp (15,40,41). It is highly toxic to guppies (39). It is slightly toxic to freshwater invertebrates (41). A 19% decrease in shell growth was observed in oysters exposed to 1.0 mg/L prometryn for 48 hours. Pink shrimp were unaffected by exposure to 1.0 mg/L prometryn for 48 hours (40). However, the compound has a 48-hour LC_{50} in the invertebrate *Daphnia* of 18.9 mg/L (15,41). The observed concentration of prometryn in bluegill and in rainbow trout is nine to ten times the ambient water concentration, indicating a low potential for bioaccumulation (41).

Effects on other organisms (non-target species)

Prometryn is nontoxic to bees and earthworms, with a reported contact LD_{50} of greater than 99 µg per bee, and a 48-hour LC_{50} of 153 mg/kg in earthworms (6,15).

Environmental fate

Breakdown in soil and groundwater

Prometryn is moderately persistent in the soil, with a field half-life of 1 to 3 months (20). It will persist longer under dry or cold conditions that are not conducive to chemical or biological activity (15,38). Following multiple annual applications of the herbicide, prometryn activity can persist for 12 to 18 months after the last application (38). Soil microorganisms readily break down prometryn in the soil (15). The amount of the herbicide evaporating from soil increases with increasing temperature and soil moisture content, but volatilization is not significant under most field conditions (15,38).

Prometryn is weakly bound to most soils and is slightly soluble in water (20). It may thus be mobile in some soils. However, it adsorbs more strongly to soils with higher proportions of clay and organic matter (38). Field leaching studies indicate that prometryn stays in the top 12 inches of treated soil (38).

Breakdown in water

No significant hydrolysis (or breakdown in water) was found when prometryn was tested over a period of 28 days in water ranging from slightly acidic to slightly alkaline and over a variety of test temperatures (38). These data indicate that prometryn is potentially persistent in the water environment.

Breakdown in vegetation

Prometryn is rapidly absorbed through both the foliage and roots of plants, and is translocated to the growing shoots (15). Removal or degradation by the plant is rapid in nonsusceptible plants, but very slow in susceptible species (15).

Physical properties

Prometryn is a colorless, crystalline solid (6).

Chemical name: N2,N4-di-isopropyl-6-methylthio-1,3,5-triazine-2,4-diyl-diamine (6)
CAS #: 7287–19–6
Molecular weight: 241.37 (6)
Water solubility: 48 mg/L @ 20°C (6)
Solubility in other solvents: s. in organic solvents, including ethanol, methanol, acetone, dichloromethane, and toluene (6)
Melting point: 118–120°C (6)
Vapor pressure: 0.13 mPa @ 20°C (6)
Partition coefficient (octanol/water): 2190 (6)
Adsorption coefficient: 400 (20)

Exposure guidelines

ADI: Not available
HA: Not available
RfD: 0.004 mg/kg/day (26)
PEL: Not available

Basic manufacturer

Ciba-Geigy Corporation
P.O. Box 18300
Greensboro, NC 27419-8300
Telephone: 800-334-9481
Emergency: 800-888-8372

8.2.7 Propazine

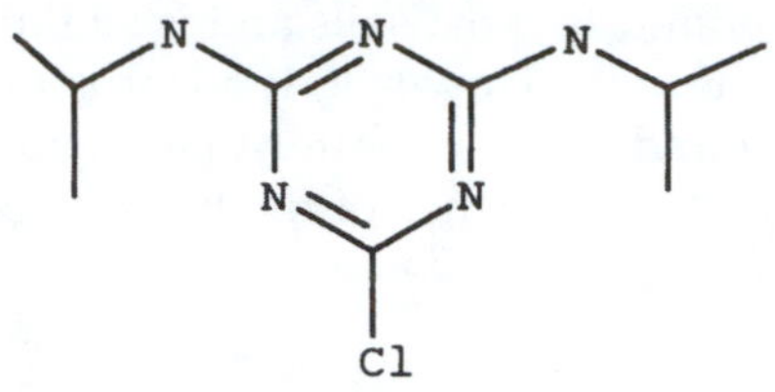

Figure 8.8 Propazine.

Trade or other names

Trade names include G-30028, Geigy 30028, Gesamil, Milocep, Milogard, Milo-Pro, Plantulin, Primatol P, Propazin, Propinex, and Prozinex.

Regulatory status

Propazine is classified as a slightly toxic compound in EPA toxicity class III and as a General Use Pesticide (GUP). Labels for products containing propazine must bear the Signal Word CAUTION.

Introduction

Propazine is an herbicide used for control of broadleaf weeds and annual grasses in sweet sorghum. It is applied as a spray at the time of planting or immediately following planting, but prior to weed or sorghum emergence. It is also used as a post-emergence selective herbicide on carrots, celery, and fennel. Propazine is available in wettable powder, liquid, and water-dispersible granular formulations.

Toxicological effects

Acute toxicity

Propazine is slightly toxic by ingestion, with reported oral LD_{50} values of 3840 mg/kg to greater than 7000 mg/kg in rats, 3180 mg/kg in mice, and 1200 mg/kg in guinea pigs (6,15). It is slightly toxic dermally; slight irritation was noted after propazine was applied to the skin of rabbits (9). Its dermal LD_{50} in rats is 10,200 mg/kg, and in rabbits is greater than 2000 mg/kg (9). Eye applications of 400 mg caused mild eye irritation in these animals (42). The inhalation LC_{50} is greater than 2.04 mg/L (6), indicating slight toxicity by this route as well.

Symptoms of exposure may include dizziness, lethargy, muscular weakness, runny nose, emaciation, diarrhea, labored breathing, and irregular breathing (43). It may be mildly irritating to the skin, eyes, and upper respiratory tract. Contact dermatitis has been reported among workers manufacturing propazine (3).

Chronic toxicity

When given daily to rabbits for 1 to 4 months, oral doses of 500 mg/kg/day propazine were reported to cause a type of anemia (3). No gross signs of toxicity or pathologic changes were evident in rats that received daily doses of 250 mg/kg propazine for 130 consecutive days (15). No clinical or physical toxic symptoms were observed in beagle dogs fed 1.25, 5, or 25 mg/kg/day of a propazine formulation in 90-day feeding studies (3).

Reproductive effects

There was an increase in the number of deaths of newborns born to female rats that were given 5 mg/kg/day propazine during 18 days of pregnancy (44). In a three-generation study with rats fed 0.15, 5, or 50 mg/kg/day, no effects on fertility, length of pregnancy, pup viability, or pup survival were observed (45). The data regarding reproductive effects are inconclusive.

Teratogenic effects

Maternal doses of 500 mg/kg/day resulted in maternal toxicity and developmental toxicity expressed as increased incidence of extra ribs, incomplete bone formation, and decreased fetal body weights in rats (45). At 50 mg/kg/day, pup body weights on day 21 of lactation were reduced, and there were pathological changes in organ weights in the second and third generations (45). These data suggest that teratogenic effects due to propazine are not likely unless the level of exposure is high.

Mutagenic effects

Propazine has shown no mutagenic effects in tests conducted on human and rat liver cells and in live hamsters (9).

Carcinogenic effects

No evidence of increased tumor frequency was detected in a 2-year study in mice fed doses up to 450 mg/kg/day propazine (9). When rats were fed 0.15, 5, or 50 mg/kg propazine each day for 2 years, there was an increase in the incidence of mammary gland tumors at the highest dose level (9). The data regarding carcinogenicity are inconclusive.

Organ toxicity

Liver damage is one of the suspected effects of propazine (42). The functioning of certain liver processes was decreased in rats given 2500 mg/kg/day propazine (9).

Fate in humans and animals

Propazine is readily absorbed and metabolized in the body (33). At 72 hours after administration of single oral doses of propazine to rats, 66% of the dose was excreted in the urine and 23% was excreted in the feces (33). This study also indicated that 77% of the dose was absorbed into the bloodstream from the gastrointestinal tract. At 8 days after the dosing, propazine or its metabolites were detected in the rats' lungs, spleen, heart, kidneys, and brain (9).

Ecological effects

Effects on birds

Propazine is practically nontoxic to slightly toxic to birds; the 8-day dietary LC_{50} is greater than 10,000 ppm in both bobwhite quail and mallard ducks (6).

Effects on aquatic organisms

Propazine is slightly toxic to fish; the 96-hour LC_{50} is 18 mg/L in rainbow trout, greater than 100 mg/L in bluegill sunfish, and greater than 32 mg/L in goldfish (6).

Effects on other organisms (non-target species)

Propazine is practically nontoxic to bees (6).

Environmental fate

Breakdown in soil and groundwater

Propazine is highly persistent in the soil environment, with reported field half-lives ranging from 35 to 231 days (20). It is broken down mainly by microbial action, and therefore will persist longer in dry, cold conditions or other conditions that inhibit microbial activity (15). Photolysis and volatilization are not important factors in propazine degradation (15).

Propazine is slightly soluble in water and is poorly bound by soils. One study found propazine to be mobile in sandy loam, loam, and clay loam soils. It was very mobile in loamy sand (45). It may contaminate groundwater, especially in areas with low organic matter and clay contents, high rainfall or irrigation rates, porous soils, and with excessive application. In 15 out of 906 groundwater samples collected from eight states, propazine was detectable. The maximum concentration found in any sample was 0.013 mg/L (45).

Breakdown in water

Propazine is resistant to breakdown by hydrolysis. After 28 days at pH 5, 60% of applied propazine remained unhydrolyzed; at pH 7, 92% remained; and at pH 9, 100% remained (4). It has been found in 33 out of 1,097 surface water samples at levels below 0.3 mg/L (9,45).

Breakdown in vegetation

Propazine is absorbed principally through plant roots, and translocated to growing shoots (meristems) and leaves of plants (15). Propazine accumulates and inhibits cell reproduction of the growth regions (meristems) in those plants that are unable to readily metabolize it; but in tolerant species, it is readily transformed via hydrolysis and ring cleavage (6).

Physical properties

Propazine is a colorless crystalline solid under normal circumstances (6).

Chemical name: 6-chloro-N2,N4-di-isopropyl-1,3,5-triazine-2,4-diyldiamine (6)
CAS #: 139–40–2
Molecular weight: 229.7 (6)
Water solubility: 5 mg/L @ 20°C (6)
Solubility in other solvents: Difficult to dissolve in organic solvents: in benzene, toluene, carbon tetrachloride, and diethyl ether (6)

Melting point: 212–214°C (6)
Vapor pressure: 0.004 mPa @ 20°C (6)
Partition coefficient (octanol/water): Not available
Adsorption coefficient: 154 (20)

Exposure guidelines

ADI: Not available
HA: 0.01 mg/L (lifetime) (9)
RfD: 0.02 mg/kg/day (26)
PEL: Not available

Basic manufacturer

Ciba-Geigy Corp.
P.O. Box 18300
Greensboro, NC 27419-8300
Telephone: 800-334-9481
Emergency: 800-888-8372

8.2.8 Simazine

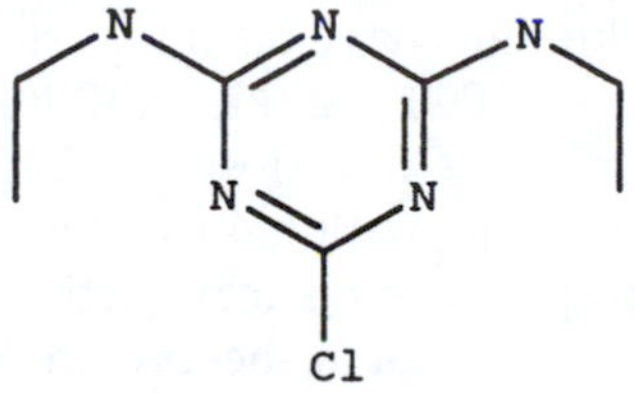

Figure 8.9 Simazine.

Trade or other names

Trade names include Aquazine, Caliber, Cekusan, Cekusima, Framed, Gesatop, Primatol S, Princep, Simadex, Simanex, Sim-Trol, Tanzine, and Totazine. This compound may also be found in formulations with other herbicides such as amitrole, paraquat dichloride, metolachlor, and atrazine.

Regulatory status

Simazine is a General Use Pesticide (GUP). It is in EPA toxicity class IV — practically nontoxic. Products containing simazine bear the Signal Word CAUTION. In November 1994, the U.S. EPA began a Special Review of simazine that could result in use restrictions or even cancellation if data warrant such action.

Introduction

Simazine is a selective triazine herbicide. It is used to control broad-leaved weeds and annual grasses in field, berry fruit, nuts, vegetable and ornamental crops, turfgrass, orchards, and vineyards. At higher rates, it is used for nonselective weed control in industrial areas. Before 1992, simazine was used to control sub-merged weeds and algae in large aquariums, farm ponds, fish hatcheries, swim-ming pools, ornamental ponds, and cooling towers. Simazine is available in wet-table powder, water-dispersible granule, liquid, and granular formulations. It may be soil-applied.

Toxicological effects

Acute toxicity

Simazine is slightly to practically nontoxic. The reported oral LD_{50} for technical simazine in rats and mice is >5,000 mg/kg (6,15); its dermal LD_{50} is 3100 mg/kg in rats and >10,000 mg/kg in rabbits (6,15). The 4-hour inhalation LC_{50} in rats is greater than 2 mg/L (6). The formulated products, in most cases, are less toxic via all routes (15).

Simazine is nonirritating to the skin and eyes of rabbits except at high doses (3). Patch tests on humans have shown that simazine is not a skin irritant, fatiguing agent, or sensitizer (3). However, rashes and dermatitis from occupational exposure to simazine have occurred (3).

The triazine herbicides disturb energy metabolism (thiamin and riboflavin func-tions). Symptoms include difficulty in walking, tremor, convulsions, paralysis, cyano-sis, slowed respiration, miosis (pinpoint pupils), gut pain, diarrhea, and impaired adrenal function (3).

No cases of poisoning in humans have been reported from ingestion of simazine (3). Rats given an oral dose of 5000 mg/kg exhibited drowsiness and irregular breathing. In another study, a single oral dose of 4200 mg/kg produced anorexia, weight loss, and some deaths in rats within 4 to 10 days (26).

For unknown reasons, sheep and cattle are especially susceptible to poisoning by simazine. Doses of 500 mg/kg were fatal in sheep with death delayed for 5 to 16 days. Symptoms exhibited by poisoned sheep included lower food intake, higher water intake, incoordination, tremors, and weakness, especially in the hindquarters (3).

Chronic toxicity

Some 90-day feeding studies showed reduced body weight at 67 to 100 mg/kg/day (10). This same effect and kidney toxicity were seen in rats at doses of 150 mg/kg/day (10). In 2-year chronic oral feeding studies in which rats were given daily dosages of 5 mg/kg/day simazine in the diet, no gross or microscopic signs of toxicity were seen (3).

When rats were given repeated doses of 15 mg/kg/day, some liver cells degen-erated during the first 3 days, but the condition did not progress. Instead, the liver adapted and the compound was metabolized (3).

Other effects observed in test animals include tremors, damage to the testes, kidneys, liver, and thyroid, disturbances in sperm production, and gene mutations (10).

Reproductive effects

No adverse effects on reproductive capacity or development were observed in a three-generation study of rats fed 5 mg/kg/day simazine (10). High rates of fetotoxicity and decreased birth weight were noted in the fetuses of pregnant rabbits fed 75 mg/kg/day (26). Reproductive effects are not likely in humans under normal circumstances.

Teratogenic effects

No dose-related teratogenic effects were observed when rabbits were given daily doses of 5, 75, or 200 mg/kg for days 7 through 19 of pregnancy (26). Chronic inhalation of a cumulative dose of 0.3 mg/L for 8 days in pregnant rats resulted in no treatment-related developmental abnormalities (10). Simazine does not appear to be teratogenic.

Mutagenic effects

Simazine has shown negative results in a variety of mutagenicity tests on bacterial cultures (10). Tests on human lung cell cultures have produced both positive and negative results (10). When injected into adult male fruit flies, simazine increased the frequency of sex-linked lethal mutations, but failed to do so when fed to larvae. Other tests for mutagenicity in fruit flies were negative (3). It is likely that simazine is either nonmutagenic or weakly mutagenic.

Carcinogenic effects

Simazine was not tumorigenic in mice at the maximum tolerated dose of 215 mg/kg/day over an 18-month period (10). In other studies, doses as low as 5 mg/kg/day produced excess tumors (thyroid and mammary) in female rats (3,10). Because of inconsistencies in the data, it is not possible to determine simazine's carcinogenic status.

Organ toxicity

Damage to the testes, kidneys, liver, and thyroid has been observed in test animals (3,10).

Fate in humans and animals

Studies in rats, goats, and sheep reveal that 60 to 70% of the ingested dose may be absorbed into the system (10), with approximately 5 to 10% distributed systemically to tissues. The remainder is eliminated via urine within 24 hours (6). Distribution led to detectable levels in red blood cells (highest), liver, kidney, fat, bone, and plasma (10).

When a cow was fed 5 ppm simazine for 3 days, no simazine was found in the cow's milk during the next 3 days. It has been reported that simazine residues were present in the urine of sheep for up to 12 days after administration of a single oral dose. The maximum concentration in the urine occurred from 2 to 6 days after administration (16).

Ecological effects

Effects on birds

Simazine is practically nontoxic to birds (6,16). The reported LD_{50} values in mallard and Japanese quail are >4600 and 1785 mg/kg, respectively (6). The acute dietary LD_{50} values in hens and pigeons are both greater than 5000 ppm (2). The 8-day dietary LC_{50} in bobwhite quail is >5260 ppm, and in mallard ducks is >10,000 ppm (6,15).

Effects on aquatic organisms

Simazine is slightly to practically nontoxic to aquatic species (6,15). The 96-hour LC_{50} for simazine is >100 mg/L (46) in rainbow trout, 100 mg/L (wettable powder) in bluegill sunfish, 0.100 mg/L in fathead minnows (46), as well as carp (2). It may be more toxic to *Daphnia* and stoneflies (46). A 96-hour LC_{50} of >3.7 mg/L is reported in oysters (15).

Effects on other organisms (non-target species)

While many mammals may be insensitive to simazine (16), sheep and cattle are especially sensitive (3). Simazine is nontoxic to bees (6,16). A soil LC_{50} in earthworms of >1000 mg/kg has been reported (16).

Environmental fate

Breakdown in soil and groundwater

Simazine is moderately persistent, with an average field half-life of 60 days (20). Soil half-lives of 28 to 149 days have been reported (20). Residual activity may remain for a year after application (2 to 4 kg/ha) in high-pH soils.

Simazine is moderately to poorly bound to soils (20). It does, however, adsorb to clays and mucks. Its low water solubility, however, makes it less mobile, limiting its leaching potential (15). Simazine has little, if any, lateral movement in soil, but can be washed along with soil particles in runoff.

Simazine is subject to decomposition by ultraviolet radiation, but this effect is small under normal field conditions. Loss from volatilization is also insignificant. In soils, microbial activity probably accounts for decomposition of a significant amount of simazine in high-pH soils. In lower pH soils, hydrolysis will occur (15).

Simazine residues have been detected in groundwater in at least 16 states. The range was from 0.00002 to 0.0034 mg/L (23).

Breakdown in water

The average half-life of simazine in ponds where it has been applied is 30 days, with the actual half-life dependent on the level of algae present, the degree of weed infestation, and other factors (15). Simazine may undergo hydrolysis at lower pH. It does not readily undergo hydrolysis in water at pH 7 (15).

Breakdown in vegetation

Plants absorb simazine mainly through the roots, with little or no foliar penetration. From the roots, it is translocated upward to the stems, leaves, and growing shoots of the plant (6,15). It acts to inhibit photosynthesis (6,15). Resistant plants readily metabolize simazine.

Plants that are sensitive to simazine accumulate it unchanged (6). It is possible that livestock or wildlife grazing on these plants could be poisoned.

Physical properties

Simazine is a white or colorless crystalline solid (6).

Chemical name: 6-chloro-N2,N4-diethyl-1,3,5-triazine-2,4-diamine (6)
CAS #: 122–34–9
Molecular weight: 201.7 (6)

Water solubility: 5 mg/L @ 20°C (6)
Solubility in other solvents: s. in methanol, chloroform and diethyl ether (6); s.s.
   in pentane (6)
Melting point: 225–227°C (6)
Vapor pressure: 0.000810 mPa @ 20°C (6)
Partition coefficient (octanol/water): 91.2 (6)
Adsorption coefficient: 130 (20)

Exposure guidelines

ADI: Not available
MCL: 0.004 mg/L (25)
RfD: 0.005 mg/kg/day (26)
PEL: Not available

Basic manufacturer

Ciba-Geigy Corporation
P.O. Box 18300
Greensboro, NC 27419-8300
Telephone: 800-334-9481
Emergency: 800-888-8372

8.2.9   Triadimefon

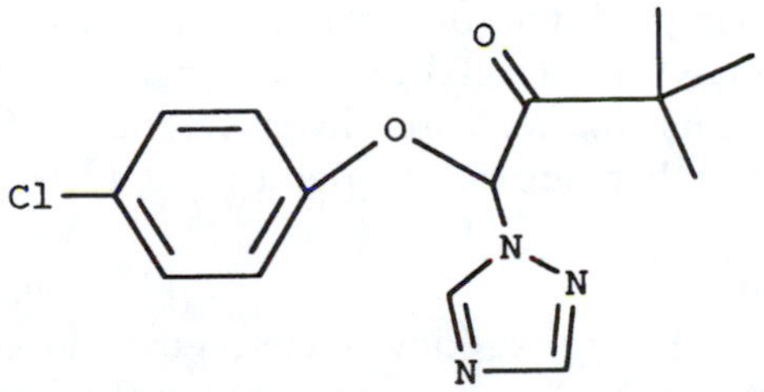

Figure 8.10 Triadimefon.

Trade or other names

Trade names for products containing triadimefon include Acizol, Amiral, Bay MEB 6447, and Bayleton (92.6% triadimefon). The compound may also be found in formulations with other fungicides such as captan, carbendazim, folpet, dodine, and propineb (1).

Regulatory status

Triadimefon is a General Use Pesticide (GUP). It is a moderately toxic compound in toxicity class II that carries the Signal Word WARNING on its label.

Introduction

Triadimefon is a systemic fungicide in the triazole family of chemicals. It is used to control powdery mildews, rusts, and other fungal pests on cereals, fruits, vegetables, turf, shrubs, and trees. It is available in wettable powder, emulsifiable concentrate, granular, and paste forms. *All data presented refer to Bayleton unless otherwise specified.*

Toxicological effects

Acute toxicity

Bayleton (92.6% triadimefon) has an acute oral LD_{50} of 300 to 600 mg/kg in rats, about 1000 mg/kg in mice, and about 500 mg/kg in rabbits and dogs (6,15). While these LD_{50} values indicate lower acute toxicity than many compounds classed as moderately toxic, triadimefon is classified this way because of its potential to cause adverse chronic effects at low to moderate dose levels. Lower potency formulations of triadimefon have lower acute toxicities (higher LD_{50} values).

Acute inhalation toxicity of the compound is moderate. The 4-hour inhalation LC_{50} is greater than 0.48 mg/L in rats and approximately the same in mice (6). Acute toxicity through skin exposure is also fairly low. The LD_{50} values for the dermal toxicity of technical triadimefon are greater than 1000 mg/kg in rats and 2000 mg/kg in rabbits (6,15).

Studies of acute effects in rats have indicated a potential to induce neurobehavioral effects (16). Data regarding eye and skin irritation are inconclusive (13).

Chronic toxicity

A number of 2-year studies have indicated that there are several toxic responses to low to moderate doses of the compound. Long-term studies of triadimefon in several species (rat, mouse,and dog) over a range of doses indicated a reduction in body weight, changes in red blood cell counts, an increase in blood cholesterol levels, and increased liver weights (16). Increased liver weights may be seen as an adaptation to toxic stress, rather than a toxic end-point related to exposure (3).

Reproductive effects

Female rats fed up to 90 mg/kg/day Bayleton over three generations showed a number of adverse effects. No effects were noted in the fetuses at maternal doses below 2.5 mg/kg/day. At the middle doses tested (around 15 mg/kg/day), the second-generation offspring experienced a decrease in weight gain. At the highest dose, the females experienced a reduction in body weight and a decrease in fertility (16). In another study conducted over two generations, the female rats showed decreased ovary weight at the 2.5-mg/kg/day dose (13). At 90 mg/kg/day, reductions in litter size, reduced offspring viability, and lower birth weight were observed in second-generation offspring (13). This evidence suggests it is unlikely that triadimefon will cause reproductive toxicity in humans under normal circumstances.

Teratogenic effects

The teratogenic potential of triadimefon is relatively low (3). Doses causing birth defects in rats were high enough to also produce maternal toxicity. Cleft palates were noted in the offspring of female rats fed moderate doses of 75 mg/kg/day for an unspecified time period. In a second study, no teratogenic effects were noted in the offspring of female rats fed 50 mg/kg/day Bayleton in the form of an emulsion. In another teratogenic study in rats, rib deformities were noted at high maternal doses of 90 mg/kg/day (13).

A study of occupationally exposed female workers showed that the highest combined dermal and inhalation level of exposure for workers was around 60 μg, which corresponds to approximately 0.008 mg/kg/shift for a 70-kg worker, a value considerably lower than the lowest dose that caused teratogenic effects in test animals (47). Thus, it is unlikely that triadimefon will cause birth defects in humans under normal circumstances.

Mutagenic effects

Six separate studies indicate that the Bayleton compound is non-mutagenic. Several other tests were inconclusive (4). It is unlikely that the compound poses a significant mutagenic risk.

Carcinogenic effects

In a 2-year dietary study with mice, the highest dose tested (600 mg/kg/day) did not produce significant increases in tumor incidence. Due to high mortality, the reliability of this data is suspect (13). Another 2-year dietary study in mice showed increased liver cell hypertrophy (which may be related to tumor formation) at doses of greater than 36 mg/kg/day in males and 6 mg/kg/day for females. Increased liver cell adenoma was detected at all levels, but carcinoma was not detected at any level in this study (13). Based on this evidence, no conclusion can be drawn about the overall carcinogenicity of triadimefon.

Organ toxicity

Triadimefon has been associated with changes in the liver, decreased kidney weights, and altered urinary bladder structure in laboratory animals exposed to 18 to 60 mg/kg/day. There is evidence that acute effects on the central nervous system may also occur (16).

Fate in humans and animals

After oral administration of a single dose of triadimefon, most of the compound was eliminated unchanged in the urine and feces within 2 to 3 days. Some breakdown of a small amount of the compound occurred in the liver. The compound has a very short residence time in the bloodstream, about $2\frac{1}{2}$ hours (6).

When applied to the skin of rats, 40% of the applied amount was excreted in urine and feces within 8 days. Additional amounts (up to 40%) were recovered unabsorbed from the skin surface and in the cage (13).

Ecological effects

Effects on birds

Triadimefon ranges from slightly toxic to practically nontoxic to birds. For instance, the compound has an LD_{50} value of greater than 4000 mg/kg in mallard ducks (6). Japanese quail are less tolerant of the compound (LD_{50} of 2000 mg/kg), and canaries are even less tolerant (LD_{50} > 1000 mg/kg) (6). Even the most tolerant species exhibited some compound-related acute toxicity such as diarrhea and regurgitation within 5 minutes of administration of the highest doses. At the lowest dose tested (500 mg/kg), no signs of diarrhea were noted (16).

Effects on aquatic organisms

The compound is slightly toxic to fish, indicating that they are more susceptible to the presence of the compound than are birds. Bluegill sunfish are the most susceptible, followed closely by goldfish, with 96-hour LC_{50} values of 11 and 10 to 50 mg/L, respectively (6). The compound is only slightly toxic to rainbow trout, with a reported LC_{50} of 14 mg/L (6).

Effects on other organisms (non-target species)

The compound is nontoxic to honeybees (6).

Environmental fate

Breakdown in soil and groundwater

Triadimefon has low to moderate persistence in soils. In a sandy loam type of soil, half of the initial amount of the compound was lost within 18 days. In loamy soil, the half-life was much shorter (about 6 days), which indicates that breakdown of the compound varies with soil type (16). Other reported soil half-lives are 14 to 60 days, with an average of 26 days (20). Triadimefon and its residues are moderately mobile and may have potential to leach to groundwater (16).

Breakdown in water

In water with a pH 3.0, 6.0, or 9.0, almost 95% of the compound remained after 28 weeks. The compound is very stable in water and does not readily undergo hydrolysis (16).

Breakdown in vegetation

In plants, a breakdown product is triadimenol (6), and translocation and metabolism may vary according to plant species. Triadimenol is of comparable toxicity to triadimefon.

Physical properties

Triadimefon consists of colorless crystals (6).

Chemical name: 1-(4-chlorophenoxy)-3,3-dimethyl-1-(1H-1,2,4-triazol-1-yl) bu-
tanone (6)
CAS #: 43121–43–3
Molecular weight: 293.76 (6)
Solubility in water: 260 mg/L @ 20°C (6)
Solubility in other solvents: m.s. in most organic solvents (1)
Melting point: 82.3°C (6)
Vapor pressure: < 0.1 mPa @ 20°C (6)
Partition coefficient (octanol/water): 1510 (6)
Adsorption coefficient: 300 (20)

Exposure guidelines

ADI: 0.03 mg/kg (22)
HA: Not available

RfD: Not available
PEL: Not available

Basic manufacturer

Miles, Inc.
8400 Hawthorn Road
P.O. Box 4913
Kansas City, MO 64120
Telephone: 816-242-2429
Emergency: 816-242-2582

References

(1) Draber, W. Structure-activity studies of photosynthesis inhibitors. In *Rational Approaches to Structure, Activity and Ecotoxicology of Agrochemicals*. Draber, W. and Fujita, T., Eds. CRC Press, Boca Raton, FL, 1992.

(2) Ware, G. W. *Fundamentals of Pesticides: A Self-Instruction Guide*, 2nd ed. Thomson Publications, Fresno, CA, 1986.

(3) Stevens, J. T. and Sumner, D. D. Herbicides. In *Handbook of Pesticide Toxicology*. Hayes, W. J., Jr. and Laws, E. R., Jr., Eds. Academic Press, New York, 1991.

(4) Gianessi, L. P. and Puffer, C. A. *The Use of Herbicides in U.S. Crop Production: Use Coefficients Listed by Active Ingredient*. Quality of the Environment Division, Resources for the Future, Washington, D.C., 1991.

(5) National Research Council. *Drinking Water and Health*. National Academy of Sciences, Washington, D.C., 1977.

(6) Kidd, H. and James, D. R., Eds. *The Agrochemicals Handbook*, 3rd ed. Royal Society of Chemistry Information Services, Cambridge, U.K., 1991 (as updated).

(7) Jordan, L. S. and Cudney, D. W. Herbicides. In *Fate of Pesticides in the Environment*. Biggar, J. W. and Seiber, J. N., Eds. University of California Agricultural Experiment Station, Davis, CA, 1987.

(8) Briggs, S. A. *Basic Guide to Pesticides: Their Characteristics and Hazards*. Hemisphere Publishing, Washington, D.C., 1992.

(9) U.S. Environmental Protection Agency. *Health Advisory Summary: Propazine*. Office of Drinking Water, Washington, D.C., 1988.

(10) U.S. Environmental Protection Agency. *Health Advisory Summary: Simazine*. Office of Drinking Water, Washington, D.C., 1988.

(11) U.S. Environmental Protection Agency. *Health Advisory Summary: Cyanazine*. Office of Drinking Water, Washington, D.C., 1988.

(12) U.S. Environmental Protection Agency. *Health Advisory Summary: Atrazine*. Office of Drinking Water, Washington, D.C., 1988.

(13) U.S. Environmental Protection Agency. *Toxicology One-Line Summary: Triadimefon*. Environmental Fate and Effects Division, Washington, D.C., 1993.

(14) U.S. Environmental Protection Agency. Amitrole: preliminary determination to terminate Special Review. *Fed. Reg.* 57, 46448–46455, 1992.

(15) Weed Science Society of America. *Herbicide Handbook*, 7th ed. Champaign, IL, 1994.

(16) U.S. National Library of Medicine. *Hazardous Substances Data Bank*. Bethesda, MD, 1995.

(17) U.S. Environmental Protection Agency. *Environmental Fate One-Line Summary: Triademifon*. Environmental Fate and Effects Division, Washington, D.C., 1988.

(18) U.S. Environmental Protection Agency. *Amitrole: Pesticide Registration Standard and Guidance Document*. Washington, D.C., 1984.

(19) Hallenbeck, W. H. and Cunningham-Burns, K. M. *Pesticides and Human Health*. Springer-Verlag, New York, 1985.

(20) Wauchope, R. D., Buttler, T. M., Hornsby A. G., Augustijn-Beckers, P. W. M., and Burt, J. P. SCS/ARS/CES pesticide properties database for environmental decisionmaking. *Rev. Environ. Contam. Toxicol.* 123, 1–157, 1992.

(21) Howard, P.H., Ed. *Handbook of Environmental Fate and Exposure Data for Organic Chemicals.* Pesticides. Lewis, Boca Raton, FL, 1989.

(22) Lu, F. C. A review of the acceptable daily intakes of pesticides assessed by the World Health Organization. *Regul. Toxicol. Pharmacol.* 21, 351–364, 1995.

(23) U.S. Environmental Protection Agency. *National Survey of Pesticides in Drinking Water Wells. Phase I Report.* Washington, D.C., 1990.

(24) U.S. Department of Interior (U.S. Geological Survey). *Spring Sampling Finds Herbicides Throughout Mississippi River and Tributaries.* Reston, VA, 1991.

(25) U.S. Environmental Protection Agency. *National Primary Drinking Water Standards (EPA 810-F94-001-A).* Office of Drinking Water, Washington, D.C., 1994.

(26) U.S. Environmental Protection Agency. *Integrated Risk Information System.* Washington, D.C., 1995.

(27) U.S. Environmental Protection Agency. *Pesticide Fact Sheet Number 41: Cyanazine.* Office of Pesticides and Toxic Substances, Washington, D.C., 1984.

(28) U.S. Environmental Protection Agency. *Cyanazine. Special Review Position. Document 1.* Washington, D.C., 1985.

(29) U.S. Environmental Protection Agency. *Health Advisory Summary: Hexazinone.* Office of Drinking Water, Washington, D.C., 1987.

(30) Food and Drug Administration. *The FDA Surveillance Index.* National Technical Information Service, Springfield, VA, 1986.

(31) U.S. Department of Agriculture (U.S. Forest Service). *Pesticide Background Statements. Vol. I. Herbicides.* Washington, D.C., 1984.

(32) E. I. DuPont de Nemours Corp. *Material Safety Data Sheet for Metribuzin Technical.* Wilmington, DE, 1991.

(33) Gosselin, R. E., Smith, R. P., and Hodge, H. C. *Clinical Toxicology of Commercial Products, Fifth Edition.* Williams and Wilkins, Baltimore, MD, 1984.

(34) U.S. Environmental Protection Agency. *Pesticide Fact Sheet Number 53: Metribuzin.* Office of Pesticides and Toxic Substances, Washington, D.C., 1985.

(35) U.S. Environmental Protection Agency. *Health Advisory Summary: Metribuzin.* Office of Drinking Water, Washington, D.C., 1988.

(36) American Conference of Governmental Industrial Hygienists, Inc. *Documentation of the Threshold Limit Values and Biological Exposure Indices,* 6th ed., Cincinnati, OH, 1991.

(37) U.S. Environmental Protection Agency. *Surveillance Index Support Document. Toxicity Summary: Prometryn.* Washington, D.C., 1982.

(38) Ciba-Geigy Corporation. *Prometryn (G 34'161): Toxicity and Safety Assessment.* Greensboro, NC, 1987.

(39) U.S. Environmental Protection Agency. *Guidance for the Reregistration of Pesticide Products Containing Prometryn as the Active Ingredient.* Washington, D.C., 1987.

(40) Pimentel, D. *Ecological Effects of Pesticides on Nontarget Species.* President's Office of Science and Technology, Washington, D.C., 1971.

(41) U.S. Environmental Protection Agency. *Pesticide Fact Sheet Number 121: Prometryn.* Office of Pesticides and Toxic Substances, Washington, D.C., 1987.

(42) National Institute for Occupational Safety and Health. *Registry of Toxic Effects of Chemical Substances.* Cincinnati, OH, 1994.

(43) Dreisbach, R. H. *Handbook of Poisoning: Prevention, Diagnosis and Treatment,* 11th ed. Lange Medical Publications, Los Altos, CA, 1983.

(44) Shepard, T. H. *Catalogue of Teratogenic Agents,* 3rd ed. Johns Hopkins University Press, Baltimore, MD, 1980.

(45) U.S. Environmental Protection Agency. *Guidance for the Reregistration of Pesticide Products Containing Propazine as the Active Ingredient.* Office of Pesticides and Toxic Substances, Washington, D.C., 1988.

(46) Johnson, W. W. and Finley, M. T. *Handbook of Acute Toxicity of Chemicals to Fish and Aquatic Invertebrates, Resource Publication 137.* U.S. Department of Interior, Fish and Wildlife Service, Washington, D.C., 1980.

(47) Maddy, K. T. *Monitoring of Occupational Explosives of Mixer/ Loaders and Applications to Bayleton 50% WP in Kern County in 1982.* California Department of Agricultural, Pest Management Division, Sacramento, CA, 1983.

chapter nine

Ureas

9.1   Class overview and general description

Background

Substituted urea compounds (compounds in which different functional groups are substituted for one or more of the hydrogens in the urea molecule) are widely used as herbicidal agents, and to a lesser degree as insecticides (1,2). The general structure of the substituted urea molecule is shown in Figure 9.1. The compound's identity, structure, and activity are defined by the substituents and their arrangement on the molecule (1,2). The substituted urea herbicides include the phenylurea herbicides (Figure 9.1A) and the sulfonylurea herbicides (Figure 9.1B), which are the two major groups of substituted urea herbicides (2). The benzoylphenylureas (Figure 9.1C) are the major class of ureas used for insect control (1). Examples of these three types of substituted urea compounds are listed in Table 9.1.

Ureas usage

Phenyl- and sulfonylureas

The phenylurea compounds include some of the most commercially important herbicides: fenuron and diuron (2). There are at least 20 different compounds within this subgroup. They are principally used for the control of annual and perennial grasses and are applied pre-emergence, i.e., before target plants have emerged (2,4). The sulfonylurea compound group is a relatively new class of herbicides first introduced in 1982, and has fewer members than the phenylurea class (2). Sulfonyl ureas are also used for control of broadleaf weed species in addition to annual and perennial grasses (2,4). They may be applied pre- or post-emergence (4). The sulfonyl urea herbicides are nearly 100 times more toxic to target plants than the older compounds (2). In addition, they are applied at relatively low application rates and have a low toxicity to humans and other animals. Thus, this class of compound is expected to see continuing research and development activity in the coming years (3).

Benzoylphenylureas

Benzoylphenylurea compounds are very useful in controlling a wide variety of insect pests, both within and outside of Integrated Pest Management systems (1). The potential for their use as insecticidal agents was first recognized in the early

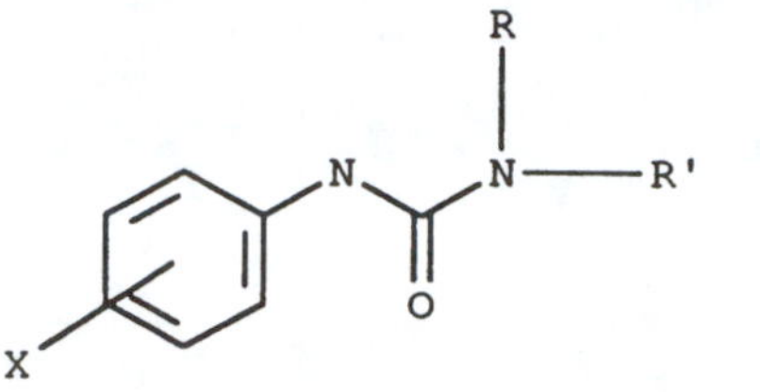

Figure 9.1 Generic structures for phenylurea (A), sulfonylurea (B), and benzoylphenyl-urea (C) compounds.

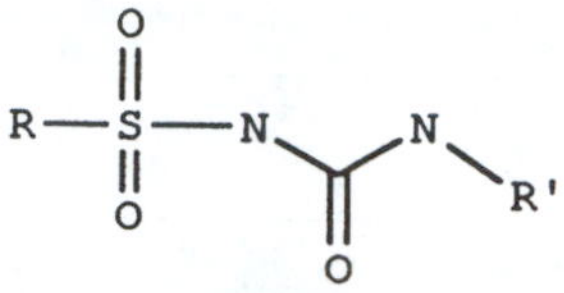

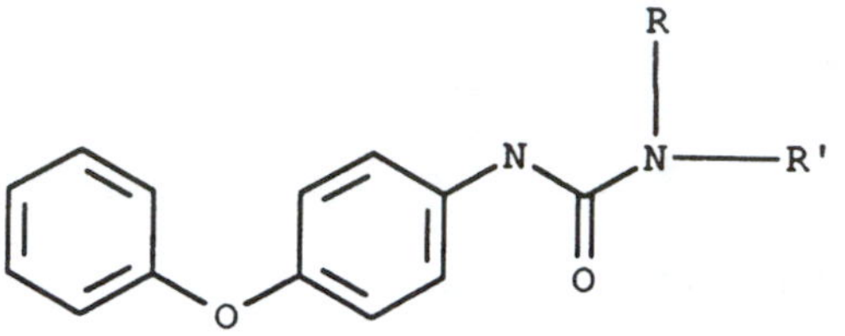

1970s by researchers in the Netherlands (1). A number of benzoylphenylurea compounds have been developed for this use, and have a number of advantages over broad-spectrum organochlorine or organophosphate insecticides (1). Chief among these are their specificity for larval and juvenile stages, very low toxicity to vertebrates, and consistent performance in the field (1). Benzoylphenyl compounds are generally expected to show low environmental persistence and relatively low or temporary impacts on most non-target invertebrate species, especially beneficial species (e.g., pollinators) (1,5,6). A notable exception to this rule are aquatic invertebrates, which may be severely affected by exposure (5,6).

Mechanism of action and toxicology: phenylureas

Mechanism of action

Phenylurea herbicides are generally well-absorbed through the roots (but not foliage) and moved through xylem to the leaves where they disrupt photosynthesis (2,4). One exception is siduron, which does not affect photosynthesis but rather root growth (2). Disruption of photosynthesis occurs by binding of the herbicide to a critical site in the Photosystem II region of the chloroplasts, shutting down CO_2 fixation and energy production (4). Indirect production of reactive lipid peroxides contributes to a loss of membrane integrity and organelle function within the cell (4). Outward signs of these processes include yellowing or blanching of the leaves (foliar chlorosis) and browning (necrosis) due to changes in the chlorophyll and cell death (4).

Acute toxicity

Phenylureas exhibit a range of acute oral toxicities, ranging from moderately to practically nontoxic; reported acute oral rodent LD_{50} values range from a low of 644 mg/kg (in the case of tebuthiuron) to greater than 5000 mg/kg (in the case of fluometuron) (4,7–13). Via the dermal route, they are generally slightly toxic, with

Table 9.1 Examples of
Substituted Urea Compounds

Phenylureas
  Benzthiazuron
  Chlorbromuron
  Chlorotoluron
  Chloroxuron
  Daimuron
  Difenoxuron
  Dimefuron
  Diuron*
  Ethidimuron
  Fenuron
  Flufenoxuron
  Fluometuron*
  Forchlorfenuron
  Isouron
  Linuron*
  Monuron
  Neburon
  Siduron
  Tebuthiuron*
Sulfonylureas
  Bensulfuron
  Chlorimuron
  Chlorsulfuron
  Primisulfuron-methyl*
  Sulfometuron-methyl*
  Triasulfuron
Benzoylphenylureas
  Chlorfluazuron
  Diflubenzuron*
  Penfluron
  Trilumuron

Note: * indicates that a profile of this
    compound is included in
    this chapter.

reported dermal LD_{50} values above 5000 mg/kg in rabbits (4,7–12). Tebuthiuron again stands out, with a reported dermal LD_{50} of greater than 200 mg/kg in rabbits, indicating moderate toxicity via this route (4). Via the inhalation route, the phenylureas are generally slightly toxic, with reported 4-hour inhalation LC_{50} values of greater than 2 mg/L in rats (4,7–13). Most of them do not cause skin sensitization in guinea pigs or skin irritation in rabbits, but most do cause slight to mild eye irritation in rabbits (4,7–12).

Linuron has shown the capacity to cause skin sensitization in guinea pigs, and tebuthiuron caused slight skin irritation in rabbits (4).

The symptoms that accompany acute exposure to phenylurea compounds vary widely. For instance, acute exposure of rats to tebuthiuron by ingestion caused a loss of appetite, a lack of energy, and muscle incoordination (13), while similar exposure of rats to fluometuron caused muscle weakness, watery eyes, extreme exhaustion, and collapse (11). Fluometuron also has caused cholinesterase depression in guinea pigs exposed to 0.6 mg/L over a 2-hour period (11). Signs of diuron exposure in rats included depression of central nervous system activity (10). Whether these differences are due to variation in reporting the symptoms or due to differences in the way the compounds act in the organisms is currently unknown (14).

Chronic toxicity

Phenylurea compounds may cause skin sensitization with repeated or prolonged exposure over extended periods of time (4,8). Anemia, effects on red blood cell levels, or other changes in blood parameters have been seen in different test animals at various dose levels. These range from 2.75 mg/kg/day in rats over a 2-year test period (linuron) to greater than 50 mg/kg/day for the same time (tebuthiuron) (4,10–17).

Reproductive effects

For most members of the class, reproductive effects have not been consistently observed. No reproductive effects were seen in any of three studies of tebuthiuron at doses up to 180 mg/kg/day during gestation of pregnant rats (4). No reproductive effects were seen on rats given linuron at 12.5 mg/kg/day in a three-generation study, nor in rats given fluometuron at doses of 50 mg/kg/day during gestation, although some maternal toxicity was seen in the latter case (4,8). Another study of fluometuron using rabbits showed increases in fetal resorptions at doses of 50 mg/kg/day and higher on days 6 to 19 of gestation (17). Some maternal toxicity was also seen at the higher dose levels in this study (17).

Teratogenic effects

Some members of the phenylurea class have shown the capacity to cause developmental effects in animal tests. Diuron has caused irregularities in skeletal formation in rats at doses above 125 mg/kg/day over days 6 to 15 of gestation, and has also caused developmental effects in offspring of rabbits given 2000 mg/kg/day during gestation (4,8). Fluometuron also caused some secondary effects in rats and rabbits receiving 100 mg/kg/day during gestation (4,8).

Both linuron and tebuthiuron have not shown adverse effects on fetal development at relevant doses. Linuron was negative for teratogenicity at doses (administered during gestation) of 25 mg/kg/day in rabbits or approximately 6 mg/kg/day in rats (4,14). Tebuthiuron did not show any teratogenic capacity at 56 mg/kg/day in rats over three generations, at 180 mg/kg/day in rats during gestation, nor at 25 mg/kg/day in rabbits (4,8).

Mutagenic effects

The vast majority of mutagenicity assays and tests for genotoxicity performed using diuron, fluometuron, linuron, and tebuthiuron have not shown them to have the potential to cause these effects (4,17,18). They have included the Ames mutagenicity assay, tests for chromosomal aberration with Chinese hamster ovary cell cultures and mouse micronuclei, and tests for DNA repair inhibition in rat liver cell lines (4,17,18).

Carcinogenic effects

Some members of the class have been linked to increased tumor production in animal systems. Fluometuron has shown mixed results, causing increased liver tumors and leukemia at doses of 87 mg/kg/day in mice over 2 years, but not in rats at any dose tested (4,17). Diuron has not caused cancer in rats at low levels of exposure (10), nor has tebuthiuron at the highest doses tested in rats and mice (4). Linuron has shown the capacity to produce increased tumors of the liver in mice at 180 mg/kg/day, and of the testes in rats at doses of 72.5 mg/kg/day (4,10,14).

Organ toxicity

Animal studies have shown the blood-forming system, liver, pancreas, spleen, and kidneys to be potentially affected by exposure to phenylurea compounds.

Fate in humans and animals

Some members of the phenylurea class (e.g., linuron and tebuthiuron) are fairly readily absorbed through the gastrointestinal tract, whereas others are rather poorly absorbed (4,7,17). Those that are not readily absorbed are excreted via feces unchanged (4,17). Those which may be absorbed and distributed systemically typically undergo rapid transformation and elimination via urine, typically within a few days (4,8). Overall, most class members possess a low potential for bioaccumulation in food animals or humans.

Ecological effects

Effects on birds

Phenylurea compounds are typically slightly to practically nontoxic to birds, with reported acute oral LD_{50} values above 2000 mg/kg for bobwhite quail in most cases (4,8). The reported acute oral LD_{50} for diuron is 1730 mg/kg in bobwhite quail (4). The reported subchronic dietary LC_{50} values in pheasants, mallards, and Japanese quail are above 5000 ppm for most members of this chemical class (4,8).

Effects on aquatic organisms

Phenylurea compounds are slightly to moderately toxic to freshwater fish, and in most cases only slightly toxic to freshwater invertebrates. Reported 48- and/or 96-hour LC_{50} values for most members of the class range from 4 to greater than 40 mg/L in such species as rainbow trout, bluegill sunfish, carp, and catfish (4,8). Diuron is more toxic to aquatic invertebrates than most members of this class, with a reported 48-hour LC_{50} of 1 to 2.5 mg/L in *Daphnia* (4,8).

Effects on other organisms (non-target species)

Phenylurea compounds are generally nontoxic to bees (4,8). Some of them (e.g., tebuthiruon) may be less selective, and may impact non-target plant species (4).

Environmental fate

Breakdown in soil and groundwater

All of the phenylureas will remain active in the soil from a few months up to a year in some cases and are broken down by the action of soil microorganisms (2). The measured field half-lives for many members of the class range from 25 to over 360 days (19). They are generally poorly bound to most soils, with reported K_{oc} values of 80 to 500, indicating a moderate to high potential for migration (19). Phenylurea compounds show little tendency to evaporate from soil, and soil pH does not appear to significantly affect their adsorption (20). Many phenylurea compounds have been detected in groundwater (20).

Breakdown in water

Phenylureas are generally stable and long-lived in aqueous systems, and will not easily undergo hydrolysis except with extreme acidic or basic conditions (3,4,20). One example is fluometuron, which has been reported to have a half-life in aqueous systems of over 100 weeks, and is stable over the pH range of 1 to 13 (11). The primary means of breakdown in aquatic systems is via microbial systems, although they may also undergo photolysis (2,8). Conditions that limit the availability of oxygen for microbial aerobic breakdown of the phenylureas will increase their longevity in the water environment.

Breakdown in vegetation

Phenylureas are readily absorbed from roots and translocated to foliage through xylem; they are not readily absorbed through the foliage (4). In tolerant plants, they are rapidly metabolized by loss of methyl groups (N-demethylation) and addition of hydroxyl groups in their place (hydroxylation), thus forming compounds that do not adversely affect the plant like the unmetabolized herbicide (4). Plants incapable of detoxifying the phenylureas, or which process them only slowly, will suffer the adverse effects the herbicides cause (4).

Mechanism of action and toxicology: sulfonylureas

Mechanism of action

Sulfonylurea herbicides are also well-absorbed and translocated through the roots, as well as through the foliage (4). Some members of the class (e.g., sulfometuron-methyl) may show lesser translocation (4). These compounds inhibit a key enzyme in the biosynthesis of necessary branched-chain amino acids (e.g., leucine, isoleucine, and valine), and thereby limit the plant's ability to construct the proteins necessary to build more cells (4). Deprived of the ability to manufacture necessary proteins, growth (meristematic) regions suffer inhibited growth and survival.

Acute toxicity

Sulfonylurea compounds are, in general, practically nontoxic via the oral route, with reported rodent LD_{50} values of greater than 5000 mg/kg for most of the compounds (4,7,8). Via the dermal route, they are generally slightly toxic, with reported dermal LD_{50} values of greater than 2000 mg/kg in rabbits (8,15). They are, in general, slightly toxic via the inhalation route in rodents, with reported 4-hour inhalation LC_{50} values of greater than 5 mg/L (4,8,15). They generally test negative for skin sensitization in guinea pigs, although sulfometuron can cause slight skin irritation in rabbits, and both primisulfuron and sulfometuron can cause slight to mild eye irritation in rabbits (4,15).

Chronic toxicity

Much like the phenylureas, sulfonylurea compounds have caused decreased body weight gain, increased liver weights, and anemia-like conditions in test animals, but at generally higher dose levels. These effects were caused by primisulfuron-methyl in dogs at doses levels of 125 mg/kg/day over 1 year, and by sulfometuron-methyl at 25 mg/kg/day in the same animals for the same length of time (15,21). In rats, these effects occurred at much higher dose levels, 180 and 375 mg/kg/day, respectively, over 90 days (15,21). Studies of primisulfuron in rats and mice over 18 months showed disorders of the teeth and bones (4).

Reproductive effects

No reproductive effects due to sulfometuron-methyl were seen in rats or rabbits at 300 mg/kg/day in separate studies (4), but 300 mg/kg/day did cause decreased fecundity (litter size) in rats in another study (4). Primisulfuron-methyl has caused effects on the testes of males rats exposed to 250 mg/kg/day over two generations (15). These effects were also observed in other chronic studies in rats. It is not clear what the overall trend for this chemical class might be.

Teratogenic effects

Sulfometuron was not observed to cause teratogenic effects in rats or rabbits at doses of 300 mg/kg/day (21), but primisulfuron-methyl did cause delayed skeletal development and/or lack of bone formation in offspring of rats given doses of 100 mg/kg/day in two separate studies (15).

Mutagenic effects

Various mutagenicity and genotoxicity assays have not shown either primisulfuron-methyl or sulfometuron-methyl to be mutagenic (4,15,21).

Carcinogenic effects

Carcinogenicity assays for sulfometuron-methyl have not shown carcinogenic activity in rats at any dose tested or mice at doses up to 120 mg/kg/day (4). Primisulfuron-methyl has shown inconsistent results in carcinogenicity assays; doses of 180 mg/kg/day produced increased liver tumors in mice over 18 months in one study, but not in another (4,15).

Organ toxicity

Animal studies have shown the liver, kidneys, spleen, and blood-forming system to be potentially affected by exposure to sulfonylurea compounds.

Fate in humans and animals

Primisulfuron is not readily absorbed via the intestines, and is generally eliminated unchanged via the feces (15). Sulfometuron is readily absorbed through the intestine, but rapidly metabolized and eliminated within 28 to 40 hours, depending on the dose (22). Neither compound is expected to bioaccumulate in animal systems.

Ecological effects

Effects on birds

Sulfonylurea compounds are practically nontoxic to wildfowl, with reported subchronic dietary LC_{50} values greater than 2000 ppm in bobwhite quail and mallard ducks (4).

Effects on aquatic organisms

Sulfonylurea compounds are only slightly toxic to freshwater fish, with reported 96-hour LC_{50} values of greater than 10 mg/L in rainbow trout and bluegill sunfish (15,23). They are practically nontoxic to *Daphnia*, with 96-hour LC_{50} values reportedly greater than 100 mg/L (15,23). Sulfometuron has been implicated in some instances of mass fish kills, but it is not clear what its actual role might have been in those cases (24).

    Effects on other organisms (non-target species)
Sulfonylurea compounds are not considered toxic to bees (4).

Environmental fate

Breakdown in soil and groundwater

Sulfonylurea compounds are of low to moderate persistence in the soil environment, with reported field half-lives of 4 to 60 days (19,25). They are readily broken down by soil microbes, and more rapidly under well-oxygenated (aerobic) conditions (4,15,24). Degradation by sunlight may play a greater role in the breakdown of sulfometuron-methyl than primisulfuron-methyl (15,24). Ground cover, pH, and soil type have all been shown to influence the rate of disappearance of sulfometuron-methyl in several field studies (24,26,27). Since photodegradation is of lesser importance for primisulfuron, ground cover may not show very much effect on its loss from soils.

    The sulfonylurea compounds are poorly bound to most soils, and may be mobile (19,25), depending on pH and water solubility (4). At pH 7, solubility ranges from 70 mg/L for primisulfuron to 27.9 mg/L for chlorsulfuron. If the water has a pH of 5, the solubility of sulfometuron-methyl decreases to 8 mg/L. The half-lives of most of the sulfonylurea compounds in soil are 1 or 2 months (4,8). Though the sulfonylurea compounds are potentially mobile within the soil, field studies show that neither primisulfuron-methyl nor sulfometuron-methyl were significantly mobile beyond a depth of 3 inches (8,20,26,27).

Breakdown in surface water

Sulfometuron is relatively short-lived in the water environment, with reported field half-lives of several days to 1 or 2 months; primisulfuron may be more persistent (15,24). Both compounds undergo aerobic breakdown, and anaerobic conditions may lengthen their residence time in water (15,24).

Breakdown in vegetation

Primisulfuron-methyl is readily absorbed by both the roots and foliage of plants, and quickly translocated to all parts of the plant (4). In resistant plants, it is rapidly metabolized, mainly by hydroxylation of the phenyl and pyrimidinyl ring structures, and subsequent linkage to glucose and elimination (4). The metabolism of sulfometuron-methyl is not well-understood (4).

Mechanism of action and toxicology: benzoylphenylureas

Mechanism of action

Benzoylphenylurea compounds inhibit the ability of insects to synthesize an integral component of their exoskeleton cuticles, chitin (1,6). These protein and chitin shells are secreted by insect epidermal cells at regulated intervals within the insect lifespan, with most chitin production occurring during early and juvenile stages of development (6). Depending on the insect species and dose rates, benzoylphenylurea inhibition of chitin synthesis results in the inability of fertilized eggs to hatch properly, the ability of hatched larvae to secrete chitin, or the ability of juvenile stages to molt properly and advance to the next life stage (6). There are several schools of thought as to how the benzoylphenylureas may act to inhibit chitin synthesis and exoskeleton formation (6). It is thought by some researchers that the most plausible mechanism is that benzoylphenylureas inhibit chitin synthetase, the enzyme responsible for stringing together the N-acetylglucosamine (amino-substituted glucose sugar molecules) building blocks into the polymer to form chitin (6).

Relatively less is known about the toxicology and environmental fate of the members of this class other than diflubenzuron (1). Insofar as the structure of a chemical compound determines its activity in biological and environmental systems, diflubenzuron can be considered representative of the toxicological and environmental properties of other members of the benzoylphenylurea chemical class. Typically, the generic chemical structure may yield reliable predictive information about members of a general chemical class, but until actual data are gathered and analyzed for a specific compound, such predictions must be considered tentative in nature.

9.2   Individual profiles

9.2.1   Diflubenzuron

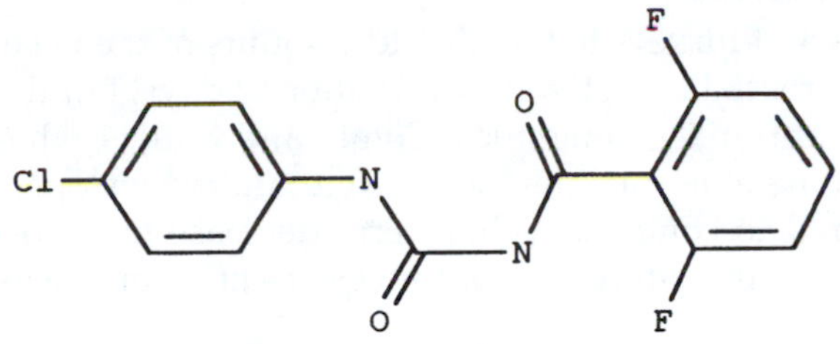

Figure 9.2 Diflubenzuron.

Trade or other names

Diflubenzuron is sold under the trade name Dimilin. Other trade names include DU112307, ENT-29054, Micromite, and OMS-1804.

Regulatory status

Some formulations of diflubenzuron may be classified as Restricted Use Pesticides (RUPs) in the U.S. RUPs may be purchased and used only by certified applicators. Diflubenzuron is classified as toxicity class III — slightly toxic. Products containing it bear the Signal Word CAUTION.

Introduction

Diflubenzuron is a benzoylphenylurea used on forest and field crops to selectively control insects and parasites. Principal target insect species are the gypsy moth, forest tent caterpiller, several evergreen-eating moths, and the boll weevil. It is also used as a larvae control chemical in mushroom operations and animal houses. Diflubenzuron is a stomach and contact poison. It acts by inhibiting the production of chitin, a compound that makes the outer covering of the insect hard and thus interferes with the formation of the insect's cuticle or shell. It is available as a suspension concentrate, wettable powder, or granules.

Toxicological effects

Acute toxicity

No overt signs of toxicity were observed in any of the acute studies conducted (16). The oral LD_{50} in rats and mice is greater than 4640 mg/kg, and the dermal LD_{50} is greater than 10,000 mg/kg in rats and greater than 4000 mg/kg in rabbits. It is nonirritating to skin and slightly irritating to eyes (8).

Chronic toxicity

Rats given moderate amounts of the compound for 2 years had enlarged spleens, while mice in a similar study had liver and spleen enlargement at slightly lower levels of exposure. In a study with cats fed over a wide range of doses for 21 days, all of the females had dose-related blood chemistry changes at low doses, and the males exhibited changes at dose levels that were slightly higher (9). The changes were reversible. The chemistry changes were associated with the formation of met-hemoglobin, a form of hemoglobin that is unable to carry oxygen.

Reproductive effects

Day-old ducks and turkeys fed moderate amounts of the pesticide in their diets for 90 days had decreased testosterone levels after 42 days, but this did not occur in chickens and pheasants in the same study. Combs and wattles, which reflect hormone activity, showed some abnormalities. Some were underdeveloped and others more developed compared to controls. A short-term decrease in testosterone levels was shown in the sexually immature rats, but no clear-cut change was shown in young bull calves (16).

A three-generation study on rats at low doses showed no effect on mating performance. It does not appear that diflubenzuron has a significant effect on reproduction (16,28).

Teratogenic effects

Diflubenzuron does not appear to be teratogenic. Newborn rats and rabbits did not develop any birth defects after their mothers were exposed to low levels of diflubenzuron (1 to 4 mg/kg/day) on days 6 to 18 of gestation (16,28).

Mutagenic effects

Extensive testing on mammalian bacterial cells shows that diflubenzuron is not mutagenic (16,28).

Carcinogenic effects

Rats fed diets containing low to moderate amounts of diflubenzuron daily for 2 years had no increase in the number of new or abnormal tissue growths or lesions. Mice fed low doses for 80 weeks showed no significant tumor development. Other studies on both species at higher levels were also negative for malignant tumors (16). Diflubenzuron does not appear to be carcinogenic.

Organ toxicity

Animal studies have shown the liver and spleen to be target organs.

Fate in humans and animals

Intestinal absorption in mammals decreases with increasing dose levels (28). For example, in rats, the total excretion in urine and bile decreased from about 50% of the dose at 4 mg/kg to only 4% at 900 mg/kg. Mice showed similar results.

A cow given 10 mg/kg orally, eliminated almost all of the product over a 4-day period. There were only minute amounts of the pesticide in the milk. The chemical is not degraded in the digestive tract, but that which is absorbed by the gut is completely broken down before excretion (16). Rabbits' skin absorbed only very small amounts, all of which was recovered in the urine.

Chickens excreted almost all of an oral dose in 13 days. Their eggs had low levels of pesticide residues (0.3 to 0.6 mg/kg) from day 9 to the end of the 9-week study. Body tissues (non-fatty) do not retain diflubenzuron (16).

Ecological effects

Effects on birds

Diflubenzuron is practically nontoxic to wild birds. Bobwhite quail and mallard ducks both have an 8-day dietary LC_{50} of greater than 4640 ppm (4,7,8).

Effects on aquatic organisms

Diflubenzuron is practically nontoxic to fish and aquatic invertebrates. The LC_{50} values (96-hour) for diflubenzuron in various fish are: bluegill sunfish, 660 mg/L; rainbow trout, 240 mg/L; saltwater minnow, 255 mg/L; and channel catfish, 180 mg/L. In oyster larvae and juveniles EC_{50} values were 130 and 250 mg/L, respectively (4,7,8). Arthropods are most susceptible in the premolting stage. For instance, fiddler crabs, exposed for as little as 1 week at levels up to 0.05 mg/L exhibited limb regeneration effects (29). Fish tissue can show some traces of the metabolites when water is contaminated with diflubenzuron; however, tissue concentrations decline steadily with time in clean water.

Effects on other organisms (non-target species)

The compound is nontoxic to bees (4).

Environmental fate

Breakdown in soil and groundwater

Diflubenzuron has a low persistence in soil. The rate of degradation in soil is strongly dependent on the particle size of the diflubenzuron (12). It is rapidly degraded by microbial processes. The half-life in soil is 3 to 4 days (19). Under field conditions, diflubenzuron has very low mobility (8,19).

Breakdown in water

In sterilized water (no microbes), there appears to be little degradation under neutral or acidic conditions. However, under field conditions, it is degraded rapidly. Residues could not be detected in field water 72 hours after an application of 110 g/ha. Other studies suggest a half-life of 1 to 3 weeks (4,7,8).

Breakdown in vegetation

Very little diflubenzuron is absorbed, metabolized, or translocated in plants. Residues on crops such as apples have a half-life of 5 to 10 weeks. The half-life in oak leaf litter is 6 to 9 months (4,7,8).

Physical properties

Diflubenzuron is a white crystalline solid (7).

Chemical name: 1-(4-chlorophenyl)-3-(2,6-difluorobenzoyl)urea (7)
CAS #: 35367-38-5
Molecular weight: 310.7 (7)
Solubility in water: 0.14 mg/L @ 20°C, insoluble (7)
Solubility in other solvents: DMSO s.; acetone s.s.; methanol s.
Melting point: 210–230°C (technical with decomposition) (7)
Vapor pressure: <0.033 mPa @ 50°C (7)
Partition coefficient (octanol/water): Not available
Adsorption coefficient: 10,000 (19)

Exposure guidelines

ADI: 0.02 mg/kg/day (30)
MCL: Not available
RfD: 0.02 mg/kg/day (31)
PEL: Not available

Basic manufacturer

Uniroyal Chemical Co., Inc.
Benson Road
Middlebury, CT 06749
Telephone: 203-573-2000
Emergency: 203-723-3670

9.2.2   *Diuron*

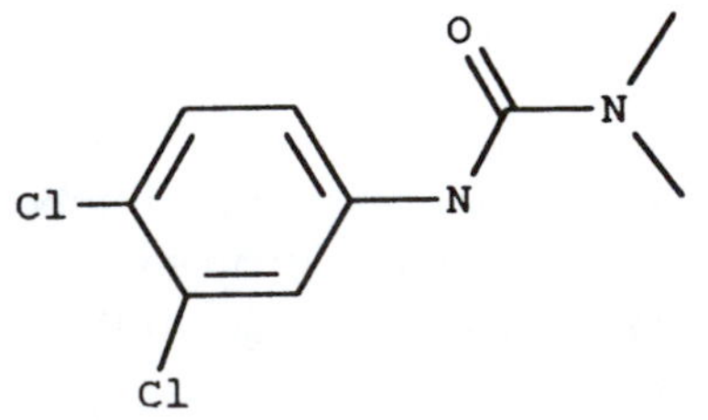

Figure 9.3 Diuron.

Trade or other names

Trade names for products containing diuron include Crisuron, Diater, Di-on, Direx, Karmex, and Unidron. It is often used in combination with other pesticides such as bromacil and hexazinone.

Regulatory status

Diuron is a General Use Pesticide (GUP). The U.S. EPA classifies it as toxicity class III — slightly toxic. However, products containing diuron bear the Signal Word WARNING because it can irritate the eyes and throat.

Introduction

Diuron is a substituted urea herbicide used to control a wide variety of annual and perennial broadleaf and grassy weeds as well as mosses. It is used on non-crop areas and many agricultural crops such as fruit, cotton, sugarcane, alfalfa, and wheat. Diuron works by inhibiting photosynthesis. It may be found in formulations as wettable powders and suspension concentrates.

Toxicological effects

Acute toxicity

Diuron is slightly toxic to mammals. The oral LD_{50} in rats is 3400 mg/kg. The dermal LD_{50} is greater than 2000 mg/kg (4,8). Some signs of central nervous system depression have been noted at high levels of diuron exposure. For humans, the only reported case of acute, oral exposure to the herbicide produced no significant symptoms or toxicity (4,8,10).

Chronic toxicity

Male rats given extremely high doses of diuron over a 2-week period showed changes in their spleen and bone marrow. Other chronic effects attributed to moderate to high doses of the pesticide over time included changes in blood chemistry, increased mortality, growth retardation, abnormal blood pigment, and anemia. When fed small amounts of diuron in food for 2 years, animal species showed no adverse effects (4,8).

Reproductive effects

Daily low doses of diuron fed to female rats through three successive generations caused significantly decreased body weight of offspring in the second and third litters. The fertility rate remained unaffected (8). It is unlikely that diuron will cause reproductive effects in humans at expected levels of exposure.

Teratogenic effects

Diuron is teratogenic at high doses. Administered to pregnant rats on days 6 through 15 of gestation, it produced no birth defects in the offspring at doses of up to 125 mg/kg/day. However, doses of 250 mg/kg/day caused wavy ribs, extra ribs, and delayed bone formation. There were also weight decreases in offspring at 500 mg/kg/day. There was no increase in the severity of the rib deformation at this higher dose (4,8). Pregnant mice given very high doses of diuron (nearly 2000 mg/kg/day) exhibited reproductive and embryotoxic effects. Developmental effects were found in their offspring (4,8).

Mutagenic effects

Diuron does not appear to be mutagenic. The majority of tests have shown that diuron does not produce mutations in animal cells or in bacterial cells (4,8).

Carcinogenic effects

Limited evidence indicates that low-level exposures to diuron does not cause cancer (10).

Organ toxicity

Low doses of diuron over extended periods of time can cause enlargement to the liver and the spleen (10).

Fate in humans and animals

Diuron is excreted in the feces and urine of test animals. Breakdown of the compound is similar in animals, plants, and soil. Cows fed very low doses of diuron in their diets had small amounts of residues in whole milk. Cattle fed small amounts accumulated low levels of diuron in fat and muscle, liver, and kidney (4,8).

Ecological effects

Effects on birds

Diuron is slightly toxic to birds. In bobwhite quail, the dietary LC_{50} is 1730 ppm. In Japanese quail and ring-necked pheasant, it is greater than 5000 ppm. The LC_{50} is approximately 5000 ppm in mallard ducks (4,8).

Effects on aquatic organisms

The LC_{50} (48-hour) values for diuron range from 4.3 to 42 mg/L in fish, and from 1 to 2.5 mg/L for aquatic invertebrates. The LC_{50} (96-hour) is 3.5 mg/L for rainbow trout (4,8). Thus, diuron is moderately toxic to fish and highly toxic to aquatic invertebrates.

Effects on other organisms (non-target species)

Diuron is nontoxic to bees (4).

Environmental fate

Breakdown in soil and groundwater

Diuron is moderately to highly persistent in soils. Residue half-lives are from 1 month to 1 year (19). Some pineapple fields contained residues 3 years after the last application. Mobility in the soil is related to organic matter and to the type of residue. The metabolites are less mobile than the parent compound (20).

In California, diuron has been found in groundwater in the 2- to 3-ppb range. It has also been found in Ontario groundwater where it has been linked with land applications (20).

Breakdown in water

Diuron is relatively stable in neutral water. Microbes are the primary agents in the degradation of diuron in aquatic environments (20).

Breakdown in vegetation

Diuron is readily absorbed through the root system of plants and less readily through the leaves and stems (4).

Physical properties

Diuron is a colorless crystalline compound in its pure form (7).

Chemical name: N-(3,4-dichlophenyl)-N,N-dimethyl urea (7)
CAS #: 330-54-1
Molecular weight: 233.10 (7)
Solubility in water: 42 mg/L @ 25°C (7)
Solubility in other solvents: s.s. in acetone, benzene, butyl stearate (7)
Melting point: 158–159°C (7)
Vapor pressure: 0.41 mPa @ 50°C (7)
Partition coefficient (octanol/water): Not available
Adsorption coefficient: 480 (19)

Exposure guidelines

ADI: Not available
HA: 0.01 mg/L (lifetime) (32)
RfD: 0.002 mg/kg/day (31)
TLV: 10 mg/m^3 (8-hour) (31)

Basic manufacturer

DuPont Agricultural Products
Walker's Mill, Barley Mill Plaza
P.O. Box 80038
Wilmington, DE 19880-0038
Telephone: 800-441-7515
Emergency: 800-441-3637

9.2.3   Fluometuron

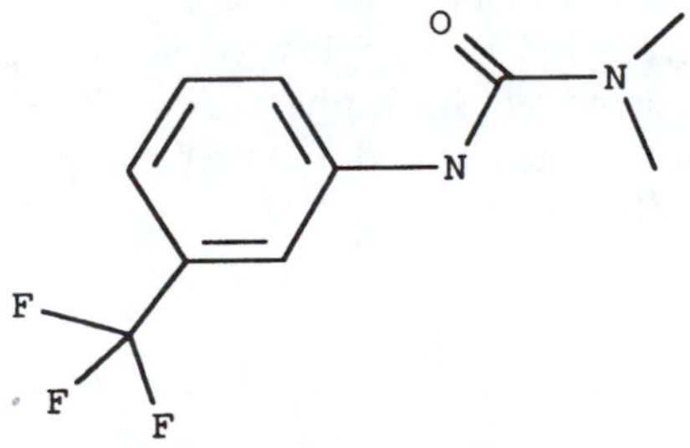

Figure 9.4 Fluometuron.

Trade or other names

Trade names include C-2059, Ciba-2059, Cotoran, Cotorex, Cottonex, Flo-Met, Higalcoton, Lanex, and Pakhtaran.

Regulatory status

Fluometuron is a practically nontoxic compound in EPA toxicity class II due to its potential to cause skin sensitization. Labels of fluometuron products must bear the Signal Word WARNING. Fluometuron is a General Use Pesticide (GUP).

Introduction

Fluometuron is a selective herbicide that acts on susceptible plants by inhibiting photosynthesis. Fluometuron is registered by the EPA exclusively for use on cotton and sugarcane. It can be applied pre-emergence, for weed control before planting, or post-emergence, after target crops and weeds come up, and may have residual activity for several months. Fluometuron is available in liquid, dry flowable, and wettable powder formulations.

Toxicological effects

Acute toxicity

Fluometuron is practically nontoxic by ingestion, with a reported oral LD_{50} of 6416 to 8900 mg/kg in rats (4,8). Via the dermal route, it is also practically nontoxic; the dermal LD_{50} is greater than 2000 mg/kg in rats, and greater than 10,000 mg/kg in rabbits (4,7). Fluometuron is a mild skin irritant and causes skin sensitization in guinea pigs. It may cause corneal opacity in test animals (33). It is irritating to the mucous membrane lining the skin, gastrointestinal tract, and respiratory system. The inhalation LC_{50} in rats is greater than 2 mg/L, indicating moderate to low toxicity by this route (4).

While there have been no reports of cases of fluometuron poisoning in humans, this herbicide is considered a mild inhibitor of cholinesterase. Cholinesterase inhibition was observed in guinea pigs exposed by inhalation to about 0.6 mg/L for 2 hours (33). Examination of rats used for LD_{50} testing revealed increased brain weight (17). Other symptoms of fluometuron poisoning in rats include muscular weakness, tearing or watery eyes, extreme exhaustion, and collapse (17).

Chronic toxicity

Rats were fed 7.5, 75, or 750 mg/kg/day for 90 days. At the highest dose, decreased body weight and congestion in the spleen, adrenals, liver, and kidneys, as well as abnormalities in red blood cells were evident (8,17). When doses of 1.5, 15, or 150 mg/kg/day were fed to puppies for 90 days, congestion of the liver, kidneys, and spleen occurred at the highest dose. No effects were seen at 15 mg/kg/day (8,17). Prolonged or repeated exposure to fluometuron may cause conjunctivitis or skin sensitization (4,8).

Reproductive effects

There were no reproductive effects due to fluometuron seen in pregnant rats given doses as high as 50 mg/kg/day during gestation, even though toxic effects in the mother were observed (4,8). Pregnant rabbits were given doses of 50, 500, or 1000 mg/kg/day by stomach tube during days 6 through 19 of gestation. An increase in the number of resorbed fetuses was found at all treatment doses. Reduction in maternal body weight and food consumption occurred at doses of 500 and 1000 mg/kg/day (17). The evidence indicates that fluometuron will not cause reproductive effects in humans at expected levels of exposure.

Teratogenic effects

Some secondary developmental effects were seen in the progeny of rats and rabbits receiving 100 mg/kg/day during gestation (4,8). These higher dose data indicate that teratogenic effects are not likely in humans at expected exposure levels.

Mutagenic effects

In various tests for mutagenicity and genotoxicity, fluometuron has not shown activity. These include the Ames mutagenicity assay, the Chinese hamster ovary cell culture assay for chromosome aberration, and DNA repair inhibition tests in rat liver and human fibroblast cell lines (4). It reportedly did show some interference with DNA synthesis in the testes of mice given a single oral dose of 2000 mg/kg (17). Based on these studies, fluometuron does not appear to be mutagenic.

Carcinogenic effects

Mice given oral doses of 87 mg/kg/day for 2 years showed evidence of liver tumors and leukemia, a condition characterized by uncontrolled growth in the number of white blood cells (4). Another study showed increased liver-cell tumor incidence in male mice, but carcinogenic effects were not observed in female mice or in rats of either sex (8,17). The available evidence is inconclusive, but suggests that carcinogenic effects in humans are not likely.

Organ toxicity

Target organs of fluometuron as determined in animal studies include brain, spleen, adrenals, liver, and kidneys, and red blood cells.

Fate in humans and animals

Fluometuron is absorbed only slowly into the body from the gastrointestinal tract. At 72 hours after rats were given oral doses of 50 mg/kg fluometuron, 15% of the dose was excreted in the urine and 49% was excreted unchanged in the feces (17). At the same time, fluometuron or its metabolites were detected in the rats' livers, kidneys, adrenal gland, pituitary gland, red blood cells, blood plasma, and spleen, with the highest concentration found in red blood cells (17).

Ecological effects

Effects on birds

Fluometuron is practically nontoxic to birds; the reported acute oral LD_{50} values for fluometuron are greater than 2150 mg/kg in bobwhite quail and 2974 mg/kg in mallard ducks (4,8). The reported 5- to 8-day dietary LC_{50} values for fluometuron were greater than 5620 ppm in bobwhite quail, 4500 ppm in mallard ducks, 3150 in ring-neck pheasant, and 4620 ppm in Japanese quail (4,8).

Effects on aquatic organisms

Fluometuron is slightly toxic to fish. The reported 96-hour LC_{50} of technical fluometuron is 30 mg/L in rainbow trout, 48 mg/L in bluegill sunfish, 170 mg/L in carp, and 55 mg/L in catfish (4,8). In catfish, the tissue concentrations in whole fish were 40 times that of the ambient water, indicating low capacity for bioaccumulation (8). The reported 48-hour LC_{50} for fluometuron in *Daphnia* (water flea) is 54 mg/L (4), indicating slight toxicity to aquatic invertebrates.

Effects on other organisms (non-target species)

Fluometuron is relatively nontoxic to bees (8).

Environmental fate

Breakdown in soil and groundwater

Fluometuron is moderately to highly persistent in the soil environment, with a reported field half-life of 12 to 171 days (19). A representative field half-life under most conditions is estimated to be 85 days (19). Breakdown in the soil environment occurs mainly through photodegradation when there is little rainfall after application, and by microbial breakdown otherwise. Fluometuron is soluble in water and poorly bound to most soils. This suggests that it would be mobile in most soils; but in field studies in California and Georgia, no residues were detected below 12 inches (4). In addition, fluometuron was not found in groundwater during a national survey (11).

Breakdown in water

Fluometuron may be highly persistent in the water environment as well. The half-life of fluometuron in water is 110 to 144 weeks (11). It is stable at pH values ranging from 1 to 13, at 20°C (3). However, exposure of 10 ppm aqueous solutions of fluometuron to natural sunlight resulted in 88% decomposition in 3 days, with a half-life of 1.2 days (4).

Breakdown in vegetation

Fluometuron is more readily absorbed by roots from soil application than by leaves from foliar application (4). The addition of a surfactant or nonphytotoxic oil to spray solutions improves the absorption of fluometuron by leaves (4). The rate at which it is absorbed, translocated, and subsequently broken down, (or metabolized) differs with various plant species (4). An understanding of these differences is important in determining the tolerance or susceptibility of plants and weeds to this chemical.

Physical properties

Fluometuron is a white to tan powder or crystalline material with an amine-like odor (7).

> Chemical name: 1,1-dimethyl-3-(a,a,a-trifluoro-m-tolyl) urea (7)
> CAS #: 2164-17-2
> Molecular weight: 232.2 970 (7)
> Water solubility: 105 mg/L @ 20°C (7)
> Solubility in other solvents: s.s. in acetone, chloroform, methanol, hexane, and organic solvents (16); s.s. in hexane (7)
> Melting point: 163–164°C (7)
> Vapor pressure: 0.067 mPa @ 20°C (7)
> Partition coefficient (octanol/water): 171 (7)
> Adsorption coefficient: 100 (19)

Exposure guidelines

> ADI: Not available
> HA: 0.09 mg/L (lifetime) (17)

RfD: 0.00025 mg/kg/day (31)
PEL: Not available

Basic manufacturer

Ciba-Geigy Corporation
P.O. Box 18300
Greensboro, NC 27419-8300
Telephone: 800-334-9481
Emergency: 800-888-8372

9.2.4   *Linuron*

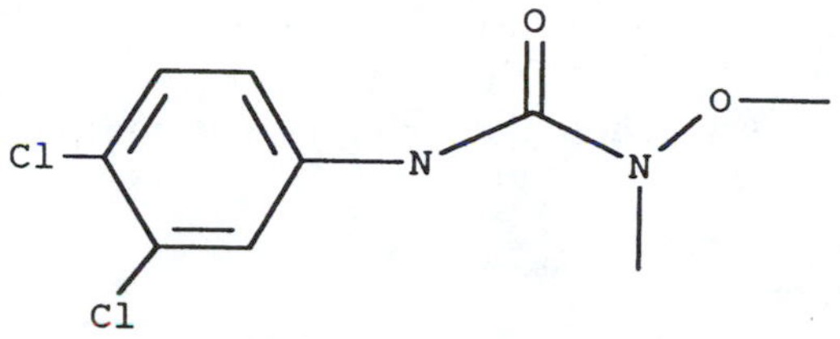

Figure 9.5 Linuron.

Trade or other names

Trade names for linuron include Afalon, Garnitan, Linex, Linorox, Linurex, Lorox, Premalin, Sarclex, and Sinuron. Linuron is often used in formulations with herbicides, insecticides, and fungicides.

Regulatory status

Linuron is a slightly toxic compound in EPA toxicity class III. It is a General Use Pesticide (GUP). Labels for products containing linuron must bear the Signal Word CAUTION.

Introduction

Linuron is a substituted urea pre- and post-emergence herbicide used to control annual and perennial broadleaf and grassy weeds on both crop and non-crop sites. It works by inhibiting photosynthesis in target weed plants. It is labeled for field and storehouse use in such crops as soybean, cotton, potato, corn, celery, parsnips, sorghum, asparagus, and carrots.

Toxicological effects

Acute toxicity

Linuron is of slight toxicity by ingestion, with reported oral LD_{50} values of 1200 to 1500 mg/kg in rats, and 2250 mg/kg in rabbits (4,8). The reported dermal LD_{50} in rabbits is greater than 5000 mg/kg (4,8). It has been reported to be a skin sensitizer in guinea pigs and an eye irritant in rabbits, but not a skin irritant in rabbits (4,8). The 4-hour inhalation LC_{50} is 6.15 mg/L, which indicates slight toxicity by this route (4).

Chronic toxicity

Skin sensitization was seen in guinea pigs repeatedly exposed (4). Alterations in red blood cells were seen in rats given 2.75 mg/kg/day over 2 years (4). Anemia was seen in dogs at doses above 6.25 mg/kg/day (4).

Reproductive effects

In a three-generation study, no reproductive effects were observed at doses of 12.5 mg/kg/day (4). These data suggest that reproductive effects are unlikely in humans at expected exposure levels.

Teratogenic effects

Pregnant rabbits fed high doses of linuron during the sensitive period of pregnancy had normal offspring at doses up to 25 mg/kg/day, even though maternal weight gain was reduced (4,14). In rats, doses of 6.25 mg/kg/day did not produce teratogenic effects. These data suggest that linuron is not likely to cause birth defects.

Mutagenic effects

Linuron caused mutations in one microbial assay (18). But in several other mutagencity and genotoxicity assays, including the Ames assay, *Escherichia coli* culture assay, Chinese hamster ovary cell culture assay, and whole animal studies, linuron showed no mutagenic or genotoxic activity (4,8,21). Thus, it appears that linuron is either nonmutagenic or slightly mutagenic.

Carcinogenic effects

Several animal studies of mice, rats, and dogs have shown that it produces nonmalignant liver and testicular tumors. In these studies, doses of 72.5 mg/kg/day in rats caused testicular adenomas, and 180 mg/kg/day in mice caused hepatocellular adenoma (4,11,14,18). These data are not sufficient to determine linuron's carcinogenicity to humans.

Organ toxicity

Rats and dogs fed linuron for 2 years had detectable residues of linuron in their blood, fat, kidney, and spleen, but these did not seem to be associated with adverse effects (4,18).

Fate in humans and animals

In rats, linuron breaks down completely after passing through the liver (4,8). It is thus unlikely to bioaccumulate in mammalian systems.

Ecological effects

Effects on birds

Linuron is slightly toxic to birds; the reported 5- to 8-day dietary LC_{50} values are greater than 5000 ppm in Japanese quail, 3000 ppm in mallard ducks, and 3500 ppm in pheasants (4,8).

Effects on aquatic organisms

Linuron is slightly toxic to fish and aquatic invertebrate species. The reported LC_{50} for linuron in trout and bluegill is 16 mg/L (8,11). The median threshold levels (i.e., levels at which adverse, sublethal effects were apparent in 50% of the test animals) are greater than 40 mg/L in crawfish and tadpoles exposed over a 48-hour period (4).

Effects on other organisms (non-target species)

Linuron is nontoxic to bees (7).

Environmental fate

Breakdown in soil and groundwater

Linuron is moderately persistent in soils, with a field half-life of 30 to 150 days in various soils and under various conditions (19,34). A representative field half-life is estimated to be approximately 60 days (12). Microbial degradation is the major process by which linuron is lost from soils; photodegradation and volatilization are not important contributors to its breakdown (4). The metabolites of linuron (3,4-dichloroaniline and carbon dioxide) are less toxic than linuron (35).

Linuron is moderately bound to soil and is soluble in water (19). Losses may occur through transport of linuron in runoff water and on suspended colloidal matter. Linuron has been found at very low concentrations in well and groundwater samples in Georgia, Missouri, Virginia, and Wisconsin (8).

Breakdown in water

Linuron is slightly to moderately soluble in water, and is not readily broken down in water (8).

Breakdown in vegetation

Linuron is more readily absorbed by roots from soil application than by leaves from foliar application (4). The rate at which it is absorbed, translocated, and subsequently broken down (or metabolized) differs with various plant species (4).

Physical properties

Linuron is an odorless, white crystalline solid (7).

Chemical name: 3-(3,4-dichlorophenyl)-1-methoxy-1-methylurea (7)
CAS #: 330-55-2
Molecular weight: 249.11 (7)
Solubility in water: 81 mg/L @ 25°C, slightly soluble (7)
Solubility in solvents: s.s. in aliphatic hydrocarbons; m.s. in ethanol; s. in acetone (7)
Melting point: 93–94°C (7)
Vapor pressure: 2 mPa @ 24°C (7)
Partition coefficient (octanol/water): 1010 (4)
Adsorption coefficient: 400 (19)

Exposure guidelines

ADI: Not available
HA: Not available
RfD: 0.002 mg/kg/day (31)
PEL: Not available

Basic manufacturer

DuPont Agricultural Products
Walker's Mill, Barley Mill Plaza
P.O. Box 80038
Wilmington, DE 19880-0038
Telephone: 800-441-7515
Emergency: 800–441–3637

9.2.5 *Primisulfuron-methyl*

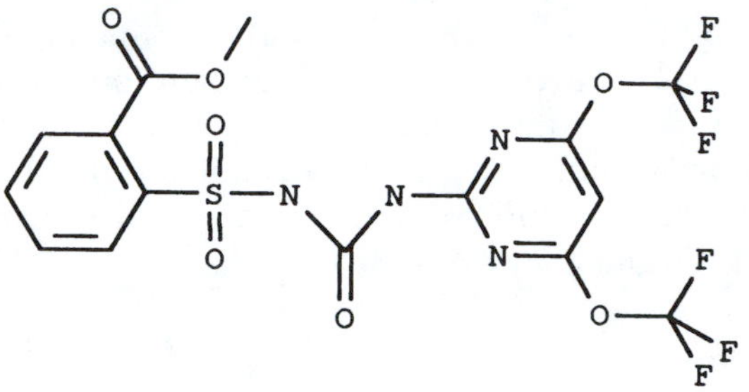

Figure 9.6 Primisulfuron-methyl.

Trade or other names

Trade names for primisulfuron-methyl include Beacon, CGAA 136872, Rifle, and Tell. Primisulfuron-methyl may be found in mixes with other herbicides such as 2,4-D, dicamba, cyanazine, bromoxynil, and atrazine.

Regulatory status

Primisulfuron-methyl is a practically nontoxic compound in EPA toxicity class IV. Labels for products containing it must bear the Signal Word CAUTION. Primisulfuron-methyl is a General Use Pesticide (GUP).

Introduction

Primisulfuron-methyl is a selective post-emergence herbicide used to control grassy and broad-leaved weeds in crop and non-crop applications. It is the methyl ester of primisulfuron, and is similar in its toxicological and environmental fate characteristics. Primisulfuron-methyl is available in wettable powders and water-dispersible granules in water-soluble packets.

Toxicological effects

Acute toxicity

Primisulfuron-methyl is practically nontoxic by ingestion, with a reported oral LD_{50} of greater than 5050 mg/kg in rats (4,8). Via the dermal route, it is moderately to practically nontoxic, with a dermal LD_{50} of greater than 2010 mg/kg in rats and rabbits (4,7). Slight skin irritation was observed in rabbits after dermal application of primisulfuron-methyl, but it did not cause skin sensitization in male guinea pigs (4,15). In rabbits, primisulfuron-methyl caused slight eye irritation that cleared within 3 days after contact (15). The 4-hour inhalation LC_{50} for primisulfuron-methyl is greater than 4.8 mg/L in rats, indicating slight toxicity by this route (4).

Chronic toxicity

Doses of 125 mg/kg/day administered in the diet to dogs over a 1-year period produced decreased body weight gain, anemia, increased liver weight, and thyroid hyperplasia (abnormal growth) (15). Rats fed dietary doses of about 180 mg/kg/day over 90 days showed effects similar to those noted in dogs, as well as spleen weight increases (24). In another study, doses of 480 mg/kg/day in rats over 18 months produced increased incidence of tooth disorders, chronic nephritis (kidney damage), and testicular atrophy (4). In two 18-month studies in mice, testicular atrophy, chronic nephritis, and increased tooth and bone disorders were seen at doses of 180 and 360 mg/kg/day (4).

Reproductive effects

Changes in the function of the testes were noted in rats fed high doses (250 mg/kg/day) of primisulfuron-methyl over two generations. There was also a decrease in the body weight of the offspring. No compound-related effects on reproduction were noted at doses below 50 mg/kg/day (15). Testicular atrophy was seen in rats in chronic studies (see above), which could result in reproductive effects. The available data suggest that reproductive effects in humans due to primisulfuron are not likely under normal circumstances.

Teratogenic effects

No teratological effects were seen in offspring of rabbits given doses up to 600 mg/kg/day. In one study of rats, delayed skeletal development and lack of ossification was seen in offspring of pregnant rats given doses of 500 mg/kg/day, while in another, 100 mg/kg/day produced incomplete ossification of the pubic bone (15). The available evidence suggests that primisulfuron-methyl is not teratogenic except at very high doses.

Mutagenic effects

Primisulfuron-methyl did not cause mutations in extensive testing. These tests included the Ames assay with and without metabolic activation, the Chinese hamster ovary cell culture, chromosomal aberration assay, and the rat liver cell unscheduled DNA synthesis assay (4,15). This indicates that this compound is not mutagenic.

Carcinogenic effects

In an 18-month study, mice fed doses of 180 mg/kg/day showed increased liver tumors, but this same dose level failed to produce the same effect in the same species in another investigation over the same time period (4,15). No carcinogenic activity was seen in rats fed up to about 450 to 500 mg/kg/day (4). The available data suggest that primisulfuron is not carcinogenic.

Organ toxicity

Target organs identified in animal studies include the liver, kidneys, spleen, and testes, as well as the skeleton.

Fate in humans and animals

Nearly all of a single dose (level not noted) of primisulfuron-methyl fed to rats was excreted unchanged in the feces and urine within 9 days (15). Relatively low concentrations were identifiable in the feces and urine of rats, goats, and chickens fed the compound (15).

Ecological effects

Effects on birds

Primisulfuron-methyl is practically nontoxic to wildfowl. Both bobwhite quail and mallard ducks show a high tolerance for the compound. The 5-day dietary LC_{50} for primisulfuron-methyl in both these species was in excess of 2150 ppm (4). In addition, mallards fed moderate amounts of the compound showed no adverse effects on reproduction at the highest dose tested (500 ppm in their diet) (15).

Effects on aquatic organisms

Primisulfuron-methyl is only slightly toxic to freshwater fish, aquatic organisms and marine (estuarine) shrimp. The reported 96-hour LC_{50} values are greater than 48 mg/L in bluegill, and greater than 13 mg/L in rainbow trout (4,15). The compound is practically nontoxic to the freshwater invertebrate *Daphnia magna* (15).

Effects on other organisms (non-target species)

Primisulfuron-methyl is nontoxic to honeybees (5,6).

Environmental fate

Breakdown in soil and groundwater

Primisulfuron-methyl is of low to moderate persistence in the soil environment, with a field half-life of 4 to 60 days. A representative value is estimated to be about 30 days. Aerobic conditions enhance the breakdown in soils (4,15). Losses due to volatilization and photodegradation are negligible (4). Acidic conditions will accelerate the breakdown process.

Primisulfuron-methyl is poorly sorbed to most soils, is soluble in water, and thus may be mobile (25). However, in field tests, no primisulfuron-methyl was detected below 9 inches of the surface (15).

Breakdown in water

Primisulfuron is resistent to hydrolysis in alkaline and neutral solutions (15). Anaerobic conditions will increase persistence. A half-life of 22 days at pH 5 has been reported (7).

Breakdown in vegetation

It is rapidly absorbed by plants and translocated throughout the plant roots and foliage (4). The herbicide works by blocking cell growth in the active growing regions of the plant (meristems) and by blocking photosynthesis (15). Primisulfuron-methyl application to crops grown previously on the same land resulted in detectable amounts in wheat, soybeans, sugar beets, corn, and lettuce (4).

Physical properties

Primisulfuron-methyl is a colorless, crystalline solid under normal conditions (7).

Chemical name: 2-[4,6-bis(difluoromethoxy)pyrimidin-2-ylcarbamoylsulfa-moyl]benzoic acid (7)
CAS #: 113036-87-6
Molecular weight: 468.3 (7)
Solubility in water: 70 mg/L @ 20°C (7)
Solubility in solvents: s.s. acetone, cyclohexane, isopropanol, methanol, and xylene (7)
Melting point: 203.1°C (7)
Vapor pressure: 0.000001 mPa (7)
Partition coefficient (octanol/water): 1.58 (7)
Adsorption coefficient: 50 (estimate) (25)

Exposure guidelines

ADI: Not available
HA: Not available
RfD: Not available
PEL: Not available

Basic manufacturer

Ciba-Geigy Corp.
P.O. Box 18300
Greensboro, NC 27419-8300
Telephone: 800-334-9481
Emergency: 800-888-8372

9.2.6   *Sulfometuron-methyl*

Trade or other names

Trade names for products containing sulfometuron-methyl include Oust Weed Killer and DPX 5648.

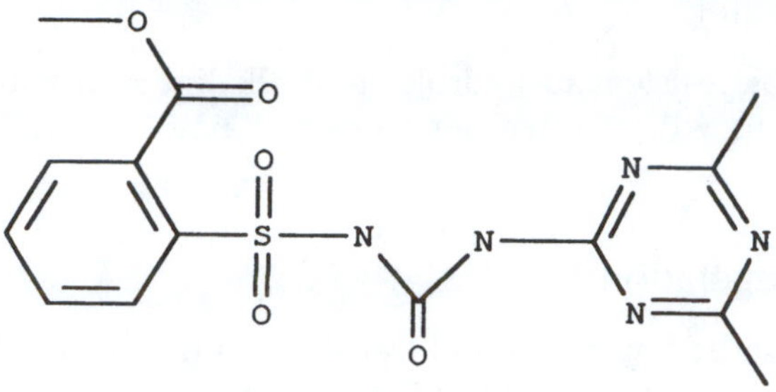

Figure 9.7 Sulfometuron-methyl.

Regulatory status

Sulfometuron-methyl is a General Use Pesticide (GUP). It is EPA toxicity class III — slightly toxic. Products containing this compound require the Signal Word CAUTION on their labels.

Introduction

Sulfometuron-methyl is a broad-spectrum sulfonylurea herbicide. It is used for the control of annual and perennial grasses and broad-leaved weeds in non-crop land. It also has forestry applications where it is used to control woody tree species. It is applied either post-emergent or pre-emergent. It works by blocking cell division in the active growing regions of stem and root tips (meristematic tissue).

Toxicological effects

Acute toxicity

Sulfometuron-methyl's acute oral toxicity is very low. The LD_{50} of sulfometuron-methyl in rats is greater than 5000 mg/kg (4,8). One study showed an LD_{50} greater than 17,000 mg/kg (8). The acute dermal toxicity of the compound is also low. The LD_{50} values for exposure through the skin ranges from over 2000 mg/kg in female rabbits to over 8000 mg/kg in male rabbits (21). The technical compound is not a skin irritant or a skin sensitizer (21). It has mild eye irritant properties in rabbits (3).

The acute inhalation LC_{50} is above 5.3 mg/L in rats, indicating its slightly toxic nature by this route (4).

Chronic toxicity

Several toxic effects were seen with chronic exposure to sulfometuron-methyl in test animals. At doses of 25 mg/kg/day, dogs experienced reduced red blood cell counts and increased liver weight (4,21). In this study, dogs were fed the compound in their food for a year.

In two other studies conducted over 90 days, rats had increased white blood cell counts (leukocytes) and anemia only at the highest dose tested (375 mg/kg/day) (4,21). In a 2-year feeding study, no effects were noted below 7.5 mg/kg/day in rats (21).

Reproductive effects

In a 90-day reproductive effects study in rats, no reproductive effects were observed at doses of 300 mg/kg/day (21). Another study in rats showed decreased fecundity and body weight at 300 mg/kg/day (4). Studies of rabbits showed no fetotoxic effects at 300 mg/kg/day, the highest dose tested (21). Reproductive effects due to sulfometuron-methyl are not likely.

Teratogenic effects

No tetatogenic effects were observed in studies of rats and rabbits at doses of 300 mg/kg/day (21).

It is unlikely that sulfometuron-methyl is teratogenic.

Mutagenic effects

The compound was not mutagenic in a variety of assays conducted on Salmonella cells and Chinese hamster ovary cells (21). It is unlikely that the compound poses a mutagenic risk.

Carcinogenic effects

No carcinogenic effects have been detected in either rats or mice exposed to sulfometuron-methyl (4,21).

Organ toxicity

As was noted above, increased liver weight may result from chronic exposure. Damage to blood-forming agents may also occur (4,21).

Fate in humans and animals

Sulfometuron-methyl is readily absorbed through the gastrointestinal tract, and is rapidly broken down and removed from the organism. Half-lives of the compound in rats ranged from 28 to 40 hours, depending on the dose (16 and 3000 mg/kg, respectively). The compound did not accumulate in rats (22).

Ecological effects

Effects on birds

Sulfometuron-methyl is practically nontoxic to birds. The acute oral LD_{50} in mallards is greater than 5000 mg/kg (4,8). An 8-day dietary study with mallard ducks and bobwhite quail also showed LC_{50} values greater than 5000 ppm for both species (4).

Effects on aquatic organisms

The compound is slightly toxic to freshwater fish. Its LC_{50} in rainbow trout and bluegill sunfish is greater than 12.5 mg/L (23). While the compound may not present a significant threat to adult aquatic organisms, the embryo hatch stage of fathead minnow may be at particular risk from the presence of the compound (23). Fish kills have been associated with sulfometuron-methyl, but other causes have not been ruled out (24).

Sulfometuron-methyl is practically nontoxic to the water flea, *Daphnia magna*. Its LC_{50} in the water flea is greater than 125 mg/L for the technical material and greater than 1000 mg/L for dispersible granules (23).

No bioaccumulation has been detected (21).

Effects on other organisms (non-target species)

No data are currently available.

Environmental fate

Breakdown in soil and groundwater

Sulfometuron-methyl is of low to moderate persistence in the soil environment. It is broken down in soil by the action of microorganisms, by the chemical action of water (hydrolysis), and through the action of sunlight (photodegradation) (4,24). Reported field half-lives of sulfometuron-methyl range from 20 to 28 days (19). In several field dissipation studies, half of the initially applied amount of the compound remained for 1 to 3 weeks, depending on soil type, vegetation cover, and pH (24,26,27). Under anaerobic soil conditions, the compound persists slightly longer although the half-life is still rather short (up to 8 weeks) (24).

Sulfometuron-methyl does not bind strongly to soil (19) and is slightly soluble in water (19), but is rapidly degraded and does not appear to pose a threat to groundwater. Field study data indicated a majority of the parent compound stays within the top 3 inches of soil (23).

Breakdown in water

In well-aerated acidic water, the compound is broken down quickly. Reported field half-lives for sulfometuron-methyl in water vary from 1 to 3 days (23) to 2 months or more (24). Photolysis is generally less important than hydrolysis in its breakdown (24). Under nonoxygenated (anaerobic) conditions in water sediments, the compound had a half-life of several months (24).

Breakdown in vegetation

Because sulfometuron-methyl is toxic to a number of plants and is nonselective, the use of the compound on non-croplands, including rights-of-way and along ditch banks, may endanger both terrestrial and aquatic plant species (4,24).

Physical properties

Sulfometuron-methyl is an off-white or colorless solid compound. The compound is odorless (7).

> Chemical name: 2-(4,6-dimethylpyrimidin-2-ylcarbamoylsulfamoyl)benzoic acid (7)
> CAS #: 74222-97-2
> Molecular weight: 364.4 (7)
> Solubility in water: 70 mg/L @ 25°C (7)
> Solubility in other solvents: s. in acetone, acetonitrile, and ethanol; s.s. in xylene; all @ 25°C (7)
> Melting point: 203–205°C (7)
> Vapor pressure: 8 mPa @ 25°C (7)
> Partition coefficient (octanol/water): 0.31 @ pH 7 (7)
> Adsorption coefficient: 78 (19)

Exposure guidelines

> ADI: Not available
> HA: Not available
> RfD: Not available
> TLV: 5 mg/m^3 (8-hour) (8)

Basic manufacturer

DuPont Agricultural Products
Walker's Mill, Barley Mill Plaza
P.O. Box 80038
Wilmington, DE 19880-0038
Telephone: 800-441-7515
Emergency: 800-441-3637

9.2.7   Tebuthiuron

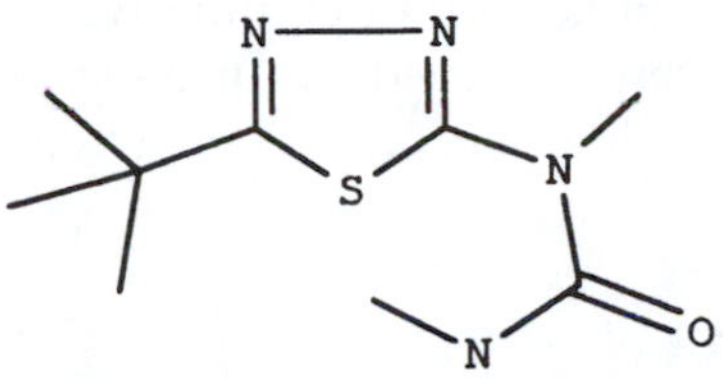

Figure 9.8 Tebuthiuron.

Trade or other names

Trade names for tebuthiuron include Brush, Bullet, Bushwacker, EL-103, Graslan, Herbec, Herbic, Perflan, Reclaim, Scrubmaster, Spike, Sprakil, and Tebusan.

Regulatory status

Tebuthiuron is a General Use Herbicide (GUP). It is EPA toxicity class III — slightly toxic. Products containing tebuthiuron must bear the Signal Word CAUTION.

Introduction

Tebuthiuron is a broad-spectrum herbicide used to control weeds in non-cropland areas, rangelands, rights-of-way, and industrial sites. It is effective on woody and herbaceous plants in grasslands and sugarcane. Weeds controlled by tebuthiuron include alfalfa, bluegrasses, chickweed, clover, dock, goldenrod, mullein, etc. Woody plants take a period of 2 to 3 years to be completely controlled.

Tebuthiuron is sprayed or spread dry on the soil surface, as granules or pellets, preferably just before or during the time of active weed growth. It is compatible with other herbicides.

Toxicological effects

Acute toxicity

Tebuthiuron has moderate to low toxicity in experimental animals when ingested. Reported oral LD_{50} values for tebuthiuron are 644 mg/kg in rats, 579 mg/kg in mice, 286 mg/kg in rabbits, greater than 200 mg/kg in cats, and greater than 500 mg/kg in dogs (4,8). Tebuthiuron is of slight to low toxicity by skin exposure. The dermal LD_{50} for tebuthiuron in rabbits is greater than 200 mg/kg (4). Neither skin irritation nor general overall intoxication were produced in rabbits that had 200 mg/kg of the material applied to their skin (4,8). Tebuthiuron did not induce sensitization or allergic reactions when tested on the skin of guinea pigs (4). Application of 67 mg herbicide in the eyes of rabbits produced short-term conjunctivitis, inflammation of the lining of the eye, but no irritation to other eye parts, the cornea, or the iris (4).

The inhalation by animals of 3.7 mg/L technical tebuthiuron for 4 hours did not cause toxicity.

Chronic toxicity

Decreases in body weight gain and red blood cell counts, along with minor effects on the pancreas, were seen in rats fed 125 mg/kg/day for 3 months (9). Exposure of rats to dietary doses of tebuthiuron as high as 80 mg/kg/day for 2 years was well tolerated, with no indication of cumulative toxicity or serious effects (4). Similarly, no toxic effects were observed in mice exposed to doses as high as 200 mg/kg/day for most of their lifetime, or in dogs given doses of 25 mg/kg/day for 1 year (4).

Reproductive effects

The reproductive capacity of rats fed dietary concentrations of tebuthiuron as high as 56 mg/kg/day was unimpaired through three successive generations, and no abnormalities were detected in either parents or offspring (4). Tebuthiuron administered to pregnant rabbits at doses as high as 25 mg/kg/day, and to rats at doses as high as 180 mg/kg/day, produced no adverse effects on either the mothers or the offspring (4). Based on these data, it is unlikely that tebuthiuron causes reproductive effects.

Teratogenic effects

No teratogenic effects were observed when rats were fed tebuthiuron at 180 mg/kg/day (4). A rabbit teratology study was also negative at 25 mg/kg/day, the highest dose tested (4). Based on these data, it is unlikely that tebuthiuron causes birth defects.

Mutagenic effects

The Ames mutagenicity asssay for tebuthiuron was negative, as were assays for structural chromosome aberrations using mouse micronuclei (4). Based on these data, it appears that tebuthiuron is mutagenic.

Carcinogenic effects

No tumor-related effects were observed in a 2-year rat feeding study at doses up to and including 80 mg/kg/day, the highest dose tested (4). A 2-year oncogenic study on mice was negative at 200 mg/kg/day, the highest dose tested (4). These data indicate that tebuthiuron is not carcinogenic.

Organ toxicity

Damage to the pancreas has been observed in animal studies as a result of exposure to tebuthiuron (36).

Fate in humans and animals

In rats, rabbits, dogs, mallards, and fish, tebuthiuron is readily absorbed into the bloodstream from the gastrointestinal tract, rapidly metabolized, and then excreted in the urine (4). Tests indicate that the herbicide is broken down and excreted within 72 hours, primarily as a variety of urinary metabolites (9).

Ecological effects

Effects on birds

Tebuthiuron is practically nontoxic to birds. The reported oral LD_{50} values are greater than 2500 mg/kg in both mallard ducks and bobwhite quail (4). A 30-day feeding of 1000 ppm tebuthiuron to hens had no effect (7).

Effects on aquatic organisms

Tebuthiuron is slightly to practically nontoxic to fish and other aquatic species (4,8). The reported 96-hour LC_{50} values are 87 to 144 mg/L in rainbow trout, and 87 to 112 mg/L in bluegill sunfish (4,8). The reported 96-hour LC_{50} values are greater than 160 mg/L in goldfish and fathead minnow (4,8). The 48-hour LC_{50} in *Daphnia magna*, an aquatic invertebrate, is 225 mg/L (4,8). The LC_{50} in fiddler crab is greater than 320 mg/L, and the LD_{50} in pink shrimp is more than 48 mg/L (4,8).

Effects on other organisms (non-target species)

Tebuthiuron is slightly toxic to bees, with a reported contact LD_{50} of 30 mg per bee (8).

Tebuthiuron may be harmful to non-target plants (4).

Environmental fate

Breakdown in soil and groundwater

Tebuthiuron is highly persistent in soil. Reported field half-lives are from 12 to 15 months in areas with over 40 inches annual rainfall, with longer half-lives expected in drier areas or in soils with high organic matter content (4). Tebuthiuron is broken down slowly in the soil through microbial degradation. Photodecomposition (or breakdown by sunlight) is negligible, as is volatilization (or evaporation from the soil surface) (4).

It is poorly bound to soil, suggesting high mobility. In field studies, however, little or no lateral movement has been seen in soils with appreciable clay or organic matter contents (4). Neither tebuthiuron nor its degradation products have been detected below the top 24 inches of soil in field studies (4). It was found in some groundwater samples in western states (e.g., Texas, California, Missouri, Oklahoma, and Washington) at levels up to 3.8 µg/L (4).

Breakdown in water

No degradation was observed in a 33-day study of photolysis of tebuthiuron in water (4).

Breakdown in vegetation

Tebuthiuron is readily absorbed through roots and translocated to other plant parts. It produces its effect by inhibiting photosynthesis, the process by which plants receive light from the sun and convert it into energy (4).

Physical properties

Tebuthiuron is an off-white to buff-colored crystalline solid with a pungent odor (7).

Chemical name: 1-(5-tert-butyl-1,3,4-thiadiazol-2-yl)-1,3-dimethylurea (7)
CAS #: 34014-18-1
Molecular weight: 228.31 (7)
Water solubility: 2500 mg/L at 25°C (7)
Solubility in other solvents: i.s. in benzene and hexane (7); s.s. in chloroform, methanol, acetone, and acetonitrile (7)
Melting point: 161.5 @ 164°C (with decomposition) (7)
Vapor pressure: 0.27 mPa @ 25°C (7)
Partition coefficient (octanol/water): 61 @ 25°C and pH 7 (7)
Adsorption coefficient: 80 (19)

Exposure guidelines

ADI: Not available
HA: 0.5 mg/L (36)
RfD: 0.07 mg/kg/day (31)
PEL: Not available

Basic manufacturer

DowElanco
9330 Zionsville Road
Indianapolis, IN 46268-1054
Telephone: 317-337-7344
Emergency: 800-258-3033

References

(1) Retnakaran, A. and Wright, J. E., Eds. Control of insect pests with benzoylphenyl urea. In *Chitin and Benzoylphenyl Ureas* (Series Entomologica Vol. 38), Wright, J. E. and Retnakaran, A., Eds. Kluwer Academic, Dordrecht, The Netherlands, 1987.

(2) Jordan, L. S. and Cudney, D. W. Herbicides. In *Fate of Pesticides in the Environment*, Biggar, J. W. and Seiber, J. N., Eds. Agricultural Experiment Station, Division of Agriculture and Natural Resources, University of California, CA, 1987.

(3) Fletcher, J. S., Pfleeger, T. G., and Ratsch, H. C. Potential environmental risks associated with the new sulfonylurea herbicides. *Environ. Sci. Technol.* 27, 2250–2252, 1993.

(4) Weed Science Society of America. *Herbicide Handbook*, 7th ed. Champaign, IL, 1994.

(5) U.S. Environmental Protection Agency. *Chemical Information Fact Sheet Number 68.1: Diflubenzuron.* Office of Pesticides and Toxic Substances, Washington, D.C., 1987.

(6) Grosscurt, A. C. and Jongsma, B. Mode of action and insecticidal properties of diflubenzuron. In *Chitin and Benzoylphenyl Ureas* (Series Entomologica Vol. 38), Wright, J. E. and Retnakaran, A., Eds. Kluwer Academic, Dordrecht, The Netherlands, 1987.

(7) Kidd, H. and James, D. R., Eds. *The Agrochemicals Handbook,* 3rd ed. Royal Society of Chemistry Information Services, Cambridge, U.K., 1991 (as updated).

(8) U.S. National Library of Medicine. *Hazardous Substances Data Bank.* Bethesda, MD, 1995.

(9) Gosselin, R. E., Smith, R. P., and Hodge, H. C. *Clinical Toxicology of Commercial Products,* 5th ed. Williams and Wilkins, Baltimore, MD, 1984.

(10) U.S. Environmental Protection Agency. *Chemical Information Fact Sheet Number 09: Diuron.* Office of Pesticides and Toxic Substances, Washington, D.C., 1983.

(11) U.S. Environmental Protection Agency. *Chemical Information Fact Sheet Number 88: Fluometuron.* Office of Pesticides and Toxic Substances, Washington, D.C., 1985.

(12) U.S. Environmental Protection Agency. *Chemical Information Fact Sheet Number 28: Linuron.* Office of Pesticides and Toxic Substances, Washington, D.C., 1984.

(13) U.S. Environmental Protection Agency. *Chemical Information Fact Sheet Number 137: Tebuthiuron.* Office of Pesticides and Toxic Substances, Washington, D.C., 1987.

(14) Wagner, S. L. *Clinical Toxicology of Agricultural Chemicals.* Oregon State University Environmental Health Sciences Center, Corvallis, OR, 1981.

(15) U.S. Environmental Protection Agency. *Chemical Information Fact Sheet Number 214: Primisulfuron-methyl.* Office of Pesticides and Toxic Substances, Washington, D.C., 1990.

(16) Food and Agriculture Organization of the United Nations. *FAO Plant Production and Protection Paper 42. Pesticide residues in food — 1981.* FAO, Geneva, Switzerland, 1981.

(17) U.S. Environmental Protection Agency. *Health Advisory Summary: Flometuron.* Office of Drinking Water, Washington, D.C., 1988.

(18) Epstein, S. S. and Legator, M. S. *The Mutagenicity of Pesticides.* MIT Press, Cambridge, MA, 1971.

(19) Wauchope, R. D., Buttler, T. M., Hornsby A. G., Augustijn-Beckers, P. W. M., and Burt, J. P. SCS/ARS/CES pesticide properties database for environmental decisionmaking. *Rev. Environ. Contam. Toxicol.* 123, 1–157, 1992.

(20) Howard, P. H., Ed. *Handbook of Environmental Fate and Exposure Data for Organic Chemicals.* Lewis, Boca Raton, FL, 1991.

(21) U.S. Environmental Protection Agency. *Toxicology One-Line Summary: Oust (MRID Nos. 244195, 245515).* Washington, D.C., 1990.

(22) E. I. DuPont de Nemours. *Metabolism of Sulfometuron Methyl in Rats.* Wilmington, DE, 1989.

(23) U.S. Environmental Protection Agency. *Oust Herbicide (Sulfometuron methyl).* Environmental Effects Branch, Washington, D.C., 1984.

(24) U.S. Environmental Protection Agency. *Pesticide Environmental Fate One-Liner Summaries: Sulfometuron methyl.* Environmental Fate and Effects Division, Washington, D.C., 1992.

(25) Augustijn-Beckers, P. W. M., Hornsby, A. G., and Wauchope, R. D. SCS/ARS/CES pesticide properties database for environmental decisionmaking. II. Additional compounds. *Rev. Environ. Contam. Toxicol.* 137, 1–82, 1994.

(26) Anderson, J. J. *Terrestrial Field Dissipation Study of 14C-DPX-T5648 in Delaware, North Carolina, and Mississippi.* E. I. DuPont de Nemours. Wilmington, DE, 1981.

(27) Trubey, R. K. *Field Soil Dissipation of Oust(R) Herbicide.* E. I. DuPont de Nemours, Wilmington, DE, 1991.

(28) Dost, F. N., Wagner, S. L., Witt, J. M., and Heumann, M. *Toxicological Evaluation of Dimilin (Diflubenzuron).* Oregon State University Extension Service, Corvallis, OR, 1985.

(29) Weis, J. S., Cohen S. R., and Kwiathowsi, J. K. Effects of diflubenzuron on limb regeneration and molting in the fiddler crab, *Uca pugilator. Aquat. Toxicol.* 10, 279–290, 1987.

(30) Lu, F. C. A review of the acceptable daily intakes of pesticides assessed by the World Health Organization. *Regul. Toxicol. Pharmacol.* 21, 351–364, 1995.

(31) U.S. Environmental Protection Agency. *Integrated Risk Information System Database,* Washington, D.C., 1995.

(32) U.S. Environmental Protection Agency. *Health Advisory Summary: Diuron.* Office of Drinking Water, Washington, D.C., 1988.

(33) Dreisbach, R. H. *Handbook of Poisoning: Prevention, Diagnosis and Treatment,* 11th ed. Lange Medical Publications, Los Altos, CA, 1983.

(34) Rao, P. S. C. and Davidson, J. M. Estimation of pesticide retention and transformation parameters required in nonpoint source pollution models. In *Environmental Impact of Nonpoint Source Pollution,* Overcash, M. R. and Davidson, J. M., Eds. Ann Arbor Science, Ann Arbor, MI, 1980.

(35) E. I. DuPont de Nemours. Product Information booklet on LOROX, Wilmington, DE, not dated.

(36) U.S. Environmental Protection Agency. *Health Advisory Summary: Tebuthiuron.* Office of Drinking Water, Washington, D.C., 1989.

chapter ten

Other pesticides

10.1 Chapter overview

The pesticide active ingredients covered in this chapter represent a wide array of compounds that do not lend themselves to easy classification. They are therefore organized according to their use; that is, fungicides, herbicides, insecticides, and others.

10.2 Fungicides

10.2.1 Benomyl

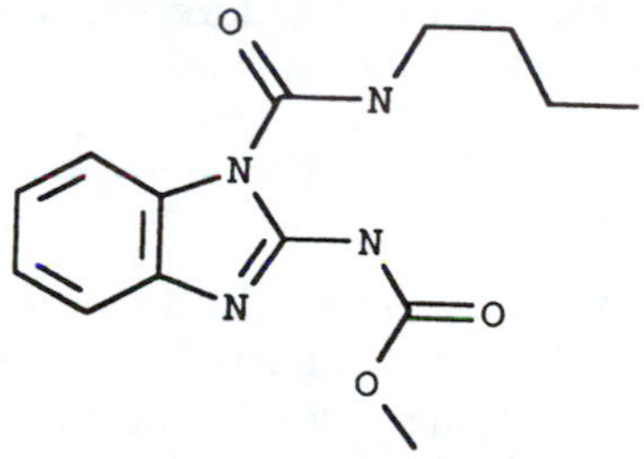

Figure 10.1 Benomyl.

Trade or other names

Commercial names for products containing benomyl include Agrocit, Benex, Benlate, Benosan, Fundazol, Fungidice 1991, and Tersan 1991. Benomyl is compatible with many other pesticides.

Regulatory status

Benomyl is a General Use Pesticide (GUP). The EPA categorizes it as toxicity class IV — practically nontoxic. Benomyl-containing products carry the Signal Word CAUTION.

Introduction

Benomyl is a systemic, benzimidazole fungicide that is selectively toxic to micro-organisms and to invertebrates, especially earthworms. It is used against a wide range of fungal diseases of field crops, fruits, nuts, ornamentals, mushrooms, and turf. Formulations include wettable powder, dry flowable powder, and dispersible granules.

Toxicological effects

Acute toxicity

Benomyl is of such a low acute toxicity to mammals that it has been impossible or impractical to administer doses large enough to firmly establish an LD_{50}. Thus, the LD_{50} is greater than 10,000 mg/kg in rats and greater than 3400 mg/kg in rabbits (using a 50% wettable powder formulation). Because of its high LD_{50}, there is a low risk for acute poisoning from this compound (1). Skin irritation may occur for workers exposed to benomyl. Skin reactions have also been seen in rats and guinea pigs. Most organisms can become sensitized to the compound as well.

Benomyl is readily absorbed into the body by inhaling the dust, but there are no reports of toxic effects to humans by this route of exposure. The inhalation LC_{50} in rats is greater than 2 mg/L (2).

Chronic toxicity

When rats were fed diets containing about 150 mg/kg/day for 2 years, no toxic effects were observed (3). Dogs fed benomyl in their diets for 3 months had no major toxic effects, but did show evidence of altered liver function at the highest dose (150 mg/kg). The damage progressed to more severely impaired liver function and liver cirrhosis after 2 years (6).

Reproductive effects

A three-generation study on rats showed no reproductive or lactational differences at a dose of 150 mg/kg/day administered in the diet (3).

In another study in rats, the testes were the most affected sites at relatively low doses of about 15 mg/kg/day. Male rats had decreased sperm counts, decreased testicular weight, and lower fertility rates. The animals recovered from these effects 70 days after feeding with the pesticide had stopped (3). Reproductive effects in humans are unlikely at expected exposure levels.

Teratogenic effects

Very high doses of benomyl can cause birth defects in test animals (4). Rats fed 150 mg/kg/day in the diet for three generations showed no birth defects. No teratogenicity was observed in another study of rats given 300 mg/kg/day on days 6 to 15 of gestation (4). At higher doses, some birth defects were noted, but they were accompanied by toxicity to the fetus (4). In another rat study where mothers were fed 1000 mg/kg/day for 4 months, the offspring showed a decrease in viability and fertility (1). These data suggest that benomyl is not likely to cause teratogenic effects under normal circumstances.

Mutagenic effects

Conflicting negative and positive results have been found in numerous mutagenicity assays. As a result, no conclusions about the mutagenicity of benomyl can be drawn (3).

Carcinogenic effects

Tumors in the livers of both male and female mice were observed in lifetime studies at doses of benomyl at 40 to 400 mg/kg/day. In a 2-year dietary study when albino rats were fed up to 2500 mg/kg/day benomyl, there were no significant adverse effects at any dose level attributable to benomyl (1).

Based on these data, it is not possible to determine the carcinogenicity of benomyl (5).

Organ toxicity

Target organs identified in animal studies included the liver and testes.

Fate in humans and animals

Benomyl's metabolism has been studied in the mouse, rat, rabbit, dog, sheep, and cow. Benomyl is rapidly broken down to carbendazim, and further to other compounds such as 5-hydroxy-2-benzimidazole carbamate (5-HBC), and then eliminated. In a rat study, benomyl, carbendazim (MBC), and 5-HBC were found in rat blood in the first 6 hours. After 18 hours, only 5-HBC was present. The urine contained about 40 to 70% of the dose, and the feces 20 to 45%. No residues were found in muscle or fat. Benomyl and its metabolites do not accumulate in tissues over long-term exposure periods (2,3). Carbendazim (MBC) and the parent compound benomyl have similar toxicological properties, but the former is not a skin sensitizer (2).

Ecological effects

Effects on birds

In bobwhite quail and mallard ducks, the 5-day dietary LC_{50} for benomyl is greater than 10,000 ppm. In redwing blackbirds, the LD_{50} value is 100 mg/kg, which indicates that benomyl is moderately toxic to this species (4).

Effects on aquatic organisms

Benomyl is highly to very highly toxic to fish. The order of susceptibility to benomyl for various fish species from least susceptible to most susceptible is: catfish, bluegill, rainbow trout, and goldfish. The LC_{50} values for the compound in fish are 0.05 to 14 mg/L in adults, and 0.006 mg/L in catfish fry (8). The main breakdown product, carbendazim, had the same order of toxicity as benomyl. Crayfish have an LC_{50} greater than 100 mg/L. The estimated bioconcentration factor (BCF) ranges from 159 in rainbow trout up to 460 in bluegill sunfish, indicating that benomyl does not tend to significantly concentrate in living tissue (8,9).

Effects on other organisms (non-target species)

A single application of benomyl to turfgrass can substantially reduce some soil-dwelling organisms. The compound is very lethal to earthworms at low concentrations over a long time period. The 7-day LC_{50} in earthworms is 1.7 mg/L, and the 14-day LC_{50} is 0.4 mg/L (6). Benomyl also decreases the mixing of soil and thatch. The effects last for up to 20 weeks (10). Benomyl is relatively nontoxic to bees (2).

Environmental fate

Breakdown in soil and groundwater

Benomyl is strongly bound to soil and does not dissolve in water in flooded rice fields to any significant extent (2,11). It is highly persistent. When applied to turf, it has a half-life of 3 to 6 months and, when applied to bare soil, the half-life is 6 to 12 months. Where four successive annual applications were applied, residues did not accumulate from one year to the next (6).

Breakdown in water

Benomyl completely degrades to carbendazim within several hours in acidic or neutral water. The half-life of carbendazim is 2 months (1).

Breakdown in vegetation

Since benomyl is a systemic fungicide, it is absorbed by plants. Once it is in the plant, it accumulates in veins and at the leaf margins (6). The metabolite carbendazim seems to be the fungicidally active agent. Benomyl residues are quite stable, with 48 to 97% remaining as the parent compound 21 to 23 days after application (6).

Physical properties

Benomyl is a tan crystalline solid compound. It has little or no odor (1).

Chemical name: methyl 1-[(butylamino)carbonyl]-H-benzimidazol-2-ylcarbamate (1)
CAS #: 17804-35-2
Molecular weight: 290.62 (1)
Solubility in water: 2 mg/L (1)
Solubility in other solvents: chloroform s.; heptane s.; ethanol s.; acetone s. (1)
Melting point: Decomposes without melting above 300°C (1)
Vapor pressure: Negligible (<1 mPa) at 20°C (1)
Partition coefficient (octanol/water): Not available
Adsorption coefficient: 1900 (11)

Exposure guidelines

ADI: 0.02 mg/kg/day (12)
HA: Not available
RfD: 0.05 mg/kg/day (13)
PEL: 5 mg/m³ (8-hour) (respirable fraction) (14)

Basic manufacturer

DuPont Agricultural Products
Walker's Mill, Barley Mill Plaza
P.O. Box 80038
Wilmington, DE 19880-0038
Telephone: 800-441-7515
Emergency: 800-441-3637

10.2.2 Captan

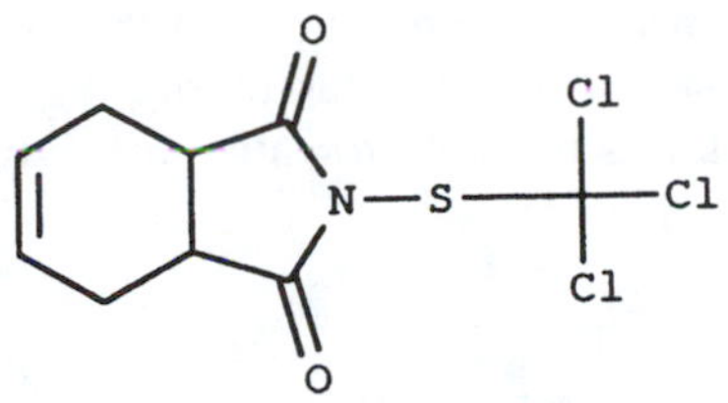

Figure 10.2 Captan.

Trade or other names

Trade names for captan include Agrox, Captal, Captec, Captol, Captonex, Clomitane, Merpan, Meteoro, Orthocide, Phytocape, Sepicap, Sorene, and Vancide 89. Captan may be found in formulations with a wide range of other pesticides.

Regulatory status

Captan is a General Use Pesticide (GUP), though most uses of the compound on food crops were canceled in the U.S. in 1989. It is categorized as toxicity class IV — practically nontoxic. However, it bears the Signal Words DANGER or CAUTION if packaged in concentrated form because it can be irritating to the skin and eyes.

Introduction

Captan is a nonsystemic phthalimide fungicide used to control diseases of many fruit, ornamental, and vegetable crops. It improves fruit finish by giving it a healthy, bright-colored appearance. It is used in agricultural production as well as by the home gardener. A major use of captan is in apple production.

Toxicological effects

Acute toxicity

The rat oral LD_{50} for captan ranges from 8400 to 15,000 mg/kg, indicating very low acute toxicity (15). The mouse LD_{50} is 7000 mg/kg. Sheep showed no effect at doses of 200 mg/kg, but experienced deaths at 250 mg/kg. The inhalation LC_{50} (2-hour) in mice is 5.0 mg/L (8). Rabbits showed little or no skin sensitization to captan, while guinea pigs were moderately sensitive (6).

Workers exposed to high concentrations of captan in air (6 mg/m³) experienced eye irritation, including burning, itching, and tearing. Skin irritation also occurred in some cases (6).

Chronic toxicity

Rats fed up to 750 mg/kg/day Orthocide for 4 weeks had decreased food intake and body weights (6). No deaths occurred in pigs given as much as 420 to 4000 mg/kg/day in the diet for 12 to 25 weeks; however cattle given six doses of 250 mg/kg experienced varied toxic effects, including death (6).

Reproductive effects

Pregnant mice exposed by inhalation to high doses of captan for 4 hours a day during days 6 to 15 of gestation showed significant mortality or weight loss. Fetal mortality accompanied these effects. Mice fed 50 mg/kg/day over three generations reproduced normally. Captan is unlikely to cause reproductive effects in humans at usual levels of exposure (6,8).

Teratogenic effects

Teratogenicity studies with rats, rabbits, hamsters, and dogs have given both negative and positive results. However, the weight of evidence suggests that captan does not produce birth defects (16).

Mutagenic effects

Although captan was mutagenic in some laboratory tests on isolated tissue cultures, the majority of evidence indicates that captan is nonmutagenic (16).

Carcinogenic effects

There is strong evidence that captan causes cancer in female mice and male rats at high doses. In addition, captan is chemically similar to two other pesticides, folpet and captafol, that have been shown to produce cancer in test animals. Tumors were associated with the gastrointestinal tract and, to a lesser degree, with the kidneys. Tumors appeared in the test animals at doses of about 300 mg/kg/day (6,8).

Organ toxicity

Most organ-specific effects are found in the kidneys of rats at and above doses of 100 mg/kg/day.

Fate in humans and animals

Studies in several animal species have shown that captan is rapidly absorbed from the gastrointestinal tract and is rapidly metabolized. Residues are excreted primarily in the urine. Rats given captan orally excreted a third in the feces and half in the urine within 24 hours.

A cow fed small amounts in its diet for 4 days had no captan in the milk at a 0.01-mg/L detection limit, nor could any be detected in the urine at a 0.1-mg/L detection limit (6).

Ecological effects

Effects in birds

Captan is practically nontoxic to birds. The LD_{50} is greater than 5000 mg/kg in mallard ducks and pheasants. The LD_{50} is 2000 to 4000 mg/kg in bobwhite quail (1). High doses administered for 90 days to chickens caused an 80% reduction in the number of eggs produced but had no effect on the fertility or hatchability of the eggs produced (6).

Effects in aquatic organisms

Captan is very highly toxic to fish. The LC_{50} (96-hour) for technical captan ranges from 0.056 mg/L in cutthroat trout and chinook salmon to 0.072 mg/L in bluegill (1).

The LC_{50} for captan in the aquatic invertebrate *Daphnia magna* is 7 to 10 mg/L, indicating that the compound is moderately toxic to this and other aquatic invertebrates (8).

Captan has a low to moderate tendency to accumulate in living tissue. Fish exposed for 3 days to concentrations that would be expected in a pond following treatment of an adjacent watershed at a rate of 1 lb/acre, had no detectable residues of captan (6). Estimates of the bioconcentration factor range from 10 to 1000 (9).

Effects on other organisms (non-target species)

Captan is not toxic to bees when used as directed (1).

Environmental fate

Breakdown in soil and groundwater

Captan has a low persistence in soil, with a half-life of 1 to 10 days in most soil environments (6). Captan was not detected in field studies of its mobility at application rates of up to 42 kg active ingredient per hectare (9).

Breakdown in water

Captan is rapidly degraded in near neutral water. Half-lives of 23 to 54 hours and 1 to 7 hours have been reported at various acidities and temperatures (6). The effective residual life in water is 2 weeks (15).

Breakdown in vegetation

Captan is taken up through leaves and roots and translocated throughout the plant. Residual fungitoxicity remains for 23 days after application on potato leaves, but residues were below the detection limit within 40 days after application (6). Some varieties of apples, pears, lettuce seeds, celery, and tomato seeds may be injured by captan at high doses (1).

Physical properties

Captan is a white to buff-colored compound in the technical form. It is a colorless crystal in its pure form (1). The technical product has a pungent smell.

Chemical name: 3a,4,7,7a-tetrahydro-2-[(trichloromethyl)thio]-1H-isoindole-1,3(2H)-dione (1)
CAS #: 133-06-2
Molecular weight: 300.61 (1)
Solubility in water: 3.3 mg/L @ 25°C (1)
Solubility in other solvents: xylene s.; acetone s.; chloroform v.s.; cyclohexanone v.s. (1)
Melting point: 178°C (1)
Vapor pressure: 1.3 mPa @ 25°C (1)
Partition coefficient (octanol/water): 610 (1)
Adsorption coefficient: 200 (11)

Exposure guidelines

ADI: 0.1 mg/kg/day (12)
MCL: Not available
RfD: 0.13 mg/kg/day (13)
TLV: 5 mg/m³ (8-hour) (17)

Basic manufacturer

Drexel Chemical Company
1700 Channel Avenue
Memphis, TN 38113
Telephone: 901-774-4370

10.2.3 Carboxin

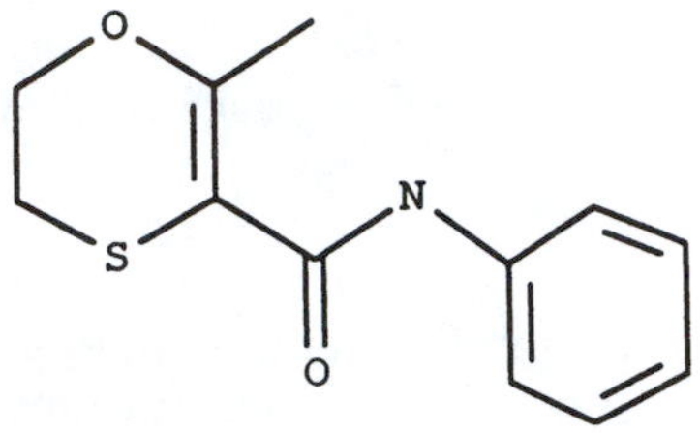

Figure 10.3 Carboxin.

Trade or other names

Trade names for products containing carboxin include Cadan, Kisvax, Kemikar, Oxalin, Padan, Sanvex, Thiobel, Vegetox, and Vitavax. It is often used in combination with other fungicides such as thiram or captan.

Regulatory status

Carboxin is classified as a General Use Pesticide (GUP) by the U.S. Environmental Protection Agency. It is in toxicity class III — slightly toxic, and products containing it carry the Signal Word CAUTION on the label.

Introduction

Carboxin is a systemic anilide fungicide. It is used as a seed treatment for control of smut, rot, and blight on barley, oats, rice, cotton, vegetables, corn, and wheat. It is also used to control fairy rings on turfgrass. Carboxin may be used to prevent the formation of these diseases or may be used to cure existing plant diseases.

Toxicological effects

Acute toxicity

Carboxin is slightly toxic. Symptoms of poisoning can include vomiting and headache. Recovery is very rapid if the exposed individual is treated quickly. The oral LD_{50} is 3820 mg/kg in rats and 3550 mg/kg in mice (18).

The compound produces very little skin irritation; however, it can seriously irritate the eyes. Acute dermal exposure results in an LD_{50} of greater than 8000 mg/kg in rabbits. The inhalation LC_{50} (1-hour) is greater than 20 mg/L in rats (18).

Chronic toxicity

Rats fed doses up to 311 mg/kg/day for 28 days showed some fluid accumulation in the liver, even at low doses (18). Another rat study showed kidney changes at somewhat higher doses (1000 mg/kg fed for 90 days) (18). A 2-year rat study with levels of up to 30 mg/kg produced no compound-related effects in physical appearance, behavior, blood chemistry, or urinalysis (18). However, there were changes in organ weights.

Male and female mice also showed liver effects after being fed high doses (912 mg/kg) of carboxin for $1\frac{1}{2}$ years (18). Beagle dogs showed no effects at the highest dose tested, 15 mg/kg for two years (18).

Reproductive effects

A three-generation study with rats showed treatment-related effects on reproductive performance at levels from 5 to 30 mg/kg/day. However, at the highest dose, there was only moderate growth suppression in nursing pups (18). It is unlikely that the compound would produce reproductive effects in humans at expected exposure levels.

Teratogenic effects

At the highest dose tested (40 mg/kg), administered to pregnant rats on days 6 to 15, there were no birth defects in the offspring (18). Pregnant rabbits treated with very high doses on days 6 to 27 of gestation had increased abortions but no fetal malformations (18). These data indicate that carboxin is not teratogenic.

Mutagenic effects

Carboxin is either a non-mutagen or is a very weak mutagen, based on information from several studies on bacteria and mammalian cells (18).

Carcinogenic effects

A 2-year study with rats fed up to 30 mg/kg/day showed no evidence of increased tumor frequency (18). Mice fed up to 900 mg/kg/day for 84 weeks had no apparent compound-related increase in tumor formation. Carboxin does not appear to cause cancer.

Organ toxicity

Animal studies have demonstrated effects in the liver and the kidneys.

Fate in humans and animals

Rats excreted almost all of a carboxin dose in 24 hours, with most excreted in urine and some in feces (15). Rabbits showed a similar excretion pattern. Carboxin is incompletely absorbed in the gut, especially in rats (18). The compound does not acumulate in animal tissues.

Only trace amounts of carboxin were found in rat tissues 48 hours after dosing (18). In milk cows fed up to 5 ppm for 10 days, less than 2% of the administered dose was found in tissues. However, significant levels were found in milk a few days after exposure. The main breakdown product is carboxin sulfoxide for which the rat oral LD_{50} is 2000 mg/kg. Carboxin sulfoxide is sold as a pesticide also and is known as oxycarboxin (19).

Ecological effects

Effects on birds

The oral LD_{50} of carboxin in chickens is very high, at 24 g/kg, indicating that it has a very low acute toxicity (8). In chronic, low-exposure experiments over $5\frac{1}{2}$ months, changes were noted in the digestive tract, cardiovascular system, and blood of chickens (20).

Effects in aquatic organisms

Carboxin is highly toxic to fish. The 96-hour LC_{50} in rainbow trout is greater than 0.1 mg/L (8). Carboxin is practically nontoxic to freshwater invertebrates. The LC_{50} is 217 mg/L in juvenile crayfish (20).

Effects on other organisms (non-target species)

The compound is nontoxic to bees (8).

Environmental fate

Breakdown in soil and groundwater

Carboxin is rapidly degraded to carboxin sulfoxide in soil. It has a low persistence, with a half-life of about 3 days in soil (11). In one study, after 7 days 95% of the parent was gone and the sulfoxide, a breakdown product, represented 31 to 45% of the amount applied (8). Minor products formed were carboxin sulfone, hydroxy carboxin, and CO_2. Carboxin does not readily adsorb to soil. Both parent compound and the sulfoxide are very mobile and could possibly leach to groundwater (18).

Breakdown in water

In water, carboxin oxidizes to the sulfoxide and sulfone within 7 days (18). This happens both under ultraviolet light and in the dark. Blue-green algae (e.g., Anabaena and Nostoc) degrade the pesticide extensively. Other algae can also break down carboxin, but not to the same extent (8).

Breakdown in vegetation

Although the distribution pattern of the parent and sulfoxide metabolite vary, carboxin is found systemically (throughout the plant) in all species of plants studied. Plants grown from treated seed had no carboxin present at 6 weeks after emergence. Carboxin sulfoxide found in plants can come either from the soil or through oxidation within the plant (8).

Physical properties

Carboxin is a colorless crystal (1).

Chemical name: A 5,6-dihydro-2-methyl-N-phenyl-1,4-oxathiin-3-carboxamide (1)
CAS #: 5234-68-4
Molecular weight: 235.31 (1)
Solubility in water: 170 mg/L @ 25°C (1)
Solubility in other solvents: acetone v.s.; benzene s.s.; methanol s.s. (1)
Melting point: Two crystal structures: 91.5-92.5°C and 98-100°C (1)
Vapor pressure: <0.025 mPa @ 25°C (1)
Partition coefficient (octanol/water): 148 (1)
Absorption coefficient: 260 (11)

Exposure guidelines

ADI: Not available
HA: 0.7 mg/L (lifetime) (18)
RfD: 0.1 mg/kg/day (13)
PEL: Not available

Basic manufacturer

Uniroyal Chemical Co., Inc.
Benson Road
Middlebury, CT 06749
Telephone: 203-573-2000
Emergency: 203-723-3670

10.2.4 Copper sulfate

$$Cu-SO_4$$

Figure 10.4 Copper sulfate.

Trade or other names

Copper sulfate is also called Agritox, Basicap, BSC Copper Fungicide, CP Basic Sulfate, and Tri-Basic Copper Sulfate. The pentahydrate form is called bluestone, blue vitriol, Salzburg vitriol, Roman vitriol, and blue copperas. Bordeaux Mixture is a combination of hydrated lime and copper sulfate. Copper sulfate is often found in combination with other pesticides.

Regulatory status

Copper sulfate is classified as a General Use Pesticide (GUP) by the U.S. Environmental Protection Agency (EPA). Copper sulfate is toxicity class I — highly toxic. It bears the Signal Words DANGER — POISON. Because of its potentially harmful effects on some endangered aquatic species, surface water use may require a permit in some places.

Introduction

Copper sulfate is a fungicide used to control bacterial and fungal diseases of fruit, vegetable, nut, and field crops. These diseases include mildew, leaf spots, blights, and apple scab. It is used as a protective fungicide (Bordeaux Mixture) for leaf application and seed treatment. It is also used as an algacide and herbicide, and to kill slugs and snails in irrigation and municipal water treatment systems. It has been used to control dutch elm disease. It is available as a dust, wettable powder, or liquid concentrate.

Toxicological effects

Acute toxicity

Copper sulfate is caustic, and acute toxicity is largely due to this property (8). There have been reports of human suicide resulting from the ingestion of gram quantities of this material (21). The lowest dose of copper sulfate that has been toxic when ingested by humans is 11 mg/kg (22). Ingestion of copper sulfate is often not toxic because vomiting is automatically triggered by its irritating effect on the gastrointestinal tract. Symptoms are severe, however, if copper sulfate is retained in the stomach, as in the unconscious victim. Some of the signs of poisoning that occurred after 1 to 12 g copper sulfate was swallowed include a metallic taste in the mouth, burning pain in the chest and abdomen, intense nausea, repeated vomiting, diarrhea, headache, sweating, shock, and discontinued urination leading to yellowing of the skin. Injury to the brain, liver, kidneys, and stomach and intestinal linings may also occur in copper sulfate poisoning (23).

Copper sulfate can be corrosive to the skin and eyes (8). It is readily absorbed through the skin and can produce a burning pain, as well as the other symptoms of poisoning resulting from ingestion. Skin contact may result in itching or eczema (8). It is a skin sensitizer and can cause allergic reactions in some individuals (24). Eye contact with this material can cause conjunctivitis, inflammation of the eyelid lining, cornea tissue deterioration, and clouding of the cornea (23).

Examination of copper sulfate-poisoned animals showed signs of acute toxicity in the spleen, liver, and kidneys (8). Injury may also occur to the brain, liver, kidneys, and gastrointestinal tract in response to overexposure to this material (22). The oral LD_{50} of copper is 472 mg/kg in rats (8).

Chronic toxicity

Vineyard sprayers experienced liver disease after 3 to 15 years of exposure to copper sulfate solution in Bordeaux Mixture (8). Long-term effects are more likely in individuals with Wilson's disease, a condition that causes excessive absorption and storage of copper (25). Chronic exposure to low levels of copper can lead to anemia (8).

The growth of rats was retarded when given dietary doses of 25 mg/kg/day copper sulfate. Dietary doses of 200 mg/kg/day caused starvation and death (8). Sheep given oral doses of 20 mg/kg/day showed blood cell and kidney damage (8). They also showed an absence of appetite, anemia, and degenerative changes (22).

Reproductive effects

Copper sulfate has been shown to cause reproductive effects in test animals. Testicular atrophy increased in birds as they were fed larger amounts of copper sulfate; sperm production was also interrupted to varying degrees (8). Reproduction and fertility was affected in pregnant rats given this material on day 3 of pregnancy (23).

Teratogenic effects

There is very limited evidence about the teratogenic effects of copper sulfate. Heart disease occurred in the surviving offspring of pregnant hamsters given intravenous copper salts on day 8 of gestation (8). These data suggest that copper sulfate is unlikely to be teratogenic in humans at expected exposure levels.

Mutagenic effects

Copper sulfate may cause mutagenic effects at high doses. At 400 and 1000 ppm, copper sulfate caused mutations in two types of microorganisms (22). Such effects are not expected in humans under normal conditions.

Carcinogenic effects

Copper sulfate at 10 mg/kg/day caused endocrine tumors in chickens given the material parenterally, that is, outside of the gastrointestinal tract through an intravenous or intramuscular injection (22). However, the relevance of these results to mammals, including humans, is not known.

Organ toxicity

Long-term animal studies indicate that the testes and endocrine glands have been affected.

Fate in humans and animals

Absorption of copper sulfate into the blood occurs primarily under the acidic conditions of the stomach. The mucous membrane lining of the intestines acts as a barrier to absorption of ingested copper (8). After ingestion, more than 99% of the copper is excreted in the feces. However, residual copper is an essential trace element that is strongly bioaccumulated (8). It is stored primarily in the liver, brain, heart, kidney, and muscles.

Ecological effects

Effects on birds

Copper sulfate is practically nontoxic to birds. It poses less of a threat to birds than to other animals. The lowest lethal dose (LD_LO) is 1000 mg/kg in pigeons and 600 mg/kg in ducks (8). The oral LD_{50} for Bordeaux Mixture in young mallards is 2000 mg/kg (27).

Effects on aquatic organisms

Copper sulfate is highly toxic to fish (28). Even at recommended rates of application, this material may be poisonous to trout and other fish, especially in soft or acid waters. Its toxicity to fish generally decreases as water hardness increases. Fish eggs are more resistant than young fish fry to the toxic effects of copper sulfate (26).

Copper sulfate is toxic to aquatic invertebrates such as crab, shrimp, and oysters. The 96-hour LC_{50} of copper sulfate to pond snails is 0.39 mg/L at 20°C. Higher concentrations of the material caused some behavioral changes, such as secretion of mucus and discharge of eggs and embryos (8).

Effects on other organisms (non-target species)

Bees are endangered by Bordeaux Mixture (1). Copper sulfate may be poisonous to sheep and chickens at normal application rates. Most animal life in soil, including large earthworms, has been eliminated by the extensive use of copper-containing fungicides in orchards (28).

Environmental fate

Breakdown in soil and groundwater

Since copper is an element, it will persist indefinitely. Copper is bound (or adsorbed) to organic materials and to clay and mineral surfaces. The degree of adsorption to soils depends on the acidity or alkalinity of the soil (8). Because copper sulfate is highly water soluble, it is considered one of the more mobile metals in soils. However, because of it binding capacity, its leaching potential is low in all but sandy soils (8).

When applied with irrigation water, copper sulfate does not accumulate in the surrounding soils. Some (60%) is deposited in the sediments at the bottom of the irrigation ditch, where it becomes adsorbed to clay, mineral, and organic particles. Copper compounds also settle out of solution (26).

Breakdown in water

As an element, copper can persist indefinitely. However, it will bind to water particulates and sediment.

Breakdown in vegetation

One of the limiting factors in the use of copper compounds is their serious potential for phytotoxicity (24). Copper sulfate can kill plants by disrupting photosynthesis. Blue-green algae in some copper sulfate-treated Minnesota lakes became increasingly resistant to the algacide after 26 years of use (28).

Physical properties

Copper sulfate crystals are blue or green-white and odorless (1).

Chemical name: copper sulfate (1)
CAS#: 7758-98-7
Molecular weight: 249.68 (pentahydrate) (1)
Water solubility: Anhydrous form; 230,500 mg/L at 25°C (1)
Solubility in other solvents: methanol s.s. (3); ethanol i.s. (1)
Melting point: Above 110°C, copper sulfate loses water of crystallization with formation of the monohydrate; above 250°C it loses all water of crystallization (1)
Vapor pressure: Nonvolatile (1)
Partition coefficient (octanol/water): Not available
Adsorption coefficent: Not available

Exposure guidelines:

ADI: Not available
MCL: Not available

RfD: Not available
PEL: 1.0 mg/m³ (8-hour) (copper dusts or mists) (14)

Basic manufacturers

CP Chemical Inc.
1 Parker Plaza
Ft. Lee, NJ 07024
Telephone: 201-944-6020

Phelps Dodge Refining Corp.
P.O. Box 20001
El Paso, TX 79998
Telephone: 800-223-8567

10.2.5 Dinocap

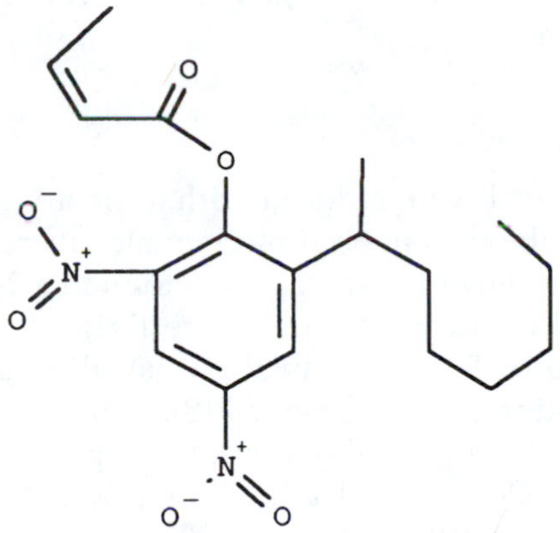

Figure 10.5 Dinocap.

Trade or other names

Trade names for products containing dinocap include Arathane, Caprane, Capryl, Cekucap 25 WP, Crotonate, Crotothane, DCPC, Dicap, Dikar (a mixture of dinocap and mancozeb), DNOPC, Ezenoan, Iscothane, Karathane, Mildane, and Mildex.

Regulatory status

Dinocap is a slightly toxic pesticide in EPA toxicity class III. Labels for products containing dinocap must bear the Signal Word CAUTION. It is a General Use Pesticide (GUP).

Introduction

Dinocap, a dinitrophenyl, was first registered in the late 1950s and has been used as a contact fungicide to control fungus and, to a lesser extent, as an acaricide for control of ticks and mites. It is applied to limit mites in apple crops, as well as foliage for control of powdery mildew on fruit, vegetable, nursery, and ornamental crops. It is available as dust, liquid concentrate, and wettable powder formulations.

Toxicological effects

Acute toxicity

Dinocap is slightly to moderately toxic by ingestion, with reported oral LD_{50} values of 980 mg/kg in rats, 2000 mg/kg in male rabbits, 53 mg/kg in mice, and 100 mg/kg in dogs (8,29,30). It is slightly toxic by skin absorption, with a reported dermal LD_{50} in rabbits of 9400 mg/kg (29). It is irritating to the skin and eyes of rabbits, and may irritate those areas as well as the eyes and the mucous membranes lining the nose, throat, and lungs in humans (8,31). The acute 4-hour inhalation LC_{50} in rats for an emulsifiable concentrate formulation is 0.36 mg/L, indicating moderate toxicity by this route.

Dinocap is included in a class of compounds that causes the following symptoms upon acute exposure: headaches, fatigue, weakness, nausea, vomiting, abdominal pain, loss of appetite, weight loss, fever, excessive sweating, rapid breathing and heart beats, shortness of breath, thirst, dehydration, heat stroke, and convulsions (29). Inhalation of dinocap can cause tightness in the chest and fluid retention in the lungs, a condition called "pulmonary edema" (29). Alcohol use may exacerbate the systemic effects of dinocap.

Chronic toxicity

Rat growth and survival were reduced with a dietary level of 125 mg/kg/day of dinocap (29). Spleen enlargement occurred in male rats receiving 125 mg/kg/day fungicide. Only male rats showed growth retardation in a 2-year study (29). Degenerative changes and cell death were seen in the livers, kidneys, and stomachs of rabbits given oral doses of 30 or 150 mg/kg/day dinocap for 90 days (8). The composition of blood and urine also changed (8).

At a dietary dose of 25 mg/kg/day dinocap, dogs showed decreased appetite and drastic weight loss, followed by death within 6 weeks. At dietary doses of 6.25 and 25 mg/kg/day, localized cell death occurred in areas of the liver (8). Neuropathy and effects on the nervous system have been hypothesized as an effect of prolonged exposure to dinocap (on the basis of its structural similarity to dinitrophenol) (32), but there is no evidence of these effects.

Greenhouse workers developed liver function abnormalities in association with exposure to the fungicide; the severity of the abnormalities varied with the length of work exposure (8).

Reproductive effects

Reproductive effects (effects on fertility or fecundity) have not been observed in animal studies (29).

Teratogenic effects

Numerous animal studies have demonstrated the teratogenic potential of dinocap (29). When pregnant rabbits were given dermal doses of 25, 50, or 100 mg/kg/day, developmental toxicity in the form of reduced fetal weight was observed at the highest dose (33). No developmental effects were seen at 50 mg/kg/day (33).

Fetal growth retardation, cleft palate, and abnormal rib formations were seen in offspring of pregnant mice exposed to dinocap during organogenesis, the organ-forming period of pregnancy (8,29). Growth retardation was seen at doses of 5 mg/kg/day, and malformations were seen at 20 mg/kg/day and higher (8,29).

Birth defects observed in the offspring of rabbits given oral or dermal doses of dinocap during pregnancy included abnormalities of the neural tube, spine, and skull at 3 mg/kg/day (34). No birth defects were discovered in rabbits given dermal doses up to and including those that cause severe skin irritation and obvious maternal poisoning (8). Following dermal applications of 100 mg/kg/day to rabbits, reduced fetal weight and an increased occurrence of skull malformations were observed (35).

Dinocap may cause teratogenic effects at very high oral doses during the critical time of pregnancy.

Mutagenic effects

No data are currently available.

Carcinogenic effects

Dinocap did not cause tumor development in mice fed the highest tolerated dose of the fungicide (32). This suggests that dinocap is not carcinogenic.

Organ toxicity

Dinocap has shown effects on the liver, kidneys and gastrointestinal tract (29,32). Effects on the nervous system are thought to be possible due to the structural similarity between dinocap and dinitrophenol (32).

Fate in humans and animals

Dinocap is readily eliminated through urine and feces in mammalian systems (8). Biological accumulation of dinocap is unlikely (29).

Ecological effects

Effects on birds

Dinocap is moderately toxic to birds; the reported 5- to 8-day dietary LC_{50} for dinocap is 790 ppm (36). Ducks fed 50 ppm of this fungicide in their food developed cataracts (32).

Effects on aquatic organisms

Dinocap is very highly toxic to fish; the reported 96-hour LC_{50} values for dinocap are 15 µg/L in rainbow trout, 33 µg/L in goldfish, and 20 µg/L in bluegill (37). The 96-hour LC_{50} is 75 µg/L in the sideswimmer (*G. fasciatus*, an invertebrate) (37).

Effects on other organisms (non-target species)

Dinocap is nontoxic to bees and other beneficial insects (8).

Environmental fate

Breakdown in soil and groundwater

Dinocap is of low persistence in the soil environment, with reported field half-lives of 4 to 6 days (11). It is highly photosensitive and is readily broken down by sunlight (11). It is also subject to microbial degradation (38).

Dinocap is moderately adsorbed to topsoils and has only slight solubility in water (11). These facts, combined with its low persistence, make it unlikely to contaminate groundwater.

Breakdown in water

Dinocap is slightly soluble in water and, if found in water, will most likely be adsorbed to suspended colloidal materials or precipitated in sediment.

Breakdown in vegetation

Dinocap is readily absorbed and translocated by treated plants. It builds up in the growing shoots and leaf tips (38). Since this material penetrates foliage rapidly, it is not likely to be washed off by rain (38). There is good crop tolerance to dinocap for the recommended uses (8). The approximate trace, or "residual," period of dinocap in plants is 1 to 2 weeks (39).

Physical properties

Dinocap is dark reddish-brown liquid (1).

Chemical name: 2,6-dinitro-4-octylphenyl crotonates; 2,4-dinitro-6-octylphenyl crotonates (1)
CAS #: 39300-45-3
Molecular weight: 364.41 (1)
Water solubility: (<0.1 mg/L); practically insoluble in water (1)
Solubility in other solvents: s. in most organic solvents such as benzene and ether (1)
Melting point: Not available
Vapor pressure: 0.0053 mPa @ 20°C (1)
Partition coefficient (octanol/water): 34,400 (1)
Adsorbtion coefficient: 550 (estimated) (11)

Exposure guidelines

ADI: 0.001 mg/kg/day (12)
MCL: Not available
RfD: Not available
PEL: Not available

Basic manufacturer

Rohm and Haas Co.
Agricultural Chemicals
Independence Mall West
Philadelphia, PA, 19106
Telephone: 215-592-3000

10.2.6 Dodine

Trade and other names

Another common name is dodine acetate. Trade names include AC 5223, Apadodine, Carpene, Curitan, Cyprex, Efuzin, Melprex, Sulgen, Syllit, Tebulan, Vandodine, and Venturol.

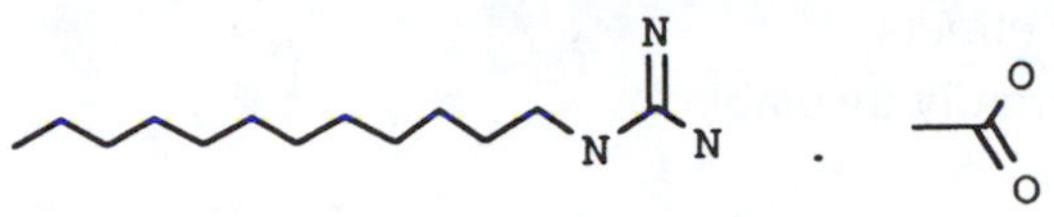

Figure 10.6 Dodine.

Regulatory status

Dodine is a General Use Pesticide (GUP). It is registered for use only in western states for peaches, pears, and cherries. It is classified as toxicity class I — highly toxic. Products containing it bear the Signal Word DANGER because of its potential to cause severe eye irritation.

Introduction

Dodine is a fungicide and bactericide used to control scab on apples, pears and pecans, brown rot on peaches, and several foliar diseases of cherries, strawberries, peaches, sycamore trees, and black walnuts. It is also used as an industrial biocide and preservative. The compound works by changing the cell walls of the fungus, causing loss of the materials from within the cell. It is available as a soluble concentrate or a wettable powder.

Toxicological effects

Acute toxicity

Because it may cause severe eye irritation, dodine is considered a highly toxic material (40). Dodine is slightly toxic via inhalation or ingestion (40). The oral LD_{50} for technical dodine in rats is 1000 mg/kg (8). The dermal LD_{50} in rats is greater than 6000 mg/kg, and in rabbits is greater than 1500 mg/kg for a single 24-hour contact, indicating slight toxicity (8). Dodine did not cause allergic skin reactions when tested on humans. However, it is an eye and skin irritant (8).

Chronic toxicity

Chronic dietary exposure in rats caused reduced weight gain in both sexes and reduced food consumption in males (8). Dogs fed dodine for 12 months exhibited structural changes in the thyroid, indicative of thyroid stimulation (40).

Reproductive effects

In a 2-year feeding study, rats given very high dietary doses of 800 mg/kg/day exhibited slight retardation of growth, but no adverse effects on reproduction or lactation (8). Offspring of mice fed dodine in the diet at doses of 74 to 89 mg/kg/day exhibited decreased numbers of pups surviving to weaning (40). Dodine is unlikely to cause reproductive effects in humans at expected exposure levels.

Teratogenic effects

No data are currently available.

Mutagenic effects

Dodine does not appear to be mutagenic. The Ames test for mutagenicity was negative on five strains of bacteria (8).

Carcinogenic effects
No data are currently available.

Organ toxicity
Long-term animal studies indicate that the thyroid was affected.

Fate in humans and animals
No data are currently available.

Ecological effects

Effects on birds

The oral LD_{50} for dodine in mallard ducks is 1142 mg/kg, suggesting that the compound is only slightly toxic to birds (8).

Effects on aquatic organisms

Dodine is highly toxic to fish (8). The 48-hour LC_{50} for dodine in harlequin fish is 0.53 mg/L (1).

Effects on other organisms (non-target species)

Dodine is nontoxic to bees (1). Its LD_{50} in honeybees is greater than 11 mg per bee for topical application (1).

Environmental fate

Breakdown in soil and groundwater

Dodine is of low persistence in soil. Its soil half-life is about 20 days. It is soluble in water but binds strongly to soil and so is unlikely to contaminate groundwater (11).

Breakdown in water

No data are currently available.

Breakdown in vegetation

No data are currently available.

Physical properties

Dodine is a colorless or white, slightly waxy crystalline solid (1).

Chemical name: 1-dodecylguanidinium acetate (1)
CAS #: 2439-10-3
Molecular weight: 287.44 (1)
Solubility in water: 630 mg/L @ 25°C (1)
Solubility in other solvents: s. in methanol and ethanol; i.s. in most organic solvents (1)
Melting point: 136°C (1)

Vapor pressure: 1300 mPa @ 20°C (1)
Partition coefficient (octanol/water): Not available
Adsorption coefficient: 100,000 (estimated) (11)

Exposure guidelines

ADI: 0.01 mg/kg/day (12)
HA: Not available
RfD: 0.004 mg/kg/day (13)
PEL: Not available

Basic manufacturer

Rhone Poulenc Ag Co.
P.O. Box 12014
2 T.W. Alexander Dr.
Research Triangle Park, NC 27709
Telephone: 919-549-2000
Emergency: 800-334-7577

10.2.7 Imazalil

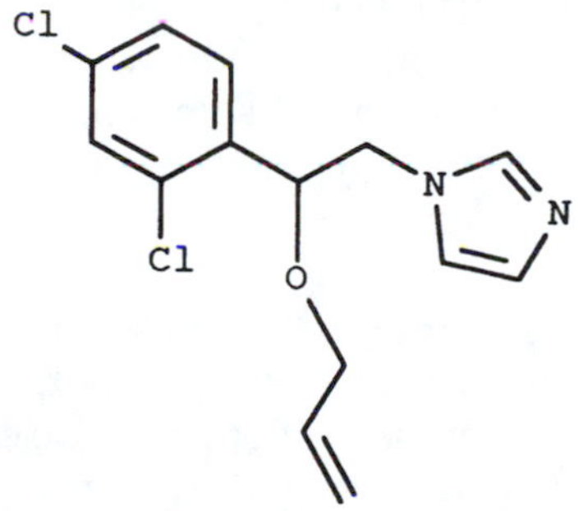

Figure 10.7 Imazalil.

Trade or other names

Trade names for products containing imazalil include Bromazil, Deccozil, Fungaflor, Freshgard, and Fungazil. The fungicide is compatible with many other types of pesticides.

Regulatory status

Imazalil is a moderately toxic compound in EPA toxicity class II. Labels for products containing it must bear the Signal Word WARNING (1). Imazalil is a General Use Pesticide (GUP).

Introduction

Imazalil is a systemic imidazole fungicide used to control a wide range of fungi on fruit, vegetables, and ornamentals, including powdery mildew on cucumber and black spot on roses. Imazalil is also used as a seed dressing and for postharvest treatment of citrus, banana, and other fruit to control storage decay. Under natural conditions, it is less likely that imazalil treatment will lead to resistant strains of fungi than as a result of treatment with other fungicides.

Toxicological effects

Acute toxicity

Imazalil is moderately toxic by ingestion, with a reported oral LD_{50} of 227 to 343 mg/kg in rats (1,8). The LD_{50} in dogs is greater than 640 mg/kg (41). The reported dermal LD_{50} is 4200 to 4880 mg/kg in rats, indicating slight toxicity (1).

Test animals have experienced symptoms such as excitation of hair follicles (goose pimples), muscle incoordination, reduced arterial tension, tremors, and vomiting (41). Contact dermatitis has been noted in some cases in sensitive individuals (41).

Chronic toxicity

Rats fed imazalil nitrate at dietary levels of up to 0.4 mg/kg/day for 14 weeks were not affected in appearance, behavior, survival, food consumption, urinalysis, or tissue composition. There were slight liver, body weight, and bilirubin changes at higher doses (41). Groups of rats fed up to 0.4 mg/kg/day for 6, 12, and 24 months did not show compound or dose-related effects on body weight gain, food consumption, appearance, behavior, or survival (41).

Similar results were found in a dog study where animals received up to 0.5 mg/kg/day for 2 years. The liver showed some slight effects at the higher doses, but all other measured and observed parameters were within normal limits (41).

Reproductive effects

In three separate three-generation rat studies at low to moderate doses of 0.4 mg/kg/day, there was a trend to a lower number of live births at the highest dose level. No differences were noted in percent of pregnancies or duration of pregnancy (3,41). These data suggest that imazalil is unlikely to cause reproductive effects under normal conditions.

Teratogenic effects

None of the rat studies mentioned above resulted in fetal abnormalities. A mouse study at doses up to 4.8 mg/kg/day was also negative. It is unlikely that imazalil is teratogenic (3,41).

Mutagenic effects

Dominant lethal mutagenic effects were not evident in male and female mice (3). Based on these data, it appears that imazalil is not mutagenic.

Carcinogenic effects

In a group of rats given imazalil for 30 months at a dose of 5.0 mg/kg/day, there were no increases in tumors compared to the controls (41). This suggests that imazalil is noncarcinogenic.

Organ toxicity

Based on animal tests, imazalil affects the nervous system and liver.

Fate in humans and animals

Imazalil is rapidly absorbed, distributed, metabolized, and excreted by rats. Following a single dose of imazalil sulfate, 90% was excreted in metabolized form within 96 hours (1). Only 3% was eliminated via the feces in nonmetabolized form, indicating almost complete absorption from the gastrointestinal tract (41). At least four metabolites are formed 48 hours after administration. Accumulation in fatty tissue did not occur (41).

Ecological effects

Effects on birds

Both the mallard duck and the Japanese quail are relatively insensitive to the fungicide. The 8-day LC_{50} values in these birds range from about 5500 to 6300 mg/kg/day (1). These values indicate that the compound is practically nontoxic to birds.

Effects on aquatic organisms

Imazalil is moderately toxic to fish. The LC_{50} for imazalil in trout is 2.5 mg/L, and in the bluegill sunfish is 3.2 mg/L (1).

Effects on other organismes (non-target species)

The compound is nontoxic to bees (1).

Environmental fate

Breakdown in soil and groundwater

Imazalil is highly persistent in the soil environment, with a reported field half-life of between 120 and 190 days (11). A representative value is estimated to be 150 days for most soils (11). It is soluble in water, but strongly bound to soils (11), and thus unlikely to pose a risk to groundwater. In a plot where seven applications were made at 14-day intervals, leaching was practically nonexistent and accumulation did not appear to be a problem (42).

Breakdown in water

In acid to neutral aqueous solutions, imazalil is stable for at least 8 weeks at 40°F. Decomposition occurs at elevated temperatures and under the influence of light (41).

Breakdown in vegetation

One week after treated barley seed was sown in soil, about 76% of the imazalil was in the adjacent soil and about 29% was in the seedcoat. After 3 weeks, only 6% was in the green plant parts. Under normal storage conditions, oranges dipped in 2000 mg active ingredient per liter and stored have residues (89%) present as the parent compound. Only a small amount of imazalil was present in the pulp, and part of this may have resulted from handling during peeling (41). Studies with apples gave similar results.

Physical properties

Imazalil is a slightly yellow to brown solidified oil (1).

Chemical name: (±)-allyl 1-(2,4-dichlorophenyl)-2-imidazol-1-ylether ester (1)
CAS #: 35554-44-0
Molecular weight: 297.18 (1)
Solubility in water: 1400 mg/L @ 20°C (1)
Solubility in other solvents: s.s. in hexane, methanol, toluene, benzene (1)
Melting point: 50°C (1)
Vapor pressure: 0.0093 mPa @ 25°C (1)
Partition coefficient (octanol/water): 6600 (1)
Adsorption coefficient: 4000 (11)

Exposure guidelines

ADI: 0.03 mg/kg/day (12)
HA: Not available
RfD: 0.013 mg/kg/day (13)
PEL: Not available

Basic manufacturer

Janssen Pharmaceutica
Plant Protection Division
1125 Trenton-Harbourton Road
Titusville, NJ 08560-1200
Telephone: 609-730-2607

10.2.8 Iprodione

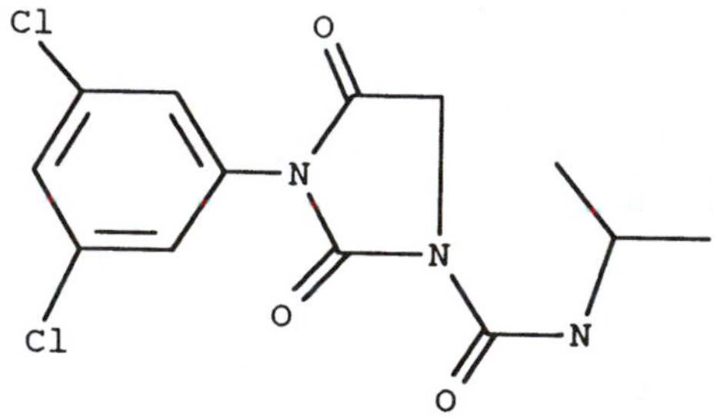

Figure 10.8 Iprodione.

Trade or other names

Trade names for commercial products containing iprodione include Chipco 26019, DOP 500F, Kidan, LFA 2043, NRC 910, Rovral, and Verisan. The compound is used in formulations with numerous other fungicides such as thiabendazole and carbendazim. It is compatible with most other pesticides.

Regulatory status

Iprodione is a slightly toxic compound and products containing it carry the Signal Word CAUTION on the label. Most products containing iprodione are General

Use Pesticides (GUPs). Chipco 26019 and Rovral (under some circumstances) may be Restricted Use Pesticides (RUPs). RUPs may be purchased and used only by certified applicators.

Introduction

Iprodione is a dicarboximide contact fungicide used to control a wide variety of crop diseases. It is used on vegetables, ornamentals, pome and stone fruit, root crops, cotton, and sunflowers to control fungal pests, and may also be used as a post-harvest fungicide and as a seed treatment. Iprodione inhibits the germination of spores and the growth of the fungal mat (mycelium).

Toxicological effects

Acute toxicity

Iprodione is slightly toxic by ingestion, with reported oral LD_{50} values of 3500 mg/kg in rats, 4000 mg/kg in mice, and greater than 4400 mg/kg in rabbits [1,8]. No dermal toxic effects were noted at doses of over 2500 mg/kg in the rat and at 1000 mg/kg in the rabbit, indicating slight toxicity by this route [1,8].

Inhalation toxicity is also low for this compound. The 4-hour inhalation LC_{50} for iprodione is greater than 3.3 mg/L in the rat [8].

Chronic toxicity

Rats given dietary doses of approximately 60 mg/kg/day over 1½ years suffered no ill effects [1,8].

Dogs fed approximately 60 mg/kg/day over 18 months also showed no adverse effects [1,8]. In another study, beagle dogs fed dietary doses of about 2.3 mg/kg/day for 1 year showed liver and kidney weight increases. At doses starting at about 1.5 mg/kg/day, the dogs had decreased prostrate weights and changes within red blood cells (damage to the hemoglobin molecules). Females also had slight decreases in uterus weights. No effects were noted below a 0.5-mg/kg/day dose [43].

Reproductive effects

Female rats fed iprodione over three successive generations showed no effects on reproduction at doses at and below 1.25 mg/kg/day [43]. Reductions in fertility and fecundity were not observed at doses of 5 mg/kg/day [43]. Based on these data, ioprodione is not likely to cause reproductive effects.

Teratogenic effects

There were no developmental effects noted in the offspring of pregnant rats receiving dietary doses of about 5.4 mg/kg/day [43]. However, the dose rate of about 120 mg/kg/day elicited unspecified developmental toxicity in the rats [43]. Rabbits did not develop any dose-related toxicity at or below 2.7 mg/kg/day iprodione, but did develop toxicity at 6 mg/kg/day [43]. It appears that iprodione is not likely to cause teratogenic effects at expected exposure levels.

Mutagenic effects

No data are currently available.

Carcinogenic effects

A 2-year feeding experiment with rats showed no increases in tumor formation or tumor precursors (neoplastic foci) at dietary doses of about 2.5 mg/kg/day (43). An 18-month study in mice showed cancer-related effects at doses up to approximately 22 mg/kg/day (43). Current evidence on the carcinogenicity of iprodione is inconclusive.

Organ toxicity

Target organs identified in animal studies include the reproductive system (prostate gland and uterus), liver, and kidneys.

Fate in humans and animals

No data are currently available (43).

Ecological effects

Effects on birds

Iprodione is slightly toxic to wildfowl. The reported acute oral LD_{50} in bobwhite quail is 930 mg/kg (1).

Effects on aquatic organisms

Iprodione is moderately toxic to fish species, with LC_{50} values ranging from 2.25 mg/L in the sunfish to 6.7 mg/L in the rainbow trout (8). Reported bioconcentration factors of 50 to 360 in carp and other fish species indicate low bioconcentration potential (8).

Effects on other organisms (non-target species)

Iprodione is nontoxic to bees (1).

Environmental fate

Breakdown in soil and groundwater

The half-life of iprodione in soil ranges from less than 7 to greater than 60 days (1,11). A representative half-life in most soils is estimated to be 14 days (11). Degradation rates vary with soil acidity, soil clay content, and history of the soil fungicide treatment. In soils that had been treated consistently with iprodione for 10 or more years, slow or little breakdown of the compound vinclozolin occurred, while in soil that had been treated with vinclozolin, rapid degradation of vinclozolin and iprodione occurred (44).

Iprodione is slightly soluble and moderately to well sorbed by most soils (11). These properties, combined with its short field half-life indicate a low potential to contaminate groundwater.

Breakdown in water

The compound breaks down very rapidly in water under aerobic conditions; the rate is less, but still rapid under near-anaerobic conditions (8). The compound is readily degraded by UV light.

Breakdown in vegetation

The compound is rapidly broken down in the plant after is taken up by the roots and translocated (1). The main metabolite in plants is 3,5-dichloroaniline (1). Iprodione alone or in combination with several other fungicides was not toxic to plants (phytotoxic) (45).

Physical properties

Iprodione is a colorless, odorless crystal (1).

Chemical name: 3-(3,5-dichlorophenyl)-N-(1-methylethyl)2,4-dioxo-1-imidazoline-carboxamide (1)
CAS #: 36734-19-7
Molecular weight: 330.17 (1)
Solubility in water: 13 mg/L @ 20°C (1)
Solubility in other solvents: v.s. in ethanol, methanol, acetonitrile, toluene, benzene, acetone, and dimethylformamide (1)
Melting point: 136°C (1)
Vapor pressure: <0.133 mPa @ 20°C (1)
Partition coefficient (octanol/water): 1260 @ 20°C (1)
Adsorption coefficient: 700 (11)

Exposure guidelines

ADI: 0.2 mg/kg/day (12)
HA: Not available
RfD: 0.04 mg/kg/day (13)
PEL: Not available

Basic manufacturer

Rhone-Poulenc Ag Co.
P.O. Box 12014
2 T.W. Alexander Dr.
Research Triangle Park, NC 27709
Telephone: 919-549-2000
Emergency: 800-334-7577

10.2.9 Metalaxyl

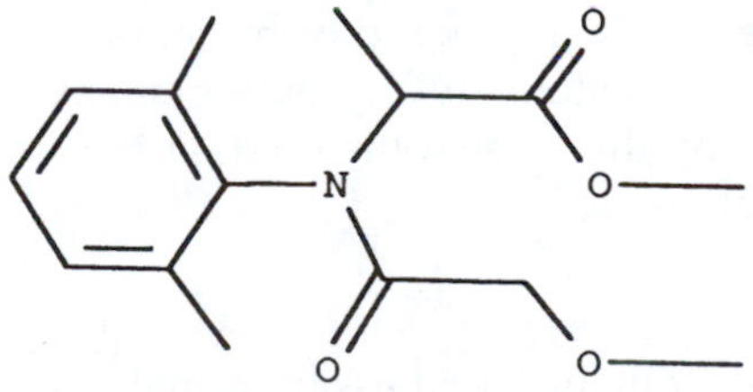

Figure 10.9 Metalaxyl.

Trade or other names

Trade names for products containing metalaxyl include Apron, Delta-Coat AD, Ridomil, and Subdue.

Regulatory status

Metalaxyl is a slightly toxic compound in EPA toxicity class III. It is a General Use Pesticide (GUP). Labels for products containing metalaxyl must bear the Signal Word CAUTION.

Introduction

Metalaxyl is a systemic benzenoid fungicide used in mixtures as a foliar spray for tropical and subtropical crops, as a soil treatment for control of soil-borne pathogens, and as a seed treatment to control downy mildews. Metalaxyl may be used on many different food crops (including tobacco), ornamentals, conifers, and turf.

Toxicological effects

Acute toxicity

The oral LD_{50} in rats is 669 mg/kg and the dermal LD_{50} is greater than 3100 mg/kg (8), indicating slight toxicity by ingestion and dermal application. Rabbits exhibited slight eye and skin irritation, but guinea pigs displayed no sensitization after metalaxyl exposure (1). No information was available regarding the inhalation toxicity of metalaxyl.

Chronic toxicity

A 90-day study of rats exposed to 0.1 to 2.5 mg/kg/day in the diet, showed some cellular enlargement in the liver at the highest dose (19). In a similar study with dogs fed diets of approximately 0.04 to 0.8 mg/kg/day for 6 months, the dogs were adversely affected by the highest dose. Manifestations included increased blood alkaline phosphatase and increased liver-to-brain weight ratio (19).

Reproductive effects

A three-generation rat study where animals were fed up to 2.5 mg/kg/day showed no compound-related maternal toxicity or reproductive effects (19). These data suggest that metalaxyl is unlikely to cause reproductive effects.

Teratogenic effects

Rats given a dosage of 120 mg/kg/day by stomach tube on days 6 to 15 of gestation exhibited no embryotoxicity or teratogenicity, nor did rabbits given a dosage of 20 mg/kg/day by the same route on days 6 to 18 (19). These data suggest that metalaxyl is not teratogenic.

Mutagenic effects

Studies including a dominant lethal assay in male mice indicate that metalaxyl has no mutagenic potential (19).

Carcinogenic effects

Available studies of the carcinogenicity of metalaxyl are inconclusive (19).

Organ toxicity

The liver is the primary target organ for metalaxyl in animal systems.

Fate in humans and animals

Studies with rats and goats showed rapid metabolism and excretion via the urine and feces (19). Metalaxyl is metabolized to a variety of products before excretion (19).

Forty-day feeding studies with dairy cattle at 15 ppm/day showed less than 0.01 ppm was stored in the muscle and fat. The liver contained 0.13 to 0.20 ppm and the kidney contained 0.26 to 0.83 ppm (19). Chickens fed for 28 days at 5 ppm in the diet had less than 0.05 ppm in the eggs, skin, fat, breast, and thigh, and less than 0.1 ppm in the liver (19).

Ecological effects

Effects on birds

Metalaxyl is reported to be practically nontoxic to birds (46).

Effects on aquatic organisms

Metalaxyl is practically nontoxic to freshwater fish. The 96-hour LC_{50} values in rainbow trout, carp, and bluegill are all above 100 mg/L (1). Freshwater aquatic invertebrates are slightly more susceptible to metalaxyl. *Daphnia magna*, a small freshwater crustacean, has an LC_{50} of 12.5 to 28 mg/L, depending on the product formulation (46). This indicates that metalaxyl is slightly toxic to this organism.

There is little tendency for metalaxyl to accumulate in the edible portion of fish. Metalaxyl did not accumulate beyond seven times the background concentration and it was quickly eliminated after exposed fish were placed in fresh (metalaxyl-free) water (46).

Effects on other organisms (non-target species)

Metalaxyl is nontoxic to bees (1).

Environmental fate

Breakdown in soil and groundwater

Under field conditions, metalaxyl has a half-life of 7 to 170 days in the soil environment (11). A representative half-life in moist soils is about 70 days (11). Increased sunlight may increase the rate of breakdown in the soil.

It is poorly sorbed by soils and highly soluble in water (11); these properties, in combination with its long persistence, pose a threat of contamination to groundwater. It readily leaches in sandy soil, although increased organic matter may decrease the rate of leaching (47). In a large-scale national survey, metalaxyl was detected in the groundwater of several states at concentrations of 0.27 µg/L to 2.3 mg/L (48).

Breakdown in water

At pH levels of 5 to 9 and temperatures of 20 to 30°C, the half-life in water was greater than 4 weeks (46). However, exposure to sunlight reduced the half-life to 1 week (46).

Breakdown in vegetation

Plants absorb foliar applications through the leaves and stems, and can translocate the compound throughout the plant. Metalaxyl is not absorbed directly from the soil by plants. The parent compound is the major residue in potato tubers and grapes, but in potato leaves and on lettuce, metabolites are the major product (7).

Physical properties

Metalaxyl is a colorless, odorless crystal (1).

Chemical name: methyl-N-(2,6-dimethylphenyl)-N-(2-xylyl)-DL-alaninate (1)
CAS #: 57837-19-1
Molecular weight: 279.34 (1)
Solubility in water: 7100 mg/L @ 20°C (1)
Solubility in other solvents: v.s. in methanol, benzene, and hexane (1)
Melting point: 71.8–72.3°C (1)
Vapor pressure: 0.293 mPa @ 20°C (1)
Partition coefficient (octanol/water): Not available
Adsorption coefficient: 50 (estimated) (11)

Exposure guidelines

ADI: 0.03 mg/kg/day (12)
HA: Not available
RfD: 0.06 mg/kg/day (13)
PEL: Not available

Basic manufacturer

Ciba-Geigy Corporation
P.O. Box 18300
Greensboro, NC 27419-8300
Telephone: 800-334-9481
Emergency: 800-888-8372

10.2.10 *Thiabendazole*

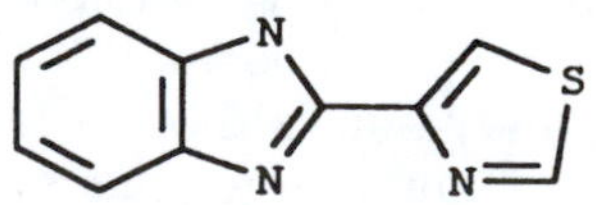

Figure 10.10 Thiabendazole.

Trade or other names

Trade names for products containing this compound include Apl-Luster, Arbotect, Mertect, Mycozol, TBZ, Tecto, and Thibenzole. The product is often used in combination with other fungicides and insecticides.

Regulatory status

Thiabendazole is a General Use Pesticide (GUP). It is in EPA toxicity class III — slightly toxic, and products containing it carry the Signal Word CAUTION on the label.

Introduction

Thiabendazole is a systemic benzimidazole fungicide used to control fruit and vegetable diseases such as mold, rot, blight, and stain. It is also active against storage diseases and Dutch Elm disease. In livestock and humans, thiabendazole is applied to treat several helminth species such as roundworms. Thiabendazole is also used medicinally as a chelating agent to bind metals. It is available as a wettable powder, suspension concentrate, flowable concentrate, and liquid.

Toxicological effects

Acute toxicity

Effects of acute overexposure to the fungicide include dizziness, anorexia, nausea, and vomiting. Other symptoms such as itching, rash, chills, and headache occur less frequently. The symptoms are brief and are related to the dose level (3).

The oral LD_{50} is 3100 to 3600 mg/kg in the rat, 1395 to 3810 mg/kg in mice, and greater than 3850 mg/kg in the rabbit (1,3). The lethal dose in sheep is 1200 mg/kg. The dermal LC_{50} in rabbits is greater than 5000 mg/kg. Thiabendazole is not a skin irritant or a sensitizer (3).

Chronic toxicity

Rats fed 200 mg/kg/day or less showed few or no growth effects. At higher doses (400 mg/kg/day), there was growth suppression. Death occurred in a few days at 1200 mg/kg/day and 30% mortality occurred within 30 days at 800 mg/kg/day. A decrease of active bone marrow at high doses was also noted (8). At doses somewhat below the LD_{50}, mice experienced significant liver, spleen, and intestinal effects.

In dogs, high daily doses (200 mg/kg/day) for 2 years produced few effects other than occasional attacks of vomiting and persistent anemia. Sheep experienced toxic depression and anorexia at very high doses (800 to 1,000 mg/kg/day). Studies on cattle, sheep, goats, swine, horses and zoo animals have shown few chronic symptoms at low doses (3).

Reproductive effects

A three-generation study in rats showed no adverse effects on reproduction at 20 to 80 mg/kg/day. However, four times this low therapeutic dose produced serious pregnancy-related disorders (eclampsia) in sheep (3).

Mice studied for five generations showed no effects at 10 mg/kg/day, but did show decreased weanling weights at 50 mg/kg/day, and decreased weanling weight and size at 250 mg/kg/day (3,8).

Reproductive effects in humans are not likely at anticipated levels of exposure.

Teratogenic effects

Pregnant rabbits fed doses of 75, 150, and 600 mg/kg/day produced pups with lower fetal weights at the highest dose tested. No birth defects were observed with thiabendazole at any dose tested (3,8).

Teratogenic effects are not likely from thiabendazole exposure.

Mutagenic effects

Several studies with bacteria have failed to produce any chromosome changes or mutations due to thiabendazole (8). It appears that the compound is not mutagenic.

Carcinogenic effects

A 2-year feeding study with rats at levels of 10 to 160 mg/kg/day produced no cancer-related effects attributable to thiabendazole (3,8). Another study conducted over 18 months at the maximum tolerated dose in mice produced no evidence of cancer-related effects (3,8). It does not appear that thiabendazole is carcinogenic.

Organ toxicity

Dogs autopsied after a 2-year feeding study had incomplete development of bone marrow, a wasting away of lymph tissue, and other abnormalities (30). Most dogs tested at about 100 mg/kg/day for 2 years developed anemia. The dogs recovered at the end of the study (3).

Fate in humans and animals

In four men given 1000 mg (approximately 14 mg/kg) thiabendazole orally, plasma concentrations peaked at 13 to 18 ppm within an hour (3). Within 4 hours, 40% of the dose was excreted and, within 24 hours, 80% was excreted, mostly in the urine as metabolites of the compound (3). Elimination is rapid in other species as well. Rats almost completely eliminate the compound after 48 hours and sheep after 96 hours (3). Metabolites are distributed throughout most body tissues in sheep, but are detectable in only a few tissues at low levels (less than 0.2 ppm) at 16 days and at very low levels (0.06 ppm or less) after 30 days (8).

Ecological effects

Effects on birds

No data are currently available.

Effects on aquatic organisms

Thiabendazole is of low toxicity to fish (8). The compound is not expected to appreciably accumulate in aquatic organisms. The bioconcentration factor for thiabendazole in whole fish is 87 times the ambient water concentrations. Fish eliminated the compound within 3 days after being placed in thiabendazole-free water (49).

Effects on other organisms (non-target species)

Earthworms are sensitive to the compound (LD_{50} = approx. 20 µg/worm), while bees are not (8). It is nontoxic to bees.

Environmental fate

Breakdown in soil and groundwater

Thiabendazole's affinity for binding to soil particles increases with increasing soil acidity. It is highly persistent. The field half-life for thiabendazole has been reported as 403 days (11). In one study, 9 months following application, most of the residues (85 to 95%) were recovered from soil. Due to its binding and slight solubility in water, it is not expected to leach readily from soil.

Breakdown in water

Thiabendazole is stable in aqueous suspension and acidic media (1). Its low water solubility will make it unlikely to be in solution, and it will most likely be bound to sediment.

Breakdown in vegetation

No metabolism was seen with seed potatoes, but photoproducts were detected on sugar beet leaves (8). Total residues in sugar beets were 78% parent compound, with the remaining 22% being benzimidazole, benzimidazole-2-carboxamide, and unidentified products. Thiabendazole is readily absorbed by roots and translocated to all parts of a plant, but predominantly to the leaf margins (8).

Physical properties

Thiabendazole is an odorless, colorless powder (1).

Chemical name: 2-(thiazol-4-yl)benzimidazole (1)
CAS #: 148-79-8
Molecular weight: 201.2 (1)
Solubility in water: <50 mg/L @ pH 5 to 12 (1)
Solubility in other solvents: s. in acetone and ethanol; s.s in benzene and chloroform (1)
Melting point: 304–305°C (1)
Vapor pressure: Negligible at room temperature (1)
Partition coefficient (octanol/water): Not available
Adsorption coefficient: 2500 (11)

Exposure guidelines

ADI: 0.1 mg/kg/day (12)
HA: Not available
RfD: 0.1 mg/kg/day (13)
PEL: Not available

Basic manufacturer

Merck Agvet
Division of Merck & Co., Inc.
P.O. Box 2000
Rahway, NJ 07065-0912
Telephone: 908-855-4277

10.2.11 Vinclozolin

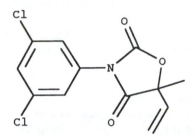

Figure 10.11 Vinclozolin.

Trade or other names

Trade names for commercial products containing vinclozolin include BAS 352F,
Drive, Ornalin, Ronilan, and Vorlan. Vinclozolin may also be used in formulations
with other fungicides such as thiram, carbendazim, chlorothalonil, maneb, and
thiophanate-methyl.

Regulatory status

Vinclozolin is a General Use Pesticide (GUP). It is toxicity class III — slightly
toxic. Products containing it carry the Signal Word CAUTION on the label.

Introduction

Vinclozolin is a nonsystemic dicarboximide fungicide used for the control of
several types of fungi in vines (such as grapes), strawberries, vegetables, fruit, and
ornamentals (1). It may also be used on turf grass. This fungicide works by inhibiting
spore germination. It is available in dust, wettable powder, smoke tablet, and water-
dispersible flowable granular forms.

Toxicological effects

Acute toxicity

Vinclozolin is practically nontoxic in experimental animals. The acute oral LD_{50}
for vinclozolin is greater than 10,000 mg/kg in rats, and around 8000 mg/kg in
guinea pigs (8).

The compound is a moderate skin irritant and will slightly irritate the membranes in the nose and throat (1). The reported dermal LD_{50} value is 72,000 mg/kg.
The 4-hour inhalation LC_{50} of a 50% concentration of vinclozolin is greater than 29

mg/L air in rats, indicating a rather low toxicity by this route of exposure. Vinclozolin is a moderate eye irritant (1,8).

Chronic toxicity

Vinclozolin was fed to dogs at relatively low levels (up to 50 mg/kg/day) for 6 months. Increases in the weight of the adrenal gland occurred in the dogs at the middle doses (7.5 mg/kg/day) for both sexes, and males had enlarged prostates. Slightly higher doses in females caused changes in the structure of the adrenal gland. Another study with dogs fed small amounts of vinclozolin showed chronic effects (unspecified) at levels of 2.5 mg/kg and above (50). Dogs are the most sensitive species of animal tested so far (13). A 2-year feeding study with rats showed reductions in body weight and changes in the blood chemistry at low doses (about 25 mg/kg/day) (13).

Male dogs experienced changes in absolute weight and fat content of the kidney at relatively low doses administered for 6 months. At slightly higher doses (15 mg/kg) for the same length of time (6 months), fat droplets appeared in the tubes within the kidney (13). A single moderate dose (about 285 mg/kg) administered by injection to male mice resulted in only minor changes in their kidneys (51).

Reproductive effects

A study that followed female rats through three successive litters showed no effects on the reproduction of those litters at doses of 72.9 mg/kg/day (13). These data suggest that vinclozolin does not cause reproductive effects.

Teratogenic effects

In one study on mice, no birth defects were noted in the offspring of pregnant females given large doses of vinclozolin (900 mg/kg/day) (13). However, the fungicide was toxic to the fetuses. In a similar study on rats, no teratogenic effects were observed at the same dose level (13).

In another study, rabbits were fed moderate amounts (up to 300 mg/kg/day) of the fungicide for an undisclosed amount of time. No effects were noted in the animals at the highest doses tested (8,13). It appears that vinclozolin is not teratogenic.

Mutagenic effects

A number of tests on the mutagenicity of vinclozolin have been negative. One of the mutation tests was run at very high doses (2000 mg/kg/day) (50). Based on the information available, it is unlikely that vinclozolin is mutagenic.

Carcinogenic effects

A 2-year study on rats showed no carcinogenic effects at the highest dose tested (219 mg/kg) (50). Another study conducted over a wide range of doses, produced some evidence of liver tumors at 219 mg/kg/day over 2 years. These data suggest that this compound is unlikely to have carcinogenic effects in humans.

Organ toxicity

Tests on dogs have shown effects on the adrenal and prostate glands.

Fate in animals and humans

Rats given a single dose of vinclozolin (level not indicated) eliminated equal portions of the breakdown products in urine and feces (8).

Ecological effects

Effects on birds

In the one bird species tested, the bobwhite quail, the LD_{50} of vinclozolin was 2510 mg/kg (8). This value indicates that the compound is practically nontoxic to this species.

Effects on aquatic organisms

Vinclozolin is only moderately toxic to freshwater fish. The LC_{50} (96-hour) for the compound is 130 mg/L in guppies (1), and 52.2 mg/L in trout (1).

Effects on other organisms (non-target species)

Vinclozolin is nontoxic to honeybees and to earthworms (1).

Environmental effects

Breakdown in soil and groundwater

Vinclozolin is of low to moderate persistence in soil. It is only partially broken down by soil microorganisms (1). Estimated half-lives of 3 days to greater than 3 weeks have been reported (52).

Field data indicate that vinclozolin will be strongly sorbed to most soils with a significant proportion of organic matter, and it is unlikely to leach significantly (52,53).

Breakdown in water

Photolysis and hydrolysis may occur, and are pH dependent, with greater photolysis and hydrolysis under neutral or slightly basic conditions (52).

Breakdown in vegetation

In its discussion of tolerance setting for vinclozolin in grapes, the EPA stated that there is no reasonable expectation for vinclozolin residues to be found in eggs, milk, meat, or poultry from its use on table grapes (50).

Physical properties

Vinclozolin is a colorless crystal with a slight aromatic odor (1).

Chemical name: (RS)-3-(3,5-dichlorophenyl)-5-methyl-5-vinyl-1,3-oxazolidine-2,4-dione (1)
CAS #: 50471-44-8
Molecular weight: 286.11 (1)
Solubility in water: 3.4 mg/L @ 20°C (1)
Solubility in other solvents: v.s. in ethanol, acetone, ethyl acetate, cyclohexane, benzene, and xylene (1)
Melting point: 108°C (1)
Vapor pressure: 0.016 mPa @ 20°C (1)

Partition coefficient (octanol/water): 1000 (1)
Adsorption coefficient: 100 (estimated) (53)

Exposure guidelines

ADI: 0.07 mg/kg/day (12)
HA: Not available
RfD: 0.025 mg/kg/day (13)
PEL: Not available

Basic manufacturer

BASF Corporation
Agricultural Products Division
P.O. Box 13528
Research Triangle Park, NC 277099-3528
Telephone: 800-669-2273
Emergency: 800-832-4357

10.3 Herbicides

10.3.1 Acifluorfen

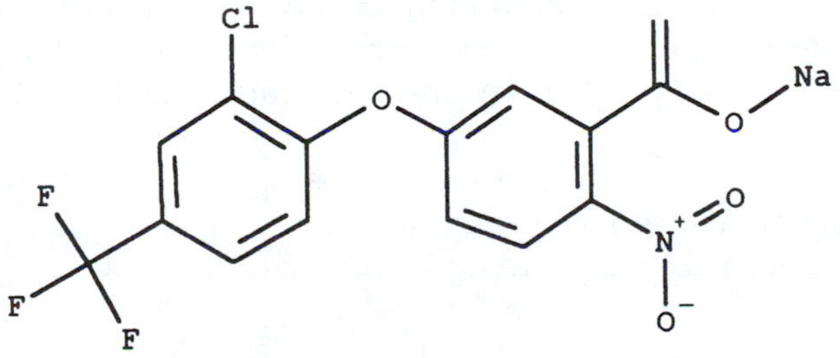

Figure 10.12 Acifluorfen.

Trade or other names

Acifluorfen and sodium acifluorfen are found in several Blazer products. Other trade names include Carbofuorfen, RH-6201, and Tackle.

Regulatory status

Acifluorfen is classified as a General Use Pesticide (GUP) by the U.S. EPA. It is toxicity class III — slightly toxic, but products containing it bear the Signal Word DANGER because it can cause serious eye injury.

Introduction

Acifluorfen is a contact, diphenolic ether herbicide used to control broadleaf weeds and grasses in soybeans, peanuts, peas, and rice. It can be applied before or after crop emergence. It is especially effective against cocklebur, velvetleaf, common lambsquarters, morning glory, and jimsonweed (2). Its use in the U.S. is estimated to be in excess of 1.4 million pounds per year.

Toxicological effects

Acute toxicity

The oral LD_{50} is 2025 mg/kg in male rats and 1370 mg/kg in female rats, indicating slight acute toxicity (1). The LD_{50} for sodium acifluorfen is 1300 mg/kg in rats (8). In rabbits, the dermal LD_{50} values of acifluorfen and formulated products are all greater than 2000 mg/kg (58). In rats, the inhalation LD_{50} (4-hour) is greater than 6.9 mg/L air (1).

Chronic toxicity

Male and female rats fed high daily doses for 4 weeks showed decreased food consumption and increased liver and kidney weights. In a 1-year study of rats fed lower doses, both sexes experienced decreased body weight and increased liver weight (54). In a 2-year study, beagle dogs fed high daily doses of acifluorfen showed irregular heart rhythms. In addition, there were some blood changes and an increase in liver and kidney weights (54).

Reproductive effects

No adverse effects were observed in rodents or their offspring when the parents were fed daily doses of acifluorfen well below lethal levels. Body weights, food consumption, fertility, and pregnancy were comparable in both treated and untreated animals (54).

However, in another rat study at higher doses, both parents and offspring suffered kidney lesions and death. This suggests that levels high enough to cause toxicity in the mother are needed to affect reproduction (54).

Teratogenic effects

Acifluorfen may have teratogenic effects at high doses. In one study, rats were given high doses of sodium acifluorfen through a stomach tube during the critical periods of pregnancy. At these doses, body weights of the fetuses were lower, and bone development was delayed (54). Teratogenic effects in humans are unlikely at expected exposure levels.

Mutagenic effects

Various mutagenesis assays of acifluorfen products on both bacteria and mammalian cells indicate that they do not cause mutations (54).

Carcinogenic effects

One study of mice fed high doses of acifluorfen for 18 months showed decreases in body weight and increases in both benign and malignant liver tumors (54). These data are not sufficient to characterize the carcinogenicity of acifluorfen (55).

Organ toxicity

In addition to being a skin and eye irritant, acifluorfen affects the weight and functions of the liver, heart, and kidneys at high doses.

Fate in humans and animals
No data are currently available.

Ecological effects

Effects on birds

Acifluorfen is practically nontoxic to mallards and is moderately toxic to bobwhite quail. The acute oral LD_{50} of acifluorfen is 2821 mg/kg in mallards and 325 mg/kg in bobwhite quail (1). The range in toxicity to these different species makes difficult any generalizations about its overall toxicity to birds.

Effects on aquatic organisms

Acifluorfen is slightly toxic to fish. The LC_{50} values for the sodium salt are 31 mg/L in bluegill and 54 mg/L in rainbow trout (56). It has a low toxicity to crustaceans. The LC_{50} (96-hour) in fiddler crabs is greater than 1000 mg/L and is 150 mg/L in freshwater clams (56).

Effects on other organisms (non-target species)

Acifluorfen is nontoxic to bees (1).

Environmental fate

Breakdown in soil and groundwater

Acifluorfen is moderately persistent in soils. In one study, acifluorfen applied to a silt loam degraded with a half-life of 59 days (54). Microbial action accounts for the majority of the compound's loss from soil. No leaching of the chemical below 3 inches was observed (54).

Breakdown in water

Acifluorfen is stable in water; no degradation was observed in laboratory studies lasting up to 28 days. However, when it is exposed to sunlight, it degrades quickly. The half-life under continuous light was 92 hours in water. When it does degrade, the primary breakdown product tends to vaporize (54).

Breakdown in vegetation

In susceptible plants, such as common cocklebur and ragweed, acifluorfen is absorbed through the leaves and roots and is translocated only slightly (57). It works by inhibiting a critical plant enzyme. In acifluorfen-resistant plants like soybeans, no acifluorfen movement from the treated leaves takes place because plants break down acifluorfen into a nontoxic form (58). High relative humidity favors herbicide penetration into the plant. High temperatures before and after spraying tend to increase susceptibility and death (59).

Physical properties

At 25°C, the acid of acifluorfen is an off-white solid, and the sodium salt is a white or brown crystalline powder (1).

Chemical name: sodium 5-(2-chloro-4-(trifluoromethyl)phenoxy)-2-nitroben-
   zoate (1)
CAS #: 62476-59-9 (sodium salt)
Molecular weight: 361.7 (1)
Solubility in water: >250,000 mg/L @ 25°C (1)
Solubility in other solvents: acetone v.s.; ethanol v.s.; xylene s.s. (1)
Melting point: 142–167°C (1)
Vapor pressure: 0.133 mPa @ 20°C (1)
Partition coefficient (octanol/water): 10–15 (58)
Adsorption coefficient: 113 (estimated) (salt) (11)

Exposure guidelines

ADI: 0.0125 mg/kg/day (12)
HA: 0.4 mg/L (longer-term) (54)
RfD: Not available
PEL: Not available

Basic manufacturer:

BASF Corp.
Agricultural Products Group
P.O. Box 13528
Research Triangle Park, NC 27709-3528
Telephone: 800-669-2273
Emergency: 800-832-4357

10.3.2 Alachlor

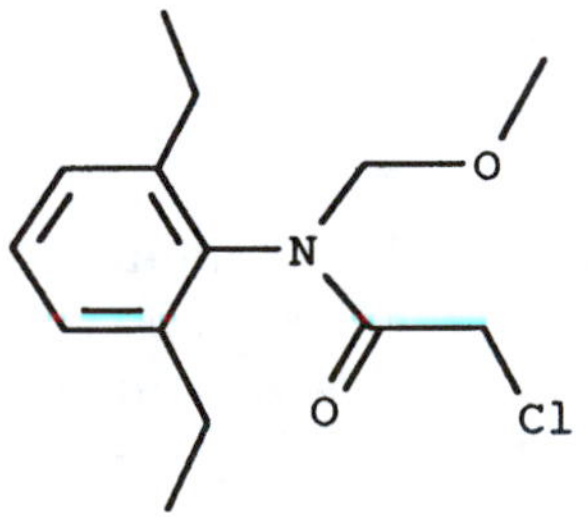

Figure 10.13 Alachlor.

Trade or other names

Trade names of commercial herbicides containing alachlor include Alanex,
Bronco, Cannon, Crop Star, Lariat, Lasso, and Partner. It mixes well with other
herbicides such as Bullet, Freedom, and Rasta, and is found in mixed formulations
with atrazine, glyphosate, trifluralin, and imazaquin.

Regulatory status

Alachlor is a Restricted Use Pesticide (RUP). RUPs may be purchased and used
only by certified applicators. The EPA categorizes it as toxicity class III — slightly

toxic. However, alachlor products bear the Signal Word DANGER on their labels because of their potential to cause cancer in laboratory animals.

Introduction

Alachlor is an aniline herbicide used to control annual grasses and broadleaf weeds in field corn, soybeans, and peanuts. It is a selective systemic herbicide, absorbed by germinating shoots and by roots. It works by interfering with a plant's ability to produce protein and by interfering with root elongation.

This compound is one of the most highly used herbicides in the U.S. Over 50 million pounds were applied annually in 1990. It is available as granules or emulsifiable concentrate.

Toxicological effects

Acute toxicity

Alachlor is a slightly toxic herbicide. The LD_{50} of alachlor in rats is between 930 and 1350 mg/kg. In the mouse, the LD_{50} is between 1910 and 2310 mg/kg (60). The dermal LD_{50} in rabbits is 13,300 mg/kg, but some of the formulated materials can be more toxic, with dermal LD_{50} values ranging from 7800 to 16,000 mg/kg (58). Skin irritation is slight to moderate. The inhalation LC_{50} in rats is reportedly greater than 23.4 mg/L for 6 hours of exposure (1,60).

Chronic toxicity

A 90-day study on rats and dogs given diets containing low to moderate amounts of alachlor (1 to 100 mg/kg/day) showed no adverse effects (58). However, a 6-month dog study showed liver toxicity at all doses above 5 mg/kg/day, and a 1-year study established that above 1 mg/kg/day, alachlor causes effects in the liver, spleen, and kidney. In 2-year rat studies, doses above 2.5 mg/kg/day caused irreversible degeneration of the iris and related eye structures (60).

Reproductive effects

High oral doses (150 or 400 mg/kg/day) fed to rats during gestation resulted in maternal and fetal toxicity, but there was no indication that reproduction was affected (60,61). Alachlor does not appear to cause reproductive effects.

Teratogenic effects

Doses of up to 150 mg/kg/day fed to rabbits on days 7 through 19 of pregnancy did not result in any birth defects (6). Similar studies in rats at doses up to 400 mg/kg/day did not result in birth defects, but toxic effects in the mothers and offspring were seen at the highest dose. These data indicate that alachlor is not likely to cause birth defects.

Mutagenic effects

Alachlor does not appear to be mutagenic. Mutagenicity assays with a variety of microbial strains at numerous concentrations of alachlor were all negative (60,61).

Carcinogenic effects

Rats given high doses of alachlor developed stomach, thyroid, and nasal turbinate tumors (60). An 18-month mouse study with doses from 26 to 260 mg/kg/day showed an increase of lung tumors at the highest dose for females but not males (60). Because of inconsistencies in these studies, the oncogenic potential of alachlor is uncertain (62).

Organ toxicity

Long-term studies in rats and dogs showed effects on the liver, spleen, and kidneys. The iris and lung have also been affected.

Fate in humans and animals

Rats given a single low dose of alachlor (14 mg/kg) eliminated most of it within the first 48 hours (60). Nearly all of the compound was eliminated in 10 days; a third to a half of the alachlor and its breakdown products were excreted in urine and nearly a half in feces. Rats absorbed close to 75% of a single low dose (14 mg/kg) through skin, while only 8 to 10% was absorbed through the skin in monkeys (60).

Ecological effects

Effects on birds

Alachlor is slightly to practically nontoxic to wildfowl. Alachlor has a 5-day dietary LC_{50} of greater than 5000 ppm in young mallard ducks and bobwhite quail (58). The LD_{50} of alachlor in other mallard ducks was greater than 2000 mg/kg (63). The LC_{50} of alachlor in pheasants is greater than 10,000 ppm (8).

Effects on aquatic organisms

Alachlor is moderately toxic to fish. The LC_{50} (96-hour) for alachlor is 2.4 mg/L in rainbow trout, 4.3 mg/L in bluegill sunfish, 6.5 mg/L in catfish, and 4.6 mg/L in carp (1,8). It is only slightly toxic to crayfish, with an LC_{50} (96-hour) of 19.5 mg/L (8,37).

The bioaccumulation factor in the channel catfish is 5.8 times the ambient water concentration, indicating that alachlor is not expected to accumulate appreciably in aquatic organisms (8).

Effects on other organisms (non-target species)

Alachlor is not toxic to bees. It is practically nontoxic to earthworms (61).

Environmental fate

Breakdown in soil and groundwater

Alachlor has a low persistence in soil, with a half-life of about 8 days (1,8). The main means of degradation is by soil microbes (58). It has moderate mobility in sandy and silty soils and thus can migrate to groundwater (60). The largest groundwater testing program for a pesticide, the National Alachlor Well Water Survey, was conducted throughout the last half of the 1980s. Over 6 million private and domestic wells were tested for the presence of alachlor. Less than 1% of all the wells had detectable levels of alachlor (64). In the wells where the compound was detected, concentrations ranged from 0.1 to 1.0 µg/L, with the majority having concentrations around 0.2 µg/L (64).

Breakdown in water

Alachlor breaks down rapidly in natural water, primarily due to the action of microorganisms. The breakdown rate is much slower in water with no oxygen (58).

Breakdown in vegetation

Absorption is primarily by germinating shoots and it is readily translocated throughout the plant (37). Higher concentrations appear in the vegetative parts than in the reproductive parts of the plant. Alachlor is rapidly metabolized to water-soluble products in plants (19). It is almost completely metabolized within 10 days (8).

Physical properties

Alachlor is a colorless to yellow crystal compound (1).

Chemical name: 2-chloro-2',6'-diethyl-N-(methoxymethyl) acetanilide (1)
CAS #: 15972-60-8
Molecular weight: 269.77 (1)
Solubility in water: 242 mg/L @ 25°C (1)
Solubility in other solvents: s. in most organic solvents (1)
Melting point: 40°C (1)
Vapor pressure: 2.9 mPa @ 25°C (1)
Partition coefficient (octanol/water): 794 (58)
Adsorption coefficient: 170 (11)

Exposure guidelines

ADI: 0.0025 mg/kg/day (12)
MCL: 0.002 mg/L (65)
RfD: 0.01 mg/kg/day (13)
PEL: Not available

Basic manufacturer

Monsanto Company
800 N. Lindbergh Blvd.
St Louis, MO 63167
Telephone: 314-694-6640
Emergency: 314-694-4000

10.3.3 *Ammonium sulfamate*

Trade or other names

Ammonium sulfamate is also called AMS. Trade names include Amcide, Amicide, Amidosulfate, Ammate, Ammate X-NI, Fyran 206k, Ikurin, Silvicide, and Sulfamate.

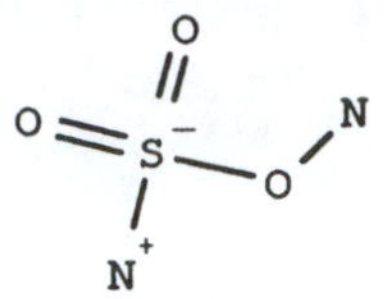

Figure 10.14 Ammonium sulfamate.

Regulatory status

Ammonium sulfamate was a General Use Pesticide (GUP) that was discontinued by DuPont in 1988. It was considered toxicity class III — slightly toxic, and products containing ammonium sulfamate had the Signal Word CAUTION.

Introduction

Ammonium sulfamate (AMS) is an herbicide used to control many types of woody plants, trees, herbaceous perennials, and annual broadleaf weeds and grasses. AMS is a contact herbicide which means that it injures only those parts of the plant to which it is applied. It is used primarily to control undesired growth along rights-of-way and for general weed and poison ivy control around homes, commercial buildings, and fruit orchards. AMS is also used as a fertilizer.

Ammonium sulfamate is applied in water solution or oil-water emulsion as a leaf or foliar spray for control of woody plants, or it is applied as crystals or concentrated solution to cuts in the bark or on freshly-cut stumps of trees to prevent resprouting.

Toxicological effects

Acute toxicity

Ammonium sulfamate is slightly toxic, with an oral LD_{50} of 3900 mg/kg in rats and 3100 mg/kg in mice (1,58). However, contact with ammonium sulfamate may cause burns to the eyes (8). The dust of ammonium sulfamate is irritating to the nose and throat and can cause coughing or difficult breathing if inhaled (8).

Chronic toxicity

In a 105-day study with rats fed 500 mg/kg/day, AMS did not cause signs of poisoning. Some inhibition of growth was seen at doses of 1000 mg/kg/day (66). There was no skin irritation, nor any signs of systemic toxicity, when 20 and 50% water-based solutions were applied to the shaved skin of rats (65).

Reproductive effects

Reproduction was not impaired when rats were given dietary doses of 17.5 or 25 mg/kg/day of AMS for 15 months (8). This suggests that AMS does not cause reproductive effects.

Teratogenic effects

No data are currently available.

Mutagenic effects

Limited data indicate that AMS in not mutagenic. The Ames/Salmonella assay was negative for AMS, indicating that it does not cause permanent changes in genetic material (8).

Carcinogenic effects

A rat study indicates that AMS is not carcinogenic at doses of 25 mg/kg/day (8). There are insufficient additional data to confidently determine the carcinogenicity status of AMS (67).

Organ toxicity

Ammonium sulfate may cause gastrointestinal tract problems if it is eaten (8).

Fate in humans and animals

AMS is readily absorbed into the bloodstream from the gastrointestinal tract (66). Following oral administration of AMS to dogs for 5 days, 80 to 84% of the dose was excreted as sulfamic acid in the urine.

Ecological effects

Effects on birds

AMS is practically nontoxic to birds. The oral LD_{50} is 3000 mg/kg in quail and 4200 mg/kg in ducks (1,22). In a 14-day feeding study, 150 and 590 mg/kg/day AMS had no effect on quail. Quail fertility was not affected when 150 mg/kg/day was mixed with their feed for two 10-day periods (58).

Effects on aquatic organisms

Ammonium sulfamate is practically nontoxic to fish. A 46% AMS solution becomes toxic in perch at 300 mg/L water (30). The LC_{50} (24-hour) of AMS is 1250 mg/L in harlequin fish (28). A concentration of 30 mg/L AMS had no effect on rainbow trout or aquatic invertebrates (8).

Effects on other organisms (non-target species)

Deer were not harmed when they were fed AMS-treated leaves (28).

Environmental fate

Breakdown in soil and groundwater

Ammonium sulfamate has a low to moderate persistence in soil and can be decomposed by soil microbes in a 6- to 8-week period (1,28). In regions with less moisture, traces of AMS and its breakdown products may remain on or in soil, water, or a crop for longer periods of time (58). Soil may become temporarily nonproductive if AMS is used at high rates (8). At dosages of 400 to 600 kg/ha, it can temporarily sterilize soil (8).

AMS does not readily bind, or adsorb, to soil particles. It dissolves readily and may leach through the soil to groundwater (8).

Breakdown in water

Ammonium sulfamate absorbs moisture readily and is highly soluble in water (8).

Breakdown in vegetation

Ammonium sulfamate is a contact herbicide, and causes injury only to the parts of the plant to which it is applied. It is absorbed rapidly through leaves, as well as through cut surfaces on wood and is translocated to other parts of the plant (58). It is nonselective and capable of killing non-target plants (58).

Physical properties

Ammonium sulfamate is a colorless, odorless, hygroscopic, water-soluble crystalline solid (1).

Chemical name: ammonium sulfamidate (1)
CAS#: 7773-06-0
Molecular weight: 114.13 (1)
Water solubility: v.s.; 684,000 mg/L @ 25°C (1)
Solubility in other solvents: i.s. in alcohol; s.s. in ethanol; s. in glycerols, glycols, and formamides (1)
Melting point: 131-132°C (1)
Vapor pressure: Negligible at room temperature (1)
Partition coefficient (octanol/water): Not available
Adsorption coefficient: 30 (estimated) (53)

Exposure guidelines

ADI: Not available
HA: 2 mg/L (lifetime) (67)
RFD: 0.2 mg/kg/day (13)
PEL: 5 mg/m3 (8-hour) (dust) (respirable fraction) (14)

Basic manufacturer

Du Pont Agricultural Products
Walker's Mill, Barley Mill Plaza
P.O. Box 80038
Wilmington, DE 19880-0038
Telephone: 800-441-7515
Emergency: 800-441-3637

10.3.4 Bentazon

Trade or other names

Trade names include Adagio, Bas 351 H, Basagran, Bendioxide, Bentazone, Entry, Leader, and Pledge. All products currently marketed in the U.S. contain the sodium salt of bentazon, referred to as sodium bentazon, as the active ingredient.

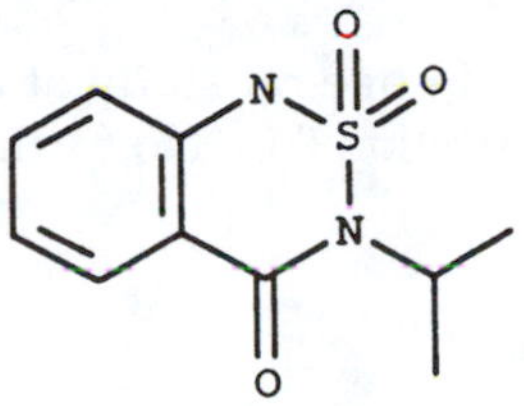

Figure 10.15 Bentazon.

Regulatory status

Bentazon is A General Use Pesticide (GUP) that is classified as toxicity class III — slightly toxic. Products containing bentazon bear the Signal Word CAUTION.

Introduction

Bentazon is a post-emergence herbicide used for selective control of broadleaf weeds and sedges (a weed) in beans, rice, corn, peanuts, mint, and others. Bentazon is a contact herbicide, which means that it causes injury only to the parts of the plant to which it is applied. It interferes with the ability of susceptible plants to use sunlight for photosynthesis. Visible injury to the treated leaf surface usually occurs within 4 to 8 hours, followed by plant death. It should not be used on blackeyed peas or garbanzo beans.

Toxicological effects

Acute toxicity

Bentazon is slightly toxic by ingestion and by dermal absorption (22). Human ingestion of high doses of this herbicide has caused vomiting, diarrhea, trembling, weakness, and irregular or difficult breathing. It is moderately irritating to the skin, eyes, and respiratory tract (68). Severe eye irritation from this material healed after 1 week (8). Symptoms that have occurred in test animals include apathy, incoordination, prostration, tremors, anorexia, vomiting, and diarrhea (58).

The LD_{50} for bentazon in cats is 500 mg/kg, in rabbits is 750 mg/kg, in mice is 400 mg/kg, and in rats is 1100 to 2063 mg/kg (1,58).

When bentazon was applied to the shaved skin of rabbits, it did not cause irritation (1,58). Its dermal LD_{50} is 4000 mg/kg in rabbits.

Chronic toxicity

Consumption of bentazon at high levels for a long time results in excessive weight loss and inflammation of the prostrate gland in animal studies (69).

Prolonged or repeated exposure of the skin or eyes to bentazon may cause dermatitis or conjunctivitis (22).

When dogs were given 2.5, 7.5, 25, or 75 mg/kg/day for 13 weeks, weight loss, general ill health, and inflammation of the prostrate occurred at the highest dose tested (69).

Reproductive effects
No data are currently available.

Teratogenic effects

Birth defects were observed in one rat study at a dose of 200 mg/kg/day, but the validity of these data is in question (70). No further data are currently available.

Mutagenic effects

No data are currently available.

Carcinogenic effects

Tumors have been seen in rats given 200 mg/kg/day of bentazon, but these results are questionable (71). No further data are currently available (1).

Organ toxicity

Animal studies have shown that the prostate gland may be affected.

Fate in humans and animals

Bentazon is rapidly absorbed and readily excreted, unchanged, in the urine. About 91% of a 0.8-mg dose administered to rats by stomach tube was excreted in the urine within 24 hours of ingestion, with less than 1% eliminated in feces. This suggests that bentazon is almost completely absorbed from the gastrointestinal tract into the bloodstream when it is ingested (69).

Ecological effects

Effects on birds

Technical and formulated bentazon are both slightly toxic to birds. The oral LD_{50} of formulated bentazon (BAS 3510H) is 2000 mg/kg in mallard ducks and 720 mg/kg in Japanese quail (58).

Effects on aquatic organisms

Bentazon is practically nontoxic to both coldwater and warmwater fish. Bentazon is slightly toxic to aquatic invertebrates. The LC_{50} (96-hour) for bentazon in rainbow trout is 510 mg/L for wettable powder. The LC_{50} (96-hour) for technical bentazon is 616 mg/L in bluegill sunfish and 190 mg/L in rainbow trout. For formulated bentazon (BAS 351-H), the LC_{50} (96-hour) in bluegills is 1060 mg/L, and in rainbow trout is 636 mg/L (1,58). The bioconcentration factor for bentazon predicted from its water solubility is 19, indicating low bioaccumulation potential.

Effects on other organisms (non-target species)

Applications of bentazon are not considered hazardous to most non-target organisms because of its generally low toxicity. Bentazon is not toxic to bees (8,70).

Environmental fate

Breakdown in soil and groundwater

Bentazon has a low persistence in soil. Its half-life is less than 2 weeks (11). Bentazon reaches undetectable levels in soil 6 weeks after its application (58,71). It is subject to breakdown by ultraviolet (UV) light from the sun and rapid degradation by soil bacteria and fungi (8).

Bentazon does not bind to soil particles and it is highly soluble in water. These characteristics usually suggest a strong potential for groundwater contamination. However, its rapid degradation is expected to prevent significant contamination of groundwater (11,72).

Based on a national survey, the EPA estimates that bentazon may be found in about 0.1% of the rural drinking water wells nationwide. Bentazon was not detected in any community water systems. It was also not detected at concentrations above 0.02 mg/L in any well (73).

Breakdown in water

Bentazon has the potential to contaminate surface water because of both its mobility in runoff water from treated crops and its pattern of use on rice, which involves either direct application to water or application to fields prior to flooding. Commercial formulations are readily water soluble (72).

Bentazon appears to be stable to hydrolysis, a chemical reaction with water. However, it has a half-life of less than 24 hours in water because it is readily broken down by sunlight (72).

Breakdown in vegetation

Bentazon is absorbed by plant leaves. In resistant (tolerant) plants, it is rapidly broken down into natural plant components (8,58).

It may be also be absorbed by the roots and translocated from the roots to other plant parts (58). The degree of translocation depends on the type of plant. Whether translocated or not, bentazon is quickly metabolized, reorganized, and incorporated into natural plant components (72).

Bentazon has little effect on germinating seeds (58). Some leaf-speckling and leaf-bronzing may occur under certain conditions of bentazon usage. Crop injury may result if bentazon is applied to crops that have been subjected to stress conditions, such as drought or widely fluctuating temperatures (8).

Physical properties

Pure bentazon is a colorless to slightly brown, odorless crystalline solid (1).

Chemical name: 3-isopropyl-1H-2,1,3-benzothiadiazin-4(3H)-one 2,2-dioxide (1)
CAS #: 25057-89-0
Molecular weight: 240.28 (1)
Solubility in water: 500 mg/L @ 20°C (1)
Solubility in other solvents: acetone v.s.; ethanols, ethyl acetates, and benzene
    s.s.; cyclohexane s.s. (1)
Melting point: 137-139°C (1)
Vapor pressure: Negligible; <0.46 mPa @ 20°C (1)
Partition coefficient (octanol/water): 0.35 (1)
Adsorption coefficient: 34 (parent acid) (11)

Exposure guidelines

ADI: 0.1 mg/kg/day (12)
HA: 0.02 mg/L (lifetime) (69)
RfD: 0.0025 mg/kg/day (13)
PEL: Not available

Basic manufacturer

BASF Corp.
Agricultural Products Group
P.O. Box 13528
Research Triangle Park, NC 27709-3528
Telephone: 800-669-2273
Emergency: 800-832-4357

10.3.5 Bromacil

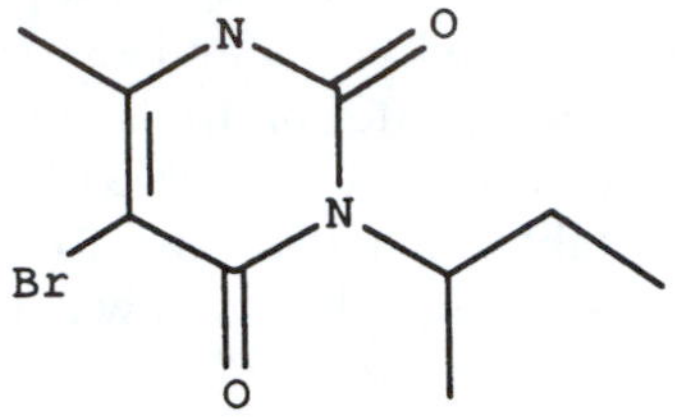

Figure 10.16 Bromacil.

Trade or other names

Trade names for bromacil include Borea, Bromax 4G, Bromax 4L, Borocil, Cynogan, Hyvar X, Hyvar XL, Isocil, Krovar, Rout, Uragan, Urox B, and Urox HX.

Regulatory status

Bromacil is classified by the U.S. Environmental Protection Agency (EPA) as a General Use Pesticide (GUP). It is in toxicity class IV — practically nontoxic in dry form, and class II — moderately toxic in the liquid form. Dry formulations containing bromacil bear the Signal Word CAUTION. Because of their irritating effects on skin, eyes, and in the respiratory tract, liquid formulations bear the Signal Word WARNING.

Introduction

Bromacil is an herbicide used for brush control on non-cropland areas. It is especially useful against perennial grasses. It is also used for selective weed control in pineapple and citrus crops. It works by interfering with photosynthesis, the process by which plants use sunlight to produce energy. Bromacil is one of a group of compounds called "uracils." It is sprayed or spread dry on the soil surface just before, or during, a period of active weed growth. It is available in granular, liquid, water-soluble liquid, and wettable powder formulations.

Toxicological effects

Acute toxicity

Liquid formulations of bromacil are moderately toxic, while dry formulations are practically nontoxic (8). The herbicide is irritating to the skin, eyes, and respiratory tract (68).

When 100 mg/kg of the herbicide was fed to dogs, it caused vomiting, watering of the mouth, muscular weakness, excitability, diarrhea, and dilation of the pupils. Rats that were fed single doses of bromacil experienced initial weight loss, paleness, exhaustion, and rapid breathing. Within 4 hours of being given 250 mg/kg, sheep became bloated and walked with stilted gaits (30). Bromacil caused mild dermal irritation when it was applied to the skin of guinea pigs. The LD_{50} is greater than 5000 mg/kg in rabbits whose skin is exposed. When bromacil was administered to the eyes of rabbits, there was irritation in the conjunctiva (the mucous membrane lining of the eye), but there was no injury to the cornea (30). The inhalation LC_{50} (4-hour) is greater than 4.8 mg/L air for rats (1).

The oral LD_{50} for bromacil in rats is 5200 mg/kg, and in mice is 3040 mg/kg (1,30).

Chronic toxicity

Enlarged livers were revealed in autopsies on rats that died after 5 days of repeated doses of bromacil at 1500 mg/kg/day (30). Sheep that died after being given 250 mg/kg/day bromacil on 4 successive days showed the following: inflammation of the mucous membrane that lines the stomach and intestines, congestion and enlargement of the liver, weakened appearance of the adrenal glands, bleeding of the heart, and swollen, bleeding lymph nodes (30).

Consumption of bromacil at high levels over a long period of time has been shown to cause damage to the testes, liver, and thyroid of laboratory animals (74).

In another study, female rats fed 62.5 mg/kg/day for 2 years, the highest dose level, exhibited decreased weight gain. No other toxic effects were observed (74). No evidence of toxicity was detected in dogs fed up to 31.2 mg/kg/day for 2 years (74). In an 18-month study in which mice were given dietary doses of 12.5, 62.5, or 250 mg/kg/day, changes in the liver and testes were observed at the 62.5-mg/kg/day dosage (74). Chickens given 500 mg/kg/day bromacil did show a decrease in weight gain (75).

Reproductive effects

Bromacil did not affect the reproduction of rats fed 12.5 mg/kg/day for three generations (8,74). This suggests that bromocil does not cause reproductive effects.

Teratogenic effects

There was no evidence of birth defects in the offspring of rats given dietary doses of 12.5 mg/kg/day bromacil, nor in rabbits that were given 7.5 mg/kg/day on days 8 through 16 of pregnancy (74). However, toxic effects and developmental abnormalities of the musculoskeletal system were seen in the embryos or fetuses of rats that inhaled very high bromacil doses of 38 mg/L for 2 hours daily, during the days 7 to 14 pregnancy (22). Toxic effects and developmental abnormalities were observed in the fetuses of pregnant rats repeatedly exposed by inhalation to bromacil (8). These data suggest that humans are unlikely to suffer teratogenic effects from bromacil under normal circumstances.

Mutagenic effects

Several mutagenic screening tests indicate that bromacil is not mutagenic (74).

Carcinogenic effects

There is limited evidence that bromacil causes cancer in animals receiving high doses over the course of their lifetimes (74). There was no evidence of carcinogenicity in rats fed 12.5 mg/kg/day bromacil for 2 years, but at 62.5 mg/kg/day, there was at slight increase in hyperplasia of the thyroid, and one rat developed benign liver tumors (17,74). An increased incidence of malignant tumors was observed in the livers of male mice given 250 mg/kg/day bromacil for 78 weeks. No effect on liver tumor incidence was observed in female mice (74). Based on these results, it is not possible to determine bromacil's carcinogenic potential.

Organ toxicity
Animal studies have shown the liver, heart, and lymph nodes are affected.

Fate in humans and animals
A number of studies show that uracils, the class of compounds to which bromacil belongs, are absorbed into the body from the gut and excreted primarily in the urine (8,74). Small amounts of bromacil were detected in the milk of lactating cows that were given 5 mg/kg in their food (30). No bromacil was found in the urine or feces of these cows (8).

Ecological effects

Effects on birds

The 8-day dietary oral LC_{50} for bromacil is over 10,000 ppm in mallards and quail (23). This indicates that it is practically nontoxic to these species.

Effects on aquatic organisms

Bromacil is slightly to practically nontoxic to fish. The 48-hour LC_{50} for bromacil in bluegill sunfish is 71 mg/L, in rainbow trout is 56 to 75 mg/L, in carp is 164 mg/L, and the 96-hour LC_{50} in fathead minnows is 182 mg/L (8). It is not toxic to aquatic invertebrates (8).

Effects on other organisms (non-target species)

Bromacil is not toxic to honeybees (1).

Environmental fate

Breakdown in soil and groundwater

Bromacil binds, or adsorbs, only slightly to soil particles, is soluble in water, and is moderately to highly persistent in soil. Its half-life is about 60 days, but may be as much as 8 months under some conditions (11). Soil persistence is correlated to the organic content of the soil (8,75). At 18 months after 22.4 kg bromacil was sprayed on abandoned field sites, residues of the herbicide were detectable, in decreasing amounts, in loamy sand, silt loam, silty clay loam, and light silty clay loam soils. Organic matter content, cation exchange capacity, and total nitrogen and soluble salt concentrations were significantly correlated with residue persistence (8).

Bromacil is expected to leach quite readily through the soil and contaminate groundwater. The amount of leaching is dependent on the soil type and the amount of rainfall or irrigation water. The potential for bromacil to leach and contaminate groundwater is greatest in sandy soils. In normal soils, it can be expected to leach to a depth of 2 to 3 feet (8).

Tests show that at increased temperatures and long exposures to sunlight, there is very little loss of bromacil from dry soil. It does not readily volatilize, nor does it break down in sunlight (8). Laboratory studies show that 5 to 30% bromacil is lost 6 to 9 weeks after application to the soil, as carbon dioxide (8).

Breakdown in water

Bromacil is estimated to have a 2-month half-life in clean river water that is low in sediment (76).

Breakdown in vegetation

In plants, bromacil is taken up rapidly by the roots and slightly absorbed through the leaves (58). When it is applied at 10 ppm, some types of algae show slowed growth, but most strains are unaffected (75). Improper application of bromacil will destroy shade trees and other desirable vegetation (76).

Physical properties

Bromacil is an odorless, white crystalline solid (1).

Chemical name: 5 bromo-6-methyl-3-(1-methylpropyl)-2,4(1H,3H)pyrimidinidi-one (1)
CAS #: 314-40-9
Molecular weight: 261.1 (1)
Solubility in water: 815 mg/L @ 25°C (1)
Solubility in other solvents: xylene v.s.; acetone v.s.; acetonitrile v.s.; ethyl alcohol v.s.; sodium hydroxide v.s. (1)
Melting point: 158–159°C (1)
Vapor pressure: 0.033 mPa @ 25°C (1)
Partition coefficient (octanol/water): not available
Adsorption coefficient: 32 (acid) (11)

Exposure guidelines

ADI: Not available
HA: 0.09 mg/L (74)
RfD: Not available
TLV: 10 mg/m³ (8-hour) (17)

Basic manufacturer

DuPont Agricultural Products
Walker's Mill, Barley Mill Plaza
P.O. Box 80038
Wilmington, DE 19880-0038
Telephone: 800-441-7515
Emergency: 800-441-3637

10.3.6 Bromoxynil

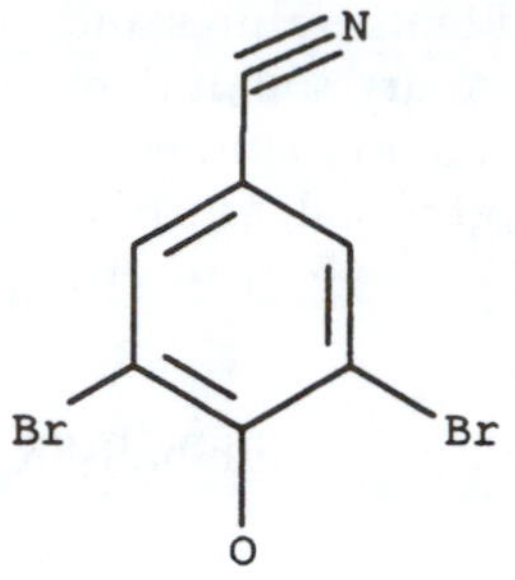

Figure 10.17 Bromoxynil.

Trade or other names

Trade names for commercial products containing bromoxynil include Brominal, Bromotril, Bronate, Buctril, Certrol B, Litarol, M&B 10064, Merit, Pardner, Sabre, and Torch. This compound may also be found in mixed formulations with other herbicides, including MCPA, linuron, dicamba, mecoprop, and 2,4-D.

Regulatory status

Bromoxynil is a Restricted Use Pesticide (RUP), and is not registered for homeowner use. RUPs may be purchased and used only by certified applicators. It is categorized as toxicity class II — moderately toxic. Products containing bromoxynil carry the Signal Word WARNING on their labels.

Introduction

Bromoxynil is a nitrile herbicide that is used for post-emergent control of annual broadleaf weeds. It is especially effective in the control of weeds in cereal, corn, sorghum, onions, flax, mint, turf, and on non-cropland. The compound works by inhibiting photosynthesis in the target plants. It is available as an emulsifiable concentrate and a suspension concentrate.

Toxicological effects

Acute toxicity

The bromoxynil product Buctril, has an oral acute LD_{50} of 779 mg/kg in the rat, and Bronate's LD_{50} is 691 mg/kg in the rat. Technical bromoxynil has an oral LD_{50} of 190 mg/kg in rats, an LD_{50} of 260 mg/kg in rabbits, and an LD_{50} of 63 mg/kg in guinea pigs, indicating moderate acute toxicity (8,22). The dermal LD_{50} of bromoxynil is greater than 2000 mg/kg in rabbits. The compound is a slight eye irritant but not a skin irritant in rabbits (1,22). However, when in contact with abraded skin, bromoxynil may produce a mild irritation (8).

Chronic toxicity

In one documented case of chronic exposure (about 1 year) of humans to bromoxynil, workers showed symptoms of weight loss, fever, vomiting, headache, and urinary problems (8).

Studies have shown that bromoxynil has no effect on rats given dietary doses of 15 and 50 mg/kg/day for 90 days (8). Doses up to 5 mg/kg/day for 2 years had no impact on blood chemistry or urine (8).

Reproductive effects

No changes in reproduction were noted in female rats fed 15 mg/kg/day bromoxynil over three generations (8). This suggests that bromoxynil does not cause reproductive effects.

Teratogenic effects

Bromoxynil is a suspected teratogen. The compound produced birth defects in rats at oral doses above 35 mg/kg. Toxic effects included abnormal rib formation and reduced fetal weight. Newborn rabbits had birth defects when bromoxynil was administered to pregnant mothers at doses above 30 mg/kg (8). In the rabbit, birth defects included changes in bone formation in the skull and hydrocephaly.

Mutagenic effects

No data are currently available.

Carcinogenic effects

Rats fed bromoxynil at low levels of 5 mg/kg and below did not develop any cancer-related effects (8).

Organ toxicity

No data were available regarding the target organs affected by bromoxynil.

Fate in animals and humans

No bromoxynil was present in the milk or feces of cows at 9 days after exposure to low doses of the herbicide. Less than 20% of the compound was excreted in urine as the parent compound (8).

Ecological effects

Effects on birds

Bromoxynil is highly toxic to pheasants (LD_{50} of 50 mg/kg) and is moderately toxic to hens (LD_{50} of 240 mg/kg), quail (LD_{50} of 100 mg/kg), and mallard ducks (LD_{50} of 200 mg/kg) (1).

Effects on aquatic organisms

Bromoxynil is very highly toxic to moderately toxic to freshwater fish; the potassium salt of bromoxynil has an LC_{50} of 5.0 mg/L in harlequin fish, 0.46 mg/L in goldfish, and 0.063 mg/L in catfish (1). Bromoxynil has an LC_{50} of 0.05 mg/L in rainbow trout (8).

Effects on other organisms (non-target species)

Bromoxynil is not toxic to bees (8).

Environmental fate

Breakdown in soil and groundwater

Bromoxynil has a low persistence in soil. In sandy soil, the half-life is about 10 days (1). Degradation in clay was slower, with half of the bromoxynil degraded to its metabolites in about a 2-week period at 25°C (8). The persistence of the compound is also slightly longer in peat field soils than in the sandy soils (77). The evidence suggests that, while bromoxynil is broken down by some soil bacteria, it may inhibit the action of other bacteria that promote the formation of nitrite by a process called "nitrification" (8).

Breakdown in water

No data are currently available.

Breakdown in vegetation

The herbicide works by disrupting the plant's ability to produce energy for cell-related activities. It is not readily translocated throughout the plant once it has been absorbed (8).

Physical properties

Bromoxynil is a colorless crystalline solid at room temperature.

Chemical name: 3,5-dibomo-4-hydroxybenzonitrile (1)
CAS #: 1689-84-5
Molecular weight: 276.93 (1)
Solubility in water: 130 mg/L @ 25°C (1)
Solubility in other solvents: methanol v.s.; ethanol v.s.; acetone v.s.; cyclohexane v.s.; tetrahydofuran v.s., all at 25°C (1)
Melting point: 194–195°C (1)
Vapor pressure: <1 mPa @ 20°C (1)
Partition coefficient (octanol/water): Not available
Adsorption coefficient: Not available

Exposure guidelines

ADI: Not available
MCL: Not available
RfD: 0.02 mg/kg/day (13)
PEL: Not available

Basic manufacturer

Rhone-Poulenc Ag Co.
P.O. Box 12014
2 T.W. Alexander Dr.
Research Triangle Park, NC 27709
Telephone: 919-549-2000
Emergency: 800-334-7577

10.3.7 Cacodylic acid

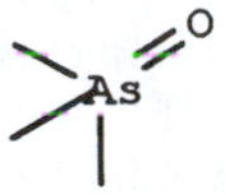

Figure 10.18 Cacodylic acid.

Trade or other names

Cacodylic acid is also known as dimethylarsinic acid. Trade names include Ansar 138, Arsan, Bolls-Eye, Broadside, Check-Mate, Cotton Aide HC, Dilic, Moncide, Montar, Phytar, Phytar 138, Phytar 600, Rad-E-Cate 25, Silvisar 510, and Sylvicor.

Regulatory status

Cacodylic acid is a General Use Pesticide (GUP). Products containing cacodylic acid are categorized toxicity class III — slightly toxic, and bear the Signal Word CAUTION.

Introduction

Cacodylic acid is an arsenical, nonselective contact herbicide that defoliates or desiccates a wide variety of plant species. It is used as a cotton defoliant, for lawn renovation, for weed control in non-crop areas such as around buildings, near perennial ornamentals, and along fence rows, and in forest management. Its phytotoxic properties are quickly inactivated upon contact with the soil. It is available in concentrated solution formulations.

Toxicological effects

Acute toxicity

Cacodylic acid is slightly toxic by inhalation and by ingestion. It is absorbed into the bloodstream more readily through inhalation than through ingestion or dermal exposure (8). Arsenate, formed from the breakdown of cacodylic acid, slows the production of energy in cells.

The symptoms of poisoning with arsenicals such as cacodylic acid include a salty taste, burning in the throat, colicky pains in the stomach and intestines, and garlicky odor of the breath, urine, and sweat. In persons exposed to sufficient arsenical dust, the onset of illness is usually characterized by difficult or short breath with pain in the chest, followed by nausea and diarrhea. Severe arsenic poisoning causes headache, dizziness, vomiting, profuse and watery diarrhea, followed by dehydration, electrolyte imbalance, gradual fall in blood pressure, stupor, convulsions, general paralysis, and possible death within 3 to 14 days (58).

Cacodylic acid may be irritating to skin and eyes (58). Product formulations may be irritating, depending on their acidity or alkalinity. Blistering or conjunctivitis may occur (8).

The oral LD_{50} for cacodylic acid in rats is 644 to 830 mg/kg (1,58). The inhalation LC_{50} for cacodylic acid in rats is 3.9 mg/L (58).

Chronic toxicity

Blanching or flushing of the skin (especially of the fingers), cirrhosis of the liver, atrophy of the bone marrow, kidney damage, and loss of sensory and motor functions have been reported. About 6 weeks after exposure to cacodylic acid, white bands may appear across the toe and finger nails (30,58).

Signs of chronic arsenate exposure include loss of appetite, weight loss, weakness, nausea, alternating diarrhea and constipation, colic, pain and tenderness in the limbs (usually starting in the fingers or toes), dermatitis, abnormal changes in skin color, loss of hair, giddiness, and headache. Prolonged exposure may cause gradual mental and physical deterioration. There may be disturbances of sight, taste, smell, and bladder function (8). Prolonged or repeated exposure of the skin to arsenicals may cause dermatitis. Prolonged eye contact may cause conjunctivitis and has caused eruptions of eyelids, conjunctiva, and even the cornea. Repeated exposure to low levels of arsenic may produce increased tolerance for arsenic (8).

Reproductive effects

Large dietary doses of cacodylic acid for extended periods of time decrease fertility. Male rats fed 226 mg/kg/day for 3 weeks showed reduced sperm production (58). Such effects are unlikely in humans at expected exposure levels.

Teratogenic effects

Oral doses of 0.35 mg/kg/day to pregnant rats produced smaller fetus size and decreased body weight. In rabbits, doses of 48 mg/kg/day decreased food intake and body weight in the offspring (58). Cacodylic acid is unlikely to cause teratogenic effects in humans at expected exposure levels.

Mutagenic effects

Cacodylic acid did not show mutagenicity in a variety of gene and chromosome assays on bacterial and mammalian cells (58).

Carcinogenic effects

No data are currently available.

Organ toxicity

Cacodylic acid affects a variety of organ systems. The kidney, liver, heart, digestive tract, and central and peripheral nervous systems are all targets (8,58).

Fate in humans and animals

Cacodylic acid is readily absorbed into the bloodstream when ingested. It is broken down in the liver. It accumulates in the skin, finger- and toenails, and hair, which also serve as means of excretion (8).

Ecological effects

Effects on birds

No data are currently available.

Effects on aquatic organisms

Cacodylic acid presents a low toxicity hazard to most fish. The LC_{50} for cacodylic acid in bluegill sunfish is 1000 mg/L (1).

Effects on other organisms (non-target species)

Cacodylic acid is not toxic to bees, with LD_{50} values reported in the 100 to 1000-ppm range (1).

Environmental fate

Breakdown in soil and groundwater

Cacodylic acid has low to moderate persistence in soil. Cacodylic acid is quickly inactivated upon contact with the soil by adsorption to soil particles and ion exchange (8,58). Soil microorganisms degrade most of the cacodylic acid in the soil (58). A breakdown product, arsenic, competes with phosphorus in the soil. It forms insoluble salts with chromium, silver, or other metals (8).

Breakdown in water

No data are currently available.

Breakdown in vegetation

Arsenic, from the metabolism of cacodylic acid, poisons cell division in plants (8).

Physical properties

Cacodylic acid is a colorless crystalline solid with an offensive odor (1).

Chemical name: dimethylarsinic acid (1)
CAS #: 75-60-5
Molecular weight: 138.0 (1)
Solubility in water: 2000 mg/L @ 25°C (1)
Solubility in other solvents: diethyl ether i.s. 3; alcohol s.s.; ethanol s.; acetic acid
    s. (1)
Melting point: 192–198°C (1)
Partition coefficient (octanol/water): Not available
Adsorption coefficient: 1000 (estimated) (8)

Exposure guidelines

ADI: Not available
MCL: Not available
RfD: 0.003 mg/kg/day (13)
PEL: Not available

Basic manufacturer

Drexel Chemical Company
1700 Channel Avenue
Memphis, TN 38113
Telephone: 901-774-4370

10.3.8 Clomazone

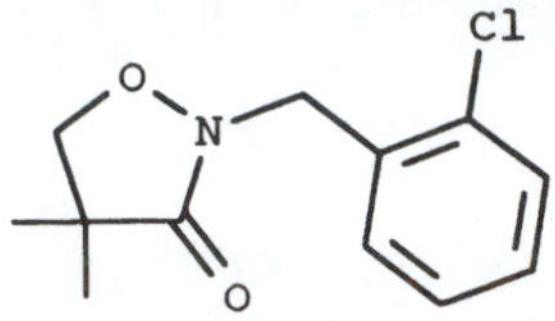

Figure 10.19 Clomazone.

Trade or other names

Trade names for clomazone include Command, Commence, Gamit, Magister, and Merit.

Regulatory status

Clomazone is a General Use Pesticide (GUP). It is classified toxicity class III — slightly toxic. Products containing clomazone bear the Signal Word CAUTION on the labels.

Introduction

Clomazone is a broad-spectrum herbicide used for control of annual grasses and broadleaf weeds in cotton, peas, pumpkins, soybeans, sweet potatoes, tobacco, winter squash, and fallow wheat fields. It can be applied early preplant, pre-emergent, or preplant-incorporated, depending on the crop, geographical area, and timing.

Toxicological effects

Acute toxicity

Clomazone is a slightly toxic material by ingestion, inhalation, and dermal exposure. The oral LD_{50} for technical clomazone is 1369 mg/kg in female rats and 2077 mg/kg in male rats (8,58). The oral LD_{50} for Command 4EC is slightly higher: 1406 mg/kg in female rats and 2343 in male rats (58). The dermal LD_{50} in rabbits is greater than 2000 mg/kg (8). The inhalation LC_{50} for technical clomazone is 4.23 mg/L in female rats and 6.52 mg/L in male rats (58). The inhalation LC_{50} for Command 4EC in rats is 4.5 to 4.7 mg/L (58).

Chronic toxicity

In 2-year feeding studies with rats and mice and a 1-year feeding study with dogs, no long-term adverse effects from Command were observed (58). In a 1-year feeding study with dogs, increased liver weight occurred at the 2.5-mg/kg/day dose (58). In 2-year feeding studies, rats fed more than 4.3 mg/kg/day exhibited elevated cholesterol levels, increased liver weights, and enlarged liver cells. Mice given doses above 15 mg/kg/day had elevated white blood cell counts (78).

Reproductive effects

In a two-generation study with rats, each generation was fed clomazone at 5, 50, 100, or 200 mg/kg/day for 11 weeks in between weaning and mating. There was no effect on reproductive performance (79). These data suggest that it does not cause reproductive effects.

Teratogenic effects

Clomazone does not appear to be teratogenic (79). No birth defects were seen in the offspring of rats given 600 mg/kg/day, the highest dose tested, nor in the offspring of rabbits given 700 mg/kg/day (78).

Mutagenic effects

Clomazone is not mutagenic. The results of several tests, including a DNA synthesis test, reverse mutation tests, and a chromosomal aberration test, were all negative (78).

Carcinogenic effects

Clomazone does not appear to be carcinogenic (80). No tumor formation occurred in mice or rats given dietary doses as high as 100 mg/kg for 2 years (58,78).

Organic toxicity

Animal studies have shown that clomazone affects the liver.

Fate in humans and animals

Metabolism studies show that 90 to 99% of the product Command administered to rats was excreted within 72 hours, and there was no significant retention of the herbicide in rat tissues (79).

Ecological effects

Effects on birds

Clomazone is practically nontoxic to birds. The oral LD_{50} for technical clomazone in bobwhite quail and mallard ducks is greater than 2510 mg/kg. The 8-day dietary LC_{50} in bobwhites and mallards is 5620 ppm (58).

Effects on aquatic organisms

Clomazone is moderately toxic to fish and aquatic invertebrates. The LC_{50} (96-hour) for technical clomazone is 19 mg/L in rainbow trout, 34 mg/L in bluegill sunfish, 6.26 mg/L in Atlantic silversides, 40.6 mg/L in sheepshead minnows, 0.566 mg/L in mysid shrimp, 5.3 mg/L in Eastern oysters, and 5.2 mg/L in *Daphnia magna* (58).

The bioconcentration factor in bluegill sunfish is 40, indicating that there is only a small potential for this compound to accumulate in aquatic organisms (80).

Effects on other organisms (non-target species)

No data are currently available.

Environmental fate

Breakdown in soil and groundwater

Clomazone is moderately persistent in soil. Microbial degradation of Command is promoted by high soil moisture, warm temperature, and by increasing the pH to 6.5. Degradation was faster in a sandy loam than in silt or clay loam soils. In field studies, the half-life of clomazone was 28 to 84 days, depending on soil type and organic matter content (58).

Clomazone is highly soluble in water, but it has a moderate tendency to adsorb to soil particles. It therefore has a low to moderate potential to contaminate groundwater (11). The product Command has low mobility in sandy loam, silt loam, and clay loam soils. It is moderately mobile in fine sand (58).

Breakdown in water

Under laboratory conditions, clomazone was not readily hydrolyzed in sterile water (81). However, clomazone is subject to photodegradation in water, with a half-life of 1.5 to 7 days reported for clomazone in solutions containing acetone (a photochemical sensitizer) (82).

Breakdown in vegetation

Clomazone inhibits the syntheses of chlorophyll and carotenoids in plants. It is absorbed by plants through the roots from the soil and by shoots. It is then translocated in the xylem and diffuses within the leaves. It does not move downward in plants or from leaf to leaf. There is no foliar absorption of clomazone. Clomazone is metabolized by plants (58).

Physical properties

Clomazone is a colorless to light brown, viscous liquid above room temperature. When cooled, it forms a white crystalline solid (1).

Chemical name: 2-(2-chlorobenzyl)-4,4-dimethyl-1,2-oxazolidin-3-one (1)
CAS #: 81777-89-1
Molecular weight: 239.7 (1)
Solubility in water: 1100 mg/L (1)
Solubility in other solvents: acetone v.s.; chloroform v.s.; cyclohexanone and methanol v.s.; toluene s. (10)
Melting point: 25°C (1)
Vapor pressure: 19.2 mPa @ 25°C (1)
Partition coefficient (octanol/water): 350 (1)
Adsorption coefficient: 300 (11)

Exposure guidelines

ADI: Not available
HA: Not available
RfD: Not available
PEL: Not available

Basic manufacturer

FMC Corporation
Agricultural Chemicals Group
1735 Market
Philadelphia, PA 19103
Telephone: 215-299-6661
Emergency: 800-331-3148

10.3.9 Dinoseb

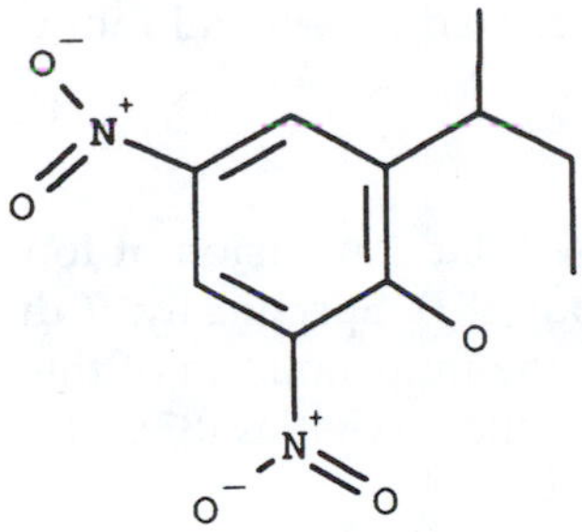

Figure 10.20 Dinoseb.

Trade or other names

Product names for pesticides containing dinoseb include Basanite, Caldon, Chemox, Chemsect DNBP, Dinitro, Dynamyte, Elgetol, Gebutox, Hel-Fire, Kiloseb, Nitropone, Premerge, Sinox General, Subitex, and Vertac Weed Killer.

Regulatory status

The use of dinoseb was canceled in the U.S. in 1986. This action was based on the potential risk of birth defects and other adverse health effects for applicators and other persons with substantial dinoseb exposure. This pesticide is not commercially available in the U.S.

Dinoseb is a highly toxic compound. Prior to its ban, it was a Restricted Use Pesticide (RUP), and products containing dinoseb were required to be labeled with a DANGER Signal Word.

Introduction

Dinoseb is a phenolic herbicide used in soybeans, vegetables, fruits and nuts, citrus, and other field crops for the selective control of grass and broadleaf weeds (e.g., in corn). It is also used as an insecticide in grapes, and as a seed-crop drying agent. It is produced in emuslifiable concentrates or as water-soluble ammonium or amine salts. *The information presented here pertains to the technical grade of dinoseb unless otherwise specified.*

Toxicological effects

Acute toxicity

Dinoseb is highly toxic by ingestion, with reported oral LD_{50} values of 25 to 58 mg/kg in rats and 25 mg/kg in guinea pigs (1,29). It is highly toxic by skin exposure, with a reported dermal LD_{50} of 80 to 200 mg/kg in rabbits, and 200 to 300 mg/kg in guinea pigs (1,29). Dinoseb did not cause skin irritation in rabbits (29).

Inhalation of dusts and sprays may be irritating to the lungs and eyes, and may cause serious illness; direct skin contact may cause irritation, yellow stains, burns, dermatitis, and more serious effects in humans (29).

In one fatal incident, a farm worker was using a backpack hand-held sprayer that leaked dinoseb onto his body and penetrated his skin (29). Symptoms in persons receiving accidental exposure include fatigue, thirst, sweating, insomnia, weight loss, headache, flushing of the face, nausea, abdominal pain, and occasional diarrhea (29). In one case, some of these persisted for several months following exposure (29).

Chronic toxicity

Dinoseb interferes with cellular conversion of food molecules (such as glucose) into usable energy for the body (83). Specifically, it disturbs the production of adenosine triphosphate (ATP) in the mitochondria of the cells, ATP being the molecule that provides energy for all cellular activities (83). This may account for many of the toxic effects caused by dinoseb.

Dietary levels of about 25 mg/kg/day caused marked food refusal and some deaths after five or more doses (29). Lower doses (5 to 20 mg/kg/day) caused statistically significant decreases in growth (29).

Some formulations of dinoseb may cause anemia-like effects, jaundice, and increased excretion of hemoglobin-related compounds (coproporphyrinuria) (29).

Reproductive effects

Dinoseb is reported to adversely affect reproduction in rats and mice at levels that are commonly found among occupational workers. Decreased sperm count and abnormal sperm shape were observed in male rats and mice after 3 weeks at low exposure levels (about 10 mg/kg/day) for 30 days (84). In a separate study, rats were exposed to relatively small quantities of dinoseb through their diet for a total of 22 weeks. Effects such as decreased fertility, slow weight gain, and poor survival of newborns appeared to be related to this pesticide (85).

Because of the adverse effects observed in laboratory animals at low chronic exposure levels, it is believed that dinoseb may cause decreased fertility or sterility in humans (83).

Teratogenic effects

Low levels of dinoseb fed to rats and rabbits caused birth defects in the fetuses of exposed females (8). When dinoseb was administered to pregnant mice, its breakdown products were found in the embryos. However, no teratogenic effects were noted.

Various tests of mice and rats fed or injected with small amounts of dinoseb (around 10 mg/kg/day) have shown maternal toxicity, decreased fetal body weights, and changes in fetal development (29). In some studies of mice, oral doses to pregnant mothers caused an increased death rate in the exposed animals but caused no fetal damage (8). Other studies indicate that dinoseb is a stronger teratogen when injected than when ingested. At low feeding levels, the compound was responsible for skeletal deformities and neurological problems in newborn rats (83).

Based on the data, dinoseb may produce teratogenic effects in humans.

Mutagenic effects

Dinoseb was not mutagenic or genotoxic in laboratory studies performed using eukaryotic cells (the type found in mammals and other higher order species) (29,83). This evidence suggests that mutagenic effects in humans due to dinoseb exposure are unlikely.

Carcinogenic effects

Dinoseb didn't cause significant increases in tumors when administered to two strains of mice at the maximum tolerated dose over a period of 18 months (29). While not carcinogenic to male mice, it was found to be carcinogenic to female mice in another study (83). The compound caused liver cancer in these animals at moderate to high doses (83). The evidence regarding the carcinogenic potential of dinoseb is currently inconclusive.

Organ toxicity

Dinoseb has the potential to damage the eye. It may also affect the immune system, liver, kidneys, and spleen (29).

Fate in humans and animals

Dinoseb is readily absorbed through the skin, gastrointestinal tract, and lung surface (29). Esters of dinoseb are rapidly transformed into dinoseb, which is the active toxicant (29). The chemical is excreted in the urine, and feces and is metabolized in the liver. Breakdown products are found in the liver, kidneys, spleen, blood, and urine (29). Dinoseb can also pass through the placenta into the fetus of experimental animals.

Ecological effects

Effects on birds

The compound is very highly toxic to birds, with reported acute oral LD_{50} values between 7 and 9 mg/kg (8); its reported 5- to 8-day dietary LC_{50} ranges from 409 ppm in quail to 515 ppm in pheasants (8). It thus has the potential to negatively impact local pheasant and songbird populations.

Effects on aquatic organisms

Dinoseb is highly toxic to fish, with reported 96-hour LC_{50} values ranging from 44 μg/L in lake trout to 118 μg/L in catfish (8). Other 96-hour LC_{50} values are 100 μg/L in coho salmon and 67 μg/L in cutthroat trout (8,37). It is more toxic to fish in acidic water than in neutral or alkaline water (37). Dinoseb has caused fish kills in small Scottish streams when washed from fields by rain (88).

The bioconcentration factor is 135 (86). Dinoseb is rapidly taken up by fish, but is is rapidly eliminated from exposed fish if placed in clean water. Dinoseb, thus, does not pose a significant risk for bioaccumulation (84).

Effects on other organisms (non-target species)

Dinoseb is toxic to bees (1).

Environmental fate

Breakdown in soil and groundwater

Dinoseb is of low persistence regardless of the form (phenolic or salt). Reported field half-lives for both types of dinoseb range from 5 to 31 days (53). An overall representative value is estimated to be 20 to 30 days in most circumstances, although persistence may be much longer in the vadose zone (9). Photodegradation and microbial breakdown may play roles in the breakdown of dinoseb in the soil environment, but volatilization should not be a significant route of loss (9).

The phenolic form of dinoseb is slightly soluble in water and moderately sorbed by most soils (53). Studies have shown soil sorption capacity to be much greater at lower pH values (9). It thus should present only a moderate risk to groundwater. On the other hand, the ammonium and amine salt forms of dinoseb are much more water soluble and much less strongly bound to soils (8). These may pose a significant risk to groundwater.

Over a 10-year period, dinoseb was found to be one of three particularly persistent contaminants in Ontario wells (8). Entry to the wells was due to spills of concentrated and dilute herbicide, drift during spraying, and storm runoff. Well water concentrations ranged from 0.05 to 5000 µg/L in these wells, and removal of dinoseb proved very difficult (9).

Breakdown in water

Photodegradation may occur in surface waters, but hydrolysis is essentially negligible (9). Dinoseb has been found in streams at about 5 µg/L (8).

Breakdown in vegetation

Dinoseb persists on treated crop soils for 2 to 4 weeks, under average conditions of use (8).

Physical properties

Dinoseb is a dark, reddish-brown solid or a dark orange viscous liquid, depending on the temperature (1).

Chemical name: 2-(*sec*-butyl)-4,6-dinitrophenol (1)
CAS #: 88-85-7
Molecular weight: 240.22 (1)
Solubility in water: 52 mg/L @ 20°C (1)
Solubility in solvents: v.s. in alcohol, ethanol, and heptane; s. in spray oil and in most organic solvents and oils (1)
Melting point: 32–42°C (1)
Vapor pressure: 6.7 mPa @ 25°C (53)
Partition coefficient (octanol/water): not available
Adsorption coefficient: 30 (estimated) (53)

Exposure guidelines

ADI: Not available
HA: Not available
RfD: 0.001 mg/kg/day (13)
PEL: Not available

Basic manufacturer

DowElanco
9330 Zionsville Road
Indianapolis, IN 46268-1054
Telephone: 317-337-7344
Emergency: 800-258-3033

10.3.10 Diquat dibromide

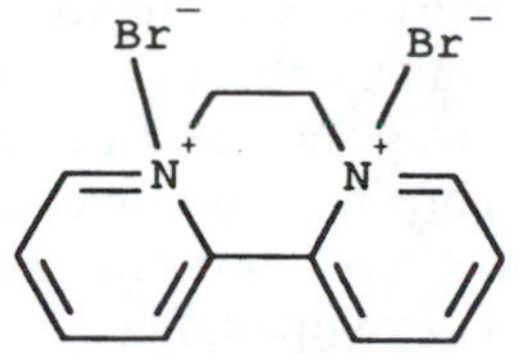

Figure 10.21 Diquat dibromide.

Trade or other names

Trade names include Aquacide, Aquakill, Dextrone, Diquat, Reglone, Reglox, Reward, Tag, Torpedo, Vegetrole, and Weedtrine-D.

Regulatory status

Diquat dibromide is a moderately toxic compound in EPA toxicity class II (1,2). It is a General Use Pesticide (GUP). Labels for products containing diquat dibromide must bear the Signal Word WARNING.

Introduction

Diquat dibromide is a nonselective, quick-acting herbicide and plant growth regulator, causing injury only to the parts of the plant to which it is applied. Diquat dibromide is referred to as a desiccant because it causes a leaf or an entire plant to dry out quickly. It is used to desiccate potato vines and seed crops, to control flowering of sugarcane, and for industrial and aquatic weed control. It is not residual; that is, it does not leave any trace of herbicide on or in plants, soil, or water.

Toxicological effects

Acute toxicity

Diquat dibromide is moderately toxic via ingestion, with reported oral LD_{50} values of 120 mg/kg in rats, 233 mg/kg in mice, 188 mg/kg in rabbits, and 187 mg/kg in guinea pigs and dogs (1,87). Cows appear to be particularly sensitive to this herbicide, with an oral LD_{50} of 30 to 56 mg/kg (17). The acute dermal LD_{50} for diquat dibromide is approximately 400 to 500 mg/kg in rabbits, indicating moderate toxicity by this route as well (58,87). A single dose of diquat dibromide was not irritating to the skin of rabbits, but repeated dermal dosing did cause mild redness, thickening, and scabbing (58). Moderate to severe eye membrane irritation occurred when diquat dibromide was administered to rabbits (88).

Ingestion of sufficient doses may cause severe irritation of the mouth, throat, esophagus, and stomach, followed by nausea, vomiting, diarrhea, severe dehydration, and alterations in body fluid balances, gastrointestinal discomfort, chest pain, diarrhea, kidney failure, and toxic liver damage (87). Skin absorption of high doses may cause symptoms similar to those that occur following ingestion (89). Very large doses of the herbicide can result in convulsions and tremors (88). Test animals (rats, mice, guinea pigs, rabbits, dogs, cows, and hens) given lethal doses of diquat dibromide showed a delayed pattern of illness, with onset approximately 24 hours following dosing, subsequent lethargy, pupil dilation, respiratory distress, weight loss, weakness, and finally, death over the course of 2 to 14 days after dosing (58,87,89).

There have been reports of workers who have had softening and color changes in one or more fingernails after contact with concentrated diquat dibromide solutions (87). In some instances, the nail was shed and did not grow in again (87). Several cases of severe eye injury in humans have occurred after accidental splashings (87). In each case, initial irritation was mild, but after several days, serious burns and sometimes scarring of the cornea developed. Direct or excessive inhalation of diquat dibromide spray mist or dust may result in oral or nasal irritation, nosebleeds, headache, sore throat, coughing, and symptoms similar to those from ingestion of diquat (87).

Chronic toxicity

Chronic effects of diquat dibromide are similar to those of paraquat (87). Cataracts, a clouding of the eyes that interferes with light entering the eye, occurred in rats and dogs given 2.5 and 5 mg/kg/day diquat dibromide, respectively (87). Cataracts increased in proportion to the dose given in test animals (cats and dogs) (17,88). Chronic exposure is necessary to produce these effects (87). Other effects on the eye (hemorrhage or retinal detachment) may occur at higher dosages (87).

Rats fed dietary doses of 2.5 mg/kg/day over 2 years did not exhibit signs of toxicity other than reduced food intake and decreased growth (17). In another study using rats, oral doses of 4 mg/kg/day over 2 years produced no behavioral or other changes in general condition (87). At this dose level, no evidence of change in the kidneys, liver, or mycocardium (heart muscle) were seen. This dosage (but not 2 mg/kg/day) caused changes in lung tissues (87).

Repeated or prolonged dermal contact may cause inflammation of the skin and, at high doses, systemic effects in other parts of the body. These may include damage to the kidneys (58). Chronic exposure may damage skin, which may increase the permeability of the skin to foreign compounds (88).

Reproductive effects

Diquat dibromide generally did not reduce fertility when tested in experimental animals (89). Rats receiving 1.25 mg/kg/day decreased their food intake and showed slowed growth, but had unchanged reproduction (89). Fertility was reduced in male mice given diquat dibromide during different stages of sperm formation (87). Neither fertility nor reproduction was affected in a three-generation study in rats given dietary doses of 12.5 or 25 mg/kg/day diquat dibromide, although some growth retardation was seen at the 25-mg/kg/day dose (87).

Based on this evidence, it is unlikely that diquat dibromide will cause reproductive effects in humans under normal circumstances.

Teratogenic effects

Offspring of pregnant rats given a fatal injected dose of 14 mg/kg diquat dibromide showed evidence of skeletal defects of the collar bone, as well as little or no ear bone formation upon examination (58,87). No deformities were found in the unborn offspring of pregnant rats injected intraperitoneally with 0.5 mg/kg/day diquat dibromide daily during organogenesis, the stage of fetal development in which organs are formed (26). Growth retardation was seen in test animals given extremely high doses of diquat. While no actual teratogenesis occurred in rats given single abdominal injections during the days 7 to 14 of pregnancy, many rats did not have normal weight gain and bone formation in the unborn was decreased (23). It is unlikely that diquat dibromide will cause teratogenic effects in humans under normal circumstances.

Mutagenic effects

There is no evidence that diquat dibromide causes permanent changes in genetic material (87). For example, no mutagenic effects were seen in mice given oral doses of 10 mg/kg/day for 5 days (23).

Carcinogenic effects

An 80-week feeding study showed that dietary doses of 15 mg/kg/day diquat dibromide did not cause tumors in rats (90). Likewise, dietary levels of 36 mg/kg/day for 2 years did not induce tumors in rats (87). Based on the evidence, it appears that diquat dibromide is not carcinogenic.

Organ toxicity

In animals, diquat dibromide may affect the gastrointestinal tract, eyes, kidneys or liver, and the lungs.

Fate in humans and animals

Absorption of diquat dibromide from the gut into the bloodstream is low (87). Oral doses are mainly metabolized within the intestines, with metabolites being excreted in the feces (30,87). Rat studies showed only a small percentage of the applied oral dose (6%) was absorbed into the bloodstream and then excreted in the urine (87).

Dermal, inhalation, or intravenous exposure results in little processing and rapid elimination in the urine (87). Following subcutaneous injection in rats, excretion of about 90% of the dose occurred in the urine on the first day and almost all of the remainder on the next day (87). Complete elimination of the herbicide was seen in urine and feces of rats within 4 days of administration of single oral doses of 5 to 10 mg/kg diquat dibromide (87).

Ecological effects

Effects on birds

Diquat dibromide ranges from slightly to moderately toxic to birds (91). The reported acute oral LD_{50} in young male mallards is 564 mg/kg (8). The oral LD_{50} for diquat dibromide is 200 to 400 mg/kg in hens (8). The 5-day dietary LC_{50} is about 1300 ppm in Japanese quail (36).

Effects on aquatic organisms

Diquat dibromide is moderately to practically nontoxic to fish and aquatic invertebrates. The 8-hour LC_{50} for diquat dibromide is 12.3 mg/L in rainbow trout and 28.5 mg/L in Chinook salmon (28). The 96-hour LC_{50} is 16 mg/L in northern pike, 20.4 mg/L in fingerling trout, 245 mg/L in bluegill, 60 mg/L in yellow perch, and 170 mg/L in black bullhead (37,92).

Research indicates that yellow perch suffer significant respiratory stress when herbicide concentrations in the water are similar to those normally present during aquatic vegetation control programs (93). There is little or no bioconcentration of diquat dibromide in fish (8).

Effects on other organisms (non-target species)

Diquat dibromide is not toxic to honeybees (1). Since diquat dibromide is a nonselective herbicide, it may present a danger to non-target plant species (91). Cows are particularly sensitive to the toxic effects of this material (17).

Environmental fate

Breakdown in soil and groundwater

Diquat dibromide is highly persistent, with reported field half-lives of greater than 1000 days (11). It is very well sorbed by soil organic matter and clay (11). Although it is water soluble (11), its capacity for strong adsorption to soil particles suggest that it will not easily leach through the soil, be taken up by plants or soil microbes, or broken down by sunlight (photochemical degradation).

Field and laboratory tests show that diquat dibromide usually remains in the top inch of soil for long periods of time after it is applied (94).

Breakdown in water

Studies on the erosion of diquat-treated soils near bodies of water indicate that diquat dibromide stays bound to soil particles, remaining biologically inactive in surface waters, such as lakes, rivers, and ponds (95). When diquat dibromide is applied to open water, it disappears rapidly because it binds to suspended particles in the water (95). Diquat dibromide's half-life is less than 48 hours in the water column, and may be on the order of 160 days in sediments due to its low bioavailability (94,95). Microbial degradation and sunlight play roles in the breakdown of the compound (95). At 22 days after a weed-infested artificial lake was treated, only 1% of the applied diquat dibromide remained in the water and 19% was adsorbed to sediments (9).

Breakdown in vegetation

Diquat dibromide is rapidly absorbed into the leaves of plants, but usually kills the plant tissues necessary for translocation too quickly to allow movement to other parts of the plant. The herbicide interferes with cell respiration, the process by which plants produce energy. Diquat dibromide is broken down on the plant surface by photochemical degradation (58). It is rapidly absorbed by aquatic weeds from the surrounding water and concentrated in the plant tissue (8). Thus, even low concentrations of the herbicide can control aquatic weeds (8).

Physical properties

Technical diquat dibromide, which is greater than 95% pure, forms white to yellow crystals (1).

Chemical name: 1,1'-ethylene-2,2'-bipyridyldiylium dibromide salt (1)
CAS #: 85-00-7
Molecular weight: 344.06 (1)
Water solubility: 700,000 mg/L @ 20°C; v.s. (1)
Solubility in other solvents: i.s. in nonpolar organic solvents such as chloroform, diethyl ether, and petroleum ether (1); s.s. in alcohol and hydroxylic solvents (1)

Melting point: Decomposes above 300°C (1)
Vapor pressure: Negligible at 20°C (1)
Partition coefficient (octanol/water): 0.000025 (1)
Adsorption coefficient: 1,000,000 (estimated) (11)

Exposure guidelines

ADI: 0.002 mg/kg/day (12)
MCL: 0.02 mg/L (65)
RfD: 0.0022 mg/kg/day (13)
TLV: 0.1 mg/m³ (8-hour) (respirable fraction) (17)

Basic manufacturer

Zeneca Ag Products
1800 Concord Pike
Wilmington, DE 19897
Telephone: 800-759-4500
Emergency: 800-759-2500

10.3.11 Glyphosate

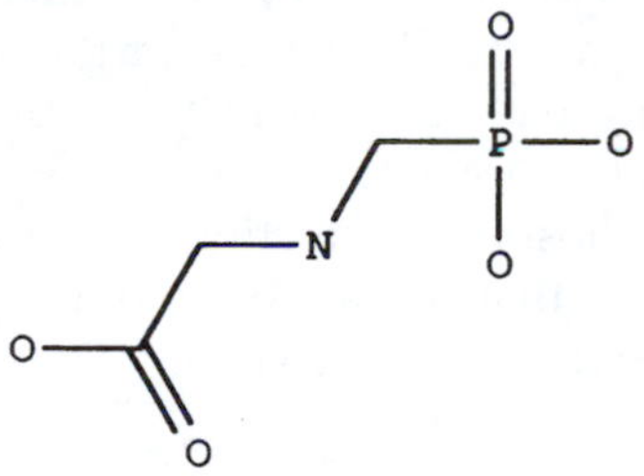

Figure 10.22 Glyphosate.

Trade or other names

Trade names for products containing glyphosate include Gallup, Landmaster, Pondmaster, Ranger, Roundup, Rodeo, and Touchdown. It may be used in formulations with other herbicides.

Regulatory status

Glyphosate acid and its salts are moderately toxic compounds in EPA toxicity class II. Labels for products containing these compounds must bear the Signal Word WARNING. Glyphosate is a General Use Pesticide (GUP).

Introduction

Glyphosate is a broad-spectrum, nonselective, systemic herbicide used for control of annual and perennial plants, including grasses, sedges, broad-leaved weeds, and woody plants. It can be used on non-croplands as well as on a great variety of crops. Glyphosate itself is an acid, but it is commonly used in salt form, most commonly the isopropylamine salt. It may also be available in acidic or trimethyl-sulfonium salt forms. It is generally distributed as water-soluble concentrates and powders. *The information presented here refers to the technical grade of the acid form of glyphosate, unless otherwise noted.*

Toxicological effects

Acute toxicity

Glyphosate is practically nontoxic by ingestion, with a reported acute oral LD_{50} of 5600 mg/kg in the rat. The toxicities of the technical acid (glyphosate) and the formulated product (Roundup) are nearly the same (58,96). The oral LD_{50} for the trimethylsulfonium salt is reported to be approximately 750 mg/kg in rats, which indicates moderate toxicity (58). Formulations may show moderate toxicity as well (LD_{50} values between 1000 and 5000 mg/kg) (58). Oral LD_{50} values for glyphosate are greater than 10,000 mg/kg in mice, rabbits, and goats (8,96).

Glyphosphate is practically nontoxic by skin exposure, with reported dermal LD_{50} values of greater than 5000 mg/kg for the acid and isopropylamine salt. The trimethylsulfonium salt has a reported dermal LD_{50} of greater than 2000 mg/kg. It is reportedly not irritating to the skin of rabbits and does not induce skin sensitization in guinea pigs (58). It does cause eye irritation in rabbits (58). Some formulations may cause much more extreme irritation of the skin or eyes (58). In a number of human volunteers, patch tests produced no visible skin changes or sensitization (58). The reported 4-hour rat inhalation LC_{50} values for the technical acid and salts were 5 to 12 mg/L (58), indicating moderate toxicity via this route. Some formulations may show high acute inhalation toxicity (58).

While it does contain a phosphatyl functional group, it is not structurally similar to organophosphate pesticides that contain organophosphate esters, and it does not significantly inhibit cholinesterase activity (1,58).

Chronic toxicity

Studies of glyphosate lasting up to 2 years have been conducted with rats, dogs, mice, and rabbits and, with few exceptions no effects were observed (96). For example, in a chronic feeding study with rats, no toxic effects were observed in rats given doses as high as 400 mg/kg/day (58). Also, no toxic effects were observed in a chronic feeding study with dogs fed up to 500 mg/kg/day, the highest dose tested (58,97).

Reproductive effects

Laboratory studies show that glyphosate produces reproductive changes in test animals very rarely and then only at very high doses (over 150 mg/kg/day) (58,96). It is unlikely that the compound would produce reproductive effects in humans.

Teratogenic effects

In a teratology study with rabbits, no developmental toxicity was observed in the fetuses at the highest dose tested (350 mg/kg/day) (97). Rats given doses up to 175 mg/kg/day on days 6 to 19 of pregnancy had offspring with no teratogenic effects, but other toxic effects were observed in both the mothers and the fetuses. No toxic effects to the fetuses occurred at 50 mg/kg/day (97).

Glyphosate does not appear to be teratogenic.

Mutagenic effects

Glyphosate mutagenicity and genotoxicity assays have been negative (58). These included the Ames test, other bacterial assays, and the Chinese hamster ovary (CHO) cell culture, rat bone marrow cell culture, and mouse dominant lethal assays (58). It appears that glyphosate is not mutagenic.

Carcinogenic effects

Rats given oral doses up to 400 mg/kg/day did not show any signs of cancer, nor did dogs given oral doses up to 500 mg/kg/day or mice fed glyphosate at doses up to 4500 mg/kg/day (58). It appears that glyphosate is not carcinogenic (97).

Organ toxicity

Some microscopic liver and kidney changes, but no observable differences in function or toxic effects, have been seen after lifetime administration of glyphosate to test animals (97).

Fate in humans and animals

Glyphosate is poorly absorbed from the digestive tract and is largely excreted unchanged by mammals. At 10 days after treatment, there were only minute amounts in the tissues of rats fed glyphosate for 3 weeks (98).

Cows, chickens, and pigs fed small amounts of glyphosate had undetectable levels (<0.05 ppm) in muscle tissue and fat. Levels in milk and eggs were also undetectable (<0.025 ppm). Glyphosate has no significant potential to accumulate in animal tissue (99).

Ecological effects

Effects on birds

Glyphosate is slightly toxic to wild birds. The dietary LC_{50} in both mallards and bobwhite quail is greater than 4500 ppm (1).

Efects on aquatic organisms

Technical glyphosate acid is practically nontoxic to fish and may be slightly toxic to aquatic invertebrates. The 96-hour LC_{50} is 120 mg/L in bluegill sunfish, 168 mg/L in harlequin, and 86 mg/L in rainbow trout (58). The reported 96-hour LC_{50} values for other aquatic species include greater than 10 mg/L in Atlantic oysters, 934 mg/L in fiddler crab, and 281 mg/L in shrimp (58). The 48-hour LC_{50} for glyphosate in *Daphnia* (water flea), an important food source for freshwater fish, is 780 mg/L (58).

Some formulations may be more toxic to fish and aquatic species due to differences in toxicity between the salts and the parent acid or to surfactants used in the formulation (58,96).

There is a very low potential for the compound to build up in the tissues of aquatic invertebrates or other aquatic organisms (96).

Effects on other organisms (non-target species)

Glyphosate is nontoxic to honeybees (1,58). Its oral and dermal LD_{50} values are greater than 0.1 mg per bee (98). The reported contact LC_{50} values for earthworms in soil are greater than 5000 ppm for both the glyphosate trimethylsulfonium salt and Roundup (58).

Environmental fate

Breakdown in soil and groundwater

Glyphosate is moderately persistent in soil, with an estimated average half-life of 47 days (58,11). Reported field half-lives range from 1 to 174 days (11). It is strongly adsorbed to most soils, even those with lower organic and clay content (11,58). Thus, even though it is highly soluble in water, field and laboratory studies show it does not leach appreciably, and has low potential for runoff (except as adsorbed to colloidal matter) (3,11). One estimate indicated that less than 2% of the applied chemical is lost to runoff (99). Microbes are primarily responsible for the breakdown of the product, and volatilization or photodegradation losses will be negligible (58).

Breakdown in water

In water, glyphosate is strongly adsorbed to suspended organic and mineral matter and broken down primarily by microorganisms (6). Its half-life in pond water ranges from 12 days to 10 weeks (97).

Breakdown in vegetation

Glyphosate may be translocated throughout the plant, including to the roots. It is extensively metabolized by some plants while remaining intact in others (1).

Physical properties

Glyphosate is a colorless crystal at room temperature (1).

Chemical name: N-(phosphonomethyl) glycine (1)
CAS #: 1071-83-6
Molecular weight: 169.08 (1)
Solubility in water: 12,000 mg/L @ 25°C (1)
Solubility in other solvents: i.s. in common organics (e.g., acetone, ethanol, and xylene) (1)
Melting point: 200°C (1)
Vapor pressure: negligible (1)
Partition coefficient: 0.0006-0.0017 (58)
Adsorption coefficient: 24,000 (estimated) (11)

Exposure guidelines

ADI: 0.3 mg/kg/day (12)
HA: 0.7 mg/L (lifetime) (98)
RfD: 0.1 mg/kg/day (13)
PEL: Not available

Basic manufacturer

Monsanto Company
800 N. Lindbergh Blvd.
St. Louis, MO 63167
Telephone: 314-694-6640
Emergency: 314-694-4000

10.3.12 Metolachlor

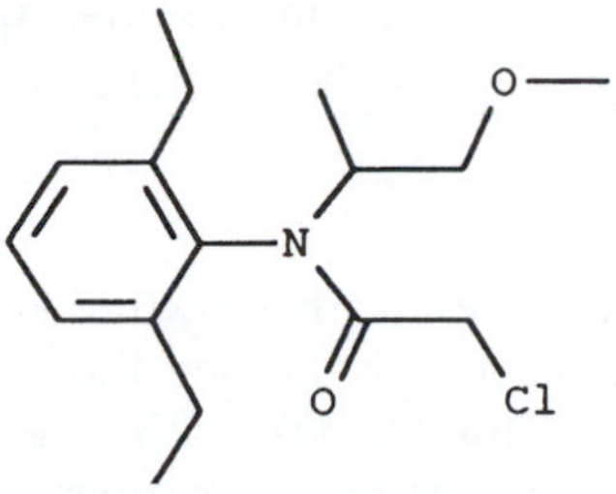

Figure 10.23 Metolachlor.

Trade or other names

Trade names for products containing metolachlor include Bicep, CGA-24705, Dual, Pennant, and Pimagram. The compound may be used in formulations with other pesticides (often herbicides that control broad-leaved weeds), including atrazine, cyanazine, and fluometuron.

Regulatory status

Metolachlor is a slightly toxic compound in EPA toxicity class III. Labels for products containing it must bear the CAUTION Signal Word. Metolachlor is in most cases a General Use Pesticide (GUP), although some products may be restricted use (RUP).

Introduction

Metolachlor is usually applied to crops before plants emerge from the soil, and is used to control certain broadleaf and annual grassy weeds in field corn, soybeans, peanuts, grain sorghum, potatoes, pod crops, cotton, safflower, stone fruits, nut trees, highway rights-of-way, and woody ornamentals. It inhibits protein synthesis; thus, high-protein crops (e.g., soy) can be adversely affected by excessive metolachlor application. Additives may be included in product formulations to help protect sensitive crops (i.e., sorghum) from injury.

Toxicological effects

Acute toxicity

Metolachlor is slightly toxic via ingestion. The reported oral LD_{50} in rats for technical grade metolachlor is from 1200 to 2780 mg/kg (1,30,100). It is slightly to practically nontoxic by skin exposure, with a reported dermal LD_{50} of greater than 2000 mg/kg (30,58). Technical metolachlor is a skin sensitizer in guinea pigs and causes slight irritation and mild eye irritation in rabbits (58). The formulated products are generally not skin sensitizers, but cause a range (slight to moderate) of skin and eye irritation in rabbits (58).

The 4-hour rat inhalation LC_{50} of greater than 4.3 mg/L indicates slight toxicity via this route (1,58). Compared to the technical grade, metholachlor formulations are generally of similar or lesser toxicity by all routes, except by inhalation; some formulated products may show higher toxicity by this route (58). However, none of the formulated products for which inhalation toxicity data are available are highly toxic by this route (58).

Human exposure most commonly occurs through skin or eye contact (101). Signs of human intoxication from metolachlor exposure include abdominal cramps, anemia, shortness of breath, dark urine, convulsions, diarrhea, jaundice, weakness, nausea, sweating, and dizziness (101).

Chronic toxicity

While metolachlor is not readily absorbed by the skin, repeated dermal exposures may create skin sensitization, especially among those who work with metolachlor (101). In rats fed metolachlor for 90 days, no effects were noted at about 90 mg/kg/day (87). In a 2-year study of mice, a similar no-effect level was found, but doses of about 300 mg/kg/day caused decreased body weight gain (58). No negative effects on mortality or organ weights were observed in male or female rats at doses of 15 mg/kg/day, but exposed females showed significantly lower weight gain and microscopic changes in their liver structure at 150 mg/kg/day (58).

Reproductive effects

In two long-term rat reproduction studies, mating, gestation, lactation, and fertility were not affected at doses of 50 mg/kg/day (58,101). However, pup weights and parental food consumption decreased at this low dose. In another 2-year rat study, metolachlor caused the wasting of testicles at doses of 150 mg/kg/day (101). In studies on mice, no effects were noted on fertility, or zygote or embryo survival rates after very high single oral doses (101). This evidence suggests that metolachlor is not likely to have an effect on reproduction in humans under normal circumstances.

Teratogenic effects

Metolachlor caused no birth defects in rats at maternal doses of 300 mg/kg/day administered during critical periods of gestation (organogenesis), although some delayed or abnormal development in offspring was seen at this dose (58). A decrease in food consumption was observed in the mother (100). In rabbits, a similar pattern of effects (no defects, but some delayed development) was also seen at doses of up to 360 mg/kg/day (58). These data indicate that teratogenic and developmental effects in humans are unlikely at expected levels of exposure.

Mutagenic effects

Metolachlor tested negative in two bacterial assays (101). Also, no mutagenicity effects were noted in a standard mouse test (101). From this evidence, it is unlikely that the compound is mutagenic.

Carcinogenic effects

Male and female mice exposed to doses up to 100 mg/kg/day for 18 to 20 months did not develop cancer (58,101), nor did male rats at doses up to 150 mg/kg/day over a 2-year period (58). Female rats given high doses for 2 years showed a significant increase in new growths, nodules, and lesions in livers at that dose (58,100). From these data, it seems unlikely that metolachlor is carcinogenic in humans.

Organ toxicity

Exposure to metolachlor can damage the liver and cause irritation of the skin, eyes, and mucous membranes. It has also caused skin sensitization in guinea pigs.

Fate in humans and animals

Studies show that orally administered metolachlor is quickly broken down into metabolites and is almost totally eliminated in the urine and feces of goats, rats, and poultry. Metolachlor itself was not detected in the urine, feces, or body tissues (101). Rats given a single oral dose of metolachlor excreted 70 to 90% of the metolachlor as metabolites within 48 hours (6).

In animals, trace amounts of metolachlor metabolites were found in kidneys, liver, blood, and milk; however, no residues were found in eggs, meat, or fat samples of laying chickens (101).

Ecological effects

Effects on birds

Metolachlor is slightly to practically nontoxic to birds. The reported oral LD_{50} is greater than 2000 mg/kg in mallard ducks and is greater than 4500 mg/kg in bobwhite quail (58). Both the mallard and the bobwhite quail show 5-day dietary LC_{50} values of greater than 10,000 ppm, also indicating very low toxicity to upland game birds and waterfowl (58,100). However, although mallard ducks showed no impairment of reproductive capabilities at high-level, long-term exposures, bobwhite quail fed a diet containing high levels of metolachlor for 17 weeks during mating, egg laying, and egg hatching produced fewer chicks (100).

Effects on aquatic organisms

Metolachlor is moderately toxic to both cold- and warmwater fish, including rainbow trout, carp, and bluegill sunfish. The reported 96-hour LC_{50} values for this compound are about 3 mg/L in rainbow trout, 5 mg/L in carp and channel catfish, and 15.0 mg/L in bluegill sunfish (58). The 48-hour LC_{50} of metolachlor in *Daphnia magna* (water flea) is 25 mg/L (58). Studies on algae and fish exposed to metolachlor in water indicate that very little is accumulated and that any accumulated material is excreted rapidly when the organisms are placed in clean water (100).

Effects on other organisms (non-target species)

Metolachlor is nontoxic to bees; its contact LC_{50} in earthworms is 140 ppm (58).

Environmental fate

Breakdown in soil and groundwater

Metolachlor is moderately persistent in the soil environment. Half-lives of 15 to 70 days in different soils have been observed (58,100). Soils with significant soil water content may show more rapid breakdown. Very little metolachlor volatilizes from the soil, and photodegradation will be a significant pathway for loss only in the top few inches (102). Breakdown is mainly dependent upon microbial activity, and thus will be temperature dependent (102). Microorganism metabolism occurs by both aerobic and anaerobic processes, and is affected by temperature, moisture, amount of leaching, soil type, nitrification, oxygen concentration, and sunlight (58,102).

Metolachlor is moderately well sorbed by most soils (58,102). Soils with higher organic matter and clay contents may sorb it better. It is slightly soluble in water (58).

Extensive leaching is reported to occur, especially in soils with low organic content (100). Metolachlor was one of four pesticides extensively studied throughout the United States in the National Alachlor Well Water Survey. This several-year project analyzed the contents of over 6 million private and domestic wells. Metolachlor was detected in about 1% of the wells (about 60,000 wells) at concentrations ranging from 0.1 to 1.0 µg/L (64). It has also been found in a number of surface water samples in 14 states, at a maximum concentration of 0.138 mg/L (9). These levels may result from runoff during spring and summer applications to fields (101).

Breakdown in water

Metolachlor is highly persistent in water over a wide range of water acidity. Its half-life at 20°C is more than 200 days in highly acid waters, and is 97 days in highly basic waters (101). Metolachlor is also relatively stable in water under natural sunlight. About 6.6% was degraded by sunlight in 30 days, a slow and minimal rate (100).

Breakdown in vegetation

Metolachlor, applied before plants emerge, is absorbed through shoots just above the seed, and may be absorbed from the soil into and through the roots (58). This chemical acts by inhibiting the production of essential plant components like chlorophylls, enzymes, and other proteins (58). Metolachlor is a growth inhibitor affecting root and shoot growth after seeds have germinated. The breakdown of metolachlor in corn, soybean, peanuts, and sorghum is similar. Residues and metabolites are found in minimal concentrations in roots, grain, and oil, but other parts of the plants may have higher levels. Some care should be exercised when crop remnants are used as forage; cotton crops may retain very high levels of residue (100).

Physical properties

Pure metolachlor is an odorless, off-white to colorless liquid at room temperature (1). In formulations, its color ranges from opaque white to tan.

> Chemical name: 2-chloro-6'-ethyl-N-(2-methoxy-1-methylethyl)acet-*o*-toluidide (1)
> CAS #: 51218-45-2
> Molecular weight: 283.80 (1)
> Solubility in water: 530 mg/L @ 20°C (1)
> Solubility in solvents: miscible with benzene, dichloromethane, hexane, methanol, octanol, xylene, toluene, dimethylformamide, ethylene dichloride, and cyclohexanone; insoluble in ethylene glycol and propylene glycol (1)
> Vapor pressure: 1.7 mPa @ 20°C (1)
> Partition coefficient (octanol/water): 2820 (1)
> Adsorption coefficient: 200 (11)

Exposure guidelines

> ADI: Not available
> HA: 0.1 mg/L (lifetime) (101)
> RfD: 0.1 mg/kg/day (13)
> PEL: Not available

Basic manufacturer

Ciba-Geigy Corp.
P.O. Box 18300
Greensboro, NC 27419-8300
Telephone: 800-334-9481
Emergency: 800-888-8372

10.3.13 Napropamide

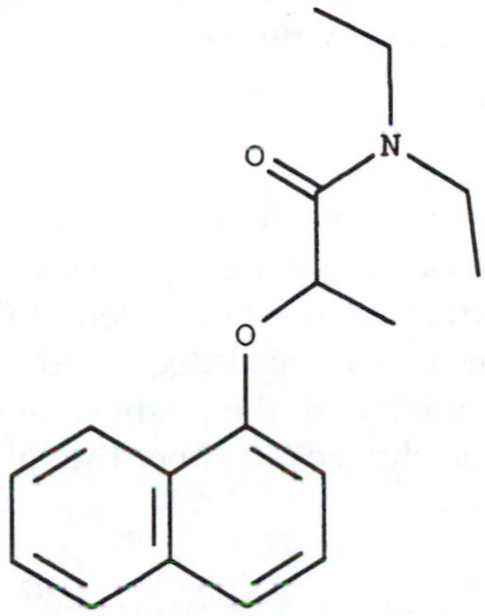

Figure 10.24 Napropamide.

Trade or other names

Trade names for products containing napropamide include Devrinol and R-7465. It may also be found in formulations with other pesticides such as monolinuron, nitralin, simazine, trifluralin, tefurthrin, and tebutam. It is compatible with many other herbicides and fungicides.

Regulatory status

Napropamide is a slightly toxic compound in EPA toxicity class III. The Signal Word CAUTION is required on the label of products containing the compound. Napropamide is a General Use Pesticide (GUP).

Introduction

Napropamide is a selective systemic amide herbicide used to control a number of annual grasses and broad-leaved weeds. Napropamide is applied to soils where vegetables, fruit trees and bushes, vines, strawberries, sunflowers, tobacco, olives, and mint or other crops are grown. The compound is absorbed by the roots and works by inhibiting root development and growth. It is available in emulsifiable concentrate, wettable powder, granules, and suspension concentrates.

Toxicological effects

Acute toxicity

Napropamide is slightly to practically nontoxic by ingestion, with reported oral LD_{50} values for the technical product of over 5000 mg/kg in male rats and 4680 mg/kg in female rats (1,103). It is slightly to practically nontoxic by skin exposure, with reported dermal LD_{50} values of greater than 4640 mg/kg in rabbits, and greater than 2000 mg/kg in guinea pigs (1,103). Data on eye and skin irritation are not available.

No studies have evaluated the acute toxicity of the technical product (95% napropamide) through inhalation exposure, but several inhalation studies conducted on Devrinol 4F (43.2% napropamide) indicate that the 4-hour inhalation LC_{50} for this formulation is greater than 0.2 mg/L in rats (13). This indicates that this formulation may be highly toxic by this route of exposure.

Toxic effects from acute exposure in rats included diarrhea, excessive tearing and urination, depression, salivation, rapid weight loss, respiratory changes, decreased blood pressure, and fluid in body cavities (13).

Chronic toxicity

Rats fed napropamide at doses up to 30 mg/kg/day for 13 weeks showed no significant effects (103). At doses of 40 mg/kg/day over the same length of time, female rats experienced a reduction in uterine weight (104,105). In a feeding study in dogs over 13 weeks, males experienced decreased liver and body weights, and some changes in blood chemistry at the highest dose tested, 100 mg/kg/day (104,105). There were no tissue changes in either the rats or dogs at doses up to 100 mg/kg/day (104,105).

Reproductive effects

One study conducted over three successive generations of rats showed a decrease in body weight gain in the fetal pups at the 100-mg/kg/day dose (13). There is not enough evidence to assess the potential of napropamide to cause reproductive effects.

Teratogenic effects

Three separate tests with rats produced conflicting results, thus making any conclusion difficult to draw. One study indicated that the compound had no effects at doses over 400 mg/kg/day in pregnant rats or in their offspring (30,101). However, two other studies showed incomplete formation of bone in the rats at doses as low as 25 mg/kg/day administered at critical times of pregnancy (organogenesis) (13).

Mutagenic effects

Three tests of the mutagenicity of napropamide all produced negative (nonmutagenic) results (30). These assays were conducted on bacterial cells and in mice. This evidence suggests that napropamide is not mutagenic.

Carcinogenic effects

Two tests, each conducted over 2 years, revealed no cancer-related changes in either rats or mice. The highest dose of napropamide in both of the studies was 100 mg/kg/day (30). In both cases, the only effects noted were decreases in body weight gains for both species. These data suggest that napropamide is not likely to be carcinogenic.

Organ toxicity

Though numerous organ-related adverse effects have been noted after acute exposure to napropamide, few have been found after chronic exposure. Only decreases in uterine and liver weights have been observed in test animals.

Fate in humans and animals

Elimination is very rapid following oral exposure to napropamide (103). Generally, most (99%) of a single dose is excreted within 4 days (103).

Ecological effects

Effects on birds

Napropamide is practically nontoxic to game birds. The compound has 5-day dietary LC_{50} values ranging from nearly 7200 ppm in the mallard duck to 5600 ppm in the bobwhite quail (1,104).

Effects on aquatic organims

The compound is slightly to moderately toxic to freshwater fish species. The LC_{50} for the compound ranges from 9 to 16 mg/L in rainbow trout, and from 20 to 30 mg/L in bluegill sunfish (1). The LC_{50} in goldfish is greater than 10 mg/L (1). Napropamide appears to be slightly toxic to aquatic invertebrates; the reported 48-hour LC_{50} for the compound in *Daphnia magna* is 14.3 mg/L (1). Tests with marine organisms indicate that the compound is slightly toxic (fiddler crab) to moderately toxic (pink shrimp and eastern oyster) to these species (104).

In one study over a 10-day period, napropamide bioaccumulated in the edible portion of the fish to 50 times the water concentration. Most of the accumulation was eliminated within 24 hours in clean water (104). This indicates that the compound is not likely to accumulate appreciably in the tissue of fish.

Effects on other organisms (non-target species)

The reported oral LD_{50} for napropamide in bees is 121 μg per bee, indicating that it is not toxic to this species (1).

Environmental fate

Breakdown in soil and groundwater

Napropamide is moderately persistent in the soil environment, with reported field half-lives of 56 to 84 days (11). A typical value for most situations is estimated to be 70 days (11). It is rapidly lost by photodegradation near the soil surface after application, and may be slowly degraded by soil fungi and bacteria (8). Little or no loss due to volatilization was seen in field studies (8).

It is moderately bound to most soils, and is slightly soluble in water (1,11). It therefore may pose a slight threat of contamination to groundwater in soils with high water tables, very low organic matter and clay content, high porosity, and high rainfall (48). As of 1988, napropamide was not found in samples from 172 different wells across the country (8).

Breakdown in water

In water, napropamide is broken down very quickly. The half-life may be as rapid as 7 minutes (104). The breakdown in water is predominantly mediated by the action of sunlight (photolysis) (104).

Breakdown in vegetation

Napropamide is typically absorbed through the roots and translocated throughout the plant in some species (e.g., tomato) but not others (e.g., corn) (1). It effectively shuts down growth by inhibiting cell division (mitosis) in the shoot regions (meristems) (1). Susceptibility is determined by rates of translocation to these regions and by the ability to detoxify the compound. In tolerant species, rapid breakdown to water-soluble compounds occurs (1).

Physical properties

Napropamide is a colorless crystal; the technical form is a brown solid (1).

Chemical name: (RS)-N,N-diethyl-2-(1-naphthyloxy)propionamide (1)
CAS #: 15299-99-7
Molecular weight: 271.36 (1)
Solubility in water: 73 mg/L @ 20°C (1)
Solubility in other solvents: v.s. in acetone, ethanol, xylene, and hexane (1)
Melting point: 74.8-75.5°C (1)
Vapor pressure: 0.53 mPa @ 25°C (1)
Partition coefficient (octanol/water): 2300 @ 25°C (1)
Adsorption coefficient: 700 (11)

Exposure guidelines

ADI: Not available
HA: Not available
RfD: 0.1 mg/kg/day (13)
PEL: Not available

Basic manufacturer

Zeneca Ag Products
1800 Concord Pike
Wilmington, DE 19897
Telephone: 800-759-4500
Emergency: 800-759-2500

10.3.14 Oryzalin.

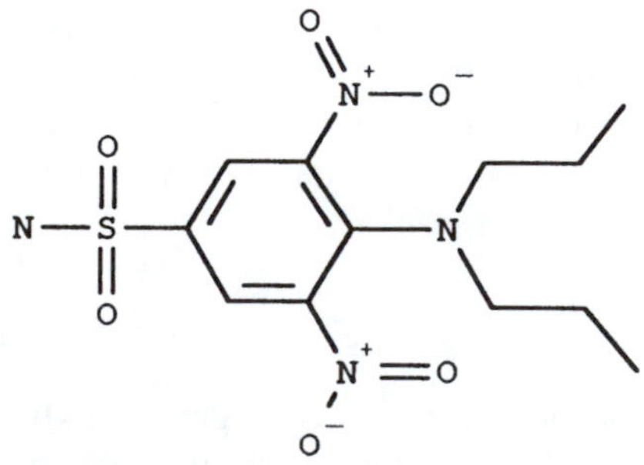

Figure 10.25 Oryzalin.

Trade or other names

Trade names for oryzalin include Dirimal, EL-119, Rycelan, Ryzelan, Ryzelon, and Surflan.

Regulatory status

Oryzalin is a slightly to practically nontoxic compound in EPA toxicity class IV. Products containing oryzalin must bear the Signal Word CAUTION. It is a General Use Pesticide (GUP)

Introduction

Oryzalin is a selective pre-emergence, surface-applied herbicide used for control of annual grasses and broadleaf weeds in fruit and nut trees, vineyards, established bermuda grass turf, and established ornamentals. It inhibits the growth of germinating weed seeds by blocking cell division in the meristems. It is available in aqueous suspension, dry flowable, and wettable powder formulations.

Toxicological effects

Acute toxicity

Oryzalin is practically nontoxic by ingestion, with reported oral LD_{50} values of greater than 5000 mg/kg in rats and mice (1,58), and greater than 1000 mg/kg in cats, dogs, and chickens (1,106). The dermal LD_{50} for technical oryzalin in rabbits is greater than 2000 mg/kg, indicating slight to practically no toxicity by this route (58). Oryzalin is reported to cause slight skin and eye irritation in the rabbit, and no skin sensitization in the guinea pig (58). It is also slightly toxic when inhaled, with a 4-hour inhalation LC_{50} of greater than 3 mg/L in rats (58). The formulated products (e.g., Surflan A.S.) may show moderate toxicity by either the oral or inhalation routes, and may show skin and eye irritation and skin sensitization properties (58). In dogs and cats, large oral doses cause nausea and vomiting (8).

Chronic toxicity

Rats fed a dietary level of about 2.5 mg/kg/day for 2 years exhibited blood changes, increased liver and kidney weights, inhibition of growth, and decreased survival (8). Repeated ingestion of large doses led to adverse changes in blood cell formation in dogs (8). Mice given dietary doses of about 200 mg/kg/day for 1 year exhibited decreased uterine and ovarian weights. Those exposed to doses of 75 mg/kg/day showed no observable effects (107).

Reproductive effects
There were no adverse effects on reproduction in a three-generation study of rats fed dietary concentrations of 12.5, 37.5, or 112.5 mg/kg/day (the highest dose tested). Fetotoxic effects appeared at 12.5 mg/kg/day (58,107).

It does not appear that oryzalin causes reproductive effects.

Teratogenic effects
There were no birth defects in the offspring of pregnant rats fed dietary concentrations as high as 112 mg/kg/day for three generations, nor in the offspring of pregnant rabbits given doses of 125 mg/kg/day, the highest dose tested (8,107). It appears that oryzalin is unlikely to cause teratogenic effects.

Mutagenic effects

Oryzalin was not mutagenic in several tests, including tests on live rats and mice and on bacterial cell cultures (107). It does not appear that oryzalin is mutagenic.

Carcinogenic effects

When oryzalin was fed to rats in doses as high as 135 mg/kg/day for 2 years, there was an increase in the incidence of thyroid, mammary, and skin tumors (107). Thyroid tumors and benign skin and mammary tumors occurred in rats fed a dietary level of 45 mg/kg/day for 2 years (107). However, there were no tumors in mice fed doses as high as 548 mg/kg/day for 2 years (107). Because of these conflicting results, it is not possible to assess the carcinogenicity of oryzalin.

Organ toxicity

Oryzalin has shown systemic effects on the thyroid, liver, and kidneys, as well as blood chemistry, in animal tests.

Fate in humans and animals

Oryzalin is moderately well-absorbed from the gastrointestinal tract, and rapidly metabolized and eliminated following absorption. When oryzalin was administered to male rats, 40% of the dose was excreted in the urine and 40% in the feces within 3 days. Similar results were obtained in tests with rabbits, a steer, and with Rhesus monkeys (106).

Ecological effects

Effects on birds

Oryzalin is slightly toxic to practically nontoxic to birds; the reported oral LD_{50} values in bobwhite quail and mallard ducks are greater than 500 mg/kg, and in chickens is 1000 mg/kg (1,58). The 5-day dietary LC_{50} values in quail and mallard ducks are greater than 5000 ppm (8).

Effects on aquatic organisms

Oryzalin is highly toxic to fish, with reported 96-hour LC_{50} values of 2.88 mg/L in bluegill sunfish, 3.26 mg/L in rainbow trout, and greater than 1.4 mg/L in goldfish fingerlings (8).

Effects on other organisms (non-target species)

The reported oral LD_{50} for oryzalin in bees is 11 μg per bee, indicating it is nontoxic to bees (1,58).

Environmental fate

Breakdown in soil and groundwater

Oryzalin is of low to moderate persistence in the field, with reported field half-lives ranging from 20 to 128 days (58,11). A representative value for soil half-life is estimated to be 20 days (11,58). Microbial degradation is mainly responsible for the breakdown of oryzalin in soils, but it may undergo photodecomposition near the soil surface (58). Volatilization is not appreciable (58).

Oryzalin is slightly soluble in water and it does not have a strong tendency to adsorb to soil particles (58,11). It is bound to a greater extent with increasing soil organic matter and clay content; in soils with low proportions of these, high water tables, and increased rainfall, oryzlin may be mobile, and thus present a risk of contamination to groundwater.

Breakdown in water

No breakdown of oryzalin by hydrolysis was observed at pH 5, 7, and 9 (8). Based on its behavior in soil, breakdown by microbial processes is probably slow in the aquatic environment due to low levels of oxygen and low microbial activity. Photodegradation may be significant in the upper portions of the water column.

Breakdown in vegetation

Oryzalin is readily absorbed via the roots, and plant metabolism of oryzalin is minimal (58).

Physical properties

Technical oryzalin is a bright yellow-orange crystalline powder (1).

Chemical name: 3,5-dinitro-N4,N4-dipropylsulfanilamide (1)
CAS #: 19044-88-3
Molecular weight: 346.36 (1)
Solubility in water: 2.5 mg/L @ pH 7 and 25°C (1)
Solubility in other solvents: v.s. in organic solvents such as acetone, methanol, and acetonitrile; s.s. in benzene and xylene; i.s. in hexane (1)
Melting point: 141-142°C (1)
Vapor pressure: <0.013 mPa @ 30°C (1)
Partition coefficient (octanol/water): 5,420 @ pH 7 (1)
Adsorption coefficient: 600 (11)

Exposure guidelines

ADI: Not available
HA: Not available
RfD: 0.05 mg/kg/day (13)
PEL: Not available

Basic manufacturer

DowElanco
9330 Zionsville Road
Indianapolis, IN 46268-1054
Telephone: 317-337-7352
Emergency: 800-258-3033

10.3.15 Oxyfluorfen

Trade or other names

Trade names for oxyfluorfen include Goal, Koltar, and RH-2915.

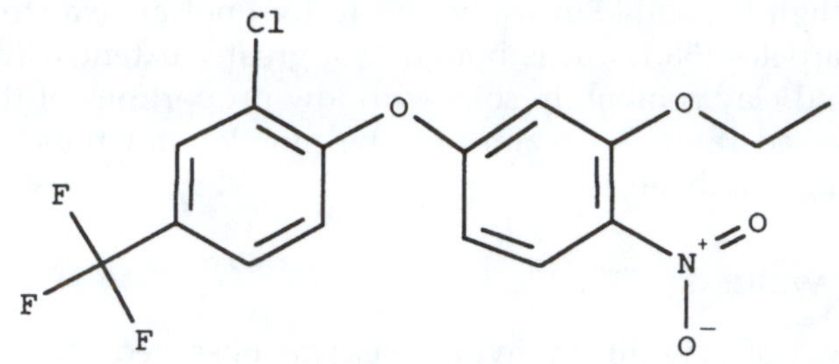

Figure 10.26 Oxyfluorfen.

Regulatory status

Oxyfluorfen is a slightly to practically nontoxic compound in EPA toxicity class III (2). Products containing oxyfluorfen must bear the Signal Word WARNING on the label because some formulated products may have higher toxicities. It is a General Use Pesticide (GUP).

Introduction

Oxyfluorfen is a selective pre- and post-emergent herbicide used to control certain annual broadleaf and grassy weeds in vegetables, fruit, cotton, and ornamentals and on non-crop areas (e.g., rail and highway rights-of-way). It is a contact herbicide and light is required for it to affect target plants. It is available in emulsifiable concentrate and granular formulations.

Toxicological effects

Acute toxicity

Oxyfluorfen is practically nontoxic by ingestion, with reported oral LD_{50} values of 5000 mg/kg in both rats and dogs, and 2700 to 5000 mg/kg in mice (1,58). The dermal LD_{50} is greater than 5000 mg/kg in both rats and rabbits, also indicating slight toxicity by this route (58). It causes no skin irritation in rabbits, no skin sensitization in guinea pigs, and moderate eye irritation in rabbits (58). However, Goal and other formulations may show severe skin and eye irritant properties and may be skin sensitizers (58). The 4-hour inhalation LC_{50} for the technical product is not available, but that for Goal 1.6E is greater than 22.64 mg/L, indicating practically no toxicity via this route (58).

Chronic toxicity

Effects on the liver have been observed in long-term feeding studies with rats, mice, and dogs (56,108).

Reproductive effects

In a developmental study with rats given doses of 10, 100, or 1000-mg/kg/day by gavage, decreased implantation, increased resorption, and lower fetal survival was seen at the 1000 mg/kg level. Toxic effects on the mothers were also seen at this dose (108). At 5 mg/kg/day, there was decreased survival of fetuses and decreased maternal and fetal weights (108). It does not appear likely that oxyfluorfen will cause reproductive effects in humans at normal levels of exposure.

Teratogenic effects

In a developmental study with rabbits, 30 mg/kg/day (the highest dose tested) produced an increase in fused sternal bones in the fetuses, as well as toxic effects on the mothers (108). These data suggest that oxyfluorfen may have teratogenic effects, but only at very high doses.

Mutagenic effects

Mutagenicity tests on rats, mice, and on bacterial cell cultures have produced mixed results. However, unscheduled DNA synthesis assays have been negative (58,108). Due to the conflicting results, it is not possible to determine the mutagenic potential of oxyfluorfen.

Carcinogenic effects

In a 20-month study with mice fed 0.3, 3, or 30 mg/kg/day, doses at and above 3 mg/kg/day produced nonsignificant increases in both benign and malignant liver tumors in male mice (58,108). No increased tumor formation was seen in female mice at any dose (58,108). No carcinogenic effects were observed in a 2-year study with rats fed doses of 2 mg/kg/day, nor in dogs at doses of 3 mg/kg/day (58,108). These data suggest that oxyfluorfen is not carcinogenic.

Organ toxicity

The liver appears to be the main target organ, based on long-term feeding studies.

Fate in humans and animals

Because oxyfluorfen is highly hydrophobic, it may have the potential to bioconcentrate in animal fatty tissues (108).

Ecological effects

Effects on birds

Oxyfluorfen is practically nontoxic to birds; the reported oral LD_{50} values are greater than 2200 mg/kg in bobwhite quail, and greater than 4000 mg/kg in mallard duck (108). The dietary 8-day dietary LC_{50} values are greater than 5000 ppm in bobwhite quail and 4000 ppm in mallard ducks (58,109). Dietary concentrations as high as 100 ppm had no effect on reproduction in mallards or bobwhite quail (109).

Effects on aquatic organisms

Oxyfluorfen is highly toxic to aquatic invertebrates, freshwater clams, oysters, aquatic plants, and fish. The reported 96-hour LC_{50} values are 200 µg/L in bluegill sunfish, 410 µg/L in rainbow trout, 400 µg/L in channel catfish, 150 µg/L in fathead minnow, and 32 µg/L in grass shrimp and oysters (58,109). Its 96-hour LC_{50} in freshwater clams is 10 µg/L; and the 96-hour LC_{50} for the product Goal 2E in *Daphnia magna*, a small freshwater crustacean, is 1500 µg/L (58,109).

Oxyfluorfen accumulated up to 13 mg/kg (13,000 µg/kg) in bluegill sunfish exposed to 10 µg/L for 40 days (108). This represents a bioconcentration factor (BCF) of 1300. The BCF in channel catfish was 700 to 5000 in one 30-day study (109). These results indicate a low to moderate potential for bioaccumulation in aquatic species.

Effects on other organisms (non-target species)

Oxyfluorfen is nontoxic to honeybees, with a reported oral LC_{50} of greater than 10,000 ppm (58).

Environmental fate

Breakdown in Soil and Groundwater

Oxyfluorfen is moderately persistent in most soil environments, with a representative field half-life of about 30 to 40 days (58,110). Oxyfluorfen is not subject to microbial degradation or hydrolysis (11,58). The main mechanism of degradation in soils may be photodegradation and evaporation/codistillation in moist soils (58,110). In laboratory studies, its soil half-life was 6 months, indicating very low rates of microbial degradation (11,58).

Oxyfluorfen is very well sorbed to most soils (11). Soil binding is highest in soils with high organic matter and clay content (11,58). Once oxyfluorfen is adsorbed to soil particles, it is not readily removed (58). It is practically insoluble in water, and therefore is unlikely to be appreciably mobile in most instances unless the sorptive capacity of the soil is exceeded. Oxyfluorfen did not leach below 4 inches in any soil except sand (11).

Breakdown in water

In water, oxyfluorfen is rapidly decomposed by light (3). Because oxyfluorfen is nearly insoluble in water and has a tendency to adsorb to soil, it will be sorbed to suspended particles or sediments (58,109).

Breakdown in vegetation

There is very little movement of oxyfluorfen within treated plants. It is not readily metabolized by plants; but since it is not readily taken up by roots, residues in plants are generally very low (58). Residues of oxyfluorfen accumulated in carrots and oats grown on previously treated fields, but not in cotton or lettuce (109).

Physical properties

Oxyfluorfen is a white to orange or red-brown crystalline solid with a smoke-like odor (1,108).

> Chemical name: 2-chloro-alpha,alpha,alpha-trifluoro-*p*-tolyl 3-ethoxy-4-nitro-
>   phenyl ether (1)
> CAS #: 42874-03-3
> Molecular weight: 361.7 (1)
> Solubility in water: 0.1 mg/L (1)
> Solubility in other solvents: v.s. in most organic solvents (e.g., acetone, cyclohex-
>   anone, isophorone) (1)
> Melting point: 84–85°C (1)
> Vapor pressure: 0.026 mPa @ 25°C (1)
> Partition coefficient (octanol/water): 29,400 (1)
> Adsorption coefficient: 100,000 (estimated) (11)

Exposure guidelines

ADI: Not available
HA: Not available
RfD: 0.003 mg/kg/day (13)
PEL: Not available

Basic manufacturer

Rohm and Haas Co.
Agricultural Chemicals
100 Independence Mall West
Philadelphia, PA 19106
Telephone: 215-592-3000

10.3.16 *Paraquat*

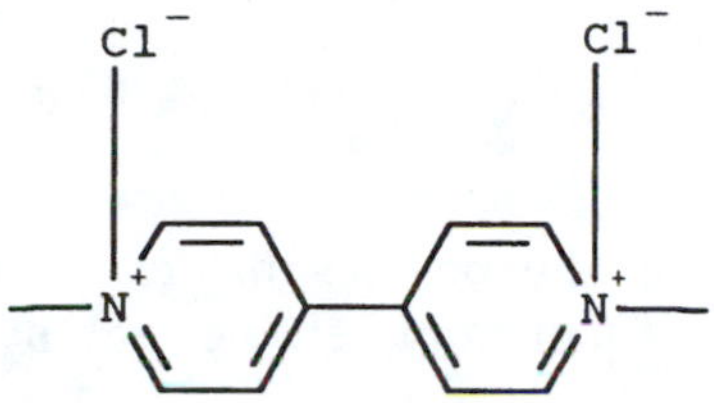

Figure 10.27 Paraquat.

Trade or other names

Product names of paraquat include Crisquat, Cyclone, Dextrone, Dexuron, Gramoxone Extra, Herbaxone, Ortho Weed, Spot Killer, and Sweep. The compound may be found in formulations with many other herbicides, including simazine and diquat dibromide.

Regulatory status

Paraquat is a highly toxic compound in EPA toxicity class I. Products containing it must be labeled with the Signal Words DANGER — POISON. Paraquat is a Restricted Use Pesticide (RUP). RUPs may be purchased and used only by certified applicators.

Introduction

Paraquat is a quaternary nitrogen herbicide widely used for broadleaf weed control. It is a quick-acting, nonselective compound that destroys green plant tissue on contact and by translocation within the plant. It has been employed for killing marijuana in the U.S. and Mexico. It is also used as a crop desiccant and defoliant, and as an aquatic herbicide.

Toxicological effects

Acute toxicity

Paraquat is highly toxic via ingestion, with reported oral LD_{50} values of 110 to 150 mg/kg in rats, 50 mg/kg in monkeys, 48 mg/kg in cats, and 50 to 70 mg/kg in cows (8,87). The toxic effects of paraquat are due to the cation, and the halogen anions have little toxic effects (87). The dermal LD_{50} in rabbits is 236 to 325 mg/kg, indicating moderate toxicity by this route (58,87). The 4-hour inhalation LC_{50} is greater than 20 mg/L for the technical grade compound (87). It causes skin and eye irritation in rabbits (severe for some of the formulated products) and also causes skin sensitization in guinea pigs in some formulations (87).

Effects due to high acute exposure to paraquat may include excitability and lung congestion, which in some cases leads to convulsions, incoordination, and death by respiratory failure (87). If swallowed, burning of the mouth and throat often occurs, followed by gastrointestinal tract irritation, resulting in abdominal pain, loss of appetite, nausea, vomiting, and diarrhea (8). Other toxic effects include thirst, shortness of breath, rapid heart rate, kidney failure, lung sores, and liver injury (32). Some symptoms may not occur until days after exposure. Persons with lung problems may be at increased risk from exposure.

Many cases of illness and/or death have been reported in humans. The estimated lethal dose (via ingestion) for paraquat in humans is 35 mg/kg (8). A maximum of 3.5 mg/hour could be absorbed through the dermal or respiratory route without damage (32).

Chronic toxicity

As indicated above, repeated exposures may cause skin irritation, sensitization, or ulcerations on contact (58,87). In animal studies, rats showed no effects after being exposed for 2 years to paraquat at doses of 1.25 mg/kg/day (8). Dogs, however, developed lung problems after being exposed for 2 years at high doses (above 34 mg/kg/day) (8).

In a study of 30 workers spraying paraquat over a 12-week period, approximately one half had minor irritation of the eyes and nose (8). Of 296 spray operators with gross and prolonged skin exposure, 55 had damaged fingernails, as indicated by discoloration, nail deformities, or loss of nails (8).

Reproductive effects

In a long-term rat study at doses up to 5 mg/kg/day, no adverse reproductive effects were reported (111). However, paraquat dichloride injected intraperitoneally at 3 mg/kg/day on days 8 to 16 of gestation increased fetal mortality in rats (8). Hens given high levels of paraquat in their drinking water for 14 days produced an increased percentage of abnormal eggs (8). Paraquat is unlikely to cause reproductive effects in humans at expected exposure levels.

Teratogenic effects

Offspring of mice dosed with high doses of paraquat during the organ-forming period of pregnancy had less-complete bone development than the mice given lower doses (111). Offspring of rats given similar treatment showed no developmental defects at any dose, but fetal and maternal body weights were lower than normal (111). Other studies of paraquat using rabbits and mice have shown no teratogenic effects (8). The weight of evidence suggests that paraquat does not cause birth defects at doses that might reasonably be encountered.

Mutagenic effects

Paraquat has been shown to be mutagenic in microorganism tests and mouse cell assays (8). It was unclear what levels of exposure were necessary to produce these effects.

Carcinogenic effects

Mice fed paraquat dichloride for 99 weeks at high levels did not show cancerous growths (112). Rats fed high doses for 113 (male) or 124 weeks (female) developed lung, thyroid, skin, and adrenal tumors (111). Thus, the evidence regarding carcinogenic effects of paraquat is inconclusive.

Organ toxicity

Paraquat affects the lungs, heart, liver, kidneys, cornea, adrenal glands, skin, and digestive system.

Fate in humans and animals

Paraquat is not readily absorbed from the stomach, and is even more slowly absorbed across the skin. Oral doses of paraquat in rats are excreted mainly in the feces, while paraquat injected into the abdomen leaves through the urine (8). In the stomach and gastrointestinal tract, paraquat metabolites may be more readily absorbed than the parent compound, but their identities and toxicities are unknown (111). Paraquat may concentrate in lung tissue, where it can be transformed into highly reactive and potentially toxic forms (87).

In one study, farm animals excreted over 90% of the administered paraquat within a few days. It was slightly absorbed and metabolized in the gastrointestinal tract. Milk and eggs contained small amounts of two paraquat metabolites (58).

Ecological effects

Effects on birds

Paraquat is moderately toxic to birds, with reported acute oral LD_{50} values of 981 and 970 mg/kg in bobwhite and Japanese quail, respectively (58). The reported 5- to 8-day dietary LC_{50} value for the compound is 4048 ppm in mallards (58).

Effects on aquatic organisms

Paraquat is slightly to moderately toxic to many species of aquatic life, including rainbow trout, bluegill, and channel catfish (8,58). The reported 96-hour LC_{50} for paraquat is 32 mg/L in rainbow trout and 13 mg/L in brown trout (58). The LC_{50} for the aquatic invertebrate *Daphnia pulex* is 1.2 to 4.0 mg/L (8). In rainbow trout exposed for 7 days to paraquat, the chemical was detected in the gut and liver, but not in the meat of the fish.

Aquatic weeds may bioaccumulate the compound. In one study, 4 days after paraquat was applied as an aquatic herbicide, sampled weeds showed significant residue levels (87). At high levels, paraquat inhibits the photosynthesis of some algae in stream waters (87).

Effects on other organisms (non-target species)

Paraquat is nontoxic to honeybees (112).

Environmental fate

Breakdown in soil and groundwater

Paraquat is highly persistent in the soil environment, with reported field half-lives of greater than 1000 days (11,58). The reported half-life for paraquat in one study ranged from 16 months (aerobic laboratory conditions) to 13 years (field study) (113). Ultraviolet light, sunlight, and soil microorganisms can degrade paraquat to products that are less toxic than the parent compound. The strong affinity for adsorption by soil particles and organic matter may limit the bioavailability of paraquat to plants, earthworms, and microorganisms (11,58). The bound residues may persist indefinitely and can be transported in runoff with the sediment.

Paraquat is not significantly mobile in most soils (8). That which does not become associated with soil particles can be decomposed to a nontoxic end-product by soil bacteria (32). Thus, paraquat does not present a high risk of groundwater contamination. Of 721 groundwater samples analyzed, only one contained paraquat, at a concentration of 20 mg/L (111).

Breakdown in water

Paraquat will be bound to suspended or precipitated sediment in the aquatic environment, and may be even more highly persistent than on land due to limited availability of oxygen. It had a half-life in a laboratory stream water column of 13.1 hours (114). In another study, paraquat dichloride was stable for up to 30 days (111). In a third study of low levels in water, paraquat had a half-life of 23 weeks (111).

Breakdown in vegetation

Paraquat dichloride droplets decompose when exposed to light after being applied to maize, tomato, and broad-bean plants. Small amounts of residues were found in potatoes treated with paraquat as a desiccant, and boiling the potatoes did not reduce the residue (8).

Physical properties

Paraquat salts are colorless, white, or pale yellow crystalline solids, which are hygroscopic and odorless (1).

Chemical name: 1,1'-dimethyl-4,4'-bipyridinium (1)
CAS #: 1910-42-5
Molecular weight: 257.2 (1)
Solubility in water: 700,000 mg/L @ 20°C (1)
Solubility in other solvents: Dichloride salt is sparingly soluble in lower alcohols (1)
Melting point: Decomposes @ 300°C (1)
Vapor pressure: Negligible @ room temperature (paraquat dichloride) (1)
Partition coefficient (octanol/water): 29,400 (58)
Adsorption coefficient: 1,000,000 (estimated) (11)

Exposure guidelines

ADI: 0.004 mg/kg/day (12)
HA: 0.03 mg/L (lifetime) (111)
RfD: 0.0045 mg/kg/day (13)
TLV: 0.1 mg/m^3 (8-hour) (respirable fraction) (17)

Basic manufacturer

Zeneca Ag Products
1800 Concord Pike
Wilmington, De 19897
Telephone: 800-759-4500
Emergency: 800-759-2500

10.3.17 Pendimethalin

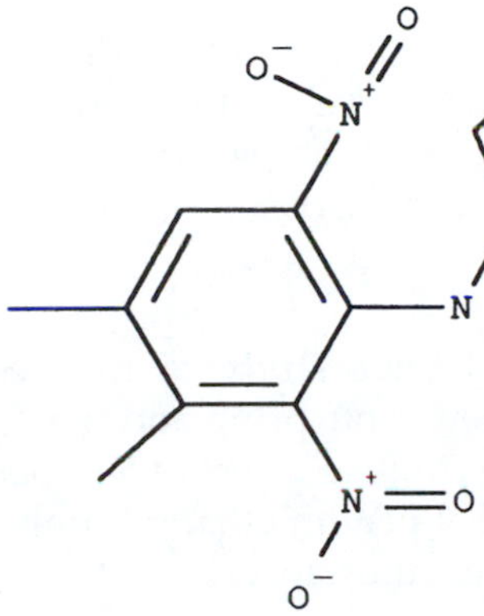

Figure 10.28 Pendimethalin.

Trade or other names

Trade names for pendimethalin include AC 92553, Accotab, Go-Go-San, Herba-dox, Penoxalin, Prowl, Sipaxol, Sovereign, Stomp, and Way-Up.

Regulatory status

Pendimethalin is a slightly toxic compound in EPA toxicity class III. Products containing pendimethalin must bear the Signal Word CAUTION or WARNING, depending on the formulation. Pendimethalin is a General Use Pesticide (GUP).

Introduction

Pendimethalin is a selective herbicide used to control most annual grasses and certain broadleaf weeds in field corn, potatoes, rice, cotton, soybeans, tobacco, peanuts, and sunflowers. It is used both pre-emergence, (i.e., before weed seeds have sprouted) and early post-emergence. Incorporation into the soil by cultivation or irrigation is recommended within 7 days following application. Pendimethalin is available in emulsifiable concentrate, wettable powder, or dispersible granule formulations.

Toxicological effects

Acute toxicity

Pendimethalin is slightly to practically nontoxic by ingestion, with reported oral LD_{50} values of 1050 mg/kg to greater than 5000 mg/kg in rats (1,58). It is slightly to practically nontoxic by skin exposure, with reported dermal LD_{50} values of greater than 2000 mg/kg (1,58). It is not a skin irritant or sensitizer in rabbits or guinea pigs, but it does cause mild eye irritation in rabbits (58). The inhalation 4-hour LC_{50} for technical pendimethalin in rats is 320 mg/L, indicating practically no toxicity via this route (58). Some formulated products (e.g., Prowl) may show slight toxicity by inhalation, and may have a greater capacity to cause skin irritation (58).

Inhalation of dusts or fumes may be mildly to moderately irritating to the linings of the mouth, nose, throat, and lungs (115).

Chronic toxicity

Increases in alkaline phosphatase level and liver weight were produced in dogs fed 50 mg/kg/day for 2 years, but not at a dose of 12.5 mg/kg/day (58,116). In a 90-day feeding study of rats, no effects were observed at doses of 40 mg/kg/day (116).

Reproductive effects

In a three-generation reproductive study of rats tested at levels up to 250 mg/kg/day, there were slightly fewer offspring and they showed decreased weight gain from weaning to maturity (116). No effects were observed at 30 mg/kg/day (58,116). This evidence suggests that pendimethalin is unlikely to cause reproductive effects in humans under normal circumstances.

Teratogenic effects

No birth defects and no toxic effects on fetuses occurred when pregnant rats were given 500 mg/kg/day, the highest dose tested. No fetotoxic or teratogenic effects were seen at the highest dose tested (60 mg/kg/day) in a teratology study with rabbits, although maternal toxicity was seen at 30 mg/kg/day (116). It does not appear that pendimethalin is teratogenic.

Mutagenic effects

Several mutagenicity studies, including tests on live animals and mammalian and bacterial cell cultures, have all indicated that pendimethalin has no mutagenic activity (116).

Carcinogenic effects

Pendimethalin did not increase tumor formation in mice given dietary doses of 75 mg/kg/day over an 18-month period (58). This evidence suggests that pendimethalin is not carcinogenic.

Organ toxicity

Chronic exposure to pendimethalin has resulted in increased liver weights in test animals.

Fate in humans and animals

Pendimethalin is largely unabsorbed from the gastrointestinal tract and is excreted unchanged in the feces (117). Pendimethalin that does become absorbed into the bloodstream from the gastrointestinal tract is rapidly metabolized in the kidneys and liver and is then excreted as metabolites via urine (117). One day after administration to rats, 90% of a 37-mg/kg dose was recovered in feces and urine.

After 4 days, this figure was 96% (117). Lower doses resulted in almost 100% excretion within 4 days. Tissue burdens of the compound were on the order of 0.3 mg/kg, with slightly higher concentrations in the body fat (117).

Ecological effects

Effects on birds

Pendimethalin is slightly toxic to birds, with an acute oral LD_{50} of 1421 mg/kg in mallard duck, and 8-day dietary LC_{50} values of greater than 3149 mg/kg in bobwhite quail and greater than 10,900 mg/kg in mallard duck (58,115).

Effects on aquatic organisms

Pendimethalin is highly toxic to fish and aquatic invertebrates. The reported 96-hour LC_{50} for pendimethalin is 199 µg/L in bluegill sunfish, 138 µg/L in rainbow trout, and 420 µg/L in channel catfish (58). The 48-hour LC_{50} in *Daphnia magna*, a small freshwater crustacean, is 280 µg/L (58,115). The bioconcentration factor for this compound in whole fish is 5100, indicating a moderate potential to acumulate in aquatic organisms (115).

Effects on other organisms (non-target species)

Pendimethalin is nontoxic to bees (1).

Environmental fate

Breakdown in soil and groundwater

Pendimethalin is moderately persistent, with a field half-life of approximately 40 days (11,58). It does not undergo rapid microbial degradation except under anaerobic conditions (58). Slight losses of pendimethalin can result from photodecomposition and volatilization.

Pendimethalin is strongly adsorbed by most soils (11,58). Increasing soil organic matter and clay is associated with increased soil binding capacity. It is practically insoluble in water, and thus will not leach appreciably in most soils, and should present a minimal risk of groundwater contamination (11,58).

Breakdown in water

Pendimethalin is stable to hydrolysis, but may be degraded by sunlight in aquatic systems (58). Pendimethalin may also be removed from the water column by binding to suspended sediment and organic matter (11,58). It is rapidly degraded under anaerobic conditions once precipitated to sediment (58).

Breakdown in vegetation

Pendimethalin is absorbed by plant roots and shoots, and inhibits cell division and cell elongation (58). Once absorbed into plant tissues, translocation is limited and pendimethalin breaks down via oxidation (58). Pendimethalin is not absorbed by the leaves of grasses, and only very small amounts are taken up by plants from the soil. Residues on crops at harvest are usually below detectable levels (0.05 ppm) (58).

Physical properties

Pendimethalin is an orange-yellow crystalline solid with a faint nutty or fruit-like odor (1).

Chemical name: N-(1-ethylpropyl)-2,6-dinitro-3,4-xylidine (1)
CAS #: 40487-42-1
Molecular weight: 281.31 (1)
Water solubility: 0.3 mg/L @ 20°C (1)
Solubility in other solvents: v.s. in most organic solvents such as acetone and xylene; s. in corn oil, isopropanol, heptane, benzene, toluene, chloroform, and dichloromethane; s.s. in petroleum ether and petrol (1)
Melting point: 54–58°C (1)
Vapor pressure: 4 mPa @ 25°C (1)
Partition coefficient (octanol/water): 152,000 (1)
Adsorption coefficient: 5000 (11)

Exposure guidelines

ADI: Not available
HA: Not available
RfD: 0.04 mg/kg/day (13)
PEL: Not available

Basic manufacturer

American Cyanamid
One Cyanamid Plaza
Wayne, NJ 07470-8426
Telephone: 201-831-2000
Emergency: 201-835-3100

10.3.18 Picloram

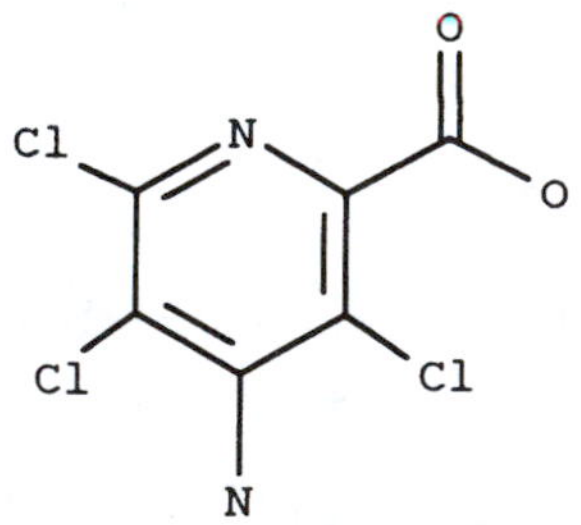

Figure 10.29 Picloram.

Trade or other names

Commercial products containing the compound include Access, Grazon, Pathway, and Tordon. It may be used in formulations with other herbicides such as bromoxynil, diuron, 2,4-D, MCPA, triclorpyr, and atrazine. It is also compatible with fertilizers.

Regulatory status

Picloram is a slightly toxic compound in EPA toxicity class III. Products containing it must bear the Signal Word CAUTION on the label. All products except for Tordon RTU and Pathway are Restricted Use Pesticides (RUPs). RUPs may be purchased and used only by certified applicators.

Introduction

Picloram, in the pyridine family of compounds, is a systemic herbicide used for control of woody plants and a wide range of broad-leaved weeds. Most grasses are resistant to picloram and thus it is used in range management programs. Picloram is formulated either as an acid (technical product), a potassium or triisopropanolamine salt, or an isooctyl ester, and is available as either soluble concentrate, pellet or granular formulations. *Information presented here pertains to the technical acid form unless otherwise indicated.*

Toxicological effects

Acute toxicity

Picloram is slightly to practically nontoxic via ingestion, with reported oral LD_{50} values of greater than 5000 to 8200 mg/kg in rats, 2000 to 4000 mg/kg in mice, and approximately 2000 mg/kg in rabbits (1,58). The reported dermal LD_{50} in rabbits is greater than 4000 mg/kg, a level that produced no mortality or toxic signs (6,58). This indicates slight toxicity via the dermal route as well. Technical picloram is reported to cause no skin and moderate eye irritation in the rabbit, and no skin sensitization in the guinea pig (1,58). Some formulations have caused mild or slight skin irritation and sensitization in test animals (58).

The technical grade is moderately toxic by inhalation, with a reported 4-hour inhalation LC_{50} of greater than 0.35 mg/L (3). Formulated products may show a lesser toxicity via this route (58).

There is no documented history of human intoxication by picloram, so symptoms of acute exposure are difficult to characterize.

Chronic toxicity

Male mice receiving picloram at dietary doses of 1000 to 2000 mg/kg/day over 32 days showed no clinical signs of toxicity nor changes in blood chemistry, but females did show decreased body weight and increased liver weights (6,8). Liver effects were also seen in rats at very high doses of 3000 mg/kg/day over an exposure period of 90 days, and above 225 mg/kg/day for 90 days (58). Dogs, sheep, and beef cattle fed low levels of picloram for a month experienced no toxic effects. The ester and triisopropanolamine salt showed low toxicity in animal tests (58). Picloram may show additive effects if mixed with other herbicides such as 2,4-D (118).

Reproductive effects

In multigenerational studies, pregnant rats exposed during critical periods of gestation to doses of about 180 mg/kg/day picloram showed no changes in fertility

(58). The fertility of pregnant mice fed 15 mg/kg/day for 4 days before and 14 days after mating was not adversely affected (8). Other studies showed no effects on fertility or fecundity in rats at doses as high as 1000 mg/kg/day (58). Picloram does not appear to cause reproductive toxicity.

Teratogenic effects

No teratogenic effects were seen in the offspring of pregnant rats exposed during gestation to 400 mg/kg/day of the acid or potassium salt, or to 1000 mg/kg/day of the ester or other salt (58). At 2000 mg/kg/day, maternal toxicity was noted but did not induce malformation in the pups (8). It appears that picloram is not teratogenic.

Mutagenic effects

One test has shown that picloram is mutagenic (to the bacterium *Saccharomyces cerevisiae*), and another test has shown that it is not mutagenic (Ames test) (118). In tests for unscheduled DNA synthesis and structural chromosome aberrations, the results were also negative (58). These data suggest that picloram is either nonmutagenic or weakly mutagenic.

Carcinogenic effects

Mice fed average doses of 18 or 30 mg/kg/day for 80 weeks and observed for another 10 weeks did not display any carcinogenic effects (8,118). Male rats fed 17.5 or about 40 mg/kg/day for 80 weeks and observed for 33 weeks showed no carcinogenicity, but females developed benign liver tumor nodules (58). Other tests have indicated an increased incidence of cancer among animals treated with picloram, but these data are difficult to interpret due to possible interference of hexachlorobenzene contaminants (8,118). These data suggest that picloram is noncarcinogenic or weakly carcinogenic.

Organ toxicity

Animal studies show the target organs for picloram to be the liver and kidneys.

Fate in humans and animals

Picloram was rapidly absorbed through the gastrointestinal tract in studies using human volunteers, and was excreted unchanged in the urine (119). Half of the product was excreted within a day or so. Skin absorption is minimal (119). Rats showed similar results, with administered doses excreted virtually unchanged in urine and feces within 48 hours (119).

Picloram does not accumulate in fat (119). No measurable residues were found in milk from cows fed small amounts of the herbicide in their diets (8). At higher levels of exposure, milk levels of picloram were low (0.05 to 0.29 ppm) and declined rapidly upon withdrawal of picloram from the diet (8).

Ecological effects

Effects on Birds

Picloram is slightly to practically nontoxic to birds; the acute oral LD_{50} is greater than 2000 to 5000 mg/kg in ducks, pheasants, and quail, with no mortality seen at even the highest levels (6).

Effects on aquatic organisms

Picloram is slightly to moderately toxic to fish and aquatic invertebrates. The reported 96-hour LC_{50} values for picloram are 19.3 mg/L in rainbow trout, 14.5 mg/L in bluegill sunfish, and 55 mg/L in fathead minnow (58). The 48-hour LC_{50} in *Daphnia* is 50 mg/L, indicating moderate toxicity (58).

Most salts are of similar or lesser toxicity, but the isooctyl ester may be highly toxic. The reported 96-hour LC_{50} for the isooctyl ester in rainbow trout is 4 mg/L, and in channel catfish is 1.4 mg/L (118). Other LC_{50} values in aquatic invertebrates ranged from 10 to 68 mg/L (8).

Picloram is not expected to accumulate appreciably in aquatic organisms; the measured bioconcentration factor in bluegill sunfish was less than 0.54 (9).

Effects on other organisms (non-target species)

The compound is nontoxic to bees (1).

Environmental fate

Breakdown in soil and groundwater

Picloram is moderately to highly persistent in the soil environment, with reported field half-lives from 20 to 300 days and an estimated average of 90 days (11). Photodegradation is significant only on the soil surface and volatilization is practically nil (58). Degradation by microorganism is mainly aerobic and dependent upon application rates (58). Increasing soil organic matter increases the sorption of picloram and the soil residence time (58).

Picloram is poorly bound to soils, although it is bound better by soils with higher proportions of soil organic matter (11). It is soluble in water and therefore may be mobile (1). These properties, combined with its persistence, mean it may pose a risk of groundwater contamination. Picloram has been detected in the groundwater of eleven states at concentrations ranging from 0.01 to 49 µg/L (9).

Breakdown in water

In laboratory studies, sunlight readily broke down picloram in water, with a half-life of 2.6 days (9,58). Herbicide levels in farm ponds were 1 mg/L directly following spraying and decreased to 0.01 mg/L within 100 days, primarily due to dilution and the action of sunlight (6).

Breakdown in vegetation

Picloram is readily absorbed by plant roots, less so by the foliage, and is readily translocated throughout plants. It remains stable and intact in plants (58).

Physical properties

Picloram is a colorless crystal (1).

Chemical name: 4-amino-3,5,6-trichloropyridine-2-carboxylic acid (1)
CAS #: 1918-02-1
Molecular weight: 241.48 (1)

Solubility in water: 430 mg/L @ 25°C (1)

Solubility in other solvents: v.s. in acetone, ethanol, benzene and dichloromethane (1)

Melting point: Decomposes @ 215°C (1)

Vapor pressure: 0.082 mPa @ 35°C (1)

Partition coefficient (octanol/water): 1.4 (58)

Adsorption coefficient: 16 (11)

Exposure guidelines

ADI: Not available

HA: 0.5 mg/L (lifetime) (120)

RfD: 0.07 mg/kg/day (13)

PEL: 5 mg/m^3 (8-hour) (respirable fraction) (14)

Basic manufacturer

DowElanco

9330 Zionsville Road

Indianapolis, IN 46268-1054

Telephone: 317-337-7344

Emergency: 800-258-3033

10.3.19 Pronamide

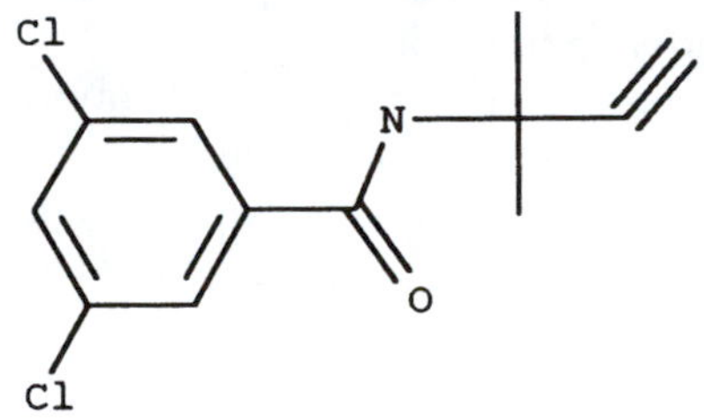

Figure 10.30 Pronamide.

Trade or other names

Pronamide is also known as propyzamide. Trade names include Benzamide, Clanex, Kerb, Propyzamide, RH-315 Rapier, and Ronamid.

Regulatory status

Pronamide is a Restricted Use Pesticide (RUP). It is EPA toxicity class III — slightly toxic. Products containing pronamid bear the Signal Word CAUTION. It may be purchased and used only by certified applicators. The U.S. Environmental Protection Agency (EPA) restricts the use of all pronamide formulations, except those in water-soluble packets, due to their potential to cause tumor growth.

Introduction

Pronamide is a selective herbicide used either before weeds emerge (pre-emergence), and/or after weeds come up (post-emergence). It controls a wide range of annual and perennial grasses, as well as certain annual broadleaf weeds. It is used primarily on lettuce and alfalfa crops, as well as on blueberries, ornamentals, fruit trees, forage legumes, and fallow lands. Pronamide is usually incorporated into the soil by cultivation, irrigation, or rain immediately following application. It is available in wettable powder and granular formulations. *Information presented here pertains to the technical product unless otherwise noted.*

Toxicological effects

Acute toxicity

Pronamide is practically nontoxic via ingestion. The reported oral LD_{50} values for pronamide range from 5620 mg/kg in female rats to 8350 mg/kg in male rats, and 10,000 mg/kg in dogs (8). Pronamide is slightly toxic by skin exposure, with a dermal LD_{50} of greater than 3160 mg/kg (1). When applied to the skin of rabbits, it produced slight local irritation, but no systemic intoxication. The 4-hour inhalation LC_{50} for pronamide is greater than 5.0 mg/L, indicating slight toxicity by this route (58).

Chronic toxicity

When dogs were fed a diet containing pronamide for 3 months, decreases in weight gain and food consumption, changes in blood chemistry, and increased liver weights were observed at doses of 15 mg/kg/day (58). In a study in rats over 3 months, similar effects were seen at doses over 10 mg/kg/day (58), and changes in thyroid, adrenal, and pituitary function were observed at 50 mg/kg/day (58). In a 2-year feeding study in dogs, the addition of pronamide to the diet at doses of 0.75, 2.5, or 7.5 mg/kg/day caused no adverse health effects at any of the doses tested (5).

Reproductive effects

When pregnant rabbits were given doses of 5, 20, or 80 mg/kg/day during days 7 to 19 of gestation (18 rabbits per dose), no effects on development or reproduction were observed at or below the 20-mg/kg dose. At 80 mg/kg, there was an increased incidence of liver lesions, one maternal death, five abortions, and a decrease in maternal and offspring weight gain (58). In a three-generation rat reproduction study, no effects on reproduction were observed at 300 ppm (15 mg/kg/day), the highest dose tested (121).

It is unlikely that pronamide will have reproductive effects except at doses high enough to cause maternal toxicity.

Teratogenic effects

No teratogenic effects were found when doses as high as 15 mg/kg/day were administered to pregnant rabbits (58,121). This evidence suggests pronamide is not teratogenic.

Mutagenic effects

Mutagenicity tests on bacteria, mammalian cell cultures, and live animals have been negative (121). It appears pronamide is not mutagenic.

Carcinogenic effects

Pronamide caused liver tumors in mice after 2 years at doses of 10 mg/kg/day and above (58). In rats, doses of 50 mg/kg/day and above produced changes in ovary and liver structure and function, as well as thyroid and testicular effects (58). These data suggest that pronamide may have carcinogenic activity at sufficient doses.

Organ toxicity

Target organs identified in animal studies include the liver, thyroid, and adrenal and pituitary glands.

Fate in humans and animals

Pronamide is not readily absorbed into the bloodstream from the gastrointestinal tracts of rats and cows. After oral doses of Kerb to rats, 54 and 0.6% of the unmetabolized Kerb was recovered in feces and urine, respectively. Unmetabolized Kerb did not appear in the urine of a cow treated orally with Kerb (121). Traces of pronamide were found in the milk of cows given feed that contained 5-ppm doses of a pronamide formulation (8). Pronamide has a low potential for bioaccumulation in animal tissues.

Ecological effects

Effects on birds

Pronamide is practically nontoxic to birds. The oral LD_{50} for pronamide in Japanese quail is 8700 mg/kg, and greater than 14,000 mg/kg in mallard ducks (58,122). The 8-day dietary LC_{50} for Kerb Technical Herbicide in bobwhite quail and mallard ducks is greater than 10,000 ppm (123).

Effects on aquatic organisms

Pronamide is practically nontoxic to warmwater fish and slightly toxic to cold-water fish. The 96-hour LC_{50} for pronamide is 100 mg/L in bluegill sunfish, 72 mg/L in rainbow trout, 350 mg/L in goldfish, 204 mg/L in harlequin fish, and 150 mg/L in guppies (8,58). The 48-hour LC_{50} for *Daphnia magna*, a small freshwater crustacean, is greater than 5.6 mg/L (58). Pronamide may be moderately toxic to aquatic invertebrates (122).

Effects on other organisms (non-target species)

Pronamide is nontoxic to honeybees (8).

Environmental fate

Breakdown in soil and groundwater

Pronamide is moderately persistent in most soils, with a reported average field half-life of 60 days (11). It is readily bound, or adsorbed, to most soils (11). Increasing soil temperature and, to a lesser extent, soil moisture and pH increases the rate of pronamide degradation in soil (121). In most soil types, there is very little movement, or leaching, of pronamide into groundwater as it is nearly insoluble in water (58). Leaching of pronamide residues in soil is most likely in soils with low organic matter content, such as loamy sands or silt loams (8).

Pronamide is inactivated by soil organic matter and will not be effective on muck, peat, or other very high organic content soils (124). Depending upon soil type and climatic conditions, persistence of pronamide may be higher. Accumulation of the herbicide from repeated annual applications to the same soil does not appear problematic.

Chemical degradation may be the main route of disappearance from the soil. Photodecomposition at the soil surface can also occur (120). A moderate amount of pronamide breakdown is carried out by soil microorganisms. The herbicide is not active against common soil microorganisms. Volatilization loss may be high under hot, dry conditions (6).

Breakdown in water

In water bodies, pronamide is stable at neutral pH. It is slowly degraded chemically, by light, and by aquatic microorganisms. Loss from volatilization is not significant (8,58). Pronamide is thought to be stable because less than 10% was hydrolyzed, or broken down in water, over a 4-week period (8). It is stable to hydrolysis between pH 4.7 and 8.8 (122).

Breakdown in vegetation

Pronamide is readily translocated from the roots to other plant parts. Absorption of pronamide through plant leaves is minimal. Pronamide is metabolized slowly by both tolerant and sensitive plants (58).

Physical properties

Pronamide is a white or off-white crystalline solid with no odor (1).

Chemical name: 3,5-dichloro-N-(1,1-dimethylpropynyl) benzamide (1)
CAS#: 23950-58-5
Molecular weight: 256.13 (1)
Water solubility: 15 mg/L @ 25°C (1)
Solubility in other solvents (@ 25°C): v.s. in dimethyl sulfoxide and methanol; s. in benzene, xylene, and carbon tetrachloride (1)
Melting point: 155–156°C (1)
Vapor pressure: 11.3 mPa @ 25°C (1)
Partition coefficient (octanol/water): 1570 (58)
Adsorption coefficient: 800 (11)

Exposure guidelines

ADI: Not available
HA: 0.05 mg/L (lifetime) (121)
RfD: 0.075 mg/kg/day (13)
PEL: Not available

Basic manufacturer

Rohm and Haas Co.
Agricultural Chemicals
100 Independence Mall West
Philadelphia, PA 19106
Telephone: 215-592-3000

10.3.20 Propanil

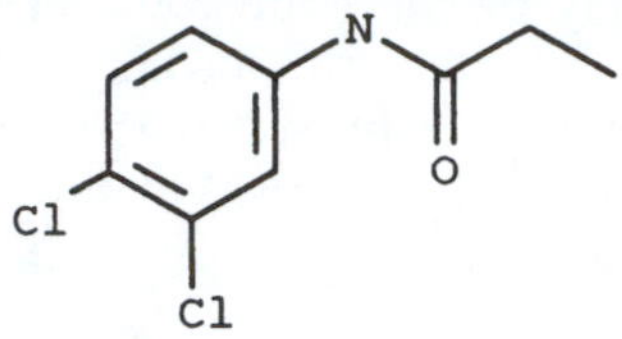

Figure 10.31 Propanil.

Trade or other names

Trade names for propanil include Arrosol, Bay 30130, Cekupropanil, Chem-Rice, DPA, DCPA, Dropaven, Erban, FW-734, Herbax, Prop-Job, Propanex, Propanilo, Riselect, S 10145, Stam, Stam 80 EDF, Stam M-4, Stampede, Strel, Supernox, Surcopur, Surpur, Vertac, Wham DF, and Wham EZ.

Regulatory status

Propanil is a General Use Pesticide (GUP). Propanil is in toxicity class II — moderately toxic, due to its potential to irritate eyes and skin. Products containing propanil bear the Signal Word WARNING.

Introduction

Propanil is an acetanilide post-emergence herbicide with no residual effect. It is used against numerous grasses and broad-leaved weeds in rice, potatoes, and wheat. It is available as emulsifiable concentrates, liquid and dry flowable, low volume, and ultra-low volume (ULV) formulations. Mixing with carbamates or organophosphorans compounds is not recommended. *The information presented here, unless otherwise noted, pertains to technical propanil.*

Toxicological effects

Acute toxicity

Propanil is moderately toxic via ingestion. The reported rat oral LD_{50} values for technical propanil range from 1080 to greater than 2500 mg/kg (8,58); and for formulations range from 500 to greater than 5000 mg/kg (58). The oral LD_{50} is 1217 mg/kg for technical propanil in dogs (87).

Technical propanil is practically nontoxic via the dermal route, with a reported LD_{50} of greater than 5000 mg/kg (58) in rabbits. The LD_{50} in rats suggests slight dermal toxicity (less than 2000 mg/kg) for one formulated product, Propanil 4 (58). However, the reported LD_{50} values in rats and rabbits indicate moderate to practically no dermal toxicity (greater than 2000 mg/kg) for four other formulated products: Wham DF, Wham EZ, Stam 80 EDF, and Stam M-4 (58).

Technical propanil is moderately toxic via the inhalation route, with a reported 4-hour LC_{50} of 1.12 mg/L (8,58). The formulated products may be less toxic, with inhalation 4-hour LC_{50} values of 2.8 mg/L for Stam M-4 and 6.1 mg/L for Stam 80 EDF (58).

Chronic toxicity

In a 2-year study, a dietary level of about 80 mg/kg/day caused a decrease in overall growth and a relative increase in the weight of the spleen and liver in female rats and of the testes in males (87). Feed consumption, growth, and hemoglobin levels in rats were reduced at daily doses of about 180 mg/kg/day over 3 weeks (87). In a 2-year study in dogs, a dietary level of about 85 mg/kg/day depressed growth in spite of increased food intake and increased relative liver weight (87). The only other change detected was a slight increase in the relative weight of the heart (87). Liver and blood changes and cyanosis were seen at 25 mg/kg/day in mice over 90 days (87).

Reproductive effects

In a three-generation study, male and female rats fed doses as high as 50 mg/kg/day propanil for 11 weeks before mating showed no effect on fertility, gestation, viability, or lactation (8). It appears that propanil does not cause reproductive effects.

Teratogenic effects

No evidence of teratogenic effects was observed in studies with rats and rabbits (58,125).

Mutagenic effects

In vitro tests of propanil, including the Ames test (with and without metabolic activation), tests on mammalian cell cultures, and cytogenetic assays on mice, failed to show mutagenic or genotoxic effects (58,125).

Carcinogenic effects

No evidence of carcinogenicity was observed in long-term studies of mice and rats (58).

Organ toxicity

Data from animal studies indicate that the most likely target organs are the liver, kidney, spleen, and possibly the testes (87). Dermatitis (rashes) and sensitization (allergies) are possible. High doses may produce anemia due to the formation of methemoglobin (87).

Fate in humans and animals

When propanil was fed to a cow for 4 days, 1.4% of the total dose was recovered in the feces, but none was detected in the urine or milk (87). This suggests that propanil is absorbed into the bloodstream through the gastrointestinal tract and that, once in the bloodstream, propanil is metabolized by the body. Propanil is lipid (fat) soluble. The liver breaks down propanil to aniline derivatives. These metabolites are responsible for the methemoglobin formation (70). Excretion is through the urine.

Ecological effects

Effects on birds

Propanil is moderately toxic to birds. The oral LD_{50} for propanil in bobwhite quail is 196 mg/kg, and in mallard ducks is 275 mg/kg (58). The 8-day dietary

LC_{50} for propanil in bobwhite quail is 2861 ppm, and in mallard ducks is 5627 ppm (58).

Effects on aquatic organisms

Propanil may be moderately to highly toxic to a wide range of aquatic species (58). The 96-hour LC_{50} for propanil is 5.4 mg/L in bluegill sunfish, 2.3 mg/L in rainbow trout, and 4.6 mg/L in sheepshead minnows (58). The 48-hour LC_{50} for propanil in *Daphnia magna*, a small freshwater crustacean, is 0.14 mg/L, and for mysid shrimp is 0.4 mg/L (58). The 96-hour LC_{50} of propanil is 5.8 mg/L in the Eastern oyster (58). The compound concentrated to levels up to 111 times the background water concentration in fathead minnows. The level of pesticide in the minnows returned to normal within 10 days after the fish were placed in a propanil-free enviroment. This indicates that the compound is not likely to concentrate appreciably in aquatic organisms.

Effects on other organisms (non-target species)

Propanil is nontoxic to honeybees, with a reported contact LC_{50} of 240 µg per bee (1).

Environmental fate

Breakdown in soil and groundwater

Propanil is of low soil persistence (11,58). The field half-life is 1 to 3 days (11,58). Propanil is soluble in water and it adsorbs only weakly to soil particles. Propanil is rapidly broken down in the soil by microorganisms, which have highest activity under warm, moist conditions. Bacteria produce by-products such as tetrachloroazobenzene and dichloroaniline (58). Its rapid breakdown in soil practically eliminates the potential for groundwater contamination (11).

Breakdown in water

Propanil will rapidly break down in water due to microbial activity, the major breakdown pathway. Reported half-lives are 2 days under aerobic conditions and 2 to 3 days under anaerobic conditions (58).

Breakdown in vegetation

Within a plant, propanil is moved from the leaves to the growing shoots, then back to other leaves. It is a contact herbicide. Resistant crop plants such as rice completely metabolize propanil (58). Carry-over of herbicidal activity to subsequent crops is not likely. Propanil may be highly phytotoxic to non-target plants if mixed with carbamates or organophosphates (1).

Physical properties

The technical product is a brownish crystalline solid with an organic acid odor (58). It is stable in emulsion concentrates, but is hydrolyzed in extremely acidic or basic conditions (58).

Chemical name: 3',4'-dichloropropionanilide (1)
CAS #: 709-98-8
Molecular weight: 218.08 (1)

Solubility in water: 225 mg/L @ 25°C (1)
Solubility in other solvents: v.s. in benzene, ethanol, acetone, and cyclohexanone;
    s. in toluene and xylene (1)
Melting point: 81-91°C (technical) (1)
Vapor pressure: 12 mPa @ 60°C (1)
Partition coefficient (octanol/water): 193 (1)
Adsorption coefficient: 149 (11)

Exposure guidelines

ADI: Not available
HA: Not available
RfD: 0.005 mg/kg/day (13)
PEL: Not available

Basic manufacturer

Rohm and Haas Co.
Agricultural Chemicals
100 Independence Mall West
Philadelphia, PA 19106
Telephone: 215-592-3000

10.3.21 Sethoxydim

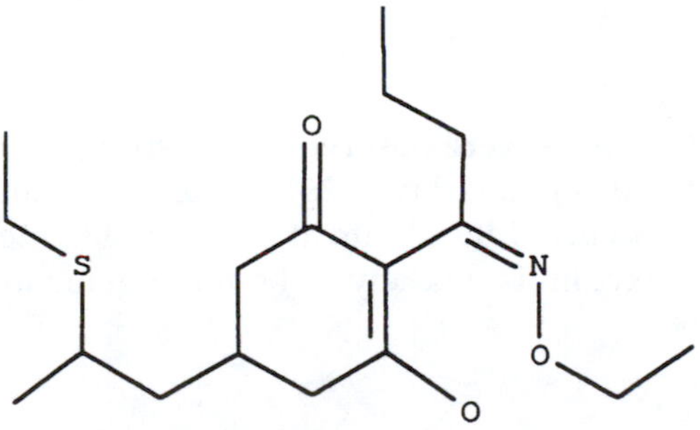

Figure 10.32 Sethoxydim.

Trade or other names

Trade names for sethoxydim include Aljaden, Alloxol S, BAS 9052H, Checkmate, Expand, Fervinal, Grasidim, Nabu, NP-55, Poast, Tritex-Extra, and Vantage.

Regulatory status

Sethoxydim is a General Use Pesticide (GUP). It is in EPA toxicity class III — slightly toxic. Products containing sethoxydim bear the Signal Word CAUTION on the label.

Introduction

Sethoxydim is a selective post-emergence herbicide used to control annual and perennial grass weeds in broad-leaved vegetable, fruit, field, and ornamental crops. It also has indoor uses. It is available in emulsifiable concentrate formulations.

Toxicological effects

Acute toxicity

Sethoxydim is slightly toxic by ingestion, and practically nontoxic by dermal absorption (58). It causes skin and eye irritation. Inhalation of dusts or vapors can cause irritation of the throat and nose (58). Other symptoms of poisoning include incoordination, sedation, tears, salivation, tremors, blood in the urine, and diarrhea (58). Sethoxydim does not cause allergic skin reactions (3).

The oral LD_{50} for sethoxydim in rats is 2600 to 3100 mg/kg (1,58). The dermal LD_{50} in rats is greater than 5000 mg/kg (1,58), and the 4-hour inhalation LC_{50} for sethoxydim in rats is greater than 6.3 mg/L (1).

Chronic toxicity

Long-term contact with sethoxydim can cause redness and swelling of the eyes or skin (8). No adverse effects were observed in mice given 2, 6, or 18 mg/kg/day for 2 years (58,126). In a 1-year dog feeding study, doses above 8.86 mg/kg/day in males and 9.41 mg/kg/day in females produced anemia (13,58).

Reproductive effects

When pregnant rabbits were fed 40, 160, or 480 mg/kg/day, decreased litter size, low fetal weights, severe maternal weight loss, increased fetal resorptions, spontaneous abortions, and maternal deaths occurred at the 480-mg/kg level (126).

Based on this study, reproductive effects are unlikely in humans at expected exposure levels.

Teratogenic effects

No developmental effects were observed in offspring of rats at maternal dose levels of 40, 100, or 250 mg/kg/day (126). Increased numbers of skeletal and visceral abnormalities occurred in rabbits at doses of 480 mg/kg/day (13). These data suggest that sethoxydim is unlikely to be teratogenic in humans at expected exposure levels.

Mutagenic effects

Several tests of the mutagenicity of sethoxydim indicate that it is not mutagenic (126).

Carcinogenic effects

No carcinogenic effects were observed at any dose level when mice were fed 6, 18, 54, or 162 mg/kg/day for 2 years (13,126). This suggests that sethoxydim is not carcinogenic.

Organ toxicity

Liver and bone marrow effects and increased thyroid weight have been reported in dogs (13).

Fate in humans and animals

Single doses of the compound fed to rats were nearly completely eliminated (98.6%) through urine and feces within 48 hours (1).

Ecological effects

Effects on birds

Sethoxydim is practically nontoxic to birds. The acute oral LD_{50} for sethoxydim in mallard ducks is greater than 2510 mg/kg (58), and in Japanese quail is greater than 5000 mg/kg (1). Its dietary LC_{50} in mallards and bobwhite quail is greater than 5620 ppm (58,127).

Effects on aquatic organisms

Sethoxydim is moderately to slightly toxic to aquatic species. A 3-hour LC_{50} of 1.5 mg/L is reported in *Daphnia* (1). In fish, 96-hour LC_{50} values range from 1.6 mg/L in carp (58), to 32 mg/L in rainbow trout, and 100 mg/L in bluegill sunfish (58).

Effects on other organisms (non-target species)

Sethoxydim has low toxicity to wildlife (8). It is nontoxic to bees (1).

Environmental fate

Breakdown in soil and groundwater

Sethoxydim is of low soil persistence. Reported field half-lives are 5 to 25 days (11,58). It has a weak tendency to adsorb to soil particles (58). Laboratory leaching tests have suggested that sethoxydim could leach in soil. However, in field tests, sethoxydim did not leach below the top 4 inches of soil, and it did not persist (127).

On soil, photodegradation of sethoxydim takes less than 4 hours (58). The product Poast photodegrades on soil surfaces with a half-life of approximately 3.7 hours (127). Disappearance of sethoxydim is primarily due to action by soil microbes.

Breakdown in water

In water, photodegradation of sethoxydim takes less than 1 hour (8). The product Poast is fairly stable to the chemical action of water (hydrolysis), with a half-life of about 40 days in a neutral solution at 25°C (127).

Breakdown in vegetation

Sethoxydim is absorbed rapidly by roots and foliage, and moves both upward and downward in plants from the point of absorption (58). Sethoxydim is rapidly detoxified in most tolerant plants (58). The product Poast accumulates in the tissues of crops planted in fields after harvest of treated crops. Measured residues were all below 0.066 ppm (11).

Physical properties

Sethoxydim is an amber-colored, oily, odorless liquid (1).

Chemical name: ($\pm$)2[1-(ethoxyimino)butyl]-5-[2-(ethylthio) propyl]-3-hydroxy-2-cyclohexen-1-one (1)
CAS #: 74051-80-2

Molecular weight: 327.5 (1)
Solubility in water: 4700 mg/L @ pH 7 and 20°C (1)
Solubility in other solvents: v.s. in methanol, hexane, and acetone (1)
Melting point: Not available
Vapor pressure: <0.1 mPa @ 20°C (1)
Partition coefficient (octanol/water): 45.1 @ pH 7 (58)
Adsorption coefficient: 100 (estimated) (11)

Exposure guidelines

ADI: Not available
HA: Not available
RfD: 0.09 mg/kg/day (13)
PEL: Not available

Basic manufacturer

BASF Corporation
Agricultural Products Group
P.O. Box 13528
Research Triangle Park, NC 27709-3528
Telephone: 800-669-2273
Emergency: 800-832-4357

10.3.22 Terbacil

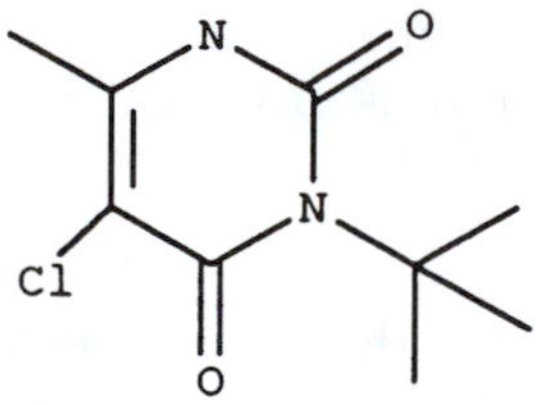

Figure 10.33 Terbacil.

Trade or other names

Trade names for terbacil include Compound 732, DuPont Herbicide 732, Geonter, and Sinbar. This compound may also be found in mixed formulations with other herbicides.

Regulatory status

Terbacil is registered by the U.S. Environmental Protection Agency (EPA) as a General Use Pesticide (GUP). The Signal Word CAUTION is required on containers of formulated terbacil. Terbacil is in EPA class IV — practically nontoxic.

Introduction

Terbacil is a selective herbicide used for control of both annual grasses, broad-leaved weeds, and some perennial weeds in sugarcane, apples, alfalfa, peaches, pecans, and mints. It is sprayed on soil surfaces preferably just before, or during, the period of active weed growth. Terbacil works in plants by inhibiting photosynthesis. It is a member of the substituted uracil chemical family. Terbacil is available in wettable powder formulations. Technical terbacil is 95% pure active material.

Toxicological effects

Acute toxicity

Terbacil has low acute toxicity (128). Clinical signs of poisoning in rats include weight loss, pallor, prostration, and rapid breathing. In dogs, a single dose of 5 mg/kg caused repeated vomiting (128). Terbacil may irritate the skin, eyes, and mucous membranes of the nose and throat. It is not a skin sensitizer (129). The oral LD_{50} of terbacil is 5000 to 7500 mg/kg in rats (1,22). The dermal LD_{50} is greater than 5000 mg/kg (the maximum feasible dose) in rabbits (1). These rabbits did not show clinical signs of toxicity, nor any obvious gross changes caused by disease. No skin irritation and only mild eye irritation were seen in rabbits at this dose (23). Similarly, there was no skin irritation or sensitization in terbacil-treated guinea pigs (23). Dogs given 5000 mg/kg terbacil exhibit vomiting and a lack of eye pupil responsiveness (128).

Chronic toxicity

No evidence of toxicity was seen in 2-year studies of rats fed doses as high as 12.5 mg/kg/day, or in dogs fed doses as high as 6.25 mg/kg/day terbacil. At 125 to 500 mg/kg/day, there was a lower rate of weight gain, liver enlargement, and other liver changes in rats. The highest dose produced a slight increase in liver weight in dogs (23).

Reproductive effects

There were no adverse effects on lactation, fertility, birth rate, pup survival, or any other aspect of reproduction in rats fed 2.5 and 12.5 mg/kg/day terbacil for three generations (23).

However, the average number of live rat fetuses per litter and the average final maternal body weight were significantly lowered in another study in the 103- and 391-mg/kg/day dosage groups (130).

Based on these data, adverse effects on reproduction in humans are not likely at expected exposure levels.

Teratogenic effects

When doses of 30, 200, or 600 mg/kg/day were administered by gavage to pregnant rabbits on days 7 to 19 of gestation, adverse effects on the fetuses appeared only at the highest dose tested. This dose also produced maternal toxicity and increased maternal mortality. No adverse effects on the mothers or the pups were observed at lower doses. In another study, pregnant rats were fed doses of 23, 103, or 391 mg/kg/day on days 6 to 15 of gestation. Abnormalities occurred in the renal pelvis, and ureter dilation was found in pups from all the treatment groups (129).

Evidence of teratogenicity is inconclusive.

Mutagenic effects

Terbacil was not mutagenic in several screening tests (129).

Carcinogenic effects

No evidence of carcinogenicity was found in rats fed 2.5, 12.5, 125, or 500 mg/kg/day terbacil for 2 years, nor in dogs fed as much as 12.5 mg/kg/day for 2 years (23). When mice were fed dietary doses of 2.5, 62.5, or 250 mg/kg for 2 years, no increased incidence of cancer was found (130). Terbacil does not appear to be carcinogenic.

Organ toxicity

Liver changes have been seen in laboratory rats exposed to high doses of terbacil (23).

Fate in humans and animals

In general, the uracil herbicides, the chemical class in which terbacil is included, are rapidly excreted in urine by mammals. This may account for their reported low toxicity (131). When given in the feed of lactating cows at 5 and 30 ppm, terbacil was excreted in the milk at levels up to 0.03 and 0.08 ppm, respectively. No herbicide was detected in the cows' urine and feces (131).

Ecological effects

Effects on birds

Terbacil is slightly toxic to birds (131). The 8-day dietary LC_{50} for terbacil is more than 56,000 ppm in Peking ducklings, and greater than 31,450 ppm in pheasant chicks (58). The LD_{50} for terbacil in quail is greater than 2250 mg/kg (129).

Effects on aquatic organisms

Terbacil is slightly to practically nontoxic to aquatic organisms. The 48-hour LC_{50} of terbacil is 86 mg/L in sunfish (1). The LC_{50} for terbacil is 102.9 mg/L in bluegill sunfish and 46.2 mg/L in rainbow trout. Terbacil is slightly toxic to freshwater invertebrates, with an LC_{50} of 65 mg/L in *Daphnia*, a small freshwater crustacean. The LC_{50} for terbacil is greater than 4.9 mg/L in marine oysters and 49 mg/L in shrimp (129). Terbacil does not bioaccumulate in bluegill sunfish (130).

Estuarine and marine organisms may be exposed to terbacil due to its use as a sugarcane herbicide (128). A study on grass shrimp with an 84.7% formulated terbacil product was sufficient to characterize the herbicide as slightly toxic to marine invertebrates. The 48-hour LC_{50} of terbacil is 1000 mg/L in fiddler crabs (1).

Effects on other organisms (non-target species)

Terbacil is nontoxic to bees (1).

Environmental fate

Breakdown in soil and groundwater

Terbacil is highly persistent in soils (11). Soil half-lives of 50 to 180 days have been reported (1,11). Data from field dissipation studies showed that terbacil persistence in soil varied with application rate, rainfall, soil type, and mobility, as well as available oxygen (129).

In most soil types, terbacil has a relatively low tendency to be adsorbed. It also is highly soluble in water. Thus, terbacil is likely to be highly mobile in soil and potentially pollute groundwater (11,131). Because of this, it should not be used on sandy or gravelly soils that have less than 1% organic matter, particularly if the water table is near the soil surface (128) and when deep tillage is used (132). Leaching may be slower in soils that are finer textured and/or have higher organic matter content (128). Terbacil was not detected in a national groundwater survey conducted by the EPA (8).

In moist soils, terbacil is subject to microbial degradation. However, data suggest that recommended rates of terbacil use may result in its persistence for more than one growing season (8).

Breakdown in water

Contamination of surface waters near terbacil-treated areas, and subsequent exposure of humans and non-target organisms, is possible due to terbacil's mobility in soil and its high water solubility (132). Terbacil is stable in water and does not readily undergo hydrolysis or photodegradation (129).

Breakdown in vegetation

At normal application rates, terbacil has residual phytotoxicity to affected species in treated soils for 1 to 2 years (128). Terbacil residues were phytotoxic to oats planted 3 years after a previous application of the herbicide. For example, in alfalfa, 12% of terbacil plus its metabolites were still found 6 to 8 months after application (8).

Terbacil is most readily absorbed through the root system of plants to which it is applied. Less is absorbed through the leaves and stems of plants. Studies of sugarcane plants indicate that terbacil is moved, or "translocated," upward into the leaves after absorption by the roots (58).

Physical properties

Terbacil is a white, crystalline, odorless solid which is noncorrosive and non-flammable (1).

Chemical name: 3-*tert*-butyl-5-chloro-6-methyluracil (1)
CAS #: 5902-51-2
Molecular weight: 216.7 (1)
Water solubility: 710 mg/L @ 25°C (1)
Solubility in other solvents: s.s. in mineral oils and aliphatic hydrocarbons; v.s. in methyl isobutyl ketone, butyl acetate, xylene, cyclohexanone, dimethylformamide, and strong aqueous alkalis (1)
Melting point: 175–177°C (1)
Vapor pressure: 0.0625 mPa @ 30°C (1)
Partition coefficient (octanol/water): 78 (58)
Adsorption coefficient: 55 (11)

Exposure guidelines

ADI: Not available
HA: 0.09 mg/L (lifetime) (130)

RfD: 0.013 mg/kg/day (13)
PEL: Not available

Basic manufacturer

DuPont Agricultural Products
Walker's Mill, Barley Mill Plaza
P.O. Box 80038
Wilmington, DE 19880-0038
Telephone: 800-441-7515
Emergency: 800-441-3637

10.3.23 Triclopyr

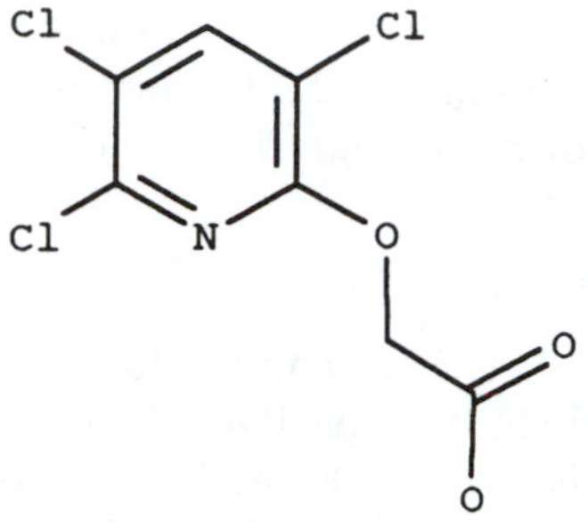

Figure 10.34 Triclopyr.

Trade or other names

Trade names for herbicides containing triclopyr include Access, Crossbow, ET, Garlon, Grazon, PathFinder, Redeem, Rely, Remedy, and Turflon. The herbicide may be mixed with picloram or with 2,4-D to extend its utility range.

Regulatory status

Some or all applications of the product Access may be classified as Restricted Use. Restricted Use Pesticides (RUPs) may be purchased and used only by certified applicators. It is toxicity class III — slightly toxic, but can cause eye irritation. The product will either have the Signal Word DANGER or CAUTION on the label, depending on the specific formulation. Products labeled DANGER include Garlon 3A, Redeem, and Turflon Amine.

Introduction

Triclopyr, a pyridine, is a selective systemic herbicide used for control of woody and broadleaf plants along rights-of-way, in forests, on industrial lands, and on grasslands and parklands. Unlike a similar product, 2,4,5-T, which has been banned in the U.S., dioxin impurities do not occur in triclopyr. Over 70,000 pounds triclopyr are used annually in the U.S. Triclopyr is commercially available mainly as a triethylamine salt or butoxyethyl ester of the parent compound.

Unless otherwise specified, data presented refer to the technical grade of the parent compound.

Toxicological effects

Acute toxicity

The oral LD_{50} of triclopyr in rats ranges from 630 to 729 mg/kg (1,6) and is over 2000 mg/kg for various amine and ester formulated products (8). Other oral LD_{50} values for triclopyr are 550 mg/kg in the rabbit and 310 mg/kg in the guinea pig (1,6). The dermal LD_{50} for the technical material in rabbits is greater than 2000 mg/kg, and greater than 4000 mg/kg for the formulations (1,6). Inhalation of triclopyr did not affect rats, but inhalation of some of the formulations did cause nasal irritation (6). A similar result was seen when rabbit eyes were exposed. The technical material had only a slight effect on rabbit eyes while some formulations caused significant eye irritation (6). These data indicate triclopyr is slightly toxic.

Chronic toxicity

Rats fed diets containing between 3 and 30 mg/kg/day triclopyr experienced no ill effects (6). Male rats fed much higher doses (100 mg/kg/day) had decreased liver and body weights and increased kidney weight (6). Male mice also showed reduced liver weight, but at 60 mg/kg/day (6). Monkeys fed smaller doses of triclopyr (20 mg/kg/day) showed no adverse effects (6).

Reproductive effects

Triclopyr fed to rabbits on days 6 to 18 of gestation at doses of 25, 50, and 100 mg/kg/day produced no effects on maternal body weight, litter size, or fetal body weight (6). A three-generation study of rats at doses of 3, 10, and 30 mg/kg/day for an 8- to 10-week period prior to breeding of each generation showed no impact of triclopyr on fertility rates (6,133). Triclopyr does not appear to cause reproductive toxicity.

Teratogenic effects

Pregnant rats given moderate to high doses of 50, 100, and 200 mg/kg/day on days 6 to 15 of gestation had offspring with mild fetotoxicity, but no birth defects (133). There were no teratogenic effects in rabbits treated on days 6 to 18 of gestation at dose rates of 10 and 25 mg/kg/day. These data suggest that triclopyr is not teratogenic.

Mutagenic effects

Triclopyr is nonmutagenic in bacterial and cytogenetic assay systems (6). A mutagenicity study using rats was weakly positive, but a negative result was found in mice, the more sensitive species (6). Based on these data, triclopyr is unlikely to be mutagenic.

Carcinogenic effects

Rats and mice fed oral doses of triclopyr at 3 to 30 mg/kg/day for 2 years showed no carcinogenic response (6,134). Even though the mice did have a high incidence of lymph cancer, this incidence was apparently characteristic of the particular strain of mice and did not represent a dose-related effect (119,134). Based on these data, triclopyr is unlikely to be carcinogenic.

Organ toxicity

Organs affected by exposure to triclopyr include the kidneys and liver (6).

Fate in humans and animals

Data from animal studies indicate that triclopyr is rapidly eliminated via the urine as the unchanged parent compound (6). At higher oral doses, some triclopyr may be eliminated through the feces as the absorption capacity of the intestine is exceeded (6,134). Reported half-lives for elimination of triclopyr from mammals are 14 hours (dog) and <24 hours (monkeys). A human elimination half-life of approximately 5 hours has been suggested (135). Minor metabolites of triclopyr may include trichloropyridinol (6).

Ecological effects

Effects on birds

Triclopyr is slightly to practically nontoxic to birds. The LD_{50} of the parent compound in the mallard duck is 1698 mg/kg, while the formulated compounds are of lower toxicity (6,58). The LC_{50} in bobwhite quail and Japanese quail fed triclopyr for 8 days are 2935 and 3278 ppm, respectively (6).

Effects on aquatic organisms

The parent compound and amine salt are practically nontoxic to fish. Triclopyr has an LC_{50} (96-hour) of 117 mg/L in rainbow trout and 148 mg/L in bluegill sunfish (6). The compound is practically nontoxic to the aquatic invertebrate *Daphnia magna*, a waterflea, with a reported LC_{50} for the amine salt of 1170 mg/L (136). The ester formulation has reported 96-hour LC_{50} values of 0.74 and 0.87 mg/L in the rainbow trout and bluegill sunfish, respectively (6,137).

The compound has little if any potential to accumulate in aquatic organisms. The bioconcentration factor for triclopyr in whole bluegill sunfish is only 1.08.

Effects on other organisms (non-target species)

The compound is nontoxic to bees (1).

Environmental fate

Breakdown in soil and groundwater

In natural soil and in aquatic environments, the ester and amine salt formulations rapidly convert to the acid, which in turn is neutralized to a relatively nontoxic salt. It is effectively degraded by soil microorganisms and has a moderate persistence in soil environments (6). The half-life in soil ranges from 30 to 90 days, depending on soil type and environmental conditions, with an average of about 46 days (137). The half-life of one of the breakdown products (trichloropyridinol) in 15 soils ranged from 8 to 279 days, with 12 of the tested soils having half-lives of less than 90 days. Longer half-lives may occur in cold or arid conditions.

Triclopyr is not strongly adsorbed to soil particles and has the potential to be mobile (6).

Breakdown in water

Triclopyr is not readily hydrolyzed at pH 5 to 9. Hydrolysis of the ester and the amine salt occurs rapidly and results in formation of triclopyr (6). Reported half-lives in water are 2.8 to 14.1 hours, depending on season and depth of water (137). The ester formulation half-life is from 12.5 to 83.4 hours (137). In water, the most important breakdown process is photolysis (137).

Breakdown in vegetation

Triclopyr is readily translocated throughout a plant after being taken up by either roots or the foliage. Cowberries contained residues of 2.4 ppm at 6 days, 0.7 to 1.1 ppm at 30 to 36 days, and 0.2 to 0.3 ppm at 92 to 98 days after application. The estimated half-life in above-ground drying foliage, as in a forest overstory, is 2 to 3 months (6).

Physical properties

Triclopyr is a fluffy, colorless solid at room temperature and is stable under normal storage conditions (1). Values presented below are for the parent acid.

Chemical name: 3,5,6-trichloro-2-pyridyloxyacetic acid (1)
CAS #: 55335-06-3
Molecular weight: 256.48 (1)
Solubility in water: 440 mg/L @ 25°C (1)
Solubility in other solvents: v.s. in acetone, acetonitrile, xylene, and benzene; s. in hexane (1)
Melting point: 148–150°C (1)
Vapor pressure: 0.168 mPa @ 25°C (1)
Partition coefficient (octanol/water): 2.64 (58)
Adsorption coefficient: 20 (amine salt, estimated); 780 (ester) (11)

Exposure guidelines

ADI: Not available
HA: Not available
RfD: 0.025 mg/kg/day (13)
PEL: Not available

Basic manufacturer

DowElanco
9330 Zionsville Road
Indianapolis, IN 46268-1054
Telephone: 317-337-7352
Emergency: 800-258-3033

10.3.24 Trifluralin

Trade or other names

Trade names trifluralin include Crisalin, Elancolan, Flurene SE, Ipersan, L-36352, M.T.F., Su Seguro Carpidor, TR-10, Trefanocide, Treficon, Treflan, Tri-4, Trifluralina 600, Triflurex Trim, and Trust. The compound may be found in formulations with other herbicides.

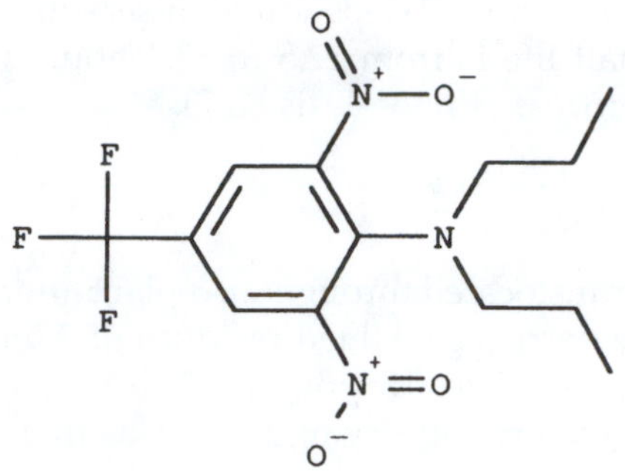

Figure 10.35 Trifluralin.

Regulatory status

Products containing trifluralin bear the Signal Words CAUTION or WARNING, depending on the type of formulation. This compound is a General Use Pesticide (GUP) in toxicity class III — slightly toxic. N-nitrosamine contaminant levels in trifluralin are required to be below 0.5 ppm, a level which the EPA believes will result in no toxic effects.

Introduction

Trifluralin is a selective, pre-emergence dinitroaniline herbicide used to control many annual grasses and broadleaf weeds in a large variety of tree fruit, nut, vegetable, and grain crops, including soybeans, sunflowers, cotton, and alfalfa. Pre-emergence herbicides are applied before weed seedlings sprout. Trifluralin should be incorporated into the soil by mechanical means within 24 hours of application. Granular formulations may be incorporated by overhead irrigation. Trifluralin is available in granular and emulsifiable concentrate formulations. The technical material is approximately 96% pure and the emulsifiable concentrate is about 45% pure.

Toxicological effects

Acute toxicity

Pure trifluralin is practically nontoxic to test animals by oral, dermal, or inhalation routes of exposure (138).

The oral LD_{50} for technical trifluralin in rats is greater than 10,000 mg/kg; in mice is greater than 5000 mg/kg; and in dogs, rabbits, and chickens is greater than 2000 mg/kg. However, certain formulated products that contain trifluralin may be more toxic than the technical material itself. For example, the oral LD_{50} for Treflan TR-10 in rats is greater than 500 mg/kg. The dermal LD_{50} for technical trifluralin in rabbits is greater than 2000 mg/kg. The 1-hour inhalation LC_{50} for technical trifluralin in rats is greater than 2.8 mg/L (58). Nausea and severe gastrointestinal discomfort may occur after eating trifluralin.

Trifluralin does not cause skin irritation. When applied to the eyes of rabbits, trifluralin produced slight irritation which cleared within 7 days (8). Skin sensitization (allergies) may occur in some individuals (8). Inhalation may cause irritation of the lining of the mouth, throat, or lungs (8).

Chronic toxicity

Prolonged or repeated skin contact with trifluralin may cause allergic dermatitis (8). The administration of 25 mg/kg/day to dogs for 2 years resulted in no observed toxicity (58). In another study of beagle dogs, toxic effects were observed at 18.75 mg/kg/day. These included decreased red blood cell counts and increases in methemoglobin, total serum lipids, triglycerides, and cholesterol (13).

Trifluralin has been shown to cause liver and kidney damage in other studies of chronic oral exposure in animals (139).

Reproductive effects

The reproductive capacity of rats fed dietary concentrations of trifluralin as high as 10 mg/kg/day was unimpaired through four successive generations. Trifluralin administered to pregnant rabbits at doses as high as 100 mg/kg/day, and to rats at doses as high as 225 mg/kg/day, produced no adverse effects on either the mothers or offspring (58).

Loss of appetite and weight loss followed by miscarriages were observed when pregnant rabbits were fed high doses of 224 or 500 mg/kg/day. Fetal weight decreased and there was an increase in the number of fetal runts at the 500-mg/kg/day dosage (8).

It is unlikely effects on reproduction will be produced in humans at expected exposure levels.

Teratogenic effects

No abnormalities were observed in the offspring of rats fed doses as high as 10 mg/kg/day for four generations (58). Studies in the rat and rabbit show no evidence that trifluralin is teratogenic. The highest doses tested in these studies were 1000 mg/kg/day in rats and 500 mg/kg/day in rabbits (138). Trifluralin does not appear to be teratogenic.

Mutagenic effects

No evidence of mutagenicity was observed when trifluralin was tested in live animals or in assays using bacterial and mammalian cell cultures (138).

Carcinogenic effects

In a 2-year study of rats fed 325 mg/kg/day, the highest dose tested, malignant tumors developed in the kidneys, bladder, and thyroid (138). However, more data are needed to characterize its carcinogenicity.

Organ toxicity

Liver, kidney, and thyroid damage appear to be the main toxic effects in chronic animal studies (139).

Fate in humans and animals

Trifluralin is not readily absorbed into the bloodstream from the gastrointestinal tract; 80% of single oral doses administered to rats and dogs was excreted in the feces (8).

Ecological effects

Effects on birds

Trifluralin is practically nontoxic to birds (63). The LD_{50} in bobwhite quail is greater than 2000 mg/kg, as it is in female mallards and pheasants (63). These values are for the technical product.

Effects on aquatic organisms

Trifluralin is very highly toxic to fish and other aquatic organisms. The 96-hour LC_{50} is 0.02 to 0.06 mg/L in rainbow trout, and 0.05 to 0.07 mg/L in bluegill sunfish (37). The 96-hour LC_{50} in channel catfish is approximately 1.4 to 3.4 mg/L (37). Variables such as temperature, pH, life stage, or size may affect the toxicity of the compound. Trifluralin is highly toxic to *Daphnia*, a species of small freshwater crustacean, with a 48-hour LC_{50} of 0.5 to 0.6 mg/L (140).

The compound shows a moderate tendency to accumulate in aquatic organisms.

Effects on other organisms (non-target species)

At exposure levels well above permissible application rates (100 mg/kg), trifluralin has been shown to be toxic to earthworms. However, permitted application rates will result in soil residues of approximately 1 ppm trifluralin, a level that had no adverse effects on earthworms (140).

It is nontoxic to bees (1).

Environmental fate

Breakdown in soil and groundwater

Trifluralin is of moderate to high persistence in the soil environment, depending on conditions. Trifluralin is subject to degradation by soil microorganisms. Trifluralin remaining on the soil surface after application may be decomposed by UV light or may volatilize. Reported half-lives of trifluralin in the soil vary from 45 to 60 days (11) to 6 to 8 months (1). After 6 months to 1 year, 80 to 90% of its activity will be gone (8).

It is strongly adsorbed on soils and nearly insoluble in water (11). Because adsorption is highest in soils high in organic matter or clay content and adsorbed herbicide is inactive, higher application rates may be required for effective weed control on such soils (8,58).

Trifluralin has been detected in nearly 1% of the 5590 wells tested. However, it has been detected at very low concentrations, typically ranging from 0.002 to 15 µg/L (8).

Breakdown in water

Trifluralin is nearly insoluble in water (1). It will probably be found adsorbed to soil sediments and particulates in the water column.

Breakdown in vegetation

Trifluralin inhibits the growth of roots and shoots when it is absorbed by newly germinated weed seedlings (58). Trifluralin residues in crop plants will occur only in root tissues that are in direct contact with contaminated soil. Trifluralin is not translocated into the leaves, seeds, or fruit of most plants. On most crops, trifluralin applied to the leaves has no effect; but on certain crops, such as tobacco and summer squash, leaf distortion may occur (58).

Physical properties

Trifluralin is an odorless, yellow-orange crystalline solid (1).

Chemical name: a,a,a-trifluoro-2,6-dinitro-N,N- dipropyl-*p*-toluidine (1)
CAS#: 1582-09-8
Molecular weight: 335.5 (1)
Solubility in water: <1 mg/L @ 27°C (1)
Solubility in other solvents: s. in organic solvents such as acetone and xylene (1)
Melting point: 48.5–49°C (1)
Vapor pressure: 13.7 mPa @ 25°C (1)
Partition coefficient (octanol/water): 118,000 @ pH 7 and 25°C (1)
Adsorption coefficient: 8000 (11)

Exposure guidelines

ADI: Not available
HA: 0.005 mg/L (lifetime) (139)
RfD: 0.0075 mg/kg/day (13)
PEL: Not available

Basic manufacturer

DowElanco
9330 Zionsville Road
Indianapolis, IN 46268-1054
Telephone: 317-337-7352
Emergency: 800-258-3033

10.4 Insecticides

10.4.1 Abamectin

Trade or other names

Abamectin is also known as avermectin B1a. Trade names include Affirm, Agri-Mek, Avermectin, Avid, MK 936, Vertimec, and Zephyr.

Regulatory status

Abamectin is a General Use Pesticide (GUP). It is classified as toxicity class IV — practically nontoxic, and has no precautionary statement on its label.

Introduction

Abamectin is a mixture of avermectins containing about 80% avermectin B1a and 20% avermectin B1b. These two components, B1a and B1b, have very similar biological and toxicological properties. The avermectins are antibiotic compounds

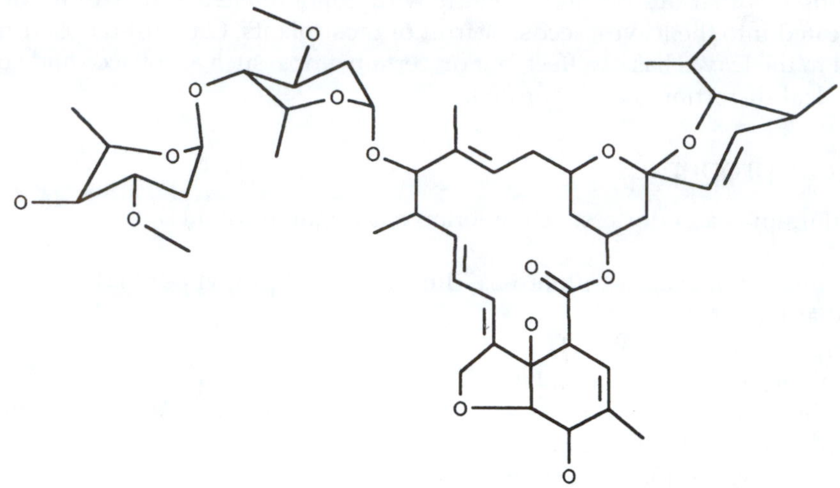

Figure 10.36 Abamectin.

derived from the soil bacterium *Streptomyces avermitilis*. Abamectin is a natural fermentation product of this bacterium. It acts as an insecticide by affecting the nervous system of and paralyzing insects. Abamectin is used to control insect and mite pests of citrus, pear, and nut tree crops, and it is used by homeowners for control of fire ants.

Toxicological effects

Acute toxicity

Abamectin is highly toxic to insects and may be highly toxic to mammals as well (141). Emulsifiable concentrate formulations may cause slight to moderate eye irritation and mild skin irritation (8). Symptoms of poisoning observed in laboratory animals include pupil dilation, vomiting, convulsions and/or tremors, and coma (141,142).

Abamectin acts on insects by interfering with the nervous system. At very high doses, it can affect mammals, causing symptoms of nervous system depression such as incoordination, tremors, lethargy, excitation, and pupil dilation. Very high doses have caused death from respiratory failure (143).

Abamectin is not readily absorbed through skin. Tests with monkeys show that less than 1% of dermally applied abamectin was absorbed into the bloodstream through the skin (141). Abamectin does not cause allergic skin reactions (142).

The oral LD_{50} for abamectin in rats is 10 mg/kg, and in mice ranges from 14 mg/kg to greater than 80 mg/kg (141,142). The oral LD_{50} for the product Avid EC in rats is 650 mg/kg (8). The dermal LD_{50} for technical abamectin in rats and rabbits is greater than 330 mg/kg (144).

Chronic toxicity

In a 1-year study with dogs given oral doses of abamectin, dogs at the 0.5- and 1-mg/kg/day doses exhibited pupil dilation, weight loss, lethargy, tremors, and recumbency (141). Similar results were seen in a 2-year study with rats fed 0.75, 1.5, or 2 mg/kg/day. Rats at all dosage levels exhibited body weight gains significantly higher than the controls. A few individuals in the high-dose group exhibited tremors (141).

When mice were fed 8 mg/kg/day for 94 weeks, the males developed dermatitis and changes in blood formation in the spleen, while females exhibited tremors and weight loss (142).

Reproductive effects

Rats given 0.40 mg/kg/day abamectin had increased stillbirths, decreased pup viability, decreased lactation, and decreased pup weights (142). These data suggest that abamectin may have the protential to cause reproductive effects at high enough doses.

Teratogenic effects

Abamectin produced cleft palate in the offspring of treated mice and rabbits, but only at doses that were also toxic to the mothers (141). There were no birth defects in the offspring of rats given up to 1 mg/kg/day (142). Abamectin is unlikely to cause teratogenic effects except at doses toxic to the mother.

Mutagenic effects

Abamectin does not appear to be mutagenic. Mutagenicity tests in live rats and mice were negative. Abamectin was shown to be nonmutagenic in the Ames test (1).

Carcinogenic effects

Abamectin is not carcinogenic in rats or mice. The rats were fed dietary doses of up to 2 mg/kg/day for 24 months, and the mice were fed up to 8 mg/kg/day for 22 months (141). These represent the maximum tolerated doses.

Organ toxicity

Animal studies indicate that abamectin may affect the nervous system.

Fate in humans and animals

Tests with laboratory animals show that ingested avermectin B1a is not readily absorbed into the bloodstream by mammals and that it is rapidly eliminated from the body within 2 days via the feces (142). Rats given single oral doses of avermectin B1a excreted 69 to 82% of the dose unchanged in the feces. The average half-life of avermectin B1a in rat tissue is 1.2 days (144). Lactating goats given daily oral doses for 10 days excreted 89% of the administered avermectin, mainly in the feces. Less than 1% was recovered in the urine (144).

Ecological effects

Effects on birds

Abamectin is practically nontoxic to birds (142). The LD_{50} for abamectin in bobwhite quail is >2000 mg/kg. The dietary LC_{50} is 3102 ppm in bobwhite quail (145). There were no adverse effects on reproduction when mallard ducks were fed dietary doses of 3, 6, or 12 ppm for 18 weeks (145).

Effects on aquatic organisms

Abamectin is highly toxic to fish and extremely toxic to aquatic invertebrates (142). Its LC_{50} (96-hour) is 0.003 mg/L in rainbow trout, 0.0096 mg/L in bluegill sunfish, 0.015 mg/L in sheepshead minnows, 0.024 mg/L in channel catfish, and

0.042 mg/L in carp. Its 48-hour LC_{50} in *Daphnia magna,* a small freshwater crustacean, is 0.003 mg/L. The 96-hour LC_{50} for abamectin is 0.0016 mg/L in pink shrimp, 430 mg/L in Eastern oysters, and 153 mg/L in blue crab (145).

While highly toxic to aquatic organisms, actual concentrations of abamectin in surface waters adjacent to treated areas are expected to be low.

Abamectin did not bioaccumulate in bluegill sunfish exposed to 0.099 µg/L for 28 days in a flow-through tank. The levels in fish were from 52 to 69 times the ambient water concentration, indicating that abamectin does not accumulate or persist in fish (145).

Effects on other organisms (non-target species)

Abamectin is highly toxic to bees, with a 24-hour contact LC_{50} of 0.002 µg per bee and an oral LD_{50} of 0.009 µg per bee (145).

Environmental fate

Breakdown in soil and groundwater

Abamectin is rapidly degraded in soil. At the soil surface, it is subject to rapid photodegradation, with reported half-lives of 8-hours to 1 day (142,145). When applied to the soil surface and not shaded, its soil half-life is about 1 week. Under dark, aerobic conditions, the soil half-life was 2 weeks to 2 months (142). Loss of abamectin from soils is thought to be due to microbial degradation. The rate of degradation was significantly decreased under anaerobic conditions (145).

Because abamectin is nearly insoluble in water and has a strong tendency to bind to soil particles, it is immobile in soil and unlikely to leach or contaminate groundwater (145). Compounds produced by the degradation of abamectin are also immobile and unlikely to contaminate groundwater (145).

Breakdown in water

Abamectin is rapidly degraded in water. After initial distribution, its half-life in artificial pond water was 4 days. Its half-life in pond sediment was 2 to 4 weeks (145). It undergoes rapid photodegradation, with a half-life of 12 hours in water (142). When tested at pH levels common to surface and groundwater (pH 5, 7, and 9), abamectin did not hydrolyze (145).

Breakdown in vegetation

Plants do not absorb abamectin from the soil (145). Abamectin is subject to rapid degradation when present as a thin film, as on treated leaf surfaces. Under laboratory conditions and in the presence of light, its half-life as a thin film was four to six hours (145).

Physical properties

Abamectin is a colorless to yellowish crystalline powder (1).

Chemical name: avermectin B1 (1)
CAS #: 71751-41-2 (avermectin B1a and avermectin B1b) (1)

Molecular weight: 873.11 (avermectin B1a); 859.08 (avermectin B1b) (1)
Solubility in water: Insoluble (1)
Solubility in other solvents: v.s. in acetone, methanol, toluene, chloroform, and
    ethanol (1)
Melting point: 150-155°C (1)
Vapor pressure: Negligible (1)
Partition coefficient (octanol/water): Not available
Adsorption coefficient: 5000 (estimated) (53)

Exposure guidelines

ADI: 0.0001 mg/kg/day (12)
HA: Not available
RfD: 0.0004 mg/kg/day (13)
PEL: Not available

Basic manufacturer

Merck Agvet
Division of Merck and Co., Inc.
P.O. Box 2000
Rahway, NJ 07065
Telephone: 908-855-4277

10.4.2 Bacillus thuringiensis

Trade or other names

Trade names include Acrobe, Bactospeine, Berliner (variety *kurstaki*), Certan
(variety *aizawai*), Dipel, Javelin, Leptox, Novabac, Teknar (variety *israelensis*), Thuricide, and Victory. *Bacillus thuringiensis* is also known as B.t.

Regulatory status

This microbial insecticide was originally registered in 1961 as a General Use
Pesticide (GUP). It is classified as toxicity class III — slightly toxic. Products containing B.t. bear the Signal Word CAUTION because of its potential to irritate eyes
and skin.

Introduction

Bacillus thuringiensis (B.t.) is a naturally occurring soil bacterium that produces
poisons which cause disease in insects. B.t. is considered ideal for pest management
because of its specificity to pests and its lack of toxicity to humans or the natural
enemies of many crop pests. There are different strains of B.t., each with specific
toxicity to particular types of insects: B.t. *aizawai* (B.t.a.) is used against wax moth
larvae in honeycombs; B.t. *israelensis* (B.t.i.) is effective against mosquitoes, blackflies,
and some midges; B.t. *kurstaki* (B.t.k.) controls various types of lepidopterous insects,
including the gypsy moth and cabbage looper. A newer strain, B.t. *san diego*, is
effective against certain beetle species and the boll weevil. To be effective, B.t. must
be eaten by insects during their feeding stage of development, when they are larvae.

B.t. is ineffective against adult insects. More than 150 insects, mostly lepidopterous larvae, are known to be susceptible in some way to B.t.

B.t. forms asexual reproductive cells, called spores, which enable it to survive in adverse conditions. During the process of spore formation, B.t. also produces unique crystalline bodies. When eaten, the spores and crystals of B.t. act as poisons in the target insects. B.t. is therefore referred to as a stomach poison. B.t. crystals dissolve in the intestine of susceptible insect larvae. They paralyze the cells in the gut, interfering with normal digestion and triggering the insect to stop feeding on host plants. B.t. spores can then invade other insect tissue, multiplying in the insect's blood, until the insect dies. Death can occur within a few hours to a few weeks of B.t. application, depending on the insect species and the amount of B.t. ingested. Typical agricultural formulations include wettable powders, spray concentrates, liquid concentrates, dusts, baits, and time-release rings.

Toxicological effects

Acute toxicity

B.t. is practically nontoxic to humans and animals. Humans exposed orally to 1000 mg/day B.t. showed no effects (146). A wide range of studies has been conducted on test animals, using several routes of exposure. The highest dose tested was 6.7×10^{11} spores per animal. The results of these tests suggest that the use of B.t. products causes few, if any, negative effects. B.t. was not acutely toxic in tests conducted on birds, dogs, guinea pigs, mice, rats, and humans. No oral toxicity was found in rats or mice fed protein crystals from B.t. var. *israelensis* (147).

The LD_{50} is greater than 5000 mg/kg for the B.t. product Javelin in rats, and greater than 13,000 mg/kg in rats exposed to the product Thuricide (147,148). Single oral dosages of up to 10,000 mg/kg did not produce toxicity in mice, rats, or dogs (148).

The dermal LD_{50} for a formulated B.t. product in rabbits is 6280 mg/kg. A single dermal application of 7200 mg/kg B.t. was not toxic to rabbits (148).

B.t. is an eye irritant; 100 g of formulated product applied in each eye of test rabbits caused continuous congestion of the iris as well as redness and swelling (149).

Very slight irritation from inhalation was observed in test animals. This may have been caused by the physical rather than the biological properties of the B.t. formulation tested (8). Mice survived 1 or more 1-hour periods of breathing mist that contained as many as 6.0×10^{10} spores B.t. per liter (143).

Chronic toxicity

No complaints were made by eight men after they were exposed for 7 months to fermentation broth, moist bacterial cakes, waste materials, and final powder created during the commercial production of B.t. (143).

Dietary administration of B.t. for 13 weeks to rats at dosages of 8400 mg/kg/day did not produce toxic effects (143).

Some reversible, abnormal redness of the skin was observed when 1 mg/kg/day formulated B.t. product was put on scratched skin for 21 days. No general, systemic poisoning was observed (8).

Reproductive effects

There is no indication that B.t. causes reproductive effects (143).

Teratogenic effects

There is no evidence indicating that formulated B.t. can cause birth defects in mammals (143,148).

Mutagenic effects

B.t. appears to have mutagenic potential in plant tissue. Thus, extensive use of B.t. on food plants might be hazardous to these crops (143). There is no evidence of mutagenicity in mammalian species.

Carcinogenic effects

Tumor-producing effects were not seen in 2-year chronic studies during which rats were given dietary doses of 8400 mg/kg/day B.t. formulation (148). It is unlikely that B.t. is carcinogenic.

Organ toxicity

There is no evidence of chronic B.t. toxicity in dogs, guinea pigs, rats, humans, or other test animals.

Fate in humans and animals

B.t. does not persist in the digestive systems of mammals that ingest it (149).

Ecological effects

Effects on birds

B.t. is not toxic to birds (8,150). The LD_{50} in bobwhite quail is greater than 10,000 mg/kg. When autopsies were performed on these birds, no pathology was attributed to B.t. Field observations of 74 bird species did not reveal any population changes after aerial spraying of B.t. formulation (148).

Effects on aquatic organisms

B.t. is practically nontoxic to fish (150). Rainbow trout and bluegills exposed for 96 hours to B.t. at concentrations of 560 and 1000 mg/L did not show adverse effects. A small marine fish (*Anguilla anguilla*) was not negatively affected by exposure to 1000 to 2000 times the level of B.t. expected during spray programs. Field observations of populations of brook trout, common white suckers, and smallmouth bass did not reveal adverse effects 1 month after aerial application of B.t. formulation (148). However, shrimp and mussels may be affected adversely (8).

Effects on other organisms (non-target species)

Applications of formulated B.t. are not toxic to most beneficial or predator insects (148). Treatment of honeycombs with B.t. var. *aizawai* does not have a detrimental effect upon bees, nor on the honey produced (151). Very high concentrations (108 spores per milliliter sucrose syrup) of B.t. var. *tenebrionis*, which is used against beetles such as the Colorado potato beetle, reduced longevity of honeybee adults, but did not cause disease (151). B.t. applied at rates used for mosquito control may cause the death of some non-target species (8).

Users of B.t. are encouraged to consult local officials or the nearest EPA regional office responsible for protecting endangered species before using B.t. products in counties where susceptible endangered species of Lepidoptera are known to be present (146). It did not have negative effects on frogs and salamanders (150).

Environmental fate

Breakdown in soil and groundwater

B.t. is a naturally occurring pathogen that readily breaks down in the environment. Due to its short biological half-life and its specificity, B.t. is less likely than chemical pesticides to cause field resistance in target insects.

B.t. is moderately persistent in soil. Its half-life in suitable conditions is about 4 months (152). B.t. spores are released into the soil from decomposing dead insects after they have been killed by it. B.t. is rapidly inactivated in soils that have a pH below 5.1 (148).

Microbial pesticides such as B.t. are classified as immobile because they do not move, or leach, with groundwater. Because of their rapid biological breakdown and low toxicity, they pose no threat to groundwater.

Breakdown in water

The EPA has not issued restrictions for the use of B.t. around bodies of water. It can be effective for up to 48 hours in water. Afterward, it gradually settles out or adheres to suspended organic matter (150).

Breakdown in vegetation

B.t. is relatively short-lived on foliage because the ultraviolet (UV) light of the sun destroys it very rapidly. Its half-life under normal sunlight conditions is 3.8 hours (153). It is not poisonous to plants and has not shown any adverse effect upon seed generation or plant vigor (150).

Physical properties

The insecticidal action of B.t. is attributed to protein crystals produced by the bacterium. The vegetative cells of B.t. are approximately 1 micron wide and 5 microns long, and are motile (146). The commercial product contains about 2.5×10^{11} viable spores per gram.

B.t. products lose some of their effectiveness when stored for more than 6 months (8). B.t. is incompatible with alkaline materials. Formulated products are not compatible with captafol, dinocap, or, under some conditions, leaf (or foliar) nutrients (8).

> Chemical name: *Bacillus thuringiensis* (1)
> CAS #: (B.t. variety *kurstaki*) 68038-71-1
> Molecular weight: Not applicable
> Water solubility: Not applicable
> Solubility in other solvents: Not applicable
> Melting point: Not applicable
> Vapor pressure: Not applicable
> Partition coefficient: Not applicable
> Adsorption coefficient: Not applicable

Exposure guidelines

ADI: Not available
MCL: Not available
RfD: Not available
PEL: Not available

Basic manufacturers

Sandoz Agro, Inc.
1300 E. Touhy Ave.
Des Plaines, IL 60018
Telephone: 708-699-1616
Emergency: 708-699-1616

Abbott Laboratories
Chem. and Agric. Prod. Div.
1401 Sheridan Rd.
North Chicago, IL 60064
Telephone: 708-937-2739

10.4.3 Hydramethylnon

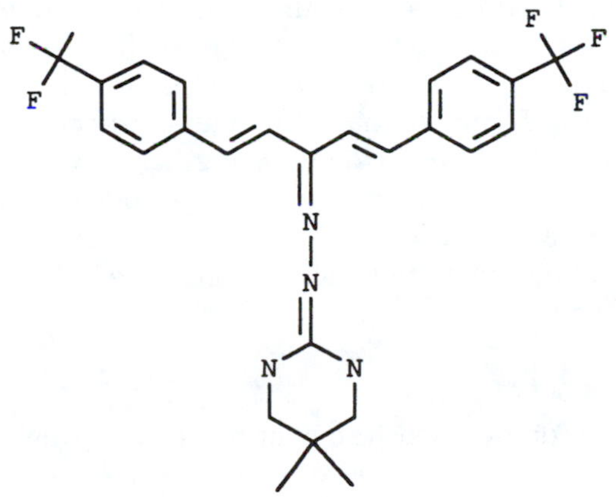

Figure 10.37 Hydramethylnon.

Trade or other names

Trade names for products containing hydramethylnon include AC 217,300, Amdro, Combat, Maxforce, and Wipeout.

Regulatory status

Hydramethylnon is a slightly toxic compound in EPA toxicity class III. Products containing hydramethylnon must bear the Signal Word CAUTION. It is a General Use Pesticide (GUP) (1).

Introduction

Hydramethylnon is a trifluoromethyl aminohydrazone insecticide used in baits to control fire ants, leafcutter ants, and cockroaches in both indoor and outdoor applications. It is available in a ready-to-use bait formulation. *The data presented here refer to the technical product unless otherwise stated.*

Toxicological effects

Acute toxicity

Hydramethylnon is slightly toxic via ingestion, with reported oral LD_{50} values of 1100 to 1300 mg/kg in rats (1). It is also slightly toxic by skin exposure, with a dermal LD_{50} in rabbits of greater than 5000 mg/kg (1). The reported 4-hour inhalation LC_{50} for hydramethylnon is greater than 5 mg/L, indicating slight toxicity by this route as well (1). It is reported not to cause skin sensitization in guinea pigs nor skin irritation in rabbits (8). It may, however, be an eye irritant in rabbits (1).

Acute exposure in humans may result in irritation of the eyes and mucous membranes of the respiratory tract.

Chronic toxicity

In a 26-week study in male and female dogs, doses of up to 3.0 mg/kg/day resulted in increased liver weights and increased liver:body weight ratios. No other effects were observable in either the structure of tissues examined, the chemistry and consistency of the blood, or the chemistry of other bodily fluids (13). A 2-year study in rats showed decreased food consumption and organ weight changes at 5.0 mg/kg/day, but not at 2.5. mg/kg/day (13). Similar effects were seen in rats over 90 days at the same doses (13). In dogs, 6 mg/kg/day caused decreased food consumption and body weight gain over a 90-day period (13). In an 18-month cancer assay, hydramethylnon at about 3.8 mg/kg/day was associated with amyloidosis, a syndrome in which abnormal protein deposition in the kidney fitration unit (glomerulus) results in damage (13).

Reproductive effects

Doses of 6 mg/kg/day caused testicular atrophy in dogs in a 90-day feeding study (13). In rats, this effect was also observed at doses of approximately 5 mg/kg/day over the same time frame (13). Testicular lesions were seen in mice at doses of approximately 3.8 mg/kg/day over 18 months. Male infertility, probably attributable to testicular effects, was seen in a three-generation rat study at 5 mg/kg/day (13). The available data suggest that reproductive effects are unlikely in humans at expected exposure levels.

Teratogenic effects

Maternal body weight gain was reduced in rats at 10 mg/kg/day, but not at 3 mg/kg/day in a teratology study (13). In the same study, maternal doses of 30 mg/kg/day were required to produce decreased fetal weights (13). In another study of potential birth defects in rabbits, reduced fetal weight gain was also noted, but at a dose of 10 mg/kg/day (13). It was not clear whether the decrements in fetal weight were severe enough to result in reduced survival rate. The available data suggest that teratogenic effects are unlikely in humans at expected exposure levels.

Carcinogenic effects

In an 18-month tumor assay in mice, no increases in tumor rates were reported at doses up to 3.8 mg/kg/day (13). The data regarding carcinogenic effects are insufficient, but suggest that hydramethylnon is not carcinogenic.

Organ toxicity

Chronic studies in several animals have shown the testis as a target organ.

Fate in humans and animals

In rats, following oral administration, hydramethylnon was rapidly eliminated in the feces and urine (8). No residues were detectable in the milk or tissues of goats at a dietary dose of 0.2 ppm in the daily diet for 8 days (8). No residues were found in the milk or tissues of cows at a dietary dose of 0.05 ppm for 21 consecutive days (1).

Ecological effects

Effects on birds

The oral LD_{50} for hydramethylnon in mallard ducks is greater than 2510 mg/kg, and in bobwhite quail is 1828 mg/kg, indicating that the compound is practically nontoxic to these species (8).

Effects on aquatic organisms

Hydramethylnon is highly to very highly toxic to fish in laboratory studies (8). The reported 96-hour LC_{50} values for hydramethylnon are 160 µg/L in rainbow trout, 100 µg/L in channel catfish, and 1700 µg/L in bluegill sunfish (1,8). The reported 72-hour LC_{50} of hydramethylnon in carp is 340 µg/L (1). The 48-hour LC_{50} for the compound in waterfleas (*Daphnia*) is 1.14 mg/L (8). Hydramethylnon accumulated in bluegill sunfish at 1300 times its concentration in surrounding waters, indicating low to moderate capacity to bioaccumulate (8). Due to its very low water solubility, it is not likely to be found in surface waters.

Effects on other organisms (non-target species)

Hydramethylnon is nontoxic to honeybees (8).

Environmental fate

Breakdown in soil and groundwater

Hydramethylnon is of low persistence in the soil environment; soil half-lives of 7 to 28 days have been reported (11,154). A representative value is estimated to be 10 days (11,154). However, a soil half-life of 18 hours has been reported, with breakdown probably due to decomposition by light and the rapid foraging of ants (11,154).

Because hydramethylnon is only slightly soluble in water and is very strongly sorbed by soil organic matter and clay particles, it is not appreciably mobile in most soils (11,154). These properties, along with its low persistence, make it unlikely to contaminate groundwater (11,154). When Amdro, a granular bait formulation, was applied to an aged soil column under laboratory conditions, 72% of the applied material remained in the treated soil after 45 days (8). Less than 0.2% was recovered in leachate. This evidence supports the conclusion that neither Amdro nor its metabolites leach (8).

Breakdown in water

The reported hydrolysis half-life for hydramethylnon is 10 to 11 days over a pH range of 7 to 8.9, and 24 to 33 days at pH 4.9 (8).

Breakdown in vegetation

Over a 90-day rotation interval, hydramethylnon did not accumulate in crop plants (154). In another study, residues in grass 4 months after treatment were less than 0.01 ppm. Negligible residues were found in radishes, barley, and French beans planted 3 months after treatment of the soil (8).

Physical properties

Hydramethylnon is an odorless, yellow to orange crystalline solid (1).

Chemical name: 5,5-dimethylperhydropyrimidin-2-one 4-trifluoromethyl-alpha-
   (4-trifluoromethylstyryl)-cinnamylidenehydrazone (1)
CAS #: 67485-29-4
Molecular weight: 494.5 (1)
Solubility in water: 0.005–0.007 mg/L @ 25°C (1)
Solubility in other solvents: v.s. in acetone, chlorobenzene, methanol, ethanol,
   and xylene (1)
Melting point: 185–190°C (1)
Vapor pressure: 0.0027 mPa @ 25°C (1)
Partition coefficient (octanol/water): 206 (1)
Adsorption coefficient: 730,000 (estimated) (11)

Exposure guidelines

ADI: Not available
HA: Not available
RfD: Not available
PEL: Not available

Basic manufacturer

American Cyanamid
One Cyanamid Plaza
Wayne, NJ 07470-8426
Telephone: 201-831-2000
Emergency: 201-835-3100

10.4.4 Methoprene

Trade or other names

Trade names for products containing methoprene include Altosid, Apex, Diacan, Dianex, Kabat, Minex, Pharorid, Precor, and ZR-515.

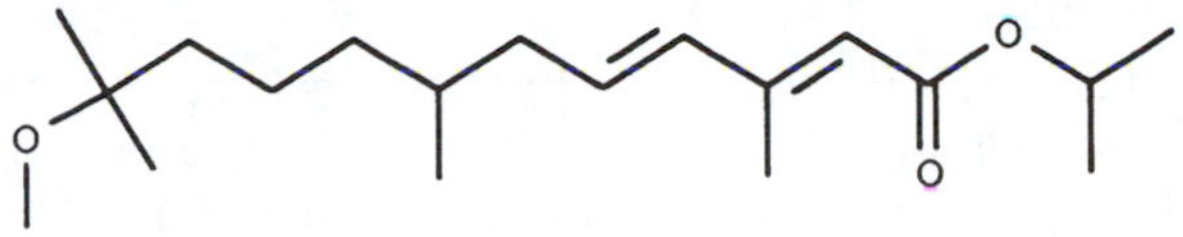

Figure 10.38 Methoprene.

Regulatory status

Methoprene is a slightly to practically nontoxic compound in EPA toxicity class IV. It is a General Use Pesticide (GUP). Labels for containers of products containing methoprene must bear the Signal Word CAUTION.

Introduction

Methoprene is a compound that mimics the action of an insect growth regulation hormone. It is used as an insecticide because it interferes with the normal maturation process. In a normal life cycle, an insect goes from egg to larva, to pupa, and eventually to adult. Methoprene artificially stunts the insect's development, making it impossible for the insect to mature to the adult stage, and thus preventing it from reproducing.

To be effective, it is essential that this growth inhibitor be administered at the proper stage of the target pest's life cycle. Methoprene is not toxic to the pupal or adult stages. Treated larvae will pupate, but adults do not hatch from the pupal stage. Methoprene is also considered a larvicide since it is effective in controlling the larval stage of insects. Methoprene is used in the production of a number of foods, including meat, milk, eggs, mushrooms, peanuts, rice, and cereals. It is also used in aquatic areas to control mosquitoes and several types of ants, flies, lice, moths, beetles, and fleas. It is available in suspension, emulsifiable and soluble concentrate formulations, as well as in briquette, aerosol, and bait form.

Toxicological effects

Acute toxicity

Methoprene is practically nontoxic when ingested or inhaled and slightly toxic by dermal absorption. The oral LD_{50} for methoprene in rats is greater than 34,600 mg/kg, and in dogs is greater than 5000 mg/kg (1). It is slightly toxic by skin exposure, with reported dermal LD_{50} values of greater than 2000 to 3000 mg/kg in rabbits (1). Methoprene is not an eye or skin irritant, and it is not a skin sensitizer (1). The inhalation LC_{50} for methoprene in rats is greater than 210 mg/L (155). No overt signs of poisoning have been reported in incidents involving accidental human exposure to methoprene (155).

Chronic toxicity

No methoprene-related effects were observed in 2-year feeding trials with rats given doses of 250 mg/kg/day, nor in mice given 30 mg/kg/day (1). Liver changes were observed in mice fed 50 to 250 mg/kg/day methoprene during an 18-month study (155). Increased liver weights occurred in rats fed 250 mg/kg/day for 90 days, but not during a 24-month feeding study in which rats were fed 125 mg/kg/day (155).

Reproductive effects

Experimental data indicate that no reproductive hazards are associated with methoprene (155). No methoprene-related effects were observed in three-generation reproduction studies in rats receiving dietary doses of 125 mg/kg/day (1).

Teratogenic effects

There have been no teratogenic effects in animals dosed with methoprene; teratogenic effects were not seen in rats at doses of about 25 mg/kg/day or in rabbits at doses of about 15 mg/kg/day (156,157). Methoprene does not appear to be teratogenic.

Mutagenic effects

Methoprene does not appear to be mutagenic. No methoprene-related mutagenic effects were observed in rats following a single dose of 2000 mg/kg (158).

Carcinogenic effects

No tumors were seen in an 18-month feeding study with mice or in a 24-month oncogenicity study with rats (156). These data suggest that methoprene is not carcinogenic.

Organ toxicity

The target organ primarily affected by methoprene after long-term exposure is the liver.

Fate in humans and animals

In mammals, methoprene is rapidly and completely broken down and excreted, mostly in the urine and feces (157). Some evidence suggests that methoprene metabolites are incorporated into natural body components (155). Methoprene is excreted unchanged in cattle feces in amounts that are sufficient to kill some larvae that breed in dung (131).

Ecological effects

Effects on birds

Methoprene is slightly toxic to birds (1,158). The reported 5- to 8-day LC_{50} values for Altosid, a methoprene formulation, are greater than 10,000 ppm in mallard ducks and bobwhite quail, and the acute oral LD_{50} for Altosid is greater than 4640 ppm in chickens (1,158). In mallards, an acute oral LD_{50} of greater than 2000 mg/kg was determined (158).

Nonlethal effects that may affect survival of the birds did appear at acute oral doses of 500 mg/kg. These effects appeared as soon as 2 hours after treatment and persisted for up to 2 days and included slowness, reluctance to move, sitting, withdrawal, and incoordination (63). These effects may decrease bird survival by making them temporarily more susceptible to predation. No effects were observed in the reproduction of bobwhite quail and mallard ducks at 30-ppm constant feeding of Altosid (158).

Effects on aquatic organisms

Methoprene is slightly to moderately toxic to fish (157). The reported 96-hour LC_{50} values for the methoprene formulation Altosid were 4.6 mg/L in bluegill sunfish, 4.4 mg/L in trout, and greater than 100 mg/L in channel catfish and largemouth bass (1,8). Methoprene residues may have a slight potential for bioconcentration in bluegill sunfish and crayfish (155).

Methoprene is very highly toxic to some species of freshwater, estuarine, and marine invertebrates. While the acute LC_{50} values are greater than 100 mg/L in freshwater shrimp, and it is greater than 0.1 mg/L in estuarine mud crabs (159).

Altosid had very little effect, if any, on exposed non-target aquatic organisms, including waterfleas, damselflies, snails, tadpoles, and mosquito fish (159).

Effects on other organisms (non-target species)

Tests with earthworms showed little if any toxic effect on contact (159). It is nontoxic to bees (1).

Environmental fate

Breakdown in soil and groundwater

Methoprene is of low persistence in the soil environment; reported field half-lives are up to 10 days (155). In sandy loam, its half-life was calculated to be about 10 days (155). When Altosid was applied at an extremely high application rate of 1 pound per acre, its half-life was less than 10 days (155). In soil, microbial degradation is rapid and appears to be the major route of its disappearance from soil (155,157). Methoprene also readily undergoes degradation by sunlight (157).

Methoprene is rapidly and tightly sorbed to most soils (155). It is slightly soluble in water (1). These properties, along with its low environmental persistence, make it unlikely to be significantly mobile. In field leaching studies, it was observed only in the top few inches of the soil, even after repeated washings with water (155,159).

Breakdown in water

Methoprene degrades rapidly in water (8). Studies have demonstrated half-lives in pond water of about 30 and 40 hours at initial concentrations of 0.001 and 0.01 mg/L, respectively (49). At normal temperatures and levels of sunlight, technical Altosid is rapidly degraded, mainly by aquatic microorganisms and sunlight (49,159).

Breakdown in vegetation

Altosid is biodegradable and nonpersistent, even in plants treated at very high rates. It has a half-life of less than 2 days in alfalfa when applied at a rate of 1 pound per acre (159). In rice, the half-life is less than 1 day (49). In wheat, its half-life was estimated to be 3 to 7 weeks, depending on the level of moisture in the plant (155). Plants grown in treated soil are not expected to contain methoprene residues.

Physical properties

Technical methoprene is a amber or pale yellow liquid with a faint fruity odor (1).

Chemical name: isopropyl(E,E)-(RS)-11-methoxy-3,7,11-trimethyldodeca-2,4-di-
    enoate (1)
CAS #: 40596-69-8
Molecular weight: 310.48 (1)
Water solubility: 1.4 mg/L @ 25°C (1)
Solubility in other solvents: Miscible in organic solvents (1)

Vapor pressure: 3.15 mPa @ 25°C (1)
Partition coefficient (octanol/water): Not available
Adsorption coefficient: Not available

Exposure guidelines

ADI: 0.1 mg/kg/day (12)
HA: Not available
RfD: Not available
PEL: Not available

Basic manufacturer

Zoecon Corp.
12005 Ford Rd., Suite 800
Dallas, TX 75234
Emergency: 708-699-1616

10.4.5 Sulfuryl fluoride

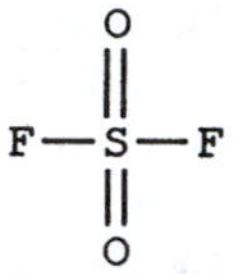

Figure 10.39 Sulfuryl fluoride.

Trade or other names

Vikane is the most common trade name for sulfuryl fluoride.

Regulatory status

All formulations of sulfuryl fluoride are Restricted Use Pesticides (RUPs) and bear the Signal Word DANGER on the product label because they pose an inhalation hazard. Sulfuryl fluoride is EPA toxicity class I — highly toxic. RUPs may be purchased and used only by certified applicators.

Introduction

Sulfuryl fluoride is an inorganic gas fumigant used in structures, vehicles, and wood products for control of drywood termites, wood-infesting beetles, and certain other insects and rodents. There are no registered uses for sulfuryl fluoride on food or feed crops.

Toxicological effects

Acute toxicity

Sulfuryl fluoride is a toxic gas that acts as a central nervous system depressant (160). Symptoms of poisoning include depression, slowed gait, slurred speech, nausea, vomiting, stomach pain, drunkenness, itching, numbness, twitching, and seizures (160,161). Inhalation of high concentrations may cause respiratory tract irritation (160) or respiratory failure (161). Skin contact with sulfuryl fluoride normally poses no hazard, but contact with liquid sulfuryl fluoride can cause pain and frostbite due to rapid vaporization (8).

Sulfuryl fluoride gas is odorless and colorless, does not cause tears or immediately noticeable eye irritation, and lacks any other warning property (162). Chloropicrin is added to products containing sulfuryl fluoride to serve as a warning indicator. Chloropicrin is a gas that causes eye and respiratory irritation and vomiting.

The oral LD_{50} for sulfuryl fluoride in rats and guinea pigs is 100 mg/kg (1). The 4-hour inhalation LC_{50} in rats is approximately 5 mg/L (1). The 1-hour LC_{50} is 12 to 15 mg/L in rats (1).

Chronic toxicity

Long-term exposure to high levels of sulfuryl fluoride may cause blood and bone effects (1). Repeated or prolonged exposure to sulfuryl fluoride may cause injury to lungs and kidneys, weakness, weight loss, anemia, bone brittleness, stiff joints, and general ill health (1). Long-term effects of sulfuryl fluoride are those of excess fluoride (160).

Rats, rabbits, guinea pigs, and female rhesus monkeys tolerated air concentrations of 417 mg/L for 7 hours per day, 5 days a week, for 6 months with no apparent adverse effects (160).

Reproductive effects

Two generations of rats were exposed to air concentrations of 21, 83, or 626 mg/L. No adverse effects on reproduction or fertility were seen at any dose. Toxic effects on the mother were accompanied by reduced pup weight at the highest doses (160). Sulfuryl fluoride does not appear to cause reproductive effects.

Teratogenic effects

There were no birth defects in the offspring of pregnant rabbits and rats exposed to air concentrations of 1063 mg/L, the highest dose tested, during days 6 to 18 (rabbits) or days 6 through 15 (rats) of pregnancy for 6 hours per day. In rabbits, both the fetuses and the mothers exhibited decreased weight gain at that dose (8). Sulfuryl fluoride does not appear to be teratogenic.

Mutagenic effects

Several tests have shown that sulfuryl fluoride is not mutagenic. It failed to produce unscheduled DNA synthesis in rat liver cells exposed to concentrations between 204 and 1020 ppm (163). Examination of bone marrow from mice exposed to air concentrations of as high as 2160 mg/L sulfuryl fluoride for 4 hours showed no mutagenic effects (164). When sulfuryl fluoride was assayed with the Ames test for mutagenic effects in bacterial cell cultures, the results were negative (165).

Carcinogenic effects

No data are currently available.

Organ toxicity

Case reports indicate that central nervous system depression and liver and kidney injuries may be possible, in addition to respiratory injury (166).

Fate in humans and animals

No data are currently available on the fate of sulfuryl fluoride in mammals. It is thought that fluoride ion may result from its metabolism (161,166).

Ecological effects

Effects on birds

Sulfuryl fluoride is a gas under normal conditions (1). It dissipates extremely rapidly after release into the environment. Exposure to birds is expected to be only at very low concentrations and of short duration.

Effects on aquatic organisms

Sulfuryl fluoride is a gas with low water solubility and will diffuse out of water into the atmosphere. Because use of sulfuryl fluoride is permitted only indoors, exposure of aquatic organisms is unlikely.

Effects on other organisms (non-target species)

No data are currently available.

Environmental fate

Environmental effects from sulfuryl fluoride are expected to be negligible because this fumigant is applied only indoors or in sealed structures (8).

Breakdown in soil and groundwater

Since sulfuryl fluoride is a gas, it will not leach or contaminate groundwater.

Breakdown in water

Sulfuryl fluoride is not readily hydrolyzed by water (160). The products of hydrolysis are sulfate and fluoride.

Breakdown in vegetation

No data are currently available.

Physical properties

Sulfuryl fluoride is a colorless, odorless gas (1).

Chemical name: sulfuryl fluoride (1)
CAS #: 2699-79-8
Molecular weight: 102.1 (1)
Solubility in water: 750 mg/L @ 25°C (1)
Solubility in other solvents: s. in ethanol, toluene, and carbon tetrachloride (1)
Melting point: –137°C (1)

Vapor pressure: 1.7 mPa @ 21.1°C (1)
Partition coefficient (octanol/water): Not available
Adsorption coefficient: Not available

Exposure guidelines

ADI: Not available
HA: Not available
RfD: Not available
PEL: 20 mg/m³ (8-hour) (14)

Basic manufacturer

DowElanco
9330 Zionsville Road
Indianapolis, IN 46268-1054
Telephone: 800-258-3033

10.5 Others

10.5.1 4-Aminopyridine

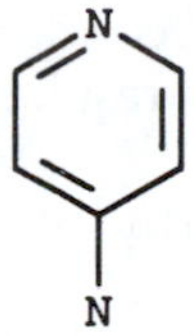

Figure 10.40 4-Aminopyridine.

Trade or other names

Common names include 4-AP, P-aminopyridine, gamma-aminopyridine, amino-4-pyridine. Trade names include Avitroland and Avitrol 200.

Regulatory status

Based on its potential hazard to fish and non-target birds, some 4-aminopyridine formulations are classified by the U.S. Environmental Protection Agency (EPA) as Restricted Use Pesticides (RUPs). RUPs may be purchased and used only by certified applicators. Grain bait formulations of 4-aminopyridine are in toxicity class III and must bear the signal word CAUTION; powder concentrate formulations are in toxicity class I and must bear the signal word DANGER.

Introduction

4-Aminopyridine, a pyridine compound, is an extremely effective bird poison. It is one of the most prominent avicides. It is registered with the EPA for use against red-winged blackbirds, blackbirds in agricultural fields, grackles, pigeons and sparrows around public buildings, and various birds around livestock feeding pens. Avitrol repels birds by poisoning a few members of a flock, causing them to become hyperactive. Their distress calls signal other birds to leave the site. Only a small number of birds need to be affected to cause alarm in the rest of the flock. After one alarming exposure, birds will usually not return to treated areas. Avitrol is available as grain baits or as a powder concentrate.

Toxicological effects

Acute toxicity

4-Aminopyridine is highly toxic to mammals. The central nervous system is strongly excited by 4-aminopyridine. Based on observations with 2-aminopyridine, a similar compound, individuals with a history of convulsive disorders may be at increased risk from exposure to 4-aminopyridine (30,31). The principal action of 4-aminopyridine in the body is to encourage message-carrying (transmitter) substances to be released throughout the nervous system, overstimulating it (68). While intended strictly for use as a bird repellent, accidental ingestion of as little as 60 mg has caused severe poisoning in adult humans (68). It is rapidly absorbed from the gastrointestinal tract (8). Poisonings are characterized by thirst, nausea, dizziness, weakness, and intense sweating, followed by impairment of normal mental functioning (toxic psychosis), lack of muscular coordination, tremors, labored breathing, and generalized seizures (167). Symptoms of Avitrol poisoning in rats, dogs, and horses include overproduction of saliva, tendency to become overstimulated, and trembling, which can progress to convulsions. Death can result from respiratory arrest or heart failure (23,30).

Skin exposure to Avitrol may lead to systemic intoxication or general overall poisoning (30).

Avitrol may contribute to the excessive formation of a substance called methemoglobin. Methemoglobin is similar to hemoglobin, the oxygen-carrying part of the blood, except that it cannot carry oxygen. When there is excess methemoglobin in the blood, oxygen cannot be transported and blood eventually becomes oxygen depleted, resulting in the condition methemoglobinemia. The LD_{50} for 4-aminopyridine is 20 to 29 mg/kg in rats and 3.7 mg/kg in dogs (8,167). It is readily absorbed through the skin (23). The dermal LD_{50} is 326 mg/kg in rabbits (23,30).

4-Aminopyridine is an eye irritant. Inflammation of the iris and conjunctivitis were noted in the eyes of albino rabbits 1 hour after 10 mg 4-aminopyridine hydrochloride were applied. These symptoms disappeared after 7 days (167).

Chronic toxicity

High dietary doses (2 to 3.25 mg/kg/day) caused increased brain weight. Brain appearance remained normal (167). However, since dietary intake is assumed to be negligible, and because significant repeated exposure is not expected to occur, the EPA has not required long-term toxicity studies of 4-aminopyridine (167).

Reproductive effects
No data are currently available.

Teratogenic effects
No data are currently available.

Mutagenic effects
No data are currently available.

Carcinogenic effects
No data are currently available.

Organ toxicity
Chronic exposure to 4-aminopyridine can cause the breakdown of proper liver and brain functioning (8). No effects were found in the blood and urine of rats and dogs.

Fate in humans and animals
4-Aminopyridine is rapidly absorbed into the bloodstream from the gastrointestinal tract (8). It is readily broken down, or metabolized, in the liver into removable compounds excreted in urine (30). After intravenous and oral doses were given to humans, 90.6 and 88.5%, respectively, were excreted in the urine (167). It does not tend to concentrate or accumulate in skin. Birds killed with Avitrol are not poisonous to predators (8,30).

Ecological effects

Effects on birds

4-Aminopyridine is highly toxic to birds. The 8-day dietary LC_{50} is 447 ppm in Japanese quail, 316 ppm in mourning doves, and 722 ppm in mallard ducks (167). Avian reproduction studies suggest ingestion of sublethal amounts of 4-aminopyridine is unlikely to cause negative effects on birds' reproductive systems (8).

There is a large potential for exposure of non-target, particularly grain-feeding birds. Migratory birds, finches, and other small seed-feeding birds may ingest lethal doses that are applied to corn and sunflower fields.

Effects on aquatic organisms

4-Aminopyridine is moderately toxic to warmwater fish. Fish become increasingly sensitive with increased exposure (167). The LC_{50} ranges from 4 (in soft water) to 2.43 mg/L (in hard water) in channel catfish. The LC_{50} is 3.40 mg/L (in soft water) and 3.20 mg/L (hard water) in bluegill (37).

Effects on other organisms (non-target species)

Endangered species may be adversely affected by 4-aminopyridine (167). There is low or nonexistent potential for secondary poisoning in animals such as cats, dogs, or birds of prey that may feed upon birds killed by Avitrol (167).

Environmental fate

Breakdown in soil and groundwater

4-Aminopyridine is readily adsorbed to soil particles and is highly persistent (167). It is broken down slowly by soil microorganisms. It is more likely to remain near the soil surface where most microbial degradation tends to occur (167).

The half-life of 4-aminopyridine in soil with oxygen ranges from 3 months in clay soil to 32 months in sandy loam soils. The rate at which 4-aminopyridine is metabolized in aerobic soil increases with greater amounts of organic matter (167).

Studies indicate that 4-aminopyridine is relatively immobile in soils. It is not expected to be present in groundwater as a result of its use on land (167).

Breakdown in water

4-Aminopyridine is not expected to be present in surface water as a result of land application of formulated products (167).

Breakdown in vegetation

Available plant metabolism data on sorghum indicate that some breakdown of 4-aminopyridine does occur, with three breakdown products; however, no metabolites were found in corn. 4-Aminopyridine is absorbed and moved from one part of a plant to another to varying degrees, depending on the manner in which it is applied. Plant uptake of 4-AP is not expected to be significant in corn and sunflowers (167).

Physical properties

Technical 4-aminopyridine is a white crystalline solid that contains about 98% active ingredient (8).

Chemical name: 4-aminopyridine (31)
CAS #: 504-24-5
Molecular weight: 94.13 (31)
Solubility in water: Soluble (31)
Solubility in other solvents: s.s. in benzene and ether (31)
Melting point: 158°C (31)
Partition coefficient (octanol/water): Not available
Adsorption coefficent: Not available

Exposure guidelines

ADI: Not available
HA: Not available
RfD: 0.00002 mg/kg/day (13)
PEL: Not available

Basic manufacturer

Avitrol Corporation
7644 East 46th St.
Tulsa, OK 74145
Telephone: 918-622-7763

10.5.2 Daminozide

Trade or other names

Trade names for products containing daminozide include Alar, Aminozide, B-995, B-Nine, Dazide, Kylar, and SADH.

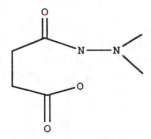

Figure 10.41 Daminozide.

Regulatory status

All use of daminozide on food crops was voluntarily canceled by the manufacturer in November 1989. It is currently registered only for use on ornamental and bedding plants. Classified toxicity III — slightly toxic, products containing daminozide bear the Signal Word CAUTION.

Introduction

Daminozide is a plant growth regulator formerly used on certain fruit crops to improve the balance between vegetative growth and fruit production, to improve fruit quality, and to synchronize fruit maturity.

Unsymmetrical dimethyl hydrazine (UDMH) is a contaminant of commercial daminozide and a metabolite of daminozide that is formed in the body, during food processing, or when spray mixes containing daminozide are left standing in the mixing tank. Commercial daminozide contains 0.005% (50 mg/L) UDMH. It is available as a water-soluble powder.

Toxicological effects

Acute toxicity

Daminozide is practically nontoxic to mammals (168). It may cause skin or eye irritation (168). Changes in liver function have been observed in animals given very high single doses (168).

The oral LD_{50} for daminozide is 8400 mg/kg in rats, and 6300/mg/kg in mice. Its dermal LD_{50} in rabbits is greater than 1600 mg/kg. The inhalation LC_{50} in rabbits is greater than 0.15 mg/L (168).

Chronic toxicity

Effects observed at higher doses included atrophy of ovaries and enlargement of the liver bile duct (hyperplasia) (169). No effects were seen in dogs fed 7.5, 75, or 187.5 mg/kg/day daminozide for 1 year (169).

Reproductive effects

Daminozide does not appear to affect reproduction. A three-generation study with rats fed 300 mg/kg/day showed no significant effects on fertility or reproductive capacity (8).

Teratogenic effects

Daminozide may be teratogenic, but only at very high doses. No birth defects occurred in the offspring of pregnant rats fed 500 mg/kg/day (8). When pregnant rats were given 1800 mg/kg/day, ossification of the bones of the sternum and spine occurred in offspring. No teratogenic or developmental effects occurred in the offspring of pregnant rabbits given 50, 150, or 300 mg/kg/day (169).

Mutagenic effects

Several *in vivo* and *in vitro* tests have shown that daminozide is not mutagenic (170).

Carcinogenic effects

Both daminozide and UDMH, a metabolite, have caused increases in the incidence of benign and malignant tumors in test animals at high doses (171). Malignant tumors were found in female rats given dietary doses of 714 and 1428 mg/kg/day (8). Malignant and benign blood vessel tumors also occurred in UDMH treated mice (168).

When rats were given UDMH in their drinking water at concentrations of 50 or 100 mg/L for 2 years, there was a significant, but slight dose-related increase in liver tumors in females, and bile duct hyperplasia and inflammation of the liver in males receiving 100 mg/L and in females receiving 50 and 100 mg/L (168).

Mice given UDMH in their drinking water for 2 years at 1.3 mg/kg/day for males and 2.6 mg/kg/day for females showed a significant increase in the incidence of lung tumors (170,171). The data suggest that daminozide is weakly or noncarcinogenic.

Organ toxicity

Some animal studies have identified the liver and lungs as target organs.

Fate in humans and animals

At 96 hours after administration of a single oral dose of 5 mg/kg to miniature swine, daminozide was detected in all body tissues at levels up to 0.073 mg/kg. The highest levels were found in the liver and kidney. About 84% of the dose was eliminated in the urine, and 1% of the dose was metabolized to UDMH (170). The majority of daminozide residues ingested by milk animals is rapidly excreted in the urine and feces (168).

Ecological effects

Effects on birds

Daminozide is practically nontoxic to birds. Its 96-hour LC_{50} in quail is 5620 mg/kg (168). The 8-day dietary LC_{50} in mallard ducks and bobwhite quail is 10,000 mg/kg (168).

Effects on aquatic organisms

Daminozide is practically nontoxic to fish. Its 96-hour LC_{50} in rainbow trout is 149 to 306 mg/L, and in bluegill is 423 to 552 mg/L (168). The 48-hour LC_{50} in *Daphnia*, a small freshwater crustacean, is 98.5 mg/L (168).

Daminozide does not bioconcentrate in fish. The concentration of this compound in the bluegill was only three times the ambient water concentration (168).

Effects on other organisms (non-target species)

Daminozide is of low toxicity to terrestrial wildlife; it is nontoxic to bees (168).

Environmental fate

Breakdown in soil and groundwater

Daminozide resists photodegradation, but it is subject to degradation by soil microorganisms (168). It has a low soil persistence. The half-life of daminozide in soil is approximately 21 days (53). In field tests, 50% of applied Alar disappeared within 1 week.

Daminozide is soluble in water and very mobile in soils (53). It appears to leach, but because it does not persist in soil; it is unlikely to contaminate groundwater (168).

Breakdown in water

In water, daminozide degrades to UDMH (168).

Breakdown in vegetation

Daminozide is rapidly absorbed through the leaves, roots, and stems of plants. It is translocated within plants and can accumulate in roots, fruits, and other plant parts (168).

Physical properties

Daminozide is a colorless to white, stable crystalline solid (1).

Chemical name: N-dimethylaminosuccinamic acid (1)
CAS #: 1596-84-5
Molecular weight: 160.2 (1)
Solubility in water: 100,000 mg/L at 25°C (1)
Solubility in other solvents: v.s. in methanol and acetone; i.s. in aliphatic hydro-
    carbons
Melting point: 157–164°C (1)
Vapor pressure: 22.7 mPa at 23°C
Partition coefficient (octanol/water): 0.031 at 21°C (1)
Adsorption coefficient: 30 (estimated) (53)

Exposure guidelines

ADI: 0.5 mg/kg/day (12)
MCL: Not available
RfD: 0.15 mg/kg/day (13)
PEL: Not available

Basic manufacturer

Uniroyal Chemical Co., Inc.
Benson Road
Middlebury, CT 06749
Telephone: 203-573-2000
Emergency: 203-723-3670

10.5.3 Diphacinone

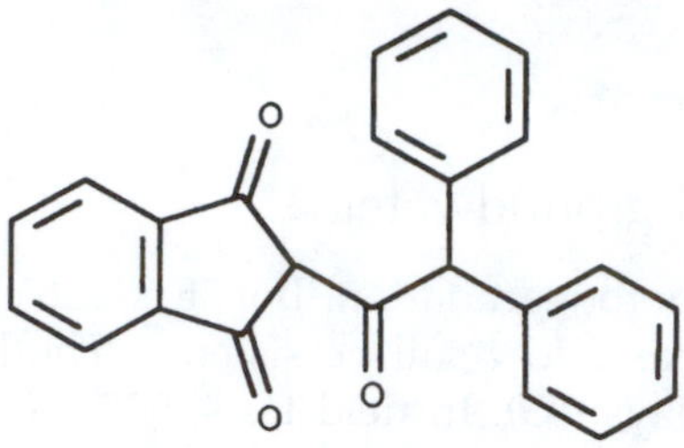

Figure 10.42 Diphacinone.

Trade or other names

Common names for diphacinone include diphacin (in Italy and Turkey), ratindan (in the former U.S.S.R.), dipazin, diphenadione, and diphenacin. Trade names include Diphacine, Ditrac, Gold Crest, Kill-Ko, P.C.Q., Promar, Ramik, Rat Killer, Rodent Cake, and Tomcat.

Regulatory status

Diphacinone is an highly toxic compound in EPA toxicity class I. All formulations of diphacinone are Restricted Use Pesticides (RUPs). RUPs may be purchased and used only by certified applicators. The Signal Word required on products containing diphacinone varies, depending on the type of formulation: DANGER applies to the technical material, WARNING applies to concentrate formulations, and CAUTION applies to bait formulations.

Introduction

Diphacinone is a rodenticide bait used for control of rats, mice, voles, and other rodents. It is available in meal, pellet, wax block, and liquid bait formulations, as well as in tracking powder and concentrate formulations. It may also be used as a anticonvulsant drug under the name of diphenadione. It has also been used success-fully in controlling vampire bats, a vector for rabies.

Toxicological effects

Acute toxicity

Diphacinone is highly toxic by ingestion, with oral LD_{50} values of 0.3 to 7 mg/kg in rats, 3.0 to 7.5 mg/kg in dogs, 14.7 mg/kg in cats, 150 mg/kg in pigs, 50 to 300 mg/kg in mice, and 35 mg/kg in rabbits (1,172). It is highly toxic by skin exposure, with reported dermal LD_{50} values of less than 200 mg/kg in rats, 340 mg/kg in mice, and greater than 3.6 mg/kg in rabbits (1,172). It is not reported to be a skin or eye irritant in rabbits, nor a skin sensitizer in guinea pigs (1,172). The 4-hour inhalation LC_{50} of diphacinone in rats of less than 2 mg/L also indicates high toxicity (1,173). Diphacinone works by inhibition of liver-synthesized coagulation proteins, leading to internal hemorrhaging (172). If the dose is sufficient, this may result in death (172). Effects exhibited in test animals during acute LD_{50} testing included labored breathing, muscular weakness, excitability, congested blood flow to the lungs, and irregular heartbeats. Other signs of poisoning include spitting of blood, bloody urine or stools, and widespread bruising or bleeding into the joints (172).

Use in humans as an anticoagulant drug has apparently been discontinued, possibly due to its structural similarity to another compound (phenindone). Phenindone was noted to pose risks of hepatitis with jaundice, damage to kidneys, severe skin irritation, and massive tissue swelling (172). The use history of diphacinone indicates that it produced no adverse health effects except occasional nausea and some hemorrhagic effects that persisted for 6 to 10 days (172). In other case reports, the anticoagulant effects are reported to persist for several weeks to months (8).

Chronic toxicity

No permanent or life-threatening effects occurred in humans on recommended dose regimes of an initial 20-mg dose (ca. 0.29 mg/kg in a 70-kg human), followed by successive 2- to 4-mg daily doses (ca. 0.03 to 0.06 mg/kg/day in a 70-kg person) for several days to weeks (172). All test animals exposed at dietary levels of 0.1 and 0.2 mg/kg/day in a 21-day study showed fatal, massive internal hemorrhaging, although at doses of 0.05 mg/kg/day, they were unaffected (172). In a 90-day study in which rats were given dietary doses of 0.002 to 0.025 mg/kg/day, single rats in each of the 0.003- and 0.013-mg/kg/day dose groups died from internal hemorrhage, but the others remained unaffected by treatment (172).

The effects due to chronic exposure are similar to those expected from acute exposure, but animal studies and human use experience suggest that there is a level of chronic exposure at which no adverse health effects may occur. The available studies also suggest that some individuals in any population may be much more succeptible to effects than others. These individuals may include those with poor nutritional status and/or Vitamin K deficiency, liver or kidney disorders, or infectious diseases (8).

Reproductive effects
No data are currently available (8).

Teratogenic effects
No data are currently available (8).

Mutagenic effects
Diphacinone was not mutagenic in the Ames test (1,172). No other data regarding mutagenic effects are currently available (1,174).

Carcinogenic effects
No data are currently available.

Organ toxicity
The principal target of diphacinone is the blood (specifically the clotting factors), but effects on the liver, kidneys, heart, and musculature have been seen, probably as secondary effects.

Fate in humans and animals
Rats eliminated 70% of the administered oral dose via the feces and 10% in the urine within 8 days (172). A similar pattern of elimination occurred in mice (172). Animal studies indicate that little metabolism takes place, and that diphacinone that

is not eliminated may concentrate to varying degrees in the liver, kidneys, and lungs (8,172). The half-life of diphacinone in humans is 15 to 20 days (8). It was determined that cattle dosed with the compound as an anti-bat measure were safe to use for dairy and/or meat production (8).

Ecological effects

Effects on birds

Diphacinone is slightly toxic to birds. The oral LD_{50} for diphacinone in mallard ducks is 3158 mg/kg (1,8), and in bobwhite quail is 1630 mg/kg (8).

Effects on aquatic organisms

Diphacinone is moderately toxic to fish species. The 96-hour LC_{50} for technical diphacinone in channel catfish is 2.1 mg/L, in bluegills is 7.6 mg/L, and in rainbow trout is 2.8 mg/L (1,174). The 48-hour LC_{50} in *Daphnia*, a small freshwater crustacean, is 1.8 mg/L (174).

Effects on other organisms (non-target species)

Studies with cattle indicate a high degree of tolerance for the compound, hence its use against vampire bats preying on cattle in Latin America (8).

Environmental fate

Breakdown in soil and groundwater

Diphacinone has a low potential to leach in soil (173).

Breakdown in water

Diphacinone is rapidly decomposed in water by sunlight (172).

Breakdown in vegetation

No data are currently available.

Physical properties

Technical diphacinone is an odorless, pale yellow powder (1).

Chemical name: 2-(diphenylacetyl)indan-1,3-dione (1)
CAS #: 82-66-6
Molecular weight: 340.38 (1)
Water solubility: 0.3 mg/L; almost
Solubility in other solvents: v.s. in acetone, acetic acid, toluene, xylene, and chloroform (1)
Melting point: 145–147°C (1)
Vapor pressure: 13.7 nPa @ 25°C (technical) (1)
Partition coefficient: Not available
Adsorbtion coefficient: Not available

Exposure guidelines

RfD: Not available
ADI: Not available
MCL: Not available
PEL: Not available

Basic manufacturer

Hacco Inc.
P.O. Box 7190
537 Atlas Ave.
Madison, WI 53707
Telephone: 608-221-6200
Emergency: 800-642-4699

10.5.4 Ethylene dibromide (dibromoethane)

Figure 10.43 Ethylene dibromide.

Trade or other names

The chemical name for EDB is 1,2-dibromoethane, and synonyms include DBE; alpha,beta-dibromoethane; dibromoethane; ethylene bromide; glycol bromide; glycol dibromide; and *sym*-dibromoethane. Product names include Bromofume, Celmide, Dibrome, Dowfume, EDB-85, Fumo-Gas, Kopfume, Nephis, and Soilfume.

Regulatory status

Ethylene dibromide (EDB) is a highly toxic compound in EPA toxicity class I. It is not registered for use in the U.S. In 1983, the EPA suspended the use of EDB as a fumigant when low-level residues were found in groundwater and some grains.

Introduction

Ethylene dibromide (EDB) is used extensively as a soil and post-harvest fumigant for crops, and as a quarantine fumigant for citrus and tropical fruits and vegetables. It also may be used as a gas in termite and Japanese beetle control, beehive and vault fumigation, and spot fumigation of milling machinery.

Toxicological effects

Acute toxicity

EDB is highly toxic via ingestion, with reported oral LD_{50} values of 108 to 146 mg/kg in rats, 250 mg/kg in mice, and 55 mg/kg in rabbits (8). The lowest fatal oral dose for humans was reportedly 90 mg/kg (8). The dermal LD_{50} was unavailable, but EDB can be rapidly absorbed through the skin (8). EDB is reportedly a severe skin and eye irritant, and can cause skin blistering with contact (8,175). The reported 2-hour inhalation LC_{50} is 3 mg/L in rats; the reported 1-hour inhalation LC_{50} is 5 mg/L, indicating high toxicity.

A woman who ingested 4.5 mL of an EDB concentrate experienced rapid breathing, vomiting, and diarrhea prior to expiring (8). Liver and kidney damage were noted in her postmortem examination (8). Administration of very high single oral doses of EDB to rats and chicks caused changes in the livers of these animals within 22 hours (176). In rats, inhalation of small concentrations of EDB vapors depresses weight gain, and slightly higher levels damage the lungs. Changes in the lungs and respiratory tract and temporary clouding of the corneas have been seen in animal studies after only 1 week of exposure to workplace levels of EDB (8). Other effects of acute exposure may include headache, depression, and nausea or loss of appetite. Four deaths have been attributed to accidental poisoning by EDB (8).

Chronic toxicity

Daily inhalation of EDB vapors for 6 to 13 weeks at levels comparable to human occupational exposures damaged the liver, kidney, and testes of rats, and changed chick livers (8).

Reproductive effects

A study of male workers exposed to low air levels of EDB at four separate manufacturing sites showed no consistent association between EDB and reproductive effects (177). However, another study of 46 men exposed to a lower average air concentration for 5 years showed adverse effects on sperm number, movement, survival, and structure (8). Bulls had abnormal sperm after 12 to 21 days of EDB exposure (8). Fetal deaths increased in offspring of adult male rats given high doses of EDB in the diet for 5 days before mating (8). Cows and ewes given lower doses did not show any effects on reproduction (8).

The available data indicate that reproductive effects may be possible in humans, but that there may be a (threshold) level below which effects are very unlikely.

Teratogenic effects

In one study, no teratogenic effects were observed in offspring of pregnant rats and mice exposed to various levels of EDB vapors for 23 hours a day during the sensitive period of gestation (178). At all doses, both rats and mice showed body weight decreases, and a significant number of deaths occurred in the adults exposed to the high doses. Since these effects occurred at doses that caused serious adverse maternal effects, it was unclear whether the observed fetal toxicity was caused simply by reduced maternal health (178). Hens fed daily diets of 1 to 2 ppm EDB had decreased egg weights (8).

The available evidence suggest that EDB does not cause teratogenic effects.

Mutagenic effects

EDB has been shown to be weakly mutagenic in several bacterial and animal test systems (8). It has not been shown to be mutagenic in humans, either in cell cultures or in the whole organism (8).

Carcinogenic effects

Rats inhaling high daily doses of EDB for 18 months developed tumors of the mammary glands, spleen, adrenals, liver, and kidney (179). When rats were subjected to airborne EDB and also given disulfiram orally over 14 months, increased tumors in the liver, kidneys, thyroids, and lungs appeared. The tumor increase was higher than that caused by EDB at the same levels over 18 months (179).

The available evidence indicates that EDB may be carcinogenic.

Organ toxicity

EDB contact may damage the lung, skin, and eyes. Acute and chronic systemic effects may be seen in the liver, kidneys, and heart, and other internal organs and systems. Lung injury can also lead to secondary effects such as pneumonia and respiratory tract infections.

Fate in humans and animals

EDB is readily and rapidly absorbed across the lung, skin, and gastrointestinal tract in liquid and vapor forms (8). Once absorbed, it is rapidly broken down, and metabolites may be found in the urine, kidneys, liver, adrenal glands, pancreas, and spleen of EDB-exposed animals shortly after application (8). About two thirds of the applied dose are excreted through urine or expired air (178). EDB has a biological half-life of less than 48 hours in rats, chicks, mice, and guinea pigs (8).

Ecological effects

Effects in birds

EDB is most likely to be in vapor form, and unless birds are in the fumigation area, during the fumigation, they are unlikely to be exposed.

Effects on aquatic organisms

EDB is slightly toxic to fish (8). The reported LC_{50} value in the shiner is 40.3 mg/L. The calculated bioconcentration factor for EDB in aquatic organisms ranges between 9.3 and 10.2, suggesting that it has a low potential for bioaccumulation (180).

Effects on other organisms (non-target species)

No data are currently available.

Environmental fate

Breakdown in soil and groundwater

The fastest degradation of EDB occurs at or near the soil surface. EDB is moderately persistent in the soil environment; a representative field half-life was estimated to be 100 days (11). In one study, after 2 months, almost all of EDB was dissipated (11,181). Sunlight readily degrades EDB. EDB near the soil surface is converted to ethylene and bromide ions, and that small percentage which remains is unchanged, possibly adsorbed to soil organic matter or clay particles or entrapped in soil micropores (181). EDB thus entrapped may be inaccessible to microbial degraders and may slowly leach to groundwater over very long periods (181). Such leaching is very slow at normal temperatures, but increases with higher temperatures (181).

Groundwater contamination of EDB has been confirmed at levels up to 0.3 mg/L (182,183). Usual levels of EDB found in groundwater were approximately 0.001 to 0.02 mg/L, similar to levels found in stored grain products (182).

Breakdown in water

EDB can be widely distributed in aqueous systems. The primary removal process for ethylene dibromide in water is evaporation (182). EDB has a half-life of slightly over 1 day in river water and about 5 days in lake water (182,184). Binding to sediment is relatively low. EDB decomposes in the presence of heat and/or light and can be slowly broken down by moisture. The major degradation products in water are also ethylene and bromide ions.

Breakdown in vegetation

Because of the inability of plants to take EDB up from soil, EDB is not likely to accumulate in plants. However, EDB's breakdown product, inorganic bromide, is taken up by plants in small amounts. Residues of EDB persist in fumigated food products for 6 to 12 weeks. EDB residues in grain and feed do not accumulate in livestock to any significant extent. Cooking may dramatically reduce the levels of EDB residues in foods (8).

Physical properties

Ethylene dibromide is a heavy, colorless liquid with a mildly sweet, chloroform odor (1).

Physical properties

Chemical name: 1,2-dibromoethane (53)
CAS #: 109-93-4
Molecular weight: 187.9 (53)
Solubility in water: 4300 mg/L at 20°C (53)
Solubility in solvents: soluble in alcohols, ethers, acetone, benzene, and most organic solvents (8)
Melting point: 9.8°C (8)
Vapor pressure: 1.5×10^6 mPa (53)
Partition coefficient (octanol/water): 53.7–61.7 (calculated) (8)
Adsorption coefficient: 34 (estimated) (53)

Exposure guidelines

ADI: 1.0 mg inorganic bromide mg/kg/day (12)
MCL: 0.00005 mg/L (65)
RfD: Not available
PEL: 154 mg/m^3 (8-hour) (14)

Basic manufacturer

United Phosphorous Ltd.
Readymoney Terrace
167 Dr. Annie Besant Rd.
Bombay 400 018 India
Telephone: 22-493-0681
Emergency: 22-493-2427

10.5.5 Metaldehyde

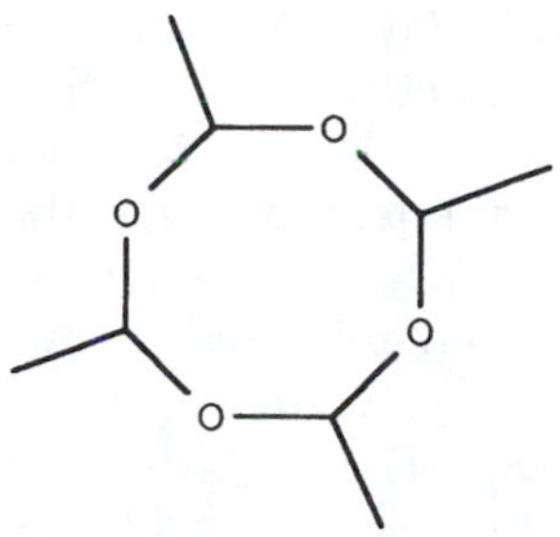

Figure 10.44 Metaldehyde.

Trade or other names

Some trade names for products containing metaldehyde include Antimilace, Antimitace, Ariotox, Cekumeta, Deadline, Halizan, Limatox, Meta, Metason, Namekil, Ortho Metaldehyde 4% Bait, Slug Death, Slug Pellets, Slug-Tox, and Slugit Pellets.

Regulatory status

Metaldehyde is a slightly toxic compound, and products containing it will be in EPA toxicity class II or III (1,2). All product labels must include the following statement on the front panel: "This pesticide may be fatal to dogs or other pets if eaten. Keep pets out of treated areas"; as well, the labels must include the signal word CAUTION or WARNING. Because of its potential short-term and long-term health effects on wildlife, metaldehyde is a Restricted Use Pesticide (RUP). RUPs may be purchased and used only by certified applicators.

Introduction

Metaldehyde is a molluscicide used in a variety of vegetable and ornamental crops in the field or greenhouse, on fruit trees, small-fruit plants, or in avocado or citrus orchards, berry plants, and banana plants. It is used to attract and kill slugs and snails. It is applied in the form of granules, sprays, dusts, pellets, or grain bait, typically to the ground around the plants or crops. It works primarily in the stomach by producing toxic effects after it is ingested by the pest. It may be formulated with or without calcium arsenate and is also available in a mixed formulation with thiram.

Toxicological effects

Acute toxicity

Metaldehyde is slightly to moderately toxic by ingestion, with reported oral LD_{50} values of 227 to 690 mg/kg in rats, 207 mg/kg in cats, 100 to 1000 mg/kg in dogs, 200 mg/kg in mice, 175 to 700 mg/kg in guinea pigs, and 290 to 1250 mg/kg in rabbits (8,185). A child died after ingesting 3000 mg (approximately 75 to 100 mg/kg for a 30- to 40-kg child) metaldehyde (186). Via the dermal route, it is also moderately toxic. The dermal LD_{50} for this molluscicide in rats is from 2275 to greater than 5000 mg/kg (8). Metaldehyde is moderately toxic by inhalation; the 4-hour inhalation LC_{50} in rats is 0.2 mg/L, and the 2-hour inhalation LC_{50} in mice is 0.35 mg/L (8).

Irritation of the skin, eye, and mucous membranes of the upper airways and gastrointestinal tract may result from contact with metaldehyde (8).

Within a few hours of accidental or intentional ingestion, the following symptoms appeared in humans: severe abdominal pain, nausea, vomiting, diarrhea, fever, convulsions, coma, and persistent memory loss. Other symptoms of high acute exposure include increased heart rate, panting, asthma attack, depression, drowsiness, high blood pressure, inability to control the release of urine and feces, incoordination, muscle tremors, sweating, excessive salivation, tearing, cyanosis, acidosis, stupor, and unconsciousness and eventual death in extreme cases (185). Kidney injury and liver cell death ("necrosis") may also occur (30,31). Mental deficiencies and memory loss from ingestion poisoning may persist for 1 year or more (185). It is thought that the formation of acetaldehyde in the gastrointestinal tract is responsible for the narcotic effects observed with metaldehyde exposure (185).

Chronic toxicity

Dosages that are not toxic when given singly do not cause illness when repeated (185). Long-term, repeated skin exposure to metaldehyde may result in dermatitis (skin inflammation) in humans (185). Prolonged eye exposure can cause conjunctivitis (8). In 2-year toxicity studies and three-generation reproductive studies in rats, changes in liver enzyme activity and increased liver and ovary weight at dietary doses of about 12.5 mg/kg/day were found (185); 50% of female rats given this dose showed paralysis (8,185). Effects on the brain (e.g., impairment of memory) may also be possible with chronic exposure at very high levels.

Reproductive effects

During a three-generation study of rats exposed to chronic ingestion of metaldehyde, adverse effects were seen on reproduction and on the survival rate of offspring (31). Doses of 50 and 250 mg/kg/day interfered with the reproduction of female rats in another three-generation test (1,31). These data suggest that metaldehyde is likely to cause reproductive effects only at high levels.

Teratogenic effects

Dietary doses of 10, 50, and 250 mg/kg metaldehyde were not teratogenic in three generations of experimental female rats (1,31). There were some increases in relative liver weights in some offspring (185). This evidence suggests that metaldehyde is unlikely to cause teratogenic effects.

Mutagenic effects

Metaldehyde has been reported to be a suspected mutagen (70). However, there was no evidence of mutagenicity when metaldehyde was tested on five strains of bacteria (185). The evidence regarding mutagenicity of metaldehyde is inconclusive.

Carcinogenic effects

Dietary doses as high as 250 mg/kg/day over a 2-year period did not increase the incidence of tumors in male and female rats (185). The study suggests that metaldehyde is not carcinogenic.

Organ toxicity

Metaldehyde or its breakdown by-products ("metabolites") may cause problems in the central nervous system by an unknown mechanism (8,185). It may also cause lesions in kidneys and the liver following systemic distribution, as well as inflammation of the skin, eye, and or mucous membranes of the airways and gastrointestinal tract with direct contact (8,185).

Fate in humans and animals

Metaldehyde is readily absorbed into the bloodstream from the gastrointestinal tract (31,185). Metaldehyde's primary decomposition product in the body is acetaldehyde (1,185). Its metabolites can cross the blood/brain barrier, as evidenced by their effect on the level of consciousness of animals (8).

Ecological effects

Effects on birds

Death of birds feeding in metaldehyde-treated areas has been reported, although the precise acute oral LD_{50} values or subchronic dietary LC_{50} values were unavailable (1). Excitability, tremors, muscle spasms, diarrhea, and/or difficult or rapid breathing was observed in poultry exposed to metaldehyde (31).

Effects on aquatic organisms

Metaldehyde is reported to be practically nontoxic to aquatic organisms (1,187).

Effects on other organisms (non-target species)

The 4% pelleted bait is reported to be toxic to wildlife (8). When used as directed, bait agents with 6% active ingredient are not toxic to bees (30). Bait pellets containing metaldehyde are attractive to dogs (11). Pets should be confined during application, and kept away from application and storage sites (187).

Environmental fate

Breakdown in soil and groundwater

Metaldehyde is of low persistence in the soil environment, with a half-life on the order of several days (8). It is weakly sorbed by soil organic matter and clay particles, and is soluble in water (1,8). Due to its low persistence, it is not a significant risk to groundwater.

Breakdown in water

Metaldehyde undergoes rapid hydrolysis to acetaldehyde, and should be of low persistence in the aquatic environment (1).

Breakdown in vegetation

Many types of flowers lose their color when they come in contact with metaldehyde dust or spray (30).

Physical properties

Metaldehyde is a white or colorless crystalline solid, with a mild characteristic odor and a powdery appearance (1).

Chemical name: r-2,c-4,c-6,c-8-tetramethyl-1,3,5,7-tetroxocane (1)
CAS #: 108-62-3
Molecular weight: 176.2 (tetramer) (1)
Water solubility: 260 mg/L at 30°C (1)
Solubility in other solvents: i.s. in acetic acid; s. in hot carbon disulfide (1); s.
in ethyl alcohol; v.s. in benzene and chloroform; s.s. in ethanol and diethyl
ether (1)
Melting point: Sublimes at approx. 112°C (1)
Vapor pressure: Negligible at room temperature (1)
Partition coefficient: Not available
Adsorption coefficient: 240 (11)

Exposure guidelines

ADI: Not available
HA: Not available
RfD: Not available
PEL: Not available

Basic manufacturer

Lonza Inc.
17-17 Route 208
Fair Lawn, NJ 07410
Telephone: 800-777-1875

10.5.6 Methyl bromide (bromomethane)

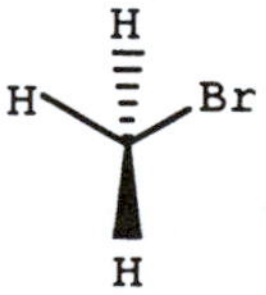

Figure 10.45 Methyl bromide.

Trade or other names

Trade or common names of methyl bromide-containing products include Brom-
o-Gas, Bromomethane, Celfume, Embafume, Haltox, MB, MeBr, Methogas, Profume,
Terr-o-Gas, and Zytox.

Regulatory status

Methyl bromide is a highly toxic compound in EPA toxicity class I. Labels for
products containing it must bear the Signal Word DANGER. Methyl bromide is a
Restricted Use Pesticide (RUP). RUPs may be purchased and used only by certified
applicators.

The EPA has expressed concerns and proposed restrictions on methyl bromide
due to concerns over its potential to destroy ozone. Ozone-depleting chemicals fall
within the scope of the Clean Air Act. Unlike FIFRA, the Clean Air Act does not
contain a risk/benefit balancing process that would allow retention of essential or

high benefit uses, nor does the listing and phase-out of ozone depleters depend on the availability of alternative products.

Introduction

Methyl bromide is chiefly used as a gas soil fumigant against insects, termites, rodents, weeds, nematodes, and soil-borne diseases. It has been used to fumigate agricultural commodities, grain elevators, mills, ships, clothes, furniture, and greenhouses. About 70% of the methyl bromide produced in the U.S. goes into pesticidal formulations.

Toxicological effects

Acute toxicity

Since bromomethane is a gas at ambient temperatures, the most significant route of exposure is inhalation (188). The reported 1-hour inhalation LC_{50} in rats is 4.5 mg/L, and the 11-hour LC_{50} in rabbits is 8 mg/L (8). Inhalation of 6 mg/L for 10 to 20 hours, or 30 mg/L for 1.5 hours, is lethal to humans (8). The compound is readily absorbed through the lung alveoli (gas exchange regions). Methyl bromide can be highly irritating to the mucous membranes of the eyes, airways, and skin with contact (17).

About 1000 human poisoning incidents caused by methyl bromide exposure have been documented, with effects ranging from skin and eye irritation to death (17). Most fatalities and injuries occurred when methyl bromide was used as a fumigant. The lowest inhalation level found to cause toxicity in humans is 0.14 mg/L in air (17). A typical delay in onset of symptoms following exposure, combined with an odor threshold (level at which most people can smell it) well above the level at which toxic effects occur, means that the victim may not realize a harmful exposure is occurring until it is too late (17).

Initial acute effects may include headache, dizziness, nausea or vomiting, chest and abdominal pain, and irritated eyes, nose, and throat (188). With sufficient exposure, symptoms of slurred speech, blurred vision, temporary blindness, mental confusion, and sweating may occur (188). More severe symptoms at even higher doses may include lung swelling; congestion; hemorrhaging of the brain, heart, and spleen; severe kidney damage; and numbness, tremors, and convulsions (188). The nervous effects observed in lab animals included degeneration of key nerve cells in various portions of the brain and peripheral nervous system (188). Death may occur from respiratory failure (188).

The rat oral LD_{50} (bromomethane administered as a liquid or in solution) is 214 mg/kg (1), also indicating moderate to high toxicity.

Chronic toxicity

Chronic exposure to methyl bromide can cause extensive damage to neurons (nerve cells) involved in cognitive processes and physical coordination or muscular control (188). These effects were seen in rats exposed to 0.51 to 1.3 mg/L, 6 hours per day, for 5 days (188). Rats exposed to 65 ppm over 4 weeks for an average of 7 hours per day for 4 to 5 days did not show neurological effects, but this level of exposure did result in severe, in some cases irreversible, neurological effects in rabbits over a similar time period (188). Exposure levels of 0.1 mg/L over 8 months (7.5 hours per day, 4 days/week) did not produce observable neurotoxicity (188). The symptoms of chronic exposure may include dizziness, vision and hearing disturbances, depression, confusion, hallucinations, euphoria, personality changes, and irritability (8). A chronic pneumonia-like syndrome may become apparent after repeated exposure to sufficient levels (8). Other targets of the fumigant identified through long-term animal studies are the heart, adrenal gland, and the testis (189).

Reproductive effects

No reproductive efffects were seen in rats exposed to up to 0.3 mg/L for 7 hours per day, 5 days a week, for 3 weeks prior to mating and during gestation (188). This suggests that methyl bromide does not cause reproductive effects.

Teratogenic effects

No teratogenic efffects were seen in rats exposed to up to 0.3 mg/L for 7 hours per day, 5 days a week, for 3 weeks prior to mating and during gestation (188). This evidence indicates that bromomethane is unlikely to cause teratogenic effects.

Mutagenic effects

Mutagenic effects were seen in mouse cell cultures, mutagenicity assays with bacteria, and in human white blood cells (190). Rat liver cells did not display increased rates of mutation after exposure to methyl bromide (190). Methyl bromide is considered to be weakly mutagenic (188).

Carcinogenic effects

In one study of industrial workers exposed to various brominated compounds, exposure to methyl bromide was suggested as the possible common factor in two fatal cases of testicular cancer, but other exposures could not be ruled out (189). In a rat study, methyl bromide given at 50 mg/kg/day by stomach tube for 90 days (gavage) induced stomach tumor increases (188,190). It appeared that the cancerous growth was due to severe localized cellular injury, with subsequent increased cell reproduction to repair tissue damage amplifying the natural incidence of mutant or abnormal cells (188). This is not likely to occur at low doses. Thus, the data are inconclusive.

Organ toxicity

Acute exposure primarily damages the lung and results in nervous system effects; chronic exposure may cause damage to the central nervous system, kidneys, and lungs. Other targets of the fumigant are the heart, nasal cavities, adrenal gland, and the testis.

Fate in humans and animals

The major route of absorption of methyl bromide vapors is through the lungs (188). Some of the compound is excreted through the lungs as unchanged methyl bromide, but a significant amount also undergoes metabolic decomposition (32). The primary breakdown products are the bromide ion and methanol, which are detectable in the blood and tissues and are excreted in the urine (32). Organic bromides (formed by reaction of bromide ion with molecular carbon centers in biomolecules) also appear in stomach fluids and mucus.

In humans, methyl bromide's half-life in blood is about 12 days (32). As a result, the toxic effects of methyl bromide can be delayed or prolonged (32). Additionally, once in a cell, this chemical inactivates many enzyme systems, so prolonged small doses can cause severe toxicity (32).

Ecological effects

Effects on birds

Bromomethane is most likely to be in vapor form, and unless birds are in the fumigation area during the fumigation, they are unlikely to be exposed.

Effects on aquatic organisms

Methyl bromide is moderately toxic to aquatic organisms. Acute toxicity in freshwater fish (bluegill sunfish) occurs at concentrations of 11 mg/L, and in salt-water fish (tidewater silversides) at about 12 mg/L (8).

Effects on other organisms (non-target species)

It is not toxic to bees (1).

Environmental fate

Breakdown in soil and groundwater

Methyl bromide quickly evaporates at temperatures ordinarily encountered in fumigating, but some may be entrapped in soil micropores following application (11). Methyl bromide is moderately persistent in the soil environment, with a field half-life of between 30 and 60 days; a representative half-life is estimated to be about 55 days (11). Transformation of methyl bromide into bromide ion increases as the amount of organic matter in the soil increases. It is soluble in water and very poorly sorbed by soils. Some leaching may occur if bromomethane is entrapped in soil micropores following fumigation; the rate of degradation for retained bromomethane in fumigated soil is 6 to 14% per day at 20°C (11).

Breakdown in water

Methyl bromide quickly evaporates at temperatures ordinarily encountered in fumigating; therefore runoff from fields into surface waters is very rare. If it does contact surface waters, the average half-life for methyl bromide under field conditions has been calculated to be 6.6 hours at 11°C (8). Another study showed the half-life in water to be 20 days at 25°C in a neutral solution (8).

Breakdown in vegetation

The amount of bromide ion (the metabolite of methyl bromide) taken up from the soil, is proportional to the protein content of the crop. Higher levels of the bromide ion will most likely be found in high-protein plants (8).

Physical properties

Methyl bromide is a colorless gas or volatile liquid, is usually odorless, but has a sweet, chloroform-like odor at high concentrations (1).

Chemical name: bromomethane (1)
CAS #: 74-83-9
Molecular weight: 94.94 (1)

Solubility in water: 13,400 mg/L at 25°C (1)
Solubility in solvents: Easily miscible with ethanol, ether, aromatic carbon disulfide, and ketones (1)
Melting point: –93.6°C (1)
Vapor pressure: 227,000,000 mPa @ 25°C (1)
Partition coefficient (octanol/water): Not available
Adsorption coefficient: 22 (11)

Exposure guidelines

ADI: 1.0 mg/kg/day (as bromide ion) (12)
HA: Not available
RfD: 0.0014 mg/kg/day (13)
PEL: 80 mg/m³ (ceiling) (14)

Basic manufacturer

Great Lakes Chemical Corporation
One Great Lakes Blvd.
P.O. Box 2200
West Lafayette, IN 47906
Telephone: 317-497-6204
Emergency: 501-862-5141

10.5.7 Tributyltin (TBT)

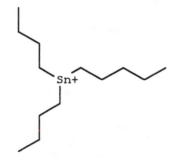

Figure 10.46 Tributyltin.

Trade or other names

Trade names for products containing tributytin include Alumacoat, Bioclean, FloTin, Fungitrol, TinSan, Ultrafresh, and Vikol.

Regulatory status

Some applications are Restricted Use, primarily those involving potential exposures to non-target aquatic organisms, such as may occur with use of TBT-containing marine paints (1). Tributyltin is in EPA toxicity class II — moderately toxic. Depending on the product, labels will be required to display the Signal Words DANGER or WARNING.

These paint uses are now regulated under the Organotin Antifouling Paint Control Act of 1988. This Act sets guidelines on the amount and rate of TBT compounds leaching from marine paints into the aquatic environment.

Introduction

The tributyltin compounds are a subgroup of the trialkyl organotin family of compounds. They are the main active ingredients in biocides used to control a broad spectrum of organisms. Uses include wood treatment and preservation, antifouling of boats (in marine paints), antifungal action in textiles and industrial water systems such as cooling tower and refrigeration water systems, wood pulp and paper mill systems, and breweries. It is also used for control of shistosomiasis in various parts of the world (188).

Tributyltin compounds are present in varying proportions in commercial products; ready-to-use wood preservatives typically contain as little as 0.3% TBT, but some products used only in manufacturing may contain as much as 48% (160).

Unless otherwise specified, all toxicity and environmental chemistry data presented are for tributyltin oxide, a representative compound.

Toxicological effects

Acute toxicity

Acute toxicity of organotin compounds is strongly influenced by the length of the alkyl chains attached to the tin (160,191). Tributyltin (TBT) is generally less toxic than trimethyl- and triethyltins (160). Generally, the toxicity of organotin compounds is influenced more by the alkyl substituents than the ionic substituent that may form the rest of the molecule (e.g., salicylate, acrylate, etc.) (191).

Tributyltin compounds are moderately toxic via both ingestion and dermal absorption. Reported oral LD_{50} values for tributyltin oxide (TBTO) range from 55 to 87 mg/kg in mice and rats (191,192). Dermal LD_{50} values are 200 mg/kg in rats and mice (191,192), and 900 mg/kg in rabbits (31,192).

The tributyltin compounds may be strongly irritating to the skin in humans, especially the hair follicles, and skin exposure may result in chemical burns in only a few minutes if the concentration of tributyltin is high enough (191). Shipyard workers exposed to TBT (occupationally exposed to dusts and vapors) developed irritated skin, dizziness, difficulty breathing, and flu-like symptoms (8). Other mucous membranes such as the eyes and nasal passages may also become irritated upon exposure.

Chronic toxicity

Although the effects of tributyltin on humans are uncertain, there have been cases of human exposure reported. Underwear treated with TBT caused severe skin irritation in wearers, and shipyard workers exposed occupationally reported reduced sense of smell, chronic headaches, and feelings of musculoskeletal stiffness (191). Prolonged exposure to organotin compounds has produced bile duct damage in several mammalian species (191), and TBT may be a potent immunotoxic agent (192). One study of male rats fed TBTO daily for 6 weeks resulted in decreased resistance to infection (193,194).

Reproductive effects

Data regarding reproductive effects are limited. In one study, increased fetal mortality was observed in rats at doses of 16 mg/kg/day (195).

Teratogenic effects

In a teratology study of tributyltin acetate on pregnant Wistar rats, a dose of 16 mg/kg/day administered on days 7 to 17 of pregnancy resulted in increased fetal mortality, incidence of cleft palate, irregularities in rib cage, and decreased fetal birth weights (195). Tributyltin was associated with offspring behavior abnormalities in another study of rats at doses up to 5 mg/kg/day (196). These effects are unlikely in humans at expected exposure levels.

Mutagenic effects

Tributyltin did not appear to be mutagenic in a large battery of mutagenicity assays, but produced DNA nucleotide base pair substitutions (point mutations) in one bacterial strain tested (192). It produced mutations in Chinese hamster ovary cells as well (192). The evidence for mutagenicity is inconclusive.

Carcinogenic effects

While one study has indicated that rats developed pituitary gland tumors after exposure to high doses of TBT, the evidence is not conclusive (8), and the carcinogenic status of TBT is still uncertain (191).

Organ toxicity

In mammals, high levels of TBTO can affect the endocrine glands, upsetting the hormone levels in the pituitary, gonad, and thyroid glands. Large doses of TBT have been shown to damage the reproductive and central nervous systems, bone structure, and the liver bile duct of mammals. TBT compounds can also damage the immune system.

Fate in humans and animals

In mice, TBTO is excreted mainly unchanged via the feces, indicating low absorption by the body. In mammalian species, tributyl compounds may be metabolized to dibutyltin derivatives and related metabolites (192). An undetermined amount of this compound is known to remain in fat, liver, kidney, and lung tissues.

Ecological effects

Effects on birds

No treatment-related mortality was observed in a 13-week study of toxic effects of TBTO in Japanese quail at dietary levels of 150 ppm; at 375 ppm, egg production, eggshell thickness, fertility, and hatchability were reduced (197). Thus, tributyltin can be considered moderately toxic to birds.

Effects on aquatic organisms

Tributyltin compounds are highly to very highly toxic to many species of aquatic organisms. TBT exposure to non-target aquatic organisms such as mussels, clams, and oysters, at low levels, may cause structural changes, growth retardation, and death (198,199).

TBT is very highly toxic to crustaceans. Lobster larvae show a nearly complete cessation of growth at just 1.0 µg/L TBT (200). Mollusks, used as indicators of TBT pollution because of their high sensitivity to these chemicals, react adversely to very low levels of TBT (0.06 to 2.3 µg/L). They release TBT very slowly from their bodies after it has been absorbed.

Imposex, the development of male characteristics in females, has been initiated by TBT exposure in several snail species. In laboratory tests, reproduction was inhibited when female snails exposed to 0.05 µg/L TBT developed male character- istics (200). Imposex was also noted in the mud snail, or dogwhelk, at less than 3 ppt TBT (200).

Oysters in France and England's marine waters are adversely affected by TBT exposure. TBT-exposed oysters have abnormal shell development, poor weight gain, brittle shells, and imposex.

TBT toxicity in the field may be substantially underestimated in laboratory studies (200). TBT binds to the sides of containers and plankton, which contributes to this underestimation of its potential toxicity. Generally, the larvae of any tested species are more sensitive to tributyltin exposure than are the adults.

TBTO has been shown to inhibit cell survival of marine unicellular algae at very low concentrations; the 72-hour EC_{50} ranges from 0.33 to 1.03 µg/L (200).

TBT is lipophilic and tends to accumulate in oysters, mussels, crustaceans, mol- lusks, fish, and algae. Freshwater species will bioaccumulate more TBT than will marine organisms. Oysters bioaccummulate TBT compounds readily, reach an equi- librium uptake soon after exposure, and are slow to release this chemical. Oysters exposed to very low TBTO concentrations bioaccumulated TBT 1000- to 6000-fold.

Juvenile Chinook salmon accumulate TBT immediately upon exposure to low TBT concentrations. TBT and its metabolite, DBT, were found in the salmon's muscle tissue (201).

Effects on other organisms (non-target species)

No data are currently available.

Environmental fate

Breakdown in soil and groundwater

Under aerobic conditions, tributyltin takes 1 to 3 months to degrade (201). But in anaerobic (airless) soils, this compound will persist for more than 2 years. Tribu- tyltin compounds may be moderately to highly persistent.

Degradation depends on temperature and the presence of microorganisms. The breakdown of TBT leads eventually to the tin ion (201). All of the breakdown products are less toxic than TBT itself.

It has not been found in groundwater.

Breakdown in water

Because of the low water solubility of TBT and other properties, it will bind strongly to suspended material such as organic material or inorganic sediments (160) and precipitate to the bottom sediment (201). Rates of sedimentation vary with location, organic content, particle size, and type of material.

Reported half-lives of the compound in freshwater are 6 to 25 days; in seawater and estuarine locations, it is 1 to 34 weeks, depending on the initial concentration

(202). Because of the low levels of UV light beyond the top-most few centimeters, it is unlikely that photolysis plays a major role in the degradation of tributyltin compounds (202).

Levels up to 0.800 μg/L have been found along the East Coast of the United States. In the Great Lakes, concentrations from 0.020 to 0.840 μg/L have been recorded. In San Diego Bay, a concentration of 1.0 μg/L TBT has been found (198).

Breakdown in vegetation

No data are currently available.

Physical properties

The physical property data presented are for tributyltin oxide, which is a slightly yellow, combustible liquid (8).

Chemical name: Bis(tributyltin)oxide (8)
CAS#: 56573-85-4
Molecular weight: 595.62 (8)
Solubility in water: ca. 4 mg/L @ 20°C and pH 7.0 (8)
Solubility in other solvents: Not available
Melting point: 53°C (8)
Vapor pressure: 0.1 mPa @ 20°C (8)
Partition coefficient (octanol/water): 5500 in 32% seawater (203)
Adsorption coefficient: Not available

Exposure guidelines

ADI: Not available
HA: Not available
RfD: 0.00003 mg/kg/day (13)
PEL: 0.1 mg/m³ (as Sn) (8-hour) (14)

Basic manufacturer

Agtrol Chemical Products
7322 Southwest Freeway
Suite 1400
Houston, TX 77074

10.5.8 Zinc phosphide

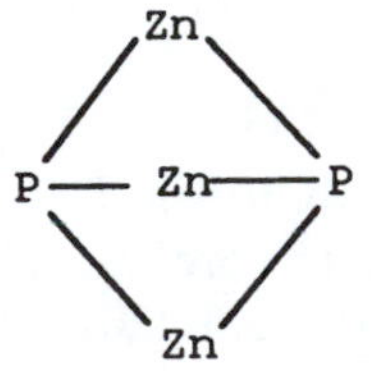

Figure 10.47 Zinc phosphide.

Trade or other names

Trade names for commercial products containing zinc phosphide include Arrex, Commando, Denkarin Grains, Gopha-Rid, Phosvin, Pollux, Ridall, Ratol, Rodenticide AG, Zinc-Tox, and ZP.

Regulatory status

Zinc phosphide is a Restricted Use Pesticide (RUP) because of its hazard to non-target organisms and its acute oral toxicity. RUPs may be purchased and used only by certified applicators. Some formulations of this rodenticide are classified as highly toxic and require the Signal Words DANGER — POISON on the label. Others are either moderately toxic or only slightly toxic, and thus require the Signal Words WARNING or CAUTION, respectively.

Introduction

Zinc phosphide is an inorganic chemical used to control rats, mice, voles, ground squirrels, prairie dogs, nutria, muskrats, feral rabbits, and gophers. It is also used as a tracking powder for the control of house mice. It is applied to crop and non-crop areas, including lawns, golf courses, highway medians, and areas adjacent to wetlands. It may be formulated as a grain-based bait, as scrap bait, or as a paste. Rodenticide baits usually contain 0.5 to 2.07% zinc phosphide, pastes contain approximately 5 to 10%.

Toxicological effects

Acute toxicity

Zinc phosphide ingested orally reacts with water and acid in the stomach and produces phosphine gas, which may account in a large part for the observed toxicity (160). Symptoms of acute zinc phosphide poisoning by ingestion include nausea, abdominal pain, tightness in the chest, excitement, agitation, and chills (160,8). Other symptoms include vomiting, diarrhea, cyanosis, rales, restlessness, and fever (8,160). The inhalation of zinc phosphide or its breadkown product (phosphine gas) may result in vomiting, diarrhea, cyanosis, rapid pulse, fever, and shock (160).

There are documented cases of adults dying from massive oral doses of 4000 to 5000 mg (approximately 55 to 70 mg/kg), although others have survived acute exposure of as high as 25,000 to 100,000 mg (approximately 350 to 1400 mg/kg) if vomiting occurred early and exposure to phosphine was limited (160).

In rats, the LD_{50} for the technical product (80 to 90% pure) is 40 mg/kg, while the LD_{50} values for lower concentration formulations are slightly higher, indicating lower acute toxicity (160). In sheep, the LD_{50} ranges from 60 to 70 mg/kg (160).

The compound is nonirritating to the skin and eyes (160).

Chronic toxicity

Rats fed zinc phosphide over a wide range of doses experienced toxic effects. Increased liver, brain, and kidney weights, and lesions on these organs, were noted in rats exposed to around 14 mg/kg/day. Body hair loss, reduction in body weight, and reduction of food intake were all noted at 3.5 mg/kg/day. The study was conducted over 13 weeks (8).

There have been no observed symptoms of chronic poisoning due to zinc phosphide exposure in humans (1). However, it has been suggested that chronic exposure to sublethal concentrations for extended periods of time may produce toxic symptoms (8).

Reproductive effects
No data are currently available.

Teratogenic effects
No data are currently available.

Mutagenic effects
No data are currently available regarding the mutagenicity of zinc phosphide. However, its metabolite, phosphine, has shown a concentration-dependent increase in chromosomal aberrations in studies using human lymphocyte cultures (8). Thus, its mutagenicity is unclear.

Carcinogenic effects
No data are currently available.

Organ toxicity
Damage to the kidneys, the liver, and the stomach have been noted in humans, but only at high acute doses of the rodenticide. Zinc phosphide reacts with water and stomach juices to release phosphine gas, which can enter the bloodstream and adversely affect the lungs, liver, kidneys, heart, and central nervous system (8).

Fate in humans and animals
Small amounts of the rodenticide fed to experimental animals may have produced an 80% absorption of zinc as well. Zinc in sufficient concentrations may have an emetic effect (8). Hypophosphite may be excreted in the urine as a metabolite of zinc phosphide (160).

There is little tendency for the compound to concentrate in living tissue, as it is readily converted to phosphine.

Ecological effects

Effects on birds

Zinc phosphide is highly toxic to wild birds. The most sensitive birds are geese (LD_{50} of 7.5 mg/kg for the white-fronted goose). Pheasants, mourning doves, quail, mallard ducks, and horned larks are also very susceptible to this compound. Blackbirds are less sensitive (8).

Effects on aquatic organisms

Zinc phosphide is highly toxic to freshwater fish. The fish species that have been evaluated include bluegill sunfish (LC_{50} of 0.8 mg/L) and rainbow trout (LC_{50} of 0.5 mg/L) (1). Carp were also found to be susceptible to zinc phosphide, especially in weakly acidic water.

Effects on other organisms (non-target species)

Zinc phosphide is also toxic to non-target mammals when ingested directly (8). Nearly 60 studies have been conducted on the toxicity of this rodenticide to wild animals.

Secondary toxicity to mammalian predators (animals eating other animals that had been exposed to the compound) from zinc phosphide is rather low, primarily because the compound does not significantly accumulate in the muscles of target species (8). Some of the toxic effects to predators have been due to the ingestion of zinc phosphide that was in the digestive tract of the target organism.

Studies on secondary organisms have focused on coyote, fox, mink, weasel, and birds of prey. Under field conditions, most of the toxic effects to non-target wildlife are due to direct exposures resulting from misuse or misapplication of this rodenticide (8).

Environmental fate

Breakdown in soil and groundwater

Zinc phosphide may be applied as an active ingredient in either bait or a dust. Under average conditions, toxic activitity persists for approximately 2 weeks (8). Soil acidity and moisture tend to accelerate the breakdown of the compound (8). Phosphine gas may be liberated as a result of this process.

Breakdown in water

No data are currently available.

Breakdown in vegetation

No data are currently available.

Physical properties

Zinc phosphide is an amorphous black-grey powder with a garlic-like odor (1). It is stable when dry and decomposes in moist air (1).

Chemical name: trizinc diphosphide (8)
CAS #: 1314-84-7
Molecular weight: 258.09 (8)
Solubility in water: Practically insoluble in water (decomposes slowly) (8)
Solubility in other solvents: Practically i.s. in alcohol; s.s. in benzene and carbon disulfide (8)
Melting point: >420°C (8)
Vapor pressure: Negligible in the dry state (as solid) (8)
Partition coefficient (octanol/water): Not available
Adsorption coefficient: Not available

Exposure guidelines

ADI: Not available
HA: Not available

RfD: 0.0003 mg/kg/day (13)
PEL: Not available

Basic manufacturer

Hacco, Inc.
P.O. Box 7190
537 Atlas Ave.
Madison, WI 53707
Telephone: 608-221-6200
Emergency: 800-642-4699

References

(1) Kidd, H. and James, D. R., Eds. *The Agrochemicals Handbook,* 3rd ed. Royal Society of Chemistry Information Services, Cambridge, U.K., 1991 (as updated).

(2) World Health Organization. *Environmental Health Criteria Number 148: Benomyl.* Geneva, Switzerland,

(3) Edwards, I. R., Ferry, D. G., and Temple, W. A. Fungicides and related compounds. In *Handbook of Pesticide Toxicology.* Hayes, W. J. and Laws, E. R., Eds. Academic Press, New York, 1991.

(4) Cummings, A. M., Ebron-McCoy, M. T., Rogers, J. M., Barbee, B. D., and Harris, S. T. Developmental effects of methyl benzimidazole carbamate following exposure during early pregnancy. *Fundam. Appl. Toxicol.* 18, 288–293, 1992.

(5) National Research Council. *Regulating Pesticides in Food: The Delaney Paradox.* National Academy Press, Washington, D.C., 1987.

(6) U.S. Department of Agriculture (U.S. Forest Service). *Pesticide Background Statements. Vol. I. Herbicides.* Washington, D.C., 1984.

(7) Food and Agriculture Organization of the United Nations. *Pesticide Residues in Food — 1983: Evaluations, FAO Plant Production and Protection Paper 61.* Geneva, Switzerland, 1983.

(8) U.S. National Library of Medicine. *Hazardous Substances Data Bank.* Bethesda, MD, 1995.

(9) Howard, P. H., Ed. *Handbook of Environmental Fate and Exposure Data for Organic Chemicals. Vol. III: Pesticides.* Lewis, Boca Raton, FL, 1991.

(10) Potter, D. A., Buxton, M. C., Redmond, C. T., Patterson, C. T., and Powell, A. J. Toxicity of pesticides to earthworms (Oligochaeta: Lumbricidae) and effects on thatch degradation in Kentucky bluegrass turf. *J. Econ. Entomol.* 83(6), 2362–2369, 1990.

(11) Wauchope, R. D., Buttler, T. M., Hornsby A. G., Augustijn-Beckers, P. W. M., and Burt, J. P. SCS/ARS/CES pesticide properties database for environmental decisionmaking. *Rev. Environ. Contam. Toxicol.* 123, 1–157, 1992.

(12) Lu, F. C. A review of the acceptable daily intakes of pesticides assessed by the World Health Organization. *Regul. Toxicol. Pharmacol.* 21, 351–364, 1995.

(13) U.S. Environmental Protection Agency. *Integrated Risk Information System Database,* Washington, D.C., 1995.

(14) U.S. Occupational Safety and Health Administration. *Permissible Exposure Limits for Air Contaminants* (29 CFR 1910. 1000, Subpart Z). U.S. Department of Labor, Washington, D.C., 1994.

(15) Chemical Information Systems, Inc. *Oil and Hazardous Materials/Technical Assistance Data System,* Baltimore, MD, 1988.

(16) U.S. Environmental Protection Agency. Captan: intent to cancel registrations; Conclusion of special reviews. *Fed. Reg.* 54, 8116–8150, 1989.

(17) American Conference of Governmental Industrial Hygienists. Inc. *Documentation of the Threshold Limit Values and Biological Exposure Indices,* 6th ed., Cincinnati, OH, 1991 (as updated).

(18) U.S. Environmental Protection Agency. *Health Advisory Summary: Carboxin.* Office of Drinking Water, Washington, D.C., 1987.

(19) Ciba-Giegy, Agricultural Division. Letter of April 20, 1992.

(20) U.S. Environmental Protection Agency. *Pesticide Abstracts*. Office of Pesticides and Toxic Substances, Management Support Division, 79–0210, 81–3526, Washington, D.C. 1968–81.

(21) National Research Council. *Drinking Water and Health*. National Adademy Press, Washington, D.C., 1977.

(22) National Institute for Occupational Safety and Health. *Registry of Toxic Effects of Chemical Substances*. Cincinnati, OH, 1981–1986.

(23) Clayton, G. D. and Clayton, F. E., Eds. *Patty's Industrial Hygiene and Toxicology, Vol. 2. Toxicology.* 3rd ed. John Wiley & Sons, New York, 1981.

(24) U.S. Environmental Protection Agency. *Guidance for Reregistration of Pesticide Products Containing Copper Sulfate. Fact Sheet Number 100.* Office of Pesticide Programs, Washington, D.C., 1986.

(25) New York State Department of Health. *Chemical Fact Sheet: Copper Sulfate.* Bureau of Toxic Substances Management. Albany, NY, 1984.

(26) Gangstad, E. O. *Freshwater Vegetation Management.* Thomson, Fresno, CA, 1986.

(27) Tucker, R. and Crabtree, D. G. *Handbook of Toxicity of Pesticides to Wildlife.* U.S. Department of Agriculture, Fish and Wildlife Service, Bureau of Sport Fisheries and Wildlife, U.S. Government Printing Office, Washington, D.C., 1970.

(28) Pimentel, D. *Ecological Effects of Pesticides on Nontarget Species.* Executive Office of the President's Office of Science and Technology, U.S. Government Printing Office, Washington, D.C., 1971.

(29) Gasiewicz, T. A. Nitro compounds and related phenolic pesticides. In *Handbook of Pesticide Toxicology.* Hayes, W. J. and Laws, E. R., Eds. Academic Press, New York, 1991.

(30) Gosselin, R. E., Smith, R. P., and Hodge, H. C. *Clinical Toxicology of Commercial Products,* 5th ed. Williams and Wilkins, Baltimore, MD, 1984.

(31) Sax, N. I. *Dangerous Properties of Industrial Materials,* 6th ed. VanNostrand Reinhold, New York, 1984.

(32) Wagner, S. L. *Clinical Toxicology of Agricultural Chemicals.* Oregon State University Environmental Health Sciences Center, Corvallis, OR, 1981.

(33) U.S. Environmental Protection Agency. Dinocap: Notice of intent to cancel registrations; Conclusion of special review. *Fed. Reg.* 54, 5908–5920, 1989.

(34) U.S. Environmental Protection Agency Health Effects Research Lab. *Postnatal Alterations in Development Resulting from Prenatal Exposure to Pesticides.* Research Triangle Park, NC, 1986.

(35) U.S. Environmental Protection Agency. *Dinocap: Special Review, Technical Support Document.* Office of Pesticide Programs, Washington, D.C., 1986.

(36) Hill, E. F. and Camardese, M. B. *Lethal Dietary Toxicities of Environmental Contaminants to Coturnix, Technical Report Number 2.* U.S. Department of Interior, Fish and Wildlife Service, Washington, D.C., 1986.

(37) Johnson, W. W. and Finley, M. T. *Handbook of Acute Toxicity of Chemicals to Fish and Aquatic Invertebrates, Resource Publications 137.* U.S. Department of the Interior, Fish and Wildlife Service, Washington, D.C., 1980.

(38) U.S. Environmental Protection Agency. *Pesticide Fact Sheet Number 65: Dinocap.* Office of Pesticides and Toxic Substances, Washington, D.C., 1978.

(39) Harding, W. C. *Pesticide Profiles.* University of Maryland Cooperative Extension Service, College Park, MD, 1979.

(40) U.S. Environmental Protection Agency. *Pesticide Fact Sheet Number 135: Dodine.* Office of Pesticides and Toxic Substances, Washington, D.C., 1987.

(41) Food and Agriculture Organization of the United Nations. Pesticide Residues in Food — 1977. *FAO Plant Production and Protection Paper 10.* FAO, Geneva, Switzerland, 1977.

(42) Food and Agriculture Organization of the United Nations. Pesticide Residues in Food — 1985. *FAO Plant Production and Protection Paper 72/2.* FAO, Geneva, Switzerland, 1985.

(43) U.S. Environmental Protection Agency. Final Rule: Pesticide Tolerance for Iprodione. *Fed. Reg.* 55, 2834–2835, 1990.

(44) Martin, C., Davet, P., Vega, D., and Cosste, C. Field effectiveness and biodegradation of cyclic imides in lettuce field soils. *Pestic. Sci.* 32(4), 427–428, 1991.

(45) Suta, V., Trandafirescu, M., Popescu, V., Voica, E., and Fugel, S. *Proc. Br. Crop Protection Conf. — Pests and Diseases.* British Crop Protection Council, Croydon, England. 1979, 103.

(46) U.S. Environmental Protection Agency. *Pesticide Fact Sheet: Metalaxyl.* Office of Pesticides and Toxic Substances, Washington, D.C., 1988.

(47) Kimmel, E. C., Casida, J. E., and Ruzo, L. O. Formamidine insecticides and chloroacetanilide herbicides: disubstituted anilines and nitrobenzenes as mammalian metabolites and bacterial mutagens. *J. Agric. Food Chem.* 34, 157–161, 1986.

(48) Williams, W. M., Holden, P. W., Parsons, D. W., and Lorber, M. N. *Pesticides in Groundwater Database, 1988 Interim Report.* U.S. Environmental Protection Agency, Office of Pesticide Programs, Washington, D.C., 1988.

(49) Menzie, C. M. *Metabolism of Pesticides, Update III. Special Scientific Report, Wildlife No. 232.* U.S. Department of the Interior, Fish and Wildlife Service, Washington, D.C., 1980.

(50) U.S. Environmental Protection Agency. Rules and Regulations. *Fed. Reg.* 52, 17954–17955, 1987.

(51) Rankin, G. O. Comparative acute renal effects of three carboximide fungicides: succinimide, vinclozolin and iprodione. *Toxicology.* 56(3), 263–272, 1989.

(52) U.S. Environmental Protection Agency. *Pesticide Environmental Fate One Liner Summaries: Vinclozolin.* Environmental Fate and Effects Division, Washington, D.C., 1991.

(53) Augustijn-Beckers, P. W. M., Hornsby, A. G., and Wauchope, R. D. SCS/ARS/CES pesticide properties database for environmental decisionmaking. II. Additional Compounds. *Rev. Environ. Contam. Toxicol.* 137, 1–82, 1994.

(54) U.S. Environmental Protection Agency. *Health Advisory Draft Report: Acifluorfen.* Office of Drinking Water, Washington, D.C., 1987.

(55) Quest, J. A., Phang, W., Hamernik, K. L., van Gemert, M., Fisher, B., Levy, R., Farber, T. M., Burnam, W. L., and Engler, R. Evaluation of the carcinogenic potential of pesticides. 1. Acifluorfen. *Regul. Toxicol. Pharmacol.* 10, 149–159, 1989.

(56) BASF Corporation. *Material Safety Data Sheet for Blazer Herbicide.* Research Triangle Park, NC, 1991.

(57) Johnson, W. O., Kollman, G. E., Swithenbank, C., and Yih, R. Y. RH-6201 (Blazer): a new broad spectrum herbicide for postemergence use in soybeans. *J. Agric. Food Chem.* 26(1), 285–286, 1978.

(58) Weed Science Society of America. *Herbicide Handbook,* 7th ed. Champaign, IL, 1994.

(59) Ritter, R. L. and Coble, H. D. Influence of temperature and relative humidity on the activity of acifluorfen. *Weed Sci.* 29, 480–485, 1981.

(60) U.S. Environmental Protection Agency. *Health Advisory: Alachlor.* Office of Drinking Water, Washington, D.C., 1987.

(61) Monsanto Company. *Material Safety Data Sheet. Alachlor Technical (94%).* St. Louis, MO, 1991.

(62) U.S. Environmental Protection Agency. *Pesticide Fact Sheet Number 97.1: Alachlor.* Office of Pesticides and Toxic Substances, Washington, D.C., 1987.

(63) Hudson, R. H., Tucker, R. K., and Haegele, M. A. *Handbook of Toxicity of Pesticides to Wildlife. Resource Publication 153.* U.S. Department of the Interior, Fish and Wildlife Service, Washington, D.C., 1984.

(64) Holden, L. and Graham, J. A. Results of the National Alachlor Well Water Survey. *Environ. Sci. Tech.* 26, 935–943, 1992.

(65) U.S. Environmental Protection Agency. *National Primary Drinking Water Standards,* 810-F-94-001A. Washington, D.C., 1994.

(66) E. I. DuPont de Nemours. *Technical Data Sheet (on Ammonium Sulfamate).* Wilmington, DE, 1972.

(67) U.S. Environmental Protection Agency. *Health Advisory: Ammonium Sulfamate.* Office of Drinking Water, Washington, D.C., 1988.

(68) Morgan, D. P. *Recognition and Management of Pesticide Poisonings,* 3rd ed. U.S. Environmental Protection Agency, U.S. Government Printing Office, Washington, D.C., 1982.

(69) U.S. Environmental Protection Agency. *Health Advisory Summary: Bentazon.* Office of Drinking Water, Washington, D.C., 1989.

(70) Hallenbeck, W. H. and Cunningham-Burns, K. M. *Pesticides and Human Health*. Springer-Verlag, New York, 1985.

(71) U.S. Environmental Protection Agency. *Guidance for the Reregistration of Pesticide Products Containing Bentazon as the Active Ingredient. EPA Case Number 182*. Office of Pesticide Programs, Washington, D.C., 1985.

(72) U.S. Environmental Protection Agency. *Chemical Fact Sheet Number 64: Bentazon and Sodium Bentazon*. Office of Pesticides and Toxic Substances, Washington, D.C., 1985.

(73) U.S. Environmental Protection Agency. *Pesticides in Ground Water Database*. Office of Pesticide Programs. (EPA 734-12-92-001) Washington, D.C., 1992.

(74) U.S. Environmental Protection Agency. *Health Advisory: Bromacil*. Office of Drinking Water, Washington, D.C., 1988.

(75) U.S. Environmental Protection Agency. *Initial Scientific and Minieconomic Review of Bromacil. 540/1-75-006*. Office of Pesticide Programs, Washington, D.C., 1975.

(76) VanDriesche, R. G. *Pesticide Profiles: Bromacil*. Cooperative Extension Service, Department of Entomology, University of Massachusetts, Amherst, MA, 1985.

(77) Smith, A. E. An analytical procedure for bromoxynil and its octanoate in soils; persistence studies with bromoxynil octanoate in combination with other herbicides in soil. *Pestic. Sci.* 11, 341–346, 1980.

(78) U.S. Environmental Protection Agency. Pesticide tolerance for 2-(2-chlorophenyl) methyl-4,4-dimethyl-3-isoxazolidinone [clomazone]. *Fed. Reg.* 56, 42574–42575, 1991.

(79) Salamon, C. M. and Borders, C. K. *2-Generation Reproduction Study in Albino Rats with FMC 57020 Technical — Final Report*. Toxigenics, Inc., Philadelphia, PA, 1984.

(80) U.S. Environmental Protection Agency. *Pesticide Fact Sheet Number 90: Command Herbicide*. Office of Pesticides and Toxic Substances, Washington, D.C., 1986.

(81) Wu, J. *Photodegradation of FMC 57020 in Water (FMC Report No. P-0869)*. Unpublished report prepared by FMC Corp., Philadelphia, PA, 1984.

(82) Dziedzic, J. E. *Hydrolysis Study (FMC Report No. P-0465)*. Unpublished report prepared by FMC Corp., Philadelphia, PA, 1982.

(83) U.S. Environmental Protection Agency. *Pesticide Fact Sheet 130: Dinoseb*. Office of Pesticides and Toxic Substances, Washington, D.C., 1986.

(84) Call, D. J., Brooke, L. T., Kent, R. S., Poirier, S. H., Knuth, M. L., and Shick, E. J. Toxicity, uptake, and elimination of the herbicides alachlor and dinoseb in freshwater fish. *J. Environ. Qual.* 13(3), 493–498, 1984.

(85) Hall, L., Linder, R., Scotti, T., Bruce, R., Moseman, R., Heiderscheit, T., Hinkle, D., Edgerton, T., Chaney, T., Goldstein, J., Gage, M., Farmer, J., Bennett, L., Stevens, J., Durham, W., and Curley, A. Subchronic and reproductive toxicity of dinoseb. *Toxicol. Appl. Pharmacol.* 45(1), 235–236, 1978.

(86) Spacie, A. and Hamelink, J. L. Alternative models for describing the bioconcentrations of organics in fish. *Environ. Toxicol. Chem.* 1, 309–320, 1982.

(87) Stevens, J. T. and Sumner, D. D. Herbicides. In *Handbook of Pesticide Toxicology*. Hayes, W. J., Jr. and Laws, E. R., Jr., Eds. Academic Press, New York, 1991.

(88) Chevron Chemical Company. *Diquat Herbicide h/a for Aquatic Plant Treatment. (Number 8616 DIQ-45.)* Agricultural Chemicals Division, San Francisco, CA, 1986.

(89) Chevron Chemical Company. *Material Safety Data Sheet: Ortho Diquat Herbicide*. Chevron Environmental Health Center, Inc., Richmond, VA, 1986.

(90) U.S. Environmental Protection Agency. Final Rule: diquat. Tolerances and exemptions from tolerances for pesticide chemicals in or on raw agricultural commodities. *Fed. Reg.* 46, 30342–30343, 1981.

(91) U.S. Environmental Protection Agency. *Guidance for Reregistration of Pesticide Products Containing as the Active Ingredient Diquat Dibromide (032201)*. Office of Pesticide Programs, Washington, D.C., 1986.

(92) Simonin, H. A. and Skea, J. C. Toxicity of diquat and cutrine to fingerling brown trout. *NY Fish & Game J.* 24(1), 37–45, 1977.

(93) Bimber, K. L. Respiratory stress in yellow perch induced by subtoxic concentrations of diquat. *Ohio J. Sci.* 76(2), 87–90, 1976.

(94) Tucker, B. V. *Diquat Environmental Chemistry*. Chevron Chemical Corporation, Ortho Agricultural Division. Richmond, VA, 1980.

(95) Gillett, J. W. The biological impact of pesticides in the environment. *Environmental Health Sciences Series No. 1*. Oregon State University, Corvallis, OR, 1970.

(96) Monsanto Company. *Toxicology of Glyphosate and Roundup Herbicide*. St. Louis, MO, 1985.

(97) U.S. Environmental Protection Agency. Pesticide tolerance for glyphosate. *Fed. Reg.* 57, 8739–8740, 1992.

(98) U.S. Environmental Protection Agency. *Health Advisory: Glyphosate*. Office of Drinking Water, Washington, D.C., 1987.

(99) Malik, J., Barry, G., and Kishore, G. Minireview: the herbicide glyphosate. *BioFactors*. 2(1), 17–25, 1989.

(100) U.S. Environmental Protection Agency. *Pesticide Fact Sheet Number 106: Metolachlor*. Office of Pesticides and Toxic Substances, Washington, D.C., 1987.

(101) U.S. Environmental Protection Agency. *Health Advisory Draft Report: Metolachlor*. Office of Drinking Water, Washington, D.C., 1987.

(102) Zimdahl, R. L. and Clark, S. K. Degradation of three acetanilide herbicides in soil. *Weed Sci.* 30, 545–548, 1982.

(103) U.S. Environmental Protection Agency. *Pesticide Environmental Fate One Liner Summaries: Devrinol*. Environmental Fate and Effects Division, Washington, D.C., 1991.

(104) U.S. Environmental Protection Agency. *Environmental Effects Branch. Chemical Profile: Napropamide*. Environmental Fate and Effects Division, Washington, D.C., 1984.

(105) U.S. Environmental Protection Agency. *Pesticide Environmental Fate One Liner Summaries: Napropamide*. Environmental Fate and Effects Division, Washington, D.C., 1991.

(106) U.S. Environmental Protection Agency. *Pesticide Fact Sheet Number 211: Oryzalin*. Office of Pesticides and Toxic Substances, Washington, D.C., 1987.

(107) U.S. Environmental Protection Agency. Pesticide tolerance for oryzalin. *Fed. Reg.* 55, 25140–25141, 1990.

(108) U.S. Environmental Protection Agency. Pesticide tolerances for oxyfluorfen. *Fed. Reg.* 57, 22202–22203, 1992.

(109) U.S. Environmental Protection Agency. Environmental Effects Branch. *Chemical Profile: Oxyfluorfen*. Washington, D.C., 1984.

(110) U.S. Environmental Protection Agency. *Pesticide Environmental Fate One Liner Summaries: Oxyfluorfen*. Environmental Fate and Effects Division, Washington, D.C., 1992.

(111) U.S. Environmental Protection Agency. *Health Advisory Draft Report: Paraquat*. Office of Drinking Water, Washington, D.C., 1987.

(112) U.S. Environmental Protection Agency. *Pesticide Fact Sheet Number 131: Paraquat*. Office of Pesticides and Toxic Substances, Washington, D.C., 1987.

(113) Rao, P. S. C. and Davidson, J. M. Estimation of pesticide retention and transformation parameters required in nonpoint source pollution models. In *Environmental Impact of Nonpoint Source Pollution*. Overcash, M. R. and Davidson, J. M., Eds. Ann Arbor Science, Ann Arbor, MI, 1980.

(114) Kosinski, J. R. and Merkle, M. G. The effect of four terrestrial herbicides on the productivity of artificial stream algal communities. *J. Environ. Qual.* 13(1), 75–82, 1984.

(115) U.S. Environmental Protection Agency. *Pesticide Fact Sheet Number 50: Pendimethalin*. Office of Pesticides and Toxic Substances, Washington, D.C., 1985.

(116) U.S. Environmental Protection Agency. Pesticide tolerance for pendimethalin. *Fed. Reg.* 52, 47734–47735, 1987.

(117) Zulalian, J. Study of the absorption, excretion, metabolism, and residues in tissues of rats treated with carbon-14-labeled pendimethalin, Prowl herbicide. *J. Agric. Food Chem.* 38, 1743–1754, 1990.

(118) National Research Council. *Drinking Water and Health. Vol. 5*. Board on Toxicology and Environmental Health Hazards, Commission on Life Sciences, Safe Drinking Water Committee, National Academy Press, Washington, D.C., 1983.

(119) Newton, M. and Dost, F. N. *Biological and Physical Effects of Forest Vegetation Management*. Washington Department of Natural Resources, Olympia, WA, 1984.

(120) U.S. Environmental Protection Agency. *Health Advisory: Picloram*. Office of Drinking Water, Washington, D.C., 1987.

(121) U.S. Environmental Protection Agency. *Health Advisory: Pronamide*. Office of Drinking Water, Washington, D.C., 1988.

(122) U.S. Environmental Protection Agency. *Pesticide Fact Sheet Number 70: Pronamide*. Office of Pesticides and Toxic Substances, Washington, D.C., 1986.

(123) Rohm and Haas Company. *Kerb Technical Herbicide (Key: 906750-0)*. Philadelphia, PA, 1991.

(124) U.S. Environmental Protection Agency. *Guidance for Reregistration of Pesticide Products Containing Pronamide as the Active Ingredient*. Office of Pesticide Programs, Washington, D.C., 1986.

(125) Rohm and Haas Company. *STAM Tech 98% DCA Herbicide (Key: 904399-2)*. Philadelphia, PA, 1991.

(126) U.S. Environmental Protection Agency. Pesticide tolerances for 2[1-(ethoxyimino)butyl]-5-[2-(ethylthio)propyl]-3-hydroxy-2-cyclohexen-1-one. *Fed. Reg.* 56, 11677–11678, 1991.

(127) U.S. Environmental Protection Agency. *EEB Chemical Profile: Sethoxydim*. Washington, D.C., 1989.

(128) U.S. Environmental Protection Agency. *Terbacil — Registration Standard*. Office of Pesticide Programs, Washington, D.C., 1982.

(129) U.S. Environmental Protection Agency. *Pesticide Fact Sheet Number 206: Terbacil*. Office of Pesticides and Toxic Substances, Washington, D.C., 1989.

(130) U.S. Environmental Protection Agency. *Health Advisory: Terbacil*. Office of Drinking Water, Washington, D.C., 1988.

(131) McEwen, F. L. and Stephenson, G. R. *The Use and Significance of Pesticides in the Environment*. John Wiley & Sons, New York, 1979.

(132) U.S. Environmental Protection Agency. *Estimating Pesticide Sorption Coefficients for Soils and Sediments*. Green, R. E. and Karickhoff, S. W. Environmental Research Laboratory, Athens, GA, 1986.

(133) Hanley, T. R., Thompson, D. J., Palmer, A. K., Bellies, R. P., and Schwetz, B. A. Teratology and reproductive studies with triclopyr in the rat and rabbit. *Fundam. Appl. Toxicol.* 4, 872–882, 1984.

(134) Dow Chemical Company. *Environmental and Toxicology Profile of Garlon Herbicides. Technical Data Sheet No. 137-1639-83*. Agricultural Products Department, Midland, MI, 1983.

(135) Carmichael, N. G. Assessment of hazards during pesticide application. *Food Add. Contam.* 6(S1): S21–S27, 1989.

(136) Gersich, F. M., Mendoza, C. G., Hopkins, D. L., and Bodner, K. M. Acute and chronic toxicity of triclopyr triethylamine salt to *Daphnia magna straus. Bull. Environ. Contam. Toxicol.* 32, 497–502, 1984.

(137) Dow Chemical Company. *Technical Information of Triclopyr, the Active Ingredient of Garlon Herbicides Technical Data Sheet No. 137-859-483*. Agricultural Products Department, Midland, MI, 1983.

(138) U.S. Environmental Protection Agency. *Guidance for the Reregistration of Pesticide Products Containing Trifluralin as the Active Ingredient*. Office of Pesticides and Toxic Substances, Washington, D.C., 1987.

(139) U.S. Environmental Protection Agency. *Health Advisory Summary: Trifluralin*. Office of Drinking Water, Washington, D.C., 1989.

(140) Mayer, F. L. and Ellersieck, M. R. *Manual of Acute Toxicity: Interpretation and Data Base for 410 Chemicals and 66 Species of Freshwater Animals. Resource Publication 160*. U.S. Department of Interior, Fish and Wildlife Service, Washington, D.C., 1986.

(141) Lankas, G. R. and Gordon, L. R. Toxicology. In *Ivermectin and Abamectin*. Campbell, W. C., Ed. Springer-Verlag, New York, 1989.

(142) U.S. Environmental Protection Agency. *Pesticide Fact Sheet Number 89.2: Avermectin B1*. Office of Pesticides and Toxic Substances, Washington, D.C., 1990.

(143) Ray, D. E. Pesticides derived from plants and other organisms. In *Handbook of Pesticide Toxicology*. Hayes, W. J., Jr. and Laws, E. R., Jr., Eds. Academic Press, New York, 1991.

(144) Thongsinthusak, T. *Estimation of Exposure of Persons in California to Pesticide Products That Contain Abamectin. HS 1567.* California Department of Food and Agriculture, Division of Pest Management, Sacramento, CA, 1990.

(145) Wislocki, P. G., Grosso, L. S., and Dybas, R. A. Environmental aspects of abamectin use in crop protection. In *Ivermectin and Abametin.* Campbell, W. C., Ed. Springer-Verlag, New York, 1989.

(146) U.S. Environmental Protection Agency. *Pesticide Fact Sheet Number 93: Bacillus thuringiensis.* Office of Pesticides and Toxic Substances, Washington, D.C., 1986.

(147) Roe, R. M. Vertebrate toxicology of the solubilized parasporal crystalline proteins of *Bacillus thuringiensis israelensis.* In *Reviews in Pesticide Toxicology 1: Toxicological Studies of Risks and Benefits.* Hodgson, E., Roe, R. M., and Motoyama, N., Eds. North Carolina State University, Raleigh, NC, 1991.

(148) Abbott Laboratories. *Toxicology Profile: Dipel, Bacillus thuringiensis Insecticide.* Chemical and Agricultural Products Division, North Chicago, IL, 1982.

(149) Spiegel, J. P. and Shadduck, J. A. Clearance of *Bacillus sphaericus* and *Bacillus thuringiensis israelensis* from mammals. *Econ. Entomol.* 83, 347–355, 1990.

(150) Agriculture Canada. *Report of New Registration: Bacillus thuringiensis SerotypeH14.* Food Protection and Inspection Branch, Ottawa, Ontario, Canada, 1982.

(151) Vandenberg, J. D. Safety of four entomopathogens for caged adult honeybees (Hymenoptera: Apidae). *Econ. Entomol.* 83(3), 756–759, 1990.

(152) Ghassemi, M. *Environmental Fates and Impact of Major Forest Use Pesticides.* U.S. Environmental Protection Agency. Washington, D.C., 1982.

(153) Dunkle, R. L. and Shasha, B. S. Response of starch-encapsulated *Bacillus thuringiensis* containing ultraviolet screens to sunlight. *Environ. Entomol.* 18(6), 1035–1041, 1989.

(154) U.S. Environmental Protection Agency. *Pesticide Environmental Fate One-Line Summary: Hydramethylnon.* Environmental Fate and Effects Division, Washington, D.C., 1992.

(155) U.S. Environmental Protection Agency. *Guidance for the Reregistration of Pesticide Products Containing Methoprene as the Active Ingredient.* Office of Pesticide Programs, Washington, D.C., 1982.

(156) U.S. Environmental Protection Agency. Methoprene: tolerances and exemptions from tolerances for pesticide chemicals in or on raw agricultural commodities. *Fed. Reg.* 46, 59248–59249, 1981.

(157) U.S. Environmental Protection Agency. *R.E.D. Facts: Methoprene.* Office of Pesticides and Toxic Substances, Washington, D.C., 1991.

(158) Zoecon Corporation. *Technical Bulletin on Altosid. Toxicological Properties.* Dallas, TX, 1974.

(159) Zoecon Corporation. *Technical Bulletin on Altosid. Environmental Properties.* Dallas, TX, 1973.

(160) Clarkson, T. W. Inorganic and organometal pesticides. In *Handbook of Pesticide Toxicology.* Hayes, W. J. and Laws, E. R., Eds. Academic Press, New York, 1991.

(161) California Dept. of Food and Agriculture. *Information on the Safe Handling of Pesticides Containing Sulfuryl Fluoride (Vikane). HS-599.* Division of Pest Management, Environmental Protection and Worker Safety, California Department of Food and Agriculture, Sacramento, CA, 1979.

(162) U.S. Environmental Protection Agency. *Pesticide Fact Sheet Number 51: Sulfuryl Fluoride.* Office of Pesticides and Toxic Substances, Washington, D.C., 1985.

(163) Gollapudi, B. B. *Evaluation of Sulfuryl Fluoride in the Rat Hepatocyte Unscheduled DNA Synthesis (UDS) Assay, Summary (Study ID TXT: K-016399-043).* Dow Elanco Company, Indianapolis, IN, 1991.

(164) Gollapudi, B. B. *Evaluation of Sulfuryl Fluoride in the Mouse Bone Marrow Micronucleus Test, Summary (Study ID TXT: K-016399-033).* Dow Elanco Company, Indianapolis, IN, 1990.

(165) Gollapudi, B. B. *Evaluation of Sulfuryl Fluoride in the Ames Salmonella/Mammalian-Microsome Bacterial Mutagenicity Assay, Summary (Study ID TXT: K-016399-037).* Dow Elanco Company, Indianapolis, IN, 1990.

(166) Breslin, W. J. *Sulfuryl Fluoride: Two Generation Inhalation Reproduction Study in Sprague-Dawley Rats, Summary.* Dow Elanco Company, Indianapolis, IN, Not Dated.

(167) U.S. Environmental Protection Agency. *Pesticide Registration Standard, 4-Aminopyridine: Avitrol.* Office of Pesticides and Toxic Substances, Washington, D.C., 1980.

(168) U.S. Environmental Protection Agency. *Chemical Fact Sheet Number 26: Daminozide.* Office of Pesticides and Toxic Substances, Washington, D.C., 1986.

(169) Uniroyal. *Database on Daminozide.* Health and Regulatory Compliance Department, Uniroyal Corporation, Middlebury, CT, 1993.

(170) U.S. Environmental Protection Agency. Daminozide: notice of final determination for non-food uses and termination of the daminozide Special Review. *Fed. Reg.* 57, 46436–46444, 1992.

(171) U.S. Environmental Protection Agency. Pesticide tolerance for daminozide. *Fed. Reg.* 54, 6392–6396, 1989.

(172) Pelfrene, A. F. Synthetic organic rodenticides. In *Handbook of Pesticide Toxicology,* Hayes, W. J. and Laws, E. R., Eds. Academic Press, New York, 1991.

(173) U.S. Environmental Protection Agency. *Pesticide Environmental Fate One-Line Summary: Diphacinone.* Environmental Fate and Effects Division, Washington, D.C., 1991.

(174) Bell Laboratories Incorporated. *Diphacinone Technical: Material Safety Data Sheet.* Bell Labs, Madison, WI, 1990.

(175) Letz, G. A., Pond S. M., Osterloh, J. D., Wade, R. L., and Becker, C. E. Two fatalities after acute occupational exposure to ethylene dibromide. *J. Am. Med. Assoc.* 252, 2428–2431, 1984.

(176) Broda, C., Nachtomi, E., and Alumot, E. Differences in liver morphology between rats and chicks treated with ethylene dibromide. *Gen. Pharmacol.* 7, 345–348, 1976.

(177) Wong, O., Utidjian, H. M., and Karten, V. S. Retrospective evaluation of reproductive performance of workers exposed to ethylene dibromide. *J. Occup. Med.* 21, 98–102, 1979.

(178) Short, R. D., Winston, J. M., Minor, J. L., Hong, C. B., Seifter, J., and Lee, C. C. Toxicity of vinylidene chloride in mice and rats and its alterations by various treatments. *Toxicol. Appl. Pharmacol.* 45, 173–182, 1978.

(179) Wong, L. C., Winston, J. M., Hong, C. B., and Plotnick, H. Carcinogenicity and toxicity of 1,2-dibromoethane in the rat. *Toxicol. Appl. Pharmacol.* 63(2), 155–165, 1982.

(180) Li, F. *Technical data submitted in support of the San Luis Drain Report of Waste Discharge.* U.S. Department of Interior, Bureau of Reclamation, Sacramento, CA, 1982.

(181) Pignatello, J. J., Sawhney, B. L., and Frink, C. R. EDB: Persistence in soil. *Science.* 236, 898–902, 1987.

(182) McConnel, J. B. *Investigation of Ethylene Dibromide (EDB) in Groundwater in Seminole County, Georgia.* U.S. Geological Survey Circular, Washington, D.C., 1984.

(183) Mackay, D. *Volatilization of Organic Pollutants from Water, 600/53-82-019.* U.S. Environmental Protection Agency, Washington, D.C., 1982.

(184) U.S. Environmental Protection Agency. *Health Advisory: Ethylene thiourea.* Office of Drinking Water, Washington, D.C., 1987.

(185) Knowles, C. O. Miscellaneous pesticides. In *Handbook of Pesticide Toxicology.* Hayes, W. J. and Laws, E. R., Eds. Academic Press, New York, 1991.

(186) U.S. Environmental Protection Agency. *Suspended, Cancelled, and Restricted Pesticides.* Office of Pesticides and Toxic Substances, Washington, D.C., 1990.

(187) U.S. Environmental Protection Agency. *Pesticide Fact Sheet Number 191: Metaldehyde.* Office of Pesticides and Toxic Substances, Washington, D.C., 1988.

(188) Gehring, P. J., Nolan, R. J., Watanabe, P. G., and Schumann, A. M. Solvents, fumigants, and related compounds. In *Handbook of Pesticide Toxocology.* Hayes, W. J., Jr. and Laws, E. R., Jr., Eds. Academic Press, New York, 1991.

(189) Mitsumori, K., Maita, K., Kosaka, T., Miyaoka, T., and Shirasu, Y. Two-year oral chronic toxicity and carcinogenicity study in rats of diets fumigated with methyl bromide. *Food Chem. Toxicol.* 28(2), 109–119, 1991.

(190) Danse, L. H., Van Velsen, F. L., and Vander Heijden, C. A. Methylbromide: Carcinogenic effects in the rat forestomach. *Toxicol. Appl. Pharmacol.* 72, 262–271, 1984.

(191) Kaloyanova, F. P. and El Batawi, M. A., Eds. In *Human Toxicology of Pesticides*. CRC Press, Boca Raton, FL, 1991.

(192) Boyer, I. J. Toxicity of dibutyltin, tributyltin and other organotin compounds to humans and experimental animals. *Toxicology.* 55(3), 253–298, 1989.

(193) Krajnc, E. I., Wester, P. W., Loeber, J. G., van Leeuwen, F. X. R., Vos, J. G., Vaessen, H. A. M. G., and van der Heijden, C. A. Toxicity of bis(tri-n-butyltin)oxide in the rat. I. Short-term effects on general parameters and on the endocrine and lymphoid systems. *Toxicol. Appl. Pharmacol.* 75, 363–386, 1984.

(194) Vos, J. G., de Klerk, A., Krajnc, E. I., Kruizinga, W., van Ommen, B., and Rozing, J. Toxicity of bis(tri-n-butyltin)oxide in the rat. II. Suppression of thymus-dependent immune responses and of parameters of nonspecific resistance after short-term exposure. *Toxicol. Appl. Pharmacol.* 75, 387–408, 1984.

(195) Noda, T., Monith, S., Yamano, T., Shinizn, M., and Sartoh, M. Teratogenicity study of tributyl-μ-acetate on rats on oral administration. *Toxicol. Lett.* 55(1), 109–115, 1991.

(196) Gardlund, A., Archer, T., Danielsson, K., Danielsson, B., Fredriksson, A., Lindqvist, N. G., Lindstrom, H., and Luthman, J. Effects of prenatal exposure to tributyltin and trihexyltin on behavior in rats. *Neurotoxicol. Teratol.* 13(1), 99–105, 1991.

(197) Solectis, R., Hilbig, V., Pfeil, R., Gericke, S., and Gottscholk, M. *Bis(tri-n-butyltin Oxide): Comparison of Effects of Single and Paired Housing on Subchronic Reproductive Toxicity Endpoints in Japanese Quail (Coturnix Coturnix Japonica) in a 13-Week Dietary Study, UBA-FB-93-025 (Original Title in German)*, Umweltbundesamt (Ministry of Environment), Federal Republic of Germany, 1992.

(198) Huggett, R. J, Unger, M. A., Seligman, P. F., and Valkis, A. D. The marine biocide tributyltin: assessing and managing the environmental risks. *Environ. Sci. Technol.* 26(2), 232–237, 1992.

(199) Michigan Department of Natural Resources. *Fact Sheet on Tributyltin Compounds*. Lansing, MI, 1987.

(200) U.S. Environmental Protection Agency. *Technical Support Document: Tributyltin*. Office of Pesticide Programs, Washington, D.C., 1985.

(201) Short, J. W. and Thrower, F. P. *Accumulation of Butyltins in Muscle Tissue of Chinook Salmon Reared in Sea Pens Treated With Tri-n-butyltin*. Northwest and Alaska Fisheries Center, National Marine Fisheries Service, U.S. National Oceanic and Atmospheric Administration, Auke Bay, AK, 1986.

(202) Clark, A. C., Steritt, R. M., and Lester, J. N. The fate of tributyltin in the aquatic environment. *Environ. Sci. Technol.* 22(6), 600–605, 1988.

(203) Laughlin, R. B., Jr., French, W., and Guard, H. E. Accumulation of bis(tributyltin) oxide by the marine mussel *Mytilis edulis*. *Environ. Sci. Technol.* 20(9), 884–890, 1986.

Section III

Basic concepts in toxicology and environmental chemistry

chapter eleven

General concepts

11.1 Pesticide regulation

Introduction

In the U.S., pesticides are regulated by a myriad of laws and agency rules. No less than 14 different Federal Acts control some aspect of the manufacture, registration, distribution, use, consumption, and disposal of pesticides. The bulk of pesticide regulation falls under the Federal Insecticide, Fungicide and Rodenticide Act (FIFRA). This legislation governs the registration, distribution, sale, and use of pesticides. The Environmental Protection Agency is responsible for the administration of this Act and for establishing rules and regulations consistent with the Act's intent.

Other Acts govern their presence in water and air, the clean-up of spills and releases, the concentrations of pesticide residues in raw and processed food, their impact on endangered species, their transportation, working conditions for manufacturers and applicators, and their disposal.

The three broad categories of concern with pesticide regulation focus on (1) the registration of new pesticides and the reregistration of existing pesticides; (2) the establishment and monitoring of pesticide levels in food products; and (3) the monitoring of pesticide levels in the environment, especially in ground and surface water.

Pesticide registration

New pesticides

The EPA is responsible under FIFRA for registering new pesticides to ensure that, when used according to label directions, they will not pose unreasonable risks to human health or the environment. FIFRA requires the EPA to balance the risks of pesticide exposure to human health and the environment against the benefits of pesticide use to society and the economy. A pesticide registration will be granted if, after careful consideration of health, economic, social and environmental costs and benefits, the benefits of the pesticide's use outweighs the costs of its use.

Pesticide registration decisions are based primarily on the EPA's evaluation of data provided by applicants. Depending on the type of pesticide, the EPA can require up to 70 different kinds of specific tests. For a major food-use pesticide, testing can cost the manufacturer many millions of dollars.

Testing is needed to determine whether a pesticide has the potential to cause adverse effects on humans, wildlife, fish, and plants. Potential human risks, which are identified using the results of laboratory tests, include acute toxic reactions, such as poisoning and skin and eye irritation, as well as possible long-term effects like cancer, birth defects, and reproductive system disorders. Data on "environmental fate" (how a pesticide behaves in the environment) are also required so that the EPA can determine, among other things, whether a pesticide poses a threat to ground or surface water.

EPA may classify a product for restricted use if it warrants special handling due to its toxicity. Restricted Use Pesticides (RUPs) may be used only by or under the supervision of certified applicators trained to handle toxic chemicals, and this classification must be shown on product labels. During registration review, the Agency may also require changes in proposed labeling, use locations, and application methods. If the pesticide is being considered for use on a food or feed crop, the applicant must petition the EPA for establishment of a food tolerance.

A brand-new active ingredient may need 6 to 9 years to move from development in the laboratory, through full completion of EPA registration requirements, to retail shelves. This time frame includes at least 2 or 3 years to obtain registration approval from the EPA.

Since 1978, when the EPA began requiring more extensive data on pesticides than in the past, over 130 brand-new chemical active ingredients have been registered; between 10 and 15 new pesticide active ingredients are registered each year.

Reregistration of existing pesticides

EPA is required by law to reregister existing pesticides that were originally registered before current scientific and regulatory standards were formally established. The reregistration process ensures that:

1. Up-to-date data sets are developed for each of these chemicals (or their registrations will be suspended or canceled).
2. Modifications are made to registrations, labels, and tolerances as necessary to protect human health and the environment.
3. Special review or other regulatory actions are initiated to deal with any unreasonable risks.

Reregistration has proved to be a massive undertaking and has proceeded slowly. Under the 1988 FIFRA amendments, the EPA must accelerate the reregistration effort so that the entire process is completed by 1997. Many pesticides are being withdrawn from manufacture and sale rather than going through the lengthy and expensive reregistration process.

Special review, cancellations, and suspensions

New data on registered products sometimes reveal the existence of a problem or a potential for hazard that was not known at the time of registration. Congress and the EPA have developed various mechanisms to reach sound scientific decisions in these situations.

Special Review: Under the law, if the EPA seeks to revoke the registration of a pesticide, the Agency must first announce its reasons and offer the registrant a formal hearing to present opposing evidence. Because the cancellation process can be very time and resource intensive, the EPA often will employ a more informal and often more productive process known as Special Review.

Special Review offers opportunities for interested parties on all sides to comment and present evidence on the risks and benefits of a pesticide. In many cases, the

Special Review results in an agreement to modify the registration to sufficiently reduce risk so that a formal hearing is no longer necessary.

Cancellation: If the Special Review process fails to resolve the issues, however, or if the EPA decides that the problem is severe enough to warrant cancellation, the EPA may issue a proposed notice of intent to cancel without holding a Special Review. The Agency is also required by FIFRA to send the proposed notice to the Scientific Advisory Panel and the U.S. Department of Agriculture (USDA), and must evaluate their comments before proceeding with a final Notice of Intent to Cancel Registration.

If no hearing is requested within 30 days of the Notice, the pesticide's registration is canceled immediately. If a hearing is requested, it is conducted in a trial-like manner before an EPA Administrative Law Judge, who issues a recommended decision to the EPA Administrator. At the end of the cancellation process, which may take 2 years or more, the decision may still be challenged in a federal court of appeals. If there is no appeal to a decision to cancel, all pertinent registrations of the pesticide are automatically canceled, and the products may no longer be sold or distributed in the U.S.

Suspension: During the entire cancellation process, the pesticide remains on the market and no regulatory restrictions are imposed on the pesticide or its use. In some cases, the EPA may believe that allowing the pesticide to stay on the market during a Special Review and/or a cancellation hearing would pose an unacceptably high risk. In such cases, the EPA may issue a suspension order that bans sale or use of the pesticide while the ultimate decision on the pesticide's status is under review.

In order to issue a suspension order, the EPA must find that use of the pesticide poses an imminent hazard. In most cases, the EPA must first offer the registrant an expedited hearing on the suspension issues. However, if the EPA finds that an emergency exists (i.e., that even during the time needed for a suspension hearing, use of the pesticide would pose unreasonable adverse effects), the Agency can ban the sale and use of a pesticide effective immediately.

Under current law, even in an emergency suspension, the EPA must assess the benefits of the pesticide as well. This provision makes emergency suspension difficult to use, and the EPA has been able to make these findings for only three pesticides: ethylene dibromide (EDB), 2,4,5-T (Silvex), and dinoseb.

Food safety and food tolerances

The food supply of the U.S. is among the safest in the world. Although many of the foods we consume may contain low levels of pesticide residues as a result of their use, numerous safeguards are built into the regulatory process to ensure that the public is protected from unreasonable risks posed by eating pesticide-treated crops and livestock.

The EPA regulates the safety of the food supply by setting tolerance levels, or maximum legal limits, for pesticides on food commodities and in animal feed available for sale in the U.S. The purpose of the tolerance program is to ensure that consumers are not exposed to unsafe levels of pesticide residues in food.

Pesticides may be registered by the EPA for use on a food or feed crop only if a tolerance (or an exemption from a tolerance) is first granted, under authority of the Federal Food, Drug and Cosmetic Act (FDCA) as amended by the Food Quality Protection Act of 1996. The EPA has approved about 300 pesticides for food uses; about 200 of them are in common use.

Setting pesticide tolerances on food

To evaluate the risks posed by pesticides in the diet, the EPA follows standard risk assessment guidelines. The Agency uses different procedures for cancer risks and non-cancer risks. Under the new Food Quality Protection Act these tolerances must result in a reasonable certainty of no harm from aggregate exposure to the pesticide. In addition, this Act requires that safety for infants and children be explicitly addressed in tolerance setting and also that the EPA develop and implement a screening program for pesticides that may be endocrine disruptors.

Non-carcinogens

For non-cancer effects, the EPA determines what the highest level of exposure to a pesticide is at which there are no observed adverse effects in animals (called the no adverse effect level, the NOEL). Then an "uncertainty factor" (usually about 100) is applied to that number to estimate the level of daily exposure to the pesticide that is acceptable for humans (that is, that would not cause any adverse effects). This level is called the Reference Dose (RfD). It was once known as the Acceptable Daily Intake (ADI).

Next, the EPA estimates people's exposure to pesticide residues in food, based on pesticide residue studies (how much residue is found on different crops after application) as well as studies of how much food people consume. This calculated value for potential human exposure to the pesticide from food is termed the theoretical maximum residue contribution (TMRC). Then, using the information on both the chemical's toxicity to humans (the RfD) and a person's potential exposure (the TMRC), the Agency sets tolerance levels that will not pose significant dietary risks to the consumer.

The EPA will usually deny a registration use if the anticipated exposure to humans from a proposed new use of a pesticide on a food crop, when added to estimated exposure from other food uses of that pesticide, exceeds the pesticide's RfD.

Carcinogens

In cases where a food-use pesticide is a carcinogen (cancer-causing agent), the EPA uses a second approach in addition to the one discussed above. The EPA assesses the cancer risk associated with exposure to the pesticide in food over the course of a person's lifetime. The EPA then determines whether that cancer risk can be considered "negligible." The EPA's pesticide program defines a negligible risk as one-in-a-million or less chance of getting cancer as a direct result of a lifetime of exposure to a particular substance. In general, the EPA will grant a tolerance and register any pesticide that poses a negligible or no-cancer risk. For pesticide residues that pose a cancer risk greater than negligible (one in a million), the EPA may register the pesticide if the benefits of its use outweigh its risks.

Monitoring residues

Pesticide tolerances, set by the EPA, are enforced by the Food and Drug Administration, which monitors all domestically produced and imported foods traveling in interstate commerce except meat, poultry, and some egg products. The FDA conducts a Total Diet Study, also known as a Market Basket Study, which measures the average American consumer's daily intake of pesticide residues from foods that are bought in typical supermarkets and grocery stores, and prepared or cooked as they would be in a household setting. The findings of the ongoing Total Diet Study show that dietary levels of most pesticides are less than 1% of the RfD. Imported foods receive special attention in the FDA's monitoring program. Above-tolerance residues in 1987 and 1988 were found in less than 1% of import samples. Even so, the FDA has tightened its import policy in the last few years: if a single shipment from a given source is found to violate U.S. tolerance regulations, all shipments from the same source are subject to automatic detention.

Monitoring meat and poultry products is conducted by the USDA's Food Safety and Inspection Service (FSIS). Each year, the FSIS conducts 10,000 to 20,000 pesticide residue analyses. Currently, fewer than 1% of these tests show illegal residues, and the violation rate has been declining steadily over the last 2 decades. State regulatory agencies are also involved in monitoring the safety of the food supply; some states have their own pesticide residue regulations for food produced and sold within state boundaries.

Environmental monitoring

Historically, environmental contamination by pesticides has been subject to less regulatory scrutiny than food supply contamination. Registration and regegistration of pesticides require the thorough testing of the fate and movement of pesticides and the effects of pesticide exposure to non-target plant and animal species. However, until recently, little regulatory emphasis was placed on the monitoring and estab-lishment of specific levels of pesticides in the environment once a pesticide was registered for use.

Requirements for monitoring are found in the Endangered Species Act, in the discharge limits for point-sources under the Clean Water Act, and in the Maximum Contaminant Levels (MCLs) in drinking water from both surface and groundwater sources.

As part of this increased emphasis, the EPA undertook a major 5-year effort to determine the extent of pesticides in the drinking water supply on a national scale.

11.2 Pesticide use in the United States

Pesticides are chemical substances used by farmers, household residents, or industry to regulate and control various kinds of pests or weeds — boll weevils, gypsy moths, corn fungi, crabgrass, bull thistles, dandelions, and the like. There are three major types of pesticides: (1) herbicides, (2) insecticides, and (3) fungicides. Herbicides are chemicals used for killing weeds or inhibiting plant growth. Insecti-cides are chemicals or mixtures of chemicals intended to prevent or destroy any insects that may destroy crops or gardens. Fungicides are chemicals used to destroy or inhibit fungi, which usually cause plant diseases.

The pesticide industry is very large; in 1992 within the U.S., $8.26 billion was spent on pesticides. Over 56.6% of the expenditures ($4.67 billion) was for herbicides; 30.0% ($2.48 billion) for insecticides; 6.4% ($525 million) for fungicides; and 7.0% ($579 million) was spent on various other pesticides.

In addition to the economic cost of the pesticides, the amounts of pesticides produced are also very large. In 1992, over a billion pounds of pesticides were sold in the U.S. About 58.7% (647 million pounds) was herbicides; 23.2% (255 million pounds) was insecticides; 10.9% (120 million pounds) was fungicides; and 7.2% (80 million pounds) was other pesticides.

From 1983 to 1993, pesticide production fluctuated from year to year, but overall grew slowly from 975 million pounds in 1983 to 1152 million pounds of pesticides in 1993, a growth rate of about 18.2%. This corresponds to an average annual growth rate of 1.8% per year during the 10-year period.

During this same period, there was moderately steady growth in herbicide production, from 587 million pounds in 1983 to 754 million pounds in 1993. The growth rate of 28.4% is nearly 50% more than the rate of overall pesticides production, corresponding to an average annual growth rate of 2.8% per year.

The annual overall production of insecticides remained constant at approximately 200 million pounds per year from 1983 to 1993.

Lastly, the number of pounds of fungicides produced increased greatly from 43 million pounds in 1983 to 78 million pounds in 1993. This represents an overall growth rate of 81.4% and an average annual growth rate of 8.4% per year.

Main uses of pesticides

Pesticides are used mainly for three purposes: (1) agriculture, (2) industry, commercial establishments, and government, and (3) home and garden.

Pesticide usage in agriculture

The number of pounds of pesticides used by the agricultural sector has remained relatively constant from 1980 to 1992. In 1980, 79% (or 846 million pounds) of pesticides was used for agriculture; while in 1992, 76% (839 million pounds) of pesticide was used in the agricultural sector.

Pesticide usage in industry, commercial establishments, and government

The number of pounds of pesticides used by industry, commercial establishments, and government increased steadily from 147 million pounds in 1980 to 193 million pounds in 1992. The moderate increase in the amount of pesticides used was due to the increased usage of herbicides and fungicides.

Pesticide usage in homes and gardens

The number of pounds of pesticides used for home and gardens decreased slightly from 82 million pounds in 1980 to 71 million pounds in 1992.

The decline was due to a reduction in the amount of insecticides used, from 42 million pounds in 1980 to 31 million pounds per year in 1992.

Most commonly used pesticides

A small number of active ingredients accounted for a large proportion of pesticide use.

For example, the combined usage of the three most commonly applied herbicides, atrazine, metolachlor, and alachlor, in 1993 was 175 to 190 million pounds. This represented about 27 to 29% of all herbicides used in the U.S.

The three most commonly used insecticides, as a group, were not as dominant as the top three herbicides. The combined usage of chlorpyrifos, diazinon, and malathion in 1993 was approximately 23 to 33 million pounds. This represents 9 to 13% of the total amount of insecticides used in the U.S.

Lastly, in 1993, the total usage of the top three fungicides combined, sulfur, chlorothalonil, and mancozeb, was approximately 49 to 60% of the total amount of fungicides applied in the U.S.

Pesticides usage by crop

The most common herbicides, triazine, metolachlor, and alachlor, were predominantly used in the planting of corn crops in the midwestern U.S., especially in the states of Illinois, Indiana, Iowa, Missouri, Nebraska, and Ohio.

Insecticides such as chlorpyrifos were predominantly used on corn, peanuts, and wheat crops.

Sulfur, the most commonly used fungicide, was primarily used on fruits, such as apples, grapes, and citrus, though it was also used extensively in peanuts in the southeastern region of the U.S. Chlorothalonil, another fungicide, was also used extensively on peanuts in the southeastern region of the U.S. as well as on tomatoes.

Conclusion

The usage of pesticides throughout the U.S. fluctuated slightly from 1983 to 1992, with an annual average growth rate of 1.8% per year, leading to an overall pesticide usage increase of approximately 18.2%.

This growth in the pesticide volume resulted mostly from an increase in fungicide usage. Fungicide usage in 1992 was nearly 100% higher than that in 1983.

Projecting the usage figures from 1983 to 1992 to the future, it is expected that pesticide usage will continue to increase slowly at a rate of about 1 to 2% per year.

11.3 Dose-response relationships in toxicology

> *"The right dose differentiates a poison and a remedy."*
>
> Paracelsus

Introduction

The science of toxicology is based on the principle that there is a relationship between a toxic reaction (the response) and the amount of poison received (the dose). An important assumption in this relationship is that there is usually a dose below which no response occurs or can be measured. A second assumption is that once a maximum response is reached, any further increases in the dose will not result in any increased effect.

True allergic reactions do not show this type of dose-response relationship. Allergic reactions are special kinds of changes in the immune system and are not really toxic responses. The difference between allergies and toxic reactions is that a toxic effect is the direct result of the toxic chemical acting on cells. Allergic responses are the result of a chemical stimulating the body to release natural chemicals which are in turn directly responsible for the effects seen. Thus, in an allergic reaction, the chemical acts merely as a trigger, not as the bullet.

For all other types of toxicity, knowing the dose-response relationship is a necessary part of understanding the cause-and-effect relationship between chemical exposure and illness. As Paracelsus once wrote, "The right dose differentiates a poison from a remedy." Keep in mind that the toxicity of a chemical is an inherent quality of the chemical and cannot be changed without changing the chemical to another form. The toxic effects on an organism are related to the amount of exposure.

Measures of exposure

Exposure to poisons can be intentional or unintentional. The effects of exposure to poisons vary with the amount of exposure, which is another way of saying "the dose." Usually, when we think of dose, we think in terms of taking one vitamin capsule a day or two aspirin every 4 hours, or something like that. Contamination of food or water with chemicals can also provide doses of chemicals each time we eat or drink. Some commonly used measures for expressing levels of contaminants are listed in Table 11.1. These measures tell us how much of the chemical is in food, water, or air. The amount we eat, drink, or breathe determines the actual dose we receive.

Concentrations of chemicals in the environment are most commonly expressed as ppm and ppb. Government tolerance limits for various poisons are often expressed using these abbreviations. Remember that these are extremely small quantities (see Table 11.1). For example, if you put 1 teaspoon of salt in 2 gallons of water, the resulting salt concentration would be approximately 1000 ppm, and it would not even taste salty!

Table 11.1 Measurements for Expressing Levels of Contaminants in Food and Water

Dose	Abbreviation	Metric equivalent	Abbreviation	Approx. amount in water
Parts per million	ppm	Milligrams per kilogram	mg/kg	1 teaspoon per 1000 gallons
Parts per billion	ppb	Micrograms per kilogram	µg/kg	1 teaspoon per 1,000,000 gallons

Dose-effect relationships

The dose of a poison is going to determine the degree of effect it produces. The following example illustrates this principle. Suppose ten goldfish are in a 10-gallon tank and we add 1 ounce shot of 100-proof whiskey to the water every 5 minutes until all the fish get drunk and swim upside down. Probably none would swim upside down after the first two or three shots. After four or five, a very sensitive fish might. After six or eight shots, another one or two might. With a dose of ten shots, five of the ten fish might be swimming upside down. After fifteen shots, there might be only one fish swimming properly and it too would turn over after seventeen or eighteen shots.

The effect measure in this example is swimming upside down. Individual sensitivity to alcohol varies, as does individual sensitivity to other poisons. There is a dose level at which none of the fish swim upside down (no observed effect). There is also a dose level at which all of the fish swim upside down. The dose level at which 50% of the fish have turned over is known as the ED_{50}, which means effective dose for 50% of the fish tested. The ED_{50} of any poison varies depending on the effect measured. In general, the less severe the effect measured, the lower the ED_{50} for that particular effect. Obviously, poisons are not tested in humans in such a fashion. Instead, animals are used to predict the toxicity that may occur in humans.

One of the more commonly used measures of toxicity is the LD_{50}. The LD_{50} (the lethal dose for 50% of the animals tested) of a poison is usually expressed in milligrams of chemical per kilogram of body weight (mg/kg). A chemical with a small LD_{50} (like 5 mg/kg) is very highly toxic. A chemical with a large LD_{50} (1000 to 5000 mg/kg) is practically nontoxic. The LD_{50} says nothing about nonlethal toxic effects though. A chemical may have a large LD_{50}, but may produce illness at very small exposure levels. It is incorrect to say that chemicals with small LD_{50} values are more dangerous than chemicals with large LD_{50} values; they are simply more acutely toxic. The danger, or risk of adverse effect of chemicals, is mostly determined by how they are used, not by the inherent toxicity of the chemical itself.

The LD_{50} values of different poisons may be easily compared; however, it is always necessary to know which specie was used for the tests and how the poison was administered (the route of exposure), since the LD_{50} of a poison may vary considerably based on the animal species and the way exposure occurs. Some poisons may be extremely toxic if swallowed (oral exposure) and not very toxic at all if splashed on the skin (dermal exposure). If the oral LD_{50} of a poison was 10 mg/kg, 50% of the animals who swallowed 10 mg/kg would be expected to die and 50% to live. The LD_{50} is determined mathematically, and in actual tests using the LD_{50}, it would be unusual to get an exact 50% response. In one test, the mortality might be 30% and in another 70%. Averaged out over many tests, the numbers would approach 50%, if the original LD_{50} determination was valid.

The potency of a poison is a measure of its strength compared to other poisons. The more potent the poison, the less it takes to kill; the less potent the poison, the more it takes to kill. The potencies of poisons are often compared using Signal Words or categories, as shown in the example in Table 11.2.

The designation toxic dose (TD) is used to indicate the dose (exposure) that will produce signs of toxicity in a certain percentage of animals. The TD_{50} is the toxic dose for 50% of the experimental subjects. The larger the TD, the less poison it takes to produce signs of toxicity. The TD does not give any information about the lethal dose because toxic effects (for example, nausea and vomiting) may not be directly related to the way that the chemical causes death. The toxicity of a chemical is an inherent property of the chemical itself. It is also true that chemicals can cause different types of toxic effects, at different dose levels, depending on the animal species tested. For this reason, when using the toxic dose designation, it is useful to precisely define the type of toxicity measured, the animal species tested, and the dose and route of administration.

Table 11.2 Toxicity Rating Scale and Labeling Requirements for Pesticides

Category	Signal word(s) required on label	LD_{50} Oral (mg/kg)	LD_{50} Dermal (mg/kg)	Probable oral label dose
I Highly toxic	DANGER — POISON (skull and cross bones)	<50	<200	A few drops to a teaspoon
II Moderately toxic	WARNING	51–500	200–2000	>1 teaspoon to 1 ounce
III Slightly toxic	CAUTION	>5000	>20,000	>1 cup
IV Practically nontoxic	None required			

Toxicity assessment is quite complex, and many factors can affect the results of toxicity tests. Some of these factors are temperature, food, light, and stressful environmental conditions. Other factors related to the animal itself include age, sex, health, and hormonal status.

The NOEL (no observable effect level) is the highest dose or exposure level of a poison that produces no noticeable toxic effect on animals. From our goldfish example, we know that there is a dose below which no effect is seen. In toxicology, residue tolerance levels of chemicals that are permitted in food or in drinking water, for instance, are usually set from 100 to 1000 times less than the NOEL to provide a wide margin of safety for humans.

The TLV (threshold limit value) for a chemical is the airborne concentration of the chemical (expressed in ppm) that produces no adverse effects in workers exposed for 8 hours per day, 5 days per week. The TLV is usually set to prevent minor toxic effects like skin or eye irritation.

Very often, people compare chemicals based on their LD_{50} values, and base decisions about the safety of a chemical on this number. This is an over-simplified approach to comparing chemicals because the LD_{50} is simply one point on the dose-response curve that reflects the potential of the compound to cause death. What is more important in assessing chemical safety is the threshold dose, and the slope of the dose-response curve, which shows how fast the response increases at the dose increases. Figure 11.1 shows examples of dose-response curves for two different chemicals having the same LD_{50}. Which of these chemicals is more toxic? Answer this question for doses below the LD_{50}, and it is chemical A that is more toxic; at the LD_{50}, they are the same, and above the LD_{50}, chemical B is more toxic. While the LD_{50} can provide some useful information, it is of limited value in risk assessment because the LD_{50} only reflects information about the lethal effects of the chemical. It is quite possible that a chemical will produce a very undesirable toxic effect (such as reproductive toxicity or birth defects) at doses that cause no deaths at all.

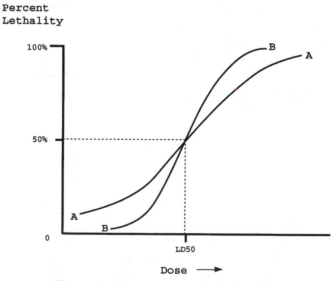

Figure 11.1 Dose-response realationships.

A true assessment of chemical toxicity involves comparisons of numerous dose-response curves covering many different types of toxic effects. The determination of which pesticides will be Restricted Use Pesticides involves this approach. Some

Restricted Use Pesticides have very large LD_{50} values (low acute oral toxicity); however, they may be very strong skin or eye irritants and thus require special handling.

The knowledge gained from dose-response studies in animals is used to set standards for human exposure and the amount of chemical residue that is allowed in the environment. As mentioned previously, numerous dose-response relationships must be determined in many different species. Without this information, it is impossible to accurately predict the health risks associated with chemical exposure. With adequate information, we can make informed decisions about chemical exposure, and work to minimize the risk to human health and the environment.

11.4 How much is a part per million?

Introduction

The health effects of any toxic substance are related to the amount of exposure, also known as the dose. The greater the dose the more severe the effects. Some chemicals can cause toxicity at very low doses, and thus it is important to be able to understand how these very small amounts are described. It is especially important to understand how low doses compare to one another and what they represent when compared to amounts of more familiar substances.

Parts per million (ppm), parts per billion (ppb), and parts per trillion (ppt), are the most commonly used terms to describe very small amounts of contaminants in our environment. But what do these terms represent? They are measures of concentration, the amount of one material within a larger amount of another material; e.g., the weight of a toxic chemical in a certain weight of food. They are expressed as concentrations rather than total amounts so we can easily compare a variety of different environmental situations. For example, scientists can measure the concentration of a chemical in the Great Lakes by looking at small samples. They do not have to measure the total amount of chemicals or water in all of the lakes.

An example might help illustrate the part per . . . idea. If you divide a pie equally into ten pieces, then each piece would be a part per ten, i.e., one tenth of the total pie. If, instead, you cut this pie into a million pieces, then each piece would be very small and would represent a millionth of the total pie or one part per million of the original pie. If you cut each of these million minute pieces into a thousand little pieces, then each of these new pieces would be one part per billion of the original pie. To give you an idea of how little this would be, a pinch of salt in 10 tons of potato chips is also one part (salt) per billion (parts chips).

In this example, the pieces of the pie were made up of the same material as the whole. However, if there was a contaminant in the pie at a level of a part per billion, one of these invisible pieces of pie would be made up of the contaminant, and the other 999,999,999 pieces would be pure pie. Similarly, one part per billion of an impurity in water represents a tiny fraction of the total amount of water. One part per billion is the equivalent of one drop of impurity in 500 barrels of water.

Comparisons and conversions

Sometimes, instead of using the part per . . . terminology, concentrations are reported in weight units; such as the weight of the impurity compared to the weight of the total. The metric system is the most convenient way to express this since metric units go by steps of ten, hundred, and thousand. For example, a milligram is a thousandth of a gram, and a gram is a thousandth of a kilogram. Thus, a milligram is a thousandth of a thousandth, or a millionth of a kilogram. A milligram is one part per million of a kilogram; thus, one part per million (ppm) is the **same** as one milligram per kilogram. Just as part per million is abbreviated as ppm, a milligram per kilogram has its own abbreviation, mg/kg. Using our abbreviations, 1 ppm equals 1 mg/kg.

Kilograms and milligrams are units of weight so they don't apply to volumes of liquids or gases. Instead of a kilogram, the unit of liquid volume most commonly used is the liter. A liter of water weighs 1 kilogram. If the contaminant is a solid, it is measured in milligrams. Thus, one part per million of a solid in a liquid can be written as a milligram per liter and abbreviated mg/L.

These are the most common units encountered. However, with the ability to detect even smaller amounts of contaminants, the terms part per billion and part per trillion are becoming more common. In the metric weight system, a microgram is a thousandth of a milligram. Since a milligram is a millionth of a kilogram, and the microgram is a thousand times smaller, it is equivalent to a billionth of a kilogram. Microgram is abbreviated μg. Thus, a part per billion of solid measure is equal to a μg/kg. Similarly, a part per billion of a solid in a liquid is equal to a μg/L.

Before going on to discuss a real example of how these measurements are used, we can compare metric weight quantities to the quantities we are most accustomed to using. A kilogram is equal to about 2 pounds. Thus, a milligram is less than a millionth of a pound. Looked at another way, it would take about five thousand milligrams (5000 mg) to make up one teaspoonful of a solid (such as salt). The unit of liquid volume, the liter, is very close to a quart. Thus, a milligram per liter is about the same as a milligram per quart. (See Table 11.3.)

Table 11.3 Metric System Quantities

For solids	
1 kilogram (kg)	= 1 million milligrams (mg)
so: 1 mg/kg	= 1 part per million
1 kilogram (kg)	= 1 billion micrograms (ug)
so: 1 μg/kg	= 1 part per billion
For liquids	
1 liter (L) of water weighs exactly 1 kg	
so: 1 mg/L	= 1 part per million and
1 μg/L	= 1 part per billion
1 kg	= about 2.2 pounds
1 L	= about 1 quart

The case of PCBs: an example

In order to appreciate how these quantities can be used in a real situation, an example is in order. In this example, we use the part per . . . terminology to compare the relative importance of PCBs in Great Lakes fish versus PCBs in Great Lakes drinking water; that is, which source might contribute most to PCB exposure of humans living in the Great Lakes states. The maximum level of PCBs legally allowed in fish sold in interstate commerce is 2 ppm (parts per million). Although there are no legally established levels for PCBs in drinking water, measurements have shown that the average PCB content of the Great Lakes drinking water is about 4 ppt (parts per trillion).

Since a part per trillion is a million times less than one part per million, the maximum allowable concentration of PCBs in fish is about a million times higher than the level of PCBs in drinking water. However, we generally consume a lot more water than fish. At the extreme, people might eat as much as a pound of fish a day or as little as one pound every 100 days (1/100 lb/day). On the other hand, people generally drink about 2 liters (equivalent to about 5 pounds) of water a day.

Thus, the consumption of water might range from about 5 to 500 times the consumption of fish. However, since there are a million times more PCBs in a pound of fish compared to a pound of water, fish can be a much larger source of PCBs than drinking water. The total amount of PCBs consumed depends mostly on the amount of fish eaten, how contaminated it is, and how it is prepared. Thus, the best way to reduce human exposure to PCBs is to reduce the levels in fish, reduce human consumption of fish with the highest contaminant levels, and prepare the consumed fish in the most appropriate manner.

Conclusion

The ability to measure concentrations of chemicals in a uniform manner provides a powerful tool for the comparison of water quality from area to area, for the establishment of water quality guidelines, or a comparison of doses of chemicals commonly found in reference works about pesticides. The use of the metric system provides an easy way to understand both liquid and solid measurements.

chapter twelve

Human health effects

12.1 Risk assessment

Background

For most of human history, concern about the toxic effects of chemicals has focused on poisons that act quickly and result in death. A well-known example is hemlock, which Socrates ingested to commit suicide. Until recently, exposure to these chemicals was not common and the risks were well known, so there was little public concern about these poisons.

In this century, however, people have become increasingly concerned with poisons, including those that cause adverse effects only after long periods of exposure. There are two main reasons for this change. One is that the average human lifespan has increased tremendously due to cures and treatments for infectious diseases. This longer lifespan has made chronic, noninfectious illness more common. The second is that the Industrial Revolution has led to new and increased uses of known chemicals and the synthesis and widespread use of newly developed chemical compounds. This tremendous increase in both the quantity and variety of chemical uses has led to greater awareness of possible adverse health effects from industrial products.

Risk assessment

One result of this attention was the establishment of the Environmental Protection Agency (EPA) in 1970 and the enactment of new legislation during the 1970s to regulate chemicals in the environment. With the passage of these laws, an important problem was how to evaluate the severity of the threat that each chemical posed under the conditions of use. This evaluation is known as risk assessment, and is based on the capacity of a chemical to cause harm (its toxicity), and the potential for humans to be exposed to that chemical in a particular situation, e.g., workplace or home.

Risk assessment did not begin in the 1970s. The safety of our food supply has been investigated since early in the 20th century. In addition, scientists in industrial toxicology laboratories had been evaluating the toxic properties of potential products as early as the 1930s. Toxic side effects of drugs had long been of concern and received increased attention in the early 1960s after the discovery that severe birth defects resulted from ingestion of a seemingly safe drug, Thalidomide. During the 1970s, risk assessment procedures for all chemicals were reevaluated, improved, and more importantly, formalized. Standardized tests were developed so consistent evaluations could be performed and the scientific basis of regulations could be more easily applied.

During this time of change, the term "risk assessment" took on a variety of meanings. However, its definition is made up of two components: toxicity (dose-response) assessment and exposure assessment. The former is a measure of the extent and type of negative effects associated with a particular level of exposure, and the latter is a measure of the extent and duration of exposure to an individual or population. For example, characterizing the risk of a pesticide to applicators requires knowing exactly what dose (amount) of the pesticide causes which effects (dose-response assessment) and the dose to which workers are exposed (exposure assessment).

Sometimes, the distinction between an exposure assessment and a dose-response assessment is forgotten and conclusions are drawn without measuring exposure levels. For example, dioxin is often referred to as the most toxic man-made chemical known based on dose-response data, and it is then concluded that it poses the greatest risk to society. However, this is not the case because the potential for dioxin exposure is usually very small.

Exposure assessment

How can exposure assessment be accomplished? There are two basic approaches: analysis of the source of exposure (e.g., levels in drinking water or workplace air), and laboratory tests (e.g., blood or urine analysis on the people thought to be exposed). Analyses of the air or water often provide the majority of usable information. These tests reveal the level of contamination in the air or water to which people are exposed. However, they only reflect the concentration at the time of testing and generally cannot be used to quantify either the type or amount of past contamination. Some estimates of past exposures may be gained from understanding how a chemical moves in the environment.

Some other types of environmental measurements may be helpful in estimating past exposure levels. For example, analyses of fish or lake sediments can provide measures of the amounts of persistent chemicals that are and were present in the water. Past levels of a persistent chemical can be estimated using the age and size of the fish, and information about how quickly the chemical accumulates in the organism. Analyses of body fluid levels of possibly exposed people provide the best measurement of direct exposure. However, they do not provide good estimates of past exposure levels because the body usually reaches a balanced state so there is no longer any change in response to continued exposure. Many chemicals are excreted from the body after exposure ceases, and a basic understanding of what happens to chemicals in the human body is often lacking for those that do persist. Thus, direct examination of a population may provide information as to whether or not exposure has occurred but not the extent, duration, or source of the exposure.

Overall, exposure assessments can be performed most reliably for recent events and much less reliably for past exposures. The difficulties in exposure assessment often make it the weak link in trying to determine the connection between an environmental contaminant and its adverse effects on human health. Although exposure assessment methods will undoubtedly improve, significant uncertainty will remain in the foreseeable future.

Dose-response assessment

Turning to the dose-response assessment, a distinction must be made between acute and chronic effects. Acute effects occur within minutes, hours, or days, while chronic effects appear only after weeks, months, or years. The quality and quantity of scientific evidence gathered is different for each type of effect and, as a result, the conclusions from the test results are also different.

Acute toxicity is easier to test or observe. Short-term animal studies provide evidence of which effects are linked with which chemicals and the levels at which these adverse effects occur. Often, some human exposure data are available as a result of accidental exposures. When these two types of evidence are available, it is usually possible to make a good estimate of the level at which a specific toxicant will cause a particular acute adverse effect in humans. This approach is the basis for the current regulation of toxic substances, especially in the occupational setting.

Chronic toxicity is much more difficult to assess. There are a variety of specific tests to ascertain adverse effects such as reproductive damage, behavioral effects, cancer, etc. It is not possible to discuss all of these, but a look at cancer assessment will reveal some of the problems inherent in long-term toxicity assessments and also focus on the health effect that seems to be of utmost concern to the general public.

In cancer assessment, it is not only the chronic nature of the disease but also the low incidence that causes difficulty. Society has decided that no more than one additional cancer in 100,000 or 1 million exposed people is acceptable, so assessment measures must be able to detect this small number. Two types of evidence are utilized to determine the chemical dose that will result in an adverse health effect. One is derived from animal experiments and the other is derived from human exposure experience.

Ideally, to detect an increase of one cancer in a million animals, millions of animals would have to be exposed to environmentally relevant amounts of the chemical. However, there are neither scientific nor economic resources to carry out this type of study. Thus, investigations are performed on smaller numbers of animals (a few hundred) who have been exposed to very large amounts of a chemical. These large amounts are necessary to produce a high enough incidence of cancer to be detectable in this small population. Thus, the results of such studies indicate the levels of a chemical that will cause cancer in a high percentage of the population.

How can this information be used to assess the chemical level that will cause one additional cancer in a million animals or, more importantly, in a million humans? Because our basic understanding is limited, mathematical models must be used to predict this level. There are a variety of possible models, and the one generally chosen is one that provides the greatest margin of safety, i.e., overestimating rather than underestimating the chemical's ability to cause cancer.

The other type of evidence utilized in chronic toxicity assessment is human exposure, better known as epidemiological evidence. With this type of study, human populations are carefully observed, and possible associations between specific chemical exposure and particular health effects are investigated. Considering the previous discussion about exposure assessment, it should be clear that this is not an easy task. It is even more difficult in cancer assessment because of the requirement of detecting very small changes in incidence, e.g., one extra cancer in a million people.

As a result, epidemiological assessments have been useful in only certain situations. One is exposure in the workplace, a place where exposure levels are usually much higher than in the environment and a place where the duration of exposure can be determined. However, even there, a sizable increase in cancer incidence is needed before a connection can be established. The conclusion that asbestos causes lung cancer is based on this type of situation. An exception to the need for a high cancer incidence is the situation where the effect is unique so that even a few cases are significant. An example of this was the observation that a small number of vinyl chloride workers developed a rare form of liver cancer. However, even with known occupational carcinogens, the question of what happens at low exposures, such as common environmental exposures, has not been fully answered.

Thus, the techniques available for assessment of chronic toxicity, especially carcinogenicity, provide rather clear evidence as to whether or not a particular chemical causes a particular effect in animals. However, there is great uncertainty about the amounts needed to produce small changes in human cancer incidence. This uncertainty, together with the difficulties in exposure assessment, make it difficult to draw definitive conclusions about the relationship between most types of environmental exposure and observed chronic health effects.

Summary

Risk assessment is a complex process that depends on the quality of scientific information that is available. It is best for assessing acute risks where effects appear soon after exposure occurs. Uncertainty becomes greater the longer the period of time between exposure and appearance of symptoms. This is largely due to increased uncertainties in exposure assessment and also the problems involved in using epidemiological or laboratory animal results in such cases. In many circumstances, these uncertainties make it impossible to come to any firm conclusions about risk. Thus, risk assessment is a process that is often useful but cannot always provide the answers that are needed.

12.2 Epidemiology

Introduction

Epidemiology is the study of diseases within human or animal populations; specifically how, when, and where they occur. Epidemiologists attempt to determine which factors are associated with diseases (risk factors), and what factors may protect people or animals against disease (protective factors). The science of epidemiology was first developed to discover and understand possible causes of contagious human diseases like smallpox, typhoid fever, and polio. It has expanded to include the study of factors associated with nontransmissible diseases like cancer, and of poisonings caused by environmental agents.

Epidemiological studies can never prove causation, which means that it cannot prove that a specific risk factor actually causes the disease being studied. Epidemiological evidence can only show that this risk factor is associated (correlated) with a higher incidence of disease in the population exposed to that risk factor. The higher the correlation, the more certain the association, but it does not prove causation.

For example, the discovery of the link between cigarette smoking and lung cancer was based on comparisons of lung cancer rates in smokers and non-smokers. The rates of lung cancer are much higher in smokers than in non-smokers. Does this prove that cigarette smoking causes lung cancer? No. To prove that cigarette smoking causes lung cancer, it was necessary to expose animals to tobacco smoke and tobacco smoke extracts. This was done under highly controlled conditions where the only difference between the controls (animals not exposed to smoke) and treated animals was the exposure to smoke. These laboratory studies proved the causal association between smoking and increased risk of cancer.

Two of the most common types of studies that epidemiologists perform are called case-control and cohort studies. A case-control study usually begins when a disease or adverse effect is noted and the causes of this disease or effect is not known. Epidemiologists then examine the histories of the cases (those showing the disease or effect) and the controls (similar individuals who are not affected) to look for differences in their pasts that might plausibly be linked to the disease or effect. For example, the history of exposure to lead can be compared between people with neurological problems and those not affected to determine if lead is a risk factor.

A cohort study is used to determine if populations with particular characteristics are more likely to develop a disease or show an effect. This is performed by following two groups of people — those with this characteristic and those without — for a period of time to see if there are any differences in the incidence of the disease or effect between the two groups. For example, a group of new workers in the lead industry without health problems and another group of workers not exposed to lead can both be followed over many years to determine if neurological problems are more common in one group than the other.

The first step in an epidemiological study is to strictly define exactly what requirements must be met in order to classify someone as having a disease or showing an effect. This seems relatively easy, and often is when the outcome is clear (a person is either dead or alive). In other instances, it can be very difficult, particularly if the experts disagree about the presence or absence of the effect. This happens often with the diagnosis of particular types of cancer. In addition, in case-control studies, it is necessary to verify that reported cases actually are cases, particularly when the study relies on personal reports and recollections about the disease made by a variety of individuals.

The strength of an epidemiological study depends on the number of individuals in each group. The more individuals that are included in the study, the more likely that a significant association will be found between the disease or effect and a risk factor. And it is just as important to determine which behavioral, environmental, and health factors will be studied as possible risk or protective factors. If inappropriate factors are chosen, and the real factors are missed, the study will not provide any useful information. For instance, an association may be found between an inappropriate factor and the disease because this inappropriate factor, which we will call Factor 1, is associated with another factor, Factor 2, which is actually related to the disease, but which was not studied. In such an instance, Factor 1 is called a *confounding variable* because it confounds the interpretation of the results of the study. Thus, it is very important that the epidemiologist choose the proper factors to study at the outset, and not study too many factors at once, since the possibility of finding confounding factors increases with the addition of more variables.

Establishing the link

Epidemiology relies heavily on statistics for establishing and quantifying the relationships between risk factors and disease, and for establishing whether or not there is an excessive amount of a particular disease occurring in a specific geographic area. Medical records can provide invaluable historical data for establishing trends in the incidence of diseases. There are vast collections of medical record information throughout the world, and sorting through the data can be a very expensive and time-consuming process. In addition, what can be gained from the records is only as good as the information they contain, and often the information is scanty or impossible to verify.

One source of information commonly used is death certificate registries, which usually contain information about the cause of death. Using information from such a registry, a study was performed that showed an unusually high incidence of deaths from lung cancer in a large agricultural valley city. There was no question that the rates of deaths from lung cancer in this city were much higher than in other cities of a similar size and location. Many possible causative agents were suggested, ranging from pesticide use to agricultural burning. Finally, it was realized that this particular city contained a renowned hospital treatment center for patients with lung cancer. As it turned out, the very high rate of cancer deaths there could be explained by the numbers of lung cancer patients who came there from all over the state for treatment and who died there.

Epidemiologists are often called upon to investigate apparent "clusters" of disease in specific geographical areas. For example, a woman in a community may have a miscarriage, and later learn that other people she knows in the neighborhood also have had miscarriages within the last couple of years. It may appear to her, her friends, and her family that a lot of miscarriages have occurred in their neighborhood. An epidemiologist may be called in to determine if the rate of miscarriages is higher than normal. The epidemiologist must interview all of the women in the community, and another similar community in a different geographical area, or select appropriate samples of women to be interviewed concerning their reproductive histories. This information will be validated through hospital records, and then analyzed and compared to other similar studies. Even in situations where the rate of disease is not higher than the normal rate, it may seem higher to the inhabitants because diseases are not evenly distributed throughout populations.

Interpreting the results

If the rate of occurrence of a disease in the general population of the U.S. is 10 per 1000, it does not mean that every group of 1000 people tested will provide ten cases of disease. Some groups may have 5 or less, others 15 or more; 10 is the mean rate. This uneven distribution is similar to the way chocolate chips cluster in cookies. Most of the time they are pretty well spread throughout; however, some cookies may have all the chips clumped together on one side. The only way to determine whether a cluster is a "real" cluster or just a "chance" cluster is to do a full-scale epidemiological study, which is an expensive and time-consuming process, and may still not able to resolve the question.

Summary

Case-control and cohort studies have been extremely valuable in discovering links between chemical exposure and disease. Perhaps the best example is the association of cigarette smoking with lung cancer and emphysema. Epidemiological studies have also been especially useful in the occupational sector where workers are exposed to a small number of chemicals, at high dosage rates, for long time periods. Epidemiological studies are occasionally relatively easy, and especially significant when they uncover a very high incidence of an unusual disease in a population. For example, the finding of a very small number (about 10) of cases of a very rare liver tumor in workers heavily exposed to vinyl chloride, was a strong signal that vinyl chloride was the causative agent. Animal studies supported this conclusion. Epidemiological studies are least powerful in studying very common diseases that occur at high incidence rates in many different populations. In such instances, it is necessary to include huge numbers of subjects in the studies. Such large studies have been undertaken in regard to cardiovascular disease, and many factors that influence the development of cardiovascular disease have been found. Such large-scale efforts have provided information about diet and exercise habits that can be used to prevent the development of cardiovascular disease.

The same is true with respect to chemically induced diseases. We all reap the benefits of epidemiological studies of workers exposed to chemicals in occupational settings. We also reap the benefits of studies done on patients who must take certain medications daily, and who sometimes develop side effects. Epidemiological studies of disease related to chemical exposure are very difficult in the general population because of the multitude of chemicals to which we are daily exposed. This is, after all, a world made up of chemicals, just as we ourselves are.

12.3 Entry and fate of chemicals in humans

Routes of entry

Chemicals, including pesticides, are widely distributed in the environment. Therefore, there are many possible sources of exposure to these chemicals for humans. Substances in ambient and indoor air may be inhaled into the lungs, while those in water may be ingested or inhaled through mist or steam (such as in the shower). Direct contact with the chemical is the most prevalent way environmental chemicals can penetrate the skin, but exposure through the skin may also occur as a result of contact with chemical contaminants in air and water (e.g., bathing or swimming).

A single chemical can enter the body through all three routes of exposure — inhalation, ingestion, and skin penetration (dermal exposure). A compound, such as chloroform, which evaporates readily and which may be found in drinking water, illustrates this point. When someone drinks this water, ingestion is the route of exposure. When it is used for showering, exposure may occur from inhalation of the steam or mist and from direct contact through the skin. Similarly, pesticide use can involve more than one route of exposure if precautions are not taken. A pesticide that is sprayed can be inhaled during use, penetrate through the skin during mixing and application, and/or be ingested through food if hands are not washed.

Absorption, distribution, and fate

Once a chemical enters the body, it is often absorbed into the bloodstream and can move throughout the body. The amount absorbed and the rate of absorption depend on the chemical and the route of exposure. This movement of the substance through the bloodstream is called "distribution." Through distribution, a chemical can come into contact with all parts of the body, not just the original site of entry. In some cases, such contact, distant from the site of entry, can lead to adverse health effects. For example, ingestion of the pesticide parathion into the stomach can lead to damage to the lungs.

Once a chemical is absorbed into the bloodstream, it can have several different fates. In many cases, it is rapidly removed from the body through the urine or feces. In other situations, it may be stored in various parts of the body, such as fat or bone, and remain in the individual for many years. A compound may also lead to a toxic effect through interaction with certain organs or tissues in the individual or with other compounds in the body. Often, a substance that is absorbed into the body interacts with particular body chemicals and is changed into one or more other chemicals. This process is called "metabolism" and the products are called "metabolites." Metabolism may lead to products that are easier for the body to excrete and so can protect the body from possible adverse effects. In other cases, however, the metabolites may be more toxic than the original chemical that was absorbed. The variety of products resulting from metabolism may have the same possible fates as the original chemical — storage, excretion, or toxicity.

Chemical properties

The particular properties of the absorbed chemical are quite critical to its fate in the body. Certain chemicals are very resistant to metabolism and readily dissolve in fat so that they tend to be stored. Dieldrin is a good example of this type of compound. Other chemicals are more rapidly metabolized and excreted, and so are gone before they can cause adverse effects. The organophosphate pesticides tend to behave this way at low doses.

An individual's characteristics

The characteristics of the individual who is exposed are also very important in the fate of the chemical. The age, sex, genetic background, previous exposures, diet, and other factors play important roles in the way that the body interacts with a chemical and in turn the potential for adverse effects. Thus, the characteristics of both the chemical and the exposed individual are important factors in determining the fate of the chemical in the body.

The time course for exposure

In the case of a single-event exposure, it is the total amount of chemical to which a person is exposed that determines the severity of the toxic effect, if any. The greater the amount of exposure, the greater the potential for adverse health effects. In some cases, this is due solely to the inherent toxicity of the chemical and, in others, also to overwhelming the ability of the body to respond. In the latter case, the body may not be able to metabolize the chemical quickly enough to prevent an increase in concentration to toxic levels. In such a situation, there is a clear threshold above which toxic signs and symptoms appear.

In the case of (repeated) multiple exposures to a chemical, it is not only the total amount of exposure, but also the rate or timing of exposure that is quite important. All processes in the body normally proceed at specific rates so that metabolism, excretion, and storage occur during a particular period of time after a chemical is absorbed. For a one-occurrence exposure, the time needed for the various processes to break down the compound completely will determine the length of the toxic response time, if any.

However, if there are repeated exposures to the same chemical, the situation is more complicated. If there is enough time between exposures so that all of the chemical from the initial exposure is excreted, and no effects persist, then each exposure is essentially independent of the previous one and can be treated as a single exposure. However, if the time between exposures is so short that some of the chemical remains from the first exposure, then a build-up of the chemical can occur. Over time, the chemical can rise to toxic levels.

The total amount of exposure can have different results, depending on whether the exposure occurred all at once or repeatedly over time (the time course of exposure). A high dose given once may have a toxic effect, while the same total dose given in small amounts over time will not. For example, drinking several ounces of alcohol at once may cause inebriation, while drinking 1 ounce every few hours may not. Also, a particular dose given a few hours apart may have an adverse effect, while the same total dose given a few days apart will not.

Summary

The possible toxic effects of exposure to a particular chemical depend on many factors. These include the characteristics of the chemical and the individual exposed, the route of exposure, the total dose, and the time course of exposure. Unfortunately, scientists have not been able to determine exactly how each of these factors will affect any specific individual so that present understanding of chemical exposures provides only general guidance. Minimizing exposure will minimize the potential for adverse effects. In addition, a general knowledge of all the contributing factors will help reveal situations that have the most potential for adverse health effects and can aid in determining the best ways to manage chemicals.

12.4 Manifestations of toxic effects

Introduction

The human body is a vastly complex biochemical organism, finely tuned, and adaptable. It contains many different regulatory systems to ensure the proper response to variations in external conditions. When it becomes too warm inside the body, the water cooling system is turned up, and more sweat is secreted by the skin. The sweat evaporates, cools the blood underneath the skin, which in turn cools the body core. The sensors in the brain detect that body temperature is within normal limits, and turn off the sweat glands. This type of regulation (known as homeostasis) occurs for all bodily processes, and usually without any awareness or thought on our part.

When external circumstances (like extreme heat or cold) or internal conditions (disease or poisoning) cannot be adjusted by normal mechanisms, the signs of discomfort and disease appear. The types of physical effects seen or felt (signs and symptoms) depend on the type of stress to which the body has been exposed. Because there are so many complex interrelationships between the systems within the body, a single change in any one system may result in numerous effects on other systems. In addition, the types of response to disease are limited; thus, signs and symptoms of disease are often quite similar for different diseases. For example, headache, fever, nausea, vomiting, and diarrhea are very common nonspecific symptoms of disease, produced by many, many agents. Because of the generality of most physiological responses to disease, many other methods have been developed to help diagnose the actual causes of disease. These methods include physical, biochemical, and immunological techniques upon which modern clinical medicine is based.

A body's homeostasis can be upset by physical, chemical, and/or biological agents that put stress on the body. The body's reaction to prolonged stress depends on the nature of the agent, the degree of stress, and the duration of stress. When the stress is too intense, and homeostasis cannot be maintained or restored, disease occurs. Poisoning by chemical agents is nothing more than chemically induced disease, and the symptoms of chemical poisoning are often the same as symptoms caused by biological agents such as bacteria or viruses. To better understand how disease is caused by exposure to toxic chemicals, we must first understand how poisons work within the body.

How poisons work

Poisons work by changing the speed of different body functions, increasing them (e.g., increasing the heart rate or sweating) or decreasing them (e.g., decreasing the rate of breathing). For example, people poisoned by parathion (an insecticide) may experience increased sweating as a result of a series of changes in the body. The first step is the biochemical inactivation of an enzyme. This biochemical change leads to a cellular change (in this case, an increase in nerve activity). The cellular change is then responsible for physiological changes, which are the symptoms of poisoning that are seen or felt in particular organ systems (in this case, the sweat glands). The basic progression from biochemical to cellular to physiological effects occurs in almost all cases of poisoning.

Depending on the specific biochemical mechanism of action, a poison may have very widespread effects throughout the body, or may cause a very limited change in physiological functioning in a particular region or organ. Parathion causes a very simple inactivation of an enzyme involved in communication between nerves. The enzyme that parathion inactivates, however, is very widespread in the body, and thus the varied effects on many body systems are easily detectable.

Toxicity

Toxicity is a general term used to indicate adverse effects produced by poisons. These adverse effects can range from slight symptoms like headaches or nausea, to severe symptoms like coma, convulsions, and death.

Toxicity is normally divided into various types, based on the number of exposures to a poison and the time it takes for toxic symptoms to develop. The two types most often referred to are acute and chronic. Acute toxicity is due to short-term exposure and happens within a relatively short period of time, whereas chronic toxicity is due to long-term exposure and happens over a longer period.

Most toxic effects are reversible and do not cause permanent damage, but complete recovery may take a long time. However, some poisons cause irreversible (permanent) damage. Poisons can affect just one particular organ system or they may produce generalized toxicity by affecting a number of systems. Usually, the type of toxicity is subdivided into categories based on the major organ systems affected. Some of these are listed in Table 12.1. Subsequent sections of this chapter more fully explain skin and nervous system effects. Another section covers the formation of tumors and cancer.

Because the body only has a certain number of responses to chemical and biological stressors, it is often difficult to sort out the signs and symptoms and determine the actual cause of human disease or illness. In many cases, it is impossible to determine whether an illness was caused by chemical exposure or by a biological agent (like a flu virus). A history of exposure to a chemical is one important clue in helping to establish the cause of illness, but such a history does not constitute conclusive evidence that the chemical was the cause. To establish this cause-effect relationship, it is important that the chemical be detected in the body (such as in the bloodstream) at levels known to cause illness. If the chemical produces a specific and easily detected biochemical effect (like the inhibition of the enzyme acetylcholinesterase), the resulting biochemical change in the body may be used as conclusive evidence.

People who handle chemicals frequently in the course of their jobs and become ill and need medical attention should tell their physicians about their occupational exposure to chemicals.

Table 12.1 General Toxicity Categories

Category	System affected	Common symptoms
Respiratory	Nose, trachea, lungs	Irritation, coughing, choking, tight chest
Gastrointestinal	Stomach, intestines	Nausea, vomiting, diarrhea
Renal	Kidney	Back pain, urinating more or less than usual
Neurological	Brain, spinal cord	Headache, dizziness, confusion, behavior, depression, coma, convulsions
Hematological	Blood	Anemia (tiredness, weakness)
Dermatological	Skin, eyes	Rashes, itching, redness, swelling
Reproductive	Ovaries, testes, fetus	Infertility, miscarriage

12.5 Toxic effects on the skin

Introduction

During mixing, loading, and application of pesticides, the skin is the most likely body surface to come into contact with the product. Many pesticides can be absorbed through the skin into the blood and cause toxic effects. The amount of pesticide absorbed through the skin (percutaneous absorption) may be enough to produce severe toxic reactions, including death. In addition, pesticides can also injure the skin directly, a process known as cutaneous toxicity. Skin irritation and skin rashes produced by irritating chemical substances are a very noticeable type of chemical toxicity. Skin infections by fungi (ringworm, athlete's foot, etc.), bacteria, or by parasites are also very common medical problems and often have the same symptoms as skin irritation caused by chemical exposure. We live in a world filled with substances that can be irritating to our skin, and we use many of these irritating substances every day without great concern because we have learned how to restrict our exposure.

The skin and the mucous membranes that cover the openings of our bodies to the external environment (such as in the nose and mouth), form protective barriers that keep water inside the body, and keep the outside materials such as bacteria, fungi, dust, dirt, etc., from coming in. The skin is really an organ of the body, and a large one at that. The skin is much more than just a simple covering, it is multilayered and, underneath the surface (which is composed of dead cells), are other layers composed of living cells which react to irritants when they get through. When an irritant reaches these sensitive live skin cells, they can only respond in a limited number of ways. One general response to any irritating chemical or physical agent (such as sunlight) is for the skin to become inflamed. Inflammation has four components: redness, pain, heat, and swelling. The degree of inflammation is directly related to the degree of chemical or physical irritation (dose-response). If the damage is great enough to cause cell death, then the response will be much more severe, and can result in areas of the skin becoming "denuded" (loss of the layers, with the deeper layers being exposed to the surface). Because the response of the skin to many different physical and chemical irritants is similar, the causes of skin irritation must usually be diagnosed by a physician who specializes in skin problems, such as a dermatologist.

One of the most common toxic affects produced by pesticides and other chemicals is cutaneous toxicity in its many forms. Cutaneous toxic reactions account for approximately one third of all pesticide-related occupational problems; however, pesticides are not the only chemicals that can cause skin toxicity. For example, allergic dermatitis produced by poison oak is the most frequent cause of temporary disability in forestry workers. Dermatitis means literally "inflammation of the skin." Most often, dermatitis is referred to as a skin "rash"; however, this term is very nonspecific. There are many different types of "rashes," and they differ in the way they appear and in how they are produced. Some of the more frequently encountered dermatitis problems in humans and food animals are primary irritant dermatitis and allergic contact dermatitis.

Primary irritant dermatitis

This type of dermatitis is caused by chemical substances that directly irritate the skin (like caustic acids or bases). The symptoms may be similar to a slight burn (redness, itching, pain) or as severe as blisters, with peeling and open wounds (ulcerations). The areas of direct contact are usually the most affected and this is one of the ways it is recognized. Treatment consists of removal of the irritant by washing and prevention of further contact with the chemical. Steroid creams (such as 0.5% hydrocortisone preparations, available without prescription) may help alleviate pain and itching. When exposure to the irritant is prevented, the irritation will not occur. Thus, the use of appropriate protective equipment can completely prevent the development of primary irritant dermatitis. (See Table 12.2.)

Table 12.2 Plants and Pesticides That May Cause
Primary Irritant Dermatitis

Pesticides		
Sulfur	Captafol	Endosulfan
Omite	Folpet	Lindane
Ziram	Toxaphene	Chloropicrin
Thiram	Methomyl	Kelthane
Zineb	Dinoseb	Triazines
Maneb	Dinitro	Benomyl
Captan	TOK	Glyphosate
Weed oil	Dacthal	
Chlorothalonil	Organophosphates	
Plants	Flowers	Trees
Tomatoes	Dieffenbachia	Rubber tree
Carrots	Castor bean	Fig tree sap
Mushroom	Daffodil	
Cucumber	Buttercup	
Parsnip	Foxglove	
Turnip	Tulip bulb	
Parsley	Narcissus bulb	
Celery		
Cowslip		
Milkweed		

Allergic contact dermatitis

The best example of allergic contact dermatitis (ACD) is poison oak and poison ivy dermatitis. This cutaneous toxic reaction is a true allergic response because the skin must be sensitized by exposure to the chemical (once or many times), and the result is a localized allergic reaction. Not everyone will develop ACD after exposure, and some workers may handle a potentially allergenic substance for years before ACD develops, or it may develop after a single exposure. The symptoms of ACD are exactly those of poison ivy or poison oak dermatitis, and vary from redness, itching, and small blisters to widespread blisters that overlap, forming very large fluid-filled blisters. Treatment involves thorough washing to remove the allergen (in the case of poison ivy and oak, it is the oils in the plants), followed by treatment to reduce the itching, pain, and swelling. A topical corticosteroid cream may be very effective in reducing the symptoms. When large blisters break or are opened, care should be taken to prevent secondary infection of these raw areas. Local anesthetic creams (containing benzocaine) should be avoided since they also can act as contact sensitizers (may also cause allergic dermatitis) and also may delay healing. (See Table 12.3.)

Table 12.3 Plants and Pesticides That May Cause Allergic Contact Dermatitis (ACD)

Pesticides		
Captan	Captafol	Benomyl
Triazines	Dichlorovos	Parathion
Malathion	Naled	Thiram
PCNB	TOK	Zineb
Maneb	Cresol	Formaldehyde
Some natural pyrethroids		
Plants	Flowers	Trees
Poison Ivy	Primrose	Cedar
Poison Oak	English ivy	Pine
Poison Sumac	Tulip bulb	Cashew
Liverwort	Chrysanthemum	
Onions	Narcissus bulb	
Garlic		
Celery		

Photosensitization dermatitis

There are two basic types of photosensitization dermatitis. The first, which is most common in non-humans, is called phototoxic photosensitization dermatitis. Phototoxicity is not very common in humans whereas, in domestic food animals, it is the most common cutaneous toxic response. Phototoxic photosensitivity occurs when a compound (a photosensitizer) is present that makes the individual sensitive to sunlight. The condition of phototoxicity occurs only after exposure to sunlight. Signs of phototoxicity are sunburn-like reactions especially in nonpigmented areas. Some of the medications and plants that cause phototoxic photosensitization in non-humans are listed in Table 12.4.

Contact phototoxic photosensitization is the most common occupational phototoxic reaction in humans, and is caused primarily by plants. This type of toxic reaction occurs when the photoactive chemical produced by the plant (or fungus) contacts the skin, is absorbed into the skin, and then is activated by sunlight. The result is the same as sunburn, and varies in intensity depending on the amount of chemical exposure and the amount of exposure to sunlight. Plants that cause photosensitization are listed in Table 12.5.

The symptoms are redness, pain, blistering, and, following recovery, hyperpigmentation of the affected area. Historians record that the ancient Egyptians knew of this reaction and used it to darken light areas of the skin. Some forms of toxic chemicals from plants are still used today to treat hypopigmentation (abnormally light pigmentation of the skin).

Table 12.4 Plants and Medications That May Cause
Photosensitization When Ingested

Plants	
Tetradymia sp. (Horsebrushes)	*Tribulus terrestris* (Goatshead)
Lecheguilla	*Hypericium* sp. (St. Johnswort)
Lantana	Blue-green algae poisoning
Kochia	
Medications	
Phenothiazines	
Sulfonamides	
Tetracyclines	

Table 12.5 Plants That Cause Photosensitization by Contact

Figs	Parsley	Carrots	Dill	Lime	Buttercup
Mustard	Klamath weed	Celery (with pink rot)			

The second type of photosensitization dermatitis is the photoallergic reaction. This disorder is very similar in appearance to allergic contact dermatitis; however, sunlight is required to initiate the process. It requires prior exposure to the chemical and sunlight, and it not very common.

Summary

Chemicals can be absorbed through skin and into the bloodstream and cause toxic effects. The chemicals may also cause reactions on the skin's surface. Some individuals may be particularly sensitive to a compound or a plant, and others may experience few or no effects from contact. Determining the cause of dermatitis may be easy or it may be difficult. If I knew I walked through poison ivy and I develop a reaction, the connection is fairly obvious. However, if I worked in the garden, or the exterminator visited a week ago, and I developed a rash, the connections are less clear.

Finding out just what has caused the dermatitis may involve "patch testing" in which the patient's skin is exposed to small patches containing dilute solutions of the suspect agents. In this way, the offending chemical can be identified and measures taken to prevent or minimize future exposure. The best way to prevent cutaneous toxicity is the appropriate and correct use of protective clothing, and the use of safe handling and application procedures.

12.6 *Effects on the nervous system: cholinesterase inhibition*

Although a variety of nervous system effects may occur from excessive exposure to pesticides, the most common problem of this type results from blockage of the action of cholinesterase in the body.

Cholinesterase is one of many important enzymes needed for the proper functioning of the nervous systems of humans, other vertebrates, and insects. Certain chemical classes of pesticides, such as organophosphates (OPs) and carbamates (CMs) work against undesirable bugs by interfering with, or "inhibiting" cholinesterase. While the effects of cholinesterase-inhibiting products are intended for insect pests, these chemicals can also be poisonous, or toxic, to humans in some situations.

Human exposure to cholinesterase-inhibiting chemicals can result from inhalation, ingestion, or eye or skin contact during the manufacture, mixing, or application of these pesticides.

Electrical switching centers, called "synapses," are found throughout the nervous systems of humans, other vertebrates, and insects. Muscles, glands, and nerve fibers called "neurons" are stimulated or inhibited by the constant firing of signals across these synapses. Stimulating signals are usually carried by a chemical called "acetylcholine." Stimulating signals are discontinued by a specific type of cholinesterase enzyme, acetylcholinesterase, which breaks down the acetylcholine. These important chemical reactions are occurring constantly at a very fast rate, with acetylcholine causing stimulation and acetylcholinesterase ending the signal. If organophosphate or carbamate insecticides are present in the synapses, however, this situation is thrown out of balance. The presence of cholinesterase-inhibiting chemicals prevents the breakdown of acetylcholine. Acetylcholine can then build up, causing a "jam" in the nervous system. Thus, when a person receives a large enough exposure to cholinesterase-inhibiting compounds, the body is unable to break down the acetylcholine.

Let us look at a typical synapse in the body's nervous system, in which a muscle is being directed by a nerve to move.

> An electrical signal, or nerve impulse, is conducted by acetylcholine across the junction between the nerve and the muscle (the synapse) stimulating the muscle to move. Normally, after the appropriate response is accomplished, cholinesterase is released, which breaks down the acetylcholine, terminating the stimulation of the muscle. The enzyme acetylcholine accomplishes this by chemically breaking down the compound into other compounds and removing them from the nerve junction. If acetylcholinesterase is unable to break down or remove acetylcholine, the muscle can continue to move uncontrollably.

Signs and symptoms of cholinesterase inhibition from exposure to CMs or OPs include the following.

1. In mild cases (within 4 to 24 hours of contact): tiredness, weakness, dizziness, nausea, and blurred vision
2. In moderate cases (within 4 to 24 hours of contact): headache, sweating, tearing, drooling, vomiting, tunnel vision, and twitching
3. In severe cases (after continued daily absorption): abdominal cramps, urinating, diarrhea, muscular tremors, staggering gait, pinpoint pupils, hypotension (abnormally low blood pressure), slow heartbeat, breathing difficulty, and possibly death, if not promptly treated by a physician

Unfortunately, some of the above symptoms can be confused with influenza (flu), heat prostration, alcohol intoxication, exhaustion, hypoglycemia (low blood sugar), asthma, gastroenteritis, pneumonia, and brain hemorrhage. This can cause problems if the symptoms of lowered cholinesterase levels are either ignored or misdiagnosed as something more or less harmful than they really are.

The types and severity of cholinesterase inhibition symptoms depend on:

1. Toxicity of the pesticide
2. Amount of pesticide involved in the exposure
3. Route of exposure
4. Duration of exposure

Although the signs of cholinesterase inhibition are similar for both carbamate and organophosphate poisoning, blood cholinesterase returns to safe levels much more quickly after exposure to CMs than after OP exposure. Depending on the degree of exposure, cholinesterase levels may return to preexposure levels after a period ranging from several hours to several days for carbamate exposure, and from a few days to several weeks for organophosphates.

When symptoms of decreased cholinesterase levels first appear, it is impossible to tell whether a poisoning will be mild or severe. In many instances, when the skin is contaminated, symptoms can quickly go from mild to severe even though the area is washed. Certain chemicals can continue to be absorbed through the skin in spite of cleaning efforts.

If someone experiences any of these symptoms, especially a combination of four or more of these symptoms during pesticide handling or through other sources of exposure, they should immediately remove themselves from possible further exposure.

Exposure to:

Carbamates and organophosphates

May result in:

Build-up of acetylcholine cholinesterase inhibition and constant firing of electrical messages with potential symptoms of twitching, trembling, paralyzed breathing, convulsions, and in extreme cases, death.

Which pesticides can inhibit cholinesterase?

Any pesticide that can bind or inhibit cholinesterase, causing the body to be unable to break down acetylcholine, is called a "cholinesterase inhibitor," or "anticholinesterase agent." The two main classes of cholinesterase-inhibiting pesticides are the organophosphates (OPs) and the carbamates (CMs).

Organophosphate insecticides include some of the most toxic pesticides. They can enter the human body through skin absorption, inhalation, or ingestion. They can affect cholinesterase activity in both red blood cells and in blood plasma, and can act directly, or in combination with other enzymes, on cholinesterase in the body. The most commonly used OPs and carbamates are included in Tables 12.6. and 12.7.

What happens as a result of overexposure to cholinesterase inhibiting pesticides?

The victim of poisoning should be transported to the nearest hospital or poison center at the first sign(s) of poisoning. Atropine and pralidoxime (2-PAM, Protopam) chloride may be given by the physician for organophosphate poisoning; atropine is the only antidote needed to treat cholinesterase inhibition resulting from carbamate exposure.

Cholinesterase testing

The following people should have their cholinesterase levels checked on a regular basis: (1) anyone that mixes, loads, applies, or expects to handle or come in contact with highly or moderately toxic organophosphate and/or carbamate pesticides (this includes anyone servicing equipment used in the process); or (2) anyone that is in contact with these chemicals for more than 30 hours within 1 month.

Every person has his/her own individual "normal" range of baseline cholinesterase values. Cholinesterase levels vary greatly within an individual, between indi-

Table 12.6 Commonly Used Organophosphate (OP) Pesticides

Acephate (Orthene)	Fensulfothion (Dasanit)
Azinphos-methyl (Guthion)	Fenthion (Baytex, Tiguvon)
Carbofuran (Furadan, F formulation)	Fonofos (Dyfonate)
Carbophenothion (Trithion)	Isofenphos (Oftanol, Amaze)
Chlorfenvinphos (Birlane)	Malathion (Cythion)
Chlorpyrifos (Dursban, Lorsban)	Methamidophos (Monitor)
Coumaphos (Co-Ral)	Methidathion (Supracide)
Crotoxyphos (Ciodrin, Ciovap)	Mevinphos (Phosdrin)
Crufomate (Ruelene)	Monocrotophos (Azodrin)
Demeton (Systox)	Naled (Dibrom)
Diazinon (Spectracide)	Oxydemeton-methyl (Metasystox-R)
Dichlorvos (DDVP, Vapona)	Parathion (Niran, Phoskil)
Dicrotophos (Bidrin)	Parathion-methyl (Bladen M)
Dimethoate (Cygon, De-Fend)	Phorate (Thimet)
Dioxathion (Delnav)	Phosalone (Zolonc)
Disulfoton (Di-Syston)	Phosmet (Irnidan, Prolate)
EPN	Phosphamidon (Dimecron)
Ethion	Temephos (Abate)
Ethoprophos (Mocap)	TEPP
Famphur	Terbufos (Counter)
Fenamiphos (Nemacur)	Tetrachlorvinphos (Rabon, Ravap)
Fenitrothion (Sumithion)	Trichlorfon (Dylox, Neguvon)

Table 12.7 Commonly Used Carbamate Pesticides

Aldicarb (Temik)	Methiocarb (Mesurol)
Bendiocarb (Ficam)	Methomyl (Lannate,
Bufencarb	Nudrin)
Carbaryl (Sevin)	Oxamyl (Vydate)
Carbofuran (Furadan)	Pirimicarb (Pirimor)
Formetanate (Carzol)	Propoxur (Baygon)

viduals, between test laboratories, and between test methods. Thus, the change in levels is the most important indicator of exposure.

Anyone anticipating exposure to organophosphate or carbamate pesticides should have their blood cholinesterase levels checked to establish "baseline values."

In general, the initial baseline test should be followed by subsequent cholinesterase testing on a regular (usually monthly) basis. However, this testing should be done weekly during the active season in individuals regularly using OPs and CMs labeled "DANGER." The test should also be repeated any time an individual becomes sick while working with OPs, or within 12 hours of his/her last exposure.

People who want to get their cholinesterase levels checked should consult with either their family or company physician for the specific requirements and procedures for cholinesterase testing in their particular state. Keep in mind that a single test method at one test laboratory should be used in the monitoring program.

A 30% drop below the individual's baseline cholinesterase means that the individual should be removed from all exposure to organophosphates and carbamates, with the individual not being allowed to return until levels return to the preexposure baseline range. Removal from exposure means avoidance of areas where the materials are handled or mixed, and avoidance of any contact with open containers or with equipment that is used for mixing, dusting, or spraying organophosphates or carbamates.

12.7 Carcinogenicity

Introduction

Although both natural and synthetic chemicals may cause a variety of toxic effects at high enough doses, the effect of most concern in the U.S. is cancer. This is not surprising considering the high incidence of this disease, the often fatal outcome, and the overall cost to society. Unfortunately, the incidence of this disease seems to increase with age so that as people live longer, there will be more and more cases of cancer in our country.

Scientists do not yet understand exactly how cancer occurs or why some chemicals seem to cause cancer and others don't. Chemicals known to cause cancer are called carcinogens, and the process of cancer development is called carcinogenesis. Up to now, scientists have identified about two dozen chemicals or occupational exposures that appear to be definitely carcinogenic to humans. Some of the most familiar are tobacco smoke and asbestos. In addition, there are a number of chemicals that cause cancer in animals and are suspected of being human carcinogens. Since not all chemicals have been tested, it is possible that the number of known and suspected human carcinogens will increase in the future.

It must be remembered, however, that as with all toxic effects, the dose or amount of exposure is critical. Just as a small enough amount of cyanide will not lead to death, smoking one cigarette will not lead to lung cancer. Thus, in order to decide on the risk that a particular carcinogen poses, it is important to determine how much of the chemical will cause how many cases of cancer in a specified population. This value can then be compared to one considered an acceptable risk: currently, the generally accepted increase in risk of cancer is one additional cancer in one million people. A few exceptions to this criterion are made in the cases of food additives where no amount of carcinogen is allowed (the Delaney Clause) and of drinking water where a goal of zero contamination for carcinogens has been set.

Carcinogen testing

Once an acceptable risk for a carcinogen has been established (e.g., by the Environmental Protection Agency for environmental contaminants), there remains the problem of determining what dose or amount of chemical will lead to this risk. There are two types of studies used to make this determination: (1) investigations of human populations (epidemiology), and (2) experiments on laboratory animals. Each of these types of studies has its advantages and disadvantages, and both have some degree of uncertainty no matter how much evidence is gathered.

Epidemiological studies in human populations

Investigations of human populations, in an attempt to establish the relationship between environmental factors and health, are called epidemiological studies. Scientists examine selected populations to single out particular exposures that might be related to toxic effects, such as in this case, cancer. Occupational groups, such as factory workers in a particular industry, are often studied for two reasons. One is that their exposures to toxic compounds are generally higher than other people's so that a higher incidence of cancer is expected, if it occurs. A higher incidence is obviously easier to detect. The second reason is that their exposure to a specific chemical is often unique and can more easily be distinguished from exposure to many other chemicals used in daily life.

From epidemiological studies of industrial worker populations, it was possible to show that asbestos is linked to lung cancer, vinyl chloride to a rare form of liver cancer, and benzene to leukemia. There have also been suggestions that exposure of farmers to pesticides might lead to cancer, but the results are not clear-cut and there is still much controversy about the epidemiological studies that have been performed on those populations. Even in clear-cut cases, it is not possible to use epidemiology to establish the exact risk of exposure to specific levels (concentrations) of these chemicals.

Laboratory animal studies

Laboratory studies have several advantages over epidemiological studies. Studies on laboratory animals are often easier to interpret because: (1) chemicals can be studied one at a time; (2) very high doses can be administered; (3) other chemicals and environmental factors can be eliminated or controlled; and (4) animals can be sacrificed during the course of the study. One main disadvantage is that it is not known how to apply these high dose results to the much lower dose exposures that occur in the real world. Equally perplexing is how animal results can be applied to human populations. In extrapolating from high to low doses and from animals to humans, regulatory agencies have taken the approach of trying to be as conservative as possible, i.e., of trying to leave a large margin of safety so that even if the studies are in error, human health (usually of the most sensitive groups in the population, such as young children and pregnant women) will be protected.

As a result, the acceptable exposure levels usually represent what is called the "worst case" situation. An assumption made in the calculation of worst-case exposure levels is that humans will be exposed to the same concentration of the chemical every day of their lives for 70 years. As a result, the governmentally established acceptable risk level does not necessarily represent the "safe level," but rather a target level with the expectation that the true risk of exposure is less than the published value.

Summary

Epidemiological investigations have been used to establish links between a particular chemical and cancer in only a few cases and cannot be employed to determine the exact levels at which cancer will occur. On the other hand, laboratory animal studies provide a way of detecting the carcinogenicity of a large number of chemicals and can provide numerical values for cancer risks. However, the relevance of the high-dose animal results to low doses or to humans is not clear.

In light of these considerations, it is not possible to determine the exact cancer risk for any human population, much less any individual. Public policy makers have tried to use worst-case analyses to be as protective of human health as possible. In addition, to minimize cancer, regulations have been designed to reduce population exposure to known human carcinogens as much as possible. In the case of known animal carcinogens, minimizing exposure is also a regulatory goal. However, since many of these chemicals are also quite beneficial to society, there are questions as to how much exposure reduction can be achieved without eliminating the benefits of these chemicals. Achieving a balance of risk and benefit is especially difficult when the uncertainties involved in determining the actual risk to humans is considered.

At present, there are a number of pesticides known to be animal carcinogens. None have been proven to be human carcinogens. Exposure to pesticides that are probable human carcinogens can be minimized through proper protective equipment and proper storage, use, and disposal of these pesticides. These measures not only protect the pesticide applicator but also the general public that consumes foods treated with pesticides or spends time in buildings treated with these chemicals.

chapter thirteen

Ecological and environmental effects

13.1 Movement of pesticides in the environment

Introduction

The widespread use and disposal of pesticides by farmers, institutions, and the general public provide many possible sources of pesticides in the environment. After release into the environment, pesticides may have many different fates. Pesticides that are sprayed can move through the air and may eventually end up in other parts of the environment, such as in soil or water. Pesticides applied directly to the soil may be washed off the soil into nearby bodies of surface water or may percolate through the soil to lower soil layers and groundwater. Pesticides injected into the soil may also be subject to the latter two fates. The application of pesticides directly to bodies of water for weed control, or indirectly as a result of leaching from boat paint, runoff from soil, or other routes, may lead not only to the build-up of pesticides in water, but may also contribute to air levels through evaporation.

This incomplete list of possibilities suggests that the movement of pesticides in the environment is very complex with transfers occurring continually among different environmental compartments. In some cases, these exchanges occur not only between areas that are close together (such as a local pond receiving some of the herbicide application on adjacent land) but may also involve transportation of pesticides over long distances. The worldwide distribution of DDT and the presence of pesticides in bodies of water far from their primary use areas are good examples of the vast potential for such movement.

While all of the above possibilities exist, this does not mean that all pesticides travel long distances or that all compounds are threats to groundwater. To understand which ones are of most concern, it is necessary to understand how pesticides move in the environment and what characteristics must be considered in evaluating contamination potential. Two things may happen to pesticides once they are released into the environment. They may be broken down, or degraded, by the action of sunlight, water or other chemicals, or microorganisms such as bacteria. This degradation process usually leads to the formation of less harmful breakdown products, but in some instances can produce more toxic products.

The second possibility is that the pesticide will be very resistant to degradation by any means and thus remain unchanged in the environment for a long period of time. The ones that are most rapidly broken down have the shortest time to move through the environment or to produce adverse effects in people or other organisms. The ones that last the longest, the so-called "persistent pesticides," can move over long distances and can build up in the environment, leading to greater potential for adverse effects.

Properties of pesticides

In addition to resistance to degradation, there are a number of other properties of pesticides that determine their behavior and fate. One is how volatile they are or, in other words, how easily they evaporate. The ones that are most volatile have the greatest potential to evaporate into the atmosphere and, if persistent, to move long distances. Another important property is solubility in water or how easily they dissolve in water. If a pesticide is very soluble in water, it is more easily carried off with rainwater as runoff or through the soil as a potential groundwater contaminant (leaching). In addition, the water-soluble pesticide is more likely to stay mixed in the surface water where it can have adverse effects on fish and other organisms. If the pesticide is very insoluble in water, it usually tends to stick to soil and also to settle to the bottom of bodies of surface water, making it less available to organisms.

Environmental characteristics

From a knowledge of these and other characteristics, it is possible to predict in a general sense how a pesticide will behave. Unfortunately, more precise prediction is not possible because the environment itself is very complex. There are, for example, huge numbers of soil types that vary with respect to the percentage of sand, organic matter, metal content, acidity, etc. All of these soil characteristics influence the behavior of a pesticide so that a pesticide that might be anticipated to contaminate groundwater in one soil may not do so in another.

Similarly, surface waters vary in their properties, such as acidity, depth, temperature, clarity (suspended soil particles or biological organisms), flow rate, and general chemistry. These properties and others all can affect pesticide movement and fate. Everyone who is familiar with the difficulty of forecasting weather knows it is partly due to problems in predicting air flow patterns. As a result, determination of pesticide distribution in the atmosphere is subject to great uncertainty.

With such great complexity, scientists cannot determine exactly what will happen to a particular pesticide once it has entered the environment. However, they can divide pesticides into general categories with regard to, for example, *persistence* and *potential* for groundwater contamination. They can also provide some idea as to where the released pesticide will most likely be found at highest levels. Thus, it is possible to gather information that can help make informed decisions about what pesticides to use in which situations and what possible risks are being faced due to a particular use.

Movement of pesticides in soil

Table 13.1 lists some of the more commonly used pesticides, with an estimate of their persistence in soil. In this table, persistence is measured as the time it takes for half of the initial amount of a pesticide to break down. Thus, if a pesticide's half-life is 30 days, half will be left after 30 days, one quarter after 60 days, one eighth after 90 days, and so on. It might seem that a short half-life would mean a pesticide would not have a chance to move far in the environment. This is generally true. However, if it is also very soluble in water and the conditions are right, it can move rapidly through certain soils. As it moves away from the surface, it moves away from the agents that are degrading it, such as sunlight and bacteria. As it gets deeper into the soil, it degrades more slowly and thus has a chance to get into groundwater. The measures of soil persistence in the table only describe pesticide behavior at or near the surface.

The downward movement of nonpersistent pesticides is not an unlikely scenario, and several pesticides with short half-lives, such as aldicarb, have been widely found in groundwater. In contrast, very persistent pesticides may have other properties that limit their potential for movement throughout the environment. Many of the chlorinated hydrocarbon pesticides are very resistant to breakdown but are also very water insoluble and tend not to move down through the soil into groundwater. They can, however, become problems in other ways since they remain on the surface for a long time where they may be subject to runoff and possible evaporation. Even if they are not very volatile, the tremendously long time that they persist can lead, over time, to measurable concentrations moving through the atmosphere and accumulating in remote areas.

Table 13.1
Pesticide Persistence in Soils

Low persistence (half-life < 30 days)	Moderate persistence (half-life 30–100 days)		High Persistence (half-life > 100 days)
Aldicarb	Aldrin	Glyphosate	Bromacil
Captan	Atrazine	Heptachlor	Chlordane
Dalapon	Carbaryl	Linuron	Lindane
Dicamba	Carbofuran	Parathion	Paraquat
Malathion	Diazinon	Phorate	Picloram
Methyl-parathion	Endrin	Simazine	TCA
Oxamyl	Fonofos	Terbacil	Trifluralin

Role of living organisms

So far, the discussion has focused on air, soil, and water. However, living organisms may also play a significant role in pesticide distribution. This is particularly important for pesticides that can accumulate in living creatures. An example of accumulation is the uptake of a very water-insoluble pesticide, such as chlordane, by a creature living in water. Since this pesticide is stored in the organism, the pesticide accumulates and levels increase over time. If this organism is eaten by a higher organism which also stores this pesticide, levels can reach much higher values in the higher organism than is present in the water in which it lives. Levels in fish, for example, can be tens to hundreds of thousands of times greater than ambient water levels of the same pesticide. This type of accumulation has a specific name. It is called "bioaccumulation."

In this regard, it should be remembered that humans are at the top of the food chain and so may be exposed to these high levels when they eat food animals that have bioaccumulated pesticides and other organic chemicals. It is not only fish but also domestic farm animals that can be accumulators of pesticides, and so care must be taken in the use of pesticides in agricultural situations.

Summary

The release of pesticides into the environment may be followed by a very complex series of events that can transport the pesticide through the air or water, into the ground, or even into living organisms. The most important route of distribution and the extent of distribution will be different for each pesticide. It will depend on the formulation of the pesticide (what it is combined with) and how and when it is released. Despite this complexity, it is possible to identify situations that can pose concern and to try to minimize them. However, there are significant gaps in the knowledge of pesticide movement and fate in the environment, and so it is best to minimize unnecessary release of pesticides into the environment. The fewer pesticides that are unnecessarily released, the safer our environment will be.

13.2 Bioaccumulation

Defining bioaccumulation

An important process by which chemicals can affect living organisms is through bioaccumulation. Bioaccumulation means an increase in the concentration of a chemical over time in a biological organism compared to the chemical's concentration in the environment. Compounds accumulate in living things any time they are taken up and stored faster than they are broken down (metabolized) or excreted. Understanding the dynamic process of bioaccumulation is very important in protecting human beings and other organisms from the adverse effects of chemical exposure, and it has become a critical consideration in the regulation of chemicals.

A number of terms are used in conjunction with bioaccumulation. **Uptake** describes the entrance of a chemical into an organism — such as by breathing, swallowing, or absorbing it through the skin — without regard to its subsequent storage, metabolism, or excretion by that organism.

Storage, a term sometimes confused with bioaccumulation, means the temporary deposit of a chemical in body tissue or in an organ. Storage is just one facet of chemical bioaccumulation. (The term also applies to other natural processes, such as the storage of fat in hibernating animals or the storage of starch in seeds.)

Bioconcentration is the specific bioaccumulation process by which the concentration of a chemical in an organism becomes higher than its concentration in the air or water around the organism. Although the process is the same for both natural and man-made chemicals, the term bioconcentration usually refers to chemicals foreign to the organism. For fish and other aquatic animals, bioconcentration after uptake through the gills (or sometimes the skin) is usually the most important bioaccumulation process.

Biomagnification describes a process that results in the accumulation of a chemical in an organism at higher levels than are found in its own food. It occurs when a chemical becomes more and more concentrated as it moves up through a food chain — the dietary linkages from single-celled plants to increasingly larger animal species.

A typical food chain includes algae eaten by a waterflea, eaten by a minnow, eaten by a trout, and finally consumed by an osprey (or human being). If each step results in increased bioaccumulation, that is, biomagnification, then an animal at the top of the food chain, through its regular diet, may accumulate a much greater concentration of chemical than was present in organisms lower in the food chain.

Biomagnification is illustrated by a study of DDT that showed that where soil levels were 10 parts per million (ppm), DDT reached a concentration of 141 ppm in earthworms and 444 ppm in robins. Through biomagnification, the concentration of a chemical in the animal at the top of the food chain may be high enough to cause death or adverse effects on behavior, reproduction, or disease resistance and thus endanger that species, even though contamination levels in the air, water, or soil are low. Fortunately, bioaccumulation does not always result in biomagnification.

The bioaccumulation process

Bioaccumulation is a normal and essential process for the growth and nurturing of organisms. All animals, including humans, daily bioaccumulate many vital nutrients, such as Vitamins A, D, and K, trace minerals, and essential fats and amino acids. What concerns toxicologists is the bioaccumulation of substances to levels in the body that can cause harm. Because bioaccumulation is the net result of the interaction of uptake, storage, and elimination of a chemical, these parts of the process will be examined further.

Uptake

Bioaccumulation begins when a chemical passes from the environment into an organism's cells. Uptake is a complex process that is still not fully understood. Scientists have learned that chemicals tend to move, or diffuse, passively from a place of high concentration to one of low concentration. The force or pressure for diffusion is called the *chemical potential*, and it works to move a chemical from outside to inside an organism.

A number of factors may increase the chemical potential of certain substances. For example, some chemicals do not mix well with water. They are called lipophilic, meaning "fat loving", or hydrophobic, meaning "water hating." In either case, they tend to move out of water and into the cells of an organism, where there are lipophilic microenvironments.

Storage

The same factors affecting the uptake of a chemical continue to operate inside an organism, hindering a chemical's return to the outer environment. Some chemicals are attracted to certain sites, and by binding to proteins or dissolving in fats, they are temporarily stored. If uptake slows or is not continued, or if the chemical is not very tightly bound in the cell, the body can eventually eliminate the chemical.

One factor important in uptake and storage is water solubility, which is the ability of a chemical to dissolve in water. Usually, compounds that are highly water soluble have a low potential to bioaccumulate and do not readily enter the cells of an organism. Once inside the organism, they are easily removed unless the cells have a specific mechanism for retaining them.

Heavy metals like mercury and certain other water-soluble chemicals are the exceptions because they bind tightly to specific sites within the body. When binding occurs, even highly water-soluble chemicals can accumulate. This is illustrated by cobalt, which binds very tightly and specifically to sites in the liver and accumulates there despite its water solubility. Similar accumulation processes occur for mercury, copper, cadmium, and lead.

Many fat-loving (lipophilic) chemicals pass into organism's cells through the fatty layer of cell membranes more easily than water-soluble chemicals. Once inside the organism, these chemicals may move through numerous membranes until they are stored in fatty tissues and begin to accumulate.

The storage of toxic chemicals in fat reserves serves to detoxify the chemical by removing it from contact with other organs. However, when fat reserves are utilized to provide energy for an organism, the materials stored in the fat may be remobilized within the organism and may again be potentially toxic. If appreciable amounts of a toxicant are stored in fat, and fat reserves are quickly used, significant toxic effects may be seen from the remobilization of the chemical.

Elimination

Another factor affecting bioaccumulation is whether an organism can break down and/or excrete a chemical. The biological breakdown of chemicals is called *metabolism*. This ability varies among individual organisms and species and also depends on the chemical's characteristics.

Chemicals that dissolve readily in fat but not in water tend to be more slowly eliminated by the body and thus have a greater potential to accumulate. Many metabolic reactions alter a chemical into more water-soluble metabolites, which are more readily excreted.

However, there are exceptions. Natural pyrethrins, insecticides that are derived from the chrysanthemum plant, are highly fat-soluble pesticides, but they are easily degraded and do not accumulate. The insecticide chlorpyrifos, which is less fat soluble but more poorly degraded, tends to bioaccumulate. Factors affecting metabolism often determine whether a chemical achieves its bioaccumulation potential in a given organism.

A dynamic equilibrium

When a chemical enters the cells of an organism, it is distributed and then excreted, stored, or metabolized. Excretion, storage, and metabolism decrease the concentration of the chemical inside the organism, increasing the potential of the chemical in the outer environment to move into the organism. During constant environmental exposure to a chemical, the amount of a chemical accumulated inside the organism and the amount leaving reach a state of dynamic equilibrium.

To understand this concept of dynamic equilibrium, imagine a tub filling with water from a faucet at the top and draining out through a pipe of smaller size at the bottom. When the water level in the tub is low, little pressure is exerted on the outflow at the bottom of the tub. As the water level rises, the pressure on the outflow increases. Eventually, the amount of water flowing out will equal the amount flowing in, and the level of the tub will not change. If the input or outflow is changed, the water in the tub adjusts to a different level.

So it is with living organisms. An environmental chemical will at first move into an organism more rapidly than it is stored, degraded, and excreted. With constant exposure, the concentration inside the organism gradually increases. Eventually, the concentration of the chemical inside the organism will reach an equilibrium with the concentration of the chemical outside the organism, and the amount of chemical entering the organism will be the same as the amount leaving. Although the amount inside the organism remains constant, the chemical continues to be taken up, stored, degraded, and excreted.

If the environmental concentration of the chemical increases, the amount inside the organism will increase until it reaches a new equilibrium. Exposure to large amounts of a chemical for a long period of time, however, may overwhelm the equilibrium (for example, overflowing the tub) and potentially cause harmful effects.

Likewise, if the concentration in the environment decreases, the amount inside the organism will also decline. If the organism moves to a clean environment, in which there is no exposure, then the chemical eventually will be eliminated from the body.

Bioaccumulation factors

This simplified explanation does not take into account all of the many factors that affect the ability of chemicals to be bioaccumulated. Some chemicals bind to specific sites in the body, prolonging their stay, whereas others move freely in and out. The time between uptake and eventual elimination of a chemical directly affects bioaccumulation. Chemicals that are immediately eliminated, for example, do not bioaccumulate.

Similarly, the duration of exposure is also a factor in bioaccumulation. Most exposures to chemicals in the environment vary continually in concentration and duration, sometimes including periods of no exposure. In these cases, an equilibrium is never achieved and the accumulation is less than expected.

Bioaccumulation varies between individual organisms as well as between species. Large, fat, long-lived individuals or species with low rates of metabolism or excretion of a chemical will tend to bioaccumulate more than small, thin, short-lived organisms. Thus, an old lake trout may bioaccumulate much more than a young bluegill in the same lake.

Summary

Bioaccumulation results from a dynamic equilibrium between exposure from the outside environment and uptake, excretion, storage, and degradation within an organism. The extent of bioaccumulation depends on the concentration of a chemical in the environment, the amount of chemical coming into an organism from the food, air, or water, and the time it takes for the organism to acquire the chemical and then excrete, store, and/or degrade it. The nature of the chemical itself, such as its solubility in water and fat, affects its uptake and storage. Equally important is the ability of the organism to degrade and excrete a particular chemical. When exposure ceases, the body gradually metabolizes and excretes the chemical.

Bioaccumulation is a normal process that can result in injury to an organism only when the equilibrium between exposure and bioaccumulation is overwhelmed. Sometimes, bioaccumulation can be a protective mechanism in which the body accumulates needed chemicals.

13.3 Ecological effects

Introduction

Chemicals released into the environment may have a variety of adverse ecological effects. Ranging from fish and wildlife kills to forest decline, ecological effects can be long-term or short-lived changes in the normal functioning of an ecosystem, resulting in biological, economic, social, and aesthetic losses. These potential effects are an important reason for the regulation of pesticides, toxic wastes, and other sources of pollution.

What is an ecosystem?

An ecosystem is the physical environment, along with the organisms (biota) inhabiting that space. Some examples of ecosystems include: a farm pond, a mountain meadow, and a rain forest.

An ecosystem follows a certain sequence of processes and events through the days, seasons, and years. These processes include not only the birth, growth, reproduction, and death of biota in that particular ecosystem, but also the interactions between species and physical characteristics of the geological environment. From these processes, the ecosystem gains a recognizable structure and function, and matter and energy are cycled and flow through the system. Over time, better adapted species will dominate. Sometimes, entire new species may be introduced and perhaps greatly alter the ecosystem.

The organization of ecosystems

The basic level of ecological organization is the individual: a single plant, or insect, or bird. The definition of ecology is based on the interactions of organisms with their environment. In the case of an individual, it would entail the relationships between that individual and numerous physical (rain, sun, wind, temperature, nutrients, etc.) and biological (plants, insects, diseases, animals, etc.) factors. The next level of organization is the population. Populations are no more than collections of individuals of the same species within an area or region. We can see populations of humans, birch trees, or sunfish in a pond. Population ecology is concerned with the interaction of the individuals with each other and with their environment.

The next, more complex level of organization is the community. Communities are made up of different populations of interacting plants, animals, and microorganisms also within some defined geographic area. Different populations within a community interact more among themselves than with populations of the same species in other communities. Therefore, there are often genetic differences between the members of two different communities. The populations in a community have evolved together, so that members of that community provide resources (such as nutrition and shelter) for each other.

The next level of organization is the ecosystem. An ecosystem consists of different communities of organisms associated within a physically defined space. For example, a forest ecosystem consists of animal and plant communities in the soil, forest floor, and forest canopy, along the stream bank and bottom, and in the stream. A stream bottom community will contain various fungi and bacteria living on dead leaves and animal wastes, protozoans and microscopic invertebrates feeding on these microbes, and larger invertebrates (worms, crayfish) and vertebrates (turtles, catfish). Each community functions somewhat separately, but is linked to the others by the forest, rainfall, and interactions between communities. For example, the stream community is heavily dependent upon leaves produced in the surrounding trees falling into the stream that feeds the microbes and other invertebrates. As another example, the rainfall and groundwater flow in a surrounding forest community greatly affects the amount and quality of water entering the stream or lake system.

Terrestrial ecosystems can be grouped into units of similar nature, termed biomes (such as a deciduous forest, grassland, coniferous forest, etc.), or into a geographic unit, termed landscapes, containing several different types of ecosystems. Aquatic ecosystems are commonly categorized on the basis of whether the water is moving (streams, river basins) or still (ponds, lakes, large lakes), and whether the water is fresh, salty (oceans), or brackish (estuaries). Landscapes and biomes (and large lakes, river basins, and oceans) are subject to global threats of pollution (acid deposition, stratospheric ozone depletion, air pollution, the greenhouse effect) and human activities (soil erosion, deforestation).

Adverse effects on ecosystems

While many natural forces — drought, fire, flood, frost, or species migration — can affect it, an ecosystem will usually continue to function in a recognizable way. For instance, a pond ecosystem may go through flood or drought but continues to be a pond. This natural resilience of ecosystems enables them to resist change and recover quickly from disruption.

On the other hand, toxic pollutants and other nonnatural phenomena can overwhelm the natural stability of an ecosystem and result in irreversible changes and serious losses, as illustrated by the following examples.

- Decline of forests, due to air pollution and acid deposition
- Loss of fish production in a stream, due to death of invertebrates from copper pollution
- Loss of timber growth, due to nutrient losses caused by mercury poisoning of microbes and soil insects
- Decline and shift in age of eagle and hawk (and other top predator) populations, due to the effects of DDT in their food supply on egg survival
- Losses of numbers of species (diversity) in ship channels subjected to repeated oil spills
- Loss of commercially valuable salmon and endangered species (bald eagle, osprey) from forest applications of DDT

Each of these pollutant-caused losses has altered ecosystem processes and components and thus affected the aesthetic and commercial value of the ecosystem.

Usually, adverse ecological effects take place over a long period of time or even at some distance from the point of chemical release. For example, DDT, though banned for use in the U.S. for over 20 years, is still entering the Great Lakes ecosystem through rainfall and dust from sources halfway around the world. The long-term effects and overall impacts of new and existing chemicals on ecosystems can only be partially evaluated by current laboratory testing procedures. Nevertheless, through field studies and careful monitoring of chemical use and biological outcome, it is possible to evaluate the short- and long-term effects of pesticides and other chemicals.

Adverse ecological effects on communities

Scientists are most concerned about the effects of chemicals and other pollutants on communities. Short-term and temporary effects are much more easily measured than long-term effects of pollutants on ecosystem communities. Understanding the impacts requires knowledge of the time course and variability of these short-term changes.

Pollutants may adversely affect communities by disrupting their normal structure and delicate interdependencies. The structure of a community includes its physical system, usually created by the plant life and geological processes, as well as the relationships between its populations of biota.

For example, a pollutant may eliminate a species essential to the functioning of the entire community; it may promote the dominance of undesirable species (weeds, trash fish); or it may simply decrease the numbers and variety of species present in the community. It may also disrupt the dynamics of the food webs in the community by breaking existing dietary linkages between species. Most of these adverse effects in communities can be measured through changes in productivity in the ecosystem. Under natural stresses (e.g., unusual temperature and moisture conditions), the community may be unable to tolerate effects of a chemical otherwise causing no harm.

An important facet of biological communities is the number and intensity of interactions between species. These interactions make the community greater than simply the sum of its parts. The community as a whole is stronger than its populations, and the ecosystem as a whole is more stable than its communities. A seriously altered interaction may adversely affect all the species dependent on it. Even so, some ecosystem properties or functions (such as nutrient dynamics) can be altered by chemicals without apparent effects on populations or communities. Thus, an important part of research in ecological effects is concerned with the relative sensitivity of ecosystems, communities, and populations to chemicals and to physical stresses.

Consider the effects of spraying an orchard with an insecticide when bees and other beneficial insects may be present and vulnerable to the toxicant. This practice is both economically and ecologically unsound, since it would deprive all plants in the area of pollinators and disrupt the dynamic control of plant pests by their natural enemies. Advanced agricultural practices, such as integrated pest management (IPM), avoid these adverse effects through appropriate timing and selection of sprays in conjunction with nonchemical approaches to insect control.

The effects of chemicals on communities can be measured through laboratory model ecosystem (microcosm) studies, through intermediate-sized model systems (mesocosms, engineered field systems, open-top plant chambers, field pens), and through full-field trials. Thus, data gathered about effects of chemicals on processes and species can be evaluated in various complex situations that reflect the real world.

Adverse effects on species

Most information on ecological effects has been obtained from studies on single species of biota. These tests have been performed in laboratories under controlled conditions and chemical exposures, usually with organisms reared in the laboratory representing inhabitants of natural systems. Most tests employ short-term, single exposures (acute toxicity assays), but long-term (chronic) exposures are used as well. Although such tests reveal which chemicals are relatively more toxic and which species are relatively more vulnerable to their effects, these tests do not disclose much about either the important interactions noted above or the role of the range of natural conditions faced by organisms in the environment.

Generally, the effects observed in these toxicity tests include reduced rates of survival or increased death rates, reduced growth and altered development, reduced reproductive capabilities, including birth defects, changes in body systems, including behavior, and genetic changes. Any of these effects can influence a species' ability to adapt and respond to other environmental stresses and community interactions.

Environmental toxicology studies performed on species in the laboratory provide the basis for much of the current regulation of pollutants and have allowed major improvements in environmental quality. However, these tests yield only a few clues as to effects on more complex systems. Long-term studies and monitoring of ecological effects of new and existing chemicals released into the environment are needed to develop an understanding of potential adverse ecological effects and their consequences.

Summary

Adverse ecological effects from environmental pollutants occur at all levels of biological organization, but most information about these effects has been obtained with single species. The effects can be global or local, temporary or permanent, short-lived (acute) or long-term (chronic). The most serious effects involve loss in production, changes in growth, development, and/or behavior, altered diversity or community structure, changes in system processes (such as nutrient cycling), and losses of valuable species. These ecological losses in turn may be biologically, economically, aesthetically, or socially important. Hence, ecological effects are of serious concern in regulating pollutants, and a variety of tests have been devised to help evaluate the potential for adverse ecological effects. Developing an understanding of how these tests and other information can be used to prevent environmental problems caused by pollutants is the basis for ecological risk assessment research.

Section IV

Appendices

Appendix A

Index of trade names

This index cross-references the trade name of the pesticides with their corresponding common name (as distinguished from their chemical name). For many of the pesticides, there are a host of trade names. This results from the number of companies producing the product, or the number of different formulations of similar products. While this list represents a great number of the current and past trade names of the pesticides covered in this book, it is not intended to be exhaustive. However, it should provide almost all of the most commonly used trade names for each pesticide.

H = herbicide   GR = growth regulator
I = insecticide   M = molluscicide
F = fungicide   A = avicide
R = rodenticide

Trade Name	Common Name	Type
58-12-315	propoxur	I
AAphytora	zineb	F
AApirol	thiram	F
AAprotect	ziram	F
AAstar	flucythrinate	I
AAtack	thiram	F
AAtrex	atrazine	H
AAvolex	ziram	F
Abat	temephos	I
Abate	temephos	I
Abater	temephos	I
Abathion	temephos	I
AC 217	hydramethylnon	I
AC 300	hydramethylnon	I
AC 3422	parathion	I
AC 3911	phorate	I
AC 5223	dodine	F
AC 52160	temephos	I
AC 92100	terbufos	I
AC 92553	pendimethalin	H
AC 222705	flucythrinate	I
Acaraben	chlorobenzilate	I
Acarin	dicofol	I
Access	triclopyr	H
Accotab	pendimethalin	H

Trade Name	Common Name	Type
Acarin	dicofol	I
Ace-Brush	triclopyr	H
Aciban	chlorpyrifos	I
Acibate	temephos	I
Acimal	malathion	I
Acinate	methomyl	I
Acioate	dimethoate	I
Acithion	ethion	I
Acitox	lindane	I
Acivap	dichlorvos (DDVP)	I
Acizol	triadimefon	F
Acme	2,4-D	H
ACP M-629	chloramben	H
ACR-2807B	fenoxycarb	GR
ACR-2913A	fenoxycarb	GR
Acrobe	Bacillus thuringiensis	I
Acuprex	zineb	F
Adagio	bentazone	H
Adion	permethrin	I
Adios	carbaryl	I
Adizon	diazinon	I
Afalon	linuron	H
Affirm	abamectin	I
Afidan	endosulfan	I

Trade Name	Common Name	Type	Trade Name	Common Name	Type
Agricorn	2,4-D, MCPA	H	Amino-4-Pyridine	4-aminopyridine	A
Agridip	coumaphos	I	Amiral	triadimefon	F
Agri-Mek	abamectin	I	Amitrol-TL	amitrole	H
Agrimet	phorate	I	Amino Triazole	amitrole	H
Agrinate	methomyl	I	Aminozide	daminozide (alar)	GR
Agritox	MCPA, chlorpyrifos	H	Amizine	amitrole, simizine	H
Agriquat	paraquat	H	Amizol	amitrole	H
Agrobac	Bacillus thuringiensis	I	Ammate	ammonium sulfamate	H
Agrocide	lindane	I	Ammate X-NI	ammonium sulfamate	H
Agrocit	benomyl	F	Ammo	cypermethrin	I
Agromethrin	cypermethrin	I	Amoxone	2,4-D	H
Agronexa	lindane	I	AMS	ammonium sulfamate	H
Agronexit	lindane	I	Amsol	2,4-D	H
Agrox	captan	F	Anelda	butylate	H
Agroxone	2,4-D	H	Anelda Plus	butylate	H
Agroxone	MCPA	H	Aneldazin	butylate	H
AGSCO 400	2,4-D	H	Anelirox	butylate	H
AGSCO MXL	MCPA	H	Ansar 138	cacodylic acid	H
Agzinphos	zinc phosphide	R	Ansar	cacodylic acid	H
Aimcocyper	cypermethrin	I	Antene	ziram	F
Aimcozeb	mancozeb	F	Anthon	trichlorfon	I
Akar	chlorobenzilate	I	Anticarie	hexachlorobenzene	I
Akar 338	chlorobenzilate	I	Antimilace	metaldehyde	M
Aktikon	atrazine	H	Antimitace	metaldehyde	M
Alanex	alachlor	H	Apadodine	dodine	F
Alanox	alachlor	H	Apavap	dichlorvos	I
Alapin	alachlor	H	Apavinphos	mevinphos	I
Alar	daminozide	GR	Apex	methoprene	I
Alatex	dalapon	H	Aphamite	parathion	I
Alazine	alachlor, atrazine	H	Apistan	fluvalinate	I
Alfadex	pyrethrin	I	Apl-Luster	thiabendazole	F
Alirox	EPTC	H	Appa	phosmet	I
Aljaden	sethoxydim	H	Apron	metalaxyl	F
Alkron	parathion	I	Aquacide	diquat dibromide	H
Alleron	parathion	I	Aqua-Kill	diquat dibromide	H
Alleviate	allethrin	I	Aqua-Kleen	2,4-D	H
Alloxol S	sethoxydim	H	Aquazine	simazine	H
Altosid	methoprene	I	Aragran	terbufos	I
Amaze	isofenphos	I	Arbex	glyphosate	H
Ambiben	chloramben	H	Arrex	zinc phosphide	R
Ambiben ds	chloramben	H	Arrivo	cypermethrin	I
Ambrocide	lindane	I	Asana	esfenvalerate	I
Ambush	permethrin	I	Arasan	thiram	F
Ambush C	cypermethrin	I	Arathane	dinocap	F
Ambushfog	permethrin	I	Arbotect	thiabendazole	F
Amcide	ammonia sulfamate	H	Arinosu-Korori	hydramethylnon	I
Amdro	hydramethylnon	I			
Amerol	amitrole	H	Ariotox	metaldehyde	I
Amiben	chloramben	H	Arpocarb	propoxur	I
Amicide	ammonia sulfamate	H			
Amidosulfate	ammonia sulfamate	H			
Aminatrix	dichlorvos	I			

Trade Name	Common Name	Type
Arrex	zinc phosphide	R
Arrivo	cypermethrin	I
Arsan	cacodylic acid	H
Arylam	carbaryl	I
Asana XL	esfenvalerate	I
Aspor	zineb	F
Assure II	quizalofop-p-ethyl	H
Astonex	diflubenzuron	I
Astro	permethrin	I
Asuntol	coumaphos	I
Atamethrin	cypermethrin	I
Atazinax	atrazine	H
Ateflox	trifluralin	H
Atgard	dichlorvos (DDVP)	I
Atra-Bor	atrazine	H
Atranex	atrazine	H
Atrataf	atrazine	H
Atratol	atrazine	H
Atred	atrazine	H
Atroban	permethrin	I
Attivar	ziram	F
Aules	thiram	F
Avadex BW	triallate	H
Avermectin	abamectin	I
Avicol	PCNB	F
Avid	abamectin	I
Avitrol	4-aminopyridine	A
Azimil	azinphos-methyl	I
Azinotox	atrazine	H
Azinugec	azinphos-methyl	I
Azolan	amitrole	H
Azole	amitrole	H
Azural	glyphosate	H
B 401	Bacillus thuringiensis	I
B 995	daminozide (alar)	GR
Bactec Bernan	Bacillus thuringiensis	I
Bactimos	Bacillus thuringiensis	I
Bactis	Bacillus thuringiensis	I
Bactospeine	Bacillus thuringiensis	I
Bactucide	Bacillus thuringiensis	I
Banel	dicamba	H
Banfel	dicamba	H
Banox	glyphosate	H
Banvel	dicamba	H
Banvel CST	dicamba	H
Banvel D	dicamba	H
Banvel XG	dicamba	H
Barrage	2,4-D	H
Barricade	cypermethrin	I

Trade Name	Common Name	Type
BAS 152J	dimethoate	I
BAS 351 H	bentazon(e)	H
BAS 352F	vinclozolin	F
BAS 90520H	sethoxydim	H
Basagran	bentazon(e)	H
Basamais	bentazon(e)	H
Basanite	dinoseb	H
Basathrin	cypermethrin	I
Basfapon	dalapon	H
Basicap	copper sulfate	F
Basinex P	dalapon	H
Basudin	diazinon	I
Batazine	simazine	H
Bay	azinphos-methyl	I
Bay 21	coumaphos	I
Bay 6159H	metribuzin	H
Bay 6443H	metribuzin	H
Bay 9010	propoxur	I
Bay 9027	azinphos-methyl	I
Bay 15922	trichlorfon	I
Bay 17147	azinphos-methyl	I
Bay 19149	dichlorvos	I
Bay 19639	disulfoton	I
Bay 70143	carbofuran	I
Bay 29493	fenthion	I
Bay 30130	propanil	H
Bay 39007	propoxur	I
Bay 68138	fenamiphos	I
Bay 94337	metribuzin	H
Bay DIC 1468	metribuzin	H
Bay L13/59	trichlorfon	I
Bay MEB 6447	triadimefon	F
Baycid	fenthion	I
Bay S276	disulfoton	I
Bay SRA 12869	isofenphos	I
Baytex	fenthion	I
Bayer 21/199	coumaphos	I
Baygon	propoxur	I
Bayleton	triadimefon	F
Baymix	coumaphos	I
Baytex	fenthion	I
Baytox	fenthion	I
BCS Copper Fungicide	copper sulfate	F
Beacon	primisulfuron-methyl	H
Beet-Kleen	chlorpropham	GR
Belmark	fenvalerate	I
Belt	chlordane	I
Bendioxide	bentazon(e)	H
Benesan	lindane	I
Benex	benomyl	F
Benexane	lindane	I

Trade Name	Common Name	Type	Trade Name	Common Name	Type
Benfos	dichlorvos (DDVP)	I	Borocil	bromacil	H
Benlate	benomyl	F	Botrilex	PCNB	F
Benomilosan	benomyl	F	Bovinos	trichlorfon	I
Benor	bromoxynil	H	Bovinox	trichlorfon	I
Benosan	benomyl	F	Bravonil	chlorpropham	GR
Bensumec	bensulide	H	Broadside	cacodylic acid	H
Bentazone	bentazon(e)	H	Bromotril	bromacil	H
Bent-cure	hexachlorobenzene	I	Brassicol	PCNB	F
Bent-no-More	hexachlorobenzene	I	Brassix	trifluralin	H
			Bravo	chlorothalonil	F
Benzamide	pronamide	H	Brifur	propoxur	I
Benzilan	chlorobenzilate	I	Briten	trichlorfon	I
Benz-o-chlor	chlorobenzilate	I	Brodan	chlorpyrifos	I
Benzulfide	bensulide	H	Bromazil	imazalil	F
Beosit	endosulfan	I	Bromax	bromacil	H
Berliner	Bacillus thuringiensis	I	Bromex	naled	I
			Brominal	bromoxynil	H
Betamec	bensulide	H	Brominex	bromoxynil	H
Betasan	bensulide	H	Brominil	bromoxynil, dicamba	H
Bi 58	dimethoate	I			
Biarbinex	heptachlor	I	Bromofume	ethylene dibromide	I
Bicep	metolachlor	H	Brom-O-Gas	methyl bromide	F
Bifex	propoxur	I	Brom-O-Sol	methyl bromide	I
Bioallethrin	allethrin	I	Bromotril	bromoxynil	H
Biobit	Bacillus thuringiensis	I	Bromoxian	bromoxynil	H
			Bronate	bromoxynil, MCPA	H
Bio Flydown	permethrin	I	Bronco	alachlor, glyphosate	H
Biothion	temephos	I	BRP	naled	I
Bithion	temephos	I	Brush	tebuthiuron	H
Blade	oxamyl	I	Bt	B.t.-aizawai	I
Bladan	parathion	I	Bt	B.t.-kurstaki	I
Bladen M	methyl parathion	I	Bt	B.t.-israelinsis	I
Bladex	cyanazine	H	Buckle	triallate	H
Blatex	hydramethylnon	I	Buctril	bromoxynil	H
Blattanex	propoxur	I	Bud Nip	chlorpropham	GR
Blazer	acifluorfen	H	Bug Master	carbaryl	I
Blotic	propetamphos	I	Buggy	glyphosate	H
Blue Copperas	copper sulfate	F	Buhach	pyrethrins	I
			Bullet	tebuthiuron	H
Bluestone	copper sulfate	F	Bushwacker	tebuthiuron	H
Blue Viking	copper sulfate	I	Butilate	butylate	H
Blue Vitriol	copper sulfate	F	Butirex	2,4-DB	H
BMP 123	Bacillus thuringiensis	I	Butormone	2,4-DB	H
			Butoxone	2,4-DB	H
BMP 144	Bacillus thuringiensis	I	Butyl-Geld	dinoseb	H
			Butyrac	2,4-DB	H
B-Nine	daminozide (alar)	GR	BW-21-Z	permethrin	I
Bolfo	propoxur	I	C 2059	fluometuron	H
Bolls-Eye	cacodylic acid	H	Cadan	carboxin	F
Bombardier	chlorothalonil	F	Caldon	dinoseb	H
Bordeaux Mixture	copper sulfate	F	Caliber	simazine	H
			Calmathion	malathion	I
Bordermaster	MCPA	H	Canogard	dichlorvos (DDVP)	I
Borea	bromacil	H	Caparol	prometryn	H
Borer-Tox	lindane	I	Capfos	fonofos	I

Trade Name	Common Name	Type	Trade Name	Common Name	Type
Command	clomazone	H	Cuberol	rotenone	I,R
Commando	zinc phosphide	R	Cudgel	fonofos	H
Commence	clomazone	H	Cuman	ziram	F
Contraven	terbufos	I	Cunitex	thiram	F
CoPilot	quizalofop-p-ethyl	H	Cupincida	heptachlor	I
Comply	fenoxycarb	GR	Cuprothex	zineb	F
Compound 732	terbacil	H	Curalan	vinclozolin	F
			Curaterr	carbofuran	I
Connonex	fluometuron	H	Curex Flea Duster	rotenone	I
Coopex	permethrin	I			
Copper Pride	copper sulfate	F	Curitan	dodine	F
Co-Ral	coumaphos	I	Cyanater	terbufos	I
Corodane	chlordane	I	Cyaforce	hydramethylnon	I
Corothion	parathion	I	Cyaforgel	hydramethylnon	I
Corsair	permethrin	I	Cybolt	flucythrinate	I
Cotnion-methyl	azinphos-methyl	I	Cyclodan	endosulfan	I
			Cyclon	hydramethylnon	I
Cotoran	fluometuron	H	Cyclone	paraquat	H
Cotorex	fluometuron	H	Cydexine	simazine	H
Cotton Aid HC	cacodylic acid	H	Cygon	dimethoate	I
			Cygon 400	dimethoate	I
Cottonex	fluometuron	H	Cymbaz	cypermethrin	I
Cotton Pro	prometryn	H	Cymbush	cypermethrin	I
Contraven	terbufos	I	Cymperator	cypermethrin	I
Corodane	chlordane	I	Cynoff	cypermethrin	I
Corothion	parathion	I	Cynogan	bromacil	H
Corozate	ziram	F	Cyper	cypermethrin	I
Counter	terbufos	I	Cypercopal	cypermethrin	I
CP 23426	triallate	H	Cyperguard 25EC	cypermethrin	I
CP 50144	alachlor	H			
CP Basic Sulfate	copper sulfate	F	Cyperhard Tech	cypermethrin	I
CP-TS 53	copper sulfate	F	Cyperkill	cypermethrin	I
CR-1693	dinocap	F	Cypermar	cypermethrin	I
Crisalamina	2,4-D	H	Cypermax	cypermethrin	I
Crisalin	trifluralin	H	Cypersect	cypermethrin	I
Crisamina	2,4-D	H	Cypersul	cypermethrin	I
Crisazina	atrazine	H	Cypertox	cypermethrin	I
Crisimia	2,4-D	H	Cypona	dichlorvos (DDVP)	I
Crisquat	paraquat	H	Cyprex	dodine	F
Crisuron	diuron	H	Cyrux	cypermethrin	I
Crittam	ziram	F	Cythion	malathion	I
Crittox MZ	mancozeb	F	Cytrol	amitrole	H
Crop Rider	2,4-D	H	2,4-D	2,4-D	H
Crop Star	alachlor	H	D 264	diazinon	I
Croptex Onyx	bromacil	H	D 735	carboxin	F
			D 1221	carbofuran	I
Crotonate	dinocap	F	DAC 893	DCPA	H
Crotothane	dinocap	F	Dacamine	2,4-D	H
Crossbow	triclopyr	H	Daconil 2787	chlorothalonil	F
Crossfire	resmethrin	I	Dacthal	DCPA	H
Cross Fire	chlorpyrifos	I	Dacthal W-75	daminozide (alar)	GR
Crunch	carbaryl	I	Dacthalor	DCPA	H
Crysthyon	azinphos-methyl	I	Dadasul	dichlorvos (DDVP)	I

Trade Name	Common Name	Type
Dalacide	dalapon	H
Dalapon 85	dalapon	H
Dalf	fenthion	I
Dalf	methyl parathion	I
Dalmation Insect Flowers	pyrethrins	I
Danex	trichlorfon	I
Daphene	dimethoate	I
Dart	glyphosate	H
Dazide	daminozide (alar)	GR
Dazzel	diazinon	I
DCMO	carboxin	F
DCMU	diuron	H
DCPC	dinocap	F
Deadline	metaldehyde	M
Decco	chlorpropham	GR
Deccozil	imazalil	F
Decemthion	phosmet	I
Decimax	diuron	H
Decofol	dicofol	I
DeDevap	dichlorvos (DDVP)	I
Ded-Weed	2,4-D	H
Ded-Weed	dalapon	H
De-Fend	dimethoate	I
Deiquat	diquat dibromide	H
Delta-Coat AD	metalaxyl	F
Demon	cypermethrin	I
Demos NF	dimethoate	I
Denapon	carbaryl	I
Denkaphon	trichlorfon	I
Denkarin Grains	zinc phosphide	R
Denkavepon	dichlorvos (DDVP)	I
Dep	trichlorfon	I
Derriban	dichlorvos (DDVP)	I
Derribanate	dichlorvos (DDVP)	I
Derrin	rotenone	I
Derringer	resmethrin	I
Derris	rotenone	I
Desormone	2,4-D	H
Destox	2,4-D	H
Detmol UA	chlorpyrifos	I
Deviban	chlorpyrifos	I
Devicarb	carbaryl	I
Devicyper	cypermethrin	I
Devigon	dimethoate	I
Devikol	dichlorvos (DDVP)	I
Devipon	dalapon	H
Devisulphan	endosulfan	I
Devithion	methyl parathion	I
Devizeb	zineb	F
Devrinol	naproamide	H
Dextrone	diquat dibromide	H

Trade Name	Common Name	Type
Dextrone X	paraquat	H
Dexuron	paraquat	H
Diacon	methoprene	I
Dianat	dicamba	H
Dianex	methoprene	I
Diater	diuron	H
Diazitol	diazinon	I
Diazol	diazinon	I
Di-Blox	diphacinone	R
Dibrom	naled	I
Dibrome	ethylene dibromide	I
Dicamba M	dicamba, MCPA	H
Dicamba P	dicamba	H
Dicambex	dicamba	H
Dicap	dinocap	F
Dicarbam	carbaryl	I
Dicaron	dicofol	I
Dicazin	atrazine, dicamba	H
Dichlorvos	DDVP	I
Dicomite	dicofol	I
Didivane	dichlorvos (DDVP)	I
Dielathion	malathion	I
Difenthos	temephos	I
Difol	dicofol	I
Difonate	fonofos	I
Digermin	trifluralin	H
Dikamin	2,4-D	H
Dikar	dinocap	F
Dikonirt	2,4-D	H
Dicotox	2,4-D, dicamba, MCPA	H
Dilic	cacodylic acid	H
Dilice	coumaphos	I
Dimate 267	dimethoate	I
Dimet	dimethoate	I
Dimetate	dimethoate	I
Dimethoate	dimethoate	I
Dimethoat Tech 95%	dimethoate	I
Dimethogen	dimethoate	I
Dimethyl Parathion	methyl parathion	I
Dimilin	diflubenzuron	I
Dinitro	dinoseb	H
Dinoxol	2,4-D	H
Di-on	diuron	H
Dipel	Bacillus thuringiensis	I
Diphacine	diphacinone	R
Diphenthos	temephos	I
Dipher	zineb	F
Dipterex	trichlorfon	I
Dipthal	triallate	H
Diquat Herbicide	diquat dibromide	H

Trade Name	Common Name	Type	Trade Name	Common Name	Type
Direx	diuron	H	Dropaven	propanil	H
Dirimal	oryzalin	H	DRP 642532	thiram	F
Disan	bensulide	H	Drupine	ziram	F
Discon-Z	zineb	F	DU 112307	diflubenzuron	I
Disonex	diazinon	I	Dual	metolachlor	H
Disyston	disulfoton	I	Dualor	metolachlor	H
Disystox	disulfoton	I	Duo-Kill	dichlorvos (DDVP)	I
Dithane	mancozeb	F	Du Pont 732	terbacil	H
Dithane 945	mancozeb	F	Du Pont 1179	methomyl	I
Dithane M-22	maneb	F	Du Pont 1991	benomyl	F
Dithane M-45	mancozeb	F	Du Pont	linuron	H
Dithane-Ultra	mancozeb	F	Herbicide 326		
Dithane Z-78	zineb	F	Du Pont	terbacil	H
Dithio-demeton	disulfoton	I	Herbicide 732		
Dithiosystox	disulfoton	I	Du Pont	bromacil	H
Ditiner	zineb	F	Herbicide 976		
Ditiozin	zineb	F			
Ditrac	diphacinone	R	Duraphos	mevinphos	I
Ditrifon	trichlorfon	I	Dursban	chlorpyrifos	I
Diumate	diuron	H	Duravos	dichlorvos (DDVP)	I
Diurex	diuron	H	Dutch-Treat	cacodylic acid	H
Diurol	amitrole	H	DW 3418	cyanazine	H
Divipan	dichlorvos (DDVP)	I	Dycarb	bendiocarb	I
DMDT	methoxychlor	I	Dyfonate	fonofos	I
DMPT	fenthion	I	Dylox	trichlorfon	I
DMU	diuron	H	Dymec	2,4-D	H
DN 289	dinoseb	H	Dynamec I	abamectin	I
DNOPC	dinocap	F	Dynamyte	dinoseb	H
Doom	dichlorvos (DDVP)	I	Dynex	diuron	H
DOP 500F	iprodione	F	Dyphonate	fonofos	I
Dormone	2,4-D	H	Dyrex	trichlorfon	I
Double R	imazalil	F	D-Z-N	diazinon	I
Dowco 179	chlorpyrifos	I	E 601	methyl parathion	I
Dowco 233	triclopyr	H	E-605	parathion	I
Dowfume	ethylene dibromide	I	E 3314	heptachlor	I
Dowicide EC-7	pentachlorophenol	I,F,H	Earthcide	PCNB	F
			EBDC	zineb	F
Dowpon	dalapon	H	ECB	chlorobenzilate	I
DPA	propanil	H	Echo	chlorpropham	GR
DPA	dalapon	H	Eclipse	fenoxycarb	GR
DPX 1410	oxamyl	I	Eclipse	glyphosate	H
DPX 3674	hexazinone	H	Ecopro	temephos	I
DPX 5648	sulfometuron-methyl	H	Ectiban	permethrin	I
			Ectrin	fenvalerate	I
DPX Y6202-31	quizalofop-p-ethyl	H	EDB-85	ethylene dibromide	F
			Edabrome	ethylene dibromide	I
Dragnet	permethrin	I	Eftol	parathion	I
Dragon	permethrin	I	Efuzin	dodine	F
Drawizon	diazinon	I	Ekatin TD	disulfoton	I
Drexel	atrazine, captan	F	Ekatox	parathion	I
Drexel	methoxychlor	I	Eksmin	permethrin	I
Drinox	heptachlor	I	EL-103	tebuthiuron	H
Drive	vinclozolin	F	EL-119	oryzalin	H

Trade Name	Common Name	Type	Trade Name	Common Name	Type
EL-4049	malathion	I	Ethyl Parathion	parathion	I
EL-12880	dimethoate	I			
Elancolan	trifluralin	H	Etilon	parathion	I
Elastrel	dichlorvos (DDVP)	I	Exagama	lindane	I
Elbanil	chlorpropham	GR	Exathion	malathion	I
Elgetol	dinoseb	H	Exetor	triclopyr	H
Embamine	2,4-D	H	Exmin	permethrin	I
Embafume	methyl bromide	F	Exotherm Termil	chlorothalonil	F
Emblem	bromoxynil	H			
Embutone	2,4-DB	H	Expand	sethoxydim	H
Embutox	2,4-DB	H	Expar	permethrin	I
Emelten-middel	temephos	I	Exporsan	bensulide	H
			Extrasim	simazine	H
Emmatos	malathion	I	Ezenosan	dinocap	F
Empal	MCPA	H	Ezy Pickin	cacodylic acid	H
Empire	chlorpyrifos	I	Faber	chlorothalonil	F
Emulsamine	2,4-D	H	Fallowmaster	dicamba	H
Endocel	endosulfan	I	Fanicide	dinoseb	H
Endocide	endosulfan	I	Far-Go	triallate	H
Endogerme	chlorpropham	GR	Farmaneb	maneb	F
Endosol	endosulfan	I	Farmco Atrazine	atrazine	H
Endosul	endosulfan	I			
Enozin	zineb	F	FB/2	diquat dibromide	H
ENT 1726	methoxychlor	I	Fecundal	imazalil	F
ENT 9932	chlordane	I	Fenamin	atrazine	H
ENT 15108	parathion	I	Fenkil	fenvalerate	I
ENT 15152	heptachlor	I	Fennotox	heptachlor	I
ENT 22374	mevinphos	I	Fenocap	dinocap	F
ENT 24105	ethion	I	Fenom	cypermethrin	I
ENT 27093	aldicarb	I	Ferkethion	dimethoate	I
ENT 27164	carbofuran	I	Fermide 850	thiram	F
Entex	fenthion	I	Fernide	thiram	F
Entry	bentazon(e)	H	Fernimine	2,4-D	H
Epigon	permethrin	I	Fernasan	thiram	F
Eptam	EPTC	H	Fernoxone	2,4-D	H
Equigard	dichlorvos (DDVP)	I	Fervinal	sethoxydim	H
Equino-Aid	trichlorfon	I	Ferxone	2,4-D	H
Equity	chlorpyrifos	I	Fesdan	phosmet	I
Eradex	chlorpyrifos	I	Fibenzol	benomyl	F
Eradicane	EPTC	H	Ficam	bendiocarb	I
Eradicane Extra	EPTC	H	Firmotox	pyrethrins	I
			Fish Tox	rotenone	I,R
Erban	propanil	H	Flectron	cypermethrin	I
Erbanil	propanil	H	Flee	permethrin	I
Erbitox	2,4-D	H	Fligene CI	cypermethrin	I
Erdex	chlorpyrifos	H	Flo-Met	fluometuron	H
Ertefon	trichlorfon	I	Flo-Pro	imazalil	F
Esgram	paraquat	H	Florasan	imazalil	F
Esbiothrin	allethrin	I	Flurene SE	trifluralin	H
Esteron	2,4-D	H	Fly-Die	dichlorvos (DDVP)	I
ET	triclopyr	H	Fly-Fighter	dichlorvos (DDVP)	I
Ethanox	ethion	I	Fly Killer-D	naled	I
Ethiol	ethion	I	Flytek	methomyl	I
Ethiosul	ethion	I	FMC 1240	ethion	I
Ethyl N	butylate	H	FMC 2070	thiram	F

Trade Name	Common Name	Type	Trade Name	Common Name	Type
FMC 5462	endosulfan	I	Fyfanon	malathion	I
FMC 9102	metiram	F	Fyran 206k	ammonium	H
FMC 10242	carbofuran	I		sulfamate	
FMC 17370	resmethrin	I	G 23992	chlorobenzilate	I
FMC 33297	permethrin	I	G 24480	diazinon	I
FMC 57020	clomazone	H	G 27692	simazine	H
Fogard	atrazine	H	G 30027	atrazine	H
Folbex	chlorobenzilate	I	G 30028	propazine	H
Folcord	cypermethrin	I	G 34161	prometryn	H
Folidol E 605	parathion	I	Gallogama	lindane	I
Folidol M	methyl parathion	I	Gallup	glyphosate	H
Folosan	PCNB	F	Gamaphex	lindane	I
Forate	phorate	I	Gamasan	lindane	I
Foray	Bacillus	I	Gamit	clomazone	H
	thuringiensis		Gamma BHC	lindane	I
Fore	mancozeb	F	Gamma-Col	lindane	I
Forlin	lindane	I	Gammalin	lindane	I
For-Mal 50	malathion	I	Gamma	lindane	I
Formec	mancozeb	F	Mean 400		
For-ester	2,4-D	H	Gamma	lindane	I
Fortrol	cyanazine	H	Mean L.O.		
Forturf	chlorothalonil	F	Gamma	lindane	I
Foschlor	trichlorfon	I	Mean Seed		
Fosdan	phosmet	I	Gamma-Up	lindane	I
Fosdrin	mevinphos	I	Gammex	lindane	I
Fosferno 50	parathion	I	Gammexane	lindane	I
Fosferno M50	methyl parathion	I	Gardentox	diazinon	I
Fostion MM	dimethoate	I	Garlon	triclopyr	H
Fostox E	parathion	I	Garnitan	linuron	H
Framed	simazine	H	Garvox	bendiocarb	I
Freshgard	imazalil	F	Gearphos	methyl parathion	I
Frumin AL	disulfoton	I	Gebutox	dinoseb	H
Fruttene	ziram	F	Geigy 338	chlorobenzilate	I
Fuching jujr	flucythrinate	I	Geigy 30028	propazine	H
Fuklasin	ziram	F	Genate	butylate	H
Fulkil	methyl parathion	I	Genate Plus	butylate	H
Fumo-Gas	ethylene dibromide	I	Genep	EPTC	H
Funconil	chlorpropham	GR	Genep Plus	EPTC	H
Fundazol	benomyl	F	Geomet	phorate	I
Fungaflor	imazalil	F	Geonter	terbacil	H
Fungazil	imazalil	F	Geophos	parathion	I
Fungostop	ziram	F	Gesagard	prometryn	H
Funomyl	benomyl	F	Gesamil	propazine	H
Furacarb	carbofuran	I	Gesaprim	atrazine	H
Furadan	carbofuran	I	Gesatop	simazine	H
Furadex	carbofuran	I	Gesfid	mevinphos	I
Furasul	carbofuran	I	GLADZ	pendimethalin	H
Furasun	carbofuran	I	Glider	glyphosate	H
Furloe	chlorpropham	GR	Glifonox	glyphosate	H
Fusilade 2000	fluazifop-p-butyl	H	Glitex	glyphosate	H
Fusilade Five	fluazifop-p-butyl	H	Glycel	glyphosate	H
Fusilade	fluazifop-p-butyl	H	Glyfonox	glyphosate	H
Super			Glyphosul	glyphosate	H
FW-293	dicofol	I	Glysate	glyphosate	H
FW-734	propanil	H	Glytex	glyphosate	H

Trade Name	Common Name	Type
Gnatrol	Bacillus thuringiensis	I
Goal	oxyfluorfen	H
Go-Go San	pendimethalin	H
Gold Crest	diphacinone	R
Gold Crest C-100	chlordane	I
Gold Crest H-60	heptachlor	I
Golden Leaf Tobacco Spray	endosulfan	I
Goldquat	paraquat	H
Gopha-Rid	zinc phosphide	R
Gordon's Amine	MCPA	H
Grain Guard	mancozeb	F
Gramevin	dalapon	H
Gramix	paraquat	H
Gramixel	paraquat	H
Gramocil	paraquat	H
Gramoxone	paraquat	H
Gramoxone Extra	paraquat	H
Grand	cypermethrin	I
Grandstand	triclopyr	H
Granutox	phorate	I
Grasidim	sethoxydim	H
Graslan	tebuthiuron	H
Grazon	picloram, triclopyr	H
Green Cross Warble Powder	rotenone	I
Green-Daisen M	mancozeb	F
Griffex	atrazine	H
Ground-Up	glyphosate	H
GS 13005	methidathion	I
Gusagrex	azinphos-methyl	I
Gusathion	azinphos-methyl	I
Guthion	azinphos-methyl	I
H 34	heptachlor	I
Halizan	metaldehyde	I
Halmark	esfenvalerate	I
Halmone	dinoseb	H
Halt	cypermethrin	I
Haltox	methyl bromide	I
Harness	bromoxynil	H
HCB	hexachlorobenzene	F
Hedonal	2,4-D	H
Hedonal M	MCPA	H
Helarion	metaldehyde	I
Hel-Fie	dinoseb	H
Heptagran	heptachlor	I
Heptamul	heptachlor	I

Trade Name	Common Name	Type
Heptox	heptachlor	I
Herbadox	pendimethalin	H
Herbax	propanil	H
Herbaxone	paraquat	H
Herbec	tebuthiuron	H
Herbic	tebuthiuron	H
Herbidal	2,4-D	H
Herbifen	2,4-D	H
Herbiflurin	trifluralin	H
Herbikill	paraquat	H
Herbizole	amitrole	H
Herb-neat	glyphosate	H
Herbolex	glyphosate	H
Herboxone	paraquat	H
Heritage	trifluralin	H
Herkol	dichlorvos (DDVP)	I
Hexasulfan	endosulfan	I
Hexathane	zineb	F
Hexathir	thiram	F
Hexavin	carbaryl	I
Hexazir	ziram	F
Hi-Dep	2,4-D	H
Higalcoton	fluometuron	H
Higalmetox	methoxychlor	I
Higalnate	molinate	H
Hildan	endosulfan	I
Hilfol	dicofol	I
Hilthion	malathion	I
Hivertox	dinoseb	H
Hockey	glyphosate	H
Hoe 02810	linuron	H
Hoe 2671	endosulfan	I
Hoe 26150	dinoseb	H
Honcho	glyphosate	H
Humextra	metolachlor	H
Hungazin	atrazine	H
Hydram	molinate	H
Hylemox	ethion	I
Hymush	thiabendazole	F
Hyvar X	bromacil	H
Hyvar XL	bromacil	H
Hy-Vic	thiram	F
ICI 0005	fluazifop-p-butyl	H
Ikurin	ammonium sulfamate	H
Imidan	phosmet	I
Imperator	permethrin, cypermethrin	I
Inakor	atrazine	H
Indothrin	permethrin	I
Inexit	lindane	I
Insectophene	endosulfan	I
Insect Powder	pyrethrins	I
Insegar	fenoxycarb	GR,I

Trade Name	Common Name	Type	Trade Name	Common Name	Type
Intox	chlordane	I	Kypchlor	chlordane	I
Invisi-Gard	propoxur	I	Kypfos	malathion	I
Ipersan	trifluralin	H	Kypman	maneb	F
Isathrine	resmethrin	I	Kypzin	zineb	F
Iscothane	dinocap	F	Kytrol	amitrole	H
Isocil	bromacil	H	L 395	dimethoate	I
Isotox	lindane	I	L-36352	trifluralin	H
Javelin	Bacillus	I	Labaycid	fenthion	I
	thuringiensis		Lanex	fluometuron	H
Jupital	chlorothalonil	F	Lannate	methomyl	I
Jureong	permethrin	I	Lanox	methomyl	I
Jury	glyphosate	H	Lantox	methomyl	I
Kabat	methoprene	I	Lariat	alachlor, atrazine	H
Kack	cacodylic acid	H	Larvakil	diflubenzuron	I
Kafil	permethrin	I	Lasso	alachlor	H
Kafil Super	cypermethrin	I	Lawn-Keep	2,4-D	H
Karamate	mancozeb	F	Lazo	alachlor	H
Karathane	dinocap	F	LE 79-519	permethrin,	I
Karbaspray	carbaryl	I		cypermethrin	
Karbofos	malathion	I	Leader	bentazon(e)	H
Karmex	diuron	H	Lebaycid	fenthion	I
Kayazinon	diazinon	I	Leivasom	trichlorfon	I
Kayazol	diazinon	I	Lentrek	chlorpyrifos	I
Kelthane	dicofol	I	Leptox	Bacillus	I
Kemikar	carboxin	F		thuringiensis	
Kemolate	phosmet	I	Lexone	metribuzin	H
Kenapon	dalapon	H	LFA 2043	iprodione	F
Kenofuran	carbofuran	I	Limatox	metaldehyde	M
Kerb	pronamide	H	Lindafor	lindane	I
Kestrel	permethrin	I	Lindagam	lindane	I
Kicker	pyrethrin	I	Lindagranox	lindane	I
Kidan	iprodione	F	Lindamul	lindane	I
Kilex	chlordane	I	Lindaterra	lindane	I
Chlordane			Linex	linuron	H
Kilex	methyl parathion	I	Linorox	linuron	H
Parathion			Lintox	lindane	I
Kil-Ko Rat	diphacinone	R	Linurex	linuron	H
Killer			Liropon	dalapon	H
Kiloseb	dinoseb	H	Litarol	bromoxynil	H
Kipsin	methomyl	I	Lock-On	chlorpyrifos	I
Kisvax	carboxin	F	Lodaco	zineb	F
Kitinex	diflubenzuron	I	Logic	fenoxycarb	GR,I
Klartan	fluvalinate	I	Lonacol	zineb	F
Knox-out	diazinon (encap.)	I	Lonacol M	maneb	F
Kobu	PCNB	F	Lorox	linuron	H
Kobutol	PCNB	F	Lorsban	chlorpyrifos	I
Koltar	oxyfluorfen	H	Loxiran	chlorpyrifos	I
Kopfume	ethylene dibromide	I	Lucanal	naled	I
Kopmite	chlorobenzilate	I	Lucaphos	dichlorvos (DDVP)	I
Kop-Thion	malathion	I	Lucathion	malathion	I
Koril	bromoxynil	H	Lypor	temephos	I
Korilene	bromoxynil	H	M 22	maneb	F
Krovar	bromacil	H	M 410	chlordane	I
Kuik	methomyl	I	Macondray	2,4-D	H
Kylar	daminozide (alar)	GR	Mafu	dichlorvos (DDVP)	I

Trade Name	Common Name	Type	Trade Name	Common Name	Type
Magister	clomazone	H	Meldane	coumaphos	I
Mahatz	chlordane	F	Melprex	dodine	F
Maizina	atrazine	H	Memilene	methomyl	I
Malamar	malathion	I	Meniphos	mevinphos	I
Malaphele	malathion	I	Menite	mevinphos	I
Malaspray	malathion	I	Mephanac	MCPA	H
Malathion	malathion	I	Mercapto- phos	fenthion	I
Malatol	malathion	I			
Malatox	malathion	I	Mercasin	prometryn	H
Malermais	atrazine	H	Mercuram	thiram	F
Malerbane	2,4-D	H	Merit	bromoxynil	H
Malixol	malathion	I	Merit	clomazone	H
Malix	endosulfan	I	Merkazin	prometryn	H
Maltox	malathion	I	Merpan	captan	F
Malmed	malathion	I	Mertect	thiabendazole	F
Mancizin	mancozeb	F	Meta	metaldehyde	M
Manco-75	mancozeb	F	Metacide	methyl parathion	I
Mancofol	mancozeb	F	Metambane	dicamba	H
Mancozan	zineb	F	Metaphos	methyl parathion	I
Mancozin	mancozeb	F	Metazintox	azinphos-methyl	I
Maneb Brestan	maneb	F	Meteoro	captan	F
			Metason	metaldehyde	I
Manebgan	maneb	F	Methavin	methomyl	I
Maneb Spritzpulver	maneb	F	Meth-O-Gas	methyl bromide	I
			Methomex	methomyl	I
Manesan	maneb	F	Methoxy- DDT	methoxychlor	I
Manex	maneb	F			
Mangavis	maneb	F	Methyl- Guthion	azinphos-methyl	I
Manox	maneb	F			
Manzate	maneb	F	Metron	methyl parathion	I
Manzate 200	mancozeb	F	Meturon	floumeturon	H
Manzeb	mancozeb	F	Mevidrin	mevinphos	I
Manzin	mancozeb	F	Mevinox	mevinphos	I
Man-Zox	maneb	F	Mevinphos	mevinphos	I
Manzi	maneb	F	Mezene	ziram	F
Manzin	mancozeb	F	MGK 264	allethrin	I
Marlate	methoxychlor	I	Micromite	diflubenzuron	I
Marmer	diuron	H	Micronyl	ziram	F
Marvex	dichlorvos (DDVP)	I	Micropearls	thiram	F
Masoten	trichlorfon	I	Micro-Tech	alachlor	H
Match	cyanazine	H	Midstream	diquat dibromide	H
Matox	hydramethylnon	I	MifaSlug	metaldehyde	I
Mavrik Aqua Flow	fluvalinate	I	Milbam	ziram	F
			Mildane	dinocap	F
Maxforce	hydramethylnon	I	Mildex	dinocap	F
Mazaline	simazine	H	Milocep	propazine	H
MB	methyl bromide	F	Milogard	propazine	F
M&B 2878	2,4-DB	H	Milo-Pro	propazine	H
M&B 10064	bromoxynil	H	Minex	methoprene	I
MC 10109	acifluorfen	H	Miracle	2,4-D	H
MCP	MCPA	H	Mirvale	chlorpropham	GR
Mebazine	atrazine	H	Mitigan	dicofol	I
Mediben	dicamba	H	Mizol	amitrole	H
Meldame	coumaphos	I	MK 936	abamectin	I

Trade Name	Common Name	Type	Trade Name	Common Name	Type
MLT	malathion	I	Niptan	EPTC	H
Mold-Ex	chlorothalonil	F	Niran	parathion,	I
Molinate	molinate	H		chlordane	I
Moltranil	molinate, propanil	H	Nitropone	dinoseb	H
MON-0573	glyphosate	H	Nitrox 80	methyl parathion	I
Moncide	cacodylic acid	H	No Bunt	hexachlorobenzene	I
Montar	cacodylic acid	H	Nogos	dichlorvos	I
M.T.F.	trifluralin	H	Nomersan	thiram	F
Multamat	bendiocarb	I	Nopcocide	chlorothalonil	F
Multimet	bendiocarb	I	N-96		
Murvin	carbaryl	I	No-Pest	dichlorvos (DDVP)	I
Muscatox	coumaphos	I	Notar	chlorothanonil	F
Muster	glyphosate	H	Novabac	Bacillus	I
Mutan	triclopyr	H		thuringiensis	
Mycozol	thiabendazole	F	Novermone	2,4-D	H
N-2790	fonofos	I	Novigam	lindane	I
NA 9184	pyrethrins	I	Noxfire	rotenone	I,R
Nabu	sethoxydim	H	Noxfish	rotenone	I,R
NAC	carbaryl	I	NP-55	sethoxydim	H
Nalkil	bromacil	H	NRC 910	iprodione	F
Namekil	metaldehyde	I	NRDC 104	resmethrin	I
Natur Gro	ryania	I	NRDC 143	permethrin	I
R-50			NRDC 149	cypermethrin	I
Natur Gro	ryania	I	Nucidol	diazinon	I
Triple Plus			Nudor	alachlor	H
NC 302D+	quizalofop-p-ethyl	H	Nudrin	methomyl	I
NC 6897	bendiocarb	I	Nurelle	cypermethrin	I
N-Disobutyl-	butylate	H	Nuvan	dichlorvos (DDVP)	I
thiocarba-			Nu-Zone	imazalil	F
mate			10ME		
Negashunt	coumaphos	I	Octachlor	chlordane	I
Neguvon	trichlorfon	I	Octa-Klor	chlordane	I
Nekol	dichlorvos (DDVP)	I	Ofirmotox	pyrethrins	I
Neocidol	diazinon	I	Oftanol	isofenphos	I
Nemacur	fenamiphos	I	Oko	dichlorvos (DDVP)	I
Nemispor	mancozeb	F	Oku	dichlorvos (DDVP)	I
Nemispot	mancozeb	F	Ole	chlorothalonil	F
Nephis	ethylene dibromide	I	Olitref	trifluralin	H
Nereb	maneb	F	OMS 771	aldicarb	I
Nespor	maneb	F	OMS 2007	flucythrinate	I
Netagrone	2,4-D	H	OR-CAL	malathion	I
Nevugon	trichlorfon	I	Stabilized		
New Kotol	lindane	I	Malathion		
Newspor	maneb	F	OR-CAL	ziram	F
Nex	carbofuran	I	Ziram 400		
Nexit	lindane	I	Ordram	molinate	H
NIA 9102	metiram	F	Ornalin	vinclozolin	F
Nialate	ethion	I	Ornamec	fluazifop-p-butyl	H
Nicouline	rotenone	I	Ornamental	chloramben	H
Nimitex	temephos	I	Weeder		
Nimitox	temephos	I	Orthocide	captan	F,H
Niomil	bendiocarb	I	Orthophos	parathion	I
Nioxyl	ziram	F	Ortho Fly	naled	I
Nipsan	diazinon	I	Killer		

Trade Name	Common Name	Type	Trade Name	Common Name	Type
Ortho Metalde- hyde 4% Bait	metaldehyde	M	Pennflo	mancozeb	F
			Penoxalin	pendimethalin	H
			Penta	pentachlorophenol	I,F,H
			Pentac Aquaflow	dienochlor	I
Ortho Weed & Spot Killer	paraquat	H	Pentacon	pentachlorophenol	I,F,H
			Pentac WP	dienochlor	I
OS-2046	mevinphos	I	Pentagen	PCNB	F
Osaquat	paraquat	H	Penwar	pentachlorophenol	I,F,H
Oust Weed Killer	sulfometuron- methyl	H	Perfekthion	dimethoate	I
			Perflan	tebuthiuron	H
Outflank	permethrin	I	Perigen	permethrin	I
Oxamimidic Acid	oxamyl	I	Perizin	coumaphos	I
			Permanone	permethrin	I
Oxatin	carboxin	F	Permasect	permethrin	I
Pacrite	imazalil	F	Permatox	pendimethalin	H
Padan	carboxin	F	Permetrina	permethrin	I
Pageant	chlorpyrifos	I	Permilan	zineb	F
Pakhtaran	fluometuron	H	Permit	permethrin	I
Pancide	azinphos-methyl	I	Persect	permethrin	I
Panthion	parathion	I	Persevtox	dinoseb	H
Parakakes	diphacinone	R	Perthrine	permethrin	I
Paramar	parathion	I	PH 60-40	diflubenzuron	I
Paraphos	parathion	I	Pharorid	methoprene	I
Parasul	methyl parathion	I	Phaser	endosulfan	I
Parataf	methyl parathion	I	Phenamiphos	fenamiphos	I
Parathene	parathion	I	Phenoxylene 50	MCPA	H
Parathion-E (encap)	parathion, azinphos-methyl	I			
			Phoschlor	trichlorfon	I
Paratox	methyl parathion	I	Phosdrin	mevinphos	I
Parawet	parathion	I	Phosfene	mevinphos	I
Parazate	zineb	F	Phoskil	parathion	I
Pardner	bromoxynil	H	Phosvin	zinc phosphide	R
Parexan	pyrethrins	I	Phosvit	dichlorvos (DDVP)	I
Partner	alachlor	H	Phytar	cacodylic acid	H
Partron M	methyl parathion	I	Phytar 138	cacodylic acid	H
Parzate	zineb	F	Phytar 600	cacodylic acid	H
Pathfinder	triclopyr	H	Phytocape	captan	F
Patrin	carbaryl	I	Phyton-27	copper sulfate	I
Pay-off (discon- tinued)	flucythrinate	I	Phytox	zineb	F
			Phytox MZ	mancozeb	F
			Phytoxone	dinoseb	H
Payze	cyanazine	H	Picket	permethrin	I
PCP	pentachlorophenol	H	Pictyl	fenoxycarb	GR,I
PCQ	diphacinone	R	Pillarben	benomyl	F
PD5	mevinphos	I	Pillarcap	captan	F
PDD 60-40-I	diflubenzuron	I	Pillarfuran	carbofuran	I
Perfekthion	dimethoate	I	Pillargon	propoxur	I
Penchloral	pentachlorophenol	I,F,H	Pillarich	chlorothalonil	F
Pennamine D	2,4-D	H	Pillarmate	methomyl	I
Pennant	metolachlor	H	Pillarquat	paraquat	H
Penncap-E	methyl parathion	I	Pillarsato	glyphosate	H
Penncap-M (encap)	methyl parathion	I	Pillartex	fenthion	I
			Pillarxone	paraquat	H
Penncozeb	mancozeb	F	Pillarzo	alachlor	H

Trade Name	Common Name	Type
Pilot Super	quizalofop-p-ethyl	H
Pimagram	metolachlor	H
PinUp	glyphosate	H
Piridane	chlorpyrifos	I
Plantgard	2,4-D	H
Planotox	2,4-D	H
Plantineb	maneb	F
Plantulin	propazine	H
Pledge	bentazon(e)	H
Plydax	terbufos	I
PMC	phosmet	I
Poast	sethoxydim	H
Point	imazalil	F
Point	permethrin	I
Polado	glyphosate	H
Policar MZ	mancozeb	F
Police	glyphosate	H
Pollux	zinc phosphide	R
Polyram	metiram	F
Polyram-Combi	metiram	F
Polyram M	maneb	F
Polyram-Ultra	thiram	F
Polytrin	cypermethrin	I
Polyram-Z	zineb	F
Pomarsol	thiram	F
Pomarsol Z Forte	ziram	F
Pondmaster	glyphosate	H
Potatoe Seed Treater M-Z	mancozeb	F
Pounce	permethrin	I
Pox	parathion	I
PP 005	fluazifop-p-butyl	H
PP 148	paraquat	H
PP 383	cypermethrin	I
PP 557	permethrin	I
Pramex	permethrin	I
Pratt	oxamyl	I
Precor	methoprene	I
Prefar	bensulide	H
Prelude	permethrin	I
Premalin	linuron	H
Premazine	simazine	H
Premerge	dinoseb	H
Prenfish	rotenone	I
Prentox DDVP	dichlorvos (DDVP)	I
Prentox	malathion	I
Prentox Methoxychlor	methoxychlor	I
Prentox Carbamate	propoxur	I

Trade Name	Common Name	Type
Prentox Fenthion 4E	fenthion	I
Pre-san	bensulide	H
Prevail	cypermethrin	I
Prevenol	chlorpropham	GR
Priltox	pentachlorophenol	I,F,H
Primatol A	atrazine	H
Primatol P	propazine	H
Primatol Q	prometryn	H
Primatol S	simazine	H
Princep	simazine	H
Prodaram	ziram	F
Proflan	trifluralin	H
Profume	methyl bromide	F
Prolan	trifluralin	H
Prolae	phosmet	I
Prolate	phosmet	I
Promar	diphacinone	R
Promet	prometryn	H
Prometrex	prometryn	H
Pronone	hexazinone	H
Pronto	trichlorfon	I
Prop Job	propanil	H
Propa	propanil	H
Propal	propanil	H
Propanae	propanil	H
Propanex	propanil	H
Propanile	propanil	H
Propanilo	propanil	H
Propax	propanil	H
Propazin	propazine	H
Propinex	propazine	H
Propogon	propoxur	I
Propyon	propoxur	I
Propyzamide	pronamide	H
Prostar	propanil	H
Protect T/O	mancozeb	F
Prowl	pendimethalin	H
Proxol	trichlorfon	I
Prozinex	propazine	H
Pryfon	isofenphos	I
Punisx	cypermethrin	I
Puralin	thiram	F
Py	pyrethrin	I
Pydrin	fenvalerate	I
Pynamin	allethrin	I
Pynosect	permethrin	I
Pynosect	resmethrin	I
Pyrenone	pyrethrin	I
Pyresin	allethrin	I
Pyrethrum	pyrethrin	I
Pyrex	pyrethrin, rotenone	I
Pyrex Insectspray	pyrethrin, malathion	I
Pyrexcel	allethrin	I

Trade Name	Common Name	Type	Trade Name	Common Name	Type
Pyrinex	chlorpyrifos	I	Rhodiasan Express	thiram	F
Pyrocide	allethrin, pyrethrin	I	Rhodiatox	parathion	I
Qamlin	permethrin	I	Rhodiacide	ethion	I
Queletox	fenthion	I	Rhodianebe	maneb	F
Quinoxone	2,4-D	H	Rhodocide	ethion	I
R-1504	phosmet	I	Rhomene	MCPA	H
R-1582	azinphos-methyl	I	Rhonox	MCPA	H
R-1608	EPTC	H	Ridall	zinc phosphide	R
R-1910	butylate	H	Ridomil	metalaxyl	F
R-4461	bensulide	H	Rifle	primisulfuron-methyl	H
R-4572	molinate	H			
R-7465	naproamide	H	RIMI 101	chlorpyrifos	I
R-23979	imazalil	F	Rion	malathion	I
Radapon	dalapon	H	Riozeb	mancozeb	F
Radazin	atrazine	H	Ripcord	cypermethrin	I
Rad-E-Cate 25	cacodylic acid	H	Riselect	propanil	H
			Ro 13-5223/000	fenoxycarb	GR
Radinex	thiabendazole	F			
Raid Flying Insect Killer	resmethrin	I	Rochlor	trichlorfon	I
			Rocyper	cypermethrin	I
Ralothrin	cypermethrin	I	Rodamine	2,4-D	H
Ramik	diphacinone	R	Rodent Cake	diphacinone	R
Rampart	carbofuran	I	Rodenticide AG	zinc phosphide	R
Rampart	phorate	I			
Ranger	glyphosate	H	Rodent Pellets	zinc phosphide	R
Rat & Mouse Blues II	diphacinone	R			
			Rodeo	glyphosate	H
Ratol	zinc phosphide	R	Roethyl-P	parathion	I
Rattler	glyphosate	H	Rofon	triadimefon	F
Rasayan-sulfan	endosulfan	I	Rogodan	dimethoate	I
			Rogodial	dimethoate	I
Ravyon	carbaryl	I	Rogor	dimethoate	I
R-Bix	paraquat	H	Rokar X	bromacil	H
RE 4355	naled	I	Roman vitriol	copper sulfate	F
Rebelate	dimethoate	I	Romethyl-P	methyl parathion	I
Reclaim	tebuthiuron	H	Romyl	benomyl	F
Redeem	triclopyr	H	Ronex	diuron	H
Reglex	diquat dibromide	H	Ronilan	vinclozolin	F
Reglon	diquat dibromide	H	ROP 500F	iprodione	F
Reglone	diquat dibromide	H	Rophosate	glyphosate	H
Reglox	diquat dibromide	H	Rotacide	rotenone	I,R
Rely	triclopyr	H	Rotalin	linuron	H
Remasan	maneb	F	Rotate	bendiocarb	I
Remedy	triclopyr	H	Rotenox	rotenone	I
Repulse	chlorothalonil	F	Rothalonil	chlorpropham	GR
Resistox	coumaphos	I	Roundup	glyphosate	H
Respond	resmethrin	I	Rout	bromacil	H
Revenge	dalapon	H	Rovral	iprodione	F
Reward	diquat dibromide	H	Roxion	dimethoate	I
Rezifilm	thiram	F	RP-26019	iprodione	F
RH-315	pronamide	H	RP-Thion	ethion	I
RH-2915	oxyfluorfen	H	Ryan 50	ryania	I
RH 6201	acifluofen	H	Ryanicide	ryania	I
Rhoden	propoxur	I			

Trade Name	Common Name	Type	Trade Name	Common Name	Type
Rycelan	oryzalin	H	Sepizin M	azinphos-methyl	I
Ryzelan	oryzalin	H	Seraphos	propetamphos	I
Ryzelon	oryzalin	H	Sevigor	dimethoate	I
Ryanodine	ryania	I	Sevimol	carbaryl	I
S 276	disulfoton	I	Sevin	carbaryl	I
S 1752	fenthion	I	Shamrox	MCPA	H
S 1844	esfenvalerate	I	Sheriff	quizalofop-p-ethyl	H
S 5602	fenvalerate	I	Shortstop	EPTC	H
S 5620A	esfenvalerate	I	Showdown	triallate	H
S 10145	propanil	H	Sialite	dinocap	F
Saber	bromoxynil	H	Sicid	rotenone	I,R
SADH	daminozide	GR	Silrifos	chlorpyrifos	I
Safer	Bacillus thuringiensis	I	Silvanol	lindane	I
			Silvapron D	2,4-D	H
Safidon	phosmet	I	Silverquat	paraquat	H
Safrotin	propetamphos	I	Silvisar 510	cacodylic acid	H
Sakkimol	molinate	H	Silvicide	ammonium sulfamate	H
Salvo	cacodylic acid, 2,4-D	H			
			Sim-Trol	simazine	H
Salzburg vitriol	copper sulfate	F	Simadex	simazine	H
			Simanex	simazine	H
SAN 52139I	propetamphos	I	Simavit	simazine	H
Sanaphen	2,4-D	H	Simazat	atrazine	H
Sandozebe	mancozeb	F	Simazip	simazine	H
Saniclor	PCNB	F	Simazol	simazine, amitrole	H
Sanmarton	fenvalerate	I	Simazon	simazine, diuron	H
Sanvex	carboxin	F	Simetrax	simazine	H
Sarclex	linuron	H	Simex	simazine, amitrole	H
Sarolex	diazinon	I	Simflow	simazine	H
Savage	2,4-D	H	Simop	simazine	H
Savit	carbaryl	I	Simosol	simazine, amitrole	H
SBP 1382	resmethrin	I	Simpar	parathion	I
Scotts Proturf	dicamba	H	Sinbar	terbacil	H
Scourge	resmethrin	I	Sinflouran	trifluralin	H
Scout	chlorpyrifos	I	Sinid	rotenone	I
Scrubmaster	tebuthiuron	H	Sinituho	pentachlorophenol	I,F,H
Scythe	paraquat	H	Sinox General	dinoseb	H
SD 15418	cyanazine	H			
Seduron	diuron	H	Sinuron	linuron	H
Seedox	bendiocarb	I	Siolcid	linuron	H
Seedoxin	bendiocarb	I	Sipaxol	pendimethalin	H
Seed Shield	mancozeb	F	Siperin	cypermethrin	I
Seed Shield Planter Box	maneb	F	Sipquat	paraquat	H
			Sitophil	dichlorvos (DDVP)	I
Sencor	metribuzin	H	Sitradol	pendimethalin	H
Sencoral	metribuzin	H	Sitrazin	simazine	H
Sencorex	metribuzin	H	Skeetal	Bacillis thuringiensis	I
Sendra	propoxur	I			
Sendran	propoxur	I	Slam	carbaryl	I
Sepicap	captan	F	Slug Death	metaldehyde	M
Sepilate	ziram	F	Slug Fest Colloidal 25	metaldehyde	M
Sepimate	ammonium sulfamate	H			
			Slug-N-Snail	metaldehyde	M
Septene	carbaryl	I	Slug Pellets	metaldehyde	M
Sepineb	zineb	H	Slug-Tox	metaldehyde	M

Trade Name	Common Name	Type	Trade Name	Common Name	Type
Slugit Pellets	metaldehyde	M	Sulfamate	ammonium sulfamate	H
SN 81742	sethoxydim	H			
Soilfume	ethylene dibromide	I	Sulfaril	carbaryl	I
Sok-Bt	Bacillis thuringiensis	I	Sulgen	dodine	F
			Sulmalthion	malathion	I
Solution	2,4-D	H	Sumi-alfa	esfenvalerate	I
Solado	glyphosate	H	Sumi-alpha	esfenvalerate	I
Solvigran	disulfoton	I	Sumibac	fenvalerate	I
Solvirex	disulfoton	I	Sumicidin	fenvalerate	I
Somonic	methidathion	I	Sumiflower	fenvalerate	I
Somonil	methidathion	I	Sumifly	fenvalerate	I
Sonic	glyphosate	H	Sumileece	fenvalerate	I
Soprathion	parathion	I	Sumipower	fenvalerate	I
Sorene	captan	F	Sumitick	fenvalerate	I
Sovereign	pendimethalin	H	Sumitox	fenvalerate	I
Spasor	glyphosate	H	Sumitox	malathion	I
SPB-1382	resmethrin	I	Sun-Bugger #4	resmethrin	I
Speedway	paraquat	H			
Spectracide	diazinon	I	Suncide	propoxur	I
Spike	tebuthiuron	H	Sunup	glyphosate	H
Spotrete	thiram	F	Super	cypermethrin	I
Spotton	fenthion	I	Superman	maneb	F
Spritz-Hormin	2,4-D	H	Supermixy	zineb	F
			Supernox	propanil	H
Sprout Nip	chlorpropham	GR	Super T	trifluralin	H
Spud Nic	chlorpropham	GR	Superzol	amitrole	H
Spur	fluvalinate	I	Supex	glyphosate	H
SR 406	captan	F	Supracide	methidathion	I
Staa-Free	bromacil	H	Supracidin	methidathion	I
Stabineb	maneb	F	Suprathion	methidathion	I
Stake	alachlor	H	Surcopur	propanil	H
Stam	propanil	H	Surflan	oryzalin	H
Stampede	propanil	H	Surpur	propanil	H
Starfire	paraquat	H	Su Seguro Carpidor	trifluralin	H
Stathion	parathion	I			
Stauffer N-2790	fonofos	I	Sutan+	butylate	H
			Sutan 6E	butylate	H
Stauffer R-1	butylate	H	Suntol	coumaphos	I
Steward	lindane	I	Swebate	temephos	I
Sting	glyphosate	H	Sweep	paraquat	H
Stinger	dimethoate	I	Swing	glyphosate	H
Stipend	chlorpyrifos	I	Syllit	dodine	F
Stirrup	glyphosate	H	Sylvicor	cacodylic acid	H
Stockade	permethrin	I	Synerol	pyrethrin	I
Stockade	cypermethrin	I	Synklor	chlordane	I
Stomoxin	permethrin	I	Syndane	chlordane	I
Stomp	pendimethalin	H	Synpran N	propanil	H
Storite	thiabendazole	F	Synpren	rotenone	I
Strel	propanil	H	Synthrin	resmethrin	I
Subdue	metalaxyl	F	Syntox	resmethrin	I
Subitex	dinoseb	H	SYS 67 Omnidel	dalapon	H
Sulban	chlorpyrifos	I			
Sulbenz	lindane	I	Tackle	acifluorfen	H
Sulfacop	copper sulfate	I	Tafethion	ethion	I

Trade Name	Common Name	Type	Trade Name	Common Name	Type
Tag	diquat dibromide	H	Thionate	endosulfan	I
Talcord	permethrin	I	Thionex	endosulfan	I
Taloberg	chlorothalonil	F	Thionic	ziram	F
Talodex	fenthion	I	Thiophos	parathion (ethyl)	I
Talon	chlorpyrifos	I	Thiosan	thiram	F
Tanzine	simazine	H	Thiosulfan	endosulfan	I
Tarene	trifluralin	H	Thiotex	thiram	F
Targa D+	quizalofop-p-ethyl	H	Thiotox	thiram	F
Targa Super	quizalofop-p-ethyl	H	Thioxamyl	oxamyl	I
Task	dichlorvos	I	Thipel	thiram	F
Taterpex	chlorpropham	GR	Thiramad	thiram	F
Tattoo	bendiocarb	I	Thirasan	thiram	F
Tau-fluvalinate	fluvalinate	I	Thirame	thiram	F
			Thiramad	thiram	F
TBZ	thiabendazole	F	Thiram Granuflo	thiram	F
TDTC Technical	triallate	H			
			Thiuramin	thiram	F
Tebulan	dodine	F	Thuricide	Bacillis thuringiensis	I
Tebuzate	thiabendazole	F			
Tecto	thiabendazole	F	Tiezene	zineb	F
Teknar	Bacillis thuringiensis	I	Tiguvon	fenthion	I
			Timbrel	triclopyr	H
Tekwaisa	methyl parathion	I	Timet	phorate	I
Tell	primisulfuron-methyl	H	Tirampa	thiram	F
			Tiuramyl	thiram	F
Temik	aldicarb	I	TMTC	thiram	F
Tender	glyphosate	H	TMTD 50 Borches	thiram	F
Tendex	propoxur	I			
Terborox	terbufos	I	Tomcat	diphacinone	R
Termi-Ded	chlordane	I	Topiclor 20	chlordane	I
Termex	chlordane	I	Toppel	cypermethrin	I
Termidan	chlordane	I	Torch	bromoxynil	H
Termiseal	chlordane	I	Tordon	picloram	F
Terrathion	phorate	I	Tornade	permethrin	I
Terraclor	PCNB	F	Tornado E	carbaryl	I
Terraguard	chlorpyrifos	I	Tornado	fluazifop-p-butyl	H
Terr-O-Gas	methyl bromide	I	Torpedo	permethrin	I
Tersan	thiram	F	Torpedo	diquat dibromide	H
Tersan 1991	benomyl	F	Totalene	trichlorfon	I
Tersan LSR	maneb	F	Totazine	simazine	H
Tercyl	carbaryl	I	Touchdown	glyphosate	H
Tetrapom	thiram	F	Torch	bromoxynil	H
TH 6040	diflubenzuron	I	Tordon	picloram	H
Thianosan	thiram	F	Torus	fenoxycarb	I
Thibenzole	thiabendazole	F	Totalene	trichlorfon	I
Thifor	endosulfan	I	Toterbane	diuron	H
ThiLor	thiram	F	Toxation	azinphos-methyl	I
Thimenox	phorate	I	Toxer Total	paraquat	H
Thimet	phorate	I	Toxichlor	chlordane	I
Thimul	endosulfan	I	Tox-R	rotenone	I,P
Thinsec	carbaryl	I	TR-10	trifluralin	H
Thiobel	carboxin	F	Tracker	dicamba	H
Thiodan	endosulfan	I	Trametan	thiram	F
Thioknock	thiram	F	Trefanocide	trifluralin	H
Thionam	ziram	F	Treficon	trifluralin	H

Trade Name	Common Name	Type
Treflan	trifluralin	H
Triangle	copper sulfate	I
Tribute	fenvalerate	I
Tributon	2,4-D	H
Tri-4	trifluralin	H
Tricarnam	carbaryl	I
Tricarbamix	ziram	F
Tricel	chlorpyrifos	I
Trichloro-phene	trichlorfon	I
Trichloro-phon	trichlorfon	I
Trifluralina 600	trifluralin	H
Triflurex	trifluralin	H
Trigard	trifluralin	H
Triherbide-CIPC	chlorpropham	GR
Trilin	trifluralin	H
Trim	trifluralin	H
Trimangol	maneb	F
Trimaran	trifluralin	H
Trimetion	dimethoate	I
Trinex	trichlorfon	I
Tripomol	thiram	F
Triscabol	ziram	F
Tristar	trifluralin	H
Tritex-Extra	sethoxydim	H
Tritisan	PCNB	F
Tritoftorol	zineb	F
Trooper	dicamba	H
Troy-Bt	Bacillus thuringiensis	I
Trust	trifluralin	H
Tuads	thiram	F
Tubatoxin	rotenone	I
Tubergran	PCNB	F
Tuff Brite	chlorpropham	GR
Tuffcide	chlorothalonil	F
Tugen	propoxur	I
Tugon	trichlorfon	I
Tulisan	thiram	F
Tumbleweed	glyphosate	H
Turcam	bendiocarb	I
Turfcide	PCNB	F
Turflon	triclopyr	H
Tycap	fonofos	I
U 46 D	2,4-D	H
U 46 M Fluid	MCPA	H
UBI C4874	quizalofop-p-ethyl	H
UC 7744	carbaryl	I
UC 21149	aldicarb	I
Ultracide	methidathion	I
Ultracidin	methidathion	I

Trade Name	Common Name	Type
Umbethion	coumaphos	I
Unden	propoxur	I
Undene	propoxur	I
Unicrop CIPC	chlorpropham	GR
Unidron	diuron	H
Unifos	dichlorvos (DDVP)	I
Unipon	dalapon	H
Uragan	bromacil	H
Urox B	bromacil	H
Urox HX	bromacil	H
Ustaad	cypermethrin	I
Uvon	prometryn	H
Vancide	ziram	F
Vancide 89	captan	F
Vancide TM	thiram	F
Vandodine	dodine	F
Vantage	sethoxydim, diphacinone	H
Vap-Malathion	malathion	I
Vapona	dichlorvos (DDVP)	I
Vaponite	dichlorvos (DDVP)	I
Varikill	fenoxycarb	I
Vault	Bacillus thuringiensis	I
Vectal	atrazine	H
Vectobac	Bacillus thuringiensis	I
Vectocide	Bacillus thuringiensis	I
Vectrin	resmethrin	I
Vegetrole	diquat dibromide	H
Vegetox	carboxin	F
Vegfru Foratox	phorate	I
Vegfru Fosmite	ethion	I
Vegfru Malatox	malathion	I
Vegiben	chloramben	H
Velpar	hexazinone	H
Velsicol 104	heptachlor	I
Velsicol 1068	chlordane	I
Velsicol 58-CS-11	dicamba	H
Vencedor	copper sulfate	I
Venceweed	2,4-DB	H
Ventene	ziram	F
Venturol	dodine	F
Verdican	dichlorvos (DDVP)	I
Verdipor	dichlorvos (DDVP)	I
Verdisol	dichlorvos (DDVP)	I
Verisan	iprodione	F

Trade Name	Common Name	Type	Trade Name	Common Name	Type
Verison	iprodione	F	Yalan	molinate	H
Vermicide Bayer 2349	trichlorfon	I	Yaltox	carbofuran	I
			Yardex	fluvalinate	I
Vertac	propanil	H	Z-C Spray	ziram	F
Vertac Weed Killer	dinoseb	H	Zeapos	atrazine	H
			Zeazin	atrazine	H
Vertimec	abamectin	I	Zebtox	zineb	F
Vidate	oxamyl	I	Zelan	MCPA	H
Vigilante	diflubenzuron	I	Zeltivar	trichlorfon	I
Vikane	sulfuryl fluoride	I	Zeltoxone	trifluralin	H
Vitavax	carboxin	F	Zephyr	abamectin	I
Vitrex	parathion	I	Zerlate	ziram	F
Vondcaptan	captan	F	Ziden	zineb	F
Vondozeb Plus	mancozeb	F	Zidil	chlorpyrifos	I
			Ziman	mancozeb	F
Volphor	phorate	I	Zimaneb	mancozeb	F
Volthion	ethion	I	Zinc Ethylenebis dithio-carbamate	zineb	F
Vonduron	diuron	H			
Vorlan	vinclozolin	F			
Vydate	oxamyl	I			
Way Up	pendimethalin	H	Zincmate	ziram	F
Weedazol	amitrole	H	Zinc metiram	metiram	F
Weed-B-Gone	2,4-D	H	Zinc-Tox	zinc phosphide	R
Weed Rhap	2,4-D, MCPA	H	Zinkcarba-mate	ziram	F
Weedar	2,4-D	H	Zinebane	zineb	F
Weedex	simazine	H	Zineber	zineb	F
Weedoff	glyphosate	H	Zinecor	zineb	F
Weedone	2,4-D, PCP	H	Zinecupryl	copper sulfate, zineb	F
Weedtox	2,4-D	H			
Weedtrine D	diquat dibromide	H	Zinosan	zineb	F
Weedtrine-II	2,4-D	H	Zinugec	zineb	F
Weedtrol	2,4-D	H	Ziram	ziram	F
Wham EZ	propanil	H	Ziramon	ziram	F
Whitmire PT-110	resmethrin	I	Ziramugec	ziram	F
			Ziramuis	ziram	F
Winner	glyphosate	H	Zirasan	ziram	F
Wipeout	hydramethylnon	I	Zirex	ziram	F
Witox	EPTC	H	Zirberk	ziram	F
WL 19805	cyanazine	H	Zithiol	malathion	I
WL 43467	cypermethrin	I	ZP	zinc phosphide	R
WL 43775	fenvalerate	I	ZR 515	methoprene	I
Wofatox	methyl parathion	I	Zytox	methyl bromide	F

Appendix B

U.S. system-metric conversions

I. Metric system (SI) prefixes

The metric system, or System Internationale (SI) as it is formally referred to, consists of basic units of length, mass, and other quantities. Basic units commonly used in the metric system are the meter, for length (abbreviated m); the gram, for mass (abbreviated g); and the liter, for liquid volume (abbreviated L). Prefixes are used to denote fractions or multiples of these basic units. For example, a nanometer (abbreviated nm) is 1×10^{-9} meter, or one-billionth of a meter, whereas a kilometer (km) is 1×10^3 meters, or 1000 meters.

Prefix	Abbreviation	Coefficient
nano-	n	10^{-9}
micro-	m or μ	10^{-6}
milli-	m	10^{-3}
centi-	c	10^{-2}
kilo-	k	10^3
mega-	M	10^6
giga-	G	10^9

II. Conversion factors

In the following sections, factors for converting from the metric system into U.S. units are presented on the left-hand side; those for converting from U.S. to metric units are on the right. For example, to convert a measurement of length expressed in centimeters (cm) to feet (ft), multiply the dimension in feet by the conversion factor 0.03281. Conversely, to convert a length measurement expressed in ft to one expressed in cm, multiply by 30.48. These conversion factors have been rounded to four significant digits where appropriate. The basic units of the U.S. system are abbreviated as follows: foot, ft; inch, in; fluid ounce, fl oz; quart, qt; gallon, gal; ounce avoirdupois, oz; pound avoirdupois, lb.

A. Length

1 cm = 0.03281 ft	1 ft = 30.48 cm
1 cm = 0.3937 in	1 in = 2.54 cm
1 m = 3.281 ft	1 ft = 0.3048 m
1 km = 0.6215	1 mi = 1.609 km

B. Area

1 m² = 10.76 ft²   1 ft² = 0.09290 m²
1 cm² = 0.1550 in²   1 in² = 6.452 cm²
1 cm² = 0.001076 ft²   1 ft² = 929.0 cm²

C. Volume

1 cm³ = 1 mL
1 cm³ = 0.03381 fl oz   1 fl oz = 29.57 cm³
1 L   = 1.057 qt   1 qt = 0.9464 L
1 L   = 0.2642 gal   1 gal = 3.785 L
1 m³ = 264.2 gal   1 gal = 0.003785 m³

D. Mass

1 g = 0.03527 oz   1 oz = 28.53 g
1 g = 0.002205 lb   1 lb = 453.6 g
1 kg = 2.205 lb   1 lb = 0.4536 kg

E. Density

1 g/cm³ = 0.5780 oz/in³   1 oz/in³ = 1.730 g/cm³

F. Temperature

To convert degrees Celsius (TC) into degrees Fahrenheit (TF):

$$TF = 1.8(TC) + 32$$

To convert degrees Fahrenheit (TF) into degrees Celsius (TC):

$$TC = (TF - 32)/1.8$$

G. Pressure

1 Pa = 0.007501 mm Hg   1 mm Hg = 133.3 Pa

H. Airborne contaminant concentrations

Most regulatory exposure standards and environmental monitoring data for airborne contaminants are expressed either as the mass of the compound (in milligrams) per volume of air (in cubic meters) of air at standard temperature and pressure (mg/m³) or as parts per million (ppm), or both. Typically, at atmospheric conditions, the corrections for temperature and pressure are not significant.

To convert the airborne concentration in mg/m³ to airborne concentration in parts per million (ppm):

$$ppm = (mg/m^3) \times 24.45/(Mol.\ Weight)$$

To convert the airborne concentration in parts per million (ppm) to airborne concentration in mg/m³:

$$mg/m^3 = (ppm) \times (Mol.\ Weight)/24.45$$

(Mol. Weight = gram molecular weight of the compound)

I. Concentrations of contaminants in water

Most regulatory exposure standards and environmental monitoring data for contaminants in water are expressed either as the mass of the compound (in milligrams) per volume of water (in liters), or mg/L. Sometimes it may be expressed as parts per million (ppm). Since the mass of one liter of water is one kilogram (1000 g), 1 mg/L is equal to 1 mg/kg in water. Thus, for environmental contaminants in water, mg/L, mg/kg, and ppm are effectively equal.

Appendix C

Glossary

Absorption: The uptake of chemicals by a cell or an organism. The movement of a chemical into or across a tissue.

Acaricide: The class of pesticides used to kill mites and ticks; also known as "miticide."

Acceptable Daily Intake (ADI): The maximum dose of a substance that is anticipated to be without health risk to humans when taken daily over the course of a lifetime. ADIs are established by the WHO.

Accumulation: The build-up of a chemical in the body due to long-term or repeated exposure.

Acetylcholinesterase: An enzyme present in nerve tissue, muscles, and red blood cells that catalyzes the hydrolysis of acetylcholine to choline and acetic acid, allowing neural transmission across synapses to occur; true cholinesterase.

Acetylcholinesterase inhibitor: A compound or group of compounds (e.g., organophosphorus compounds) that block the action of the enzyme acetylcholinesterase, interfering with the transmission of impulses between nerve cells.

ACGIH: American Conference of Governmental Industrial Hygienists; establishes exposure limits for workers.

Acidic: Sour; having a pH of less than 7.

Active ingredient (a.i.): a chemical compound in a pesticide formulation that produces the pesticidal effect in the target species.

Acute: Single or short-term exposure; used to describe brief exposures and effects that appear promptly after exposure.

Additive effect: Combined effect of two or more chemicals equal to the sum of their individual effects.

Adenoma: A benign tumor of glandular origin.

Adjuvant: Substance added to a formulated pesticide product to act as a wetting or spreading agent, sticker, penetrant, or emulsifier to enhance the effectiveness of the product.

Adsorption: The process by which chemicals are bound to a solid surface, especially soil particles.

ADI — see Acceptable Daily Intake.

Adsorption coefficient (K_{oc}): A measure of a material's tendency to adsorb to soil particles. High K_{oc} values indicate a tendency for the material to be adsorbed by soil particles rather than remain dissolved in the soil solution. Strongly adsorbed molecules will not leach or move unless the soil particle to which they are adsorbed moves (as in erosion). K_{oc} values of less than 500 indicate little or no adsorption and a potential for leaching.

$$K_{oc} = \frac{\text{conc. adsorbed}/\text{conc. dissolved}}{\%\ \text{organic carbon in the soil}}$$

Aerobic: Requiring oxygen or free air; conditions in which oxygen is present.

Alkaline: Basic; having a pH of greater than 7.

Ambient: Environmental or surrounding conditions.

Anaerobic: A process that does not require oxygen or free air; conditions in which oxygen is absent.

Antagonism: The combined action of two or more substances to produce an effect less than the sum of their individual effects; the opposite of synergism.

Aquatic invertebrates: Organisms that do not have a spinal column and live in water; includes insects, crayfish, and mites.

Aqueous: Watery; pertaining to water.

Arthropods: Organisms such as insects, arachnids (spiders and mites), and crustaceans that lack backbones (invertebrates).

Assay: Analysis to determine the presence, absence, or quantity of a particular chemical or effect.

Atrophy: The wasting away or reduction in the size of a cell, tissue, or organ(s).

Avicide: The class of pesticides used to kill birds.

Basic: Alkaline; having a pH of greater than 7.

Benign tumor: A noncancerous tumor.

Bioaccumulation: The uptake and concentration of a substance in plants or animals.

Biocide: A material that has the capacity to kill forms of life.

Bioconcentration: The accumulation of a chemical in tissues of an organism (such as fish) to levels that are greater than the level in the medium (such as water) in which the organism resides.

Bioconcentration Factor (BCF): A measure of the tendency for a chemical to accumulate. The ratio of the concentration of a substance in a living organism (mg/kg) to the concentration of that substance in the surrounding environment (mg/L for aquatic systems).

$$BCF = \frac{\text{Concentration in organism, mg/kg}}{\text{Environmental concentration, mg/L}}$$

Biodegradation: Breakdown of a substance into more elementary compounds by the action of microorganisms such as bacteria.

Biomagnification: Process by which substances such as pesticides or heavy metals become more concentrated with each succeeding step up the food chain.

Bone marrow: The soft tissue contained inside the bone.

Broad-spectrum pesticide: A pesticide that kills a wide range of pest species.

Cancer: A disease characterized by the rapid and uncontrolled growth of aberrant cells into malignant tumors.

Carcinogen: Any substance capable of producing cancer.

Carcinogenicity: The ability of a substance to produce cancer.

Carcinoma: A malignant tumor of epithelial origin.

Cardiovascular system: The heart and blood vessels.

Carrier: Material added to an active ingredient to facilitate its preparation, storage, shipment, or use.

CAS No.: Chemical Abstracts Service Registry Number. The CAS No. is a unique number to specify each chemical (also CAS #).

CAUTION: see "Signal Word."

Central nervous system (CNS): Portion of the nervous system that consists of the brain and spinal cord.

Certified applicator: A person who has been trained and certified as a pesticide applicator by a state agency.

Chemical class: A group of chemical compounds containing common structural elements.

Cholinergic: Resembling acetylcholine, especially in physiological action.

Cholinesterase: An enzyme in the body necessary for proper nerve function.

Chromosomal aberration: An irregularity in the number or structure of chromosomes.

Chromosome: Rod-like structure in the nucleus of a cell that contains the genes responsible for heredity.

Chronic: Occurring over a long period of time; used to describe exposures and effects that occur only after a long exposure.

Clay: A soil component consisting of very fine particles (<0.002 mm diameter). Clay particles provide large surface area for adsorption of molecules. Clay soils provide the most resistance to leaching.

Clinical studies: Studies of humans under controlled conditions.

Common names: Common names for pesticides are established or accepted by the following organizations: ANSI, American National Standards Institute; BSI, British Standards Institution; ISO, International Organization for Standardization; WSSA, Weed Science Society of America. A common name generally is not accepted for use on EPA registered labels or in CFR regulations until it has been accepted by ANSI.

Concentration: The amount of pesticide or other chemical in a quantity of liquid, solid, or gas, expressed as lb/gal, mL/L, etc.

Congenital: Existing at birth.

Conjunctivitis: Inflammation of the membrane that lines the inner surface of the eyelids.

Contact dermatitis: Skin swelling due to either acute irritation from short-term contact with a substance, or from chronic sensitization that develops from long-term skin contact with an irritating substance.

Contact poison: A poison that affects the target organisms through physical contact rather than through ingestion or inhalation (animals) or translocation (plants). Compare with "Stomach poison."

Cornea: The transparent front portion of the eyeball.

Cost/benefit analysis: A quantitative evaluation of the costs that would be incurred versus the overall benefits to society of a proposed action such as the establishment of an acceptable dose of a toxic chemical.

Cytotoxicity: The capacity of a material to interfere with normal cell function.

DANGER: see Signal Word.

Degradate: A compound that results from the degradation of another compound.

Degradation: Chemical or biological breakdown of a complex compound into simpler compounds. Common processes of degradation are hydrolysis, photolysis, and microbial breakdown.

Dermal: Of the skin; through or by the skin.

Diffusion: The movement of suspended or dissolved particles from a more concentrated region to a less concentrated region. Diffusion tends to distribute particles uniformly throughout the available volume.

Dipterous: Of or pertaining to an order of insects that includes flies and mosquitoes.

Distribution: The movement of a chemical from the blood to other tissues.

Dominant lethal assay: A mutagenesis test used to assess the ability of a chemical to produce genetic damage. Male animals are treated with a test substance acutely (single dose) or over the entire period of sperm production. These males are then mated with females, which are examined for the number of total implantations and viable fetuses.

Dose: The amount of a compound that is ingested, applied to the skin, intravenously injected, or otherwise delivered to the organism. Dose is often expressed in milligrams of chemical per kilogram of body weight (mg/kg) in cases of single administration, and milligrams per kilogram per day (mg/kg/day) in cases of chronic exposures or long-term animal studies. Note that dose is the amount that is applied to the organism, rather than the concentration in food, air, or water.

Dose-Response: A quantitative relationship between the dose of a chemical and the degree/severity of an effect caused by the chemical.

Dust: A pesticide formulation that consists of an active ingredient impregnated on a finely ground carrier such as clay, talc, or calcium carbonate.

Ecology: The study of the interrelationships between living organisms and their environment, both physical and biological.

Ecosystem: The interacting system of a biological community and its nonliving environment.

Elimination: The removal of a chemical from the body, primarily in the urine, feces, or exhaled air.

Embryo: The developing animal during pregnancy, especially during the early stages.

Embryonic: Pertaining to embryos.

Embryotoxicity: A toxic effect on the embryo.

Emulsifiable Concentrate (EC): A pesticide formulation consisting of an active ingredient and an emulsifying agent in an organic solvent. The solvent is usually not soluble in water. When an EC product is mixed with water prior to application, the resulting mix is a dispersion of fine, oily particles in water.

Environmental fate: The destiny of a chemical after release to the environment; involves considerations such as transport through air, soil, and water, bioconcentration, and degradation.

Enzyme: A protein, synthesized by a cell, that acts as a catalyst in a specific chemical reaction.

EPA: U.S. Environmental Protection Agency.

Epidemiology: The study of the incidence and distribution of disease and/or toxic effects in a population.

Excretion: The removal of a chemical from the body in urine, feces, or expired air.

Exposure: Contact with a chemical. Some common routes of exposure are dermal (skin), oral (by mouth), and inhalation (breathing). Also refers to the concentration in the environment or food to which the organism is exposed.

FAO: Food and Agriculture Organization of the United Nations.

FDA monitoring: The collection and analysis for pesticide residues carried out by the Food and Drug Administration.

Fecundity: The number of offspring produced by an animal.

Fertility: The ability of an organism to produce viable gametes and successfully engage in sexual reproduction.

Fetal resorption: Following in utero fetal death, the partial or complete dissolution of fetal tissue and maternal absorption of the resulting products.

Fetotoxicity: A compound-induced toxic effect on the fetus.

Fetus: Unborn or unhatched vertebrate, especially one that has attained the basic structure of its kind; in humans, a fetus is the developing organism from approximately 3 months after conception until birth.

Flash point: The lowest temperature at which a liquid gives off ignitable vapors.

Food chain: A sequence of species in which each species serves as a food source for the next species. Food chains usually begin with species that consume plant material (herbivores) and proceed to larger and larger carnivores. Ex: grasshopper eaten by snake eaten by owl.

Formulated pesticide product: A mixture of one or more pesticidal active ingredients with inert ingredients such as carriers and diluents; a packaged, ready-for-sale pesticide product. Types of formulations include liquids, dusts, emulsifiable concentrates, ultra-low volume, and granular products.

Fungicide: A class of pesticides used to kill fungi, primarily those that cause diseases of plants.

Gastrointestinal tract (GI tract): The entire digestive canal from mouth to anus.

Gavage: Forced feeding by stomach tube.

General Use Pesticide (GUP): A pesticide that may be purchased and used by individuals without any special certification or licensing. Pesticides classified by the EPA as GUP have not been shown to pose undue hazards to applicators, the general public, or to the environment. Compare with "Restricted Use Pesticide."

Genotoxicity: Any toxic modification or alteration of the structure or function of genetic material.

Gestation: The duration of pregnancy. In the human, gestation is normally 9 months.

Goiter: Enlargement of the thyroid gland.

Granular formulation: A dry, ready-to-use pesticide product that consists of an active ingredient mixed with or impregnated into a carrier such as coarse particles of clay.

Groundwater: Water located in saturated zones below the soil surface. Wells and springs are fed by groundwater.

GUP: see General Use Pesticide.

HA: see Lifetime Health Advisory.

Half-life: The time required for half of the residue to lose its analytical identity whether through dissipation, decomposition, metabolic alteration, or other factors. The half-life concept may be applied to residues in crops, soil, water, animals, or specific tissues.

Hazard: The potential that the use of a product would result in an adverse effect on humans or the environment in a given situation.

Hectare: A metric measure of area equal to 2.47 acres.

Hemotoxicity: A toxic effect on blood components or properties such as changes in hemoglobin, pH, electrolytes, or protein.

Henry's Law Constant: A parameter used in evaluating air exposure pathways. Values for Henry's Law Constant (H) are calculated using the following equation and the values for solubility, vapor pressure, and molecular weight:

$$H \text{ (atm-m}^3/\text{mole)} = \frac{\text{Vapor pressure (atm)} \times \text{Mole Wt. (g/mole)}}{\text{Water solubility (g/m}^3)}$$

Hepatoma: A malignant tumor occurring in the liver.

Herbicide: A pesticide used for killing or inhibiting plant growth. A weed or grass killer.

Hormone: A chemical substance secreted in one part of an organism and transported to another part of that organism where it has a specific effect.

Hydrology: The study of the properties, distribution, behavior, and effects of water on the earth's surface.

Hydrolysis: A chemical reaction (either spontaneous or enzyme-catalyzed) wherein a molecule reacts with water, resulting in the breakdown of the molecule. Literally, "cleavage by water."

Hydrolyze: To subject to hydrolysis; to undergo hydrolysis.

Hypoxia: A deficiency of oxygen.

Inert ingredient: An ingredient in a formulated pesticide product that will not prevent, destroy, repel, or mitigate any pest and which is intentionally included in the product. Includes carriers and materials that dilute the active ingredient.

Inhalation: Drawing of air into the lungs.

Insecticide: A class of pesticides used to kill insects.

Intake: Amount of material inhaled, ingested, or absorbed dermally during a specified period of time.

Intraperitoneal (IP): The introduction by injection of a substance into the peritoneal cavity, which is comprised of the abdominal and pelvic spaces and contains the large internal organs.

In vitro studies: Studies of chemical effects conducted in tissues, cells, or subcellular extracts from an organism (i.e., not in the living organism).

In vivo studies: Studies of chemical effects conducted in intact living organisms.

Invertebrates: Organisms that lack a spinal column.

Irreversible: Permanent, incurable.

Iris: The colored circular portion of the eyeball.

Isomers: Two or more chemical compounds having the same basic structure but different properties.

K_d: Soil-water adsorption coefficient, calculated by using measurements of pesticide distribution between soil and water.

K_{oc}: see "Adsorption coefficient."

K_{ow}: see "Octanol-water partition coefficient."

Larva: The immature form and stage of development of some insect species.

Larvicide: A class of pesticides used to kill insect larvae. Usually refers to chemicals used for controlling mosquito larvae, but also to chemicals for controlling caterpillars on crops.

Latency: Time from the first exposure to a chemical until the appearance of a toxic effect.

LC: Lethal Concentration.

LC_{50}: The concentration of toxicant necessary to kill 50% of the organisms being tested. It is usually expressed in units of parts per million (ppm) for dietary studies, as mg/L or mg/m^3 for airborne toxicants such as dusts or vapors, and as mg/L for toxicants dissolved in water.

LC_{50}/96 hr: The concentration at which the test compound is lethal to 50% of the test organisms in a test of 96 hours duration.

LC_{Lo}: Lethal concentration, low. The lowest concentration to cause death in test animals.

LD: Lethal dose.

LD_{50}: The amount of a chemical that is lethal to one half (50%) of the experimental animals exposed to it. LD_{50} values are usuallly expressed as the weight of the chemical per unit of body weight (mg/kg). It may be fed (oral LD_{50}), applied to the skin (dermal LD_{50}), or administered in the form of vapors (inhalation LD_{50}).

LD_{Lo}: Lethal dose, low. The lowest dose that causes death in test animals.

Leaching: The movement of a pesticide chemical or another substance downward through soil as a result of water movement, potentially causing contamination of groundwater resources.

Lethality: Capacity for producing death.

LHA: see "Lifetime Health Advisory."

Lifetime Health Advisory (HA or LHA): A nonenforceable guideline set by the EPA and used to evaluate the health significance of a contaminant in drinking water when no Maximum Contaminant Level has been set by the EPA for that contaminant. HAs are set at a concentration of contamination that can be consumed daily in drinking water over the course of a lifetime with no adverse health effects.

LOAEL: Lowest-Observed-Adverse-Effect-Level; the lowest dose in an experiment that produced an observable adverse effect.

Loam: A soil of intermediate texture containing moderate amounts of sand, silt, and clay.

Malignant tumor: A cancerous tumor.

Mammals: The class of organisms that have backbones (vertebrates), have hair, and suckle their young.

MCL: Maximum Contaminant Level. The highest amount of a contaminant allowed by the EPA in water supplied by a municipal water system; also referred to as "drinking water standard."

Mean: The average value of a variable over a range of observations.

Median: The middle value of an array of values arranged in numerical order; that value of a variable which corresponds to the 50th percentile of a standard normal distribution.

Metabolism: The process of chemical change by which energy is provided in living cells.

Metabolite: Any product of metabolism, especially a transformed chemical.

mg/kg: Milligrams per kilogram.

mg/kg/day: Milligrams per kilogram per day.

mg/L: Milligrams per liter; equivalent to ppm for water.

µg/L: Micrograms per liter; equivalent to parts per billion (ppb) for water.

mg/m^3: Milligrams of material per cubic meter (of air or water).

µg/m^3: Microgram per cubic meter (µg/m^3) (of air or water).

Molluscicide: A class of pesticides used to kill mollusks, primarily snails and slugs.

Monitoring: Measuring concentrations of substances in environmental media or in human or other biological tissues.

MOS: Margin of Safety. Refer to Uncertainty factor.

MTD: Maximum Tolerated Dose; the highest dose of a chemical that does not alter the life span or severely affect the health of an animal.

Mucous membranes: Any tissue lining body cavities and canals that comes in contact with the air and is kept moist by secretions of various types of glands (e.g., inside the mouth).

Musculoskeletal system: Composed of, or pertaining to the muscles and the skeleton.

Mutagen: An agent that causes a permanent genetic change in a cell other than that which occurs normally.

Mutagenicity: The ability of a substance to produce a detectable and heritable change in genetic material, which may be transmitted to the progeny of affected individuals through germ cells (germinal mutation) or from one cell generation to another within the individual (somatic mutation).

Mutation: An alteration in genetic structure.

Necrosis: Death of cells or tissue.

NEL: The No-Effect-Level is the concentration at or below which there will be no defined effect, either deleterious or beneficial, on populations exposed to the pollutant in question.

Nervous system: Includes the brain and all the nerves.

Neural tube: A hollow tube in vertebrate embryos that changes and eventually forms the brain and spinal cord.

Neurotoxicity: The ability of a substance to destroy nerve tissues or affect behavior.

Neurotoxin: Any substance that is capable of destroying or adversely affecting nerve tissue.

NIOSH: National Institute of Occupational Safety and Health; a branch of the U.S. Department of Health and Human Services. Responsible for recommending limits for workplace exposures.

NOAEL: No-Observed-Adverse-Effect-Level; the highest dose in an experiment that does not produce an observable adverse effect.

NOEL: No-Observed-Effect-Level; the dosage or exposure level at which no effect(s) can be detected.

Nonvolatile: Having very little or no capacity to vaporize or become a gas.

Nonsystemic: Not capable of affecting the entire system; limited to particular areas of a plant or animal.

Non-target vegetation: Vegetation that is not expected or not planned to be affected by an herbicide application.

Non-target species: Any plant, animal, or pest that is not intended to be affected by a pesticide application.

Octanol–water partition coefficient (K_{ow}): A measurement of how a chemical is distributed at equilibrium between octanol and water. It is used often in the assessment of environmental fate and transport for organic chemicals. Additionally, K_{ow} is a key variable used in the estimation of other properties.

Oncogen: Any substance capable of inducing tumors; a carcinogen.

Oncogenicity: The ability of a substance to produce either benign or malignant tumors.

OP: Organophosphate Pesticide.

Oral: Of the mouth; through or by the mouth.

Organic Matter (OM): Soil particles created by the decomposition of plant tissues. Soil OM has a high adsorptive capacity (higher than clay) and offers resistance to leaching of soil water and dissolved molecules.

Organogenesis: The time period during embryonic development in which all major organs and organ systems are formed. During this period, the embryo is most susceptible to factors interfering with development.

OSHA: Occupational Safety and Health Administration, a branch of the U.S. Department of Labor.

PEL: Permissible Exposure Level; OSHA workplace air standard.

Pesticide: EPA defines a pesticide as, "... any substance or mixture of substances intended for preventing, destroying, repelling, or mitigating any pest, and any substance or mixture of substances intended for use as a plant regulator, defoliant or dessicant." The following is a list of pesticides as classified by their target species:

Pesticide	Target species
Acaricide	Mites, ticks
Algaecide	Algae
Attractant	Insects, birds, other vertebrates

Avicide	Birds
Bactericide	Bacteria
Defoliant	Unwanted plant leaves
Desiccant	Unwanted plant tops
Fungicide	Fungi
Growth regulator	Insect and plant growth
Herbicide	Weeds
Insecticide	Insects
Miticide	Mites
Molluscicide	Snails, slugs
Nematicide	Nematodes
Piscicide	Fish
Predacide	Vertebrates
Repellents	Insects, birds, other vertebrates
Rodenticide	Rodents
Silvicide	Trees and woody vegetation
Slimicide	Slime molds
Sterilants	Insects, vertebrates

pH: Hydrogen ion concentration; used to express the degree of acidity or alkalinity of a material.

Photolysis: Breaking of chemical bonds by light energy; in most environmental applications, energy from light in the UV region of sunlight.

Phototoxicity: Toxicity resulting from sequential exposure to a photosensitizing agent and sunlight.

Physiological: Having to do with the mechanics of body function.

Post-emergence: After plants have emerged from the ground.

Potency: A measure of the relative strength of a chemical.

Potentiation: The ability of a substance to increase the toxic effect(s) of another compound.

ppb: Parts per billion; a measure of concentration. May represent the concentration of a residue in soil, water, or whole animals. For example, 1 ppb is equivalent to 1 second in 32 years.

ppm: Parts per million; a measure of concentration. May represent the concentration of a residue in soil, water, or whole animals. For example, 1 ppm is equivalent to 1 minute in 2 years.

Pre-emergence: Before plants have emerged from the ground.

Pupil: The opening in the colored portion (iris) of the eye, through which light passes.

Rate: The amount of active ingredient or acid equivalent applied per unit area or other treatment unit.

Reentry interval: The period of time designated by Federal law between the application of pesticides to crops, and the entrance of workers into the fields without protective clothing.

Reference Dose (RfD): The dose of a compound below which it is believed that chronic effects due to exposure to the compound will be unlikely in almost all individuals within the exposed population.

Registration standard: A document produced by the EPA that details the various conditions which chemical registrants must meet in order to reregister pesticide products containing active ingredients. Through its Registration Standards program, the EPA is reexamining, by current scientific standards, the health and environmental safety of approximately 600 active ingredients contained in some 45,000 currently registered products.

Renal: Pertaining to the kidney.

Reproductive effects: Adverse effects on the reproductive processes that result in significantly reduced fertility or fecundity in exposed animals.

Residue: That quantity of pesticide, its degradation products, and/or its metabolites remaining on or in the soil, plant parts, animal tissues, whole organisms, and surfaces.

Resorption of fetuses: The disappearance of part, or all, of a fetus caused by biochemical reactions that may involve dissolution, absorption, and/or other actions.

Respiratory system/tract: All the passages through which air is taken in and out of the body with breathing, including the nose, trachea, larynx, and lungs, where an exchange of oxygen and carbon dioxide takes place.

Restricted Use Pesticide (RUP): A designation given to a pesticide by the EPA that restricts purchase and use of that pesticide to certified applicators. The designation is given to pesticides that pose a potential hazard to applicators, the general public, or to the environment.

RfD: see Reference Dose.

Risk assessment: A qualitative or quantitative evaluation of the environmental and/or health risk resulting from exposure to a chemical or physical agent (pollutant); combines exposure assessment results with toxicity assessment results to estimate risk.

Risk: The potential for realization of unwanted negative consequences or events.

Rodenticide: A class of pesticides that kills rodents, especially rats and mice.

RUP: see "Restricted Use Pesticide."

Sand: A soil component consisting of coarse particles (0.05 to 2.0 mm diameter). Sand particles provide relatively little surface area for adsorption of molecules. Soil water and dissolved materials move/leach easily through sandy soils.

Selective herbicide: One that kills specific undesirable plants, sparing desirable plants; this is done through different types of toxic action or by the manner in which the material is used (its formulation, dosage, timing, placement, etc.).

Sensitization: The development of a hypersensitive or allergic reaction upon repeated exposure to a substance. The reaction may be immediate or delayed and may be of short-term or chronic duration.

Signal Word: One of three words (CAUTION, DANGER, WARNING) that must appear in large, bold letters on a pesticide label. Signal words are determined by the EPA toxicity class (I–IV), which is related to the level of acute toxicity posed by the pesticide product, or its ability to cause eye or skin irritation or sensitization.

Silt: A soil component consisting of moderately sized particles (0.002 to 0.05 mm diameter). Silt particles provide a moderate amount of surface area for adsorption of molecules, providing some resistance to leaching of charged molecules.

Soil mobility: Movement of a compound through soil from the treated area by leaching, volatilization, adsorption and desorption, or dispersal by water. Leaching is of particular concern because of the potential for contamination of groundwater.

Soil organic matter: see "Organic Matter."

Solubility: The concentration of a substance that dissolves in a given solvent. High solubility: readily dissolves. Low solubility: does not dissolve very well.

Solvent solubility: Concentration of a material that dissolves in a given solvent.

Sorption: A surface phenomenon that may be either absorption or adsorption, or a combination of the two; often used when the specific mechanism is not known.

Special Review: An intensive analysis of all the data on a chemical, including its risks and benefits. The initiation of a Special Review process is an announcement of the EPA's concern about the safety of a pesticide's use. Existing pesticides suspected of posing unreasonable risks to human health, non-target organisms, or the environment are referred to EPA for Special Review. This process may or may not lead to the cancellation of a material.

Spleen: An organ near the stomach or intestines of most vertebrates; the site of final destruction of blood cells, blood storage, and production of lymphocytes.

Statistically significant: Probably caused by something other than mere chance.

STEL: Short-Term Exposure Limit. The maximum allowable concentration, or ceiling, not to be exceeded at any time during a 15-minute exposure period up to four times per day. The level to which workers may be exposed continuously for 15 minutes without adverse effects.

Sterility: Total inability to reproduce.

Stomach poison: A pesticide that must be eaten in order to kill or poison an organism. Compare with "Contact poison."

Subchronic: Intermediate between acute and chronic toxicities; subchronic toxicity studies involve repeated daily exposures of animals to a chemical for part (not exceeding 10%) of a lifespan. In rodents, this period extends up to 90 days of exposure.

Subcutaneous: Under the skin.

Surface water: Water at the soil surface in open bodies such as streams, rivers, ponds, lakes, and oceans.

Susceptibility: Capacity to be affected by chemical treatment.

Synergism: An interaction of two or more chemicals that results in an effect that is greater than the sum of their effects taken independently.

Systemic: Affecting the body generally; distributed throughout the body.

Systemic insecticide: Capable of being absorbed by plants into the plant sap, or by animals into the bloodstream, without undue harm to the plant or animal; capable of poisoning insects feeding on the plant juice or animal blood. For example, a systemic insecticide can be applied to the soil, enter the roots of the plant, travel to the leaves, and kill insects feeding on the leaves.

Systemic toxicity: Poisoning of the whole system or organism, rather than poisoning that affects, for example, a single organ.

$T_{1/2}$: see "Half-life."

Target: The target species is the organism that the pesticide is intended to control. Conversely, the non-target species are those that, because they are either beneficial or harmless, are not to be killed by the pesticide.

TD: Toxic dose; the dose of a chemical that produces signs of toxicity.

TD_{Lo}: The lowest dose that produces signs of toxicity.

Teratogen: Any substance capable of producing structural abnormalities of prenatal origin, present at birth or manifested shortly thereafter.

Teratogenic: The ability to produce birth defects. A teratogenic agent has the ability to induce or increase the incidence of congenital malformations, i.e., permanent structural or functional deviations arising during gestation.

Teratogenicity: The ability of a substance to produce irreversible birth defects or anatomical or functional disorders as a result of an effect on the developing embryo or fetus.

Testicular: Referring to the testes, a pair of male reproductive glands.

Threshold: The lowest dose of a chemical at which a specified measurable effect is observed and below which it is not observed.

Thymus: An immune system gland located at the base of the neck in young vertebrates; it tends to disappear or become nonfunctional in adults.

Thyroid gland: An endocrine gland that lies in front of the trachea, or wind pipe.

Tissue: A group of similar cells.

TLV: Threshold Limit Value. The highest allowable air concentration of a chemical in which workers may work for many years (8 hours a day, 40 hours per week) without negative health effects. Expressed as milligrams per cubic meter of air (mg/m³).

TLV-TWA: Threshold Limit Value-Time Weighted Average. The time-weighted average concentration for a normal 8-hour workday and a 40-hour work week, to which nearly all workers may be repeatedly exposed without adverse effect.

Tolerance: (1) A legal limit, established by the EPA, for the maximum amount of a pesticide residue that may be present in or on raw agricultural commodities, processed foods, and processed animal feeds. Temporary tolerances, which cover residues resulting from an experimental use, generally expire after 1 year. (2) Capacity to withstand pesticide treatment without adverse effect on normal growth and function.

Toxic: Capable of producing adverse health effects in a given species or range of species.

Toxicity: (1) The capacity or property of a substance to cause adverse effects. (2) The specific quantity of a substance that may be expected, under specific conditions, to do damage to a specific living organism.

Toxicity class: A regulatory designation defined by EPA according to the acute toxicity of the pesticide product through the oral, dermal, and inhalation routes, as well as its capacity to cause eye or skin irritation or sensitization.

Translocation: Transport of a substance through a plant from the site of absorption to other parts of the plant.

Tumor: An abnormal mass of cells produced by unregulated overgrowth of cells, and which has no physiologic function. Tumors may be benign or malignant.

TWA: The Time-Weighted Average concentration is the average exposure concentration based on the duration of exposure to airborne concentration as it varies during an 8-hour workday.

Ultra-Low Volume (ULV): A liquid pesticide product that is applied undiluted at a rate of $\frac{1}{2}$ gallon per acre or less.

Uncertainty factor: A number (equal to or greater than 1) used to divide NOAEL or LOAEL values derived from measurements in animals or small groups of humans, in order to estimate "safe" values for the whole human population; also called Margin of Safety.

Vapor pressure: A relative measure of the volatility of a chemical in its pure state; the pressure exerted by a gas that is in equilibrium with its solid or liquid form.

Volatile: Capable of vaporizing or evaporating readily.

Volatilization: The process by which a compound enters the vapor phase from liquid form.

WARNING: see "Signal Word."

Water solubility: The maximum concentration of a chemical that dissolves in pure water at a specific temperature and pH.

WHO: World Health Organization.

Index